DEPARTMENT OF ECONOMIC AND SOCIAL AFFAIRS
Statistics Division

ST/ESA/STAT/SER.J/58

2014 Energy Statistics Yearbook

United Nations
New York, 2016

The Department of Economic and Social Affairs of the United Nations Secretariat is a vital interface between global policies in the economic, social and environmental spheres and national action. The Department works in three main interlinked areas: (i) it compiles, generates and analyses a wide range of economic, social and environmental data and information on which States Members of the United Nations draw to review common problems and to take stock of policy options; (ii) it facilitates the negotiations of Member States in many intergovernmental bodies on joint courses of action to address ongoing or emerging global challenges; and (iii) it advises interested Governments on the ways and means of translating policy frameworks developed in United Nations conferences and summits into programmes at the country level and, through technical assistance, helps build national capacities.

NOTE

Symbols of United Nations documents are composed of capital letters combined with figures. Mention of such a symbol indicates reference to a United Nations document.

General Disclaimer

The designations employed and the presentation of material in this publication do not imply the expression of any opinion whatsoever on the part of the Secretariat of the United Nations concerning the legal status of any country, territory, city or area, or of its authorities, or concerning the delimitation of its frontiers or boundaries.

Where the designation "country or area" appears in the headings of tables, it covers countries, territories, cities or areas. In prior issues of this publication, where the designation "country" appears in the headings of tables, it should be interpreted to cover countries, territories, cities or areas.

ST/ESA/STAT/SER.J/58

UNITED NATIONS PUBLICATION
Sales number: E.17.XVII.5

ISBN: 978-92-1-161616-3
eISBN: 978-92-1-060078-1
ISSN 0256-6400

CONTENTS

Page

Conversion factors

TABLES

Summary Tables

Solid Fuels

Liquid fuels

Gaseous fuels

Electricity and heat

Nuclear fuels

Energy resources

FIGURES

Page

Total energy supply and final consumption

Solid Fuels

Liquid Fuels

Gaseous fuels

Electricity and Heat

Statistics compiled as of December 2016

INTRODUCTION

The *Energy Statistics Yearbook 2014* is a comprehensive collection of international energy statistics prepared by the United Nations Statistics Division. It is the fifty-eighth in a series of annual compilations which commenced under the title *World Energy Supplies in Selected Years, 1929-1950.*[1] It updates the statistical series shown in the previous issue. Supplementary series of monthly data on production of energy may be found in the *Monthly Bulletin of Statistics.*[2]

The principal objective of the *Yearbook* is to provide a global framework of comparable data on long-term trends in the supply of both primary and secondary forms of energy. Data for each energy product and aggregated data for the total energy supply are shown for individual countries and territories, and are summarized into regional and world totals. The data are compiled primarily from the annual energy questionnaire distributed by the United Nations Statistics Division and supplemented by official national statistical publications, as well as publications from international and regional organizations. Where official data are not available or are inconsistent, estimates are made by the Statistics Division based on governmental, professional or commercial materials. Estimates include, but are not limited to, extrapolated data based on partial year information, use of annual trends, trade data based on partner country reports, breakdowns of aggregated data and analysis of current energy events.

This issue of the *Yearbook* contains data in original and common energy units (tons of coal equivalent and joules) for the years 2011-2014.[3]

This publication presents energy statistics consistent with the International Recommendations for Energy Statistics (IRES), which were endorsed by the United Nations Statistical Commission in 2011. IRES contains the Standard International Energy product Classification (SIEC), the first definitive standard classification for energy products. It has been built on a set of internationally harmonized definitions of energy products developed by the Intersecretariat Working Group on Energy Statistics (InterEnerStat) as mandated by the United Nations Statistical Commission. The adoption of SIEC as an international standard classification for energy products represents a significant step forward for energy statistics at the international level. SIEC not only provides a unified set of product definitions, but also uses a standard coding scheme, a common hierarchy of categories, and provides links to other internationally agreed product classifications, such as the Central Product Classification (CPC) and the Harmonized Commodity Description and Coding System (HS).

Within this methodological framework:

a) Table 1 shows a more historical perspective of total energy supply, with data for selected years going back to 1990.
b) Table 2 provides more detail on the components of total energy supply, namely primary production (broken down by type), trade, international bunkers and stock changes. Total energy supply is also shown on a per capita basis.
c) Table 3 provides more detail on total final consumption of energy products, broken down by consuming sector (including non-energy use), and also shows total final consumption per capita.

Prior to the 2012 edition, this publication contained a table 4 which contrasted with the first three tables of the former publication in grouping together both "commercial" and "non-commercial" energy. This distinction assumed that energy products such as biomass are not bought and sold, but rather collected and consumed with no commercial transaction. While this can often be the case, this distinction is somewhat arbitrary and not necessarily relevant for measuring national or global energy use patterns. IRES therefore makes no distinction between commercial and non-commercial energy for presentational purposes (although they may of course be collected differently), and so they are included together.

By referring to previous volumes of the publication, time series can be established from 1950 to the present for the shown products. In addition to the basic tables showing production, trade, stock changes, international bunkers,

[1] Statistical Papers, Series J, No. 1 (United Nations publication, Sales No. 1952.XVII.3).

[2] Statistical Papers, Series Q, *Monthly Bulletin of Statistics*, (United Nations publication).

[3] Table 1 shows data back to 1990 to give a more historical perspective.

supply and consumption, information is included on various other topics such as:

(a) The principal importers and exporters of hard coal, crude oil and natural gas for the years 2013 and 2014;

(b) The capacity of oil refineries and electricity generating plants by type;

(c) The ratio of crude oil and natural gas liquids reserves to crude oil and natural gas liquids production (R/P ratio);

(d) The electricity generated from different sources (including both electricity-only and combined heat and power (CHP) plants) such as from combustible fuels, nuclear, hydro, wind, and solar;

(e) The heat produced (from both heat-only and CHP plants) from combustible fuels, geothermal, solar thermal, nuclear, chemical heat, heat pumps and electric boilers, and direct use of solar thermal and geothermal.

The information contained in the *Yearbook* is also available in electronic format.[4] Requests for information should be directed to United Nations Publications at: order@un.org.

Acknowledgement is due to the following specialized and intergovernmental agencies whose publications have been utilized in supplementing our statistics: Food and Agriculture Organization of the United Nations (FAO), International Atomic Energy Agency (IAEA), International Energy Agency of the Organisation for Economic Co-operation and Development (IEA/OECD), International Sugar Organization (ISO), Organization of Arab Petroleum Exporting Countries (OAPEC), Organization of the Petroleum Exporting Countries (OPEC), Organización Latinoamericana de Energía (OLADE), Statistical Office of the European Union (Eurostat), World Energy Council (WEC). Acknowledgement is also made to governmental, energy and statistical authorities of the Member States which have been cooperative in providing data.

The annual energy data are being collected and processed by the Industrial and Energy Statistics Section of UNSD, headed by Mr. Ralf Becker. The processing of the energy data and preparation for publication was carried out by Mr. Leonardo Souza, Mr. Alexander Blackburn, Mr. Man Soni, Ms. Costanza Giovannelli, Ms. Peng Guo and Mr. Graham Osborn. The table design and the update of the yearbook content were carried out also with the assistance of Mr. Harpal Shergill.

Enquiries, comments and suggestions for improving this publication are welcome and should be addressed to: energy_stat@un.org.

[4] For details, see http://unstats.un.org/unsd/energy/yearbook

ABBREVIATIONS AND SYMBOLS

The following symbols and abbreviations have been used:

cSt	centistoke
m^3	cubic metre
°C	degree Celsius
GJ	gigajoule (10^9 joules)
GWh	gigawatt-hour (10^6 kWh)
IRES	International Recommendations for Energy Statistics
kcal	kilocalorie
kg	kilogramme
kg/cm^2	kilogramme per square centimetre
kJ	kilojoule (10^3 joules)
kWh	kilowatt-hour
LPG	liquefied petroleum gas
mg/g	milligramme per gramme
MJ	megajoule (10^6 joules)
MW	megawatt
NGL	natural gas liquids
PJ	petajoule (10^{15} joules)
t	metric ton
TCE	ton of coal equivalent
TJ	terajoule (10^{12} joules)
*	estimate by the United Nations Statistics Division
..	data not applicable or not available

ENERGY PRODUCT DEFINITIONS

This section provides definitions of energy products utilized in the publication. These definitions correspond to SIEC and as such, some definitions have cha nged since the 2011 edition of the Yearbook. The nature of these definition changes is often cosmetic in nature, with wording differences outnumbering true methodological differences. However, some historical data were revised to reflect changes in definition.

Some tables may include names or concepts not included in these product definitions. See the specific table's notes for how its product definitions relate to SIEC definitions.

SOLID FOSSIL FUELS

Hard coal – Coals with a gross calorific value (moist, ash-free basis) which is not less than 24 MJ/kg or which is less than 24 MJ/kg provided that the coal has a vitrinite mean random reflectance greater than or equal to 0.6 per cent. Hard coal comprises anthracite and bituminous coals.

Brown coal – Coals with a gross calorific value (moist, ash-free basis) less than 24 MJ/ kg and a vitrinite mean random reflectance less than 0.6 per cent. Brown coal comprises sub-bituminous coal and lignite.

Peat – A solid fuel formed from the partial decomposition of dead vegetation under conditions of high humidity and limited air access (initial stage of coalification). It is available in two forms for use as a fuel, sod peat and milled peat. Only peat used as fuel is included.

Patent fuel – A composition fuel made by moulding hard coal fines into briquette shapes with the addition of a binding agent such as pitch.

Brown coal briquettes – A composition fuel made of brown coal produced by briquetting under high pressure with or without the addition of a binding agent. Either sub-bituminous coal or lignite may be used, including dried lignite fines and dust.

Peat products – This includes products such as peat briquettes derived directly or indirectly from sod peat and milled peat.

Coal coke – This group includes the solid, cellular, infusible material remaining after the carbonisation of certain coals. Various cokes are defined according to the type of coal carbonised and their conditions of carbonisation or use: coke oven coke, gas coke, coke breeze and semi cokes. For collection purposes coke breeze and semi cokes are included with coke oven coke.

Coal tar – The liquid by-product of the carbonization of coal in coke ovens. Coal tar may be separated by distillation into several liquid products which may be used for pharmaceutical or wood preservative purposes.

Other coal products – coal products not elsewhere specified (if any).

Oil shale – A sedimentary rock which contains organic matter in the form of kerogen. Kerogen is a waxy hydrocarbon-rich material regarded as a precursor of petroleum. Oil shale may be burned directly or processed by heating to extract shale oil. Note that this product was formerly included under Lignite by the IEA.

LIQUID FOSSIL FUELS

Conventional crude oil – A mineral oil of fossil origin extracted by conventional means from underground reservoirs, and comprises liquid or near-liquid hydrocarbons and associated impurities such as sulphur and metals. Conventional crude oil exists in the liquid phase under normal surface temperature and pressure, and usually flows to the surface under the pressure of the reservoir. This is termed "conventional" extraction. Crude oil includes condensate from condensate fields, and "field" or "lease" condensate extracted with the crude oil.

The various crude oils may be classified according to their sulphur content ("sweet" or "sour") and API gravity ("heavy" or "light"). There are no rigorous specifications for the classifications but a heavy crude oil may be assumed to have an API gravity of less than 20º and a sweet crude oil may be assumed to have less than 0.5% sulphur content. For simplicity this publication often refers to conventional crude oil as simply crude oil.

Natural gas liquids (NGL) – Natural gas liquids are a mixture of ethane, propane, butane (normal and iso), (iso) pentane and a few higher alkanes collectively referred to as pentanes plus.

NGL are produced in association with oil or natural gas. They are removed in field facilities or gas separation plants before sale of the gas. All of the components of NGL except ethane are either liquid at the surface or are liquefied for disposal.

The definition given above is the most commonly used. However, there is some use of terms based on the vapour pressure of the components which are liquid at the surface or can be easily liquefied. The three resulting groups are in order of increasing vapour pressure: condensates, natural gasoline and liquefied petroleum gas.

NGL may be distilled with crude oil in refineries, blended with refined oil products or used directly. NGL differs from LNG (liquefied natural gas) which is obtained by liquefying natural gas from which the NGL has been removed.

Additives and oxygenates – Compounds added to or blended with oil products to modify their properties (octane, cetane, cold properties, etc.). Examples are: (a) oxygenates such as alcohols (methanol, ethanol) and ethers [MTBE (methyl tertiary butyl ether), ETBE (ethyl tertiary butyl ether), TAME (tertiary amyl methyl ether)]; (b) esters (e.g., rapeseed or dimethylester, etc.); and (c) chemical compounds (such as TML, TEL and detergents). Some additives/oxygenates may be derived from biomass while others may be of fossil hydrocarbon origin.

Before the 2012 publication, additives and oxygenates were classified under Crude petroleum/Conventional crude oil.

Other hydrocarbons – This includes non-conventional oils and hydrogen. Non-conventional oils refer to oils obtained by non-conventional production techniques, that is oils which are extracted from reservoirs containing extra heavy oils or oil sands which need heating or treatment (e.g., emulsification) in situ before they can be brought to the surface for refining/processing. They also include the oils extracted from oil sands, extra heavy oils, coal and oil shale which are at, or can be brought to, the surface without treatment and require processing after mining (ex situ processing). Non-conventional oils may also be produced from natural gas.

The oils may be divided into two groups: (i) oils for transformation (e.g., synthetic crudes extracted from extra heavy oils, oil sands, coal and oil shale); and (ii) oils for direct use (e.g., emulsified oils such as orimulsion and GTL liquids).

Before the 2012 publication, other hydrocarbons were classified under Crude petroleum/Conventional crude oil.

Oil products – Products obtained from crude oil, non-conventional oils or gases from oil and gas fields. They may be produced through the refining of conventional crude and non-conventional oils or during the separation of natural gas from gases extracted from oil or gas fields. This category includes Aviation gasoline, Motor gasoline, Gasoline-type jet fuel, Kerosene-type jet fuel, Other kerosene, Gas/diesel oil, Fuel oil, LPG, Feedstocks, Naphtha, White spirit and SBP, Lubricants, Bitumen, Paraffin waxes, Petroleum coke, Refinery gas, Ethane, and Other oil products.

Aviation gasoline – Gasoline prepared especially for aviation piston engines with additives which assure performance under flight conditions. Aviation gasolines are predominantly alkylates (obtained by combining C4 and C5 isoparaffins with C3, C4 and C5 olefins) with the possible addition of more aromatic components including toluene. The distillation range is 25ºC to 170ºC.

Motor gasoline – A mixture of some aromatics (e.g., benzene and toluene) and aliphatic hydrocarbons in the C5 to C12 range. The distillation range is 25ºC to 220ºC.

Additives are blended to improve octane rating, improve combustion performance, reduce oxidation during storage, maintain cleanliness of the engine and improve capture of pollutants by catalytic converters in the exhaust system. Motor gasoline may also contain biogasoline products when blended.

Gasoline-type jet fuel – Light hydrocarbons

for use in aviation turbine power units, distilling between 100ºC and 250ºC. They are obtained by blending kerosene and gasoline or naphtha in such a way that the aromatic content does not exceed 25 per cent in volume, and the vapour pressure is between 13.7 kPa and 20.6 kPa. Gasoline-type jet fuel is also known as "aviation turbine fuel".

Kerosene :-Mixtures of hydrocarbons in the range C9 to C16 and distilling over the temperature interval 145ºC to 300°C, but not usually above 250ºC and with a flash point above 38ºC.

The chemical compositions of kerosenes depend on the nature of the crude oils from which they are derived and the refinery processes that they have undergone. Kerosenes obtained from crude oil by atmospheric distillation are known as straight-run kerosenes. Such streams may be treated by a variety of processes to produce kerosenes that are acceptable for blending as jet fuels.
Kerosenes are primarily used as jet fuels. They are also used as domestic heating and cooking fuels, and as solvents. Kerosenes may include components or additives derived from biomass when blended.

Kerosene-type jet fuel – A blend of kerosenes suited to flight conditions with particular specifications, such as freezing point.

The specifications are set down by a small number of national standards committees, most notably ASTM (U.S.), MOD (UK), GOST (Russia).

Other kerosene – Kerosene which is used for heating, cooking, lighting, solvents and internal combustion engines. Other names for this product are burning oil, vaporizing oil, power kerosene and illuminating oil.

Gas oil/diesel oil – Gas oils are middle distillates, predominantly of carbon number range C11 to C25 and with a distillation range of 160ºC to 420°C. The principal marketed products are fuels for diesel engines (diesel oil), heating oils and marine fuel. Gas oils are also used as middle distillate feedstock for the petrochemical industry and as solvents. Also included here is heavy gasoil, which is a mixture of predominantly gas oil and fuel oil which distills in the range of approximately 380ºC to 540ºC.

Fuel oil – Comprises residual fuel oil and heavy fuel oil. Residual fuel oils have a distillation range of 350ºC to 650ºC and a kinematic viscosity in the range 6 to 55 cSt at 100ºC. Their flash point is always above 60ºC and their specific gravity is above 0.95. Heavy fuel oil is a general term describing a blended product based on the residues from various refinery processes. Other names commonly used to describe fuel oil include: bunker fuel, bunker C, fuel oil No. 6, industrial fuel oil, marine fuel oil and black oil.

Residual and heavy fuel oil are used in medium to large industrial plants, marine applications and power stations in combustion equipment such as boilers, furnaces and diesel engines. Residual fuel oil is also used as fuel within the refinery.

Liquefied petroleum gases (LPG) – LPG refers to liquefied propane (C3H8) and butane (C4H10) or mixtures of both. Commercial grades are usually mixtures of the gases with small amounts of propylene, butylene, isobutene and isobutylene stored under pressure in containers.

The mixture of propane and butane used varies according to purpose and season of the year. The gases may be extracted from natural gas at gas separation plants or at plants re-gasifying imported liquefied natural gas. They are also obtained during the refining of crude oil. LPG may be used for heating and as a vehicle fuel.

See also the definition for natural gas liquids. Certain oil field practices also use the term LPG to describe the high vapour pressure components of natural gas liquids.

Ethane – A naturally straight-chain hydrocarbon (C_2H_6). Ethane is obtained at gas separation plants or from the refining of crude oil. It is a valuable feedstock for petrochemical manufacture.

Refinery gas – Includes a mixture of non-condensable gases mainly consisting of hydrogen, methane, ethane and olefins obtained during distillation of crude oil or treatment of oil products (e.g., cracking) in refineries or from nearby petrochemical plants. It is used mainly as a fuel within the refinery.

Feedstocks – This includes refinery feedstocks, i.e. oils or gases from crude oil refining or the processing of hydrocarbons in the petrochemical industry which are destined for further processing in the refinery excluding blending. Typical feedstocks include naphthas, middle distillates, pyrolysis gasoline and heavy oils from vacuum distillation and petrochemical plants.

Naphtha – Light or medium oils distilling between 30°C and 210°C which do not meet the specification for motor gasoline. Different naphthas are distinguished by their density and the content of paraffins, isoparaffins, olefins, naphthenes and aromatics. The main uses for naphthas are as feedstock for high octane gasolines and the manufacture of olefins in the petrochemical industry.

White spirit and SBP industrial spirits – White spirit and special boiling point industrial spirits (SBP) are refined distillate intermediates with a distillation in the naphtha/kerosene range. They are mainly used for non-fuel purposes and sub-divided as: (a) white spirit - an industrial spirit with a flash point above 30°C and a distillation range of 135°C to 200°C; and (b) industrial spirit (SBP) - light oils distilling between 30°C and 200°C.

There are 7 or 8 grades of industrial spirits, depending on the position of the cut in the distillation range. The grades are defined according to the temperature difference between the 5 per cent and 90 per cent volume distillation points (which is not more than 60°C). White spirit and Industrial spirits are mostly used as thinners and solvents.

Lubricants – Oils, produced from crude oil, for which the principal use is to reduce friction between sliding surfaces and during metal cutting operations.

Lubricant base stocks are obtained from vacuum distillates which result from further distillation of the residue from atmospheric distillation of crude oil. The lubricant base stocks are then further processed to produce lubricants with the desired properties.

Bitumen – A solid, semi-solid or viscous hydrocarbon with a colloidal structure, being brown to black in color. It is obtained as a residue in the distillation of crude oil and by vacuum distillation of oil residues from atmospheric distillation. It should not be confused with the non-conventional primary extra heavy oils which may also be referred to as bitumen.

In addition to its major use for road pavements, bitumen is also used as an adhesive, a waterproofing agent for roof coverings and as a binder in the manufacture of patent fuel. It may also be used for electricity generation in specially designed power plants.

Bitumen is also known in some countries as asphalt but in others asphalt describes the mixture of bitumen and stone aggregate used for road pavements.

Paraffin waxes – Residues extracted when dewaxing lubricant oils. The waxes have a crystalline structure which varies in fineness according to the grade, and are colourless, odourless and translucent, with a melting point above 45°C.

Paraffin waxes are also known as "petroleum waxes".

Petroleum coke – Petroleum coke is a black solid obtained mainly by cracking and carbonizing heavy hydrocarbon oils, tars and pitches. It consists mainly of carbon (90 to 95 per cent) and has a low ash content.

The two most important categories are "green coke" and "calcined coke".

Green coke (raw coke) is the primary solid carbonization product from high boiling hydrocarbon fractions obtained at temperatures below 630°C. It contains 4-15 per cent by weight of matter that can be released as volatiles during subsequent heat treatment at temperatures up to approximately 1330°C.

Calcined coke is a petroleum coke or coal-derived pitch coke obtained by heat treatment of green coke to about 1330°C. It will normally have a hydrogen content of less than 0.1 per cent by weight.

In many catalytic operations (e.g., catalytic cracking) carbon or catalytic coke is deposited on the catalyst, thus deactivating it. The catalyst is reactivated by burning off the coke which is used as a fuel in the refining process. The coke is not recoverable in a concentrated form.

Other oil products – Products (including

partly refined products) from the refining of crude oil and feedstocks which are not specified above. These products will include basic chemicals and organic chemicals destined for use within the refinery or for sale to or processing in the chemical industry such as propylene, benzene, toluene and xylene.

GASEOUS FOSSIL FUELS

Natural gas – A mixture of gaseous hydrocarbons, primarily methane, but generally also including ethane, propane and higher hydrocarbons in much smaller amounts and some non-combustible gases such as nitrogen and carbon dioxide.

The majority of natural gas is separated from both "non-associated" gas originating from fields producing hydrocarbons only in gaseous form, and "associated" gas produced in association with crude oil.

The separation process produces natural gas by removing or reducing the hydrocarbons other than methane to levels which are acceptable in the marketable gas. The natural gas liquids (NGL) removed in the process are distributed separately.

Natural gas also includes methane recovered from coal mines (colliery gas) or from coal seams (coal seam gas) and shale gas. When distributed it may also contain methane from anaerobic fermentation or the methanation of biomass.

Natural gas may be liquefied (LNG) by reducing its temperature in order to simplify storage and transportation when production sites are remote from centres of consumption and pipeline transportation is not economically practicable.

Gas works gas (and other manufactured gases for distribution) – This includes gases obtained from the carbonisation or gasification of carbonaceous material of fossil or biomass origins in Gas Works. The gases comprise: (a) gases obtained from carbonisation or gasification of coals, cokes, biomass or waste; and (b) substitute natural gas (a methane-rich gas) made from synthesis gas.

Synthesis gas is a mixture of mainly hydrogen and carbon monoxide obtained by cracking hydrocarbons with high temperature steam.

Coke oven gas – A gas produced from coke ovens during the manufacture of coke oven coke.

Blast furnace gas – The by-product gas of blast furnace operation consisting mainly of nitrogen, carbon dioxide and carbon monoxide. The gas is recovered as it leaves the furnace. Its calorific value arises mainly from the carbon monoxide produced by the partial combustion of coke and other carbon bearing products in the blast furnace. It is used to heat blast air and as a fuel in the iron and steel industry. It may also be used by other nearby industrial plants. Note that where carbonised biomass (e.g, charcoal or animal meal) is used in blast furnaces, part of the carbon supply may be considered renewable.

Other recovered gases - Combustible gases of solid carbonaceous origin recovered from manufacturing and chemical processes not elsewhere defined. Examples of fuel gas production from metals and chemicals processing are in the production of zinc, tin, lead, ferroalloys, phosphorus and silicon carbide.

BIOFUELS AND WASTE

Fuelwood, wood residues and by-products – Fuelwood or firewood (in log, brushwood, pellet or chip form) obtained from natural or managed forests or isolated trees. Also included are wood residues used as fuel and in which the original composition of wood is retained. Charcoal and black liquor are excluded.

Bagasse – The fuel obtained from the fibre which remains after juice extraction in sugar cane processing. It is often used as a fuel within the sugar milling industry.

Charcoal – The solid residue from the carbonisation of wood or other vegetal matter through slow pyrolysis.

Animal waste – Excreta of animals, meat and fish residues which, when dry, are used directly as a fuel.

This excludes waste used in anaerobic fermentation plants. Fuel gases from these plants are included under biogases.

Vegetal waste – Refers to SIEC's 'Other vegetal materials and residues', which are solid primary biofuels not specified elsewhere, including straw, vegetable husks, ground nut shells, pruning brushwood, olive pomace and other wastes arising from the maintenance, cropping and processing of plants.

Municipal waste – Household waste and waste from companies and public services that resembles household waste and which is collected at installations specifically designed for the disposal of mixed wastes with recovery of combustible liquids, gases or heat.

Municipal wastes can be divided into renewable and non-renewable fractions.

Industrial waste – Non-renewable waste which is combusted with heat recovery in plants other than those used for the incineration of municipal waste.

Examples are used tires, specific residues from the chemical industry and hazardous wastes from health care. Combustion includes co-firing with other fuels.

The renewable portions of industrial waste combusted with heat recovery are classified according to the biofuels which best describe them.

Black liquor – The alkaline-spent liquor obtained from the digesters during the production of sulphate or soda pulp required for paper manufacture. The lignin contained in the liquor burns to release heat when the concentrated liquor is sprayed into a recovery furnace and heated with hot gases at 900ºC. Black liquor is used as a fuel in the pulping process.

Biogasoline – Liquid fuels derived from biomass and used in spark-ignition internal combustion engines.

Common examples are: bioethanol (including both hydrous and anhydrous ethanol); biomethanol; biobutanol; bio ETBE (ethyl-tertio-butyl-ether); and bio MTBE (methyl-tertio-butyl-ether). Biogasoline may be blended with petroleum gasoline or used directly in engines. The blending may take place in refineries or at or near the point of sale.

Biodiesels – Liquid biofuels derived from biomass and used in diesel engines.

Biodiesels obtained by chemical modification are a linear alkyl ester made by transesterification of vegetable oils or animal fats with methanol. The transesterification distinguishes biodiesel from straight vegetable and waste oils. Biodiesel has a flash point of around 150°C and a density of about 0.88 kg/litre. Biological sources of biodiesel include, but are not limited to, vegetable oils made from canola (rapeseed), soybeans, corn, oil palm, peanut or sunflower. Some liquid biofuels (straight vegetable oils) may be used without chemical modification and their use usually requires modification of the engine.

A further category of diesel fuels can be produced by a range of thermal processes (including for example gasification followed by Fischer Tropsch synthesis, pyrolysis followed by hydrogenation, or conversion of sugar to hydrocarbons using microorganisms (e.g. yeast)). A wide range of biomass feedstocks, including cellulosic materials and algal biomass could be used in such processes.

Biodiesels may be blended with petroleum diesel or used directly in diesel engines.

Bio jet kerosene – Liquid biofuels derived from biomass and blended with or replacing jet kerosene. Bio jet kerosene can be produced by a range of thermal processes (including for example gasification followed by Fischer-Tropsch synthesis, pyrolysis followed by hydrogenation, or conversion of sugar to hydrocarbons using microorganisms (e.g. yeast). A wide range of biomass feedstocks, including cellulosic materials and algal biomass could be used in such processes.

Other liquid biofuels – Includes liquid biofuels not elsewhere specified.

Biogases – Gases arising from the anaerobic fermentation of biomass and the gasification of solid biomass (including biomass in wastes).

The biogases from anaerobic fermentation are composed principally of methane and carbon dioxide and comprise landfill gas, sewage sludge gas and other biogases from anaerobic fermentation.

Biogases can also be produced from

thermal processes (by gasification or pyrolysis) of biomass and are mixtures containing hydrogen and carbon monoxide (usually known as syngas) along with other components. These gases may be further processed to modify their composition and can be further processed to produce substitute natural gas.

The gases are divided into two groups according to their production: biogases from anaerobic fermentation and biogases from thermal processes.

Biogases are used mainly as a fuel but can be used as a chemical feedstock.

ELECTRICITY AND HEAT

Electricity –The transfer of energy through the physical phenomena involving electric charges and their effects when at rest and in motion.

Electricity can be generated through different processes such as: the conversion of energy contained in falling or streaming water, wind or waves; the direct conversion of solar radiation through photovoltaic processes in semiconductor devices (solar cells); or by the combustion of fuels.

Electricity production in this Yearbook refers to gross production, which is the sum of the electrical energy production by all the generating units/installations concerned (including pumped storage) measured at the output terminals of the main generators.

Editorial note: as of the 2012 edition of the Yearbook, Table 1 changed in scope as described in the introduction. The total energy supply figure is now consistent with energy balances as defined within IRES and published in the *Energy Balances* and *Electricity Profiles*[5] that are complementary to this publication. As such, see the changes in presentation for electricity from non-combustible fuels described below under General Notes.

Primary electricity refers to electrical energy from hydro, solar photovoltaics, wind, tide, wave and ocean sources.

Secondary electricity is defined as electricity derived from a product already accounted for within energy statistics. This includes electricity generated from combustible fuels, in addition to electricity coming from solar heat, nuclear heat, geothermal heat and heat from chemical sources.

Main activity producers – These are enterprises which produce electricity or heat as their principal activity. Formerly known as public utilities, these enterprises may be privately or publicly owned companies.

Autoproducers (electricity) – These are enterprises which produce electricity but for whom the production is not their principal activity.

Autoproducers (heat) – These are enterprises which produce heat for sale but for whom the production is not their principal activity. Deliveries of fuels for heat generated by an establishment for its own purposes are classified within the part of final consumption where they are consumed, and not as transformation inputs.

Heat – The energy obtained from the translational, rotational and vibrational motion of the constituents of matter, as well as changes in its physical state.

Heat can be produced by different production processes, both primary and secondary. It is usually sold in the form of steam or hot water, and for the purposes of energy statistics it is important to note that heat only represents quantities of energy for sale (or generated and consumed directly by direct use of solar thermal or geothermal). This means that any quantities "generated" at an end-use site (for example through the combustion of coal to power a boiler, or an electric filament heating a kettle) are not considered within the scope of heat production.

In Tables 1 and 2, which follow energy balance methodology, the heat column includes primary heat from sources such as nuclear, solar thermal and geothermal which are used to generate electricity. These quantities are not taken into account in Table 34.

Uranium production – Comprises the U content of uranium ores and concentrates intended for treatment for uranium recovery.

[5] Statistical Papers, Series W (United Nations publication).

ENERGY FLOW DEFINITIONS

Production is defined as the capture, extraction or manufacture of fuels or energy in forms which are ready for general use. In energy statistics, two types of production are distinguished, primary and secondary.

Primary production is the capture or extraction of fuels or energy from natural energy flows, the biosphere and natural reserves of fossil fuels within the national territory in a form suitable for use. Inert matter removed from the extracted fuels and quantities reinjected, flared or vented are not included. The resulting products are referred to as "primary" products.

Secondary production is the manufacture of energy products through the process of transformation of primary fuels or energy. The quantities of secondary fuels reported as production include quantities lost through venting and flaring during and after production. In this manner, the mass, energy and carbon within the primary source(s) from which the fuels are manufactured may be balanced against the secondary fuels produced. Fuels, electricity and heat produced are usually sold but may be partly or entirely consumed by the producer (but see the definition of autoproducer heat plants for an important exception.)

When considering complete energy balances (such as in Table 1, where total energy supply is displayed), or indeed adding together a mix of primary and secondary fuels (like Table 4), the distinction between primary and secondary production is crucial. Within a complete energy balance (or a coal table showing both primary and secondary energy), only primary energy is shown to avoid double counting of quantities transformed into secondary products. For the product-specific tables however, this distinction is not necessary as double counting does not apply, and so primary and/or secondary production is included under production to give a complete idea of the "supply" of that product. See individual table notes for details.

Imports of energy products comprise all fuel and other energy products entering the national territory. Goods simply being transported through a country (goods in transit) and goods temporarily admitted are excluded but re-imports, which are domestic goods exported but subsequently readmitted, are included. The bunkering of fuel outside the reference territory by national merchant ships and civil aircraft engaged in international travel is excluded from imports. Fuels delivered to national merchant ships and civil aircraft which are outside of the national territory and are engaged in international travel should be classified as "International Marine" or "Aviation Bunkers", respectively, in the country where such bunkering is carried out. Note that the "country of origin" of energy products should be recorded as a country from which goods were imported.

Exports of energy products comprise all fuel and other energy products leaving the national territory with the exception that exports exclude quantities of fuels delivered for use by merchant (including passenger) ships and civil aircraft, of all nationalities, during international transport of goods and passengers. Goods simply being transported through a country (goods in transit) and goods temporarily withdrawn are excluded but re-exports, foreign goods exported in the same state as previously imported, are included. Fuels delivered to foreign merchant ships and civil aircraft engaged in international travel are classified as "International Marine" or "Aviation Bunkers", respectively. Note that "country of destination" of energy products (that is country of the last known destination as it is known at the time of exportation) should be recorded as a country to which these products are exported to.

For electricity, trade data include "goods in transit", i.e. electricity transmitted through the country from one neighbour to another, as there is no practical way of discerning which quantities are re-exported and which are consumed by the transit country.

International Marine Bunkers are quantities of fuels delivered to merchant (including passenger) ships, of any nationality, for consumption during international voyages transporting goods or passengers. International voyages take place when the ports of departure and arrival are in different national territories. Fuels delivered for consumption by ships during domestic transportation, fishing or military uses are not included here. For the purposes of energy statistics International Marine Bunkers are not included in exports.

International Aviation Bunkers are quantities of fuels delivered to civil aircraft, of any nationality, for consumption during international flights transporting goods or passengers. International flights take place when the ports of departure and arrival are in different national territories. Fuels delivered for consumption by aircraft undertaking domestic or military flights are not included here. For the purposes of energy statistics International Aviation Bunkers are not included in exports.

Stocks – For the purposes of energy statistics, stocks are quantities of energy products that are held on the national territory and can be used to: (a) maintain service under conditions where supply and demand are variable in their timing or amount due to normal market fluctuations, or (b) supplement supply in the case of a supply disruption. Stocks used to manage a supply disruption may be called "strategic" or "emergency" stocks and are often held separately from stocks designed to meet normal market fluctuations. **Stock changes** are defined as the increase (stock build) or decrease (stock draw) in the quantity of stock over the reporting period and are thus calculated as the difference between the closing and opening stocks.

Energy supply – is defined as energy production plus imports minus exports minus international marine bunkers minus international aviation bunkers minus stock changes. Depending on the table in this Yearbook, production can refer to either primary, secondary, or both primary and secondary production (see the note under production above).

Energy industries own use refers to consumption of fuels and energy for the direct support of the production, and preparation for use of fuels and energy. Quantities of fuels which are transformed into other fuels or energy are not included here but within the transformation use. Neither are quantities which are used within parts of the energy industry not directly involved in the activities listed in the definition. These quantities are reported within final consumption.

Final Consumption refers to all fuel and energy that is delivered to users for both their energy and non-energy uses which do not involve a transformation process.

CAPACITIES AND OTHER DEFINITIONS

CAPACITIES

Refinery capacity is the theoretical maximum capacity of crude oil refineries available for operation at the end of the reference year.

Net maximum electrical capacity is the maximum active power that can be supplied continuously, with all plants running, at the point of outlet (i.e., after taking the power supplies for the station auxiliaries and allowing for the losses in those transformers considered integral to the station). This assumes no restriction of interconnection to the network. Does not include overload capacity that can only be sustained for a short period of time (e.g., internal combustion engines momentarily running above their rated capacity). The net maximum electricity-generating capacity represents the sum of all individual plants' maximum capacities available to run continuously throughout a prolonged period of operation in a day.

ENERGY RESOURCES AND RESERVES[6]

Hard Coal, Brown coal and Peat

Proved amount in place is the resource remaining in known deposits that has been carefully measured and assessed as exploitable under present and expected local economic conditions with existing available technology.

Proved recoverable reserves are the tonnage within the proved amount in place that can be recovered in the future under present and expected local economic conditions with existing available technology.

Estimated additional amount in place is the indicated and inferred tonnage additional to the proved amount in place that is of foreseeable economic interest. It includes estimates of amounts which could exist in unexplored extensions of known deposits or in undiscovered deposits in known coal-bearing areas, as well as amounts inferred through knowledge of favourable geological conditions. Speculative amounts are not included.

[6] *World Energy Resources - 2016.*
World Energy Council (WEC)

Crude Oil and Natural Gas Liquids

Proved recoverable reserves are the quantity within the proved amount in place that can be recovered in the future under present and expected local economic conditions with existing available technology. Proved amount in place is the resource remaining in known natural reservoirs that has been carefully measured and assessed as exploitable under present and expected local economic conditions with existing available technology.

The **ratio of crude oil reserves to production (R/P ratio)** is used to show the length of time those reserves would last in years if production continued at the then-current level and there were no further increases in the proved recoverable reserves. The ratio is calculated by dividing the proved recoverable crude oil and NGL reserves (which for the most part refer to reserves at the end of 2014) by the production. These R/P ratios should be viewed with caution. The definition used for proved recoverable reserves is very restrictive and confined to those known reserves which can be recovered with reasonable certainty under existing economic conditions. The R/P ratios, therefore, can frequently give a very pessimistic impression of the expected life of a country's reserves. In addition, for some of those countries whose R/P ratios appear very large, it can reasonably be assumed that the figures for proved recoverable reserves include some unproved reserves.

Oil Shale and Bituminous Sands

Proved recoverable reserves are the amount, expressed as tonnage of recoverable synthetic oil that has been both carefully measured and has also been assessed as exploitable under present and expected local economic conditions with existing available technology.

Natural Gas

Proved recoverable reserves are the volume within the proved amount in place that can be recovered in the future under present and expected local economic conditions with existing available technology.

Uranium

Reasonably assured resources refer to recoverable uranium that occurs in known mineral deposits of delineated size, grade and configuration such that the quantities which could be recovered within the given production cost ranges with currently proven mining and processing technology can be specified. Estimates of tonnage and grade are based on specific sample data and measurements of the deposits and on knowledge of deposit characteristics.

Inferred Resources refer to recoverable uranium (in addition to reasonably assured resources) that is inferred to occur, based on direct geological evidence, in extensions of well-explored deposits and in deposits in which geological continuity has been established, but where specific data and measurements of the deposits and knowledge of their characteristics are considered to be inadequate to classify the resource as reasonably assured resources.

Hydropower

Gross theoretical capability is the annual energy potentially available in the country if all natural flows were turbined down to sea level or to the water level of the border of the country (if the water course extends into another country) with 100% efficiency from the machinery and driving water-works. Unless otherwise stated, the figures have been estimated on the basis of atmospheric precipitation and water runoff.

GENERAL NOTES

Calculation of primary energy equivalent for electricity and heat

As mentioned previously, the publication now treats electricity and heat consistently with IRES energy balances. Notably, this leads to the following notional efficiencies applying to electricity and heat generation (to back calculate the primary energy):

- 100% for electricity from hydro, wind, tide, wave, ocean and solar PV;
- 33% for electricity from nuclear heat or solar thermal;
- 100% for solar thermal heat;
- 50% for heat produced from geothermal heat;
- 10% for electricity from geothermal heat.

Consumption

Consumption throughout the *Yearbook* is defined consistently with the definition of final consumption in IRES, namely the sum of flows reflecting energy consumption by energy consumers, as well as non-energy use of energy products. As such, the publication should no longer have the same issues with negative consumption figures that previous versions did, when consumption was calculated based on supply flows and as such negative figures could appear for countries that exported secondary products.

Negative Supply

Negative supply can represent a residual or statistical difference between production and exports when a particular product is mainly exported, or when a transferred product is subsequently exported.

Time Period

The period to which the data refer is the calendar year, with the exception of the data of the following countries which refer to the fiscal year:

Afghanistan and Iran (Islamic Rep. of) - beginning 21 March of the year stated;

Australia, Bangladesh, Bhutan, Egypt (for the latter two, electricity only), Nepal - ending June of the year stated;

Pakistan - starting July of the year stated

India and Myanmar - beginning April of the year stated.

Trade data

Figures displayed for international trade by partner country in Tables 6, 15 and 28 are based on data from the International Energy Agency and from the United Nations Statistics Division COMTRADE (Commodity Trade Statistics). The total of all exports will likely differ from the figures given in other tables because not all of the importers are included.

Per Capita Data

Data on a *per capita* basis use population data from the United Nations Population Division as a denominator. These indicators may sometimes show differences for certain years compared to previous versions of this publication, due to revisions made by the Population Division rather than changes to the underlying energy data.

Coverage

As Table 1 contains data back to 1990, data are included for former countries such as the Union of Soviet Socialist Republics, and older versions of countries such as Ethiopia which have since had territories split from them.

COUNTRY NOTES

Australia – Data exclude the overseas territories.

Bonaire, St Eustatius, Saba – Data prior to 2012 are included in Netherlands Antilles (former).

China – Data exclude Hong Kong and Macao Special Administrative Regions (Hong Kong SAR and Macao SAR) and Taiwan Province. Data for other kerosene include kerosene-type jet fuel until 2013.

China, Hong Kong SAR – Data on kerosene-type jet fuel include aviation gasoline and other kerosene.

China, Macao SAR – For confidentiality reasons, some data on coal and coal products, petroleum products and biomass are not available.

Curaçao – Data prior to 2012 are included in Netherlands Antilles (former).

Czechoslovakia – prior to 1993, refers to the former Czechoslovakia composed of Czechia and Slovakia.

Denmark – Data exclude Greenland and the Danish Faroes.

Egypt – Data for lubricants are included with other oil products for 2011-2013.

France – Data include Monaco, and exclude the following overseas departments and territories: Guadeloupe, French Guiana, Martinique, New Caledonia, French Polynesia, Reunion, Saint Barthélemy, St. Pierre and Miquelon, and Wallis and Futuna.

French Guiana – Electricity consumption series from 1990 was updated based on new data.

Gambia – Data for Other kerosene have been reclassified as kerosene-type jet fuel.

Germany – Through the accession of the **German Democratic Republic** to the **Federal Republic of Germany** with effect from 3 October 1990, the two German States united to form one sovereign state. Data for 1990 are shown separately for the former states.

Greenland – Data for kerosene-type jet fuel include other kerosene.

Indonesia – Data include Timor-Leste until 2001.

Italy – Data include San Marino and the Holy See.

Japan – Data include Okinawa.

Kuwait – The data for crude oil production include 50 per cent of the output of the Neutral Zone.

Madagascar – Data for bitumen include petroleum coke.

Netherlands – Data exclude Suriname and the Netherlands Antilles (former).

Netherlands Antilles (former) – The country, comprising Bonaire, Curaçao, Saba, St. Eustatius and Sint Marteen (Dutch part), was dissolved on 10 October 2010. For the purposes of this yearbook, data up to 2011 refer to the former territory of the Netherlands Antilles as it was known prior to 10 October 2010.

Pacific Islands – Until 1991, Trust Territory comprising Federated States of Micronesia, Marshall Islands, Northern Mariana Islands and Palau.

Portugal – Data include the Azores and Madeira.

Saudi Arabia – Data for crude oil production include 50 per cent of the output of the Neutral Zone.

Serbia – Data exclude Kosovo from 2000 onwards.

Sint Maarten (Dutch part) – Data prior to 2012 are included in Netherlands Antilles (former).

South Sudan – Data prior to 2012 are included in Sudan (former).

Spain – Data include the Canary Islands.

Sudan – Data prior to 2012 are included in Sudan (former).

Sudan (former) – South Sudan became an independent state on 9 July 2011. This country represents the whole territory of Sudan as it was on 1 January 2011 for data up to and including 2011.

Switzerland – Data include Liechtenstein for oil statistics.

Ukraine – The Government of Ukraine has informed the United Nations that it is not in a position to provide statistical data concerning the Autonomous Republic of Crimea and the city of Sevastopol. For confidentiality reasons, data on the following products (mainly on production) may not be available or may be included with other products: lignite (included with peat), patent fuel (included with peat briquettes), aviation gasoline, kerosene-type jet fuel, other kerosene, petroleum coke, paraffin waxes and white spirit (the latter three included with other oil products).

Union of Soviet Socialist Republics – Referred to as USSR, comprised up to 1990 Armenia, Azerbaijan, Belarus, Estonia, Georgia, Kazakhstan, Kyrgyzstan, Latvia, Lithuania, Republic of Moldova, Russian Federation, Tajikistan, Turkmenistan, Ukraine and Uzbekistan.

United Kingdom – Shipments of coal and oil to Jersey, Guernsey and the Isle of Man from the United Kingdom are not classed as exports. Supplies of coal and oil to these islands are, therefore, included as part of UK supply. Exports of natural gas to the Isle of Man are included with the exports to Ireland.

United States – Includes the 50 states and the District of Columbia. Oil statistics as well as coal trade statistics also include Puerto Rico, Guam, the U.S. Virgin Islands, American Samoa, Johnston Atoll, Midway Islands, Wake Island and the Northern Mariana Islands.

Yemen – On 22 May 1990 **Democratic Yemen** and **Yemen Arab Republic** merged to form a single State. Since that date they have been represented as one Member with the name 'Yemen'. Data for 1990 are shown separately for the former states.

Yugoslavia – Data until 1991 refer to the former Socialist Federal Republic of Yugoslavia, namely Bosnia and Herzegovina, Croatia, The Former Yugoslav Republic of Macedonia, Montenegro, Slovenia and Serbia (including Kosovo).

UNITS OF MEASUREMENT AND CONVERSION

Metric units of measurement are used throughout the *Yearbook*. The following tables provide the necessary factors for converting energy data from one measurement system (mass, volume or heat) to another. Unless otherwise stated in the Table Notes, in the case of solid fuels, liquid fuels and gases, the conversions are based on the net calorific value.

The comparison between different fuels is presented in metric tons of coal equivalent (TCE, Table 5), and terajoules (Tables 1, 2, 3 and 12) on the basis of the heat energy which may be obtained from each of them. One TCE is defined as 0.0293076 TJ, or 29307.6 MJ. This is derived from the standard coal equivalency of 7000 calories per gram.

The procedure to convert from original units to the common unit terajoule is as follows:

- Data in original (mass) units multiplied by the factors in Table I will give data in terajoules.

Further, to convert electricity data to terajoules, the following energy equivalence is applied: 1 GWh = 3.6 TJ.

Data on natural gas are collected in terajoules on a gross calorific value basis by convention. When comparing natural gas with other energy sources, it is more appropriate to compare useful energy, and thus the data have been converted to be on a net calorific basis by multiplying them by 0.9.

Table I

NET CALORIFIC VALUES OF ENERGY PRODUCTS
GJ/ton

PRODUCTION, EXPORTS AND CHANGES IN STOCKS

HARD COAL	2011	2012	2013	2014
Standard factor	**25.8**	**25.8**	**25.8**	**25.8**
Argentina	24.7	24.7	24.7	24.7
Australia	27.1	27.0	26.9	26.9
Bangladesh	20.9	20.9	20.9	20.9
Botswana	23.6	23.6	23.6	23.6
Brazil	18.6	18.6	18.6	18.6
Bulgaria	17.4	17.2	20.0	27.2
Canada	24.7	24.7	25.1	25.0
Chile	15.7	17.4	22.0	22.0
China	20.6	19.9	19.9	20.1
Colombia	27.2	27.2	27.2	27.2
Czechia	26.7	26.9	27.8	27.0
Egypt	25.5	25.5	25.7	25.7
France	26.0	26.0	26.0	26.0
Georgia	25.0	25.0	28.0	25.1
Germany	29.3	27.4	27.7	27.6
India	18.9	18.9	17.0	16.8
Indonesia	26.1	25.8	25.8	25.8
Italy	26.6	26.6	26.6	26.6
Kazakhstan	18.6	18.6	18.6	18.6
Korea, Democratic People's Republic of	26.2	26.2	26.3	26.3
Korea, Republic of	19.3	18.6	18.6	18.6
Kyrgyzstan	18.6	18.6	18.6	18.6
Malaysia	26.4	26.4	26.4	26.4
Mexico	28.9	29.0	28.9	28.9
Mongolia	28.5	28.5	28.5	28.5
Mozambique	26.2	26.9	26.9	26.9
New Zealand	29.7	29.7	29.9	29.9
Niger	16.0	16.0	16.0	16.0
Norway	28.1	28.1	28.1	28.1
Pakistan	19.8	18.8	18.8	18.8
Peru	26.7	27.2	27.1	26.1
Poland	23.7	23.7	23.9	23.8
Russian Federation	26.1	26.0	26.2	26.2
South Africa	23.6	23.6	23.7	23.7
Spain	18.6	18.4	19.5	19.2
Sri Lanka	26.4	29.3	29.3	29.3
Tajikistan	18.6	18.6	18.6	18.6
Turkey	24.9	23.7	25.7	25.5
Ukraine	23.5	24.9	24.8	24.4
United Kingdom	24.4	24.5	24.4	24.4
United States	26.6	26.5	26.3	26.3
Uzbekistan	18.6	18.6	18.6	18.6
Viet Nam	23.4	23.4	23.4	23.4
Venezuela (Bolivarian Republic of)	30.6	30.6	30.6	30.6
Zambia	24.7	24.7	24.7	24.7
Zimbabwe	27.0	27.0	27.0	27.0

Table I (continued)

NET CALORIFIC VALUES OF ENERGY PRODUCTS
GJ/ton

PRODUCTION, EXPORTS AND CHANGES IN STOCKS

BROWN COAL	2011	2012	2013	2014
Standard factor	**14.0**	**14.0**	**14.0**	**14.0**
Albania	9.8	9.8	9.8	9.8
Australia	13.0	13.0	12.4	12.1
Bosnia and Herzegovina	13.5	13.6	13.6	13.5
Bulgaria	7.0	7.0	7.0	6.8
Canada	17.1	17.0	16.5	16.9
Czechia	12.3	12.3	12.4	12.3
Georgia	16.7	16.7	16.7	17.2
Germany	9.0	9.0	9.1	9.1
Greece	5.4	5.4	5.2	5.3
Hungary	7.2	7.2	7.1	7.0
India	9.5	9.5	9.5	9.5
Indonesia	20.0	22.0	22.0	20.9
Kazakhstan	14.7	14.7	14.7	14.7
Korea, Democratic People's Republic of	17.6	17.6	17.6	17.6
Kyrgyzstan	14.7	14.7	14.7	14.7
Mexico	19.4	19.9	19.9	20.0
Mongolia	14.4	14.4	14.4	14.4
Montenegro	9.2	9.2	9.2	9.2
New Zealand	19.9	19.9	19.8	19.6
Philippines	22.1	22.1	22.1	22.1
Poland	8.4	8.3	8.3	8.2
Romania	7.9	7.8	7.9	7.9
Russian Federation	14.9	14.9	14.9	14.9
Serbia	8.0	8.0	8.0	8.0
Slovakia	10.6	10.4	10.4	11.1
Slovenia	11.2	10.7	11.6	11.0
Spain	13.4	13.8	13.3	13.8
Tajikistan	14.7	14.7	14.7	14.7
Thailand	12.1	12.1	12.1	10.7
T.F. Yugoslav Republic of Macedonia	7.2	7.1	6.6	6.4
Turkey	9.3	8.7	10.4	10.0
United States	18.3	18.2	18.1	18.1
Uzbekistan	14.7	14.7	14.7	14.7

COKE OVEN COKE	2011	2012	2013	2014
Standard factor	**28.2**	**28.2**	**28.2**	**28.2**
Argentina	28.5	28.5	28.5	28.5
Australia	27.0	27.0	27.0	27.0
Austria	29.0	29.0	29.0	29.0
Belgium	29.3	29.3	29.3	29.3
Bosnia and Herzegovina	26.9	26.9	26.9	26.9
Brazil	30.5	30.5	30.5	30.5

Table I (continued)

NET CALORIFIC VALUES OF ENERGY PRODUCTS
GJ/ton

PRODUCTION, EXPORTS AND CHANGES IN STOCKS

COKE OVEN COKE	2011	2012	2013	2014
Canada	27.5	27.5	27.5	27.5
Chile	28.6	28.6	28.6	28.6
China	28.4	28.4	28.4	28.4
Colombia	20.1	20.1	20.1	20.1
Czechia	28.2	28.4	28.5	28.5
Egypt	27.2	27.2	27.2	27.2
Estonia	28.5	28.5	28.5	28.5
Finland	29.3	29.3	29.3	29.3
France	28.0	28.0	28.0	28.0
Germany	28.7	28.7	28.7	28.7
Hungary	29.7	29.5	29.2	29.5
Italy	29.0	29.0	29.0	29.0
Japan	29.4	29.4	29.2	29.2
Kazakhstan	25.1	25.1	25.1	25.1
Korea, Republic of	29.3	28.9	28.9	28.9
Mexico	26.5	26.5	28.7	28.7
Netherlands	28.5	28.5	28.5	28.5
New Zealand	29.5	29.5	29.5	29.5
Other Asia	28.2	28.2	28.2	28.2
Poland	27.8	28.1	27.9	27.8
Russian Federation	29.0	29.0	29.0	29.0
Slovakia	28.3	28.4	28.1	27.9
South Africa	26.5	26.5	26.5	26.5
Spain	26.8	26.8	26.8	26.8
Sweden	28.1	28.1	28.1	28.1
Turkey	27.5	28.2	29.7	27.6
Ukraine	25.1	25.1	25.1	25.1
United Kingdom	29.8	29.8	29.8	29.8
United States	28.8	28.8	28.9	28.9
Zimbabwe	25.1	25.1	25.1	25.1

PATENT FUEL	2011	2012	2013	2014
Standard factor	**20.7**	**20.7**	**20.7**	**20.7**
France	32.0	32.0	32.0	32.0
Germany	31.4	31.4	31.4	31.4
Hungary	22.0	22.0	22.0	22.0
Korea, Republic of	19.3	18.6	18.6	18.6
Russian Federation	29.0	29.0	29.0	29.0
United Kingdom	31.0	31.0	31.1	31.1

BROWN COAL BRIQUETTES	2011	2012	2013	2014
Standard factor	**20.7**	**20.7**	**20.7**	**20.7**
Australia	21.0	21.0	21.0	21.0
Belgium	20.0	20.7	20.7	20.7
Bulgaria	13.1	12.5	10.2	12.2

Table I (continued)

NET CALORIFIC VALUES OF ENERGY PRODUCTS
GJ/ton

PRODUCTION, EXPORTS AND CHANGES IN STOCKS

BROWN COAL BRIQUETTES	2011	2012	2013	2014
Czechia	21.6	22.0	20.7	21.1
Germany	21.3	21.1	21.3	21.4
Serbia	17.6	17.6	17.6	17.7

PEAT	2011	2012	2013	2014
Standard factor	**9.8**	**9.8**	**9.8**	**9.8**
Austria	8.8	8.8	8.8	8.8
Belarus	10.2	10.2	10.5	10.5
Burundi	14.7	14.7	14.7	14.7
Estonia	10.3	10.0	10.4	10.4
Falkland Islands (Malvinas)	9.5	9.5	9.5	9.5
Finland	10.2	10.1	10.2	10.0
Ireland	8.6	9.1	8.1	8.8
Russian Federation	10.0	10.0	10.0	10.0
Sweden	12.5	12.5	12.5	12.5
Ukraine	9.7	10.0	9.7	9.7

OIL SHALE	2011	2012	2013	2014
Standard factor	**8.9**	**8.9**	**8.9**	**8.9**
Israel	2.9	2.9	2.9	2.9

Table I (continued)

NET CALORIFIC VALUES OF ENERGY PRODUCTS
GJ/ton

IMPORTS

HARD COAL	2011	2012	2013	2014
Albania	27.2	27.2	27.2	27.2
Argentina	30.1	30.1	30.1	28.4
Australia	27.1	27.0	26.9	27.9
Austria	28.3	28.2	28.5	28.6
Azerbaijan	27.7	27.7	27.7	27.7
Bangladesh	20.9	20.9	20.9	20.9
Belarus	25.0	26.1	25.2	25.6
Belgium	27.5	27.4	27.4	27.5
Bosnia and Herzegovina	30.7	30.7	30.7	30.7
Brazil	30.5	30.5	30.5	30.5
Bulgaria	25.6	25.9	26.0	27.2
Canada	26.3	26.7	28.0	27.9
Chile	23.5	23.9	25.0	25.0
China	21.4	16.2	15.8	17.9
China, Hong Kong SAR	26.4	26.4	26.4	26.4
Croatia	24.4	24.7	25.2	24.8
Czechia	26.7	26.6	26.4	25.4
Denmark	24.4	24.2	24.5	24.7
Egypt	25.5	28.2	25.7	25.7
Estonia	27.2	27.2	27.2	27.2
Finland	25.7	26.0	25.9	25.9
France	27.2	27.2	27.4	27.7
Georgia	25.0	25.0	28.0	25.1
Germany	26.3	26.4	26.9	27.0
Greece	23.9	27.2	27.2	26.6
Hungary	31.2	32.1	32.1	29.2
Iceland	28.1	28.1	28.1	28.1
India	26.3	26.3	26.3	26.3
Indonesia	26.7	28.2	28.2	28.2
Iran (Islamic Republic of)	26.9	25.8	25.8	28.1
Ireland	25.7	25.0	25.7	25.5
Israel	25.1	25.1	25.1	25.0
Italy	27.6	27.5	26.3	26.0
Japan	25.9	25.6	25.8	25.9
Jordan	26.7	26.7	26.7	26.7
Kazakhstan	18.6	18.6	18.6	18.6
Korea, Democratic People's Republic of	25.8	26.7	25.8	25.8
Korea, Republic of	25.8	25.6	25.6	25.5
Kyrgyzstan	18.6	18.6	20.9	20.9
Latvia	26.2	26.3	26.3	24.2
Lebanon	27.7	27.7	27.7	27.7
Lithuania	25.1	25.1	25.1	25.1
Luxembourg	24.8	24.7	24.8	24.7
Malaysia	26.4	26.4	26.4	26.4
Mauritius	26.0	26.0	26.0	25.8
Mexico	26.3	26.1	26.6	26.5
Moldova, Republic of	18.6	23.7	24.5	24.1

Table I (continued)

NET CALORIFIC VALUES OF ENERGY PRODUCTS
GJ/ton

IMPORTS

HARD COAL	2011	2012	2013	2014
Morocco	27.6	27.6	27.6	27.6
Namibia	22.7	22.7	22.7	22.7
Nepal	25.1	25.1	25.1	25.1
Netherlands	25.4	25.4	25.0	25.1
New Zealand	28.2	28.1	28.2	28.4
Norway	28.1	28.1	28.1	28.1
Other Asia	27.4	27.4	27.4	27.4
Pakistan	29.1	27.6	27.6	27.6
Peru	25.0	25.0	25.0	25.8
Poland	24.4	24.6	24.9	25.0
Portugal	25.0	25.2	24.9	24.8
Romania	26.5	26.5	26.3	25.8
Russian Federation	25.4	25.3	25.3	25.2
Serbia	23.0	23.0	23.3	25.6
Senegal	25.9	25.9	25.9	25.9
Slovakia	28.0	28.1	28.2	28.7
Slovenia	25.3	25.4	26.2	25.2
South Africa	31.0	31.0	31.0	31.0
Spain	24.4	24.0	24.4	23.9
Sri Lanka	26.4	29.3	29.3	29.3
Sweden	28.8	28.6	28.6	28.8
Switzerland	25.5	25.5	25.5	25.5
T.F. Yugoslav Republic of Macedonia	24.0	25.5	26.3	25.8
Tajikistan	18.6	18.6	18.6	18.6
Thailand	26.4	26.4	26.4	26.4
Turkey	27.1	27.5	27.4	26.9
Ukraine	25.5	27.3	25.5	26.7
United Kingdom	26.1	25.8	25.9	26.0
United States	26.2	26.1	26.1	26.2
Uruguay	29.3	25.8	25.8	25.8
Viet Nam	23.4	23.4	23.4	23.4
Zambia	24.7	24.7	24.7	24.7
Zimbabwe	27.0	27.0	27.0	27.0

BROWN COAL	2011	2012	2013	2014
Albania	9.8	9.8	9.8	9.8
Austria	15.6	19.4	20.0	20.4
Bosnia and Herzegovina	12.0	9.3	11.9	10.3
Cambodia	19.9	19.9	19.9	19.9
Canada	18.2	18.2	17.4	17.9
Croatia	17.6	17.1	15.4	15.9

Table I (continued)

NET CALORIFIC VALUES OF ENERGY PRODUCTS
GJ/ton

IMPORTS

BROWN COAL	2011	2012	2013	2014
Czechia	18.7	13.1	11.6	11.1
France	17.0	17.0	17.0	17.0
Germany	16.7	16.7	16.7	17.7
Greece	4.2	4.0	4.1	4.0
Hungary	17.7	17.5	16.9	17.4
Israel	20.4	20.8	20.8	20.8
Italy	18.6	18.8	18.8	18.8
Kazakhstan	14.7	14.7	14.7	14.7
Korea, Republic of	20.9	21.4	21.4	21.4
Mexico	14.1	14.1	13.9	13.9
Montenegro	9.2	9.2	9.2	9.2
Netherlands	20.0	20.0	20.0	20.0
New Zealand	20.6	20.6	20.6	20.5
Other Asia	18.9	18.9	18.9	18.9
Philippines	22.1	22.1	22.1	22.1
Poland	8.4	8.5	8.4	8.2
Romania	24.6	24.5	21.5	18.2
Russian Federation	14.9	14.9	14.9	14.9
Serbia	17.0	17.0	17.0	17.0
Singapore	9.7	9.7	9.7	9.7
Slovakia	14.7	13.5	14.3	15.0
Slovenia	18.4	18.4	17.4	18.5
Switzerland	20.1	20.1	20.1	23.6
T.F. Yugoslav Republic of Macedonia	18.6	20.0	14.4	17.0
Ukraine	10.0	10.0	10.0	10.0
United States	18.5	17.9	17.8	19.2
Uzbekistan	14.7	14.7	14.7	14.7

COKE OVEN COKE	2011	2012	2013	2014
Australia	27.0	27.0	27.0	27.0
Austria	29.0	29.0	29.0	29.0
Belarus	29.0	29.0	29.0	29.0
Belgium	29.3	29.3	29.3	29.3
Brazil	30.5	30.5	30.5	30.5
Canada	27.5	27.5	27.5	27.5
Chile	28.6	28.6	28.6	28.6
China	28.4	28.4	28.4	28.4
Croatia	29.3	29.3	29.3	29.3
Czechia	27.4	28.5	25.8	26.2
Denmark	29.3	29.3	29.3	29.3
Egypt	27.2	27.2	27.2	27.2
Finland	29.3	29.3	29.3	29.3
France	28.0	28.0	28.0	28.0
Germany	28.7	28.7	28.7	28.7

Table I (continued)

NET CALORIFIC VALUES OF ENERGY PRODUCTS
GJ/ton

IMPORTS

COKE OVEN COKE	2011	2012	2013	2014
Hungary	29.4	29.6	29.2	29.5
Iceland	26.7	26.7	26.7	26.7
Italy	29.0	29.0	29.0	29.0
Japan	29.4	29.4	29.2	29.2
Kazakhstan	25.1	25.1	25.1	25.1
Korea, Republic of	29.3	28.9	28.9	28.9
Luxembourg	28.5	28.5	28.5	28.5
Mexico	26.5	26.5	27.8	27.8
Netherlands	28.5	28.5	28.5	28.5
Norway	28.5	28.5	28.5	28.5
Other Asia	28.2	28.2	28.2	28.2
Peru	26.8	28.2	28.2	28.2
Poland	28.0	28.0	28.0	28.0
Portugal	29.6	29.6	29.5	29.5
Romania	26.4	26.4	26.4	26.3
Slovakia	28.2	28.0	27.5	27.8
Slovenia	29.5	29.6	29.9	30.1
Spain	26.8	26.8	26.8	26.8
Sweden	28.1	28.1	28.1	28.1
Switzerland	25.5	25.5	25.5	25.5
T.F. Yugoslav Republic of Macedonia	25.2	25.6	25.9	25.5
Turkey	27.9	28.2	29.7	27.6
United Kingdom	29.8	29.8	29.8	29.8
United States	28.8	28.8	28.9	28.9
Uruguay	28.5	28.5	28.5	28.5

PATENT FUEL	2011	2012	2013	2014
Austria	31.0	31.0	31.0	31.0
Belgium	29.3	30.5	30.5	30.5
France	32.0	32.0	32.0	32.0
Germany	31.4	31.4	31.4	31.4
Netherlands	29.3	29.3	29.3	29.3
Poland	23.2	23.2	23.2	23.2
Slovakia	28.0	28.0	28.0	28.0
United Kingdom	31.0	31.0	31.1	31.1

BROWN COAL BRIQUETTES	2011	2012	2013	2014
Austria	19.3	19.3	19.3	19.8
Belgium	20.0	20.7	20.7	20.7
Czechia	21.6	22.0	20.7	21.1
Denmark	18.3	18.3	18.3	18.3
Germany	21.3	20.9	20.1	25.6
Hungary	23.0	19.8	19.8	19.8
Ireland	19.8	19.8	19.8	19.8
Luxembourg	22.2	22.2	22.2	22.2

Table I (continued)

NET CALORIFIC VALUES OF ENERGY PRODUCTS
GJ/ton

IMPORTS

BROWN COAL BRIQUETTES	2011	2012	2013	2014
Netherlands	20.0	20.0	20.0	20.0
Poland	17.5	17.6	17.6	17.5
Serbia	15.9	15.9	15.9	15.8
Slovakia	17.0	17.0	17.0	17.7

PEAT	2011	2012	2013	2014
Finland	10.2	10.1	10.2	10.0
Romania	8.8	8.8	8.8	8.8
Russian Federation	10.0	10.0	10.0	10.0
Sweden	12.9	12.9	12.9	12.9
T.F. Yugoslav Republic of Macedonia	9.6	9.1	9.1	9.1
Ukraine	9.3	10.0	9.3	9.7

OTHER PRODUCTS

Conventional crude oil	42.3
Natural gas liquids (NGL)	44.2
Aviation gasoline	44.3
Motor gasoline	44.3
Gasoline-type jet fuel	44.3
Kerosene-type Jet fuel	44.1
Other kerosene	43.8
Gas oil/diesel oil	43.0
Fuel oil	40.4
Liquefied petroleum gases (LPG)	47.3
Ethane	46.4
Refinery gas	49.5
Feedstocks	43.0
Naphtha	44.5
White spirit and SBP industrial spirit	40.2
Lubricants	40.2
Bitumen	40.2
Paraffin waxes	40.2
Petroleum coke	32.5
Other oil products	40.2
Fuelwood, wood residues and by-products	12.6
Bagasse	7.7
Peat products	9.8
Charcoal	29.5

Table I (continued)

STANDARD HEAT VALUES OF GASES
kJ/m^3

Natural gas	39021
Coke oven gas	17590
Gas works gas	17590
Blast furnace gas	4000

Table II

SELECTED CONVERSION FACTORS FOR CRUDE OIL AND OIL PRODUCTS

(To convert from metric tons into the following units, multiply by the factor in the appropriate column)

Commodity	Litre	U.S. Gallon	Imperial Gallon	Barrel	Barrels / day	Cubic Metre
Conventional crude oil (average specific gravity)	1164	308	256	7.32	0.02005	1.164
Natural gas liquids (NGL)	1653	437	364	10.40	0.02849	1.653
Additives and oxygenates	1033	315	262	7.50	0.02055	1.033
Other hydrocarbons	1177	311	259	7.40	0.02027	1.177
Aviation gasoline	1370	362	301	8.62	0.02362	1.370
Motor gasoline	1351	357	297	8.50	0.02329	1.351
Gasoline-type jet fuel	1317	348	290	8.28	0.02270	1.317
Kerosene-type jet fuel	1235	326	272	7.77	0.02129	1.235
Other kerosene	1235	326	272	7.77	0.02129	1.235
Gas oil/diesel oil	1149	304	253	7.23	0.01981	1.149
Fuel oil	1053	278	232	6.62	0.01814	1.053
Liquefied petroleum gases (LPG)	1852	489	407	11.65	0.03192	1.852
Ethane	2679	708	589	16.85	0.04616	2.679
Refinery gas	1272	336	280	8.00	0.02192	1.272
Feedstocks	1177	311	259	7.40	0.02027	1.177
Naphtha	1389	367	306	8.74	0.02395	1.389
White spirit and SBP industrial spirits	1235	326	272	7.77	0.02129	1.235
Lubricants	1111	294	244	6.99	0.01915	1.111
Bitumen	962	254	212	6.05	0.01658	0.962
Paraffin waxes	1250	330	275	7.86	0.02153	1.250
Petroleum coke	877	232	193	5.52	0.01512	0.877
Other oil products	1099	290	241	6.91	0.01893	1.099

TABLES

Table 1

Total energy supply
Petajoules

Table Notes

For tables 1 to 3

Data on individual energy commodities are aggregated and presented in this table. As discussed in the introduction, this table follows the principles of energy balances as recommended in IRES.

Instead of breaking energy supply down by flow, Table 1 gives a longer historical perspective back to 1990. As such, historical countries that existed in 1990 are included here. Due to the break-up of the Former Union of Soviet Socialist Republics, there is a reallocation between Europe and Asia between 1991 and 1992.

Please refer to the Definitions Section on pages ix to xv for the appropriate product description/ classification.

Figure 1: Total energy supply by region, 1990

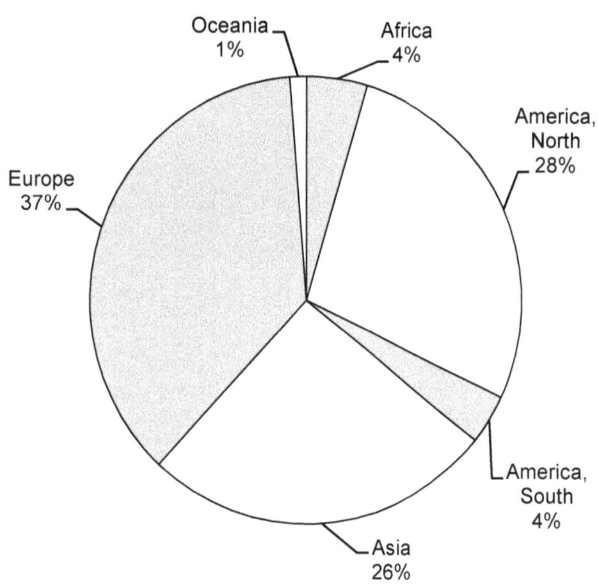

Figure 2: Total energy supply by region, 2014

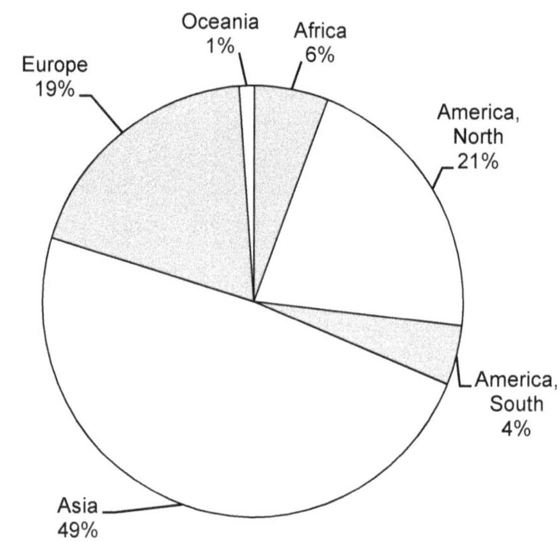

Figure 3: Total energy supply by region, 1990-2014

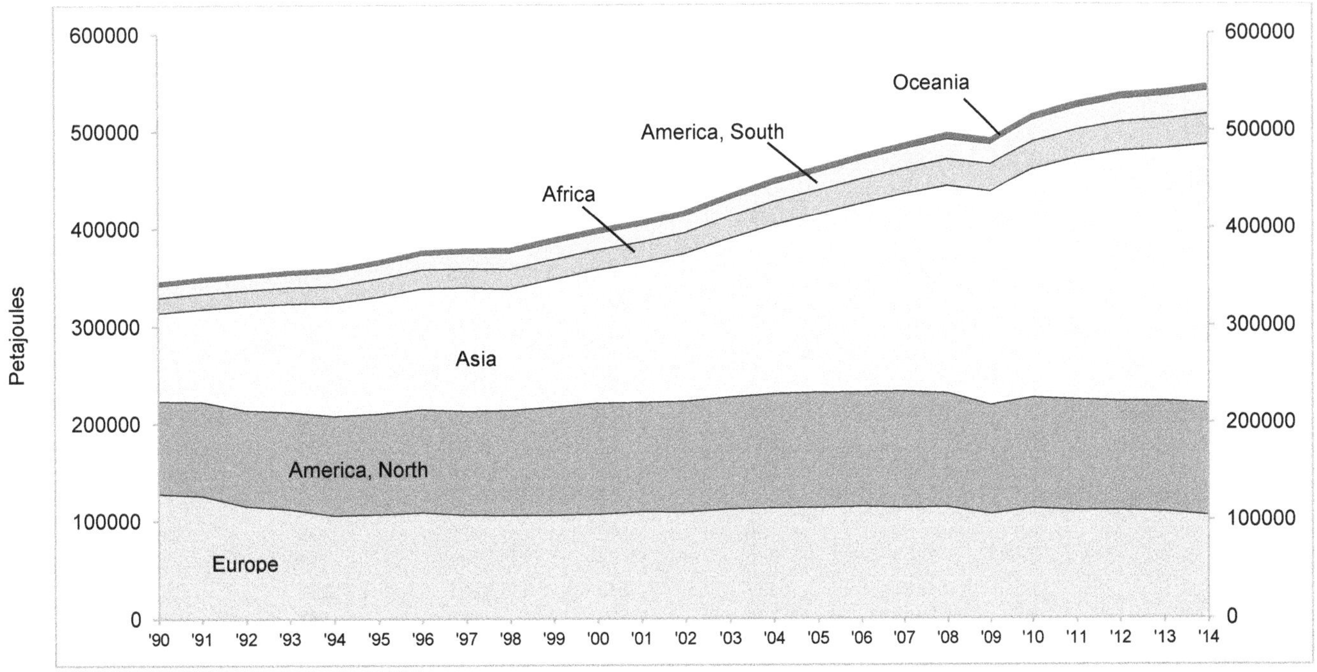

Table 1

Total Energy Supply

Petajoules

Country or area	1990	1995	2000	2005	2010	2013	2014
World	**346,313**	**368,048**	**400,413**	**463,050**	**516,833**	**541,770**	**547,533**
Africa	**15,556**	**18,759**	**20,886**	**24,442**	**28,218**	**30,501**	**31,211**
Algeria	1,075	1,270	1,170	1,583	1,641	1,959	2,123
Angola	238	343	308	363	509	581	613
Benin	74	89	83	105	153	170	179
Botswana	39	44	57	65	76	83	98
Burkina Faso	93	104	83	114	143	164	162
Burundi	58	66	55	81	89	60	62
Cabo Verde	*2	*3	*4	8	9	9	9
Cameroon	193	253	264	295	291	307	318
Central African Rep.	31	31	22	22	23	23	23
Chad	45	50	57	66	73	77	79
Comoros	2	3	3	*4	*5	*5	5
Congo	36	44	40	45	70	109	109
Côte d'Ivoire	204	245	278	418	425	566	580
Dem. Rep. of the Congo	477	584	635	825	831	1,151	1,202
Djibouti	*6	*7	8	*9	11	*11	*11
Egypt	1,271	1,605	2,288	2,742	3,242	3,428	3,259
Equatorial Guinea	5	6	9	52	69	69	72
Eritrea	..	43	30	32	31	33	34
Ethiopia	..	918	1,022	1,132	1,293	1,394	1,428
Ethiopia, incl. Eritrea	543	..	..	..	..	..	..
Gabon	50	56	62	72	93	107	109
Gambia	7	7	9	10	12	13	14
Ghana	187	277	271	251	267	367	372
Guinea	105	126	128	135	147	144	145
Guinea-Bissau	20	21	23	26	28	29	30
Kenya	289	337	388	443	*793	*868	909
Lesotho	*36	*41	*48	47	52	58	*59
Liberia	38	30	49	63	75	83	85
Libya	468	653	685	727	860	805	743
Madagascar	85	107	118	127	150	166	167
Malawi	64	65	85	90	98	101	102
Mali	43	48	57	60	64	66	73
Mauritania	*23	*26	29	36	45	*51	*51
Mauritius	33	41	49	57	63	63	65
Mayotte	*1	*2	*2	*3	4	*4	*4
Morocco	352	428	477	603	758	791	800
Mozambique	248	264	292	364	421	469	549
Namibia	..	38	42	53	65	59	74
Niger	62	73	81	*77	70	87	*91
Nigeria	2,780	3,082	3,599	4,402	5,015	5,606	5,637
Réunion	24	31	42	49	58	57	58
Rwanda	35	54	54	71	84	*95	*96
Sao Tome and Principe	*1	1	*2	*2	2	*3	*3
Senegal	*81	*87	96	114	165	155	165
Seychelles	2	*3	8	10	6	6	7
Sierra Leone	50	47	55	56	60	67	69
Somalia	68	77	90	107	126	128	129
South Africa	4,179	4,750	5,044	5,465	6,323	6,287	6,568
South Sudan	..	..	..	..	..	28	29
Sudan	..	..	..	..	..	617	626
Sudan (former)	451	506	556	654	711	..	..
Swaziland	20	21	41	40	44	48	50
Togo	55	71	89	99	131	135	139
Tunisia	192	220	293	349	427	430	435
Uganda	281	312	338	376	425	454	466
United Rep. of Tanzania	206	477	572	722	854	1,002	1,046

Total Energy Supply

Petajoules

Country or area	1990	1995	2000	2005	2010	2013	2014
Zambia	222	244	260	305	348	398	415
Zimbabwe	402	422	438	414	392	454	464
America, North	**95,554**	**103,281**	**114,127**	**118,427**	**113,970**	**113,417**	**115,244**
Anguilla	*1	*1	*1	*2	*2	*2	*2
Antigua and Barbuda	*4	*4	*5	*6	*8	*8	*8
Aruba	*23	*23	*26	30	*28	*12	*12
Bahamas	*27	*23	*23	*25	24	40	35
Barbados	16	13	16	18	20	19	17
Belize	8	10	11	11	13	13	13
Bermuda	*7	*8	*7	*8	*9	7	9
Bonaire, St Eustatius, Saba	..	..	..	..	..	*5	*5
British Virgin Islands	*1	*1	*1	*2	*2	*2	*3
Canada	8,752	9,684	10,504	11,229	10,896	11,168	11,530
Cayman Islands	4	*5	*7	7	8	8	8
Costa Rica	68	110	137	177	203	208	212
Cuba	628	452	462	412	566	524	531
Curaçao	..	..	..	..	..	77	85
Dominica	*1	1	2	2	*2	*2	*2
Dominican Republic	143	236	283	266	298	307	295
El Salvador	92	135	166	189	178	167	170
Greenland	8	*7	*8	9	11	9	9
Grenada	2	2	3	3	4	5	4
Guadeloupe	18	*21	26	32	*33	*37	*37
Guatemala	195	252	307	365	391	452	517
Haiti	64	71	84	142	159	171	174
Honduras	122	142	157	170	191	217	224
Jamaica	113	130	148	151	103	113	104
Martinique	21	*21	*22	29	27	*30	*31
Mexico	5,179	5,498	6,356	7,434	7,413	7,976	7,812
Montserrat	0	0	0	*1	1	1	1
Neth. Antilles (former)	82	*95	83	84	69	..	..
Nicaragua	86	102	107	119	125	148	154
Panama	62	62	104	120	143	161	136
Puerto Rico	*1	0	13	24	27	56	60
Sint Maarten (Dutch part)	..	..	..	..	..	*11	*11
St. Kitts-Nevis	*2	*2	*3	*3	3	*3	*3
St. Lucia	2	5	5	*5	6	6	6
St. Pierre-Miquelon	1	1	*1	*1	*1	*1	*1
St. Vincent-Grenadines	*1	*2	2	*3	3	3	*3
Trinidad and Tobago	251	258	414	678	845	824	822
Turks and Caicos Islands	0	*1	*1	*2	*3	*3	*3
United States	79,568	85,901	94,630	96,666	92,155	90,619	92,198
America, South	**12,401**	**14,527**	**16,804**	**18,897**	**22,192**	**23,739**	**24,404**
Argentina	1,928	2,194	2,556	2,919	3,307	3,379	3,638
Bolivia (Plur. State of)	105	149	206	220	263	319	341
Brazil	5,960	7,062	8,020	9,252	10,916	12,149	12,549
Chile	587	772	1,073	1,178	1,278	1,613	1,505
Colombia	983	1,104	1,103	1,180	1,390	1,511	1,505
Ecuador	255	339	328	475	562	623	663
Falkland Is. (Malvinas)	1	*1	*1	*1	*1	*1	*1
French Guiana	9	*9	8	10	*11	*11	*13
Guyana	24	29	32	29	32	34	35
Paraguay	*84	118	153	180	234	237	247
Peru	396	444	542	565	773	830	914
Suriname	30	34	36	26	40	31	34
Uruguay	96	107	119	126	173	197	199
Venezuela (Bolivar. Rep.)	1,944	2,167	2,629	2,737	3,211	2,803	2,760

Table 1

Total Energy Supply

Petajoules

Country or area	1990	1995	2000	2005	2010	2013	2014
Asia	**90,689**	**120,093**	**136,481**	**182,822**	**234,205**	**259,164**	**265,722**
Afghanistan	*46	*29	25	36	137	158	154
Armenia	..	68	84	105	*119	138	139
Azerbaijan	..	536	485	573	486	567	584
Bahrain	213	257	324	323	517	556	563
Bangladesh	785	921	1,001	1,189	1,485	1,623	1,687
Bhutan	37	39	44	48	59	63	63
Brunei Darussalam	107	71	73	76	136	127	149
Cambodia	121	119	142	144	222	250	267
China	30,390	39,163	42,461	68,833	101,525	118,296	119,578
China, Hong Kong SAR	316	398	570	579	550	585	599
China, Macao SAR	15	18	24	28	33	33	36
Cyprus	54	69	87	92	103	82	83
Georgia	..	59	93	135	140	167	188
India	10,370	13,035	16,135	22,809	28,903	33,195	35,327
Indonesia	3,759	4,587	4,970	7,087	8,330	8,966	8,801
Iran (Islamic Rep. of)	2,819	3,887	5,386	7,221	8,579	9,205	9,958
Iraq	546	1,072	995	1,488	1,437	2,049	2,054
Israel	480	649	763	767	966	963	945
Japan	18,318	20,696	21,678	21,774	20,863	18,982	18,477
Jordan	133	172	205	289	303	328	350
Kazakhstan	..	2,185	1,560	2,352	3,363	3,550	3,352
Korea, Dem.Ppl's.Rep.	1,391	921	826	896	795	459	505
Korea, Republic of	3,851	6,083	7,854	8,764	10,441	11,007	11,164
Kuwait	746	835	824	1,133	1,350	1,475	1,427
Kyrgyzstan	..	101	101	114	115	167	159
Lao People's Dem. Rep.	55	59	68	74	75	79	78
Lebanon	*113	173	207	206	262	289	306
Malaysia	773	1,768	1,958	2,632	2,955	3,417	3,535
Maldives	2	4	6	9	13	16	19
Mongolia	116	89	87	104	168	444	245
Myanmar	449	498	538	619	656	683	805
Nepal	154	297	349	388	446	483	542
Oman	153	236	322	445	778	1,018	1,014
Other Asia	1,947	2,623	3,413	4,242	4,651	4,517	4,601
Pakistan	1,290	1,722	2,082	2,642	2,946	3,056	3,101
Philippines	1,039	1,264	1,551	1,469	1,572	1,780	1,899
Qatar	199	553	584	728	1,165	1,395	1,837
Saudi Arabia	2,875	3,516	4,487	6,047	7,707	7,985	8,876
Singapore	656	640	756	805	1,075	1,125	1,151
Sri Lanka	187	225	296	324	360	388	414
State of Palestine	*20	*22	38	58	54	61	69
Syrian Arab Republic	476	585	729	740	911	530	454
Tajikistan	..	95	91	99	94	106	119
Thailand	1,784	2,635	3,075	4,067	4,945	5,359	5,643
Timor-Leste	..	..	..	*4	*4	*7	*8
Turkey	2,191	2,598	3,202	3,549	4,419	4,874	5,052
Turkmenistan	..	574	625	805	951	1,100	1,123
United Arab Emirates	791	1,108	1,719	1,842	2,587	2,805	3,326
Uzbekistan	..	1,788	2,130	2,050	1,809	1,799	1,829
Viet Nam	791	908	1,262	1,747	2,318	2,510	2,759
Yemen	..	142	197	274	326	344	309
Yemen Arab Rep. (former)	*47	..	..	..	..	..	..
Yemen, Dem. (former)	*84	..	..	..	..	..	..
Europe	**127,755**	**106,665**	**106,646**	**112,731**	**111,737**	**108,471**	**104,490**
Albania	92	47	65	91	91	102	103
Andorra	*7	7	9	10	10	9	9
Austria	1,038	1,119	1,197	1,421	1,426	1,391	1,346

Total Energy Supply

Petajoules

Country or area	1990	1995	2000	2005	2010	2013	2014
Belarus	..	1,043	1,029	1,125	1,155	1,142	1,162
Belgium	2,005	2,237	2,426	2,436	2,524	2,332	2,203
Bosnia and Herzegovina	..	56	179	206	269	266	325
Bulgaria	1,181	966	773	826	741	703	745
Croatia	..	325	349	407	390	352	335
Czechia	..	1,753	1,741	1,885	1,860	1,751	1,725
Czechoslovakia (former)	2,960	..	..	..	..	..	..
Denmark	725	811	785	786	812	729	673
Estonia	..	217	196	220	236	258	254
Faeroe Islands	10	*8	*10	*10	*9	*10	*9
Finland	1,135	1,170	1,348	1,436	1,527	1,386	1,408
France	9,381	9,924	10,537	11,332	10,929	10,584	10,149
German Dem. R. (former)	*3,676	..	..	..	..	..	..
Germany	..	14,037	14,034	14,009	13,656	13,261	12,776
Germany, Fed. R. (former)	11,391	..	..	..	..	..	..
Gibraltar	2	4	5	6	7	8	8
Greece	902	963	1,148	1,280	1,169	989	983
Guernsey	..	..	..	1	1	0	1
Hungary	1,211	1,093	1,049	1,155	1,079	947	959
Iceland	87	90	139	158	278	309	312
Ireland	416	446	573	610	608	546	535
Italy	6,366	6,714	7,230	7,847	7,324	6,530	6,168
Jersey	..	1	1	3	3	3	3
Latvia	..	197	157	190	203	196	196
Liechtenstein	..	..	..	..	3	3	3
Lithuania	..	376	306	362	286	282	287
Luxembourg	143	133	142	185	178	168	161
Malta	28	29	28	37	35	32	33
Montenegro	..	..	..	42	48	42	40
Netherlands	2,723	3,053	3,103	3,352	3,447	3,180	3,004
Norway	876	964	1,075	1,117	1,425	1,365	1,199
Poland	4,406	4,256	3,780	3,886	4,251	4,122	3,941
Portugal	720	840	1,026	1,104	975	888	874
Republic of Moldova	..	183	66	89	91	93	95
Romania	2,620	1,925	1,532	1,626	1,474	1,342	1,331
Russian Federation	..	26,720	25,994	27,338	28,932	30,634	29,886
Serbia	..	..	..	672	648	618	550
Serbia and Montenegro	..	522	556	..	..	..	..
Slovakia	..	746	744	778	737	700	658
Slovenia	..	254	269	306	306	287	280
Spain	3,744	4,191	5,074	5,915	5,337	4,841	4,732
Sweden	1,967	2,089	1,971	2,137	2,112	2,050	1,998
Switzerland	1,017	1,003	1,041	1,080	1,091	1,113	1,043
T.F.Yug.Rep. Macedonia	..	151	169	163	121	113	110
Ukraine	..	7,073	5,574	5,872	5,491	4,844	4,431
United Kingdom	8,527	8,925	9,216	9,219	8,442	7,951	7,448
USSR (former)	56,587	..	..	..	..	..	..
Yugoslavia, SFR (former)	1,809	..	..	..	..	..	..
Oceania	**4,358**	**4,723**	**5,468**	**5,732**	**6,512**	**6,479**	**6,462**
Australia	3,650	3,928	4,540	4,762	5,431	5,333	5,252
Cook Islands	0	*1	*1	1	*1	*1	*1
Fiji	21	*23	22	24	22	*23	*23
French Polynesia	*7	7	9	12	13	*12	*12
Kiribati	0	0	0	*1	*1	*1	*1
Marshall Islands	..	*1	*1	*1	*1	*1	*1
Micronesia (Fed. States of)	..	*2	*2	2	2	*2	*2
Nauru	*2	*1	*1	*1	*1	*1	*1
New Caledonia	21	27	30	37	46	56	56
New Zealand	555	629	747	747	834	871	931

Table 1

Total Energy Supply

Petajoules

Country or area	1990	1995	2000	2005	2010	2013	2014
Pacific Islands (former)	*3	..	..	..	..	..	..
Palau	..	*3	*4	*4	*4	*4	*4
Papua New Guinea	88	89	99	126	141	160	*161
Samoa	*3	*3	*3	*4	*5	5	*5
Solomon Islands	*5	*5	*5	*6	*6	*6	*6
Tonga	1	1	1	*2	2	2	2
Vanuatu	1	1	*2	2	3	3	3

Table 2

Components of total energy supply

Petajoules and Gigajoules per capita

Table Notes

Included in the **Primary energy production** for **Solids** are hard coal, lignite, peat, oil shale, primary solid biofuels and wastes; **Liquids** are comprised of crude oil, natural gas liquids, other hydrocarbons, additives and oxygenates, and liquid biofuels; **Gas** comprises natural gas and biogas; **Electricity** is comprised of primary electricity generation from hydro, wind, tide, wave and solar sources; and **Heat** comprises primary heat from nuclear, geothermal, chemical and solar sources.

Stock changes, trade and International bunkers refer to all primary and secondary forms of energy.

Supply is calculated as primary energy production plus import minus exports minus International bunkers minus stock changes.

Please refer to the Definitions Section on pages ix to xv for the appropriate product description/ classification

Figure 4: Total per capita energy supply by region, selected years

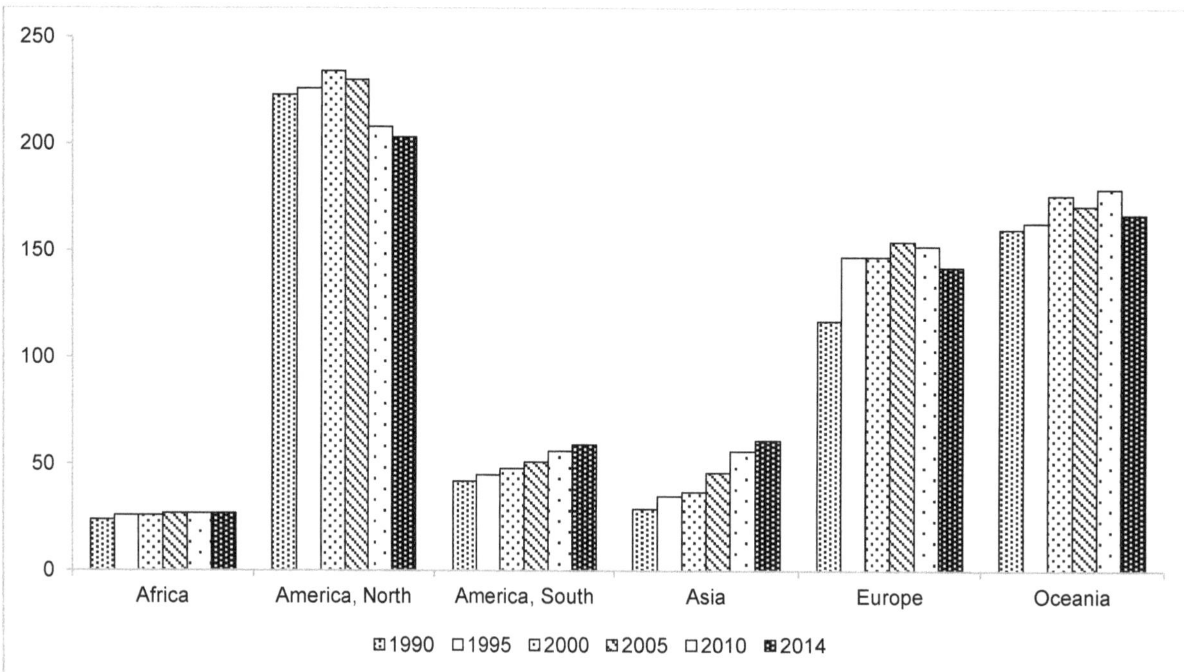

Figure 5: Production of solids by region, 2014

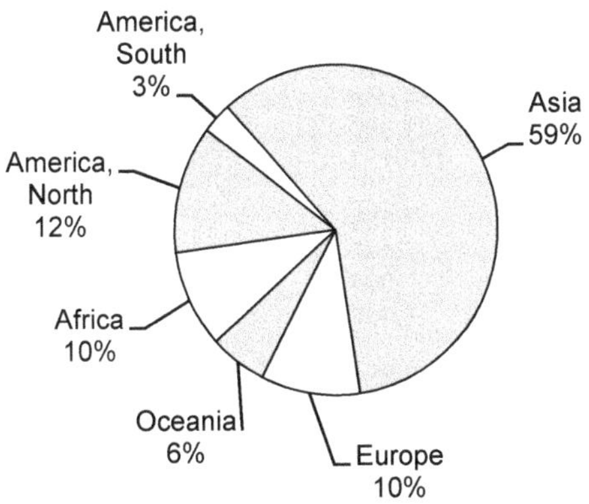

Figure 6: Production of liquids by region, 2014

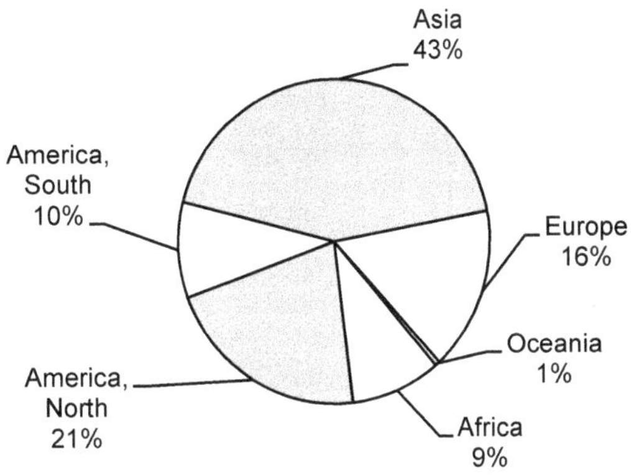

Figure 7: Production of gases by region, 2014

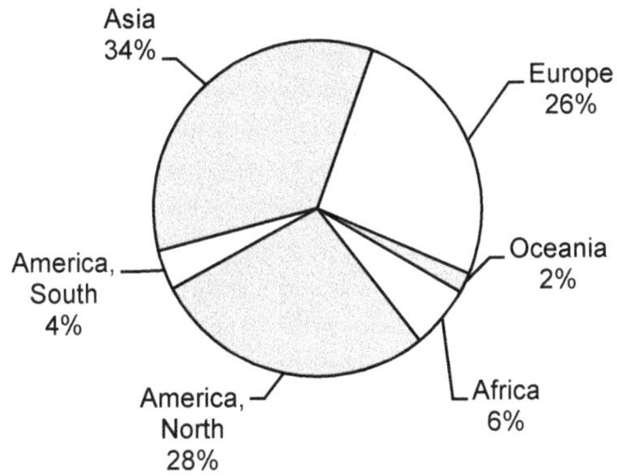

Figure 8: Primary electricity production by region, 2014

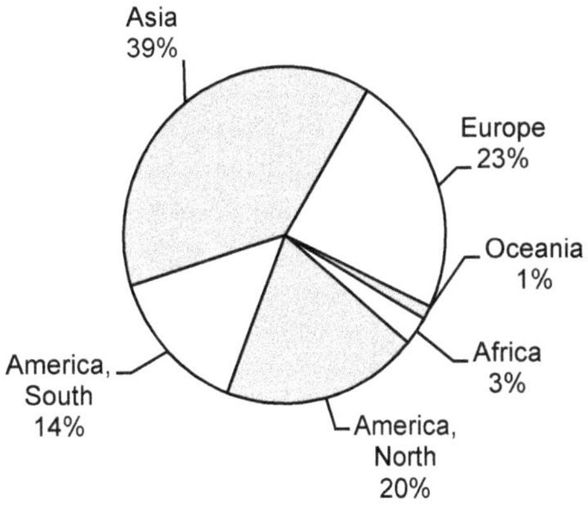

Table 2

Components of Total Energy Supply

Petajoules and Gigajoules per capita

Country or area	Year	Primary energy production					
		Total	Solids	Liquids	Gases	Electricity	Heat
World	**2011**	**545,218**	**208,757**	**172,880**	**118,363**	**14,299**	**30,919**
	2012	**554,058**	**212,090**	**176,590**	**120,268**	**15,404**	**29,706**
	2013	**560,493**	**214,201**	**177,877**	**121,890**	**16,492**	**30,032**
	2014	**567,417**	**214,774**	**181,441**	**123,123**	**17,241**	**30,837**
Africa	**2011**	**44,681**	**19,166**	**17,767**	**7,129**	**411**	**207**
	2012	**47,122**	**19,892**	**19,174**	**7,447**	**415**	**194**
	2013	**45,731**	**20,187**	**17,869**	**7,006**	**442**	**227**
	2014	**45,612**	**20,566**	**16,850**	**7,460**	**471**	**265**
Algeria	2011	6,008	1	3,091	2,914	2	..
	2012	5,928	1	2,888	3,037	2	..
	2013	5,676	1	2,789	2,885	1	..
	2014	5,900	0	2,959	2,940	1	..
Angola	2011	3,770	283	3,447	26	14	..
	2012	3,967	288	3,639	26	14	..
	2013	4,001	294	3,658	32	17	..
	2014	3,902	300	3,559	25	18	..
Benin	2011	89	89	..	..	..	..
	2012	91	91	..	..	..	..
	2013	94	94	..	..	..	..
	2014	96	96	..	..	..	..
Botswana	2011	25	25	..	..	..	..
	2012	41	41	..	..	0	..
	2013	42	42	..	..	0	..
	2014	47	47	..	..	0	..
Burkina Faso	2011	120	119	..	..	0	..
	2012	121	121	..	..	0	..
	2013	123	123	..	..	0	..
	2014	123	122	..	..	0	..
Burundi	2011	87	86	..	..	0	..
	2012	54	53	..	..	0	..
	2013	56	56	..	..	0	..
	2014	56	56	..	..	1	..
Cabo Verde	2011	2	1	..	..	0	..
	2012	2	1	..	..	0	..
	2013	2	2	..	..	0	..
	2014	2	2	..	..	0	..
Cameroon	2011	341	190	127	8	16	..
	2012	352	195	131	9	17	..
	2013	376	200	142	16	17	..
	2014	408	205	161	23	18	..
Central African Rep.	2011	19	18	..	..	*1	..
	2012	19	18	..	..	*1	..
	2013	19	18	..	..	*1	..
	2014	19	18	..	..	*1	..
Chad	2011	309	66	243	..	..	..
	2012	288	67	220	..	..	..
	2013	275	69	207	..	..	..
	2014	288	70	218	..	..	..
Comoros	2011	2	2	..	..	..	..
	2012	3	3	..	..	..	..
	2013	3	3	..	..	..	..
	2014	3	3	..	..	..	..
Congo	2011	712	56	648	5	3	..
	2012	676	59	606	7	4	..
	2013	612	61	541	8	3	..
	2014	648	62	574	8	3	..

Components of Total Energy Supply

Petajoules and Gigajoules per capita

Imports	Exports	International bunkers		Stock changes	Supply	
		Aviation	Marine		Total	Per capita
211,558	**209,143**	**6,832**	**8,581**	**2,871**	**529,349**	**75**
213,606	**213,766**	**6,944**	**7,752**	**795**	**538,409**	**76**
216,504	**219,681**	**7,068**	**8,024**	**454**	**541,770**	**75**
215,540	**217,039**	**7,138**	**8,116**	**3,131**	**547,533**	**75**
5,487	**20,470**	**295**	**231**	**201**	**28,971**	**27**
6,051	**22,710**	**320**	**263**	**-19**	**29,898**	**27**
6,460	**21,192**	**300**	**247**	**-49**	**30,501**	**27**
6,852	**20,479**	**296**	**270**	**208**	**31,211**	**27**
114	4,375	20	11	1	1,715	47
221	4,231	22	11	-5	1,890	50
240	3,920	23	11	3	1,959	51
153	3,856	18	11	44	2,123	55
150	3,375	9	6	1	529	24
197	3,559	8	15	3	580	26
203	3,605	9	10	0	581	25
241	3,496	10	15	8	613	25
84	9	7	..	0	157	16
88	10	7	..	0	163	16
94	10	8	..	0	170	16
102	11	8	..	0	179	17
50	3	1	..	0	71	34
52	3	1	..	18	73	34
53	2	1	..	10	83	38
50	5	0	..	-7	98	44
31	..	*2	..	..	149	9
36	0	*2	..	..	156	9
44	0	*2	..	..	164	10
41	0	*2	..	..	162	9
4	..	0	..	0	91	9
4	0	0	..	0	58	6
4	0	0	..	0	60	6
6	0	0	..	0	62	6
13	..	2	3	..	10	21
11	..	2	2	..	9	17
11	..	2	2	..	9	17
12	..	2	3	..	9	17
80	134	3	2	0	281	13
83	139	3	2	0	291	13
87	151	3	2	0	307	14
89	173	3	2	..	318	14
*5	0	*1	..	..	23	5
*6	0	*1	..	..	23	5
*6	0	*1	..	..	23	5
*6	0	*1	..	..	23	5
*5	239	*1	..	..	74	6
*5	217	*1	..	..	75	6
*5	202	*1	..	..	77	6
*5	*213	*1	..	..	79	6
*2	..	0	..	..	4	6
*2	..	0	..	..	5	6
*3	..	0	..	..	*5	*7
*2	..	0	..	..	5	6
12	594	2	..	35	93	22
9	577	2	..	0	106	25
13	527	2	..	-12	109	25
14	537	2	..	14	109	24

Table 2

Components of Total Energy Supply

Petajoules and Gigajoules per capita

Country or area	Year	Primary energy production					
		Total	Solids	Liquids	Gases	Electricity	Heat
Côte d'Ivoire	2011	514	382	71	55	6	..
	2012	523	394	62	60	6	..
	2013	540	413	53	69	6	..
	2014	540	423	40	69	7	..
Dem. Rep. of the Congo	2011	902	822	51	..	28	..
	2012	1,112	1,034	51	..	27	..
	2013	1,149	1,071	48	..	30	..
	2014	1,179	1,103	45	..	32	..
Djibouti	2011	3	3	..	..	..	..
	2012	3	3	..	..	..	..
	2013	3	3	..	..	..	..
	2014	3	3	..	..	..	..
Egypt	2011	3,716	218	1,455	1,990	54	..
	2012	3,669	220	1,485	1,912	53	..
	2013	3,448	222	1,464	1,709	53	..
	2014	3,509	221	1,300	1,931	56	..
Equatorial Guinea	2011	761	4	538	219	0	..
	2012	819	4	586	228	*1	..
	2013	801	4	575	221	*1	..
	2014	841	4	604	232	*1	..
Eritrea	2011	25	25	..	..	0	..
	2012	25	25	..	..	0	..
	2013	26	26	..	..	0	..
	2014	27	27	..	..	0	..
Ethiopia	2011	1,236	1,213	0	..	23	0
	2012	1,259	1,231	0	..	27	1
	2013	1,285	1,252	0	..	31	1
	2014	1,308	1,273	0	..	35	1
Gabon	2011	596	51	529	13	3	..
	2012	604	52	535	13	3	..
	2013	558	53	491	11	3	..
	2014	559	*53	491	12	3	..
Gambia	2011	6	6	..	..	..	..
	2012	7	7	..	..	..	..
	2013	7	7	..	..	..	..
	2014	7	7	..	..	..	..
Ghana	2011	299	128	144	..	27	..
	2012	347	143	175	..	29	..
	2013	401	149	223	..	30	..
	2014	410	152	225	2	30	..
Guinea	2011	112	110	..	0	1	..
	2012	112	111	..	0	2	..
	2013	113	111	..	0	2	..
	2014	113	112	..	0	*2	..
Guinea-Bissau	2011	24	24	..	..	..	..
	2012	24	24	..	..	..	..
	2013	25	25	..	..	..	..
	2014	25	25	..	..	..	..
Kenya	2011	*662	*598	..	..	12	52
	2012	*690	*621	..	..	15	55
	2013	*719	*638	..	..	16	64
	2014	750	632	..	..	13	105
Lesotho	2011	30	29	..	..	2	..
	2012	30	29	..	..	2	..
	2013	31	29	..	..	2	..
	2014	31	29	..	..	2	..

Components of Total Energy Supply

Petajoules and Gigajoules per capita

Imports	Exports	International bunkers		Stock changes	Supply	
		Aviation	Marine		Total	Per capita
103	128	2	0	-1	487	24
151	144	3	1	-2	528	25
159	132	2	1	-2	566	26
149	106	3	2	-3	580	26
39	52	6	..	..	882	13
40	51	6	..	..	1,095	16
57	48	7	..	..	1,151	16
78	45	10	..	..	1,202	16
10	*1	4	*1	*-2	10	12
*13	*1	*4	*1	0	*11	*13
*13	*1	*4	*1	0	*11	*13
*13	*1	*4	*1	0	*11	*13
493	629	34	16	-6	3,536	42
560	733	36	15	-5	3,450	40
681	666	25	10	0	3,428	39
531	735	30	64	-48	3,259	36
*10	*698	*2	..	..	72	96
*10	*751	*2	..	..	76	98
*10	*741	*2	..	..	69	86
*10	*778	*2	..	..	72	87
7	..	0	..	0	32	7
7	..	0	..	0	33	7
7	..	0	..	0	33	7
8	..	0	..	0	34	7
102	1	12	..	-2	1,327	15
110	2	14	..	0	1,352	15
129	3	15	..	1	1,394	15
142	4	17	..	1	1,428	15
17	502	3	8	1	99	63
22	513	3	8	-1	102	63
27	467	3	8	0	107	65
28	468	3	8	..	109	64
7	0	*1	..	..	13	7
7	0	*1	..	..	13	7
7	0	*1	..	..	13	7
8	0	*1	..	..	14	7
188	195	6	3	0	283	11
177	192	6	3	0	323	13
198	220	6	2	5	367	14
202	230	5	0	4	372	14
38	0	*1	..	..	149	13
36	0	*1	..	..	148	13
32	0	*1	..	..	144	12
*33	0	*1	..	..	145	12
*5	0	0	..	..	28	17
*5	0	0	..	..	29	17
*5	0	0	..	..	29	17
*5	0	*1	..	..	30	17
199	8	30	1	-1	*823	*20
184	3	30	1	0	*841	*20
231	57	24	1	-1	*868	*20
591	422	9	1	0	909	20
*26	0	0	..	..	56	28
*27	0	0	..	..	58	28
*27	0	0	..	..	58	28
*28	0	0	..	..	59	28

Table 2

Components of Total Energy Supply

Petajoules and Gigajoules per capita

Country or area	Year	Primary energy production					
		Total	Solids	Liquids	Gases	Electricity	Heat
Liberia	2011	66	66	..	..	..	..
	2012	69	69	..	..	..	..
	2013	71	71	..	..	..	..
	2014	73	73	..	..	..	..
Libya	2011	1,352	7	1,077	269	..	..
	2012	3,590	6	3,166	417	..	..
	2013	2,635	6	2,191	438	..	..
	2014	1,505	6	1,073	426	..	..
Madagascar	2011	124	121	..	..	2	..
	2012	124	121	..	..	3	..
	2013	127	124	..	..	3	..
	2014	129	126	..	..	3	..
Malawi	2011	87	81	..	..	6	..
	2012	88	81	..	..	7	..
	2013	88	81	..	..	7	..
	2014	89	82	..	..	*7	..
Mali	2011	53	50	..	..	3	..
	2012	53	51	..	..	3	..
	2013	53	49	..	..	4	..
	2014	55	52	..	..	*3	..
Mauritania	2011	33	17	16	..	..	..
	2012	31	18	14	..	..	..
	2013	31	18	13	..	..	..
	2014	31	18	13	..	..	..
Mauritius	2011	11	11	..	..	0	..
	2012	10	10	..	..	0	..
	2013	11	10	..	0	0	..
	2014	10	10	..	..	0	..
Mayotte	2011	0	0	..	..	0	0
	2012	0	0	..	..	0	0
	2013	0	0	..	..	0	0
	2014	0	0	..	..	0	0
Morocco	2011	76	63	0	2	10	..
	2012	73	62	0	2	9	..
	2013	81	63	0	3	15	..
	2014	78	62	0	4	13	..
Mozambique	2011	548	369	1	118	61	..
	2012	671	483	1	132	55	..
	2013	734	531	2	147	53	..
	2014	779	558	2	160	59	..
Namibia	2011	19	13	..	..	6	0
	2012	18	13	..	..	5	0
	2013	19	13	..	..	6	0
	2014	19	14	..	..	5	0
Niger	2011	62	58	4	..	0	..
	2012	88	61	27	..	0	..
	2013	101	64	38	..	0	..
	2014	*101	*65	36	..	0	..
Nigeria	2011	10,811	4,281	5,198	1,311	21	..
	2012	11,097	4,522	5,145	1,409	20	..
	2013	10,656	4,549	4,817	1,271	19	..
	2014	10,851	4,549	4,832	1,451	19	..
Réunion	2011	8	5	..	0	2	1
	2012	8	5	..	0	3	1
	2013	9	5	..	0	3	1
	2014	8	5	..	0	2	1

Components of Total Energy Supply

Petajoules and Gigajoules per capita

Imports	Exports	International bunkers		Stock changes	Supply	
		Aviation	Marine		Total	Per capita
13	..	0	*1	..	79	19
15	..	0	*1	..	82	20
13	..	0	*1	..	83	19
12	..	0	*1	..	85	19
165	809	2	3	133	569	90
206	3,074	9	14	-17	716	114
236	2,235	7	10	-185	805	129
341	1,020	5	4	74	743	119
33	0	3	*1	-1	154	7
39	0	3	*1	1	158	7
42	0	2	*1	0	166	7
*42	0	*2	*1	0	167	7
13	0	0	..	..	99	7
12	0	0	..	..	100	6
13	0	0	..	..	101	6
13	0	0	..	..	102	6
*16	*1	*1	..	..	66	4
*15	*1	*1	..	..	66	4
*15	*1	*1	..	..	66	4
*21	*1	*1	..	..	73	4
31	6	1	..	*11	47	13
34	15	1	..	*-1	51	13
*34	*13	1	..	0	*51	*13
*34	*13	1	..	..	*51	*13
67	..	5	12	-1	62	49
68	..	5	11	1	62	49
71	..	5	11	2	63	50
70	..	5	12	*-2	65	51
4	..	0	0	0	4	19
*4	..	0	0	..	*4	*19
*4	..	0	0	..	*4	*19
*4	..	0	0	..	*4	*19
752	25	28	*1	-4	778	24
824	23	41	*1	10	824	25
841	58	25	6	42	791	23
869	50	28	6	63	800	23
68	163	3	..	11	440	18
66	245	3	..	42	447	17
88	280	3	..	70	469	18
110	300	5	..	35	549	20
47	4	*1	..	0	61	27
54	4	*2	..	0	66	29
45	4	*1	..	0	59	25
61	5	2	..	..	74	31
14	0	1	..	2	73	4
4	7	1	..	0	84	5
4	16	1	..	1	87	5
4	*12	1	..	*1	*91	*5
341	5,803	8	16	20	5,305	32
380	5,930	8	17	-65	5,588	33
361	5,377	15	15	3	5,606	32
349	5,560	13	16	-26	5,637	32
56	..	*3	0	1	60	71
52	..	*2	0	-2	59	70
50	..	*2	0	0	57	67
51	..	*2	0	-1	58	68

Table 2

Components of Total Energy Supply

Petajoules and Gigajoules per capita

Country or area	Year	Primary energy production					
		Total	Solids	Liquids	Gases	Electricity	Heat
Rwanda	2011	*78	*78	..	0	1	..
	2012	*81	*80	..	0	1	..
	2013	*83	*83	..	0	1	..
	2014	*84	*83	..	0	1	..
Sao Tome and Principe	2011	1	1	..	..	0	..
	2012	1	1	..	..	0	..
	2013	1	1	..	..	0	..
	2014	1	1	..	..	0	..
Senegal	2011	90	87	..	1	..	1
	2012	92	89	..	1	..	1
	2013	75	73	..	2	0	1
	2014	77	75	..	2	0	1
Seychelles	2011	0	0	..	..	..	..
	2012	0	0	..	..	..	..
	2013	0	0	..	..	0	..
	2014	0	0	..	..	0	..
Sierra Leone	2011	52	52	..	..	1	..
	2012	52	52	..	..	0	..
	2013	53	52	..	..	0	..
	2014	53	53	..	..	0	..
Somalia	2011	128	128	..	..	..	..
	2012	128	128	..	..	..	..
	2013	128	128	..	..	..	..
	2014	129	129	..	..	..	..
South Africa	2011	6,881	6,654	24	47	8	149
	2012	6,995	6,804	14	40	5	132
	2013	6,999	6,783	12	43	5	156
	2014	7,102	6,888	13	36	11	153
South Sudan	2011	..	..	..	..	..	..
	2012	74	8	66	..	0	..
	2013	217	8	210	..	0	..
	2014	337	8	329	..	0	..
St. Helena and Depend.	2011	0	0	..	..	0	..
	2012	0	0	..	..	0	..
	2013	0	0	..	..	0	..
	2014	0	0	..	..	0	..
Sudan	2011	..	..	..	..	..	..
	2012	619	379	216	..	24	..
	2013	679	387	261	..	30	..
	2014	682	396	254	..	32	..
Sudan (former)	2011	1,467	487	957	..	23	..
	2012	..	..	..	..	..	..
	2013	..	..	..	..	..	..
	2014	..	..	..	..	..	..
Swaziland	2011	34	33	..	..	1	..
	2012	37	36	..	..	1	..
	2013	38	37	..	..	1	..
	2014	39	38	..	..	1	..
Togo	2011	102	101	..	..	1	..
	2012	105	104	..	..	1	..
	2013	108	107	..	..	0	..
	2014	111	110	..	..	0	..
Tunisia	2011	312	39	146	122	1	4
	2012	309	40	145	119	1	5
	2013	298	40	134	118	2	5
	2014	274	40	120	108	2	4

Components of Total Energy Supply

Petajoules and Gigajoules per capita

Imports	Exports	International bunkers		Stock changes	Supply	
		Aviation	Marine		Total	Per capita
*10	0	*1	..	..	*88	*8
11	0	*1	..	..	*91	*8
12	0	*1	..	..	*95	*9
13	0	*1	..	..	*96	*8
*2	..	0	..	..	*2	*14
*2	..	0	..	..	*3	*15
*2	..	0	..	..	*3	*14
*2	..	0	..	..	*3	*14
97	7	9	4	-4	172	13
90	4	9	3	-2	168	12
100	7	11	3	-2	155	11
110	8	12	3	-1	165	11
12	..	2	6	..	5	50
14	..	1	7	..	6	64
14	..	1	7	..	6	59
15	..	*1	*7	..	7	72
11	..	0	..	..	63	11
13	..	0	..	..	65	11
15	..	0	..	..	67	11
17	..	0	..	..	69	11
*11	9	2	*1	..	128	13
*11	8	2	*1	..	128	13
*11	8	2	*1	..	128	12
*11	8	2	*1	..	129	12
1,373	1,764	39	131	7	6,313	121
1,428	1,964	38	145	6	6,269	119
1,450	1,966	38	140	18	6,287	118
1,449	1,848	37	109	-12	6,568	122
..	..	..	..	..	..	..
20	66	1	..	..	27	2
21	209	2	..	..	28	2
22	328	2	..	..	29	2
0	..	..	..	..	0	37
0	..	..	..	..	0	41
0	..	..	..	..	0	40
0	..	..	..	..	0	*41
..	..	..	..	..	..	..
48	67	10	1	0	590	16
49	103	*8	1	*-1	617	16
46	90	11	1	..	626	16
49	786	14	1	..	716	15
..	..	..	..	..	..	..
..	..	..	..	..	..	..
..	..	..	..	..	..	..
*16	7	0	..	0	43	36
*16	*6	0	..	0	46	38
*18	*8	0	..	0	48	38
17	6	0	..	0	50	40
30	..	3	1	-3	131	20
28	..	3	1	-2	131	19
30	..	3	1	-2	135	20
31	..	3	1	-2	139	20
236	129	10	1	3	405	38
281	153	12	1	1	422	39
281	134	12	0	2	430	39
309	134	12	1	2	435	39

Table 2

Components of Total Energy Supply

Petajoules and Gigajoules per capita

Country or area	Year	Primary energy production: Total	Solids	Liquids	Gases	Electricity	Heat
Uganda	2011	396	390	..	..	*6	..
	2012	398	392	..	..	*6	..
	2013	401	394	..	..	*7	..
	2014	409	400	..	..	9	..
United Rep. of Tanzania	2011	814	777	..	30	7	..
	2012	843	803	..	34	6	..
	2013	889	849	..	34	6	..
	2014	935	894	..	32	9	..
Zambia	2011	332	291	..	..	41	..
	2012	346	301	..	..	44	..
	2013	362	314	..	..	48	..
	2014	374	324	..	..	51	..
Zimbabwe	2011	373	354	..	..	19	..
	2012	353	334	..	..	19	..
	2013	400	382	..	..	18	..
	2014	482	462	1	..	20	..
America, North	**2011**	**103,580**	**28,679**	**29,448**	**31,148**	**3,227**	**11,078**
	2012	**105,022**	**26,760**	**32,045**	**32,144**	**3,183**	**10,891**
	2013	**108,320**	**26,375**	**34,907**	**32,428**	**3,333**	**11,277**
	2014	**114,496**	**26,838**	**38,609**	**34,226**	**3,407**	**11,417**
Anguilla	2011	0	0	..	..	..	..
	2012	0	0	..	..	..	..
	2013	0	0	..	..	..	..
	2014	0	0	..	..	..	..
Antigua and Barbuda	2011	..	..	..	..	..	..
	2012	..	..	..	..	..	..
	2013	..	..	..	..	..	..
	2014	..	..	..	..	..	..
Aruba	2011	*5	0	*4	..	0	..
	2012	*4	0	*3	..	0	..
	2013	*1	0	0	..	0	..
	2014	*1	0	0	..	0	..
Bahamas	2011	0	0	..	..	..	..
	2012	0	0	..	..	..	..
	2013	0	0	..	..	..	..
	2014	0	0	..	..	..	..
Barbados	2011	3	1	2	1	..	..
	2012	3	1	2	1	..	..
	2013	3	1	1	1	..	..
	2014	2	1	1	0	..	..
Belize	2011	14	4	9	0	*1	..
	2012	12	4	6	0	*1	..
	2013	10	4	5	0	*1	..
	2014	9	4	4	0	*1	..
Bermuda	2011	*1	*1	..	..	..	..
	2012	*1	*1	..	..	..	..
	2013	*1	*1	..	..	..	..
	2014	*1	*1	..	..	..	..
Bonaire, St Eustatius, Saba	2011	..	..	..	..	..	..
	2012	..	..	..	..	..	..
	2013	..	..	..	..	..	..
	2014	0	..	..	..	0	..
British Virgin Islands	2011	0	0	..	..	..	..
	2012	0	0	..	..	..	..
	2013	0	0	..	..	..	..
	2014	0	0	..	..	..	..

Components of Total Energy Supply

Petajoules and Gigajoules per capita

Imports	Exports	International bunkers		Stock changes	Supply	
		Aviation	Marine		Total	Per capita
53	0	*4	..	..	445	13
50	0	*4	..	..	443	13
57	0	*4	..	..	454	12
63	1	5	..	..	466	12
79	..	5	2	..	887	19
96	..	5	2	..	932	19
121	..	6	2	..	1,002	20
119	..	6	2	..	1,046	20
38	2	1	..	0	366	26
45	6	2	..	-1	383	26
44	7	2	..	-1	398	26
42	6	2	..	-6	415	26
54	10	0	..	..	417	29
58	9	0	..	..	402	28
67	13	0	..	..	454	30
59	10	0	..	67	464	30
37,925	**25,159**	**1,162**	**1,449**	**-7**	**113,742**	**206**
34,618	**25,911**	**1,164**	**913**	**-175**	**111,827**	**200**
32,472	**26,744**	**1,183**	**991**	**-1,542**	**113,417**	**201**
31,018	**27,256**	**1,214**	**913**	**887**	**115,244**	**203**
*2	..	..	..	..	*2	*144
*2	..	..	..	..	*2	*140
*2	..	..	..	..	*2	*136
*2	..	..	..	..	*2	*136
*10	0	2	0	..	*7	*84
*10	0	2	0	..	*8	*85
*10	0	*2	0	..	*8	*84
*10	0	*2	0	..	*8	*84
*459	*432	*3	0	..	*28	*271
*122	*105	*4	0	..	*17	*168
*16	0	*4	0	..	*12	*119
*16	0	*4	0	..	*12	*120
35	0	*4	*4	..	27	74
38	0	*4	*5	..	29	77
54	*2	*3	*5	*4	40	107
38	0	*2	*5	*-4	35	90
25	2	*6	..	0	20	72
24	2	*5	..	0	20	70
24	1	*6	..	0	19	69
17	1	*1	..	0	17	59
9	8	0	..	0	14	43
7	6	0	..	0	13	37
8	4	1	..	0	13	38
8	4	1	..	0	13	36
7	..	*1	0	..	7	105
8	..	*1	0	..	7	115
7	..	*1	0	..	7	108
9	..	*1	0	..	9	136
..	..	..	..	..	..	..
*14	..	0	*8	..	*5	*215
*14	..	0	*9	..	*5	*213
*14	..	0	*9	..	*5	*208
*2	..	..	..	..	*2	*89
*2	..	..	..	..	*2	*87
*2	..	..	..	..	*2	*85
*3	..	..	..	..	*3	*85

Table 2

Components of Total Energy Supply

Petajoules and Gigajoules per capita

Country or area	Year	Primary energy production					
		Total	Solids	Liquids	Gases	Electricity	Heat
Canada	2011	17,041	1,946	7,134	5,559	1,390	1,013
	2012	17,507	1,943	7,682	5,445	1,411	1,026
	2013	18,309	2,032	8,204	5,474	1,480	1,119
	2014	19,276	2,036	8,834	5,777	1,464	1,165
Cayman Islands	2011	..	..	..	..	..	..
	2012	..	..	..	..	..	..
	2013	..	..	..	..	..	..
	2014	..	..	..	..	..	..
Costa Rica	2011	105	31	1	..	27	46
	2012	106	27	1	..	28	50
	2013	110	28	1	0	26	55
	2014	110	28	..	0	27	55
Cuba	2011	206	42	128	36	0	..
	2012	206	42	127	36	0	..
	2013	202	41	123	37	0	..
	2014	212	47	123	42	0	..
Curaçao	2011	..	..	..	..	..	..
	2012	0	..	..	..	0	..
	2013	0	..	..	..	0	..
	2014	0	..	..	..	0	..
Dominica	2011	0	0	..	..	0	..
	2012	0	0	..	..	0	..
	2013	0	0	..	..	0	..
	2014	0	0	..	..	0	..
Dominican Republic	2011	27	21	..	..	5	..
	2012	29	23	..	..	7	..
	2013	29	22	..	..	8	..
	2014	28	22	..	..	5	..
El Salvador	2011	94	32	..	..	7	55
	2012	91	29	..	..	7	55
	2013	86	23	..	1	6	56
	2014	86	24	..	0	6	56
Greenland	2011	1	0	..	..	1	..
	2012	1	0	..	..	1	..
	2013	1	0	..	..	1	..
	2014	2	0	..	..	1	..
Grenada	2011	0	0	..	..	..	..
	2012	0	0	..	..	..	..
	2013	0	0	..	..	..	..
	2014	0	0	..	..	..	..
Guadeloupe	2011	*4	*1	..	..	*1	*2
	2012	4	*1	..	..	1	2
	2013	5	*1	..	..	1	3
	2014	5	*1	..	..	1	3
Guatemala	2011	302	253	25	..	15	9
	2012	309	259	24	..	16	9
	2013	310	264	21	..	17	8
	2014	327	280	21	..	17	9
Haiti	2011	135	135	..	..	0	..
	2012	140	139	..	..	1	..
	2013	140	139	..	..	1	..
	2014	136	135	..	..	0	..
Honduras	2011	101	91	..	..	11	..
	2012	104	93	..	..	11	..
	2013	106	95	..	..	11	..
	2014	105	95	..	..	11	..

Components of Total Energy Supply

Petajoules and Gigajoules per capita

Imports	Exports	International bunkers		Stock changes	Supply	
		Aviation	Marine		Total	Per capita
3,881	9,778	40	21	-110	11,193	324
3,844	10,300	42	17	-75	11,066	317
3,576	10,809	34	18	-145	11,168	317
3,423	10,952	30	15	171	11,530	324
9	..	0	..	..	8	149
8	..	*1	..	..	8	134
8	..	0	..	..	8	131
8	..	0	..	..	8	130
105	3	0	0	1	205	45
105	6	0	0	-2	207	44
108	3	7	0	0	208	44
111	2	7	0	1	212	44
342	..	*6	*1	..	541	48
343	..	*6	*1	..	542	48
333	..	*9	*1	..	524	46
330	..	*9	*1	..	531	47
..	..	..	..	..	..	..
486	332	3	65	..	87	570
496	351	3	65	..	77	499
524	370	3	66	..	85	545
*2	0	0	0	..	*2	*28
*2	0	0	0	..	*2	*29
*2	0	0	0	..	*2	*28
*2	0	0	0	..	*2	*29
295	..	19	..	3	300	30
308	..	21	..	0	316	31
297	..	19	..	0	307	30
291	..	21	..	3	295	28
89	4	5	..	-5	180	30
87	0	4	..	-3	177	29
88	1	6	..	0	167	27
91	2	7	..	-1	170	28
11	..	..	2	-1	11	200
9	..	..	1	0	9	166
9	..	..	1	0	9	165
8	..	..	1	0	9	154
4	..	0	..	0	4	36
4	..	0	..	0	4	39
5	..	0	..	0	5	43
4	..	0	..	0	4	35
*35	..	5	..	..	*34	*75
*37	..	*5	..	..	*36	*78
*37	..	*5	..	..	*37	*80
*37	..	*5	..	..	*37	*80
164	31	3	13	1	419	28
164	27	2	13	4	427	28
168	30	2	14	-20	452	29
225	35	1	15	-15	517	32
30	..	1	..	..	164	16
33	..	2	..	..	170	17
35	..	4	..	-1	171	16
39	..	1	..	0	174	16
122	15	2	0	-5	212	28
131	18	2	0	0	214	28
134	20	2	0	0	217	28
133	20	3	..	-9	224	28

Table 2

Components of Total Energy Supply

Petajoules and Gigajoules per capita

Country or area	Year	Primary energy production					
		Total	Solids	Liquids	Gases	Electricity	Heat
Jamaica	2011	7	6	..	..	1	..
	2012	7	6	..	..	1	..
	2013	7	6	..	..	1	..
	2014	8	7	..	..	1	..
Martinique	2011	*1	*1	..	..	0	..
	2012	*1	*1	..	..	0	..
	2013	*1	*1	..	..	0	..
	2014	*1	*1	..	..	0	..
Mexico	2011	9,091	762	6,079	1,765	137	349
	2012	8,886	677	6,062	1,709	128	311
	2013	8,843	694	5,983	1,697	116	353
	2014	8,514	689	5,771	1,562	164	329
Montserrat	2011	..	..	..	..	..	..
	2012	..	..	..	..	..	..
	2013	..	..	..	..	..	..
	2014	..	..	..	..	..	..
Neth. Antilles (former)	2011	0	0	..	..	..	..
	2012	..	..	..	..	..	..
	2013	..	..	..	..	..	..
	2014	..	..	..	..	..	..
Nicaragua	2011	66	54	..	..	2	10
	2012	78	57	..	..	3	19
	2013	89	61	..	..	4	24
	2014	91	62	..	..	4	24
Panama	2011	27	12	..	..	15	..
	2012	31	12	..	..	19	..
	2013	31	13	..	..	19	..
	2014	31	12	..	..	19	..
Puerto Rico	2011	1	..	..	..	1	..
	2012	0	..	..	..	0	..
	2013	1	..	..	..	1	..
	2014	1	..	..	..	1	..
Sint Maarten (Dutch part)	2011	..	..	..	..	..	..
	2012	..	..	..	..	..	..
	2013	..	..	..	..	..	..
	2014	..	..	..	..	..	..
St. Kitts-Nevis	2011	0	0	..	..	0	..
	2012	0	0	..	..	0	..
	2013	0	0	..	..	0	..
	2014	0	0	..	..	0	..
St. Lucia	2011	0	0	..	..	..	..
	2012	0	0	..	..	..	..
	2013	0	0	..	..	..	..
	2014	0	0	..	..	..	..
St. Pierre-Miquelon	2011	0	..	..	..	0	..
	2012	0	..	..	..	0	..
	2013	0	..	..	..	0	..
	2014	0	..	..	..	0	..
St. Vincent-Grenadines	2011	0	0	..	..	0	..
	2012	0	0	..	..	0	..
	2013	0	0	..	..	0	..
	2014	0	0	..	..	0	..
Trinidad and Tobago	2011	1,697	0	267	1,430	..	..
	2012	1,670	0	232	1,438	..	..
	2013	1,673	0	229	1,444	..	..
	2014	1,663	0	227	1,436	..	..

Components of Total Energy Supply

Petajoules and Gigajoules per capita

Imports	Exports	International bunkers		Stock changes	Supply	
		Aviation	Marine		Total	Per capita
124	9	8	3	0	110	40
119	9	8	3	1	105	38
122	8	8	2	-1	113	41
113	7	9	1	1	104	37
41	*6	*4	*2	..	30	77
41	*6	*4	*2	..	30	77
*41	*6	*4	*2	..	*30	*76
*43	*6	*4	*3	..	*31	*79
2,184	3,370	113	39	-1	7,753	64
2,353	3,122	120	39	14	7,944	65
2,350	3,048	126	34	9	7,976	64
2,474	2,961	134	35	45	7,812	62
1	..	0	0	0	1	114
1	..	0	0	0	1	118
1	..	0	0	0	1	146
1	..	0	0	0	1	133
523	336	4	96	..	87	410
..	..	..	..	..	..	..
..	..	..	..	..	..	..
..	..	..	..	..	..	..
69	1	1	..	3	129	22
62	1	1	..	0	138	24
61	1	1	..	-1	148	25
64	1	1	..	-1	154	26
259	0	17	139	-23	152	41
272	0	19	*141	-10	152	41
285	0	22	134	-1	161	42
269	0	26	122	16	136	35
26	..	..	..	..	27	7
48	..	..	..	..	48	13
55	..	..	..	..	56	15
58	..	..	..	..	60	16
..	..	..	..	..	..	..
*21	..	*2	*8	..	*11	*311
*21	..	2	*8	..	*11	*306
*21	..	*2	*8	..	*11	*303
3	..	0	..	0	3	62
3	..	0	..	0	3	60
*3	..	0	..	..	*3	*59
*3	..	0	..	..	*3	*59
6	..	..	0	..	6	33
6	..	..	0	..	6	33
6	..	..	0	..	6	33
6	..	..	0	..	6	33
*1	..	..	0	..	*1	*160
*1	..	..	0	..	*1	*158
*1	..	..	0	..	*1	*162
*1	..	..	0	..	*1	*174
3	0	0	..	..	3	28
4	0	0	..	..	4	35
3	0	0	..	..	3	29
*3	0	0	..	..	*3	*29
192	1,035	13	16	-6	831	622
142	977	7	19	0	809	603
177	1,005	12	23	-13	824	611
126	945	9	19	-6	822	607

Table 2

Components of Total Energy Supply

Petajoules and Gigajoules per capita

Country or area	Year	Primary energy production					
		Total	Solids	Liquids	Gases	Electricity	Heat
Turks and Caicos Islands	2011	0	0	..	..	..	..
	2012	0	0	..	..	..	..
	2013	0	0	..	..	..	..
	2014	0	0	..	..	..	..
United States	2011	74,650	25,286	15,799	22,358	1,612	9,595
	2012	75,832	23,446	17,904	23,515	1,547	9,419
	2013	78,361	22,948	20,340	23,774	1,640	9,660
	2014	83,887	23,394	23,626	25,407	1,682	9,777
America, South	**2011**	**30,354**	**6,132**	**16,875**	**4,525**	**2,582**	**238**
	2012	**30,633**	**6,421**	**16,685**	**4,748**	**2,535**	**244**
	2013	**30,904**	**6,660**	**16,993**	**4,547**	**2,466**	**238**
	2014	**31,217**	**6,547**	**17,240**	**4,720**	**2,476**	**234**
Argentina	2011	3,248	97	1,507	1,461	114	69
	2012	3,160	101	1,453	1,430	107	69
	2013	2,981	96	1,393	1,304	120	67
	2014	3,167	116	1,477	1,362	150	62
Bolivia (Plur. State of)	2011	706	34	103	561	8	..
	2012	804	37	118	641	8	..
	2013	910	37	135	730	9	..
	2014	962	39	145	771	8	..
Brazil	2011	10,220	2,798	5,143	558	1,552	169
	2012	10,305	2,886	5,097	635	1,513	173
	2013	10,360	2,961	5,091	709	1,431	167
	2014	10,948	2,972	5,668	754	1,388	166
Chile	2011	412	257	25	53	77	0
	2012	544	405	21	43	74	1
	2013	626	495	21	35	73	2
	2014	540	397	20	30	90	3
Colombia	2011	5,107	2,499	1,981	451	176	..
	2012	5,283	2,581	2,051	479	172	..
	2013	5,697	2,654	2,414	469	160	..
	2014	5,380	2,584	2,187	447	162	..
Ecuador	2011	1,114	30	1,019	24	40	0
	2012	1,141	32	1,034	31	44	0
	2013	1,188	30	1,082	36	40	0
	2014	1,247	33	1,136	37	42	0
Falkland Is. (Malvinas)	2011	0	0	..	..	0	..
	2012	0	0	..	..	0	..
	2013	0	0	..	..	0	..
	2014	0	0	..	..	0	..
French Guiana	2011	3	1	..	..	2	..
	2012	3	1	..	..	2	..
	2013	3	1	..	..	2	..
	2014	3	1	..	..	2	..
Guyana	2011	10	10	..	..	..	..
	2012	8	8	..	..	..	..
	2013	7	7	..	..	..	..
	2014	7	7	..	..	..	..
Paraguay	2011	337	127	3	..	207	..
	2012	344	124	3	..	217	..
	2013	340	119	4	..	217	..
	2014	325	122	4	..	199	..
Peru	2011	885	109	285	413	78	0
	2012	899	100	286	432	79	1
	2013	954	120	307	444	80	2
	2014	1,020	122	318	497	81	2

Components of Total Energy Supply

Petajoules and Gigajoules per capita

Imports	Exports	International bunkers		Stock changes	Supply	
		Aviation	Marine		Total	Per capita
*3	0	..	..	..	*3	*84
*3	0	..	..	..	*3	*86
*3	0	..	..	..	*3	*85
*3	0	..	..	..	*3	*86
28,853	10,128	904	1,112	135	91,224	292
25,757	11,000	900	589	-104	89,204	283
23,911	11,454	901	672	-1,374	90,619	286
22,488	11,948	929	612	688	92,198	289
5,517	**12,826**	**255**	**347**	**63**	**22,380**	**56**
5,811	**12,565**	**290**	**318**	**-143**	**23,415**	**58**
6,239	**12,779**	**291**	**320**	**15**	**23,739**	**58**
6,367	**12,420**	**298**	**314**	**149**	**24,404**	**59**
543	319	31	56	29	3,356	81
600	294	31	70	-17	3,382	80
704	191	38	72	4	3,379	79
818	251	38	64	-6	3,638	85
37	460	2	..	6	275	27
42	525	2	..	1	319	31
41	631	2	..	-1	319	31
46	664	3	..	0	341	32
2,890	1,611	91	179	57	11,172	56
2,950	1,462	94	159	-91	11,631	57
3,200	1,199	98	143	-30	12,149	59
3,305	1,422	98	143	41	12,549	61
1,057	26	18	21	3	1,401	81
1,074	36	29	7	-5	1,551	89
1,115	67	26	9	26	1,613	92
1,083	73	27	8	10	1,505	85
110	3,745	*50	*13	-2	1,410	30
135	3,896	*54	*7	0	1,461	31
158	4,240	44	34	26	1,511	32
199	3,867	46	37	124	1,505	31
243	733	15	20	-2	591	39
250	774	15	16	-2	588	38
285	819	16	15	0	623	40
334	883	17	16	2	663	42
*1	..	..	..	..	*1	*255
*1	..	..	..	..	*1	*259
*1	..	..	..	..	*1	*258
*1	..	..	..	..	*1	*257
*10	..	*2	..	..	*12	*48
*9	..	*2	..	..	*11	*45
*10	..	*2	..	..	*11	*45
*12	..	*2	..	..	*13	*50
25	..	0	..	0	34	46
28	..	1	..	0	36	47
27	..	0	..	0	34	45
28	..	0	..	0	35	46
69	172	1	..	-5	237	38
73	177	1	..	2	237	37
68	175	1	..	-5	237	37
78	153	2	..	1	247	38
309	388	33	7	-15	781	26
330	411	29	9	-55	835	28
345	475	32	..	-38	830	27
312	370	35	..	12	914	30

Table 2

Components of Total Energy Supply

Petajoules and Gigajoules per capita

Country or area	Year	Primary energy production					
		Total	Solids	Liquids	Gases	Electricity	Heat
Suriname	2011	44	2	39	0	3	..
	2012	45	2	39	0	4	..
	2013	42	2	39	0	2	..
	2014	44	2	40	0	3	..
Uruguay	2011	83	59	1	..	24	..
	2012	83	62	1	..	20	..
	2013	95	63	2	..	30	..
	2014	113	73	2	..	37	..
Venezuela (Bolivar. Rep.)	2011	8,184	109	6,770	1,003	301	..
	2012	8,013	81	6,581	1,056	295	..
	2013	7,699	73	6,505	820	301	..
	2014	7,460	80	6,243	823	314	..
Asia	**2011**	**250,767**	**123,357**	**77,729**	**39,447**	**4,642**	**5,591**
	2012	**254,438**	**125,985**	**78,004**	**40,217**	**5,480**	**4,752**
	2013	**257,360**	**127,602**	**77,749**	**41,144**	**6,097**	**4,768**
	2014	**259,583**	**127,151**	**78,153**	**42,456**	**6,627**	**5,195**
Afghanistan	2011	62	53	..	6	3	..
	2012	56	47	..	6	3	..
	2013	59	51	..	5	*3	..
	2014	63	55	..	5	*3	..
Armenia	2011	55	19	..	..	9	28
	2012	52	19	..	..	8	25
	2013	47	14	..	..	8	25
	2014	48	14	..	..	7	27
Azerbaijan	2011	2,525	4	1,937	575	10	..
	2012	2,458	4	1,842	606	7	..
	2013	2,486	6	1,846	629	5	..
	2014	2,459	6	1,787	662	5	..
Bahrain	2011	862	0	418	444	..	..
	2012	828	0	382	446	..	..
	2013	911	0	432	479	..	..
	2014	961	0	443	519	..	..
Bangladesh	2011	1,307	592	10	701	3	..
	2012	1,347	598	10	736	3	..
	2013	1,410	601	11	794	4	..
	2014	1,438	610	11	814	3	..
Bhutan	2011	73	48	..	..	25	..
	2012	72	48	..	..	25	..
	2013	75	48	..	..	28	..
	2014	75	49	..	..	25	..
Brunei Darussalam	2011	780	0	332	448	0	..
	2012	773	0	333	440	0	..
	2013	710	0	282	427	0	..
	2014	679	0	264	415	0	..
Cambodia	2011	159	159	..	..	0	..
	2012	165	163	..	..	2	..
	2013	171	167	..	..	4	..
	2014	178	172	..	..	7	..
China	2011	96,925	80,834	8,827	3,562	2,769	933
	2012	99,224	81,867	9,056	3,764	3,485	1,052
	2013	101,306	82,878	9,195	4,186	3,841	1,205
	2014	101,394	81,669	9,273	4,572	4,448	1,431
China, Hong Kong SAR	2011	..	..	..	..	..	..
	2012	..	..	..	..	..	..
	2013	..	..	..	..	..	..
	2014	..	..	..	..	..	..

Components of Total Energy Supply

Petajoules and Gigajoules per capita

Imports	Exports	International bunkers		Stock changes	Supply	
		Aviation	Marine		Total	Per capita
18	25	*2	*1	0	34	65
21	24	*2	*2	0	38	72
21	29	*2	*1	0	31	59
20	28	*2	*1	0	34	62
121	2	4	16	-6	189	56
146	4	4	12	8	202	60
109	2	3	10	-7	197	58
103	6	3	9	-1	199	58
86	5,343	6	35	-2	2,887	98
151	4,961	26	37	17	3,123	105
156	4,951	26	37	39	2,803	93
31	4,702	26	36	-33	2,760	90
92,536	**86,361**	**2,674**	**4,129**	**2,283**	**247,855**	**59**
97,106	**86,890**	**2,740**	**4,031**	**1,127**	**256,756**	**60**
101,705	**90,985**	**2,874**	**4,225**	**1,817**	**259,164**	**60**
103,352	**88,862**	**2,893**	**4,067**	**1,391**	**265,722**	**61**
125	1	0	..	..	185	6
114	1	0	..	..	169	6
101	*1	0	..	..	158	5
*93	*2	0	..	..	*154	*5
90	9	*2	..	0	135	45
103	9	*2	..	0	145	49
99	7	*2	..	0	138	46
98	5	*2	..	..	139	46
2	1,989	19	3	-11	527	57
2	1,869	16	4	13	558	60
6	1,893	17	3	11	567	60
12	1,879	15	3	-9	584	61
483	809	26	..	-1	511	392
489	796	25	..	3	494	370
480	817	18	..	0	556	412
462	848	17	..	-5	563	413
232	8	15	3	-18	1,531	10
236	6	14	4	-34	1,594	10
235	5	14	4	0	1,623	10
258	4	14	4	-13	1,687	11
8	21	0	..	..	59	81
8	19	0	..	..	61	82
9	21	0	..	..	63	83
8	19	0	..	..	63	82
11	619	5	4	2	162	406
16	621	4	3	1	160	396
13	588	4	3	0	127	310
16	538	3	3	3	149	357
75	..	2	..	..	232	16
80	..	2	..	..	242	16
82	..	3	..	..	250	17
92	..	3	..	..	267	17
19,085	1,948	248	369	2,026	111,417	83
20,010	1,658	262	360	1,324	115,631	85
21,360	1,884	280	312	1,894	118,296	87
22,203	1,980	307	289	1,443	119,578	87
1,328	39	306	384	25	574	81
1,274	43	302	347	9	573	81
1,291	32	306	361	6	585	82
1,249	25	318	316	-10	599	83

Table 2

Components of Total Energy Supply

Petajoules and Gigajoules per capita

Country or area	Year	Primary energy production					
		Total	Solids	Liquids	Gases	Electricity	Heat
China, Macao SAR	2011	*2	*2	..	..	..	..
	2012	*2	*2	..	..	..	..
	2013	*2	*2	..	..	..	..
	2014	*2	*2	..	..	..	..
Cyprus	2011	4	0	0	0	0	3
	2012	4	0	0	0	1	3
	2013	5	0	0	0	1	3
	2014	5	1	0	0	1	3
Georgia	2011	50	19	2	0	28	0
	2012	48	20	2	0	26	0
	2013	63	30	2	0	30	1
	2014	61	28	2	0	30	1
India	2011	23,002	18,708	1,816	1,672	458	349
	2012	23,320	19,223	1,801	1,429	512	355
	2013	22,401	18,384	1,789	1,216	643	370
	2014	23,103	19,253	1,728	1,151	581	390
Indonesia	2011	17,971	12,513	2,102	2,975	45	337
	2012	18,812	13,585	2,025	2,817	46	339
	2013	19,776	14,866	1,972	2,537	61	339
	2014	19,481	14,702	1,930	2,421	68	360
Iran (Islamic Rep. of)	2011	14,464	87	9,046	5,283	44	4
	2012	12,506	112	6,818	5,511	46	20
	2013	12,544	114	6,794	5,533	54	49
	2014	13,291	*110	6,895	6,186	51	48
Iraq	2011	5,813	2	5,586	210	15	..
	2012	6,423	2	6,199	206	16	..
	2013	6,507	2	6,249	239	17	..
	2014	6,744	2	6,500	231	11	..
Israel	2011	197	2	1	147	1	46
	2012	136	1	1	87	1	46
	2013	273	1	3	221	2	46
	2014	313	1	3	260	3	46
Japan	2011	2,128	410	27	134	334	1,222
	2012	1,168	410	25	128	314	291
	2013	1,159	455	22	116	348	218
	2014	1,114	469	21	108	401	115
Jordan	2011	9	3	0	6	0	..
	2012	8	3	0	5	0	..
	2013	7	3	0	4	0	..
	2014	7	3	0	4	0	..
Kazakhstan	2011	6,961	2,134	3,410	1,388	28	..
	2012	7,030	2,212	3,375	1,415	28	..
	2013	7,198	2,198	3,483	1,489	28	..
	2014	7,091	2,092	3,443	1,525	30	..
Korea, Dem.Ppl's.Rep.	2011	843	796	..	..	48	..
	2012	849	800	..	..	49	..
	2013	853	803	..	..	50	..
	2014	871	825	..	..	47	..
Korea, Republic of	2011	1,941	181	33	26	23	1,678
	2012	1,908	196	34	26	22	1,631
	2013	1,799	206	31	27	27	1,508
	2014	2,023	236	42	21	25	1,699
Kuwait	2011	6,366	0	5,883	483	..	..
	2012	7,136	0	6,591	545	..	..
	2013	7,113	0	6,555	558	..	..
	2014	6,941	0	6,427	514	..	..

Components of Total Energy Supply

Petajoules and Gigajoules per capita

Imports	Exports	International bunkers		Stock changes	Supply	
		Aviation	Marine		Total	Per capita
32	..	..	..	0	34	62
32	..	..	..	0	34	61
32	..	..	..	0	33	58
36	..	..	3	0	36	62
112	..	13	8	-5	100	90
111	..	12	8	1	94	83
98	..	10	10	1	82	71
96	..	10	10	-2	83	72
114	6	2	..	0	157	37
121	4	2	..	0	164	40
113	5	4	..	0	167	41
136	5	4	..	0	188	47
10,610	2,683	208	0	199	30,522	24
12,380	2,889	208	0	-28	32,630	26
13,929	3,016	170	53	-103	33,195	26
14,876	*2,599	*167	*54	-169	35,327	27
2,018	9,773	30	8	-187	10,365	42
2,208	10,382	32	8	-72	10,669	43
2,389	13,082	39	15	62	8,966	36
2,520	12,848	57	12	282	8,801	35
578	6,226	51	93	-77	8,750	116
251	3,679	50	79	-107	9,055	119
315	3,506	52	126	-30	9,205	119
436	3,608	55	149	-43	9,958	127
411	4,525	35	7	-6	1,663	52
533	5,052	14	7	11	1,871	57
572	4,991	17	7	13	2,049	60
649	5,247	23	8	61	2,054	58
986	207	36	13	-37	965	128
1,098	164	35	13	11	1,012	132
1,055	286	47	10	21	963	123
1,008	322	47	5	3	945	119
18,243	607	256	163	42	19,303	152
18,778	583	268	161	51	18,882	149
18,976	742	272	159	-20	18,982	149
18,309	658	266	141	-118	18,477	146
294	0	5	0	-12	309	46
337	0	6	1	7	332	47
316	0	*6	0	-10	328	45
354	0	6	0	5	350	47
440	*3,959	7	1	-74	3,508	212
550	*4,215	9	0	107	3,249	193
641	4,202	7	0	80	3,550	208
308	4,020	10	..	16	3,352	193
42	293	..	..	..	592	24
69	315	..	..	..	603	24
44	439	..	..	..	459	18
44	411	..	..	..	505	20
11,787	2,293	168	350	44	10,874	220
12,002	2,459	168	341	-49	10,992	222
12,147	2,380	177	333	49	11,007	221
12,181	2,472	178	339	50	11,164	223
118	5,041	30	43	-18	1,388	429
91	5,584	35	45	-21	1,584	463
75	5,656	32	46	-21	1,475	410
120	5,578	31	47	-22	1,427	380

Table 2

Components of Total Energy Supply

Petajoules and Gigajoules per capita

Country or area	Year	Primary energy production					
		Total	Solids	Liquids	Gases	Electricity	Heat
Kyrgyzstan	2011	68	13	3	1	51	..
	2012	73	18	3	1	51	..
	2013	74	22	4	1	47	..
	2014	80	28	3	1	48	..
Lao People's Dem. Rep.	2011	78	66	..	..	13	..
	2012	78	66	..	..	*13	..
	2013	77	62	..	..	15	..
	2014	72	57	..	..	15	..
Lebanon	2011	9	5	..	..	3	1
	2012	9	5	..	..	4	1
	2013	10	5	..	..	5	1
	2014	7	5	..	..	1	1
Malaysia	2011	3,394	118	1,272	1,977	27	..
	2012	3,369	115	1,291	1,931	33	..
	2013	3,529	117	1,240	2,133	39	..
	2014	3,738	105	1,303	2,281	49	..
Maldives	2011	0	0	..	..	0	..
	2012	0	0	..	..	0	..
	2013	0	0	..	..	0	..
	2014	0	0	..	..	0	..
Mongolia	2011	860	847	13	..	..	..
	2012	780	761	18	..	..	..
	2013	799	774	26	..	..	..
	2014	677	640	37	..	..	..
Myanmar	2011	987	459	39	462	27	..
	2012	954	465	36	424	28	..
	2013	973	470	36	435	32	..
	2014	1,078	477	34	535	32	..
Nepal	2011	326	310	..	3	13	..
	2012	365	348	..	4	13	..
	2013	409	391	..	7	11	..
	2014	451	431	..	7	13	..
Oman	2011	2,894	0	1,840	1,053	..	..
	2012	3,014	0	1,917	1,097	..	..
	2013	3,150	0	1,964	1,186	..	..
	2014	3,101	0	1,966	1,134	..	..
Other Asia	2011	562	68	4	11	19	459
	2012	551	65	5	14	26	441
	2013	562	64	5	12	27	454
	2014	566	66	3	12	23	462
Pakistan	2011	2,195	612	149	1,275	103	57
	2012	*2,216	613	167	*1,279	107	49
	2013	*2,193	632	191	*1,199	116	55
	2014	*2,202	661	182	*1,189	115	55
Philippines	2011	882	311	39	140	35	358
	2012	969	394	35	134	37	369
	2013	936	394	36	123	36	346
	2014	991	408	49	130	34	371
Qatar	2011	8,802	0	3,169	5,634	..	..
	2012	9,184	0	3,269	5,915	..	..
	2013	9,271	0	3,252	6,019	..	..
	2014	9,166	0	3,205	5,961	..	..
Saudi Arabia	2011	24,664	..	22,110	2,554	0	..
	2012	26,010	..	23,236	2,773	0	..
	2013	25,573	..	22,770	2,803	0	..
	2014	25,904	..	22,993	2,911	0	..

Components of Total Energy Supply

Petajoules and Gigajoules per capita

Imports	Exports	International bunkers		Stock changes	Supply	
		Aviation	Marine		Total	Per capita
86	15	2	..	-3	139	25
116	11	1	..	4	174	31
108	10	2	..	2	167	29
101	10	0	..	11	159	27
*9	*14	0	..	..	73	11
*9	*14	0	..	..	73	11
*11	9	0	..	..	79	12
*12	5	0	..	..	78	12
262	..	10	1	..	260	57
293	..	9	1	..	292	59
291	..	11	1	..	289	55
311	..	10	1	..	306	55
1,732	1,920	76	8	54	3,068	107
1,864	2,085	75	2	-76	3,146	108
2,094	2,035	102	14	54	3,417	116
1,895	2,019	108	8	-37	3,535	118
16	..	1	*2	..	14	42
21	..	3	*2	..	16	46
22	..	*5	*2	..	16	45
29	..	*8	*2	..	19	54
44	484	..	..	171	249	90
51	455	..	..	74	302	108
56	383	..	..	29	444	155
53	437	..	..	47	245	84
65	362	1	0	-8	697	13
61	365	1	0	2	646	12
81	369	2	0	1	683	13
154	457	2	0	-31	805	15
67	0	*4	..	..	388	14
69	0	*4	..	..	430	16
78	0	*4	..	..	483	17
97	0	*6	..	0	542	19
98	2,037	17	23	-2	916	285
99	2,076	18	32	2	985	278
113	2,197	16	36	-4	1,018	261
85	2,105	22	48	-2	1,014	239
4,696	566	87	66	-25	4,565	196
4,734	694	91	48	6	4,446	191
4,812	719	97	49	-8	4,517	194
4,964	716	106	47	60	4,601	197
862	39	9	4	0	3,005	17
859	32	8	4	*-7	3,039	17
913	42	9	4	*-5	3,056	17
949	42	9	4	*-6	3,101	17
907	144	48	9	8	1,580	17
953	126	48	8	14	1,726	18
1,001	128	49	7	-27	1,780	18
1,115	194	46	4	-38	1,899	19
4	7,419	64	..	-7	1,331	699
0	7,539	79	..	-13	1,579	783
5	7,778	88	..	15	1,395	664
1	7,246	55	..	29	1,837	846
433	17,223	89	129	257	7,399	257
707	18,407	92	137	-243	8,324	282
1,035	18,517	82	138	-114	7,985	264
1,009	17,892	101	130	-87	8,876	287

Table 2

Components of Total Energy Supply

Petajoules and Gigajoules per capita

Country or area	Year	Primary energy production					
		Total	Solids	Liquids	Gases	Electricity	Heat
Singapore	2011	26	26	..	..	..	..
	2012	26	26	..	..	..	..
	2013	27	27	..	..	..	..
	2014	27	27	..	..	..	..
Sri Lanka	2011	180	164	..	..	17	..
	2012	173	161	..	..	12	..
	2013	184	158	..	..	26	..
	2014	179	161	..	..	17	..
State of Palestine	2011	9	4	..	..	..	5
	2012	8	3	..	..	..	5
	2013	8	3	..	..	..	5
	2014	9	4	..	..	..	5
Syrian Arab Republic	2011	1,056	0	777	267	12	..
	2012	590	0	360	219	12	..
	2013	304	0	113	180	11	..
	2014	237	0	59	166	11	..
Tajikistan	2011	65	5	1	1	58	..
	2012	71	8	1	0	61	..
	2013	73	10	1	0	61	..
	2014	76	17	1	0	58	..
Thailand	2011	2,831	1,124	580	*1,098	29	0
	2012	3,041	1,147	641	*1,223	30	0
	2013	3,060	1,175	642	1,217	26	0
	2014	3,338	1,235	835	1,242	26	0
Timor-Leste	2011	178	1	177	..	..	..
	2012	169	1	168	..	..	..
	2013	168	1	167	..	..	..
	2014	143	1	142	..	..	..
Turkey	2011	1,345	897	101	29	205	113
	2012	1,282	799	100	27	229	126
	2013	1,310	795	103	27	241	144
	2014	1,303	812	107	26	177	181
Turkmenistan	2011	2,735	0	477	2,258	0	..
	2012	2,852	0	497	2,355	0	..
	2013	3,208	0	533	2,675	..	..
	2014	3,270	0	541	2,730	..	..
United Arab Emirates	2011	8,151	..	6,314	1,837	..	..
	2012	8,470	..	6,563	1,907	..	..
	2013	8,681	..	6,764	1,917	..	..
	2014	8,642	..	6,737	1,905	..	..
Uzbekistan	2011	2,400	57	157	2,150	37	..
	2012	2,382	57	140	2,145	40	..
	2013	2,268	61	132	2,033	42	..
	2014	2,339	66	124	2,105	43	..
Viet Nam	2011	2,805	1,704	655	298	149	..
	2012	2,856	1,614	722	329	192	..
	2013	2,878	1,604	721	342	210	..
	2014	2,977	1,646	751	359	222	..
Yemen	2011	765	4	422	339	..	..
	2012	619	5	340	274	..	..
	2013	759	5	382	373	..	..
	2014	668	5	312	351	..	..
Europe	**2011**	**101,798**	**21,085**	**29,970**	**33,951**	**3,243**	**13,549**
	2012	**102,440**	**22,214**	**29,635**	**33,619**	**3,608**	**13,363**
	2013	**102,878**	**21,853**	**29,434**	**34,392**	**3,948**	**13,251**
	2014	**100,312**	**21,322**	**29,691**	**31,838**	**4,037**	**13,425**

Components of Total Energy Supply

Petajoules and Gigajoules per capita

Imports	Exports	International bunkers		Stock changes	Supply	
		Aviation	Marine		Total	Per capita
6,782	3,873	*252	1,747	-23	959	185
6,489	3,667	*282	1,728	15	823	155
6,588	3,478	*295	1,728	-11	1,125	208
6,678	3,535	*297	1,717	4	1,151	209
224	3	14	12	-4	380	19
245	0	17	11	8	381	19
220	6	15	11	-16	388	19
246	1	20	14	-24	414	20
50	0	..	..	0	59	14
50	0	..	..	0	58	13
53	0	..	..	0	61	14
57	0	..	..	-2	69	15
174	307	1	40	45	837	41
119	38	1	21	0	650	33
272	26	1	19	0	530	27
254	25	1	10	0	454	24
28	2	1	..	..	90	12
33	3	1	..	..	99	12
38	4	2	..	..	106	13
51	5	3	..	..	119	14
2,571	491	*147	0	*-55	4,819	72
2,799	552	*147	0	*-26	5,167	77
2,851	522	161	31	-163	5,359	79
2,894	445	160	41	-57	5,643	83
*4	*177	0	..	..	*5	*4
*4	*168	0	..	..	*5	*5
*6	*167	0	..	..	*7	*6
*7	*142	0	..	..	*8	*7
3,726	354	50	6	-55	4,716	64
4,108	362	44	8	57	4,918	66
4,027	344	58	38	23	4,874	64
4,228	329	112	42	-4	5,052	65
0	1,678	21	..	..	1,036	203
0	1,762	18	..	..	1,072	207
0	2,089	20	..	..	1,100	210
0	2,127	21	..	..	1,123	212
1,725	6,255	*290	*619	..	2,712	310
1,778	6,432	*300	*639	..	2,877	321
1,924	6,767	*338	*696	..	2,805	310
1,865	6,344	231	607	..	3,326	366
47	464	..	..	..	1,983	70
48	402	..	..	..	2,027	71
47	516	..	..	..	1,799	62
48	558	..	..	..	1,829	62
548	847	24	10	*36	2,438	27
545	834	29	6	*82	2,450	27
479	722	34	4	*86	2,510	27
506	630	34	4	*56	2,759	30
148	631	3	4	..	276	11
159	522	2	4	..	250	10
199	606	5	4	..	344	13
178	529	4	4	..	309	12
67,674	**54,703**	**2,269**	**2,381**	**289**	**109,829**	**149**
67,573	**55,732**	**2,246**	**2,178**	**-115**	**109,972**	**149**
67,018	**56,941**	**2,228**	**2,200**	**57**	**108,471**	**147**
65,355	**55,985**	**2,234**	**2,508**	**452**	**104,490**	**142**

Table 2

Components of Total Energy Supply

Petajoules and Gigajoules per capita

Country or area	Year	Primary energy production: Total	Solids	Liquids	Gases	Electricity	Heat
Albania	2011	62	9	38	1	15	..
	2012	70	9	44	1	17	..
	2013	85	9	51	1	25	..
	2014	84	8	58	1	17	..
Andorra	2011	0	0	..	..	0	..
	2012	0	0	..	..	0	..
	2013	1	0	..	..	0	..
	2014	1	0	..	..	0	..
Austria	2011	480	224	52	65	131	9
	2012	530	233	50	71	168	9
	2013	507	229	46	58	165	9
	2014	505	220	54	58	164	9
Belarus	2011	174	93	72	8	0	..
	2012	173	93	71	8	0	..
	2013	167	88	71	8	1	..
	2014	155	76	71	8	0	..
Belgium	2011	667	107	17	5	13	525
	2012	587	103	18	7	19	441
	2013	621	101	18	8	24	470
	2014	520	89	21	9	28	374
Bosnia and Herzegovina	2011	193	177	..	..	16	..
	2012	188	173	..	..	15	..
	2013	192	166	..	..	26	..
	2014	252	230	..	..	21	..
Bulgaria	2011	517	304	2	15	14	182
	2012	491	282	1	13	19	176
	2013	444	248	3	10	25	158
	2014	474	261	4	7	26	176
Croatia	2011	192	59	29	84	19	1
	2012	176	61	28	69	19	1
	2013	186	62	27	64	33	1
	2014	182	58	27	62	35	1
Czechia	2011	1,343	979	21	18	16	308
	2012	1,363	970	21	25	17	330
	2013	1,258	853	19	33	19	335
	2014	1,221	820	21	34	16	330
Denmark	2011	855	102	466	251	35	1
	2012	779	97	423	220	37	1
	2013	692	97	368	184	42	1
	2014	666	93	344	179	49	1
Estonia	2011	211	209	..	0	1	..
	2012	213	211	..	0	2	..
	2013	237	234	..	0	2	..
	2014	242	239	..	0	2	..
Faeroe Islands	2011	0	..	..	..	0	..
	2012	0	..	..	..	0	..
	2013	0	..	..	..	0	..
	2014	1	..	..	..	1	..
Finland	2011	716	404	10	2	47	253
	2012	720	391	12	2	63	252
	2013	759	432	15	4	49	259
	2014	757	425	16	4	52	259
France	2011	5,629	461	131	36	214	4,788
	2012	5,565	503	140	35	281	4,606
	2013	5,638	544	144	30	331	4,589
	2014	5,694	489	148	19	311	4,727

Components of Total Energy Supply

Petajoules and Gigajoules per capita

Imports	Exports	International bunkers		Stock changes	Supply	
		Aviation	Marine		Total	Per capita
62	29	0	0	0	95	33
58	40	0	0	0	88	30
79	51	0	1	10	102	35
83	49	0	1	13	103	36
9	0	0	..	..	9	111
9	0	0	..	..	9	115
8	0	0	..	..	9	118
8	0	0	..	..	9	119
1,291	297	30	1	61	1,382	164
1,322	416	29	1	24	1,382	163
1,212	328	28	1	-28	1,391	164
1,184	275	28	1	38	1,346	158
1,774	743	0	..	-26	1,231	130
1,945	829	0	..	9	1,280	135
1,657	671	4	..	7	1,142	120
1,703	686	5	..	5	1,162	122
3,286	1,239	63	285	12	2,353	214
3,193	1,242	58	256	-24	2,249	203
3,290	1,243	55	260	21	2,332	209
3,306	1,316	57	228	23	2,203	196
150	51	0	..	-2	294	77
131	38	0	..	3	278	73
129	50	0	..	5	266	70
133	61	0	..	-1	325	85
497	199	7	3	1	804	109
505	220	7	3	-1	767	105
504	228	7	4	6	703	97
495	227	7	3	-14	745	103
301	109	4	1	-2	381	89
287	105	5	..	-6	359	84
285	116	5	..	-2	352	82
270	119	5	..	-8	335	79
883	379	13	..	47	1,788	170
806	350	13	..	28	1,779	169
838	348	12	..	-16	1,751	166
872	343	12	..	13	1,725	164
665	713	36	29	-7	749	134
663	683	36	22	-17	718	128
791	686	35	26	6	729	130
738	644	38	31	18	673	119
78	42	1	8	0	238	179
131	80	2	17	12	233	176
106	66	1	18	1	258	195
116	84	2	14	5	254	193
*9	..	0	1	..	*8	*168
*9	..	0	1	..	*9	*177
*10	..	0	1	..	*10	*202
*9	..	0	1	..	*9	*182
1,132	330	27	8	16	1,467	272
1,015	342	26	5	-55	1,416	261
1,100	410	27	5	30	1,386	254
1,100	393	27	4	25	1,408	257
6,673	1,377	249	104	58	10,516	166
6,546	1,305	242	98	-71	10,538	166
6,472	1,228	240	90	-32	10,584	166
6,177	1,350	242	76	54	10,149	158

Table 2

Components of Total Energy Supply

Petajoules and Gigajoules per capita

Country or area	Year	Primary energy production: Total	Solids	Liquids	Gases	Electricity	Heat
Germany	2011	5,145	2,682	287	673	310	1,193
	2012	5,126	2,727	272	669	356	1,102
	2013	5,032	2,637	274	659	381	1,081
	2014	5,004	2,633	282	599	407	1,083
Gibraltar	2011	..	..	..	..	..	..
	2012	..	..	..	..	..	..
	2013	..	..	..	..	..	..
	2014	..	..	..	..	..	..
Greece	2011	403	355	8	3	29	9
	2012	437	379	9	4	36	9
	2013	389	317	9	4	51	9
	2014	368	304	8	4	43	9
Guernsey	2011	..	..	..	..	..	..
	2012	..	..	..	..	..	..
	2013	..	..	..	..	..	..
	2014	..	..	..	..	..	..
Hungary	2011	451	134	47	91	3	175
	2012	442	131	55	76	4	176
	2013	426	132	50	68	3	172
	2014	423	131	48	63	4	176
Iceland	2011	260	0	..	0	45	215
	2012	281	0	..	0	44	236
	2013	283	0	..	0	46	237
	2014	284	0	..	0	46	238
Ireland	2011	70	41	1	10	18	0
	2012	53	24	1	9	17	0
	2013	94	66	1	8	19	0
	2014	84	54	1	7	21	1
Isle of Man	2011	0	0	..	..	0	..
	2012	0	0	..	..	0	..
	2013	0	0	..	..	0	..
	2014	0	0	..	..	0	..
Italy	2011	1,340	285	263	336	239	216
	2012	1,466	387	254	344	267	214
	2013	1,542	396	266	341	321	217
	2014	1,539	360	279	327	346	227
Jersey	2011	1	1	..	..	..	..
	2012	*1	*1	..	..	..	..
	2013	*1	*1	..	..	..	..
	2014	*1	*1	..	..	..	..
Latvia	2011	100	86	2	1	11	..
	2012	109	90	3	2	14	..
	2013	98	82	3	3	11	..
	2014	108	95	3	3	8	..
Liechtenstein	2011	1	1	..	..	0	0
	2012	1	1	..	..	0	0
	2013	1	1	..	..	0	0
	2014	1	1	..	..	0	0
Lithuania	2011	64	42	8	0	3	10
	2012	65	42	9	0	3	10
	2013	69	46	9	1	4	9
	2014	74	50	8	1	4	11
Luxembourg	2011	5	4	..	1	1	0
	2012	5	4	..	1	1	0
	2013	6	4	..	1	1	0
	2014	6	5	..	1	1	0

Components of Total Energy Supply

Petajoules and Gigajoules per capita

Imports	Exports	International bunkers: Aviation	International bunkers: Marine	Stock changes	Supply: Total	Supply: Per capita
10,009	1,695	330	114	54	12,960	161
10,053	1,723	352	106	-8	13,005	162
10,640	1,983	359	96	-27	13,261	165
10,282	2,076	345	96	-8	12,776	158
181	..	0	174	..	7	228
183	..	0	175	..	7	231
168	..	0	160	..	8	245
128	..	0	119	..	8	259
1,255	423	32	116	-44	1,131	101
1,374	545	28	95	18	1,124	101
1,310	618	29	90	-28	989	89
1,390	665	33	79	-2	983	89
1	..	..	..	..	1	18
1	..	..	..	..	1	18
0	..	..	..	..	0	6
1	..	..	..	..	1	9
739	186	10	..	-56	1,050	105
726	202	7	..	-30	989	99
727	221	7	..	-22	947	95
810	211	8	..	55	959	97
34	..	6	3	0	285	889
34	..	6	2	0	306	945
34	..	7	1	1	309	951
36	..	8	1	0	312	952
596	69	29	4	14	550	118
570	76	24	4	-31	549	118
572	53	27	6	34	546	117
546	56	31	5	2	535	114
0	0	..	..	..	0	2
1	0	..	..	..	1	11
0	0	..	..	..	0	*4
0	0	..	..	..	0	*3
7,158	1,187	140	105	-20	7,086	119
6,888	1,307	133	104	6	6,804	114
6,262	1,076	129	92	-23	6,530	109
5,770	929	132	81	-1	6,168	103
3	..	..	..	..	3	33
2	..	..	..	..	3	32
2	..	..	..	..	3	30
2	..	..	..	..	3	28
171	54	5	9	10	193	93
176	59	5	10	11	200	98
183	68	5	10	2	196	98
159	74	5	10	-17	196	99
2	..	..	..	..	3	77
2	..	..	..	..	3	78
2	..	..	..	..	3	79
2	..	..	..	..	3	74
590	350	2	6	0	296	97
582	344	3	5	-4	299	99
588	371	3	4	-2	282	95
529	312	3	0	1	287	98
198	10	17	..	0	176	339
194	10	16	..	1	173	325
186	7	16	..	0	168	308
181	8	17	..	1	161	290

Table 2

Components of Total Energy Supply

Petajoules and Gigajoules per capita

Country or area	Year	Primary energy production					
		Total	Solids	Liquids	Gases	Electricity	Heat
Malta	2011	0	..	0	0	0	0
	2012	0	..	0	0	0	0
	2013	0	..	0	0	0	0
	2014	1	..	0	0	0	0
Montenegro	2011	30	26	..	..	4	..
	2012	29	24	..	..	5	..
	2013	32	23	..	..	9	..
	2014	29	23	..	..	6	..
Netherlands	2011	2,689	110	85	2,430	19	46
	2012	2,701	110	111	2,417	19	44
	2013	2,894	107	132	2,599	22	33
	2014	2,447	116	148	2,113	24	47
Norway	2011	8,303	103	4,004	3,758	438	1
	2012	8,560	98	3,726	4,220	516	0
	2013	8,134	113	3,547	4,004	468	1
	2014	8,204	101	3,624	3,978	497	3
Poland	2011	2,855	2,621	45	167	20	2
	2012	2,986	2,733	56	171	24	2
	2013	2,969	2,699	69	168	30	2
	2014	2,819	2,545	71	165	36	3
Portugal	2011	231	130	14	2	76	10
	2012	197	116	11	2	59	8
	2013	241	122	11	3	94	10
	2014	250	122	12	3	102	11
Republic of Moldova	2011	10	9	1	0	0	..
	2012	10	10	0	0	0	..
	2013	12	11	1	0	0	..
	2014	15	14	0	0	0	..
Romania	2011	1,162	429	184	363	58	128
	2012	1,142	426	174	365	53	125
	2013	1,090	350	181	361	72	127
	2014	1,109	342	176	368	96	127
Russian Federation	2011	54,534	7,174	21,713	23,148	597	1,902
	2012	55,216	8,043	21,984	22,642	597	1,950
	2013	56,263	8,020	22,113	23,583	652	1,895
	2014	54,829	8,240	22,299	21,675	632	1,984
Serbia	2011	466	372	45	17	31	0
	2012	450	349	49	18	33	0
	2013	474	368	51	18	37	0
	2014	393	286	49	19	40	0
Slovakia	2011	259	60	8	6	15	169
	2012	261	60	7	8	16	170
	2013	269	63	7	7	20	172
	2014	264	63	7	8	17	170
Slovenia	2011	161	77	0	2	13	69
	2012	150	72	0	2	15	61
	2013	151	73	0	2	17	59
	2014	154	60	..	1	23	71
Spain	2011	1,332	334	39	13	291	654
	2012	1,374	331	32	14	282	715
	2013	1,420	308	52	22	363	675
	2014	1,432	301	68	16	358	689
Sweden	2011	1,375	435	21	5	261	654
	2012	1,494	461	26	5	310	692
	2013	1,450	452	17	6	256	718
	2014	1,428	440	10	6	270	701

Components of Total Energy Supply

Petajoules and Gigajoules per capita

Imports	Exports	International bunkers		Stock changes	Supply	
		Aviation	Marine		Total	Per capita
97	0	4	56	2	35	85
93	0	4	50	1	37	90
97	6	4	50	4	32	78
95	8	5	51	-2	33	78
19	3	0	..	..	45	73
16	2	1	..	..	43	69
14	3	1	..	..	42	67
16	4	1	..	0	40	64
7,483	6,275	151	616	-61	3,191	191
7,955	6,741	145	566	-4	3,208	192
8,156	7,139	148	542	40	3,180	189
8,242	7,002	154	534	-4	3,004	178
314	7,397	18	16	25	1,160	234
294	7,565	17	14	10	1,248	249
356	7,090	21	15	-2	1,365	269
294	7,292	22	6	-20	1,199	233
2,086	606	21	7	45	4,261	110
1,980	656	23	6	171	4,110	106
1,922	802	23	6	-61	4,122	107
1,953	778	25	6	21	3,941	102
931	145	39	24	5	949	90
938	179	40	25	0	890	85
968	270	40	27	-16	888	85
916	240	42	26	-16	874	84
85	1	1	..	1	93	23
82	1	1	..	0	91	22
85	2	1	..	1	93	23
82	1	1	..	0	95	23
502	181	6	0	-31	1,508	75
503	169	5	1	2	1,469	74
439	191	6	2	-11	1,342	68
469	244	8	3	-9	1,331	68
1,284	25,193	276	126	-83	30,305	212
1,160	24,870	284	76	69	31,078	217
1,148	26,077	261	215	225	30,634	214
1,143	25,062	243	631	150	29,886	208
243	38	2	0	-5	673	75
201	33	2	0	11	605	67
198	51	2	0	0	618	69
203	51	3	1	-7	550	62
688	216	2	..	13	717	132
629	205	2	..	-4	688	127
649	223	2	..	-7	700	129
616	207	1	..	15	658	121
209	61	1	1	0	307	149
219	66	1	2	4	296	143
209	72	1	3	-3	287	139
223	96	1	2	-2	280	135
5,113	714	156	360	-25	5,240	112
5,252	1,055	155	352	-142	5,207	112
5,001	1,249	154	303	-125	4,841	104
5,175	1,339	157	327	52	4,732	102
1,351	560	32	72	-6	2,068	219
1,329	686	30	71	-42	2,079	218
1,201	506	34	67	-6	2,050	213
1,332	648	31	72	11	1,998	206

Table 2

Components of Total Energy Supply

Petajoules and Gigajoules per capita

Country or area	Year	Primary energy production					
		Total	Solids	Liquids	Gases	Electricity	Heat
Switzerland	2011	513	90	0	3	117	302
	2012	530	96	0	4	140	290
	2013	540	97	0	4	141	298
	2014	552	94	0	4	140	314
T.F.Yug.Rep. Macedonia	2011	73	67	0	..	5	0
	2012	64	60	0	..	4	0
	2013	57	51	0	..	6	0
	2014	53	48	0	..	4	0
Ukraine	2011	3,513	1,698	145	650	38	981
	2012	3,552	1,741	145	645	40	980
	2013	3,578	1,771	134	671	54	947
	2014	3,203	1,427	119	629	36	992
United Kingdom	2011	5,413	592	2,212	1,785	78	747
	2012	4,880	572	1,902	1,549	95	762
	2013	4,576	478	1,745	1,461	126	765
	2014	4,482	460	1,713	1,468	151	691
Oceania	**2011**	**14,039**	**10,338**	**1,090**	**2,162**	**193**	**256**
	2012	**14,403**	**10,818**	**1,049**	**2,092**	**183**	**261**
	2013	**15,299**	**11,524**	**924**	**2,373**	**206**	**271**
	2014	**16,196**	**12,349**	**898**	**2,423**	**224**	**302**
Australia	2011	13,202	10,102	993	2,008	88	12
	2012	13,538	10,585	931	1,925	85	12
	2013	14,429	11,297	814	2,197	108	13
	2014	15,282	12,137	779	2,232	121	13
Cook Islands	2011	..	..	..	..	..	..
	2012	..	..	..	..	..	..
	2013	..	..	..	..	..	..
	2014	..	..	..	..	..	..
Fiji	2011	7	5	..	..	2	..
	2012	6	4	..	..	2	..
	2013	8	6	..	..	2	..
	2014	8	6	..	..	1	..
French Polynesia	2011	*1	0	..	..	*1	..
	2012	*1	0	..	..	*1	..
	2013	*1	0	..	..	*1	..
	2014	*1	0	..	..	*1	..
Kiribati	2011	0	0	..	..	..	..
	2012	0	0	..	..	..	..
	2013	0	0	..	..	..	..
	2014	0	0	..	..	..	..
Marshall Islands	2011	..	..	..	..	..	..
	2012	0	..	..	..	0	..
	2013	0	..	..	..	0	..
	2014	0	..	..	..	0	..
Micronesia (Fed. States of)	2011	0	0	..	..	0	..
	2012	0	0	..	..	0	..
	2013	0	0	..	..	0	..
	2014	0	0	..	..	0	..
Nauru	2011	0	..	..	..	0	..
	2012	0	..	..	..	0	..
	2013	0	..	..	..	0	..
	2014	0	..	..	..	0	..
New Caledonia	2011	2	..	..	..	2	0
	2012	2	..	..	..	2	0
	2013	2	..	..	..	2	0
	2014	1	..	..	..	1	0

Components of Total Energy Supply

Petajoules and Gigajoules per capita

Imports	Exports	International bunkers		Stock changes	Supply	
		Aviation	Marine		Total	Per capita
735	135	65	0	-8	1,056	133
746	135	66	0	7	1,068	133
775	140	66	0	-5	1,113	136
702	144	67	0	0	1,043	126
76	17	0	..	1	131	63
73	12	0	..	-2	126	61
63	8	0	..	-2	113	54
66	8	1	..	0	110	53
2,414	427	11	..	265	5,225	115
1,950	332	13	..	21	5,135	113
1,668	343	*5	..	54	4,844	107
1,443	278	*6	..	-68	4,431	98
6,297	3,255	480	129	37	7,809	124
6,749	3,108	465	111	-78	8,024	126
6,913	2,944	461	106	26	7,951	124
6,356	2,703	460	98	130	7,448	116
2,420	**9,625**	**177**	**43**	**41**	**6,573**	**178**
2,447	**9,957**	**184**	**48**	**120**	**6,540**	**174**
2,610	**11,040**	**193**	**42**	**155**	**6,479**	**170**
2,595	**12,037**	**204**	**44**	**43**	**6,462**	**167**
1,941	9,463	138	24	34	5,484	243
1,975	9,772	143	31	143	5,423	237
2,079	10,855	151	25	145	5,333	229
2,063	11,853	162	29	49	5,252	222
*1	..	0	..	..	*1	*49
*1	..	0	..	..	*1	*49
*1	..	0	..	..	*1	*48
*1	..	0	..	..	*1	*48
17	0	0	*2	..	21	24
16	0	*1	*2	..	20	23
*18	0	*1	*2	..	*23	*26
*18	0	*1	*2	..	*23	*26
14	..	*1	*2	..	12	46
14	..	*1	*2	..	12	44
14	..	*1	*2	..	12	44
13	..	*1	*2	..	12	43
*1	..	0	..	..	*1	*8
*1	..	0	..	..	*1	*8
*1	..	0	..	..	*1	*8
1	..	0	..	..	1	8
*1	..	..	..	..	*1	*28
1	..	..	..	..	1	27
*1	..	..	..	..	*1	*28
*1	..	..	..	..	*1	*27
*2	..	0	..	..	*2	*18
*2	..	0	..	..	*2	*20
*2	..	0	..	..	*2	*21
*2	..	0	..	..	*2	*22
*1	..	0	0	0	*1	*55
*1	..	0	0	0	*1	*55
*1	..	0	0	0	*1	*61
*1	..	0	0	0	*1	*63
49	0	*1	0	2	48	191
46	0	*1	0	-1	48	188
56	..	0	0	2	56	219
55	..	0	0	0	56	215

Table 2

Components of Total Energy Supply

Petajoules and Gigajoules per capita

Country or area	Year	Primary energy production					
		Total	Solids	Liquids	Gases	Electricity	Heat
New Zealand	2011	741	169	97	149	97	229
	2012	739	167	86	162	90	233
	2013	739	160	76	171	90	242
	2014	783	143	85	187	96	273
Niue	2011	0	0	..	..	0	0
	2012	0	0	..	..	0	0
	2013	0	0	..	..	0	0
	2014	0	0	..	..	0	0
Palau	2011	..	..	..	..	..	..
	2012	..	..	..	..	..	..
	2013	..	..	..	..	..	..
	2014	..	..	..	..	..	..
Papua New Guinea	2011	81	57	0	6	3	15
	2012	112	57	32	5	4	15
	2013	114	57	34	5	4	15
	2014	*114	57	*34	*5	*4	*15
Samoa	2011	*1	*1	..	..	0	..
	2012	*2	*1	..	..	0	..
	2013	*2	*1	..	..	0	..
	2014	*2	*1	..	..	0	..
Solomon Islands	2011	*3	*3	..	..	..	..
	2012	*3	*3	..	..	..	..
	2013	*3	*3	..	..	..	..
	2014	*3	*3	..	..	..	..
Tonga	2011	0	0	..	..	..	..
	2012	0	0	..	..	..	..
	2013	0	0	..	..	0	..
	2014	0	0	..	..	0	..
Tuvalu	2011	..	..	..	..	..	..
	2012	..	..	..	..	..	..
	2013	..	..	..	..	..	..
	2014	..	..	..	..	..	..
Vanuatu	2011	1	1	0	..	0	..
	2012	1	1	0	..	0	..
	2013	1	1	0	..	0	..
	2014	1	1	0	..	0	..
Wallis and Futuna Is.	2011	..	..	..	..	..	..
	2012	..	..	..	..	..	..
	2013	..	..	..	..	..	..
	2014	..	..	..	..	..	..

Components of Total Energy Supply

Petajoules and Gigajoules per capita

Imports	Exports	International bunkers		Stock changes	Supply	
		Aviation	Marine		Total	Per capita
307	163	33	14	5	834	189
306	153	34	12	-22	868	196
327	136	36	12	11	871	195
325	134	35	12	-3	931	207
0	..	0	..	..	0	56
0	..	0	..	..	0	58
0	..	0	..	..	0	61
0	..	0	..	..	0	62
*4	..	*1	..	..	*4	*172
*4	..	*1	..	..	*3	*169
*4	..	*1	..	..	*4	*170
*4	..	*1	..	..	*4	*172
71	0	*2	*1	0	149	21
68	32	*2	0	0	145	20
95	49	*2	0	-2	160	22
*97	*50	*3	0	*-3	*161	*22
3	..	0	..	0	4	23
*3	..	0	..	..	*4	*23
3	..	0	..	0	5	24
*3	..	0	..	..	*5	*24
*3	0	0	0	..	*6	*11
*3	0	0	0	..	*6	*11
*3	0	0	0	..	*6	*11
*3	0	0	0	..	*6	*11
2	..	0	0	..	2	15
2	..	0	0	..	2	15
2	..	0	0	..	2	15
2	..	0	0	..	2	15
0	..	..	..	..	0	10
0	..	..	..	..	0	*13
0	..	..	..	..	0	*13
0	..	..	..	..	0	*13
2	0	0	0	0	3	12
2	0	0	0	0	3	11
2	0	0	0	0	3	10
3	0	0	0	0	3	12
0	..	0	..	..	0	28
0	..	0	..	..	0	26
0	..	0	..	..	0	26
0	..	0	..	..	0	26

Table 3

Total final consumption and its components

Petajoules and Gigajoules per capita

Table Notes

Total final consumption refers to all fuel and energy that is delivered to users for both their energy and non-energy uses which do not involve a transformation process.

Non-energy use consists of the use of energy products as raw materials for the manufacture of products outside the scope of energy statistics, as well as for direct uses that do not involve using the products as a source of energy, nor as a transformation input. Examples are lubrication, sealing, preservation, road surfacing, and use as a solvent.

Industry refers to the use of energy products for energy purposes by economic units belonging to manufacturing, construction and non-fuel mining industries, namely the following: Iron and steel, Chemical and petrochemical, Non-ferrous metals, Non-metallic minerals; Transport equipment; Machinery; Mining and quarrying; Food and Tobacco; Paper, pulp and print; Wood and wood products (other than pulp and paper); Textile and leather; Construction; and Industries not elsewhere specified. It excludes the use of energy products for transport.

Transport refers to the consumption of fuels and electricity by any economic entity to transport goods or persons between points of departure and destination within the national territory. As such, it excludes international aviation and shipping, which are included under international marine and aviation bunkers respectively.

Other refers to energy consumed by consumers not classified under Industry. It includes Households, Commerce and public services, Agriculture, Forestry, Fishing, and Not elsewhere specified.

Please refer to the Definitions Section on pages ix to xv for the appropriate product description/ classification.

Figure 9: World total final consumption by component, 2014

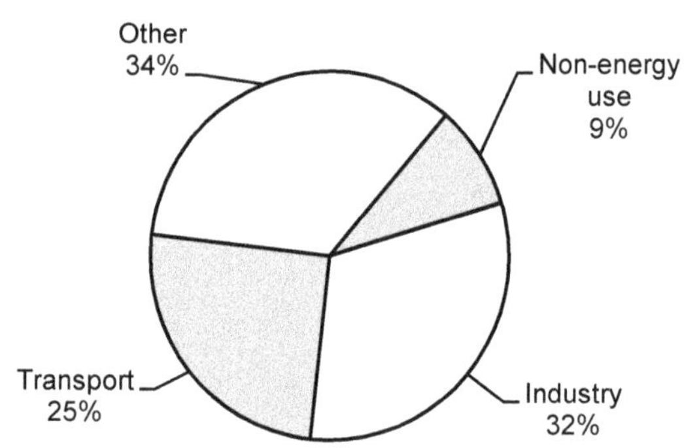

Figure 10: Total final consumption by region, 1993-2014

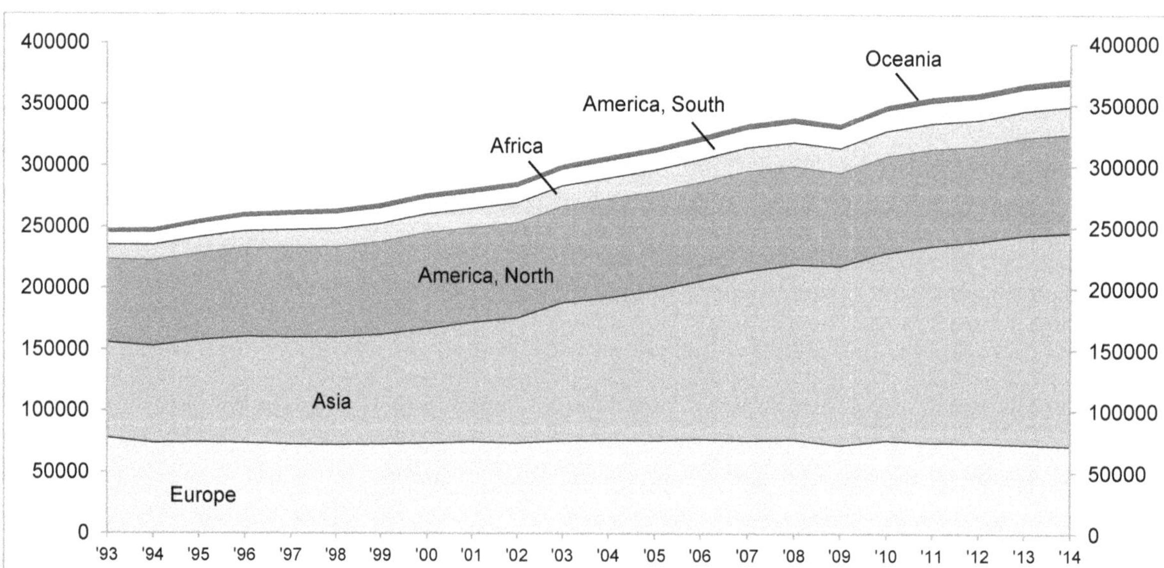

Total final consumption and its components

Petajoules and gigajoules per capita

Country or area	Year	Total final consumption	Non-energy use	Energy use: Industry	Energy use: Transport	Energy use: Other	TFC per capita
World	**2011**	**356,472**	**32,851**	**113,766**	**85,518**	**124,338**	**51**
	2012	**359,846**	**33,065**	**113,511**	**87,712**	**125,558**	**51**
	2013	**367,579**	**33,659**	**115,621**	**91,141**	**127,159**	**51**
	2014	**371,589**	**33,922**	**116,760**	**93,440**	**127,467**	**51**
Africa	**2011**	**20,793**	**771**	**3,697**	**3,430**	**12,896**	**19**
	2012	**21,204**	**773**	**3,698**	**3,597**	**13,135**	**19**
	2013	**22,011**	**789**	**3,990**	**3,793**	**13,439**	**20**
	2014	**22,097**	**784**	**3,682**	**4,020**	**13,611**	**19**
Algeria	2011	1,150	80	225	457	388	31
	2012	1,268	116	218	503	432	34
	2013	1,327	109	223	521	475	35
	2014	1,458	155	227	603	473	37
Angola	2011	*430	5	50	98	*277	*20
	2012	*442	10	48	98	*285	*19
	2013	*471	12	39	112	*308	*20
	2014	*479	11	38	117	*313	*20
Benin	2011	*138	..	3	45	*90	*14
	2012	*143	..	3	47	*93	*14
	2013	*150	..	3	51	*96	*15
	2014	*159	..	5	54	*100	*15
Botswana	2011	65	1	18	28	18	31
	2012	66	1	18	28	19	31
	2013	68	1	18	29	20	31
	2014	60	1	14	31	15	27
Burkina Faso	2011	*130	0	*3	*19	*107	*8
	2012	*136	0	*3	*24	*109	*8
	2013	*144	0	*4	*29	*110	*8
	2014	*139	0	*4	*24	*112	*8
Burundi	2011	*89	1	*1	3	*85	*9
	2012	*56	0	*1	3	*52	*6
	2013	*53	1	*1	2	*49	*5
	2014	*55	0	*1	5	*49	*5
Cabo Verde	2011	7	0	0	4	3	14
	2012	7	0	0	4	3	14
	2013	7	0	0	3	3	13
	2014	6	0	0	3	3	13
Cameroon	2011	*252	1	14	38	*199	*12
	2012	*260	1	15	40	*204	*12
	2013	*268	1	16	42	*209	*12
	2014	*277	1	16	45	*215	*12
Central African Rep.	2011	*18	0	0	*2	*15	*4
	2012	*18	0	0	*2	*15	*4
	2013	*18	0	0	*2	*15	*4
	2014	*18	0	0	*2	*15	*4
Chad	2011	*62	0	5	*1	*56	*5
	2012	*63	0	5	*1	*57	*5
	2013	*64	0	6	*1	*58	*5
	2014	*66	0	*6	*1	*59	*5
Comoros	2011	*3	..	..	*1	*3	*4
	2012	*3	..	..	0	*3	*4
	2013	*3	..	..	*1	*3	*5
	2014	*3	..	..	*1	*3	*4

Table 3

Total final consumption and its components

Petajoules and gigajoules per capita

Country or area	Year	Total final consumption	Non-energy use	Energy use			TFC per capita
				Industry	Transport	Other	
Congo	2011	*74	1	2	22	*48	*18
	2012	*76	1	3	22	*50	*18
	2013	*82	1	3	26	*52	*19
	2014	*84	1	2	28	*53	*19
Côte d'Ivoire	2011	263	1	19	21	223	13
	2012	288	1	21	33	234	14
	2013	308	3	27	36	243	14
	2014	322	2	28	40	253	15
Dem. Rep. of the Congo	2011	848	0	182	29	636	12
	2012	695	0	130	29	535	10
	2013	796	0	133	45	617	11
	2014	895	0	141	62	692	12
Djibouti	2011	*6	0	0	1	*5	*7
	2012	*7	0	0	1	*5	*8
	2013	*7	0	0	*1	*5	*8
	2014	*7	0	0	*1	*5	*8
Egypt	2011	2,459	345	646	470	998	29
	2012	2,418	313	639	458	1,008	28
	2013	2,554	*317	865	426	947	29
	2014	2,245	290	609	531	815	25
Equatorial Guinea	2011	*70	0	*58	*6	*6	*93
	2012	*74	0	*62	*6	*6	*96
	2013	*67	0	*55	*6	*6	*84
	2014	*69	0	*58	*6	*6	*84
Eritrea	2011	*21	0	0	2	*18	*4
	2012	*21	0	1	2	*19	*4
	2013	*22	0	1	2	*19	*4
	2014	*22	0	1	2	*19	*4
Ethiopia	2011	*1,224	3	43	41	*1,137	*14
	2012	*1,250	3	42	46	*1,158	*14
	2013	*1,288	3	51	54	*1,179	*14
	2014	*1,316	4	56	58	*1,198	*14
Gabon	2011	*85	1	25	10	*49	*54
	2012	*87	0	26	10	*51	*54
	2013	*93	0	28	12	*53	*56
	2014	*93	0	28	11	*53	*55
Gambia	2011	*10	..	*2	*1	*7	*6
	2012	*10	..	*2	*1	*7	*6
	2013	*10	..	*2	*1	*7	*5
	2014	*11	..	*3	*1	*7	*6
Ghana	2011	240	*3	42	89	106	10
	2012	276	*3	50	107	117	11
	2013	282	*3	49	*109	122	11
	2014	281	*3	48	*110	120	10
Guinea	2011	*133	*1	*16	*14	*103	*12
	2012	*132	*1	*13	*14	*103	*11
	2013	*129	*1	*10	*15	*103	*11
	2014	*130	*1	*11	*15	*103	*11
Guinea-Bissau	2011	*26	*1	0	*1	*24	*16
	2012	*27	*1	0	*1	*25	*16
	2013	*27	*1	0	*1	*25	*15
	2014	*28	*1	0	*1	*25	*15

Total final consumption and its components

Petajoules and gigajoules per capita

Country or area	Year	Total final consumption	Non-energy use	Energy use: Industry	Energy use: Transport	Energy use: Other	TFC per capita
Kenya	2011	464	10	68	63	325	11
	2012	473	12	57	66	338	11
	2013	485	2	62	79	342	11
	2014	498	2	50	103	343	11
Lesotho	2011	*54	..	*19	6	*29	*26
	2012	*55	..	*19	6	*29	*27
	2013	*55	..	*20	6	*30	*27
	2014	*56	..	*20	*6	*30	*26
Liberia	2011	*69	0	..	*4	*64	*17
	2012	*73	0	..	*5	*68	*17
	2013	*74	0	..	*8	*67	*17
	2014	*74	0	..	*5	*69	*17
Libya	2011	339	49	26	189	76	54
	2012	387	51	23	234	79	62
	2013	444	74	46	246	78	71
	2014	388	30	32	263	64	62
Madagascar	2011	*118	0	*7	*12	*98	*5
	2012	*122	*1	*10	*12	*98	*5
	2013	*120	*2	*15	*13	*91	*5
	2014	*122	*2	*14	*13	*94	*5
Malawi	2011	*63	0	*12	*9	*42	*4
	2012	*63	0	*11	*9	*43	*4
	2013	*64	0	*11	*9	*43	*4
	2014	*65	0	*12	*10	*43	*4
Mali	2011	*57	..	*3	*5	*50	*4
	2012	*59	..	*3	*5	*51	*4
	2013	*57	..	3	*5	*49	*3
	2014	*60	..	*4	*5	*51	*3
Mauritania	2011	*37	0	7	*12	*18	*10
	2012	*40	0	8	*13	*18	*11
	2013	*40	0	*8	*13	*19	*10
	2014	*41	0	*8	13	*20	*10
Mauritius	2011	36	1	9	18	9	29
	2012	36	1	8	18	9	28
	2013	37	1	8	18	9	29
	2014	37	1	8	19	9	30
Mayotte	2011	3	..	..	1	1	12
	2012	3	..	..	*1	1	12
	2013	3	..	..	*1	1	12
	2014	3	..	..	*1	1	12
Morocco	2011	605	21	156	*151	276	18
	2012	633	14	179	*151	289	19
	2013	605	19	128	205	252	18
	2014	611	18	125	208	260	18
Mozambique	2011	*349	..	77	27	*246	*14
	2012	*321	..	80	26	*215	*12
	2013	*412	..	84	27	*301	*16
	2014	*465	..	99	53	*312	*17
Namibia	2011	*55	1	*7	22	*25	*25
	2012	*60	1	*9	22	*27	*26
	2013	*55	*1	7	*20	*26	*23
	2014	67	1	9	26	31	28

Table 3

Total final consumption and its components

Petajoules and gigajoules per capita

Country or area	Year	Total final consumption	Non-energy use	Energy use			TFC per capita
				Industry	Transport	Other	
Niger	2011	67	0	3	7	57	4
	2012	74	0	4	9	61	4
	2013	80	0	4	12	63	4
	2014	*81	0	*4	*14	*63	*4
Nigeria	2011	*4,648	40	*412	351	*3,844	*28
	2012	*4,860	47	*425	362	*4,026	*29
	2013	*4,900	50	*415	340	*4,095	*28
	2014	*4,871	57	*307	302	*4,204	*27
Réunion	2011	40	*1	3	24	11	48
	2012	40	*1	4	23	12	47
	2013	39	*1	4	22	12	46
	2014	40	*1	4	23	12	47
Rwanda	2011	*72	0	*6	*5	*60	*7
	2012	*75	*1	*6	*6	*62	*7
	2013	*77	*1	*6	*7	*64	*7
	2014	*78	*1	*6	*7	*65	*7
Sao Tome and Principe	2011	*2	0	0	0	*1	*10
	2012	*2	0	0	0	*1	*11
	2013	*2	0	0	0	*1	*10
	2014	*2	0	0	0	*1	*10
Senegal	2011	112	2	18	30	63	8
	2012	112	2	17	28	64	8
	2013	107	2	17	31	57	8
	2014	112	2	18	33	58	8
Seychelles	2011	3	0	1	*1	1	33
	2012	*5	0	1	*1	*3	*49
	2013	*4	0	1	*1	*2	*42
	2014	*5	0	1	*2	*3	*54
Sierra Leone	2011	*52	0	*1	*6	*46	*9
	2012	*54	0	*1	*6	*47	*9
	2013	*56	0	*2	*7	*48	*9
	2014	*58	0	*2	*8	*48	*9
Somalia	2011	*95	..	*2	2	*92	*10
	2012	*96	..	*2	2	*92	*10
	2013	*96	..	*2	2	*92	*9
	2014	*98	..	*2	2	*94	*9
South Africa	2011	2,918	177	1,083	726	932	56
	2012	2,963	165	1,084	726	988	56
	2013	3,100	158	1,150	792	1,000	58
	2014	3,096	172	1,165	754	1,004	57
South Sudan	2011	..	..	..	..	..	..
	2012	23	..	0	13	9	2
	2013	22	..	0	13	9	2
	2014	23	..	0	13	9	2
St. Helena and Depend.	2011	0	..	..	0	0	18
	2012	0	..	..	0	0	23
	2013	0	..	..	0	0	20
	2014	0	..	..	0	0	*20
Sudan	2011	..	..	..	..	..	..
	2012	418	9	60	116	233	11
	2013	428	10	62	112	244	11
	2014	434	10	68	108	248	11

Total final consumption and its components

Petajoules and gigajoules per capita

Country or area	Year	Total final consumption	Non-energy use	Energy use			TFC per capita
				Industry	Transport	Other	
Sudan (former)	2011	*489	10	*74	120	*285	*10
	2012	..	..	..	..	..	..
	2013	..	..	..	..	..	..
	2014	..	..	..	..	..	..
Swaziland	2011	*32	..	*15	*8	*10	*27
	2012	*35	..	*15	*8	*11	*28
	2013	*37	..	*17	*8	*12	*30
	2014	38	0	18	8	*11	30
Togo	2011	84	0	4	20	59	13
	2012	82	0	3	17	62	12
	2013	84	0	3	18	63	12
	2014	87	0	3	18	65	12
Tunisia	2011	291	6	78	92	115	27
	2012	303	7	85	91	120	28
	2013	304	8	88	86	124	28
	2014	316	9	93	89	126	28
Uganda	2011	*405	0	*19	23	*363	*12
	2012	*407	0	*20	25	*362	*12
	2013	*416	0	*28	27	*361	*11
	2014	*424	1	*32	29	*363	*11
United Rep. of Tanzania	2011	773	1	114	52	606	16
	2012	813	2	121	65	624	17
	2013	859	2	131	83	644	17
	2014	902	2	142	82	677	17
Zambia	2011	284	2	87	11	184	20
	2012	306	3	101	13	189	21
	2013	318	3	105	14	196	21
	2014	327	3	110	16	198	21
Zimbabwe	2011	*373	1	33	31	*309	*26
	2012	*391	1	39	35	*316	*27
	2013	*391	1	28	39	*323	*26
	2014	*395	1	30	36	*329	*26
America, North	**2011**	**79,077**	**7,266**	**15,280**	**30,388**	**26,143**	**143**
	2012	**77,817**	**6,639**	**14,987**	**31,018**	**25,173**	**139**
	2013	**79,635**	**6,627**	**15,225**	**31,162**	**26,620**	**141**
	2014	**80,889**	**6,511**	**15,281**	**31,793**	**27,304**	**142**
Anguilla	2011	*1	..	..	*1	*1	*104
	2012	*1	..	..	*1	*1	*103
	2013	*1	..	..	*1	*1	*99
	2014	*1	..	..	*1	*1	*99
Antigua and Barbuda	2011	*4	0	0	*3	*1	*51
	2012	*5	0	0	*3	*1	*52
	2013	*5	0	0	*3	*1	*52
	2014	*5	0	0	*3	*1	*52
Aruba	2011	*6	0	..	*3	*3	*62
	2012	*6	0	..	*3	*3	*62
	2013	*6	0	..	*3	*3	*62
	2014	*6	0	..	*3	*3	*62
Bahamas	2011	19	0	3	7	9	51
	2012	23	*1	4	7	11	62
	2013	*34	0	*9	*10	*15	*89
	2014	*26	0	*8	*8	*10	*68

Table 3

Total final consumption and its components

Petajoules and gigajoules per capita

Country or area	Year	Total final consumption	Non-energy use	Energy use			TFC per capita
				Industry	Transport	Other	
Barbados	2011	12	..	2	6	5	44
	2012	12	..	2	6	5	44
	2013	12	..	1	6	4	41
	2014	12	..	2	6	*4	41
Belize	2011	12	0	*3	6	3	36
	2012	11	0	*3	4	3	31
	2013	11	0	*3	5	3	32
	2014	11	0	*3	5	3	31
Bermuda	2011	5	0	0	*2	2	75
	2012	*6	0	0	*4	2	*97
	2013	4	0	0	*2	2	65
	2014	5	0	0	*2	3	76
Bonaire, St Eustatius, Saba	2011	..	..	..	..	..	..
	2012	*4	*1	*1	*2	*1	*180
	2013	*4	*1	*1	*2	*1	*178
	2014	*4	*1	*1	*2	*1	*174
British Virgin Islands	2011	*1	..	0	*1	*1	*52
	2012	*1	..	0	*1	*1	*51
	2013	*1	..	0	*1	*1	*50
	2014	*2	..	0	*1	*1	*51
Canada	2011	8,186	971	1,911	2,475	2,829	237
	2012	8,263	1,048	1,921	2,526	2,768	237
	2013	8,356	954	1,984	2,601	2,817	237
	2014	8,415	911	2,012	2,639	2,854	236
Cayman Islands	2011	5	0	1	3	1	87
	2012	5	0	1	2	1	80
	2013	5	0	1	2	1	80
	2014	5	0	1	2	1	77
Costa Rica	2011	152	2	35	72	44	33
	2012	152	2	36	74	40	33
	2013	147	2	37	67	40	31
	2014	148	3	37	68	40	31
Cuba	2011	303	5	167	30	101	27
	2012	*304	5	*175	29	95	*27
	2013	*296	*5	*164	30	97	*26
	2014	*278	*4	*157	28	88	*24
Curaçao	2011	..	..	..	..	..	..
	2012	28	4	6	14	3	184
	2013	28	4	6	15	3	183
	2014	29	4	7	15	3	184
Dominica	2011	*2	..	0	*1	0	*22
	2012	*2	..	0	*1	0	*22
	2013	*2	..	0	*1	0	*22
	2014	*2	..	0	*1	0	*22
Dominican Republic	2011	212	14	51	84	63	21
	2012	224	15	54	89	66	22
	2013	215	14	56	79	66	21
	2014	203	10	55	69	69	20
El Salvador	2011	107	1	27	43	36	18
	2012	99	2	23	38	37	16
	2013	93	1	18	38	36	15
	2014	96	1	19	39	36	16

Total final consumption and its components

Petajoules and gigajoules per capita

Country or area	Year	Total final consumption	Non-energy use	Energy use: Industry	Energy use: Transport	Energy use: Other	TFC per capita
Greenland	2011	8	0	1	2	6	150
	2012	8	0	1	2	6	149
	2013	8	0	1	2	6	148
	2014	8	0	1	2	5	139
Grenada	2011	3	..	0	1	1	25
	2012	3	..	0	1	1	25
	2013	3	..	0	1	1	25
	2014	3	..	0	1	1	26
Guadeloupe	2011	*19	*1	*1	*10	*7	*41
	2012	*19	*1	*1	*10	7	*41
	2013	*19	*1	*1	*11	7	*41
	2014	*19	*1	*1	*11	7	*41
Guatemala	2011	*316	2	33	76	*206	*21
	2012	*322	2	33	77	*211	*21
	2013	*340	2	44	84	*211	*22
	2014	400	1	80	90	227	25
Haiti	2011	124	0	11	15	98	12
	2012	122	0	9	13	99	12
	2013	124	0	10	12	102	12
	2014	129	0	13	16	100	12
Honduras	2011	*173	..	36	44	*94	*23
	2012	*188	..	47	46	*95	*24
	2013	*186	..	44	46	*97	*24
	2014	*176	..	25	48	*103	*22
Jamaica	2011	78	1	34	28	15	28
	2012	74	1	33	26	15	27
	2013	81	1	40	26	*14	29
	2014	77	1	32	30	*14	28
Martinique	2011	*18	0	0	*10	*7	*45
	2012	*18	0	0	*11	*7	*46
	2013	*18	0	0	*11	*7	*46
	2014	*19	0	0	*11	*7	*47
Mexico	2011	5,013	277	1,489	2,142	1,104	42
	2012	4,973	220	1,478	2,153	1,122	41
	2013	4,991	229	1,518	2,130	1,115	40
	2014	4,937	256	1,405	2,137	1,139	39
Montserrat	2011	0	0	0	0	0	81
	2012	0	0	0	0	0	84
	2013	*1	0	0	0	0	*113
	2014	*1	0	0	0	0	*100
Neth. Antilles (former)	2011	42	6	10	21	5	198
	2012	..	..	..	..	..	..
	2013	..	..	..	..	..	..
	2014	..	..	..	..	..	..
Nicaragua	2011	88	1	13	23	51	15
	2012	92	1	13	25	53	16
	2013	94	1	13	24	55	16
	2014	99	1	14	26	57	16
Panama	2011	122	..	36	51	35	33
	2012	130	..	41	52	37	35
	2013	132	..	41	52	39	35
	2014	137	..	40	55	41	35

Table 3

Total final consumption and its components

Petajoules and gigajoules per capita

Country or area	Year	Total final consumption	Non-energy use	Energy use			TFC per capita
				Industry	Transport	Other	
Puerto Rico	2011	72	..	10	..	62	19
	2012	70	..	10	..	60	19
	2013	71	..	9	..	62	19
	2014	71	..	8	..	63	19
Sint Maarten (Dutch part)	2011	..	..	..	..	..	..
	2012	*9	*1	*2	*5	*1	*260
	2013	*9	*1	*2	*5	1	*256
	2014	*10	*1	*2	*5	*1	*253
St. Kitts-Nevis	2011	2	0	0	1	1	34
	2012	2	0	0	1	1	33
	2013	*2	0	0	*1	*1	*33
	2014	*2	0	0	*1	*1	*33
St. Lucia	2011	*4	0	0	*2	2	*21
	2012	*4	0	0	*2	2	*21
	2013	*4	0	0	*2	2	*21
	2014	*4	0	0	*2	2	*21
St. Pierre-Miquelon	2011	*1	..	..	0	0	*108
	2012	*1	..	..	0	0	*106
	2013	*1	..	..	0	0	*108
	2014	*1	..	..	*1	0	*117
St. Vincent-Grenadines	2011	*2	0	0	*2	1	*22
	2012	*3	0	0	*2	1	*25
	2013	*2	0	0	*1	1	*21
	2014	*2	0	0	*1	*1	*23
Trinidad and Tobago	2011	564	423	79	44	18	423
	2012	561	406	94	43	18	418
	2013	564	404	94	45	22	418
	2014	575	399	111	44	20	424
Turks and Caicos Islands	2011	*1	..	..	0	*1	*33
	2012	*1	..	..	0	*1	*34
	2013	*1	..	..	0	*1	*33
	2014	*1	..	..	0	*1	*33
United States	2011	63,394	5,560	11,329	25,177	21,329	203
	2012	62,066	4,929	10,998	25,743	20,396	197
	2013	63,761	5,006	11,126	25,845	21,784	201
	2014	64,972	4,914	11,245	26,417	22,395	203
United States Virgin Is.	2011	3	..	1	..	2	27
	2012	3	..	1	..	1	25
	2013	2	..	1	..	1	23
	2014	2	..	1	..	1	22
America, South	**2011**	**17,678**	**1,009**	**6,486**	**5,943**	**4,240**	**44**
	2012	**18,276**	**1,015**	**6,466**	**6,406**	**4,390**	**45**
	2013	**18,334**	**982**	**6,419**	**6,429**	**4,504**	**45**
	2014	**18,686**	**966**	**6,345**	**6,797**	**4,577**	**45**
Argentina	2011	2,451	189	697	756	810	59
	2012	2,508	202	680	756	870	60
	2013	2,436	206	686	648	896	57
	2014	2,609	204	722	737	946	61
Bolivia (Plur. State of)	2011	217	2	45	86	84	22
	2012	245	2	50	92	101	24
	2013	253	2	55	102	94	24
	2014	268	2	56	109	102	25

Total final consumption and its components

Petajoules and gigajoules per capita

Country or area	Year	Total final consumption	Non-energy use	Energy use			TFC per capita
				Industry	Transport	Other	
Brazil	2011	9,525	674	3,652	3,303	1,895	47
	2012	9,793	664	3,649	3,524	1,956	48
	2013	9,966	648	3,623	3,715	1,979	49
	2014	10,146	636	3,562	3,867	2,081	49
Chile	2011	1,045	30	398	302	316	61
	2012	1,042	31	379	310	322	60
	2013	1,112	26	417	338	330	63
	2014	1,037	27	434	328	247	58
Colombia	2011	949	*24	246	350	329	20
	2012	1,013	*20	238	422	332	22
	2013	1,050	15	266	402	368	22
	2014	1,115	17	312	447	339	23
Ecuador	2011	473	20	97	184	172	31
	2012	492	22	105	192	174	32
	2013	514	21	107	205	180	33
	2014	570	21	113	220	216	36
Falkland Is. (Malvinas)	2011	*1	..	0	0	0	*230
	2012	*1	..	0	0	0	*234
	2013	*1	..	0	0	0	*233
	2014	*1	..	0	0	0	*232
French Guiana	2011	9	0	0	*4	5	36
	2012	9	0	0	*4	5	36
	2013	9	0	0	*4	5	35
	2014	9	0	0	*4	5	35
Guyana	2011	26	0	9	8	9	34
	2012	28	0	8	9	10	36
	2013	26	0	7	9	9	34
	2014	27	0	7	10	10	35
Paraguay	2011	212	1	67	71	72	34
	2012	211	1	70	68	72	33
	2013	212	1	70	71	69	33
	2014	222	1	73	75	73	34
Peru	2011	624	11	178	245	191	21
	2012	650	30	180	262	179	22
	2013	705	16	197	281	211	23
	2014	684	14	192	269	209	22
Suriname	2011	22	1	5	7	9	41
	2012	25	1	5	8	11	47
	2013	23	1	5	8	10	44
	2014	24	1	5	8	10	45
Uruguay	2011	159	2	54	46	57	47
	2012	162	3	54	47	58	48
	2013	170	4	57	50	59	50
	2014	179	3	68	50	58	52
Venezuela (Bolivar. Rep.)	2011	1,965	55	1,038	581	291	67
	2012	2,098	39	1,046	712	300	70
	2013	1,858	40	929	596	293	61
	2014	1,794	38	801	673	282	58
Asia	**2011**	**160,757**	**15,741**	**68,721**	**25,198**	**51,097**	**38**
	2012	**164,525**	**16,706**	**68,489**	**26,809**	**52,521**	**39**
	2013	**170,785**	**17,457**	**70,924**	**29,968**	**52,437**	**40**
	2014	**174,454**	**17,538**	**72,495**	**30,926**	**53,494**	**40**

Table 3

Total final consumption and its components

Petajoules and gigajoules per capita

Country or area	Year	Total final consumption	Non-energy use	Energy use			TFC per capita
				Industry	Transport	Other	
Afghanistan	2011	177	0	*49	108	*20	6
	2012	*162	0	*43	*96	*23	*5
	2013	149	0	*46	81	*22	5
	2014	*146	0	*49	72	*25	*5
Armenia	2011	*103	*2	*15	*22	*64	*35
	2012	*109	*2	*17	*23	*68	*37
	2013	*103	*2	*16	*22	*63	*34
	2014	*103	*1	*17	*22	*62	*34
Azerbaijan	2011	318	26	37	85	170	34
	2012	327	27	49	93	158	35
	2013	345	32	49	105	158	36
	2014	363	31	51	109	171	38
Bahrain	2011	213	71	60	42	40	163
	2012	220	79	53	46	42	165
	2013	242	71	80	47	44	179
	2014	254	69	89	48	48	186
Bangladesh	2011	1,203	66	227	125	785	8
	2012	1,228	63	244	126	795	8
	2013	1,237	64	257	121	795	8
	2014	1,283	58	265	129	831	8
Bhutan	2011	*58	0	6	5	*47	*79
	2012	*60	0	7	5	*48	*80
	2013	*61	0	7	*6	*48	*81
	2014	*62	0	8	*5	*48	*81
Brunei Darussalam	2011	72	35	7	18	12	181
	2012	79	40	8	19	13	195
	2013	46	8	6	18	13	111
	2014	60	22	5	19	14	144
Cambodia	2011	201	1	32	46	122	14
	2012	211	1	35	49	127	14
	2013	219	1	38	50	131	15
	2014	231	0	37	54	140	15
China	2011	67,548	5,851	38,002	7,599	16,095	50
	2012	68,737	6,246	37,214	8,394	16,882	51
	2013	72,483	6,622	38,782	9,023	18,056	53
	2014	74,768	6,938	39,834	9,347	18,648	55
China, Hong Kong SAR	2011	304	1	32	66	205	43
	2012	306	1	30	68	207	43
	2013	308	2	25	81	200	43
	2014	314	1	26	81	206	43
China, Macao SAR	2011	22	0	2	4	15	40
	2012	24	0	2	5	17	42
	2013	25	0	*2	5	18	44
	2014	27	0	3	5	19	47
Cyprus	2011	70	3	9	31	27	63
	2012	64	2	7	29	26	57
	2013	58	1	8	26	23	51
	2014	58	1	9	25	23	51
Georgia	2011	133	13	32	34	54	32
	2012	138	10	32	36	60	33
	2013	151	12	27	43	70	37
	2014	169	13	26	52	77	42

Total final consumption and its components

Petajoules and gigajoules per capita

Country or area	Year	Total final consumption	Non-energy use	Energy use: Industry	Energy use: Transport	Energy use: Other	TFC per capita
India	2011	*21,325	1,193	8,090	1,006	*11,036	*17
	2012	*21,874	1,356	8,202	1,016	*11,301	*17
	2013	24,045	1,524	9,630	3,101	*9,790	19
	2014	*25,349	1,384	10,794	*3,346	*9,825	*20
Indonesia	2011	*7,732	349	*3,039	1,614	*2,730	*32
	2012	*8,037	343	*3,100	1,807	*2,786	*32
	2013	7,209	294	2,460	1,718	*2,737	29
	2014	*6,679	347	*1,621	1,758	*2,953	*26
Iran (Islamic Rep. of)	2011	7,117	997	*1,669	1,695	2,756	95
	2012	7,078	941	*1,726	1,769	2,642	93
	2013	7,456	1,065	*1,722	1,829	2,840	97
	2014	7,637	909	1,844	1,993	2,890	98
Iraq	2011	931	26	158	457	290	29
	2012	964	32	167	494	271	29
	2013	1,097	62	245	490	301	32
	2014	900	39	199	402	260	26
Israel	2011	588	65	56	217	250	78
	2012	615	64	64	237	250	80
	2013	582	77	87	219	199	74
	2014	604	79	100	234	191	76
Japan	2011	12,708	1,586	3,886	3,035	4,200	100
	2012	12,674	1,608	3,786	3,104	4,176	100
	2013	12,732	1,601	3,694	3,094	4,343	100
	2014	12,428	1,527	3,713	2,989	4,199	98
Jordan	2011	206	5	40	85	75	30
	2012	217	5	29	107	77	31
	2013	219	5	37	109	68	30
	2014	235	7	43	108	76	32
Kazakhstan	2011	1,800	23	999	208	570	109
	2012	1,753	35	957	221	540	104
	2013	1,797	34	1,021	208	534	105
	2014	1,537	31	690	206	609	88
Korea, Dem.Ppl's.Rep.	2011	501	..	328	18	155	20
	2012	512	..	337	18	157	21
	2013	386	..	238	20	127	15
	2014	426	..	270	18	137	17
Korea, Republic of	2011	6,805	1,716	1,997	1,235	1,857	138
	2012	6,992	1,859	1,992	1,269	1,873	141
	2013	7,080	1,878	2,002	1,315	1,884	142
	2014	7,193	1,976	2,076	1,336	1,804	144
Kuwait	2011	602	131	165	163	143	186
	2012	673	160	186	175	152	197
	2013	642	118	192	180	152	179
	2014	709	116	244	184	165	189
Kyrgyzstan	2011	117	3	15	39	59	21
	2012	148	3	25	56	64	26
	2013	144	3	15	56	70	25
	2014	131	2	26	34	69	22
Lao People's Dem. Rep.	2011	*70	0	*6	*3	*62	*11
	2012	*71	0	*5	*3	*63	*11
	2013	*72	0	6	*3	*63	*11
	2014	*72	0	7	*3	*63	*11

Table 3

Total final consumption and its components

Petajoules and gigajoules per capita

Country or area	Year	Total final consumption	Non-energy use	Energy use			TFC per capita
				Industry	Transport	Other	
Lebanon	2011	158	2	27	72	57	34
	2012	181	3	25	76	77	37
	2013	174	2	26	72	73	33
	2014	197	2	28	78	90	35
Malaysia	2011	1,892	402	490	644	357	66
	2012	1,996	421	565	645	365	69
	2013	2,109	349	547	834	379	72
	2014	2,142	323	531	908	380	72
Maldives	2011	*12	0	0	*10	*2	*36
	2012	*14	0	0	*12	*2	*40
	2013	*13	0	0	*11	*2	*38
	2014	*16	0	0	*14	*2	*45
Mongolia	2011	125	..	23	16	86	45
	2012	136	..	26	19	90	48
	2013	142	..	26	18	99	50
	2014	135	..	24	20	90	46
Myanmar	2011	*602	9	79	56	*457	*12
	2012	*607	10	79	41	*476	*12
	2013	626	13	79	44	490	12
	2014	*681	7	98	63	*513	*13
Nepal	2011	*372	..	32	*32	*308	*14
	2012	*413	..	31	*36	*346	*15
	2013	*465	..	36	*41	*389	*17
	2014	*522	..	*45	*47	*430	*19
Oman	2011	622	98	304	137	82	194
	2012	705	97	362	156	90	199
	2013	731	88	373	172	98	187
	2014	780	96	404	174	106	184
Other Asia	2011	2,738	754	984	510	491	118
	2012	2,732	794	956	499	483	117
	2013	2,824	859	981	499	485	121
	2014	2,845	875	968	505	497	122
Pakistan	2011	2,331	138	739	512	941	13
	2012	2,336	121	*712	527	976	13
	2013	2,353	135	*705	537	975	13
	2014	2,398	134	*722	558	983	13
Philippines	2011	982	10	248	348	377	10
	2012	1,090	12	242	366	470	11
	2013	1,157	19	263	385	490	12
	2014	1,212	27	273	402	510	12
Qatar	2011	545	156	178	139	72	286
	2012	659	180	253	146	79	327
	2013	667	181	247	161	78	317
	2014	772	213	279	194	86	355
Saudi Arabia	2011	5,124	1,028	1,846	1,530	721	178
	2012	5,533	1,145	1,929	1,670	789	188
	2013	5,525	1,080	1,908	1,728	809	183
	2014	5,887	1,155	2,041	1,821	869	191
Singapore	2011	689	250	235	112	92	133
	2012	678	240	234	109	95	128
	2013	711	259	249	106	97	132
	2014	725	276	247	103	98	132

Total final consumption and its components

Petajoules and gigajoules per capita

Country or area	Year	Total final consumption	Non-energy use	Energy use: Industry	Energy use: Transport	Energy use: Other	TFC per capita
Sri Lanka	2011	330	6	76	98	151	16
	2012	328	4	79	99	146	16
	2013	334	*6	79	*105	144	16
	2014	340	*2	83	*110	144	16
State of Palestine	2011	42	1	*2	13	26	10
	2012	49	0	*5	15	29	11
	2013	51	1	3	19	28	12
	2014	65	1	3	33	29	14
Syrian Arab Republic	2011	521	57	131	164	168	25
	2012	400	41	101	128	130	20
	2013	356	32	88	120	117	18
	2014	299	30	71	91	107	16
Tajikistan	2011	81	0	28	5	48	10
	2012	89	1	24	13	52	11
	2013	96	1	21	13	61	12
	2014	107	1	16	22	69	13
Thailand	2011	*2,855	229	*867	*924	*835	*43
	2012	*3,069	271	*920	*995	*883	*46
	2013	3,305	464	993	957	892	49
	2014	3,256	423	1,008	939	886	48
Timor-Leste	2011	*3	..	..	*1	*2	*3
	2012	*4	..	..	*1	*3	*4
	2013	*5	..	..	*1	*4	*4
	2014	*5	..	..	*1	*4	*5
Turkey	2011	3,346	259	1,013	630	1,444	46
	2012	3,550	299	1,018	731	1,502	47
	2013	3,535	298	985	831	1,420	46
	2014	3,551	231	1,036	873	1,411	46
Turkmenistan	2011	677	..	47	155	476	133
	2012	704	..	51	160	493	136
	2013	729	..	53	173	503	139
	2014	750	..	56	182	512	141
United Arab Emirates	2011	*2,089	*80	*1,236	485	288	*239
	2012	*2,267	*78	*1,411	477	300	*253
	2013	*2,390	*93	*1,441	537	320	*264
	2014	2,355	72	1,391	543	350	259
Uzbekistan	2011	1,445	78	317	130	920	51
	2012	1,491	80	331	129	952	52
	2013	1,272	66	284	113	809	44
	2014	1,290	66	292	111	821	44
Viet Nam	2011	2,061	17	829	423	791	23
	2012	2,052	30	789	416	816	23
	2013	2,079	35	798	406	839	23
	2014	2,164	50	788	449	877	23
Yemen	2011	192	1	34	94	63	8
	2012	168	1	29	77	61	7
	2013	247	1	44	116	86	10
	2014	222	1	41	103	77	8
Europe	**2011**	**74,158**	**7,833**	**18,389**	**19,024**	**28,912**	**101**
	2012	**73,990**	**7,703**	**18,666**	**18,340**	**29,281**	**100**
	2013	**72,660**	**7,558**	**17,789**	**18,215**	**29,098**	**99**
	2014	**71,265**	**7,844**	**17,685**	**18,323**	**27,413**	**97**

Table 3

Total final consumption and its components

Petajoules and gigajoules per capita

Country or area	Year	Total final consumption	Non-energy use	Energy use: Industry	Energy use: Transport	Energy use: Other	TFC per capita
Albania	2011	84	2	16	33	33	29
	2012	81	1	16	31	32	28
	2013	92	6	15	35	37	32
	2014	94	3	20	35	35	33
Andorra	2011	9	0	0	5	4	105
	2012	9	0	0	5	4	109
	2013	8	0	0	4	4	112
	2014	8	0	0	4	4	113
Austria	2011	1,124	68	321	337	397	133
	2012	1,134	75	316	336	407	134
	2013	1,158	73	318	353	414	136
	2014	1,118	82	310	349	377	131
Belarus	2011	873	164	200	166	343	92
	2012	962	228	200	171	362	101
	2013	830	94	192	183	361	87
	2014	853	141	189	174	349	90
Belgium	2011	1,677	303	435	390	548	152
	2012	1,687	311	441	362	574	152
	2013	1,760	332	453	359	617	158
	2014	1,689	354	447	370	518	150
Bosnia and Herzegovina	2011	139	8	28	48	55	36
	2012	133	7	28	44	54	35
	2013	123	3	31	40	50	32
	2014	184	3	32	42	107	48
Bulgaria	2011	402	22	112	117	151	55
	2012	400	20	108	124	149	55
	2013	381	20	108	112	142	53
	2014	393	22	109	125	137	55
Croatia	2011	314	26	54	81	153	73
	2012	298	23	48	80	147	69
	2013	294	24	47	80	143	69
	2014	280	24	46	79	131	66
Czechia	2011	1,072	107	298	252	415	102
	2012	1,062	110	287	245	419	101
	2013	1,057	106	281	243	427	100
	2014	1,047	119	285	252	391	99
Denmark	2011	601	12	101	186	301	108
	2012	576	11	95	171	298	103
	2013	569	12	90	169	299	101
	2014	543	11	88	171	274	96
Estonia	2011	120	3	25	32	61	90
	2012	122	4	24	32	63	92
	2013	126	7	28	31	60	96
	2014	121	5	23	31	62	92
Faeroe Islands	2011	7	0	1	3	4	148
	2012	8	0	1	3	4	157
	2013	9	0	1	3	5	182
	2014	8	0	1	3	4	165
Finland	2011	952	47	347	185	372	176
	2012	958	44	335	181	398	177
	2013	931	48	331	180	371	171
	2014	1,033	49	433	176	375	188

Total final consumption and its components

Petajoules and gigajoules per capita

Country or area	Year	Total final consumption	Non-energy use	Energy use			TFC per capita
				Industry	Transport	Other	
France	2011	6,351	606	1,105	1,866	2,775	100
	2012	6,529	606	1,104	1,865	2,954	103
	2013	6,658	573	1,183	1,856	3,045	104
	2014	6,249	592	1,090	1,867	2,700	97
Germany	2011	9,233	946	2,349	2,299	3,640	115
	2012	9,289	922	2,305	2,280	3,782	115
	2013	9,503	918	2,302	2,319	3,963	118
	2014	9,126	934	2,291	2,354	3,547	113
Gibraltar	2011	6	1	..	5	1	193
	2012	6	1	..	5	1	195
	2013	7	1	..	5	1	206
	2014	7	1	..	5	1	219
Greece	2011	801	37	139	284	341	72
	2012	720	30	126	238	325	65
	2013	642	27	119	238	258	58
	2014	651	30	129	239	253	59
Guernsey	2011	1	..	0	..	1	22
	2012	*1	..	0	..	*1	*21
	2013	*1	..	0	..	*1	*21
	2014	*1	..	0	..	*1	*20
Hungary	2011	745	86	109	170	380	75
	2012	688	81	98	164	345	69
	2013	696	63	157	149	327	70
	2014	707	71	165	167	304	72
Iceland	2011	106	1	55	12	39	331
	2012	109	0	57	11	41	338
	2013	114	0	59	12	43	350
	2014	115	1	60	12	43	351
Ireland	2011	435	14	93	152	177	94
	2012	427	12	91	148	176	91
	2013	430	13	91	150	176	92
	2014	424	9	93	156	165	91
Isle of Man	2011	2	..	0	..	2	24
	2012	2	..	0	..	2	28
	2013	2	..	0	..	2	27
	2014	2	..	0	..	2	26
Italy	2011	5,388	388	1,213	1,639	2,148	90
	2012	5,285	332	1,174	1,545	2,234	88
	2013	5,096	267	1,090	1,517	2,222	85
	2014	4,910	303	1,061	1,571	1,975	82
Jersey	2011	7	..	2	3	2	76
	2012	7	..	2	3	2	76
	2013	8	..	2	3	3	76
	2014	7	..	2	3	2	70
Latvia	2011	175	4	34	41	96	85
	2012	182	4	38	39	100	89
	2013	174	4	36	40	94	86
	2014	175	4	37	42	92	88
Liechtenstein	2011	*3	..	..	..	*3	*78
	2012	*3	..	..	..	*3	*79
	2013	3	..	..	..	3	80
	2014	3	..	..	..	3	74

Table 3

Total final consumption and its components

Petajoules and gigajoules per capita

Country or area	Year	Total final consumption	Non-energy use	Energy use			TFC per capita
				Industry	Transport	Other	
Lithuania	2011	246	51	39	62	93	80
	2012	251	52	42	63	94	83
	2013	237	42	41	63	91	80
	2014	244	45	41	70	89	84
Luxembourg	2011	166	1	29	98	37	320
	2012	163	2	27	94	40	306
	2013	160	2	25	92	41	294
	2014	154	2	26	89	38	276
Malta	2011	17	0	2	8	7	41
	2012	18	0	2	8	8	42
	2013	18	0	2	8	8	43
	2014	18	0	2	8	9	44
Montenegro	2011	32	..	12	9	12	52
	2012	31	..	10	9	12	50
	2013	28	..	8	8	12	45
	2014	28	..	6	8	14	44
Netherlands	2011	2,557	586	591	504	877	153
	2012	2,571	602	574	482	913	154
	2013	2,575	603	564	469	938	153
	2014	2,392	600	553	439	800	142
Norway	2011	860	96	252	199	313	174
	2012	862	95	242	196	330	172
	2013	871	97	243	203	327	171
	2014	847	100	243	205	299	164
Poland	2011	2,863	217	597	732	1,318	74
	2012	2,831	212	585	702	1,332	73
	2013	2,806	215	614	664	1,313	73
	2014	2,707	224	601	669	1,213	70
Portugal	2011	778	74	224	254	225	74
	2012	707	56	198	234	220	67
	2013	697	58	194	230	215	67
	2014	696	61	186	231	218	67
Republic of Moldova	2011	84	1	9	26	47	20
	2012	82	2	10	23	47	20
	2013	85	2	10	24	48	21
	2014	86	2	10	25	49	21
Romania	2011	998	75	282	209	432	50
	2012	1,011	75	266	227	444	51
	2013	957	63	247	220	426	48
	2014	952	66	255	223	409	48
Russian Federation	2011	19,305	2,830	5,387	4,145	6,944	135
	2012	19,384	2,836	6,009	3,944	6,595	135
	2013	18,349	2,990	5,201	3,940	6,218	128
	2014	19,116	3,207	5,267	3,938	6,703	133
Serbia	2011	400	35	109	84	173	44
	2012	355	14	98	75	169	40
	2013	360	23	98	82	157	40
	2014	338	22	80	84	151	38
Slovakia	2011	465	53	138	110	163	86
	2012	433	41	139	99	154	80
	2013	448	45	136	99	168	83
	2014	416	39	140	94	143	77

Total final consumption and its components

Petajoules and gigajoules per capita

Country or area	Year	Total final consumption	Non-energy use	Energy use: Industry	Energy use: Transport	Energy use: Other	TFC per capita
Slovenia	2011	215	5	52	80	79	105
	2012	211	5	50	81	75	102
	2013	206	5	50	77	74	100
	2014	199	6	51	76	65	96
Spain	2011	3,763	286	891	1,374	1,212	81
	2012	3,588	253	869	1,266	1,201	77
	2013	3,446	212	869	1,191	1,173	74
	2014	3,339	173	843	1,198	1,125	72
Sweden	2011	1,395	81	480	333	501	147
	2012	1,392	76	471	326	518	146
	2013	1,369	87	465	323	494	142
	2014	1,346	83	450	334	479	139
Switzerland	2011	808	24	155	252	377	101
	2012	835	22	156	254	403	104
	2013	855	22	157	253	422	105
	2014	785	22	150	251	364	95
T.F.Yug.Rep. Macedonia	2011	79	3	24	20	32	38
	2012	75	2	21	19	33	36
	2013	73	2	21	22	28	35
	2014	72	2	19	22	28	35
Ukraine	2011	3,183	253	1,091	531	1,308	70
	2012	3,073	256	1,039	483	1,294	68
	2013	2,959	207	927	498	1,328	66
	2014	2,599	145	879	434	1,140	58
United Kingdom	2011	5,315	316	987	1,690	2,323	84
	2012	5,438	277	966	1,674	2,522	86
	2013	5,490	292	980	1,670	2,548	86
	2014	5,182	289	970	1,696	2,227	81
Oceania	**2011**	**4,009**	**231**	**1,192**	**1,535**	**1,051**	**108**
	2012	**4,035**	**230**	**1,204**	**1,544**	**1,057**	**107**
	2013	**4,155**	**246**	**1,273**	**1,573**	**1,063**	**109**
	2014	**4,199**	**280**	**1,272**	**1,581**	**1,066**	**108**
American Samoa	2011	0	..	0	..	0	9
	2012	0	..	0	..	0	9
	2013	0	..	0	..	0	9
	2014	0	..	0	..	0	9
Australia	2011	3,278	189	965	1,297	827	145
	2012	3,298	183	980	1,306	828	144
	2013	3,386	192	1,028	1,331	834	145
	2014	3,384	206	1,009	1,332	836	143
Cook Islands	2011	*1	..	..	*1	0	*34
	2012	*1	..	..	*1	0	*35
	2013	*1	..	..	*1	0	*35
	2014	*1	..	..	*1	0	*35
Fiji	2011	*19	0	*8	*7	3	*22
	2012	*18	0	*7	*7	3	*20
	2013	*20	0	*9	*7	3	*22
	2014	*20	0	*9	*7	3	*23
French Polynesia	2011	*8	0	..	*5	*3	*30
	2012	*8	0	..	*5	3	*29
	2013	*8	0	..	*5	3	*28
	2014	*8	0	..	*5	*3	*27

Table 3

Total final consumption and its components

Petajoules and gigajoules per capita

Country or area	Year	Total final consumption	Non-energy use	Energy use			TFC per capita
				Industry	Transport	Other	
Guam	2011	6	..	..	..	6	36
	2012	6	..	..	..	6	35
	2013	6	..	..	..	6	34
	2014	*6	..	..	..	*6	*34
Kiribati	2011	*1	..	..	0	0	*6
	2012	*1	..	..	0	0	*6
	2013	*1	..	..	0	0	*6
	2014	*1	..	..	0	0	*6
Marshall Islands	2011	0	..	..	..	0	*5
	2012	0	..	..	..	0	5
	2013	0	..	..	..	0	*5
	2014	0	..	..	..	0	*5
Micronesia (Fed. States of)	2011	*1	0	0	*1	0	*13
	2012	*1	0	0	*1	0	*14
	2013	*2	0	0	*1	0	*16
	2014	*2	0	0	*1	0	*17
Nauru	2011	0	..	..	0	0	*35
	2012	0	..	..	0	0	*35
	2013	0	..	..	0	0	*39
	2014	0	..	..	0	0	*42
New Caledonia	2011	*33	*4	*17	*8	*3	*131
	2012	*33	*4	*17	*8	*3	*130
	2013	35	4	19	9	3	136
	2014	*37	*4	*20	*9	*3	*142
New Zealand	2011	534	37	167	192	138	121
	2012	543	42	167	190	144	122
	2013	560	49	177	192	142	125
	2014	600	68	193	197	143	134
Niue	2011	0	..	0	0	0	41
	2012	0	..	0	0	0	43
	2013	0	..	0	0	0	44
	2014	0	..	0	0	0	45
Northern Mariana Islands	2011	*1	..	..	..	*1	*21
	2012	*1	..	..	..	*1	*22
	2013	*1	..	..	..	*1	*22
	2014	*1	..	..	..	*1	*21
Palau	2011	*3	0	0	*3	0	*141
	2012	*3	0	0	*3	0	*139
	2013	*3	0	0	*3	0	*138
	2014	*3	0	0	*3	0	*140
Papua New Guinea	2011	*111	..	*35	*16	*60	*16
	2012	*109	..	31	19	*60	*15
	2013	*121	..	40	20	*61	*17
	2014	*122	..	*41	*21	*61	*16
Samoa	2011	4	0	0	2	2	20
	2012	*4	0	0	*2	*2	*20
	2013	4	0	0	2	2	20
	2014	*4	0	0	*2	*2	*20
Solomon Islands	2011	*5	0	0	*1	*4	*9
	2012	*5	0	0	*1	*4	*9
	2013	*5	0	0	*1	*4	*9
	2014	*5	0	0	*1	*4	*9

Total final consumption and its components

Petajoules and gigajoules per capita

Country or area	Year	Total final consumption	Non-energy use	Energy use: Industry	Energy use: Transport	Energy use: Other	TFC per capita
Tonga	2011	1	0	..	1	0	12
	2012	*1	0	..	*1	0	*12
	2013	*1	0	..	0	*1	*12
	2014	*1	0	..	*1	*1	*12
Tuvalu	2011	0	..	..	0	0	*7
	2012	0	..	..	0	0	*7
	2013	0	..	..	0	0	*7
	2014	0	..	..	0	0	*8
Vanuatu	2011	*2	0	..	*1	*1	*10
	2012	*2	0	..	*1	*1	*9
	2013	*2	0	..	*1	*1	*8
	2014	*3	0	..	*2	*1	*11
Wallis and Futuna Is.	2011	0	..	..	0	0	14
	2012	0	..	..	0	0	*13
	2013	0	..	..	0	0	*13
	2014	0	..	..	0	0	*13

Table 4

Production, trade and supply of coal

Thousand metric tons of coal equivalent and kilograms per capita

Table Notes

Production includes hard coal, brown coal, oil shale and peat. Production from other sources of hard coal and brown coal is also included, representing recovered slurries, middlings, and other low- grade coal products which cannot be classified according to type of coal and which includes coal recovered from waste piles and other waste receptacles.

Imports, **Exports** and **Changes in stocks** refer to hard coal, brown coal, oil shale, peat, coke oven coke, gas coke, coal tar, patent fuel, brown coal briquettes, peat products and other coal products.

Please refer to the Definitions Section on pages ix to xv for the appropriate product description/ classification.

Figure 11: Production of coal by region, 1990-2014

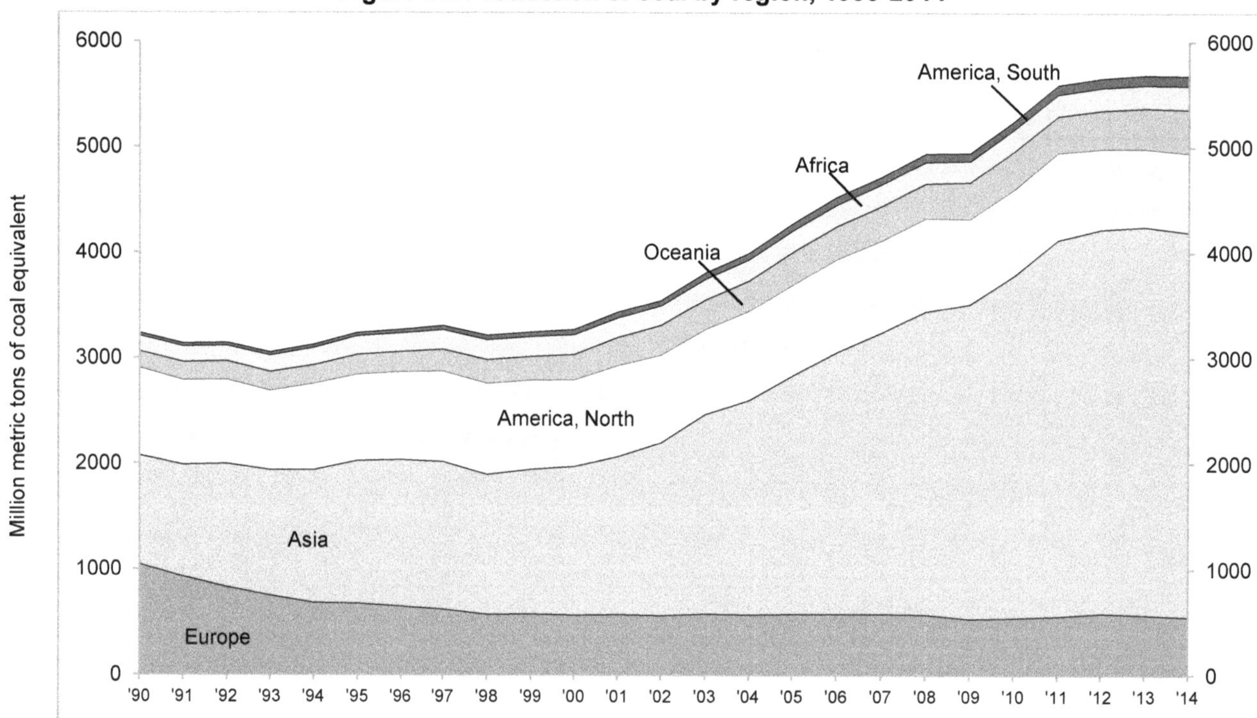

Production, trade and supply of coal

Thousand metric tons of coal equivalent and kilograms per capita

Country or area	Year	Production	Imports	Exports	Changes in stocks	Supply	
						Total	Per capita
World	**2011**	**5,587,415**	**1,006,685**	**995,500**	**56,783**	**5,541,817**	**790**
	2012	**5,652,924**	**1,063,009**	**1,072,387**	**21,997**	**5,621,549**	**792**
	2013	**5,682,830**	**1,124,823**	**1,222,899**	**35,665**	**5,549,088**	**773**
	2014	**5,675,350**	**1,173,118**	**1,235,551**	**58,172**	**5,554,746**	**765**
Africa	**2011**	**207,937**	**11,962**	**56,360**	**347**	**163,192**	**152**
	2012	**216,005**	**12,940**	**64,756**	**2,393**	**161,796**	**147**
	2013	**217,432**	**9,525**	**64,616**	**2,800**	**159,541**	**141**
	2014	**223,986**	**11,605**	**60,685**	**3,315**	**171,591**	**148**
Algeria	2011	..	421	..	-1	422	12
	2012	..	405	..	-13	418	11
	2013	..	260	..	-33	293	8
	2014	..	246	..	28	218	6
Benin	2011	..	..	..	..	..	..
	2012	..	..	..	..	..	..
	2013	..	..	..	..	..	..
	2014	..	62	..	..	62	6
Botswana	2011	634	..	96	0	539	258
	2012	1,184	..	87	607	490	230
	2013	1,205	..	54	337	813	374
	2014	1,378	..	156	-226	1,448	652
Burundi	2011	4	..	..	..	4	0
	2012	10	..	..	..	10	1
	2013	8	..	..	..	8	1
	2014	6	..	..	..	6	1
Egypt	2011	8	*886	*269	-16	*640	*8
	2012	18	*895	*247	-13	*679	*8
	2013	3	516	199	10	310	4
	2014	0	682	119	..	563	6
Ethiopia	2011	18	124	..	-7	150	2
	2012	..	197	..	..	197	2
	2013	..	256	..	..	256	3
	2014	..	283	..	..	283	3
Kenya	2011	..	337	..	..	337	8
	2012	..	302	..	..	302	7
	2013	..	334	..	..	334	8
	2014	..	*414	..	..	*414	*9
Lesotho	2011	..	*590	..	..	*590	*290
	2012	..	*599	..	..	*599	*291
	2013	..	*607	..	..	*607	*292
	2014	..	*616	..	..	*616	*292
Madagascar	2011	..	134	..	..	134	6
	2012	..	232	..	..	232	10
	2013	..	375	..	..	375	16
	2014	..	338	..	..	338	14
Malawi	2011	73	15	*7	..	81	5
	2012	*66	13	*7	..	*72	*5
	2013	59	21	*7	..	73	4
	2014	56	17	*7	..	66	4
Mauritius	2011	..	585	..	17	568	453
	2012	..	646	..	48	598	475
	2013	..	627	..	-3	630	498
	2014	..	680	..	26	654	516

Table 4

Production, trade and supply of coal

Thousand metric tons of coal equivalent and kilograms per capita

Country or area	Year	Production	Imports	Exports	Changes in stocks	Supply	
						Total	Per capita
Morocco	2011	..	5,022	..	-322	5,343	162
	2012	..	5,667	..	155	5,512	164
	2013	..	4,224	..	-63	4,287	126
	2014	..	6,147	..	380	5,767	167
Mozambique	2011	552	..	145	360	46	2
	2012	4,158	..	2,714	1,429	15	1
	2013	5,502	..	3,549	1,939	14	1
	2014	5,814	..	4,296	1,208	310	11
Namibia	2011	..	26	..	..	26	12
	2012	..	104	..	..	104	45
	2013	..	40	..	..	40	17
	2014	..	*26	..	..	*26	*11
Niger	2011	144	..	..	5	140	8
	2012	137	..	..	-8	145	8
	2013	132	..	..	4	128	7
	2014	140	..	..	*5	135	7
Nigeria	2011	28	..	..	..	28	0
	2012	42	..	..	..	42	0
	2013	39	..	..	..	39	0
	2014	40	..	..	..	40	0
Réunion	2011	..	660	..	59	602	719
	2012	..	604	..	-21	625	741
	2013	..	597	..	-1	598	704
	2014	..	560	..	-3	563	658
Senegal	2011	..	355	..	..	355	27
	2012	..	309	..	..	309	22
	2013	..	309	..	..	309	22
	2014	..	325	..	..	325	22
South Africa	2011	203,921	2,528	55,513	252	150,684	2,885
	2012	208,589	2,740	61,315	208	149,805	2,835
	2013	207,194	968	60,283	609	147,270	2,757
	2014	210,645	902	55,809	-398	156,135	2,893
Swaziland	2011	*123	*96	*95	0	*124	*102
	2012	*220	*85	*133	0	*173	*140
	2013	*199	*129	*199	0	*129	*103
	2014	230	61	117	0	175	138
Tunisia	2011	..	15	..	..	15	1
	2012	..	0	..	..	0	0
	2013	..	0	..	..	0	0
	2014	..	0	..	..	0	0
Uganda	2011	..	0	..	..	0	0
	2012	..	0	..	..	0	0
	2013	..	151	..	..	151	4
	2014	..	*151	..	..	*151	*4
United Rep. of Tanzania	2011	71	..	..	..	71	2
	2012	70	..	..	..	70	1
	2013	75	..	..	..	75	1
	2014	217	..	..	..	217	4
Zambia	2011	0	115	12	..	103	7
	2012	43	107	15	..	135	9
	2013	148	*63	*36	..	176	12
	2014	134	*54	*3	..	185	12

Production, trade and supply of coal

Thousand metric tons of coal equivalent and kilograms per capita

Country or area	Year	Production	Imports	Exports	Changes in stocks	Supply	
						Total	Per capita
Zimbabwe	2011	2,360	52	224	..	2,188	153
	2012	1,468	36	238	..	1,266	87
	2013	2,869	48	290	..	2,627	176
	2014	5,327	42	177	2,296	2,895	190
America, North	**2011**	**828,998**	**29,907**	**115,293**	**4,108**	**739,503**	**1,339**
	2012	**766,894**	**26,238**	**131,259**	**2,515**	**659,358**	**1,182**
	2013	**742,560**	**24,706**	**126,860**	**-26,109**	**666,515**	**1,183**
	2014	**753,666**	**26,701**	**107,849**	**3,865**	**668,652**	**1,176**
Bahamas	2011	..	*1	..	..	*1	*4
	2012	..	*1	..	..	*1	*4
	2013	..	*1	..	..	*1	*4
	2014	..	*1	..	..	*1	*4
Bermuda	2011	..	0	..	..	0	1
	2012	..	0	..	..	0	1
	2013	..	0	..	..	0	1
	2014	..	0	..	..	0	1
Canada	2011	48,082	9,062	27,629	53	29,462	854
	2012	47,914	9,186	29,332	-176	27,944	801
	2013	50,057	7,975	33,501	-647	25,178	715
	2014	49,845	7,778	29,484	435	27,705	778
Costa Rica	2011	..	90	..	..	90	20
	2012	..	93	..	..	93	20
	2013	..	93	..	..	93	20
	2014	..	93	..	..	93	20
Cuba	2011	..	5	..	..	5	0
	2012	..	6	..	..	6	1
	2013	..	12	..	..	12	1
	2014	..	2	..	..	2	0
Dominican Republic	2011	..	777	..	80	697	69
	2012	..	769	..	-6	775	76
	2013	..	748	..	14	734	71
	2014	..	919	..	-11	930	89
Guadeloupe	2011	..	*176	..	..	*176	*384
	2012	..	*176	..	..	*176	*382
	2013	..	*176	..	..	*176	*380
	2014	..	*176	..	..	*176	*378
Guatemala	2011	..	503	..	69	434	29
	2012	..	508	..	70	438	29
	2013	..	533	..	95	438	28
	2014	..	634	..	0	634	40
Honduras	2011	..	228	..	-19	247	32
	2012	..	249	..	0	249	32
	2013	..	254	..	0	254	32
	2014	..	192	..	..	192	24
Jamaica	2011	..	63	..	-7	70	25
	2012	..	83	..	18	64	23
	2013	..	84	..	5	79	28
	2014	..	83	..	-9	92	33
Mexico	2011	14,813	6,958	240	766	20,764	173
	2012	11,350	6,415	215	-604	18,153	149
	2013	10,967	7,153	5	-330	18,444	149
	2014	11,074	7,277	4	273	18,074	144

Table 4

Production, trade and supply of coal

Thousand metric tons of coal equivalent and kilograms per capita

Country or area	Year	Production	Imports	Exports	Changes in stocks	Supply	
						Total	Per capita
Panama	2011	..	285	..	..	285	77
	2012	..	441	..	..	441	118
	2013	..	454	..	..	454	119
	2014	..	456	..	..	456	118
United States	2011	766,103	11,760	87,424	3,167	687,271	2,200
	2012	707,630	8,311	101,712	3,212	611,017	1,941
	2013	681,536	7,223	93,354	-25,247	620,652	1,957
	2014	692,746	9,089	78,362	3,177	620,297	1,942
America, South	**2011**	**85,991**	**31,079**	**76,645**	**48**	**40,376**	**101**
	2012	**88,895**	**29,505**	**79,604**	**-1,020**	**39,815**	**98**
	2013	**94,276**	**31,378**	**82,192**	**731**	**42,731**	**104**
	2014	**91,899**	**34,247**	**79,712**	**-6**	**46,440**	**112**
Argentina	2011	76	1,767	58	7	1,778	43
	2012	80	1,631	117	-36	1,630	39
	2013	70	889	71	-9	896	21
	2014	70	2,153	56	-27	2,193	51
Brazil	2011	3,494	20,996	45	279	24,166	121
	2012	4,199	18,837	0	-669	23,705	117
	2013	5,454	20,775	0	303	25,926	127
	2014	5,037	23,101	0	394	27,744	135
Chile	2011	351	7,792	0	285	7,858	457
	2012	422	8,499	0	72	8,849	509
	2013	2,273	8,826	945	583	9,571	545
	2014	3,081	8,531	1,681	279	9,652	543
Colombia	2011	79,710	0	74,487	27	5,196	112
	2012	82,727	0	78,378	0	4,349	93
	2013	85,292	0	80,334	11	4,947	104
	2014	82,250	..	76,743	0	5,507	115
Falkland Is. (Malvinas)	2011	*4	..	..	..	*4	*1,480
	2012	*4	..	..	..	*4	*1,476
	2013	*4	..	..	..	*4	*1,469
	2014	*4	..	..	..	*4	*1,462
Paraguay	2011	..	1	..	..	1	0
	2012	..	1	..	..	1	0
	2013	..	0	..	..	0	0
	2014	..	0	..	..	0	0
Peru	2011	167	520	158	-549	1,078	36
	2012	211	533	159	-387	972	32
	2013	195	884	154	-158	1,083	35
	2014	206	459	261	-652	1,056	34
Uruguay	2011	..	3	..	..	3	1
	2012	..	3	..	..	3	1
	2013	..	3	..	..	3	1
	2014	..	2	..	..	2	1
Venezuela (Bolivar. Rep.)	2011	2,190	..	1,897	..	293	10
	2012	1,251	..	950	..	301	10
	2013	988	..	687	..	300	10
	2014	1,251	..	970	..	282	9
Asia	**2011**	**3,568,708**	**681,730**	**322,566**	**48,273**	**3,879,600**	**920**
	2012	**3,640,927**	**729,645**	**347,496**	**12,914**	**4,010,162**	**941**
	2013	**3,679,199**	**784,661**	**448,750**	**51,820**	**3,963,290**	**921**
	2014	**3,645,518**	**828,822**	**438,635**	**38,566**	**3,997,139**	**919**

Production, trade and supply of coal

Thousand metric tons of coal equivalent and kilograms per capita

Country or area	Year	Production	Imports	Exports	Changes in stocks	Supply	
						Total	Per capita
Afghanistan	2011	1,303	197	46	..	1,454	50
	2012	1,092	197	46	..	1,243	42
	2013	1,186	*197	*50	..	1,333	43
	2014	1,336	*197	*57	..	1,476	47
Armenia	2011	0	3	0	..	3	1
	2012	0	2	0	..	2	1
	2013	0	1	0	..	1	0
	2014	0	1	1	..	1	0
Azerbaijan	2011	..	6	..	-2	8	1
	2012	..	5	..	-1	6	1
	2013	..	5	..	0	5	0
	2014	..	2	..	-1	3	0
Bangladesh	2011	476	660	..	67	1,069	7
	2012	596	714	..	20	1,290	8
	2013	610	726	..	-75	1,412	9
	2014	676	705	..	59	1,322	8
Bhutan	2011	96	..	65	..	31	43
	2012	87	..	41	..	46	62
	2013	68	..	5	..	64	85
	2014	107	..	23	..	84	110
Cambodia	2011	..	13	..	..	13	1
	2012	..	14	..	..	14	1
	2013	..	65	..	..	65	4
	2014	..	332	..	..	332	22
China	2011	2,642,128	162,323	13,488	39,132	2,751,831	2,041
	2012	2,673,376	159,397	7,273	15,127	2,810,372	2,073
	2013	2,702,652	176,778	9,633	49,416	2,820,381	2,070
	2014	2,659,489	178,067	12,196	30,480	2,794,881	2,041
China, Hong Kong SAR	2011	..	11,274	0	697	10,577	1,501
	2012	..	11,114	0	702	10,412	1,466
	2013	..	11,674	2	572	11,100	1,549
	2014	..	12,407	0	778	11,629	1,609
Cyprus	2011	..	0	..	-11	11	10
	2012	..	0	..	0	0	0
	2013	..	0	..	0	0	0
	2014	..	4	..	1	4	3
Georgia	2011	202	199	9	0	391	93
	2012	241	197	4	0	434	105
	2013	231	229	4	9	447	109
	2014	175	241	1	-7	422	104
India	2011	362,878	94,714	1,908	160	455,523	365
	2012	375,310	133,896	2,869	-7,534	513,870	407
	2013	342,209	155,438	1,416	-4,246	500,476	391
	2014	365,990	193,601	807	-2,870	561,654	434
Indonesia	2011	348,350	*95	243,988	999	103,458	423
	2012	384,941	133	273,978	-140	111,236	448
	2013	427,577	617	374,099	690	53,405	213
	2014	418,681	2,437	366,932	9,351	44,834	176
Iran (Islamic Rep. of)	2011	1,111	1,270	297	..	2,084	28
	2012	993	938	320	..	1,612	21
	2013	1,054	897	291	..	1,660	22
	2014	910	556	195	1	1,270	16

Table 4

Production, trade and supply of coal

Thousand metric tons of coal equivalent and kilograms per capita

Country or area	Year	Production	Imports	Exports	Changes in stocks	Supply	
						Total	Per capita
Israel	2011	42	10,652	..	-232	10,925	1,445
	2012	42	11,923	..	-308	12,273	1,595
	2013	42	10,933	..	926	10,049	1,285
	2014	40	9,402	..	85	9,356	1,178
Japan	2011	..	155,747	1,000	125	154,621	1,215
	2012	..	161,685	1,488	-85	160,283	1,261
	2013	..	174,513	1,209	13	173,291	1,365
	2014	..	169,745	498	42	169,205	1,334
Jordan	2011	..	0	..	..	0	0
	2012	..	0	..	..	0	0
	2013	..	454	..	..	454	63
	2014	..	527	..	..	527	71
Kazakhstan	2011	72,708	771	19,221	300	53,958	3,259
	2012	75,376	921	20,417	1,800	54,080	3,215
	2013	74,913	854	21,181	1,045	53,541	3,131
	2014	71,343	1,054	19,300	188	52,907	3,046
Korea, Dem.Ppl's.Rep.	2011	25,602	192	9,991	..	15,804	642
	2012	25,750	1,183	10,734	..	16,199	654
	2013	25,852	246	14,974	..	11,124	447
	2014	26,572	292	14,018	..	12,846	513
Korea, Republic of	2011	1,369	113,548	..	223	114,694	2,324
	2012	1,331	108,685	..	-58	110,074	2,219
	2013	1,154	110,541	..	254	111,441	2,236
	2014	1,111	113,603	..	-1,367	116,082	2,318
Kyrgyzstan	2011	430	655	19	0	1,066	192
	2012	603	897	8	0	1,492	264
	2013	738	907	97	224	1,323	230
	2014	941	920	152	41	1,668	285
Lao People's Dem. Rep.	2011	391	..	*195	..	196	31
	2012	389	..	195	..	194	30
	2013	285	..	*144	..	140	21
	2014	106	..	0	..	106	16
Lebanon	2011	..	236	..	..	236	51
	2012	..	236	..	..	236	48
	2013	..	189	..	..	189	36
	2014	..	236	..	..	236	42
Malaysia	2011	2,626	19,043	202	531	20,936	733
	2012	2,658	20,315	332	75	22,566	778
	2013	2,606	19,870	467	152	21,857	742
	2014	2,420	19,577	163	23	21,811	729
Mongolia	2011	28,670	0	16,072	5,838	6,759	2,450
	2012	25,796	0	14,896	2,497	8,403	2,992
	2013	26,213	1	12,154	1,022	13,038	4,560
	2014	21,641	1	13,741	1,531	6,370	2,189
Myanmar	2011	305	*8	..	..	313	6
	2012	408	6	..	..	415	8
	2013	448	37	..	..	485	9
	2014	537	0	..	..	537	10
Nepal	2011	11	408	..	..	419	15
	2012	8	305	..	..	313	11
	2013	12	380	..	..	392	14
	2014	12	680	..	..	692	25

Production, trade and supply of coal

Thousand metric tons of coal equivalent and kilograms per capita

Country or area	Year	Production	Imports	Exports	Changes in stocks	Supply	
						Total	Per capita
Other Asia	2011	..	58,643	96	82	58,465	2,514
	2012	..	56,023	110	-647	56,560	2,427
	2013	..	58,007	13	434	57,560	2,466
	2014	..	57,867	68	-560	58,359	2,498
Pakistan	2011	2,441	4,028	..	..	6,469	37
	2012	2,040	3,499	..	..	5,539	31
	2013	2,207	2,941	..	..	5,148	28
	2014	2,279	2,603	..	..	4,882	26
Philippines	2011	5,189	8,466	2,063	354	11,238	119
	2012	5,541	9,175	2,392	568	11,755	122
	2013	5,347	11,076	2,565	-638	14,497	149
	2014	5,730	11,728	4,348	-2,374	15,484	156
Singapore	2011	..	9	0	..	9	2
	2012	..	35	0	..	35	7
	2013	..	380	4	..	376	70
	2014	..	565	11	..	555	101
Sri Lanka	2011	..	684	..	257	427	21
	2012	..	962	..	239	724	35
	2013	..	1,131	..	370	761	37
	2014	..	1,607	..	145	1,461	71
Syrian Arab Republic	2011	..	4	..	..	4	0
	2012	..	1	..	..	1	0
	2013	..	1	..	..	1	0
	2014	..	1	..	..	1	0
Tajikistan	2011	142	10	14	..	138	18
	2012	249	15	1	..	263	33
	2013	323	7	..	..	329	41
	2014	549	8	..	..	557	67
Thailand	2011	8,836	14,696	5	375	23,151	346
	2012	7,485	16,559	0	-330	24,373	363
	2013	7,288	16,890	8	-459	24,629	365
	2014	6,603	19,155	21	3,070	22,668	335
Turkey	2011	25,486	22,190	0	-450	48,126	655
	2012	22,271	27,803	6	-62	50,130	670
	2013	22,388	25,466	13	1,331	46,510	610
	2014	23,147	27,702	114	-578	51,313	662
United Arab Emirates	2011	..	637	4	..	633	73
	2012	..	1,930	52	..	1,878	210
	2013	..	2,376	6	..	2,370	262
	2014	..	2,627	1	..	2,626	289
Uzbekistan	2011	1,930	50	20	..	1,961	70
	2012	1,935	65	20	..	1,980	69
	2013	2,072	10	20	..	2,062	71
	2014	2,253	..	21	..	2,233	76
Viet Nam	2011	35,987	146	13,862	-175	22,446	251
	2012	32,409	656	12,314	1,050	19,701	218
	2013	31,724	659	10,393	779	21,211	232
	2014	32,869	206	*5,967	528	26,580	288
Yemen	2011	..	153	..	..	153	6
	2012	..	157	..	..	157	6
	2013	..	163	..	..	163	6
	2014	..	166	..	..	166	6

Table 4

Production, trade and supply of coal

Thousand metric tons of coal equivalent and kilograms per capita

Country or area	Year	Production	Imports	Exports	Changes in stocks	Supply	
						Total	Per capita
Europe	**2011**	**553,026**	**251,101**	**158,665**	**2,070**	**643,392**	**874**
	2012	**581,004**	**264,123**	**168,494**	**843**	**675,790**	**917**
	2013	**566,435**	**273,535**	**188,650**	**1,180**	**650,141**	**881**
	2014	**549,170**	**270,637**	**201,370**	**10,223**	**608,213**	**824**
Albania	2011	2	210	..	..	212	73
	2012	2	246	..	..	248	86
	2013	3	215	..	..	218	76
	2014	0	279	..	..	279	96
Andorra	2011	..	0	..	..	0	2
	2012	..	0	..	..	0	*2
	2013	..	0	..	..	0	*2
	2014	..	0	..	..	0	*3
Austria	2011	0	4,470	5	-510	4,976	591
	2012	0	4,775	3	149	4,624	547
	2013	0	4,473	2	-265	4,736	558
	2014	0	4,386	1	25	4,361	512
Belarus	2011	980	181	115	139	908	96
	2012	930	458	132	108	1,148	121
	2013	812	584	55	145	1,196	126
	2014	513	747	39	-8	1,228	129
Belgium	2011	..	6,093	1,041	49	5,002	455
	2012	..	5,537	1,226	-265	4,576	413
	2013	..	5,411	922	-215	4,704	422
	2014	5	5,430	705	116	4,614	411
Bosnia and Herzegovina	2011	5,817	1,482	632	-56	6,723	1,754
	2012	5,676	1,114	424	98	6,268	1,637
	2013	5,451	1,245	429	51	6,216	1,626
	2014	5,383	1,489	528	28	6,317	1,655
Bulgaria	2011	8,887	2,943	44	86	11,700	1,591
	2012	8,015	2,220	48	187	10,001	1,369
	2013	6,839	1,440	21	-199	8,457	1,166
	2014	7,327	1,397	50	-469	9,144	1,270
Croatia	2011	..	992	1	-16	1,007	234
	2012	..	793	1	-109	902	210
	2013	..	1,072	10	97	964	226
	2014	..	866	13	-71	924	217
Czechia	2011	29,995	3,042	6,719	416	25,902	2,459
	2012	29,580	2,552	5,617	1,811	24,704	2,343
	2013	25,390	2,717	5,511	-782	23,378	2,217
	2014	24,200	3,970	4,922	385	22,864	2,169
Denmark	2011	..	5,130	0	506	4,623	829
	2012	..	3,310	9	-224	3,525	629
	2013	..	4,110	43	-419	4,485	797
	2014	..	3,627	44	175	3,408	603
Estonia	2011	5,802	62	58	-12	5,818	4,381
	2012	5,765	76	43	366	5,432	4,103
	2013	6,322	55	42	3	6,332	4,797
	2014	6,469	76	42	127	6,376	4,844
Finland	2011	2,453	6,595	8	429	8,610	1,596
	2012	1,492	3,825	49	-1,416	6,683	1,232
	2013	2,573	4,871	71	-107	7,480	1,372
	2014	2,290	5,219	74	1,062	6,373	1,163

Production, trade and supply of coal

Thousand metric tons of coal equivalent and kilograms per capita

Country or area	Year	Production	Imports	Exports	Changes in stocks	Supply	
						Total	Per capita
France	2011	132	14,673	137	-22	14,690	232
	2012	257	15,755	226	-550	16,337	257
	2013	278	16,796	181	-895	17,788	278
	2014	266	13,128	52	75	13,268	207
Germany	2011	67,370	46,952	1,376	1,657	111,290	1,384
	2012	67,995	47,549	1,628	-266	114,182	1,419
	2013	64,364	53,493	1,649	-478	116,686	1,448
	2014	63,047	53,136	2,149	216	113,818	1,411
Greece	2011	10,721	338	8	-215	11,267	1,010
	2012	11,492	275	12	135	11,621	1,046
	2013	9,612	333	13	-39	9,971	902
	2014	9,120	285	8	-156	9,553	868
Hungary	2011	2,351	1,867	373	-136	3,981	399
	2012	2,295	1,815	386	-120	3,843	386
	2013	2,302	1,599	595	-57	3,363	339
	2014	2,269	1,512	606	-1	3,175	321
Iceland	2011	..	127	..	-2	130	404
	2012	..	139	..	6	132	409
	2013	..	139	..	0	139	428
	2014	..	124	..	0	124	379
Ireland	2011	1,086	2,023	16	289	2,805	603
	2012	451	1,911	18	-994	3,338	715
	2013	1,845	2,120	18	1,037	2,909	623
	2014	1,387	1,736	14	233	2,876	615
Italy	2011	83	22,109	273	-770	22,690	380
	2012	73	22,826	304	-619	23,214	388
	2013	66	18,847	243	-660	19,330	323
	2014	78	18,777	338	-158	18,675	312
Latvia	2011	0	166	8	1	158	77
	2012	3	135	10	-3	131	64
	2013	3	109	9	-9	113	56
	2014	2	70	5	-18	85	43
Lithuania	2011	17	376	25	32	335	109
	2012	24	319	29	-11	325	108
	2013	34	408	37	29	375	126
	2014	40	294	8	10	316	109
Luxembourg	2011	..	83	..	..	83	160
	2012	..	77	..	..	77	145
	2013	..	67	..	..	67	124
	2014	..	76	..	..	76	136
Montenegro	2011	620	3	17	..	606	973
	2012	561	0	17	..	544	872
	2013	532	1	7	..	525	841
	2014	520	1	9	..	512	819
Netherlands	2011	..	21,422	10,964	125	10,333	619
	2012	..	25,451	16,022	-1,945	11,374	679
	2013	..	35,932	23,397	1,343	11,193	666
	2014	..	40,947	27,319	1,141	12,486	740
Norway	2011	1,329	1,122	1,442	-198	1,207	244
	2012	1,178	1,149	1,222	-64	1,169	233
	2013	1,779	1,023	1,984	-290	1,107	218
	2014	1,606	1,172	1,542	22	1,214	236

Table 4

Production, trade and supply of coal

Thousand metric tons of coal equivalent and kilograms per capita

Country or area	Year	Production	Imports	Exports	Changes in stocks	Supply	
						Total	Per capita
Poland	2011	79,688	12,645	12,233	773	79,327	2,055
	2012	82,619	8,741	12,237	5,580	73,542	1,905
	2013	81,642	9,211	15,559	-1,819	77,113	1,997
	2014	77,220	9,174	14,030	770	71,594	1,854
Portugal	2011	..	3,069	..	-87	3,156	299
	2012	..	4,332	..	139	4,193	399
	2013	..	3,612	..	-173	3,785	362
	2014	..	3,694	..	-119	3,813	367
Republic of Moldova	2011	..	131	..	6	126	31
	2012	..	150	..	-6	155	38
	2013	..	217	..	8	209	51
	2014	..	120	..	-10	130	32
Romania	2011	9,619	1,586	33	-521	11,692	581
	2012	9,066	1,765	12	-10	10,829	543
	2013	6,652	1,505	6	-37	8,188	414
	2014	6,356	1,386	15	-423	8,149	415
Russian Federation	2011	234,639	27,505	113,164	-168	149,149	1,041
	2012	263,451	26,093	118,693	490	170,361	1,189
	2013	263,248	25,039	127,480	3,617	157,191	1,096
	2014	271,061	22,723	140,360	2,429	150,996	1,053
Serbia	2011	11,178	1,145	11	-114	12,427	1,377
	2012	10,397	586	8	76	10,900	1,213
	2013	10,958	441	54	22	11,324	1,267
	2014	8,161	695	27	-94	8,923	1,003
Slovakia	2011	859	4,562	227	-111	5,305	980
	2012	811	4,552	102	312	4,949	914
	2013	835	4,093	110	-136	4,954	914
	2014	828	4,171	87	2	4,909	905
Slovenia	2011	1,715	369	2	-2	2,083	1,012
	2012	1,562	430	1	-18	2,009	974
	2013	1,536	368	6	-13	1,911	926
	2014	1,169	346	1	7	1,507	729
Spain	2011	3,784	13,588	1,098	-1,516	17,790	381
	2012	3,515	18,502	1,596	-1,489	21,911	470
	2013	2,518	11,538	603	-2,196	15,649	337
	2014	2,326	13,600	960	-1,359	16,325	353
Sweden	2011	314	3,383	25	117	3,555	376
	2012	201	2,471	22	-462	3,113	326
	2013	266	2,636	26	-283	3,160	328
	2014	192	2,873	30	30	3,004	310
Switzerland	2011	..	153	..	-37	190	24
	2012	..	177	..	5	173	21
	2013	..	186	..	10	176	22
	2014	..	157	..	-43	199	24
T.F.Yug.Rep. Macedonia	2011	2,015	200	8	101	2,106	1,019
	2012	1,780	203	8	-23	1,998	966
	2013	1,504	144	0	0	1,648	795
	2014	1,407	165	0	11	1,561	752
Ukraine	2011	56,096	11,236	7,686	1,984	57,662	1,268
	2012	57,628	14,180	7,419	2,551	61,838	1,364
	2013	58,440	12,984	8,950	2,684	59,790	1,324
	2014	46,225	14,820	6,926	2,006	52,113	1,158

Production, trade and supply of coal

Thousand metric tons of coal equivalent and kilograms per capita

Country or area	Year	Production	Imports	Exports	Changes in stocks	Supply	
						Total	Per capita
United Kingdom	2011	15,474	29,068	918	-147	43,770	693
	2012	14,184	39,631	971	-2,574	55,419	872
	2013	10,629	44,502	615	1,205	53,311	834
	2014	9,704	37,970	469	4,282	42,923	667
Oceania	**2011**	**342,755**	**907**	**265,970**	**1,938**	**75,753**	**2,047**
	2012	**359,200**	**558**	**280,777**	**4,353**	**74,628**	**1,985**
	2013	**382,928**	**1,017**	**311,833**	**5,243**	**66,870**	**1,751**
	2014	**411,111**	**1,107**	**347,300**	**2,208**	**62,710**	**1,618**
Australia	2011	338,620	230	263,786	1,850	73,215	3,248
	2012	355,096	123	278,539	4,862	71,818	3,135
	2013	379,014	86	309,693	5,100	64,308	2,763
	2014	407,766	180	345,530	2,298	60,118	2,545
New Caledonia	2011	..	547	..	70	477	1,910
	2012	..	435	..	-42	476	1,881
	2013	..	550	..	45	505	1,969
	2014	..	589	..	0	589	2,268
New Zealand	2011	4,134	130	2,184	18	2,061	468
	2012	4,104	1	2,238	-468	2,334	526
	2013	3,914	381	2,140	98	2,057	461
	2014	3,345	338	1,770	-90	2,003	446

Table 5

Production, trade and supply of hard coal

Thousand metric tons and kilograms per capita

Table Notes

Production includes production from other sources as defined in the previous table.

Data for China include brown coal.

Please refer to the Definitions Section on pages ix to xv for the appropriate product description/ classification.

Figure 12: Production of hard coal by region, 1990-2014

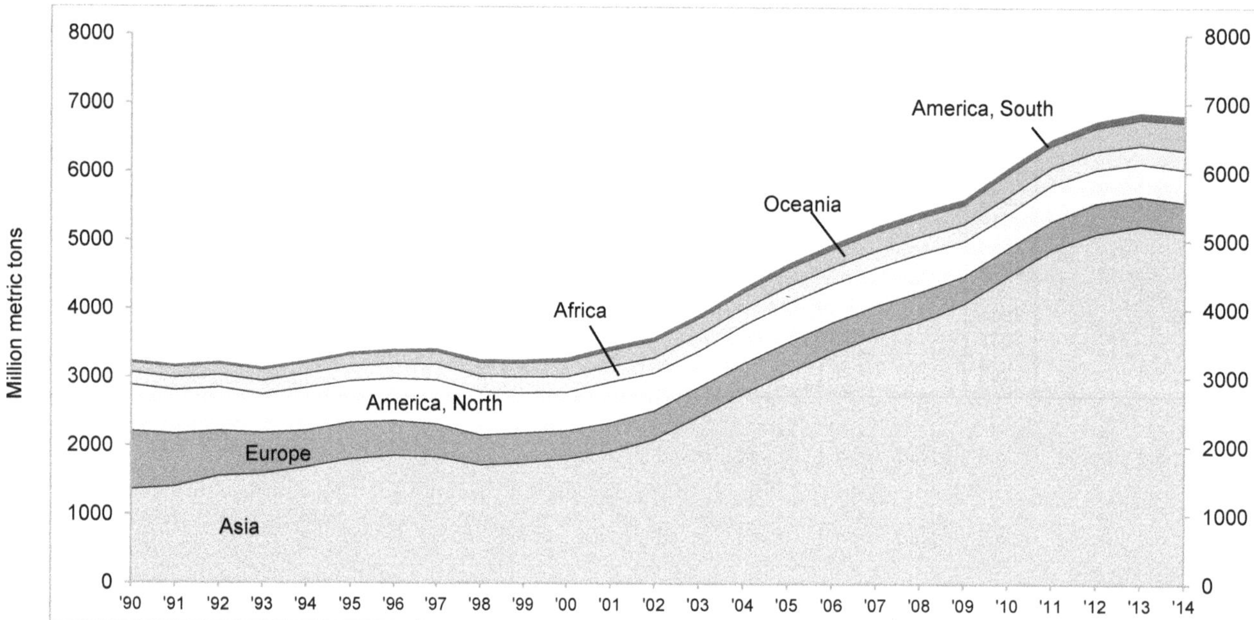

Figure 13: Major hard coal producing countries, 2014

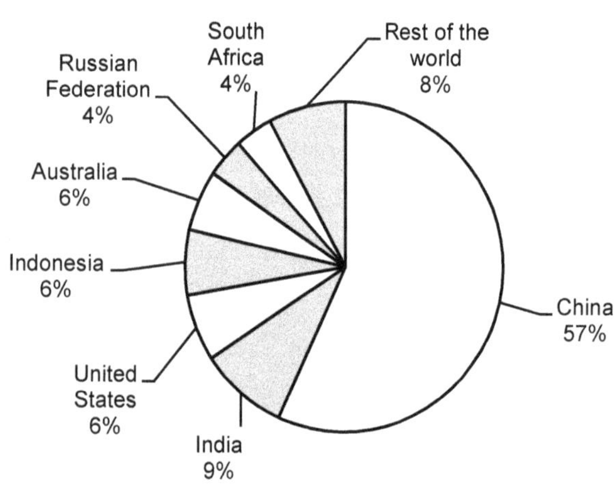

Figure 14: Major hard coal consuming countries, 2014

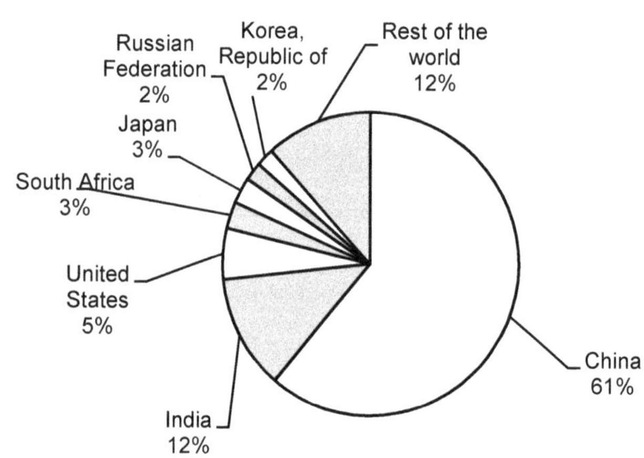

Production, trade and supply of hard coal

Thousand metric tons and kilograms per capita

Country or area	Year	Production	Imports	Exports	Changes in stocks	Supply	
						Total	Per capita
World	**2011**	**6,492,681**	**1,110,663**	**1,082,303**	**69,038**	**6,452,002**	**920**
	2012	**6,748,197**	**1,244,904**	**1,169,715**	**38,056**	**6,785,330**	**956**
	2013	**6,877,142**	**1,328,377**	**1,342,175**	**38,498**	**6,824,846**	**950**
	2014	**6,830,636**	**1,337,991**	**1,344,021**	**79,382**	**6,745,224**	**928**
Africa	**2011**	**257,354**	**12,180**	**69,275**	**460**	**199,800**	**187**
	2012	**266,945**	**13,133**	**79,300**	**2,749**	**198,028**	**180**
	2013	**267,727**	**9,790**	**78,869**	**3,219**	**195,430**	**173**
	2014	**275,353**	**11,869**	**74,042**	**3,472**	**209,708**	**181**
Algeria	2011	..	0	..	0	0	0
	2012	..	0	..	0	0	0
	2013	..	0	..	0	0	0
	2014	..	20	..	-1	21	1
Benin	2011	..	..	..	..	..	..
	2012	..	..	..	..	..	..
	2013	..	..	..	..	..	..
	2014	..	70	..	..	70	7
Botswana	2011	788	..	119	0	669	320
	2012	1,471	..	108	754	609	286
	2013	1,496	..	67	419	1,010	464
	2014	1,712	..	194	-281	1,799	810
Egypt	2011	9	*987	..	..	*996	*12
	2012	21	*905	..	..	*926	*11
	2013	3	416	..	..	419	5
	2014	0	434	..	..	434	5
Ethiopia	2011	21	141	..	-8	170	2
	2012	..	224	..	..	224	2
	2013	..	291	..	..	291	3
	2014	..	321	..	..	321	3
Kenya	2011	..	383	..	..	383	9
	2012	..	343	..	..	343	8
	2013	..	379	..	..	379	9
	2014	..	*470	..	..	*470	*10
Lesotho	2011	..	*670	..	..	*670	*330
	2012	..	*680	..	..	*680	*331
	2013	..	*690	..	..	*690	*331
	2014	..	*700	..	..	*700	*332
Madagascar	2011	..	153	..	..	153	7
	2012	..	263	..	..	263	12
	2013	..	426	..	..	426	19
	2014	..	384	..	..	384	16
Malawi	2011	83	17	*8	..	92	6
	2012	*75	15	*8	..	*82	*5
	2013	67	24	*8	..	83	5
	2014	64	20	*8	..	75	5
Mauritius	2011	..	660	..	19	641	512
	2012	..	729	..	54	675	536
	2013	..	708	..	-3	711	563
	2014	..	772	..	29	743	586
Morocco	2011	..	5,326	..	-341	5,667	171
	2012	..	6,010	..	164	5,846	174
	2013	..	4,480	..	-67	4,547	134
	2014	..	6,519	..	403	6,116	177

Table 5

Production, trade and supply of hard coal

Thousand metric tons and kilograms per capita

Country or area	Year	Production	Imports	Exports	Changes in stocks	Supply	
						Total	Per capita
Mozambique	2011	617	..	162	403	52	2
	2012	4,530	..	2,957	1,557	16	1
	2013	5,992	..	3,865	2,112	15	1
	2014	6,332	..	4,679	1,315	338	12
Namibia	2011	..	34	..	..	34	15
	2012	..	134	..	..	134	58
	2013	..	52	..	..	52	22
	2014	..	*33	..	..	*33	*14
Niger	2011	264	..	..	8	256	15
	2012	251	..	..	-15	265	15
	2013	242	..	..	7	235	13
	2014	256	..	..	*9	247	13
Nigeria	2011	32	..	..	..	32	0
	2012	48	..	..	..	48	0
	2013	44	..	..	..	44	0
	2014	46	..	..	..	46	0
Réunion	2011	..	750	..	67	684	817
	2012	..	686	..	-23	710	842
	2013	..	678	..	-2	679	800
	2014	..	636	..	-4	639	748
Senegal	2011	..	402	..	..	402	30
	2012	..	349	..	..	349	25
	2013	..	350	..	..	350	25
	2014	..	367	..	..	367	25
South Africa	2011	252,757	2,390	68,807	312	186,028	3,561
	2012	258,575	2,591	76,009	258	184,899	3,499
	2013	256,282	915	74,565	753	181,879	3,405
	2014	260,540	853	69,029	-492	192,856	3,573
Swaziland	2011	*140	*109	*108	0	*141	*116
	2012	*250	*97	*151	0	*196	*159
	2013	*226	*146	*226	0	*146	*117
	2014	216	52	133	0	136	107
Tunisia	2011	..	17	..	..	17	2
	2012	..	0	..	..	0	0
	2013	..	0	..	..	0	0
	2014	..	0	..	..	0	0
Uganda	2011	..	0	..	..	0	0
	2012	..	0	..	..	0	0
	2013	..	172	..	..	172	5
	2014	..	*172	..	..	*172	*5
United Rep. of Tanzania	2011	81	..	..	..	81	2
	2012	79	..	..	..	79	2
	2013	85	..	..	..	85	2
	2014	246	..	..	..	246	5
Zambia	2011	0	85	14	..	71	5
	2012	51	67	15	..	103	7
	2013	176	*12	*39	..	149	10
	2014	159	..	..	..	159	10
Zimbabwe	2011	2,562	56	57	..	2,562	180
	2012	1,594	39	52	..	1,580	108
	2013	3,114	52	99	..	3,067	206
	2014	5,783	46	..	2,493	3,335	219

Production, trade and supply of hard coal

Thousand metric tons and kilograms per capita

Country or area	Year	Production	Imports	Exports	Changes in stocks	Supply	
						Total	Per capita
America, North	**2011**	**531,254**	**29,397**	**124,102**	**12,813**	**423,736**	**767**
	2012	**489,680**	**25,760**	**140,753**	**-2,290**	**376,977**	**676**
	2013	**479,607**	**23,737**	**134,366**	**-15,169**	**384,147**	**682**
	2014	**485,757**	**25,700**	**116,710**	**9,140**	**385,607**	**678**
Canada	2011	34,621	8,560	32,610	1,886	8,685	252
	2012	35,375	8,863	34,648	-29	9,619	276
	2013	38,282	6,263	38,989	-1,564	7,120	202
	2014	35,950	6,124	34,364	245	7,465	210
Costa Rica	2011	..	8	..	..	8	2
	2012	..	3	..	..	3	1
	2013	..	3	..	..	3	1
	2014	..	3	..	..	3	1
Cuba	2011	..	2	..	..	2	0
	2012	..	4	..	..	4	0
	2013	..	13	..	..	13	1
	2014	..	1	..	..	1	0
Dominican Republic	2011	..	882	..	91	791	79
	2012	..	874	..	-6	880	87
	2013	..	850	..	16	833	81
	2014	..	1,043	..	-13	1,056	101
Guadeloupe	2011	..	*200	..	..	*200	*436
	2012	..	*200	..	..	*200	*434
	2013	..	*200	..	..	*200	*431
	2014	..	*200	..	..	*200	*429
Guatemala	2011	..	571	..	78	493	33
	2012	..	577	..	79	498	32
	2013	..	606	..	108	497	32
	2014	..	720	..	0	720	45
Honduras	2011	..	112	..	..	112	15
	2012	..	110	..	..	110	14
	2013	..	112	..	..	112	14
	2014	..	119	..	..	119	15
Jamaica	2011	..	72	..	-8	80	29
	2012	..	94	..	21	73	26
	2013	..	95	..	6	89	32
	2014	..	94	..	-10	105	38
Mexico	2011	5,800	7,404	242	186	12,776	106
	2012	2,096	6,790	217	-367	9,036	74
	2013	2,122	7,489	4	108	9,499	77
	2014	1,952	7,518	4	-20	9,486	76
Panama	2011	..	237	..	..	237	64
	2012	..	323	..	..	323	86
	2013	..	341	..	..	341	90
	2014	..	354	..	..	354	92
United States	2011	490,833	11,349	91,250	10,580	400,352	1,282
	2012	452,209	7,922	105,888	-1,987	356,230	1,132
	2013	439,203	7,766	95,373	-13,843	365,439	1,152
	2014	447,855	9,523	82,342	8,938	366,098	1,146
America, South	**2011**	**94,374**	**29,988**	**81,366**	**540**	**42,456**	**106**
	2012	**97,943**	**29,059**	**84,412**	**-707**	**43,297**	**107**
	2013	**104,717**	**30,204**	**87,190**	**1,163**	**46,569**	**114**
	2014	**102,132**	**33,059**	**84,414**	**276**	**50,501**	**122**

Table 5

Production, trade and supply of hard coal

Thousand metric tons and kilograms per capita

Country or area	Year	Production	Imports	Exports	Changes in stocks	Supply	
						Total	Per capita
Argentina	2011	90	1,718	29	0	1,779	43
	2012	95	1,586	78	-14	1,617	38
	2013	83	864	27	17	903	21
	2014	83	2,220	..	..	2,303	54
Brazil	2011	5,505	18,007	71	574	22,867	114
	2012	6,617	16,486	0	-515	23,618	117
	2013	8,594	18,041	0	534	26,101	128
	2014	7,936	20,351	0	460	27,827	135
Chile	2011	654	9,665	0	531	9,788	569
	2012	712	10,359	0	199	10,872	625
	2013	3,028	10,301	1,259	743	11,327	644
	2014	4,104	9,978	2,041	422	11,619	654
Colombia	2011	85,842	0	79,273	0	6,569	142
	2012	89,092	0	83,295	0	5,797	124
	2013	91,854	0	85,117	0	6,737	142
	2014	88,578	..	81,190	..	7,388	155
Paraguay	2011	..	1	..	..	1	0
	2012	..	2	..	..	2	0
	2013	..	0	..	..	0	0
	2014	..	0	..	..	0	0
Peru	2011	183	595	174	-565	1,169	39
	2012	227	624	128	-377	1,100	36
	2013	211	995	128	-131	1,209	40
	2014	231	507	253	-606	1,091	35
Uruguay	2011	..	2	..	..	2	1
	2012	..	3	..	..	3	1
	2013	..	3	..	..	3	1
	2014	..	3	..	..	3	1
Venezuela (Bolivar. Rep.)	2011	2,100	..	1,819	..	281	10
	2012	1,200	..	911	..	289	10
	2013	947	..	659	..	288	10
	2014	1,200	..	930	..	270	9
Asia	**2011**	**4,872,292**	**773,732**	**360,012**	**50,711**	**5,235,301**	**1,242**
	2012	**5,105,822**	**896,554**	**389,681**	**31,429**	**5,581,265**	**1,310**
	2013	**5,216,567**	**973,825**	**508,739**	**44,926**	**5,636,727**	**1,309**
	2014	**5,132,232**	**982,351**	**483,146**	**52,198**	**5,579,239**	**1,283**
Afghanistan	2011	1,480	..	..	..	1,480	51
	2012	1,240	..	..	..	1,240	42
	2013	1,347	..	..	..	1,347	44
	2014	1,517	..	..	..	1,517	48
Armenia	2011	..	3	0	..	3	1
	2012	..	3	0	..	3	1
	2013	..	1	0	..	1	0
	2014	..	1	..	..	1	0
Azerbaijan	2011	..	6	..	-2	8	1
	2012	..	5	..	-1	6	1
	2013	..	5	..	0	5	1
	2014	..	2	..	-1	3	0
Bangladesh	2011	667	924	..	94	1,497	10
	2012	835	1,000	..	28	1,807	12
	2013	855	1,000	..	-105	1,960	12
	2014	947	988	..	83	1,852	12

Production, trade and supply of hard coal

Thousand metric tons and kilograms per capita

Country or area	Year	Production	Imports	Exports	Changes in stocks	Supply	
						Total	Per capita
Bhutan	2011	109	..	73	..	36	49
	2012	99	..	47	..	52	70
	2013	78	..	5	..	73	96
	2014	122	..	27	..	95	125
China	2011	3,764,435	222,358	14,665	41,548	3,930,580	2,915
	2012	3,945,128	288,411	9,275	37,720	4,186,544	3,089
	2013	3,974,322	327,018	7,508	43,684	4,250,148	3,119
	2014	3,873,919	291,220	5,742	41,062	4,118,335	3,007
China, Hong Kong SAR	2011	..	12,529	0	775	11,754	1,669
	2012	..	12,351	0	780	11,571	1,629
	2013	..	12,974	2	636	12,336	1,722
	2014	..	13,789	0	865	12,924	1,788
Cyprus	2011	..	0	..	-12	12	11
	2012	..	0	..	0	0	0
	2013	..	0	..	0	0	0
	2014	..	5	..	1	4	3
Georgia	2011	..	81	7	0	74	18
	2012	..	76	4	0	71	17
	2013	..	82	4	0	77	19
	2014	..	127	2	*-3	129	32
India	2011	539,950	102,853	2,032	26	640,745	514
	2012	557,707	145,790	2,825	-11,890	712,562	564
	2013	565,766	168,440	2,188	-7,535	739,553	578
	2014	612,435	212,103	1,238	-4,269	827,569	639
Indonesia	2011	346,943	41	267,787	1,064	78,133	319
	2012	385,899	78	304,051	-217	82,143	331
	2013	458,463	509	424,325	860	33,787	134
	2014	435,743	2,273	410,083	10,695	17,238	68
Iran (Islamic Rep. of)	2011	1,164	14	308	..	870	12
	2012	1,041	91	333	..	799	10
	2013	1,104	48	305	..	847	11
	2014	958	140	205	..	893	11
Israel	2011	..	12,310	..	-271	12,581	1,663
	2012	..	13,816	..	-360	14,176	1,842
	2013	..	12,643	..	1,083	11,560	1,479
	2014	..	11,021	..	100	10,921	1,376
Japan	2011	..	175,423	2	..	175,421	1,379
	2012	..	183,861	3	..	183,858	1,446
	2013	..	195,609	2	..	195,607	1,540
	2014	..	188,069	1	..	188,068	1,483
Jordan	2011	..	0	..	..	0	0
	2012	..	0	..	..	0	0
	2013	..	331	..	..	331	46
	2014	..	474	..	..	474	64
Kazakhstan	2011	108,082	152	30,064	474	77,696	4,693
	2012	112,779	206	30,322	2,848	79,815	4,745
	2013	112,884	228	31,762	1,608	79,742	4,663
	2014	107,091	15	28,471	61	78,574	4,523
Korea, Dem.Ppl's.Rep.	2011	25,500	0	11,178	..	14,322	581
	2012	25,800	1,089	12,011	..	14,878	601
	2013	26,600	63	16,703	..	9,960	400
	2014	27,090	113	15,610	..	11,593	463

Production, trade and supply of hard coal

Thousand metric tons and kilograms per capita

Country or area	Year	Production	Imports	Exports	Changes in stocks	Supply	
						Total	Per capita
Korea, Republic of	2011	2,084	124,935	..	340	126,679	2,567
	2012	2,094	121,434	..	-92	123,620	2,492
	2013	1,815	124,704	..	400	126,119	2,530
	2014	1,748	126,167	..	-2,151	130,066	2,597
Kyrgyzstan	2011	109	1,020	0	0	1,129	203
	2012	153	1,399	0	0	1,552	275
	2013	251	1,271	144	354	1,024	178
	2014	298	1,276	236	-57	1,395	239
Lao People's Dem. Rep.	2011	167	..	..	..	167	26
	2012	130	..	..	..	130	20
	2013	104	..	..	..	104	16
	2014	106	..	..	..	106	16
Lebanon	2011	..	250	..	..	250	54
	2012	..	250	..	..	250	51
	2013	..	200	..	..	200	38
	2014	..	250	..	..	250	45
Malaysia	2011	2,916	21,145	224	590	23,247	814
	2012	2,951	22,558	369	83	25,057	863
	2013	2,894	22,063	518	169	24,270	824
	2014	2,687	21,738	181	25	24,219	810
Mongolia	2011	26,962	0	16,476	6,017	4,469	1,620
	2012	23,124	0	15,155	1,716	6,253	2,227
	2013	23,793	1	12,439	945	10,410	3,641
	2014	19,215	1	13,986	1,705	3,525	1,211
Myanmar	2011	351	*10	..	..	361	7
	2012	471	8	..	..	479	9
	2013	535	47	..	..	582	11
	2014	642	0	..	..	642	12
Nepal	2011	13	476	..	..	489	18
	2012	9	356	..	..	365	13
	2013	14	443	..	..	457	16
	2014	14	794	..	..	808	29
Other Asia	2011	..	53,644	..	152	53,492	2,300
	2012	..	48,544	..	-714	49,258	2,114
	2013	..	52,633	..	879	51,754	2,218
	2014	..	52,534	..	-578	53,112	2,273
Pakistan	2011	3,613	4,057	..	..	7,670	44
	2012	3,179	3,710	..	..	6,889	39
	2013	3,438	3,119	..	..	6,557	36
	2014	3,551	2,760	..	..	6,311	34
Singapore	2011	..	7	1	..	6	1
	2012	..	40	0	..	40	7
	2013	..	431	5	..	427	79
	2014	..	642	12	..	630	114
Sri Lanka	2011	..	760	..	286	475	23
	2012	..	962	..	239	724	35
	2013	..	1,131	..	370	761	37
	2014	..	1,607	..	145	1,461	71
Tajikistan	2011	176	15	22	..	169	22
	2012	318	24	2	..	340	43
	2013	482	10	..	..	492	61
	2014	822	12	..	..	834	101

Production, trade and supply of hard coal

Thousand metric tons and kilograms per capita

Country or area	Year	Production	Imports	Exports	Changes in stocks	Supply	
						Total	Per capita
Thailand	2011	..	16,213	6	*-124	16,331	244
	2012	..	18,360	0	*-44	18,404	274
	2013	..	18,726	9	186	18,531	275
	2014	..	21,243	23	3,733	17,487	258
Turkey	2011	2,528	23,678	0	-26	26,232	357
	2012	2,292	29,195	7	20	31,460	420
	2013	1,968	26,633	5	418	28,178	370
	2014	1,833	29,816	64	121	31,464	406
United Arab Emirates	2011	..	654	4	..	650	74
	2012	..	2,132	58	..	2,074	232
	2013	..	2,656	7	..	2,649	293
	2014	..	2,939	1	..	2,938	323
Uzbekistan	2011	60	..	..	..	60	2
	2012	60	..	..	..	60	2
	2013	199	..	..	..	199	7
	2014	408	..	..	..	408	14
Viet Nam	2011	44,984	0	17,163	-219	28,040	314
	2012	40,512	627	15,219	1,313	24,607	272
	2013	39,655	621	12,808	974	26,494	290
	2014	41,086	42	*7,265	660	33,203	359
Yemen	2011	..	174	..	..	174	7
	2012	..	178	..	..	178	7
	2013	..	185	..	..	185	7
	2014	..	189	..	..	189	7
Europe	**2011**	**414,265**	**264,570**	**160,888**	**2,457**	**515,490**	**700**
	2012	**445,917**	**279,851**	**171,842**	**1,768**	**552,158**	**749**
	2013	**432,675**	**290,136**	**194,720**	**-1,319**	**529,410**	**718**
	2014	**426,207**	**284,261**	**208,946**	**11,779**	**489,743**	**664**
Andorra	2011	..	0	..	..	0	2
	2012	..	0	..	..	0	*2
	2013	..	0	..	..	0	*3
	2014	..	0	..	..	0	*4
Austria	2011	..	3,388	1	-441	3,828	454
	2012	..	3,642	2	142	3,498	414
	2013	..	3,182	1	-306	3,487	411
	2014	..	3,157	0	31	3,126	367
Belarus	2011	..	116	2	23	91	10
	2012	..	415	1	56	358	38
	2013	..	586	0	43	543	57
	2014	..	781	5	32	744	78
Belgium	2011	..	5,937	667	126	5,144	467
	2012	..	5,374	627	-319	5,066	457
	2013	..	5,284	598	-311	4,997	448
	2014	5	5,226	557	190	4,484	399
Bosnia and Herzegovina	2011	..	1,388	1	110	1,277	333
	2012	..	1,058	5	16	1,037	271
	2013	..	1,164	1	58	1,105	289
	2014	..	1,408	6	63	1,339	351
Bulgaria	2011	18	3,300	39	124	3,155	429
	2012	15	2,440	44	171	2,240	307
	2013	17	1,547	3	-198	1,759	243
	2014	35	1,410	41	-444	1,848	257

Table 5

Production, trade and supply of hard coal

Thousand metric tons and kilograms per capita

Country or area	Year	Production	Imports	Exports	Changes in stocks	Supply	
						Total	Per capita
Croatia	2011	..	1,104	0	-17	1,121	261
	2012	..	875	0	-127	1,002	234
	2013	..	1,182	11	113	1,058	248
	2014	..	962	14	-80	1,028	242
Czechia	2011	11,423	2,310	6,279	12	7,442	706
	2012	12,319	1,747	5,050	1,796	7,220	685
	2013	8,744	2,010	4,818	-1,121	7,057	669
	2014	8,825	3,208	4,364	298	7,371	699
Denmark	2011	..	6,137	0	612	5,525	991
	2012	..	3,978	11	-271	4,238	757
	2013	..	4,895	52	-501	5,344	950
	2014	..	4,287	52	207	4,028	713
Estonia	2011	..	67	0	-3	70	53
	2012	..	82	1	17	64	48
	2013	..	59	0	-2	61	46
	2014	..	82	0	4	78	59
Finland	2011	..	6,970	..	967	6,003	1,113
	2012	..	3,909	..	-693	4,602	848
	2013	..	5,067	..	-769	5,836	1,070
	2014	..	5,439	..	881	4,558	832
France	2011	149	14,427	61	-32	14,547	230
	2012	290	15,856	177	-854	16,823	265
	2013	313	16,959	191	-1,248	18,329	287
	2014	300	12,853	0	-58	13,211	206
Germany	2011	12,960	47,845	204	1,680	58,921	733
	2012	11,558	49,034	347	-155	60,400	751
	2013	8,260	54,319	251	-416	62,744	779
	2014	8,337	53,740	208	139	61,730	765
Greece	2011	..	395	7	8	380	34
	2012	..	276	9	-84	351	32
	2013	..	344	11	31	302	27
	2014	..	310	9	27	274	25
Hungary	2011	..	1,560	0	-110	1,670	167
	2012	..	1,487	0	-17	1,504	151
	2013	..	1,319	0	-13	1,332	134
	2014	..	1,361	0	20	1,341	136
Iceland	2011	..	111	..	6	105	327
	2012	..	104	..	-1	105	325
	2013	..	120	..	0	120	369
	2014	..	100	..	0	100	306
Ireland	2011	..	2,289	9	267	2,013	433
	2012	..	2,219	12	-231	2,438	522
	2013	..	2,383	13	237	2,133	457
	2014	..	1,974	13	-68	2,029	434
Italy	2011	92	23,261	3	-885	24,235	406
	2012	80	24,062	58	-662	24,746	414
	2013	73	19,637	0	-866	20,576	344
	2014	86	19,453	1	-90	19,628	328
Latvia	2011	..	183	7	4	172	83
	2012	..	144	6	-1	139	68
	2013	..	119	8	-10	121	60
	2014	..	85	6	-23	102	51

Production, trade and supply of hard coal

Thousand metric tons and kilograms per capita

Country or area	Year	Production	Imports	Exports	Changes in stocks	Supply	
						Total	Per capita
Lithuania	2011	..	356	18	36	302	98
	2012	..	320	23	0	297	98
	2013	..	417	35	29	353	119
	2014	..	300	0	-7	307	105
Luxembourg	2011	..	89	..	..	89	171
	2012	..	82	..	..	82	154
	2013	..	75	..	..	75	138
	2014	..	85	..	..	85	153
Netherlands	2011	..	24,463	12,605	181	11,677	700
	2012	..	28,996	18,370	-2,154	12,780	763
	2013	..	41,769	27,009	1,791	12,969	772
	2014	..	47,261	31,256	1,395	14,610	866
Norway	2011	1,386	709	1,504	-187	778	157
	2012	1,229	723	1,273	-75	754	150
	2013	1,855	644	2,068	-297	728	143
	2014	1,675	761	1,608	25	803	156
Poland	2011	76,448	14,955	7,007	869	83,527	2,164
	2012	79,813	10,165	7,070	6,838	76,070	1,970
	2013	77,017	10,515	10,846	-2,097	78,783	2,040
	2014	73,271	10,417	8,956	1,173	73,559	1,905
Portugal	2011	..	3,595	..	-102	3,697	350
	2012	..	5,035	..	162	4,873	463
	2013	..	4,246	..	-202	4,448	425
	2014	..	4,371	..	-141	4,512	434
Republic of Moldova	2011	..	207	..	9	198	49
	2012	..	184	..	-7	191	47
	2013	..	259	..	9	250	61
	2014	..	146	..	-12	158	39
Romania	2011	..	300	36	14	250	12
	2012	..	261	8	-22	275	14
	2013	..	237	6	7	224	11
	2014	..	153	5	-13	161	8
Russian Federation	2011	219,288	30,344	123,762	-189	126,059	880
	2012	252,062	28,602	130,268	552	149,844	1,046
	2013	252,307	27,873	138,979	4,066	137,135	957
	2014	264,042	25,291	153,100	2,390	133,843	933
Serbia	2011	..	186	1	-11	196	22
	2012	..	173	0	11	162	18
	2013	..	101	3	-5	103	12
	2014	..	160	0	-8	168	19
Slovakia	2011	..	3,972	..	-86	4,058	750
	2012	..	4,166	..	249	3,917	723
	2013	..	3,839	..	-78	3,917	723
	2014	..	3,784	..	-64	3,848	710
Slovenia	2011	..	24	2	5	17	8
	2012	..	24	1	7	16	8
	2013	..	23	7	7	9	4
	2014	..	22	1	14	7	3
Spain	2011	4,262	16,168	1,190	-2,137	21,377	458
	2012	3,910	22,414	1,876	-2,135	26,583	570
	2013	2,542	13,662	677	-3,410	18,937	408
	2014	2,669	16,394	1,284	-1,436	19,215	415

Table 5

Production, trade and supply of hard coal

Thousand metric tons and kilograms per capita

Country or area	Year	Production	Imports	Exports	Changes in stocks	Supply	
						Total	Per capita
Sweden	2011	..	3,076	1	32	3,043	322
	2012	..	2,281	1	-496	2,776	291
	2013	..	2,540	1	-293	2,832	294
	2014	..	2,774	1	101	2,672	275
Switzerland	2011	..	99	..	-45	144	18
	2012	..	141	..	3	138	17
	2013	..	144	..	7	137	17
	2014	..	39	..	-52	91	11
T.F.Yug.Rep. Macedonia	2011	..	2	0	0	2	1
	2012	..	2	0	0	2	1
	2013	..	6	0	0	6	3
	2014	..	6	0	0	6	3
Ukraine	2011	69,687	12,709	6,991	2,453	72,952	1,604
	2012	67,674	14,764	6,114	3,018	73,306	1,618
	2013	68,779	14,208	8,537	3,128	71,322	1,579
	2014	55,315	14,695	7,034	2,404	60,572	1,346
United Kingdom	2011	18,552	32,528	491	-836	51,425	814
	2012	16,967	44,816	488	-2,966	64,261	1,011
	2013	12,768	49,402	594	1,298	60,278	942
	2014	11,647	41,765	425	4,881	48,106	748
Oceania	**2011**	**323,142**	**795**	**286,661**	**2,057**	**35,219**	**952**
	2012	**341,891**	**547**	**303,726**	**5,107**	**33,605**	**894**
	2013	**375,848**	**685**	**338,292**	**5,678**	**32,563**	**853**
	2014	**408,956**	**751**	**376,763**	**2,518**	**30,426**	**785**
Australia	2011	320,811	195	284,510	1,981	34,515	1,531
	2012	339,614	96	301,516	5,273	32,921	1,437
	2013	373,568	53	336,196	5,611	31,814	1,367
	2014	407,007	111	375,044	2,466	29,608	1,253
New Caledonia	2011	..	566	..	72	494	1,978
	2012	..	450	..	-43	493	1,948
	2013	..	570	..	47	523	2,039
	2014	..	610	..	0	610	2,349
New Zealand	2011	2,331	34	2,151	4	210	48
	2012	2,277	1	2,210	-123	191	43
	2013	2,280	62	2,096	20	226	51
	2014	1,949	30	1,719	52	208	46

Table 6

International trade of hard coal
(Principal importers/exporters)

Thousand metric tons

2013

Importers	Exporters					
	World	Australia	Canada	Colombia	Indonesia	Kazakhstan
Austria	3,182	..	..	10	..	..
Belgium	5,284	1,306	..	339	..	..
Brazil	18,041	3,073	1,685	2,747	..	0
Canada	6,263	..	..	1,348	..	..
Chile	10,301	1,760	516	5,871	..	..
China	327,018	73,839	*10,012	503	*60,925	257
China, Hong Kong SAR	12,974	451	..	..	*11,213	..
Denmark	4,895	..	..	1,914	..	..
Finland	5,067	206	368	..	..	105
France	16,959	3,031	34	2,517	..	..
Germany	54,319	4,566	1,176	8,131	..	..
India	168,440	33,379	1,360	68	*87,069	..
Israel	12,643	..	..	5,229	..	..
Italy	19,637	744	1,136	1,928	3,205	36
Japan	195,609	124,393	*9,081	212	37,509	..
Korea, Republic of	124,704	49,950	*10,307	324	*30,573	..
Malaysia	22,063	3,168	..	..	*13,810	..
Mexico	7,489	1,104	278	861	..	..
Morocco	4,480	..	..	..	..	..
Netherlands	41,769	2,225	469	15,784	14	48
Pakistan	3,119	208	..	..	887	..
Poland	10,515	1,013	120	60	..	182
Portugal	4,246	..	..	3,390	..	..
Russian Federation	27,873	..	..	45	..	22,787
Slovakia	3,839	..	..	..	..	..
Spain	13,662	1,006	59	2,519	3,392	6
Thailand	18,726	3,832	0	..	13,217	..
Turkey	26,633	905	339	7,158	147	..
Ukraine	14,208	..	326	73	..	565
United Kingdom	49,402	2,147	561	11,494	..	18
United States	7,766	..	911	5,966	605	..

Table 6

International trade of hard coal
(Principal importers/exporters)

Thousand metric tons

2013

Exporters					
Mongolia	Poland	Russian Federation	South Africa	United States	Viet Nam
..	751	38	..	680	..
..	70	1,170	318	1,655	..
..	..	1,329	808	7,764	..
..	..	0	..	4,719	..
..	..	..	..	2,154	..
*11,817	..	*24,494	12,743	7,149	*10,359
..	..	145	..	524	..
..	560	1,826	317	37	..
..	302	3,590	..	411	..
..	1,355	2,803	2,159	4,383	..
..	4,480	11,829	3,132	*8,824	..
..	..	461	21,237	3,556	51
..	..	2,621	3,848	..	..
..	..	4,410	2,156	5,456	..
..	..	12,583	477	*6,115	1,006
..	..	*13,243	165	*5,151	*1,011
..	..	438	1,797	..	206
..	..	..	..	*4,410	..
..	387	589	453	2,803	..
..	..	7,386	4,906	*8,444	2
..	..	..	1,129	..	..
..	..	6,541	..	521	..
..	..	11	164	507	..
1	..	..	..	0	..
..	700	986	..	303	..
..	17	2,352	1,552	1,741	..
..	..	4	..	1	91
..	214	8,590	3,347	4,029	..
..	..	10,288	..	2,626	..
..	529	20,249	484	*10,732	..
..	..	33	..	..	..

Table 6

International trade of hard coal
(Principal importers/exporters)
Thousand metric tons

2014

Importers	Exporters					
	World	Australia	Canada	Colombia	Indonesia	Kazakhstan
Austria	3,157	236	..	70	..	..
Belgium	5,226	1,287	..	377	..	..
Brazil	20,351	3,747	2,263	5,051	..	72
Canada	6,124	..	..	965	..	..
Chile	9,978	823	272	6,112	..	..
China	291,220	95,640	*6,193	106	*41,834	38
China, Hong Kong SAR	13,789	532	..	..	12,610	..
Denmark	4,287	..	..	1,833	..	..
Finland	5,439	0	599	5	..	183
France	12,853	2,883	29	1,737	..	..
Germany	53,740	2,339	*2,782	7,528	..	..
India	212,103	38,533	1,711	0	*100,866	..
Israel	11,021	..	..	5,891	..	..
Italy	19,453	591	723	2,333	3,304	566
Japan	188,069	119,675	*7,981	170	33,395	..
Korea, Republic of	126,167	54,431	*9,166	0	29,606	..
Malaysia	21,738	4,062	..	..	12,085	..
Mexico	7,518	2,437	158	449	..	..
Morocco	6,519	..	..	..	..	..
Netherlands	47,261	2,404	25	*12,170	0	0
Pakistan	2,760	109	..	..	953	..
Poland	10,417	1,224	0	107	..	65
Portugal	4,371	..	..	3,851	..	..
Russian Federation	25,291	..	..	48	..	22,864
Slovakia	3,784	..	91	..	..	..
Spain	16,394	916	1	5,611	3,936	13
Thailand	21,243	4,076	0	..	16,520	..
Turkey	29,816	631	492	9,375	..	..
Ukraine	14,695	300	281	61	..	657
United Kingdom	41,765	1,249	434	*9,197	..	0
United States	9,523	..	839	7,526	713	..

International trade of hard coal (Principal importers/exporters)

Thousand metric tons

2014

Exporters					
Mongolia	Poland	Russian Federation	South Africa	United States	Viet Nam
..	887	14	..	542	..
..	0	1,308	464	1,225	..
..	..	1,043	998	*5,886	..
..	..	0	..	5,041	..
..	..	..	..	2,696	..
*13,706	..	25,237	5,757	1,428	*5,328
..	..	573	..	0	..
..	363	1,690	309	2	..
..	176	3,814	..	602	..
..	7	2,636	2,682	2,186	..
..	3,291	13,900	5,089	*8,545	..
..	..	1,124	26,973	4,160	27
..	..	2,460	2,496	..	..
..	..	4,001	1,772	5,220	..
..	..	14,886	133	*5,956	*507
..	..	17,394	323	3,894	*1,063
..	..	1,417	1,760	..	69
..	..	..	..	4,445	..
..	588	1,748	949	2,218	..
..	..	8,086	10,290	*9,168	10
..	..	..	*1,641	..	..
..	..	6,506	..	379	..
..	..	0	156	288	..
20	..	..	..	0	..
..	532	904	..	537	..
..	22	2,222	1,555	1,227	..
..	..	0	..	0	180
..	67	8,645	3,985	4,325	..
..	..	*10,261	299	2,573	..
..	212	17,677	136	*9,345	..
..	..	39	..	..	..

Table 7

Production, trade and supply of brown coal and oil shale

Thousand metric tons and kilograms per capita

Table Notes

Data refer to brown coal and oil shale.

Please refer to the Definitions Section on pages ix to xv for the appropriate product description/ classification.

Figure 15: Production of brown coal and oil shale by region, 1990-2014

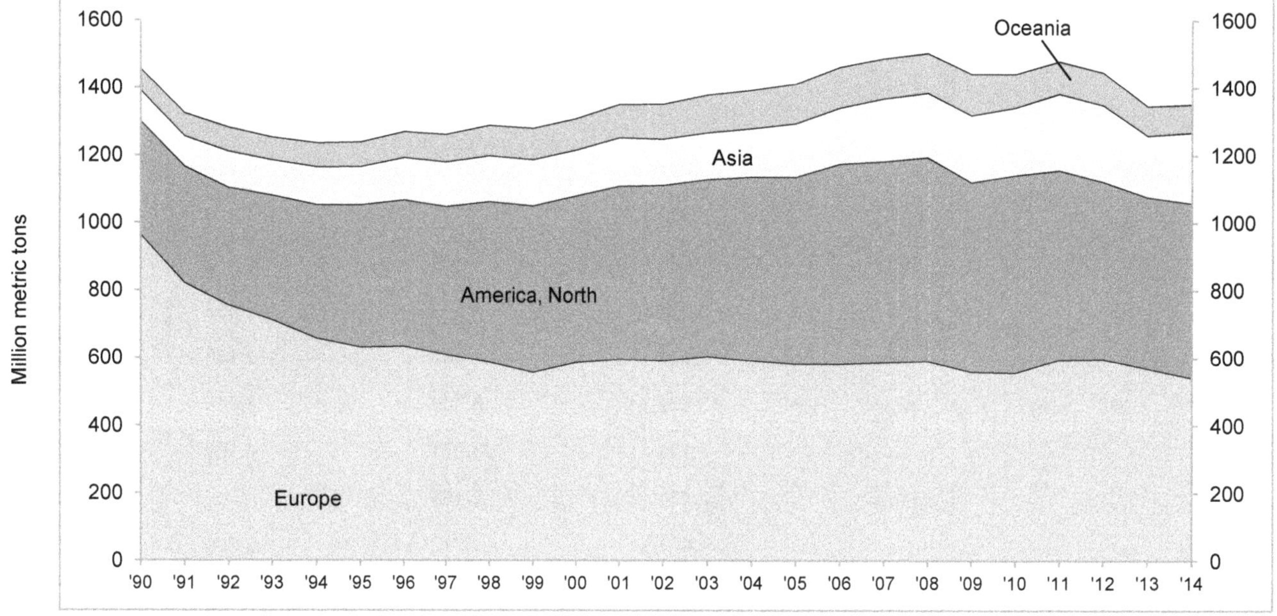

Figure 16: Major brown coal producing countries, 2014

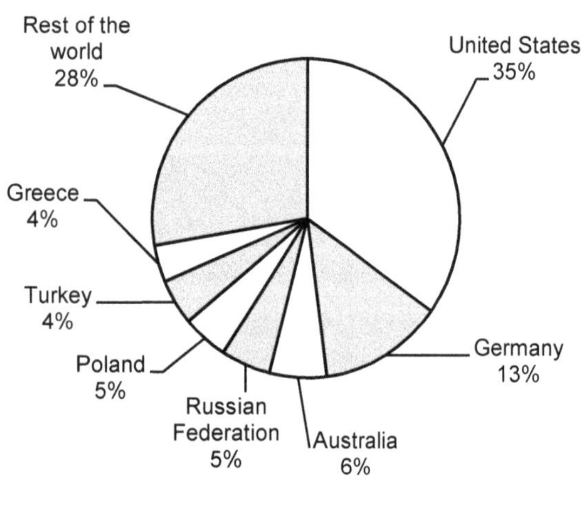

Figure 17: Major brown coal consuming countries, 2014

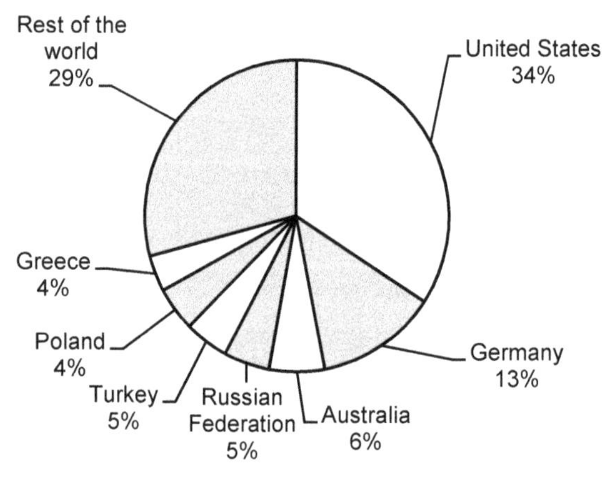

Production, trade and supply of brown coal and oil shale

Thousand metric tons and kilograms per capita

Country or area	Year	Production	Imports	Exports	Changes in stocks	Supply	
						Total	Per capita
World	**2011**	**1,478,692**	**36,372**	**20,740**	**-13,138**	**1,507,461**	**215**
	2012	**1,446,218**	**38,955**	**26,648**	**12,662**	**1,445,863**	**204**
	2013	**1,345,180**	**38,706**	**21,493**	**-20,657**	**1,383,051**	**193**
	2014	**1,351,709**	**43,807**	**28,350**	**-16,483**	**1,383,648**	**190**
Africa	**2011**	..	..	..	..	..	..
	2012	..	..	..	..	..	..
	2013	..	..	..	..	..	..
	2014	**84**	**32**	..	..	**115**	**0**
Swaziland	2011	..	..	..	..	..	..
	2012	..	..	..	..	..	..
	2013	..	..	..	..	..	..
	2014	84	32	..	..	115	91
America, North	**2011**	**561,299**	**1,767**	**6,229**	**-11,563**	**568,400**	**1,029**
	2012	**524,842**	**1,343**	**8,344**	**7,785**	**510,056**	**914**
	2013	**508,151**	**2,604**	**11,483**	**-20,653**	**519,925**	**923**
	2014	**516,845**	**2,472**	**5,988**	**-6,823**	**520,152**	**915**
Canada	2011	32,493	1,240	176	-2,053	35,610	1,032
	2012	31,121	954	158	-109	32,026	918
	2013	30,626	2,289	111	1,099	31,705	900
	2014	33,068	1,695	101	466	34,196	961
Mexico	2011	13,718	3	..	879	12,842	107
	2012	13,656	3	..	-355	14,014	115
	2013	13,065	3	..	-642	13,710	111
	2014	13,435	3	..	430	13,008	104
United States	2011	515,088	524	6,053	-10,389	519,948	1,664
	2012	480,065	386	8,186	8,249	464,016	1,474
	2013	464,460	312	11,372	-21,110	474,510	1,496
	2014	470,342	774	5,887	-7,719	472,948	1,481
Asia	**2011**	**225,718**	**28,510**	**11,661**	**853**	**241,714**	**57**
	2012	**226,323**	**31,207**	**14,450**	**2,411**	**240,668**	**56**
	2013	**181,378**	**29,916**	**5,961**	**-204**	**205,536**	**48**
	2014	**209,290**	**33,927**	**16,618**	**-6,723**	**233,322**	**54**
Bangladesh	2011	..	0	..	..	0	0
	2012	..	0	..	..	0	0
	2013	..	30	..	..	30	0
	2014	..	..	..	..	..	..
Cambodia	2011	..	19	..	..	19	1
	2012	..	21	..	..	21	1
	2013	..	96	..	..	96	6
	2014	..	490	..	..	490	32
Cyprus	2011	..	1	..	..	1	1
	2012	..	1	..	..	1	1
	2013	..	1	..	..	1	1
	2014	..	0	..	..	0	0
Georgia	2011	353	..	..	0	353	84
	2012	422	..	..	0	422	102
	2013	404	..	..	-1	405	99
	2014	299	..	..	0	299	74
India	2011	42,332	0	0	441	41,891	34
	2012	46,598	0	107	442	46,049	36
	2013	44,271	0	2	367	43,902	34
	2014	48,257	1	3	-1,316	49,571	38

Table 7

Production, trade and supply of brown coal and oil shale

Thousand metric tons and kilograms per capita

Country or area	Year	Production	Imports	Exports	Changes in stocks	Supply	
						Total	Per capita
Indonesia	2011	*57,575	..	*8,000	..	*49,575	*203
	2012	*59,724	..	*8,000	..	*51,724	*209
	2013	*31,284	..	0	..	*31,284	*125
	2014	*48,946	..	*8,000	..	*40,946	*161
Israel	2011	416	165	..	..	581	77
	2012	423	154	..	..	577	75
	2013	421	172	..	..	593	76
	2014	396	..	..	..	396	50
Jordan	2011	..	..	..	..	..	..
	2012	..	..	..	..	..	..
	2013	..	..	..	..	..	..
	2014	..	62	..	..	62	8
Kazakhstan	2011	8,368	3	286	56	8,029	485
	2012	7,748	7	2,339	-98	5,514	328
	2013	6,690	1	2,049	45	4,597	269
	2014	6,894	..	2,474	205	4,215	243
Korea, Dem.Ppl's.Rep.	2011	4,684	..	..	..	4,684	190
	2012	4,488	..	..	..	4,488	181
	2013	3,342	..	..	..	3,342	134
	2014	3,741	..	..	..	3,741	149
Korea, Republic of	2011	..	4,215	..	..	4,215	85
	2012	..	2,834	..	..	2,834	57
	2013	..	1,803	..	..	1,803	36
	2014	..	4,865	..	..	4,865	97
Kyrgyzstan	2011	722	16	38	..	700	126
	2012	1,011	21	15	..	1,017	180
	2013	1,157	4	12	..	1,149	200
	2014	1,505	21	5	154	1,367	234
Lao People's Dem. Rep.	2011	512	..	*409	..	103	16
	2012	575	..	409	..	166	26
	2013	404	..	*302	..	102	15
	2014	26	..	0	..	26	4
Mongolia	2011	5,068	..	152	-8	4,924	1,785
	2012	6,802	..	368	1,690	4,743	1,689
	2013	6,330	..	155	212	5,964	2,086
	2014	6,073	..	328	-252	5,997	2,061
Myanmar	2011	47	..	..	..	47	1
	2012	62	..	..	..	62	1
	2013	42	..	..	..	42	1
	2014	50	..	..	..	50	1
Other Asia	2011	..	12,945	..	-64	13,009	559
	2012	..	16,085	..	82	16,003	687
	2013	..	13,318	..	-834	14,152	606
	2014	..	13,247	..	55	13,192	565
Philippines	2011	6,882	10,963	2,736	469	14,639	155
	2012	7,349	11,896	3,173	754	15,317	160
	2013	7,092	14,415	3,401	-847	18,952	194
	2014	7,600	15,182	5,767	-3,148	20,163	203
Tajikistan	2011	61	..	..	..	61	8
	2012	94	..	..	..	94	12
	2013	34	..	..	..	34	4
	2014	56	..	..	..	56	7

Production, trade and supply of brown coal and oil shale

Thousand metric tons and kilograms per capita

Country or area	Year	Production	Imports	Exports	Changes in stocks	Supply	
						Total	Per capita
Thailand	2011	21,327	..	..	1,309	20,018	299
	2012	18,066	..	..	-299	18,365	273
	2013	17,591	..	..	-1,512	19,103	283
	2014	18,042	..	..	-409	18,451	272
Turkey	2011	73,588	..	..	-1,350	74,938	1,019
	2012	69,169	..	..	-160	69,329	926
	2013	58,425	..	..	2,366	56,059	735
	2014	63,416	..	..	-2,011	65,427	844
United Arab Emirates	2011	..	83	..	..	83	10
	2012	..	58	..	..	58	6
	2013	..	56	..	..	56	6
	2014	..	59	..	..	59	6
Uzbekistan	2011	3,784	100	39	..	3,845	137
	2012	3,793	130	39	..	3,884	136
	2013	3,891	20	40	..	3,871	133
	2014	3,989	..	41	..	3,948	134
Europe	**2011**	**595,069**	**5,957**	**2,842**	**-2,449**	**600,633**	**816**
	2012	**597,317**	**6,405**	**3,854**	**2,971**	**596,896**	**810**
	2013	**568,443**	**5,729**	**4,049**	**84**	**570,039**	**773**
	2014	**541,674**	**6,935**	**5,722**	**-2,723**	**545,611**	**739**
Albania	2011	7	..	..	..	7	2
	2012	7	..	..	..	7	2
	2013	9	..	..	..	9	3
	2014	0	..	..	..	0	0
Austria	2011	..	112	6	1	105	12
	2012	..	90	0	0	90	11
	2013	..	85	0	0	85	10
	2014	..	98	0	0	98	12
Bosnia and Herzegovina	2011	12,628	29	488	-443	12,612	3,291
	2012	12,195	20	508	79	11,628	3,037
	2013	11,765	15	370	-94	11,504	3,009
	2014	11,651	20	425	-392	11,638	3,049
Bulgaria	2011	37,110	..	86	28	36,996	5,030
	2012	33,412	..	91	351	32,970	4,514
	2013	28,615	..	80	-184	28,719	3,960
	2014	31,268	2	51	-221	31,440	4,366
Croatia	2011	..	76	..	0	76	18
	2012	..	46	..	-4	50	12
	2013	..	54	..	-2	56	13
	2014	..	44	..	-2	46	11
Czechia	2011	46,639	76	1,144	1,200	44,371	4,212
	2012	43,533	338	1,203	230	42,438	4,024
	2013	40,385	312	1,111	650	38,936	3,692
	2014	38,177	1,387	905	329	38,330	3,636
Estonia	2011	18,734	..	..	-5	18,739	14,110
	2012	18,796	..	..	1,270	17,526	13,237
	2013	20,511	..	..	25	20,486	15,519
	2014	20,995	..	..	367	20,628	15,672
France	2011	..	82	..	..	82	1
	2012	..	76	..	..	76	1
	2013	..	146	..	..	146	2
	2014	..	177	..	..	177	3

Table 7

Production, trade and supply of brown coal and oil shale

Thousand metric tons and kilograms per capita

Country or area	Year	Production	Imports	Exports	Changes in stocks	Supply	
						Total	Per capita
Germany	2011	176,502	0	5	193	176,304	2,192
	2012	185,432	0	276	-20	185,176	2,301
	2013	182,696	18	179	44	182,491	2,265
	2014	178,178	13	1,171	64	176,956	2,194
Greece	2011	58,666	111	10	-1,211	59,978	5,378
	2012	62,956	137	18	1,165	61,910	5,573
	2013	53,924	97	17	-382	54,386	4,920
	2014	50,845	26	..	-1,007	51,878	4,716
Hungary	2011	9,555	311	1	-79	9,944	996
	2012	9,290	295	71	-388	9,902	994
	2013	9,558	245	220	-338	9,921	1,000
	2014	9,551	181	384	-34	9,382	949
Italy	2011	..	210	..	..	210	4
	2012	..	439	..	..	439	7
	2013	..	500	..	..	500	8
	2014	..	478	..	..	478	8
Lithuania	2011	..	40	6	3	31	10
	2012	..	1	0	-15	16	5
	2013	..	4	0	0	4	1
	2014	..	1	0	0	1	0
Montenegro	2011	1,973	9	53	..	1,929	3,097
	2012	1,786	1	55	..	1,732	2,776
	2013	1,692	3	23	..	1,672	2,677
	2014	1,655	3	28	..	1,630	2,607
Netherlands	2011	..	21	0	-4	25	1
	2012	..	31	0	2	29	2
	2013	..	41	0	17	24	1
	2014	..	41	..	10	31	2
Poland	2011	62,841	76	145	64	62,708	1,625
	2012	64,280	147	134	138	64,155	1,662
	2013	65,849	195	218	-108	65,934	1,707
	2014	63,877	176	303	-95	63,845	1,653
Romania	2011	35,512	692	1	-1,606	37,809	1,880
	2012	33,945	1,023	17	134	34,817	1,746
	2013	24,721	867	1	-80	25,667	1,297
	2014	23,565	931	37	-1,615	26,074	1,327
Russian Federation	2011	76,352	2,048	831	0	77,569	542
	2012	77,299	1,673	1,422	0	77,550	541
	2013	73,680	1,530	1,775	123	73,312	511
	2014	68,893	1,552	2,404	785	67,256	469
Serbia	2011	41,105	635	34	-371	42,077	4,663
	2012	38,234	479	26	73	38,614	4,299
	2013	40,297	306	53	230	40,320	4,511
	2014	30,011	460	14	-233	30,690	3,451
Slovakia	2011	2,376	602	..	-173	3,151	582
	2012	2,292	675	..	21	2,946	544
	2013	2,353	437	..	60	2,730	504
	2014	2,188	474	..	211	2,451	452
Slovenia	2011	4,501	503	..	-5	5,009	2,433
	2012	4,278	604	..	-53	4,935	2,392
	2013	3,876	532	..	-34	4,442	2,151
	2014	3,108	466	..	1	3,573	1,729

Production, trade and supply of brown coal and oil shale

Thousand metric tons and kilograms per capita

Country or area	Year	Production	Imports	Exports	Changes in stocks	Supply	
						Total	Per capita
Spain	2011	2,359	..	..	-450	2,809	60
	2012	2,271	..	..	80	2,191	47
	2013	1,826	..	..	153	1,673	36
	2014	1,230	..	..	-940	2,170	47
Switzerland	2011	..	72	..	3	69	9
	2012	..	53	..	3	50	6
	2013	..	64	..	6	58	7
	2014	..	134	..	3	131	16
T.F.Yug.Rep. Macedonia	2011	8,209	252	32	406	8,023	3,884
	2012	7,311	277	33	-95	7,650	3,697
	2013	6,686	278	2	-2	6,964	3,360
	2014	6,482	271	0	46	6,708	3,232
Oceania	**2011**	**96,606**	**138**	**9**	**21**	**96,714**	**2,613**
	2012	**97,736**	**0**	**0**	**-506**	**98,242**	**2,613**
	2013	**87,208**	**458**	**0**	**115**	**87,551**	**2,293**
	2014	**83,816**	**441**	**22**	**-214**	**84,449**	**2,179**
Australia	2011	93,992	..	..	..	93,992	4,170
	2012	95,086	..	..	..	95,086	4,150
	2013	84,862	..	..	..	84,862	3,647
	2014	81,787	..	..	..	81,787	3,462
New Zealand	2011	2,614	138	9	21	2,722	618
	2012	2,650	0	0	-506	3,156	711
	2013	2,346	458	0	115	2,689	602
	2014	2,029	441	22	-214	2,662	592

Table 8

Production, trade and supply of coke

Thousand metric tons and kilograms per capita

Table Notes

Data comprise coal tar, gas coke and coke oven coke.

Please refer to the Definitions Section on pages ix to xv for the appropriate product description/ classification.

Figure 18: Per capita supply of coke by region, 2014

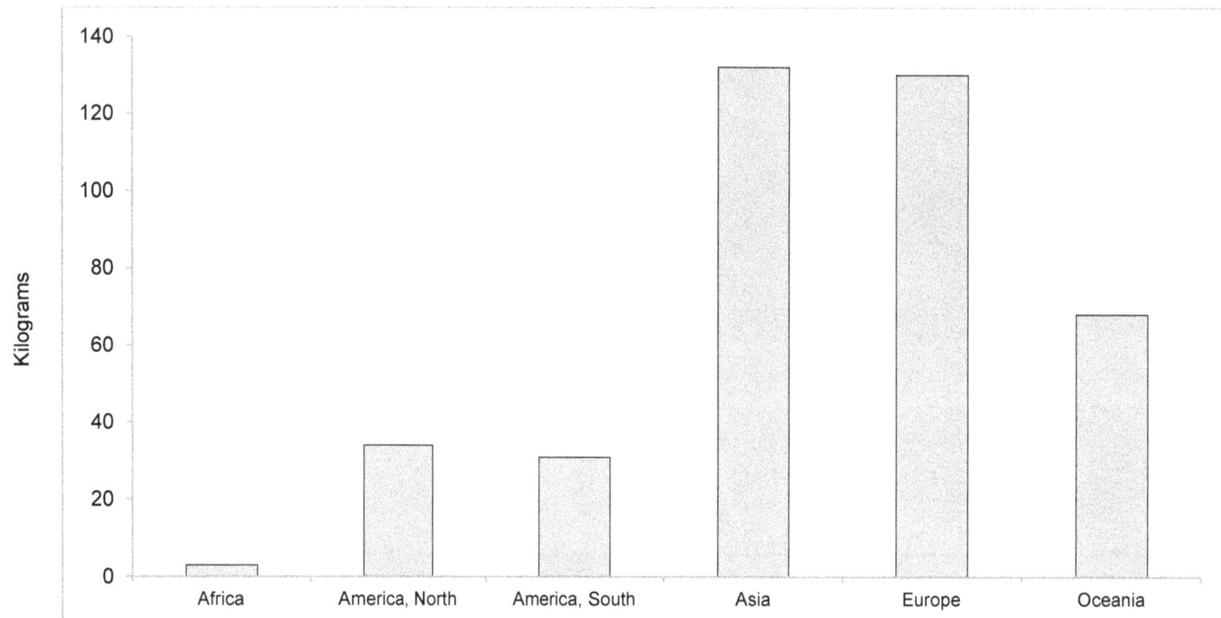

Figure 19: Major coke producing countries, 2014

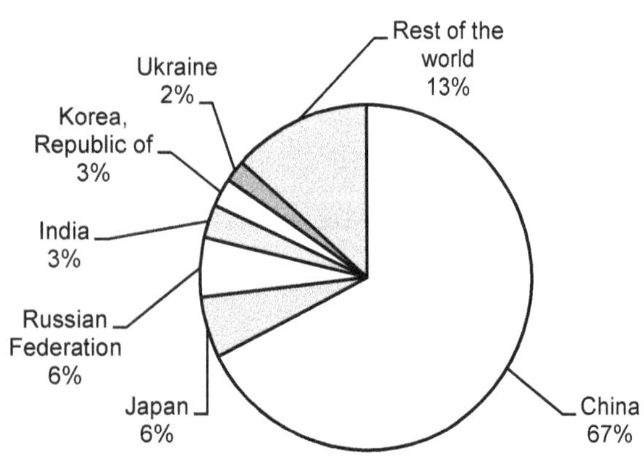

Figure 20: Major coke consuming countries, 2014

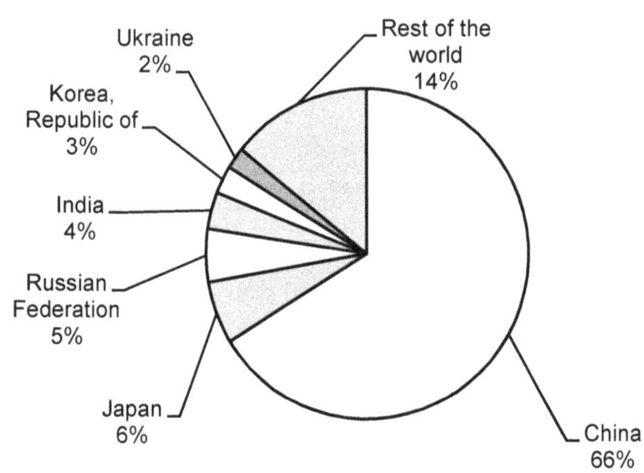

Production, trade and supply of coke

Thousand metric tons and kilograms per capita

Country or area	Year	Production	Imports	Exports	Changes in stocks	Supply	
						Total	Per capita
World	**2011**	**655,656**	**24,005**	**23,623**	**10,336**	**645,702**	**92**
	2012	**655,223**	**22,931**	**22,884**	**-11,740**	**667,010**	**94**
	2013	**715,595**	**26,530**	**24,977**	**21,131**	**696,017**	**97**
	2014	**711,170**	**28,801**	**28,112**	**2,366**	**709,494**	**98**
Africa	**2011**	**3,201**	**511**	**490**	**-18**	**3,240**	**3**
	2012	**3,172**	**499**	**489**	**-27**	**3,209**	**3**
	2013	**3,125**	**487**	**448**	**-23**	**3,187**	**3**
	2014	**2,798**	**616**	**338**	**30**	**3,046**	**3**
Algeria	2011	..	438	..	-1	439	12
	2012	..	421	..	-13	434	12
	2013	..	270	..	-34	304	8
	2014	..	236	..	30	206	5
Egypt	2011	*843	*28	*290	-17	*598	*7
	2012	*786	*26	*266	-14	*560	*7
	2013	*577	*162	214	11	*514	*6
	2014	377	324	128	..	573	6
South Africa	2011	2,072	..	..	..	2,072	40
	2012	2,070	..	..	..	2,070	39
	2013	2,218	..	..	..	2,218	42
	2014	2,111	..	..	..	2,111	39
Zambia	2011	..	45	0	..	45	3
	2012	..	52	2	..	50	3
	2013	..	*55	*3	..	*52	*3
	2014	..	*56	*3	..	*53	*3
Zimbabwe	2011	286	..	200	..	86	6
	2012	316	..	221	..	95	7
	2013	330	..	231	..	99	7
	2014	310	..	207	..	103	7
America, North	**2011**	**19,061**	**2,581**	**931**	**-346**	**21,057**	**38**
	2012	**18,941**	**2,391**	**912**	**-220**	**20,640**	**37**
	2013	**18,746**	**1,579**	**790**	**322**	**19,213**	**34**
	2014	**18,349**	**1,881**	**919**	**-115**	**19,426**	**34**
Bermuda	2011	..	0	..	..	0	1
	2012	..	0	..	..	0	1
	2013	..	0	..	..	0	1
	2014	..	0	..	..	0	1
Canada	2011	2,804	656	50	-365	3,775	109
	2012	2,949	554	27	-94	3,570	102
	2013	2,480	664	29	79	3,036	86
	2014	2,219	975	61	-47	3,180	89
Costa Rica	2011	..	87	..	..	87	19
	2012	..	94	..	..	94	20
	2013	..	94	..	..	94	20
	2014	..	94	..	..	94	20
Cuba	2011	..	3	..	..	3	0
	2012	..	2	..	..	2	0
	2013	..	1	..	..	1	0
	2014	..	1	..	..	1	0
Honduras	2011	..	134	..	-20	154	20
	2012	..	158	..	0	158	20
	2013	..	161	..	0	161	21
	2014	..	91	..	..	91	11

Table 8

Production, trade and supply of coke

Thousand metric tons and kilograms per capita

Country or area	Year	Production	Imports	Exports	Changes in stocks	Supply	
						Total	Per capita
Mexico	2011	2,268	336	1	..	2,603	22
	2012	2,228	391	1	..	2,618	21
	2013	2,368	374	1	..	2,741	22
	2014	2,383	501	..	..	2,884	23
Panama	2011	..	79	..	..	79	21
	2012	..	163	..	..	163	43
	2013	..	160	..	..	160	42
	2014	..	150	..	..	150	39
United States	2011	13,989	1,286	880	39	14,356	46
	2012	13,764	1,029	884	-126	14,035	45
	2013	13,898	125	760	243	13,020	41
	2014	13,747	69	858	-68	13,026	41
America, South	**2011**	**13,189**	**2,189**	**1,314**	**-74**	**14,139**	**35**
	2012	**12,826**	**1,647**	**1,602**	**-439**	**13,311**	**33**
	2013	**13,027**	**1,967**	**1,980**	**-53**	**13,068**	**32**
	2014	**13,290**	**1,836**	**2,221**	**-85**	**12,990**	**31**
Argentina	2011	1,276	..	35	7	1,234	30
	2012	765	..	53	-25	737	18
	2013	722	..	50	-24	696	16
	2014	840	..	58	-28	810	19
Brazil	2011	9,683	2,142	..	-82	11,907	59
	2012	9,683	1,591	..	-328	11,602	57
	2013	9,393	1,896	..	-34	11,323	55
	2014	9,496	1,818	..	98	11,216	54
Chile	2011	563	33	..	0	596	35
	2012	473	55	..	-47	575	33
	2013	451	34	..	26	459	26
	2014	459	5	153	-39	350	20
Colombia	2011	1,667	..	1,279	39	350	8
	2012	1,905	..	1,507	0	399	9
	2013	2,461	..	1,893	17	551	12
	2014	2,495	..	1,973	0	522	11
Peru	2011	..	14	0	-38	52	2
	2012	..	1	42	-39	-2	0
	2013	..	37	37	-38	38	1
	2014	..	13	37	-116	92	3
Asia	**2011**	**513,704**	**7,312**	**5,222**	**10,304**	**505,490**	**120**
	2012	**516,646**	**7,851**	**3,887**	**-11,039**	**531,650**	**125**
	2013	**579,327**	**10,422**	**6,290**	**20,647**	**562,812**	**131**
	2014	**576,646**	**10,590**	**9,558**	**2,251**	**575,428**	**132**
Afghanistan	2011	48	205	48	..	205	7
	2012	48	205	48	..	205	7
	2013	*52	*205	*52	..	*205	*7
	2014	*59	*205	*59	..	*205	*6
China	2011	434,330	116	3,297	10,290	420,859	312
	2012	438,315	76	1,020	-10,767	448,137	331
	2013	483,478	35	4,673	20,340	458,501	337
	2014	479,809	2	8,507	2,360	468,943	342
Georgia	2011	..	129	3	..	126	30
	2012	..	131	0	..	131	32
	2013	..	151	0	10	141	35
	2014	..	131	0	-4	135	33

Production, trade and supply of coke

Thousand metric tons and kilograms per capita

Country or area	Year	Production	Imports	Exports	Changes in stocks	Supply	
						Total	Per capita
India	2011	9,965	2,365	618	..	11,712	9
	2012	9,271	3,007	1,050	..	11,228	9
	2013	26,007	4,171	154	..	30,024	23
	2014	22,290	3,294	102	..	25,482	20
Indonesia	2011	*16	*60	..	..	*76	0
	2012	*18	*60	..	..	*78	0
	2013	23	130	..	..	153	1
	2014	27	257	81	..	203	1
Iran (Islamic Rep. of)	2011	714	1,307	3	..	2,018	27
	2012	543	892	2	..	1,433	19
	2013	997	889	0	..	1,886	24
	2014	1,124	438	..	1	1,561	20
Japan	2011	40,619	845	996	125	40,343	317
	2012	41,245	956	1,481	-85	40,805	321
	2013	41,184	2,163	1,214	13	42,120	332
	2014	40,627	3,337	500	42	43,422	342
Jordan	2011	..	0	..	..	0	0
	2012	..	0	..	..	0	0
	2013	..	159	..	..	159	22
	2014	..	68	..	..	68	9
Kazakhstan	2011	2,663	785	20	-33	3,461	209
	2012	2,569	918	27	51	3,409	203
	2013	2,379	827	23	3	3,180	186
	2014	2,698	1,218	15	55	3,846	221
Korea, Dem.Ppl's.Rep.	2011	..	200	..	..	200	8
	2012	..	200	..	..	200	8
	2013	..	198	..	..	198	8
	2014	..	200	..	..	200	8
Korea, Republic of	2011	15,807	406	..	..	16,213	328
	2012	15,229	367	..	..	15,596	314
	2013	15,246	364	..	..	15,610	313
	2014	19,099	405	..	..	19,504	390
Other Asia	2011	5,446	103	100	-20	5,469	235
	2012	5,152	235	114	-33	5,306	228
	2013	5,678	175	14	155	5,684	244
	2014	6,277	173	71	-57	6,436	275
Pakistan	2011	193	..	..	..	193	1
	2012	185	..	..	..	185	1
	2013	77	..	..	..	77	0
	2014	81	..	..	..	81	0
Philippines	2011	..	208	..	..	208	2
	2012	..	214	..	..	214	2
	2013	..	216	..	..	216	2
	2014	..	292	..	..	292	3
Singapore	2011	..	4	..	..	4	1
	2012	..	0	..	..	0	0
	2013	..	0	..	..	0	0
	2014	..	0	..	..	0	0
Syrian Arab Republic	2011	..	4	..	..	4	0
	2012	..	1	..	..	1	0
	2013	..	1	..	..	1	0
	2014	..	1	..	..	1	0

Table 8

Production, trade and supply of coke

Thousand metric tons and kilograms per capita

Country or area	Year	Production	Imports	Exports	Changes in stocks	Supply	
						Total	Per capita
Thailand	2011	..	108	..	-58	166	2
	2012	..	36	..	-173	209	3
	2013	..	38	..	0	38	1
	2014	..	38	..	-146	184	3
Turkey	2011	3,903	309	0	0	4,212	57
	2012	4,072	378	0	-32	4,482	60
	2013	4,205	531	8	126	4,602	60
	2014	4,555	352	61	..	4,846	63
United Arab Emirates	2011	..	8	0	..	8	1
	2012	..	16	1	..	16	2
	2013	..	2	0	..	2	0
	2014	..	1	0	..	1	0
Viet Nam	2011	..	151	137	..	14	0
	2012	..	159	144	..	15	0
	2013	..	168	152	..	16	0
	2014	..	178	161	..	17	0
Europe	**2011**	**102,514**	**11,358**	**15,151**	**470**	**98,251**	**133**
	2012	**100,065**	**10,506**	**15,153**	**-15**	**95,432**	**129**
	2013	**97,740**	**12,034**	**14,490**	**239**	**95,045**	**129**
	2014	**96,670**	**13,797**	**14,230**	**285**	**95,953**	**130**
Austria	2011	1,371	1,140	0	-81	2,592	308
	2012	1,361	1,202	0	13	2,550	302
	2013	1,403	1,305	1	29	2,678	316
	2014	1,375	1,232	1	-9	2,615	307
Belarus	2011	..	83	..	1	82	9
	2012	..	90	..	4	86	9
	2013	..	81	..	-4	85	9
	2014	..	65	..	-2	67	7
Belgium	2011	1,991	203	334	-71	1,931	175
	2012	1,906	214	560	37	1,523	137
	2013	1,744	236	354	73	1,553	139
	2014	1,491	232	176	-63	1,610	143
Bosnia and Herzegovina	2011	887	18	442	36	427	111
	2012	700	0	199	48	453	118
	2013	759	22	279	37	465	122
	2014	908	9	354	156	407	107
Bulgaria	2011	0	62	0	-6	68	9
	2012	0	66	0	1	65	9
	2013	0	71	0	0	71	10
	2014	0	90	0	9	81	11
Croatia	2011	..	27	1	-2	28	7
	2012	..	29	1	0	28	7
	2013	..	27	1	1	25	6
	2014	..	28	1	-2	29	7
Czechia	2011	2,791	817	536	-105	3,177	302
	2012	2,677	708	484	71	2,830	268
	2013	2,682	733	466	8	2,941	279
	2014	2,737	597	529	-27	2,832	269
Denmark	2011	..	21	..	-3	24	4
	2012	..	21	..	0	21	4
	2013	..	19	..	0	19	3
	2014	..	14	..	1	13	2

Production, trade and supply of coke

Thousand metric tons and kilograms per capita

Country or area	Year	Production	Imports	Exports	Changes in stocks	Supply	
						Total	Per capita
Estonia	2011	24	..	25	-1	0	0
	2012	25	..	23	2	0	0
	2013	22	..	20	-3	5	4
	2014	25	..	26	-1	0	0
Finland	2011	854	444	6	44	1,248	231
	2012	894	320	46	22	1,146	211
	2013	909	373	70	31	1,181	217
	2014	925	398	73	16	1,234	225
France	2011	3,043	1,209	87	7	4,158	66
	2012	3,303	933	72	217	3,947	62
	2013	3,438	826	12	222	4,030	63
	2014	3,385	808	54	132	4,007	62
Germany	2011	7,990	3,713	177	20	11,506	143
	2012	8,050	3,254	218	-121	11,207	139
	2013	8,273	3,489	302	-71	11,531	143
	2014	8,770	3,520	491	71	11,728	145
Hungary	2011	1,119	9	372	-1	757	76
	2012	1,086	7	369	-3	727	73
	2013	978	9	546	40	401	40
	2014	967	44	514	-12	509	51
Iceland	2011	..	23	..	-9	32	100
	2012	..	43	..	8	35	108
	2013	..	27	..	0	27	83
	2014	..	31	..	0	31	95
Italy	2011	4,788	33	273	33	4,515	76
	2012	4,184	0	254	-19	3,949	66
	2013	2,415	929	246	127	2,971	50
	2014	1,920	1,215	341	-77	2,871	48
Latvia	2011	..	3	..	0	3	1
	2012	..	6	..	0	6	3
	2013	..	1	..	-1	2	1
	2014	..	0	..	0	0	0
Lithuania	2011	..	18	..	..	18	6
	2012	..	19	..	..	19	6
	2013	..	19	..	..	19	6
	2014	..	17	..	1	16	5
Luxembourg	2011	..	1	..	..	1	2
	2012	..	1	..	..	1	2
	2013	..	1	..	..	1	2
	2014	..	1	..	..	1	2
Netherlands	2011	2,120	183	55	-30	2,278	136
	2012	1,983	261	108	-82	2,218	132
	2013	2,071	182	323	-205	2,135	127
	2014	2,103	500	608	-60	2,055	122
Norway	2011	..	455	0	-19	474	96
	2012	..	469	2	8	459	91
	2013	..	417	1	-5	421	83
	2014	..	455	..	-2	457	89
Poland	2011	9,782	148	6,860	56	3,014	78
	2012	9,269	148	6,769	22	2,626	68
	2013	9,754	182	6,969	-79	3,046	79
	2014	9,952	195	7,037	-164	3,274	85

Table 8

Production, trade and supply of coke

Thousand metric tons and kilograms per capita

Country or area	Year	Production	Imports	Exports	Changes in stocks	Supply	
						Total	Per capita
Portugal	2011	..	3	..	0	3	0
	2012	..	1	..	0	1	0
	2013	..	0	..	-1	1	0
	2014	..	..	..	..	..	..
Republic of Moldova	2011	..	0	..	..	0	0
	2012	..	1	..	..	1	0
	2013	..	0	..	..	0	0
	2014	..	0	..	..	0	0
Romania	2011	0	802	0	-110	912	45
	2012	0	747	0	-29	776	39
	2013	0	716	0	-24	740	37
	2014	0	738	0	27	711	36
Russian Federation	2011	36,112	160	2,530	..	33,742	236
	2012	36,404	563	2,361	..	34,606	242
	2013	36,449	162	2,522	..	34,089	238
	2014	39,518	150	2,528	155	36,985	258
Serbia	2011	..	640	0	0	640	71
	2012	..	167	1	12	154	17
	2013	..	184	0	-16	200	22
	2014	..	290	0	1	289	32
Slovakia	2011	1,680	468	236	35	1,877	347
	2012	1,620	218	106	58	1,674	309
	2013	1,596	161	115	-79	1,721	318
	2014	1,617	210	91	-11	1,747	322
Slovenia	2011	..	31	..	-4	35	17
	2012	..	30	..	-5	35	17
	2013	..	31	..	-6	37	18
	2014	..	32	..	-5	37	18
Spain	2011	2,052	165	374	51	1,792	38
	2012	1,807	168	461	-208	1,722	37
	2013	1,826	190	168	-1	1,849	40
	2014	1,605	248	130	26	1,697	37
Sweden	2011	1,223	214	25	89	1,323	140
	2012	1,115	92	22	22	1,163	122
	2013	1,079	98	26	3	1,148	119
	2014	1,113	98	30	-72	1,253	129
Switzerland	2011	..	20	..	..	20	3
	2012	..	21	..	..	21	3
	2013	..	19	..	..	19	2
	2014	..	17	..	..	17	2
T.F.Yug.Rep. Macedonia	2011	..	42	0	1	41	20
	2012	..	14	0	0	13	6
	2013	..	3	0	0	3	1
	2014	..	2	0	1	2	1
Ukraine	2011	20,473	177	2,351	11	18,288	402
	2012	19,788	501	2,577	-7	17,719	391
	2013	18,383	702	1,982	58	17,045	377
	2014	14,473	1,636	1,158	..	14,951	332
United Kingdom	2011	4,214	26	467	528	3,245	51
	2012	3,893	192	520	-86	3,651	57
	2013	3,959	819	87	105	4,586	72
	2014	3,786	925	88	196	4,427	69

Production, trade and supply of coke

Thousand metric tons and kilograms per capita

Country or area	Year	Production	Imports	Exports	Changes in stocks	Supply	
						Total	Per capita
Oceania	**2011**	**3,987**	**54**	**515**	**0**	**3,526**	**95**
	2012	**3,572**	**37**	**841**	**0**	**2,768**	**74**
	2013	**3,630**	**41**	**979**	**0**	**2,692**	**71**
	2014	**3,417**	**81**	**846**	**0**	**2,652**	**68**
Australia	2011	3,509	54	515	0	3,048	135
	2012	3,086	37	841	0	2,282	100
	2013	3,139	41	979	0	2,201	95
	2014	2,914	81	846	0	2,149	91
New Zealand	2011	478	..	..	..	478	109
	2012	486	..	..	..	486	110
	2013	491	..	..	..	491	110
	2014	503	..	..	..	503	112

Table 9

Production, trade and supply of patent fuel

Thousand metric tons and kilograms per capita

Table Notes

Data refer to patent fuel.

Please refer to the Definitions Section on pages ix to xv for the appropriate product description/classification.

Figure 21: Major patent fuel producing countries, 2014

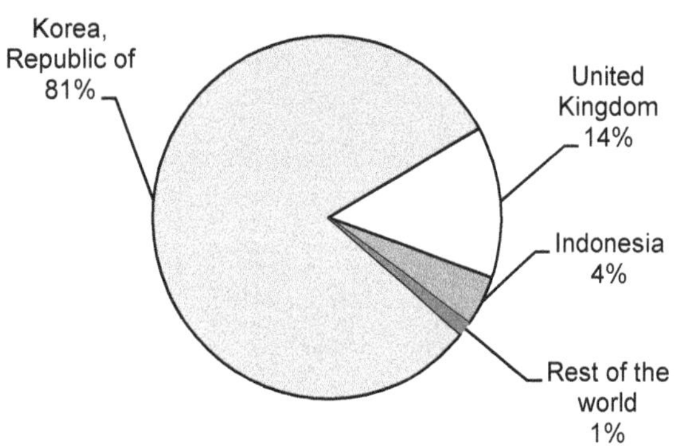

Figure 22: Major patent fuel consuming countries, 2014

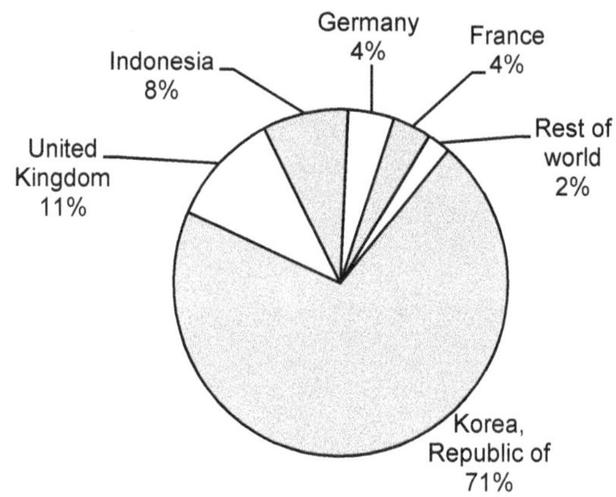

Production, trade and supply of patent fuel

Thousand metric tons and kilograms per capita

Country or area	Year	Production	Imports	Exports	Changes in stocks	Supply	
						Total	Per capita
World	**2011**	**2,276**	**434**	**65**	**90**	**2,555**	**0**
	2012	**2,260**	**283**	**49**	**61**	**2,433**	**0**
	2013	**2,353**	**276**	**177**	**-73**	**2,525**	**0**
	2014	**2,019**	**250**	**40**	**-78**	**2,307**	**0**
Africa	**2011**	**5**	**..**	**..**	**..**	**5**	**0**
	2012	**4**	**..**	**..**	**..**	**4**	**0**
	2013	**3**	**..**	**..**	**..**	**3**	**0**
	2014	**3**	**..**	**..**	**..**	**3**	**0**
Niger	2011	*5	..	..	..	*5	0
	2012	*4	..	..	..	*4	0
	2013	*3	..	..	..	*3	0
	2014	*3	..	..	..	*3	0
America, North	**2011**	**..**	**2**	**..**	**..**	**2**	**0**
	2012	**..**	**2**	**..**	**..**	**2**	**0**
	2013	**..**	**2**	**..**	**..**	**2**	**0**
	2014	**..**	**2**	**..**	**..**	**2**	**0**
Bahamas	2011	..	*2	..	..	*2	*5
	2012	..	*2	..	..	*2	*5
	2013	..	*2	..	..	*2	*5
	2014	..	*2	..	..	*2	*5
Asia	**2011**	**1,977**	**0**	**0**	**73**	**1,904**	**0**
	2012	**1,998**	**0**	**0**	**73**	**1,925**	**0**
	2013	**2,014**	**3**	**135**	**-96**	**1,978**	**0**
	2014	**1,720**	**3**	**0**	**-96**	**1,819**	**0**
Indonesia	2011	*155	0	0	*73	*82	0
	2012	*165	0	0	*73	*92	0
	2013	96	3	135	*-96	60	0
	2014	91	3	0	-96	190	1
Korea, Republic of	2011	1,822	..	..	..	1,822	37
	2012	1,833	..	..	..	1,833	37
	2013	1,917	..	..	..	1,917	38
	2014	1,629	..	..	..	1,629	33
Kyrgyzstan	2011	0	..	..	..	0	0
	2012	0	..	..	..	0	0
	2013	1	..	..	..	1	0
	2014	..	..	..	..	..	..
Europe	**2011**	**294**	**432**	**65**	**17**	**644**	**1**
	2012	**258**	**281**	**49**	**-12**	**502**	**1**
	2013	**336**	**271**	**42**	**23**	**542**	**1**
	2014	**296**	**245**	**40**	**18**	**483**	**1**
Austria	2011	..	0	..	0	0	0
	2012	..	7	..	0	7	1
	2013	..	13	..	4	9	1
	2014	..	9	..	3	6	1
Belgium	2011	..	23	19	2	2	0
	2012	..	13	13	-2	2	0
	2013	..	14	9	3	2	0
	2014	..	10	7	..	3	0
Bulgaria	2011	..	..	..	..	..	..
	2012	..	..	..	..	..	..
	2013	..	..	..	..	..	..
	2014	19	..	..	1	18	2

Table 9

Production, trade and supply of patent fuel

Thousand metric tons and kilograms per capita

Country or area	Year	Production	Imports	Exports	Changes in stocks	Supply	
						Total	Per capita
France	2011	0	81	0	0	81	1
	2012	0	89	0	0	89	1
	2013	0	70	0	0	70	1
	2014	0	86	0	0	86	1
Germany	2011	..	275	1	..	274	3
	2012	..	133	1	..	132	2
	2013	..	135	1	..	134	2
	2014	..	103	1	..	102	1
Hungary	2011	5	..	..	2	3	0
	2012	0	..	..	-3	3	0
	2013	0	..	..	-1	1	0
	2014	0	..	..	-1	1	0
Netherlands	2011	..	0	..	..	0	0
	2012	..	1	1	..	0	0
	2013	..	0	0	..	0	0
	2014	..	..	..	..	..	..
Poland	2011	..	20	13	..	7	0
	2012	..	12	2	..	10	0
	2013	..	12	2	..	10	0
	2014	3	14	8	..	9	0
Slovakia	2011	..	12	..	..	12	2
	2012	..	11	..	..	11	2
	2013	..	12	..	..	12	2
	2014	..	9	..	..	9	2
United Kingdom	2011	289	21	32	13	265	4
	2012	258	15	32	-7	248	4
	2013	336	15	30	17	304	5
	2014	274	14	24	15	249	4

Production, trade and supply of brown coal briquettes and peat products

Thousand metric tons and kilograms per capita

Table Notes

Data refer to brown coal briquettes (BKB) and peat products.

Please refer to the Definitions Section on pages ix to xv for the appropriate product description/classification.

Figure 23: Production of BKB and peat products by region, 1990-2014

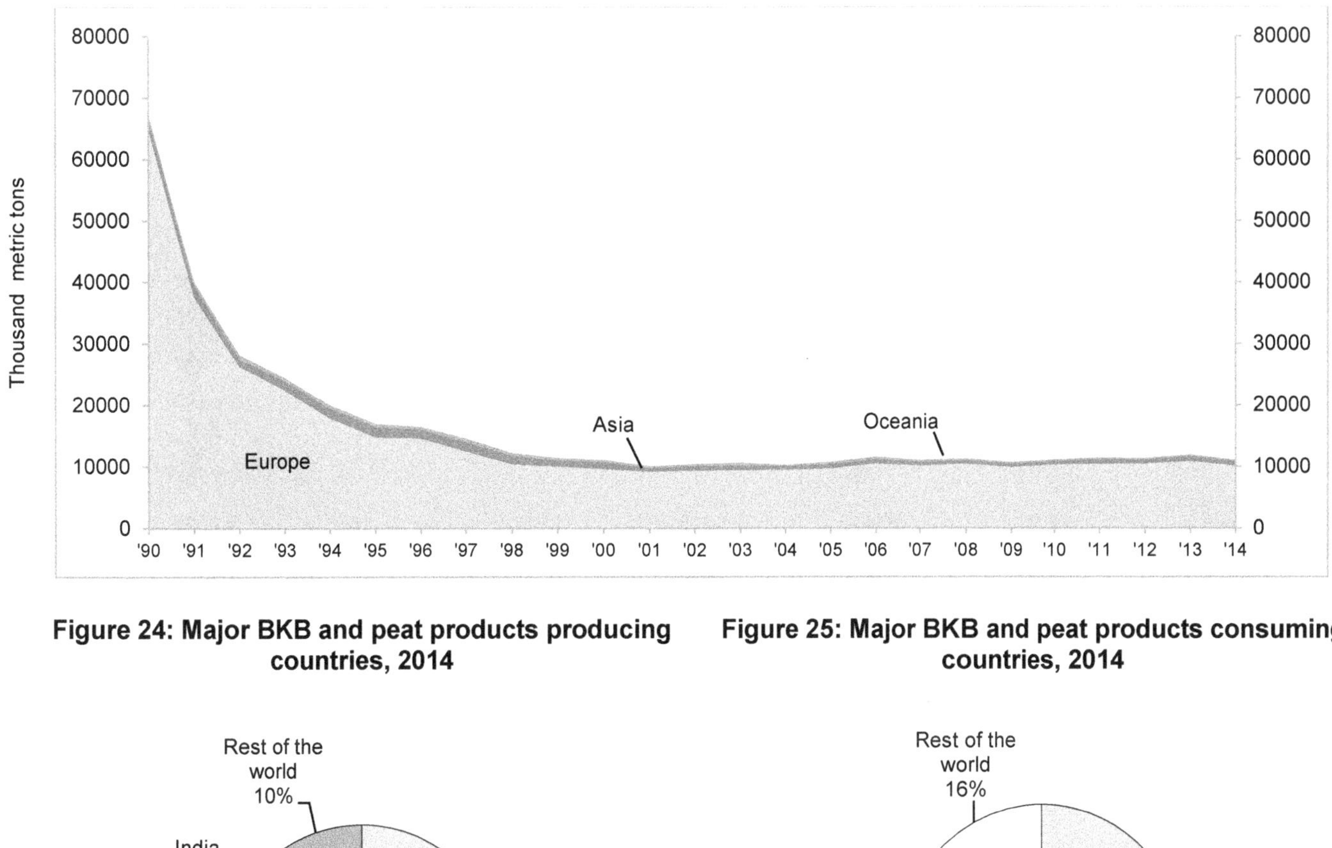

Figure 24: Major BKB and peat products producing countries, 2014

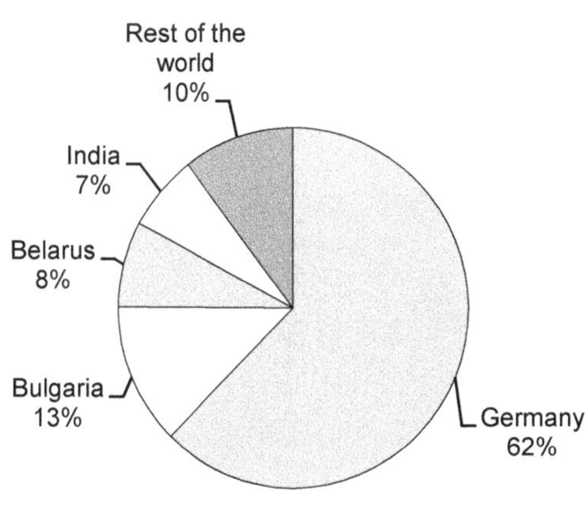

Figure 25: Major BKB and peat products consuming countries, 2014

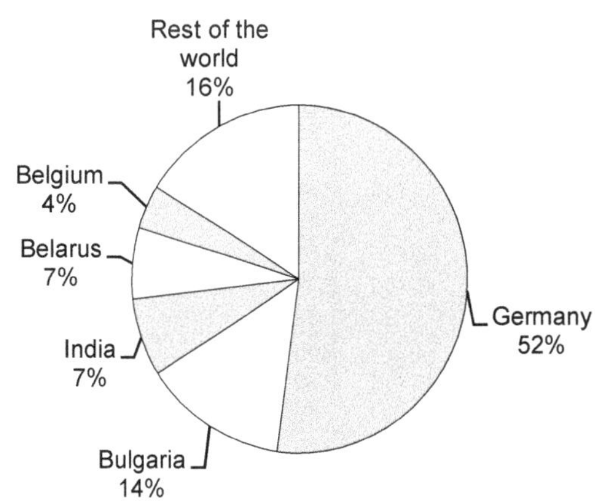

Table 10

Production, trade and supply of brown coal briquettes and peat products

Thousand metric tons and kilograms per capita

Country or area	Year	Production	Imports	Exports	Changes in stocks	Supply	
						Total	Per capita
World	**2011**	**11,365**	**1,220**	**1,944**	**8**	**10,633**	**2**
	2012	**11,310**	**1,271**	**1,970**	**131**	**10,480**	**1**
	2013	**11,881**	**1,168**	**1,894**	**-92**	**11,247**	**2**
	2014	**11,018**	**1,280**	**1,849**	**23**	**10,426**	**1**
America, North	**2011**	**..**	**0**	**..**	**..**	**0**	**0**
	2012	**..**	**4**	**..**	**..**	**4**	**0**
	2013	**..**	**4**	**..**	**..**	**4**	**0**
	2014	**..**	**3**	**..**	**..**	**3**	**0**
Mexico	2011	..	0	..	..	0	0
	2012	..	4	..	..	4	0
	2013	..	4	..	..	4	0
	2014	..	3	..	..	3	0
Asia	**2011**	**745**	**0**	**..**	**..**	**745**	**0**
	2012	**669**	**0**	**..**	**..**	**669**	**0**
	2013	**811**	**0**	**..**	**..**	**811**	**0**
	2014	**754**	**0**	**..**	**..**	**754**	**0**
India	2011	745	..	..	..	745	1
	2012	669	..	..	..	669	1
	2013	811	..	..	..	811	1
	2014	*754	..	..	..	*754	*1
Europe	**2011**	**10,407**	**1,220**	**1,918**	**-15**	**9,724**	**13**
	2012	**10,413**	**1,267**	**1,965**	**125**	**9,590**	**13**
	2013	**10,846**	**1,164**	**1,789**	**-19**	**10,240**	**14**
	2014	**10,051**	**1,277**	**1,843**	**-21**	**9,506**	**13**
Albania	2011	..	297	..	..	297	103
	2012	..	348	..	..	348	121
	2013	..	304	..	..	304	105
	2014	..	394	..	..	394	137
Austria	2011	..	20	1	-8	27	3
	2012	..	23	1	0	22	3
	2013	..	17	0	0	17	2
	2014	..	15	0	0	15	2
Belarus	2011	1,362	..	339	97	926	98
	2012	1,392	..	393	74	925	97
	2013	1,117	..	166	85	866	91
	2014	852	..	105	53	694	73
Belgium	2011	..	433	90	0	343	31
	2012	..	402	93	-2	311	28
	2013	..	328	..	-1	329	29
	2014	..	426	..	1	425	38
Bulgaria	2011	856	..	..	27	829	113
	2012	995	..	..	2	993	136
	2013	1,576	..	..	-57	1,633	225
	2014	1,392	..	..	-36	1,428	198
Czechia	2011	0	162	8	3	151	14
	2012	0	173	11	-3	165	16
	2013	0	159	18	0	141	13
	2014	..	148	10	-3	141	13
Denmark	2011	..	1	..	..	1	0
	2012	..	1	..	..	1	0
	2013	..	1	..	..	1	0
	2014	..	0	..	..	0	0

Production, trade and supply of brown coal briquettes and peat products

Thousand metric tons and kilograms per capita

Country or area	Year	Production	Imports	Exports	Changes in stocks	Supply	
						Total	Per capita
Estonia	2011	75	..	78	-16	13	10
	2012	52	..	41	-2	13	10
	2013	56	..	53	-10	13	10
	2014	62	..	50	..	12	9
Finland	2011	46	..	0	..	46	9
	2012	28	..	0	..	28	5
	2013	18	..	0	..	18	3
	2014	9	..	..	..	9	2
Germany	2011	6,922	83	1,372	-137	5,770	72
	2012	6,786	58	1,392	5	5,447	68
	2013	6,973	73	1,460	-39	5,625	70
	2014	6,883	78	1,522	-6	5,445	68
Hungary	2011	..	11	..	..	11	1
	2012	..	7	..	..	7	1
	2013	..	6	..	..	6	1
	2014	..	4	..	..	4	0
Ireland	2011	192	22	22	15	177	38
	2012	197	25	21	-19	220	47
	2013	240	46	20	37	229	49
	2014	207	28	6	16	213	46
Lithuania	2011	7	68	2	12	61	20
	2012	5	76	13	-5	73	24
	2013	7	86	8	9	76	26
	2014	6	57	12	-14	65	22
Luxembourg	2011	..	9	..	..	9	17
	2012	..	9	..	..	9	17
	2013	..	4	..	..	4	7
	2014	..	4	..	..	4	7
Netherlands	2011	..	80	..	..	80	5
	2012	..	77	..	..	77	5
	2013	..	45	..	..	45	3
	2014	..	25	..	..	25	1
Poland	2011	..	16	4	..	12	0
	2012	..	10	0	..	10	0
	2013	..	37	0	..	37	1
	2014	..	70	..	..	70	2
Russian Federation	2011	42	..	1	0	41	0
	2012	82	..	0	0	82	1
	2013	74	..	2	0	72	1
	2014	60	..	2	..	58	0
Serbia	2011	568	9	1	-7	583	65
	2012	519	18	0	59	478	53
	2013	516	7	62	-35	496	55
	2014	271	4	39	-40	276	31
Slovakia	2011	..	9	..	0	9	2
	2012	..	40	..	16	24	4
	2013	..	51	..	-12	63	12
	2014	..	23	..	-7	30	6
Ukraine	2011	337	..	..	-1	338	7
	2012	357	..	..	0	357	8
	2013	269	..	..	4	265	6
	2014	309	1	97	15	198	4

Table 10

Production, trade and supply of brown coal briquettes and peat products

Thousand metric tons and kilograms per capita

Country or area	Year	Production	Imports	Exports	Changes in stocks	Supply	
						Total	Per capita
Oceania	**2011**	**213**	**..**	**26**	**23**	**164**	**4**
	2012	**228**	**..**	**5**	**6**	**217**	**6**
	2013	**224**	**..**	**105**	**-73**	**192**	**5**
	2014	**213**	**..**	**6**	**44**	**163**	**4**
Australia	2011	213	..	26	23	164	7
	2012	228	..	5	6	217	9
	2013	224	..	105	-73	192	8
	2014	213	..	6	44	163	7

Production, trade and supply of peat

Thousand metric tons and kilograms per capita

Table Notes

Data refer to peat.

Please refer to the Definitions Section on pages ix to xv for the appropriate product description/classification.

Figure 26: World production of peat, 1990-2014

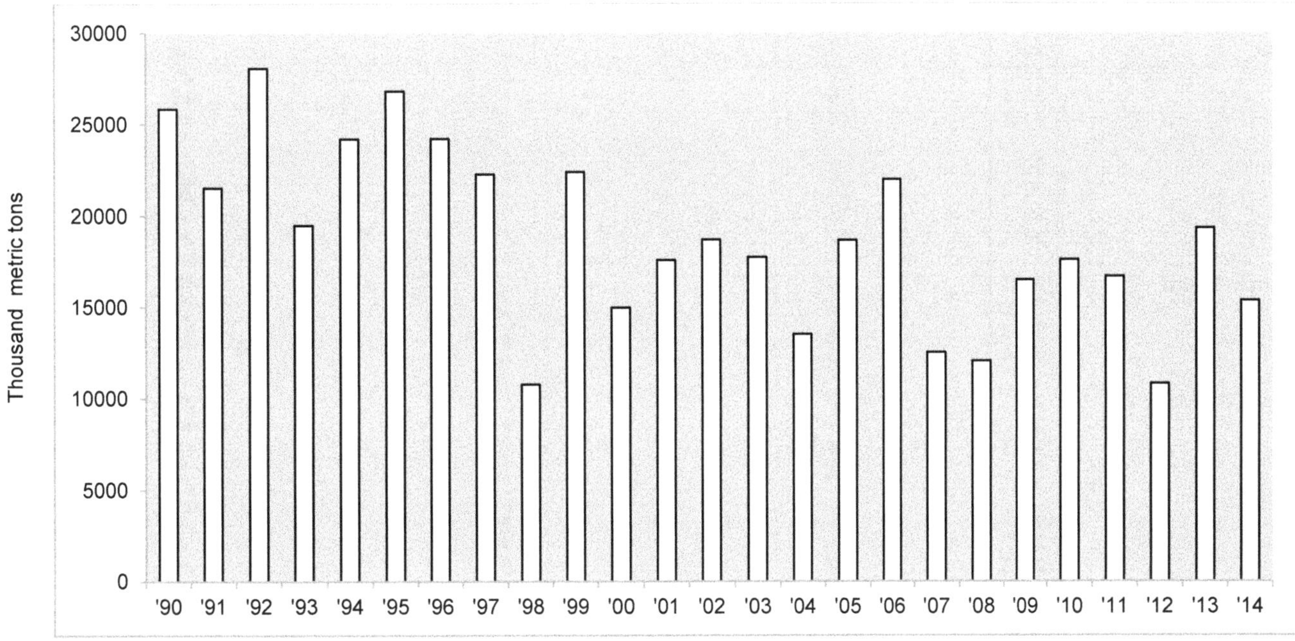

Figure 27: Major peat producing countries, 2014

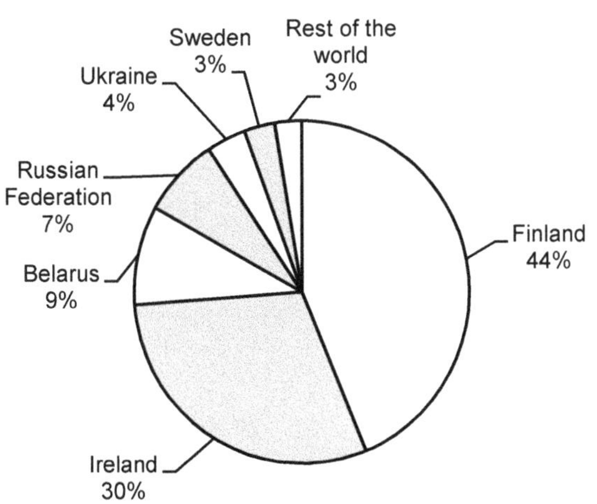

Figure 28: Major peat consuming countries, 2014

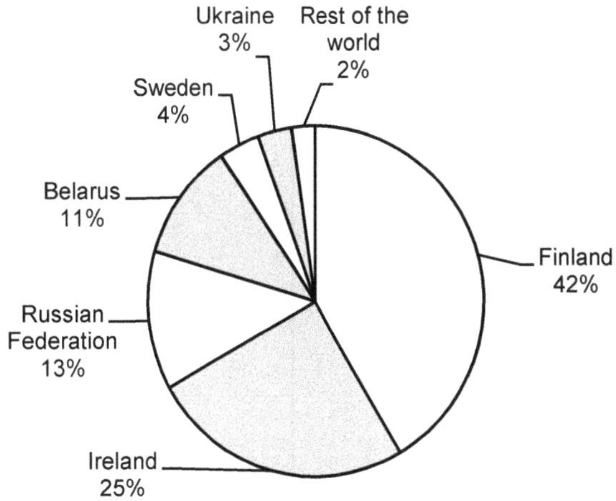

Table 11

Production, trade and supply of peat

Thousand metric tons and kilograms per capita

Country or area	Year	Production	Imports	Exports	Changes in stocks	Supply	
						Total	Per capita
World	**2011**	**16,657**	**557**	**177**	**-919**	**17,956**	**3**
	2012	**10,814**	**493**	**152**	**-4,969**	**16,123**	**2**
	2013	**19,315**	**262**	**134**	**4,470**	**14,972**	**2**
	2014	**15,337**	**230**	**226**	**930**	**14,411**	**2**
Africa	**2011**	**8**	**..**	**..**	**..**	**8**	**0**
	2012	**20**	**..**	**..**	**..**	**20**	**0**
	2013	**16**	**..**	**..**	**..**	**16**	**0**
	2014	**11**	**..**	**..**	**..**	**11**	**0**
Burundi	2011	8	..	..	..	8	1
	2012	20	..	..	..	20	2
	2013	16	..	..	..	16	1
	2014	11	..	..	..	11	1
America, South	**2011**	**13**	**..**	**..**	**..**	**13**	**0**
	2012	**13**	**..**	**..**	**..**	**13**	**0**
	2013	**13**	**..**	**..**	**..**	**13**	**0**
	2014	**13**	**..**	**..**	**..**	**13**	**0**
Falkland Is. (Malvinas)	2011	*13	..	..	..	*13	*4,553
	2012	*13	..	..	..	*13	*4,541
	2013	*13	..	..	..	*13	*4,519
	2014	*13	..	..	..	*13	*4,497
Asia	**2011**	**1**	**4**	**1**	**..**	**5**	**0**
	2012	**1**	**4**	**1**	**..**	**4**	**0**
	2013	**0**	**4**	**0**	**..**	**4**	**0**
	2014	**1**	**5**	**2**	**..**	**5**	**0**
Armenia	2011	*1	0	1	..	0	0
	2012	*1	0	1	..	0	0
	2013	0	1	0	..	0	0
	2014	*1	0	2	..	0	0
Georgia	2011	..	1	..	..	1	0
	2012	..	1	..	..	1	0
	2013	..	0	..	..	0	0
	2014	..	2	..	..	2	0
Viet Nam	2011	..	3	..	..	3	0
	2012	..	3	..	..	3	0
	2013	..	3	..	..	3	0
	2014	..	3	..	..	3	0
Europe	**2011**	**16,635**	**552**	**176**	**-919**	**17,930**	**24**
	2012	**10,780**	**488**	**151**	**-4,969**	**16,086**	**22**
	2013	**19,286**	**258**	**134**	**4,470**	**14,940**	**20**
	2014	**15,311**	**225**	**224**	**930**	**14,382**	**19**
Austria	2011	1	..	..	..	1	0
	2012	1	..	..	..	1	0
	2013	1	..	..	..	1	0
	2014	1	..	..	..	1	0
Belarus	2011	2,823	..	..	247	2,576	272
	2012	2,679	..	..	86	2,593	273
	2013	2,269	..	..	233	2,036	214
	2014	1,433	..	..	-144	1,577	166
Estonia	2011	323	..	23	-5	305	230
	2012	166	..	17	-108	257	194
	2013	263	..	13	10	240	182
	2014	261	..	0	36	225	171

Production, trade and supply of peat

Thousand metric tons and kilograms per capita

Country or area	Year	Production	Imports	Exports	Changes in stocks	Supply	
						Total	Per capita
Finland	2011	7,054	105	5	-1,331	8,485	1,573
	2012	4,330	92	9	-2,385	6,798	1,253
	2013	7,388	39	3	1,558	5,866	1,076
	2014	6,722	49	2	787	5,982	1,092
Ireland	2011	3,708	..	..	170	3,538	760
	2012	1,452	..	..	-2,544	3,996	856
	2013	6,657	..	..	2,938	3,719	796
	2014	4,604	..	..	956	3,648	780
Latvia	2011	1	0	5	-8	4	2
	2012	9	0	13	-7	3	1
	2013	10	4	4	2	8	4
	2014	5	0	0	2	3	2
Lithuania	2011	42	..	12	-12	42	14
	2012	61	..	13	6	42	14
	2013	84	..	12	4	68	23
	2014	100	..	9	48	43	15
Romania	2011	1	40	2	0	39	2
	2012	2	0	0	0	2	0
	2013	1	39	2	0	38	2
	2014	2	34	3	0	33	2
Russian Federation	2011	1,431	18	55	0	1,394	10
	2012	1,162	6	49	0	1,119	8
	2013	1,522	6	50	-222	1,700	12
	2014	1,149	9	66	-755	1,847	13
Sweden	2011	736	360	..	..	1,096	116
	2012	472	363	..	..	835	87
	2013	624	144	..	..	768	80
	2014	450	110	..	..	560	58
T.F.Yug.Rep. Macedonia	2011	..	5	0	..	5	3
	2012	..	4	0	..	4	2
	2013	..	0	0	..	0	0
	2014	..	0	0	..	0	0
Ukraine	2011	515	24	74	20	445	10
	2012	446	23	50	-17	436	10
	2013	467	26	50	-53	496	11
	2014	584	23	144	0	463	10

Production of biofuels and wastes

Terajoules

Table Notes

The conversion of fuelwood (solid volume, 9.135 TJ/m^3) is based on an average 20-30% moisture content of wood from an unspecified species.

The methodology used for estimating fuel bagasse production for certain countries is derived from the work done by the Economic Commission for Latin America and the Caribbean (ECLAC). Although several different methodologies have been devised, the one selected assumes a yield of 3.26 tons of fuel bagasse at 50% humidity per ton of cane sugar produced. This methodology was used for instances where bagasse production data were derived from sugar production figures extracted from the Sugar Yearbook of the International Sugar Organization (London).

Biogasoline includes Other liquid biofuels.

The total column excludes charcoal as it is the only secondary energy product in the table.

Please refer to the Definitions Section on pages ix to xv for the appropriate product description/classification.

Figure 29: Production of fuelwood by region, 1990-2014

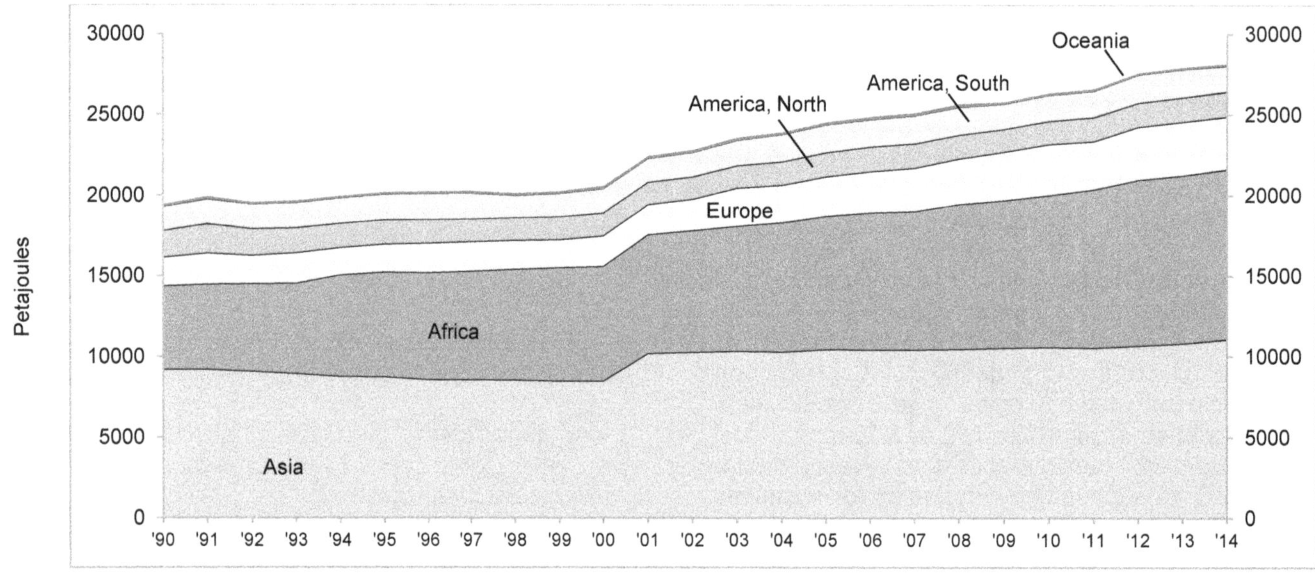

Figure 30: Production of charcoal by region, 1990-2014

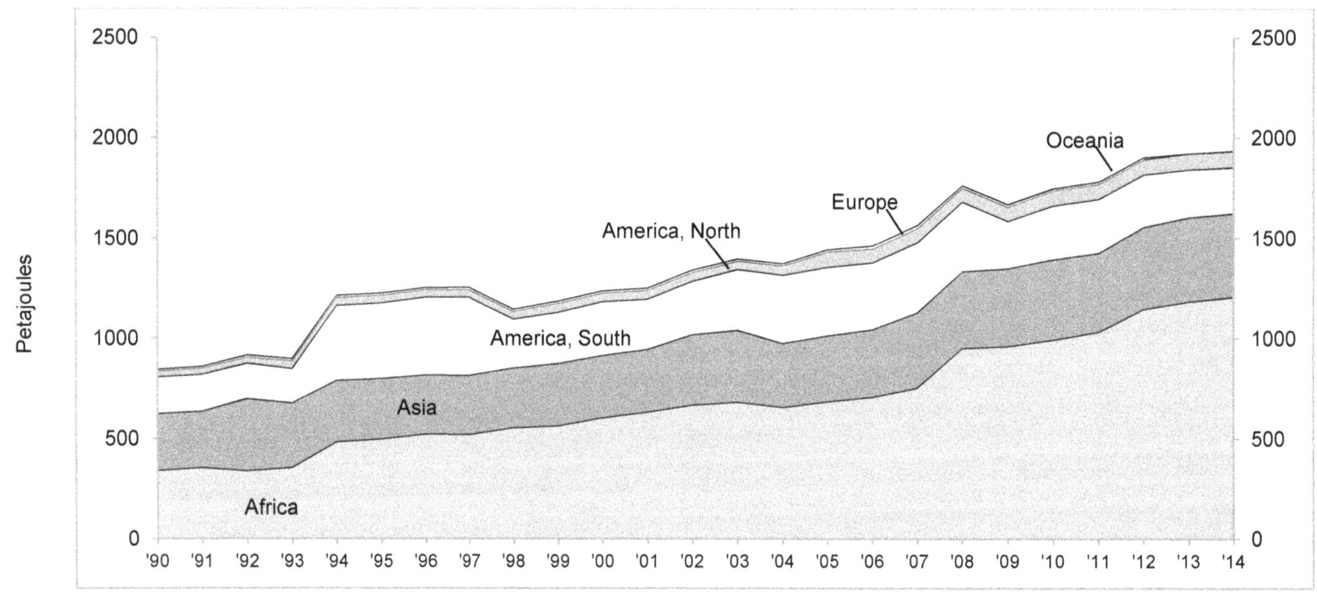

Figure 31: Production of total solid biofuels by region, 2014

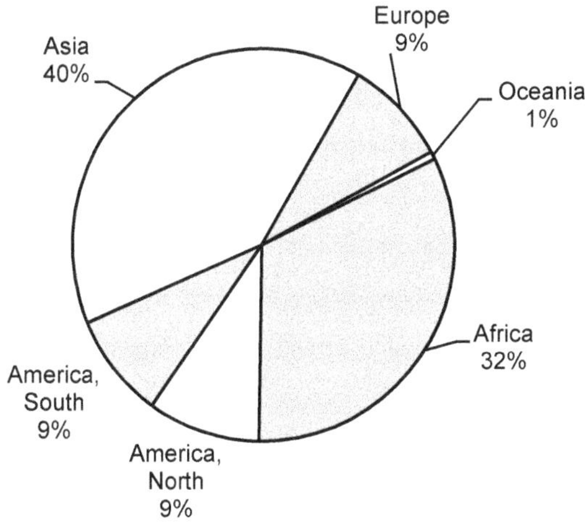

Figure 32: Production of biogases by region, 2014

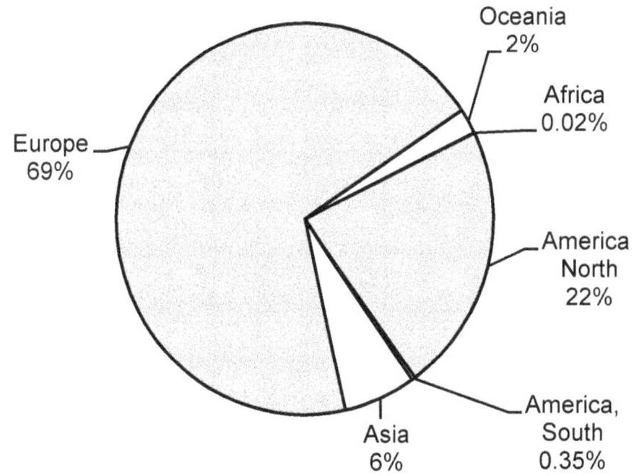

Figure 33: Production of bagasse by region, 2014

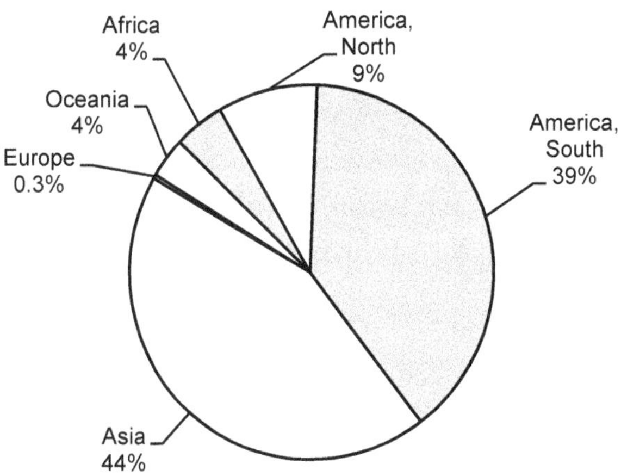

Figure 34: Production of liquids biofuels by region, 2014

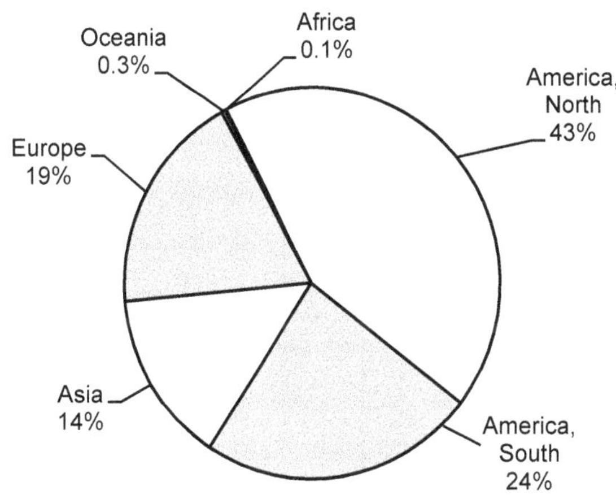

Figure 35: Production of energy from wastes by region, 2014

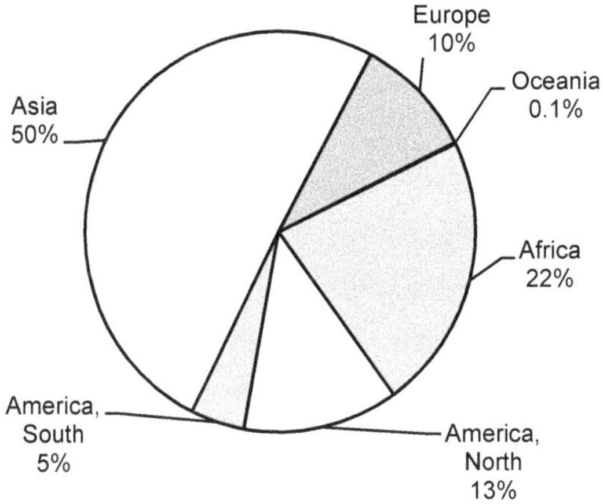

Table 12

Production of biofuels and wastes

Terajoules (TJ)

Country or area	Year	Solid biofuels					
		Fuelwood	Animal waste	Vegetal waste	Bagasse	Black liquor	Charcoal
World	**2011**	**26,593,984**	**957,307**	**9,284,522**	**3,311,085**	**1,638,656**	**1,780,435**
	2012	**27,593,627**	**977,792**	**9,417,034**	**3,521,582**	**1,541,452**	**1,902,002**
	2013	**27,929,406**	**1,054,880**	**9,682,747**	**3,726,762**	**1,709,822**	**1,930,985**
	2014	**28,148,008**	**1,076,776**	**9,904,781**	**3,789,362**	**1,851,589**	**1,947,654**
Africa	**2011**	**9,832,865**	**147,435**	**2,875,341**	**191,548**	**4,000**	**1,032,220**
	2012	**10,273,596**	**150,944**	**2,937,537**	**174,661**	**3,950**	**1,145,989**
	2013	**10,433,662**	**154,551**	**3,008,932**	**192,502**	**4,000**	**1,181,616**
	2014	**10,545,164**	**158,206**	**3,086,034**	**187,247**	**4,000**	**1,206,222**
Algeria	2011	674	..	..	..	..	..
	2012	652	..	..	..	..	..
	2013	918	..	..	..	..	..
	2014	251	..	..	..	..	..
Angola	2011	171,352	..	111,378	..	..	36,020
	2012	174,762	..	113,594	..	..	33,040
	2013	178,186	..	115,819	..	..	33,689
	2014	181,598	..	118,037	..	..	34,338
Benin	2011	88,096	..	386	226	..	7,434
	2012	90,537	..	397	226	..	7,641
	2013	92,987	..	408	226	..	7,847
	2014	95,473	..	419	252	..	8,054
Botswana	2011	6,270	..	..	..	..	2,146
	2012	6,299	..	..	..	..	2,164
	2013	6,328	..	..	..	..	2,181
	2014	6,328	..	..	..	..	2,181
Burkina Faso	2011	118,417	..	..	934	..	17,777
	2012	120,080	..	..	934	..	18,275
	2013	121,779	..	..	1,035	..	18,786
	2014	121,779	..	..	633	..	18,786
Burundi	2011	85,842	..	..	521	..	1,770
	2012	52,462	..	..	583	..	1,770
	2013	54,874	..	..	650	..	7,936
	2014	54,801	..	..	542	..	7,788
Cabo Verde	2011	1,439	..	..	..	..	21
	2012	1,473	..	..	..	..	24
	2013	1,508	..	..	..	..	26
	2014	1,501	..	..	..	..	19
Cameroon	2011	183,993	..	6,187	..	..	6,461
	2012	188,698	..	6,344	..	..	6,638
	2013	193,507	..	6,505	..	..	6,785
	2014	198,403	..	6,670	..	..	6,962
Central African Rep.	2011	18,270	..	..	..	..	5,774
	2012	18,270	..	..	..	..	5,880
	2013	18,270	..	..	..	..	5,988
	2014	18,270	..	..	..	..	6,097
Chad	2011	65,626	..	..	747	..	12,181
	2012	66,686	..	..	806	..	12,493
	2013	67,769	..	..	770	..	12,812
	2014	68,869	..	..	829	..	13,142
Comoros	2011	2,491	..	..	..	..	1,204
	2012	2,556	..	..	..	..	1,236
	2013	2,621	..	..	..	..	1,271
	2014	2,689	..	..	..	..	1,307
Congo	2011	37,561	..	18,780	..	..	6,166
	2012	39,381	..	19,691	..	..	5,782
	2013	40,389	..	20,187	..	..	5,900
	2014	41,393	..	20,697	..	..	6,077

Production of biofuels and wastes

Terajoules (TJ)

Non-biomass wastes		Liquid biofuels		Biogases	Total
Industrial waste	Municipal waste	Biogasoline	Biodiesel		
775,563	**2,442,373**	**1,780,526**	**804,403**	**640,510**	**48,228,928**
799,187	**2,565,761**	**1,795,523**	**888,619**	**733,154**	**49,833,732**
815,199	**2,732,544**	**1,929,788**	**970,380**	**841,672**	**51,393,201**
820,461	**2,851,887**	**2,029,816**	**1,084,151**	**911,884**	**52,468,715**
20,850	..	**2,975**	..	**80**	**13,075,093**
20,850	..	**3,028**	..	**121**	**13,564,688**
20,850	..	**3,109**	..	**428**	**13,818,033**
20,805	..	**4,368**	..	**159**	**14,005,983**
..	..	..	..	..	674
..	..	..	..	..	652
..	..	..	..	..	918
..	..	..	..	..	251
..	..	..	..	..	282,730
..	..	..	..	..	288,356
..	..	..	..	..	294,005
..	..	..	..	..	299,635
..	..	..	..	..	88,708
..	..	..	..	..	91,160
..	..	..	..	..	93,621
..	..	..	..	..	96,144
..	..	..	..	..	6,270
..	..	..	..	..	6,299
..	..	..	..	..	6,328
..	..	..	..	..	6,328
..	..	..	..	..	119,351
..	..	..	..	..	121,014
..	..	..	..	..	122,813
..	..	..	..	..	122,412
..	..	..	..	..	86,362
..	..	..	..	..	53,046
..	..	..	..	..	55,524
..	..	..	..	..	55,343
..	..	..	..	..	1,439
..	..	..	..	..	1,473
..	..	..	..	..	1,508
..	..	..	..	..	1,501
..	..	..	..	..	190,180
..	..	..	..	..	195,042
..	..	..	..	..	200,012
..	..	..	..	..	205,073
..	..	..	..	..	18,270
..	..	..	..	..	18,270
..	..	..	..	..	18,270
..	..	..	..	..	18,270
..	..	..	..	..	66,373
..	..	..	..	..	67,491
..	..	..	..	..	68,539
..	..	..	..	..	69,697
..	..	..	..	..	2,491
..	..	..	..	..	2,556
..	..	..	..	..	2,621
..	..	..	..	..	2,689
..	..	..	..	..	56,341
..	..	..	..	..	59,072
..	..	..	..	..	60,576
..	..	..	..	..	62,090

Table 12

Production of biofuels and wastes

Terajoules (TJ)

Country or area	Year	Solid biofuels					
		Fuelwood	Animal waste	Vegetal waste	Bagasse	Black liquor	Charcoal
Côte d'Ivoire	2011	380,476	..	1,364	..	..	37,377
	2012	392,717	..	1,412	..	..	38,586
	2013	411,252	..	1,450	..	..	41,359
	2014	421,878	..	1,489	..	..	42,392
Dem. Rep. of the Congo	2011	806,022	..	16,426	..	..	22,214
	2012	1,023,871	..	10,467	..	..	104,725
	2013	1,060,330	..	10,804	..	..	108,472
	2014	1,096,788	..	5,824	..	..	112,189
Djibouti	2011	3,303	..	..	..	..	1,437
	2012	3,355	..	..	..	..	1,460
	2013	3,408	..	..	..	..	1,484
	2014	3,461	..	..	..	..	1,509
Egypt	2011	160,370	..	34,235	23,386	..	40,480
	2012	160,781	..	35,000	23,386	..	40,849
	2013	161,199	..	35,798	25,023	..	41,223
	2014	161,625	..	36,601	23,160	..	41,595
Equatorial Guinea	2011	4,083	..	..	..	..	266
	2012	4,083	..	..	..	..	280
	2013	4,083	..	..	..	..	289
	2014	4,083	..	..	..	..	295
Eritrea	2011	20,980	2,466	1,395	..	..	4,071
	2012	21,427	2,519	1,425	..	..	4,160
	2013	21,896	2,574	1,456	..	..	4,248
	2014	22,382	2,631	1,488	..	..	4,337
Ethiopia	2011	937,337	134,928	132,256	8,108	..	112,885
	2012	949,733	138,376	135,636	6,950	..	115,678
	2013	962,335	141,839	139,031	9,267	..	118,540
	2014	975,143	145,441	142,561	9,946	..	121,472
Gabon	2011	47,528	..	2,879	612	..	608
	2012	48,691	..	2,949	579	..	620
	2013	49,854	..	3,019	612	..	631
	2014	*49,786	..	3,048	598	..	643
Gambia	2011	6,422	..	..	..	..	1,779
	2012	6,513	..	..	..	..	1,817
	2013	6,595	..	..	..	..	1,856
	2014	6,687	..	..	..	..	1,894
Ghana	2011	127,881	..	..	..	..	36,391
	2012	143,472	..	..	..	..	43,493
	2013	148,795	..	..	..	..	46,540
	2014	151,939	..	..	..	..	50,744
Guinea	2011	109,713	..	..	453	..	10,502
	2012	110,196	..	..	453	..	10,733
	2013	110,689	..	..	453	..	10,969
	2014	111,209	..	..	453	..	11,209
Guinea-Bissau	2011	24,092	..	..	..	..	1,935
	2012	24,438	..	..	..	..	1,981
	2013	24,788	..	..	..	..	2,027
	2014	25,141	..	..	..	..	2,075
Kenya	2011	*555,572	..	27,622	14,765	..	178,283
	2012	*578,693	..	28,378	14,057	..	183,699
	2013	*600,973	..	22,353	15,057	..	188,664
	2014	593,629	..	23,393	14,640	..	188,422
Lesotho	2011	19,181	*8,140	*1,325	..	..	2,903
	2012	19,253	*8,160	*1,330	..	..	2,956
	2013	19,326	*8,180	*1,335	..	..	3,009
	2014	19,398	*8,190	*1,340	..	..	3,065

Production of biofuels and wastes

Terajoules (TJ)

Non-biomass wastes		Liquid biofuels		Biogases	Total
Industrial waste	Municipal waste	Biogasoline	Biodiesel		
..	..	..	..	..	381,840
..	..	..	..	..	394,129
..	..	..	..	..	412,702
..	..	..	..	..	423,367
..	..	..	..	..	822,448
..	..	..	..	..	1,034,338
..	..	..	..	..	1,071,134
..	..	..	..	..	1,102,612
..	..	..	..	..	3,303
..	..	..	..	..	3,355
..	..	..	..	..	3,408
..	..	..	..	..	3,461
..	..	..	..	..	217,992
..	..	..	..	..	219,168
..	..	..	..	..	222,020
..	..	..	..	..	221,386
..	..	..	..	..	4,083
..	..	..	..	..	4,083
..	..	..	..	..	4,083
..	..	..	..	..	4,083
..	..	..	..	..	24,841
..	..	..	..	..	25,371
..	..	..	..	..	25,926
..	..	..	..	..	26,501
..	..	161	..	..	1,212,790
..	..	134	..	..	1,230,829
..	..	161	..	..	1,252,632
..	..	188	..	..	1,273,279
..	..	..	..	..	51,019
..	..	..	..	..	52,219
..	..	..	..	..	53,485
..	..	..	..	..	*53,432
..	..	..	..	..	6,422
..	..	..	..	..	6,513
..	..	..	..	..	6,595
..	..	..	..	..	6,687
..	..	..	..	..	127,881
..	..	..	..	..	143,472
..	..	..	..	..	148,795
..	..	..	..	..	151,939
..	..	..	..	..	110,166
..	..	..	..	..	110,649
..	..	..	..	..	111,142
..	..	..	..	..	111,663
..	..	..	..	..	24,092
..	..	..	..	..	24,438
..	..	..	..	..	24,788
..	..	..	..	..	25,141
..	..	..	..	..	*597,959
..	..	..	..	..	*621,128
..	..	..	..	..	*638,383
..	..	..	..	..	631,662
..	..	..	..	..	28,646
..	..	..	..	..	28,743
..	..	..	..	..	28,841
..	..	..	..	..	28,928

Table 12

Production of biofuels and wastes

Terajoules (TJ)

Country or area	Year	Solid biofuels					
		Fuelwood	Animal waste	Vegetal waste	Bagasse	Black liquor	Charcoal
Liberia	2011	66,265	..	..	..	..	7,177
	2012	68,588	..	..	..	..	7,445
	2013	70,992	..	..	..	..	7,722
	2014	73,480	..	..	..	..	8,010
Libya	2011	6,523	..	..	..	..	..
	2012	6,466	..	..	..	..	..
	2013	6,395	..	..	..	..	..
	2014	6,333	..	..	..	..	..
Madagascar	2011	119,669	..	..	1,636	..	35,252
	2012	119,166	..	..	1,762	..	34,989
	2013	121,733	..	..	1,938	..	46,021
	2014	124,318	..	..	1,440	..	46,021
Malawi	2011	49,928	..	..	7,678	..	14,804
	2012	50,500	..	..	8,056	..	15,167
	2013	51,094	..	..	7,761	..	15,538
	2014	51,705	..	..	7,976	..	15,919
Mali	2011	49,192	..	..	805	..	4,160
	2012	49,740	..	..	805	..	4,160
	2013	48,637	..	..	805	..	5,065
	2014	50,860	..	..	1,259	..	5,045
Mauritania	2011	17,148	..	..	..	..	5,623
	2012	17,532	..	..	..	..	5,773
	2013	17,925	..	..	..	..	5,927
	2014	18,325	..	..	..	..	6,086
Mauritius	2011	183	..	..	10,525	..	18
	2012	183	..	..	9,969	..	18
	2013	174	..	..	9,738	..	18
	2014	164	..	..	9,336	..	18
Mayotte	2011	242	..	..	..	..	..
	2012	242	..	..	..	..	..
	2013	242	..	..	..	..	..
	2014	*242	..	..	..	..	..
Morocco	2011	61,769	..	..	1,690	..	3,180
	2012	61,579	..	..	809	..	3,221
	2013	61,390	..	..	1,367	..	3,263
	2014	61,202	..	..	670	..	3,304
Mozambique	2011	325,576	..	17,096	9,807	..	13,688
	2012	333,814	..	17,529	9,985	..	20,503
	2013	342,172	..	17,968	9,645	..	20,414
	2014	357,972	..	18,798	10,572	..	21,093
Namibia	2011	11,317	..	1,383	..	..	4,189
	2012	11,580	..	1,415	..	..	4,248
	2013	11,858	..	1,449	..	..	3,983
	2014	12,458	..	1,449	..	..	4,278
Niger	2011	52,805	432	335	..	..	1,236
	2012	56,570	434	442	..	..	1,325
	2013	58,776	500	378	..	..	1,376
	2014	*60,071	*500	*405	..	..	*1,420
Nigeria	2011	2,184,881	..	2,094,664	755	..	102,749
	2012	2,369,125	..	2,151,823	273	..	118,148
	2013	2,336,770	..	2,210,364	273	..	110,419
	2014	2,278,015	..	2,269,954	158	..	113,398
Réunion	2011	925	..	..	4,082	..	435
	2012	923	..	..	4,095	..	435
	2013	922	..	..	3,920	..	434
	2014	921	..	..	4,145	..	434

Production of biofuels and wastes

Terajoules (TJ)

Non-biomass wastes		Liquid biofuels		Biogases	Total
Industrial waste	Municipal waste	Biogasoline	Biodiesel		
..	..	..	..	..	66,265
..	..	..	..	..	68,588
..	..	..	..	..	70,992
..	..	..	..	..	73,480
..	..	..	..	..	6,523
..	..	..	..	..	6,466
..	..	..	..	..	6,395
..	..	..	..	..	6,333
..	..	..	..	..	121,305
..	..	..	..	..	120,928
..	..	..	..	..	123,671
..	..	..	..	..	125,758
*20,850	..	..	..	..	78,456
*20,850	..	..	..	..	79,406
*20,850	..	..	..	..	79,705
*20,805	..	..	..	..	80,486
..	..	..	..	..	49,997
..	..	..	..	..	50,545
..	..	..	..	..	49,443
..	..	..	..	..	52,119
..	..	..	..	..	17,148
..	..	..	..	..	17,532
..	..	..	..	..	17,925
..	..	..	..	..	18,325
..	..	..	..	..	10,708
..	..	..	..	..	10,152
..	..	..	..	240	10,151
..	..	..	..	..	9,500
..	..	..	..	..	242
..	..	..	..	..	242
..	..	..	..	..	242
..	..	..	..	..	*242
..	..	..	..	..	63,459
..	..	..	..	..	62,388
..	..	..	..	..	62,757
..	..	..	..	..	61,872
..	..	..	..	..	352,479
..	..	..	..	..	361,328
..	..	..	..	..	369,785
..	..	..	..	..	387,342
..	..	..	..	..	12,700
..	..	..	..	..	12,995
..	..	..	..	..	13,307
..	..	..	..	..	13,907
..	..	..	..	..	53,572
..	..	..	..	..	57,446
..	..	..	..	..	59,654
..	..	..	..	..	*60,976
..	..	..	..	..	4,280,300
..	..	..	..	..	4,521,221
..	..	..	..	..	4,547,407
..	..	..	..	..	4,548,127
..	..	..	..	80	5,086
..	..	..	..	121	5,140
..	..	..	..	188	5,030
..	..	..	..	159	5,225

Table 12

Production of biofuels and wastes

Terajoules (TJ)

Country or area	Year	Solid biofuels					
		Fuelwood	Animal waste	Vegetal waste	Bagasse	Black liquor	Charcoal
Rwanda	2011	*77,355	..	..	226	..	*16,449
	2012	*79,803	..	..	226	..	*16,968
	2013	*82,368	..	..	151	..	*17,865
	2014	*83,129	..	..	252	..	*17,995
Sao Tome and Principe	2011	987	..	..	..	..	266
	2012	996	..	..	..	..	268
	2013	1,005	..	..	..	..	271
	2014	1,005	..	..	..	..	271
Senegal	2011	83,078	..	1,195	2,863	..	16,550
	2012	85,358	..	611	3,072	..	16,992
	2013	69,442	..	1,021	2,055	..	14,219
	2014	71,649	..	1,069	2,152	..	14,662
Seychelles	2011	29	..	..	..	..	..
	2012	29	..	..	..	..	..
	2013	29	..	..	..	..	..
	2014	29	..	..	..	..	..
Sierra Leone	2011	51,330	..	..	201	..	11,564
	2012	51,704	..	..	201	..	11,830
	2013	52,097	..	..	201	..	12,125
	2014	52,471	..	..	286	..	12,420
Somalia	2011	127,494	..	..	579	..	35,022
	2012	127,494	..	..	579	..	35,022
	2013	127,495	..	..	579	..	35,022
	2014	128,241	..	..	579	..	33,515
South Africa	2011	548,742	..	82,828	46,130	..	20,877
	2012	557,140	..	84,096	49,497	..	20,945
	2013	565,815	..	85,405	59,279	..	21,013
	2014	574,809	..	86,763	53,287	..	21,081
South Sudan	2011	..	..	..	..	..	..
	2012	8,064	..	..	..	..	325
	2013	7,826	..	..	..	..	325
	2014	8,138	..	..	..	..	325
St. Helena and Depend.	2011	5	..	..	..	..	..
	2012	5	..	..	..	..	..
	2013	6	..	..	..	..	..
	2014	*5	..	..	..	..	..
Sudan	2011	..	..	..	..	..	..
	2012	339,332	..	39,640	..	..	44,929
	2013	346,699	..	40,634	..	..	45,725
	2014	354,540	..	41,639	..	..	47,082
Sudan (former)	2011	421,711	..	46,833	18,101	..	56,552
	2012	..	..	..	..	..	..
	2013	..	..	..	..	..	..
	2014	..	..	..	..	..	..
Swaziland	2011	9,847	..	..	15,630	*4,000	1,292
	2012	9,987	..	..	15,390	*3,950	1,325
	2013	10,131	..	..	17,290	*4,000	1,361
	2014	10,277	..	..	17,336	*4,000	1,397
Togo	2011	100,480	..	955	..	..	18,231
	2012	103,262	..	1,225	..	..	18,733
	2013	106,063	..	1,258	..	..	19,234
	2014	108,910	..	1,292	..	..	19,765
Tunisia	2011	33,907	1,469	3,963	..	..	5,753
	2012	34,058	1,455	4,075	..	..	5,871
	2013	34,214	1,458	4,088	..	..	6,018
	2014	34,384	1,444	4,074	..	..	6,136

Production of biofuels and wastes

Terajoules (TJ)

Non-biomass wastes		Liquid biofuels		Biogases	Total
Industrial waste	Municipal waste	Biogasoline	Biodiesel		
..	..	..	..	..	*77,581
..	..	..	..	..	*80,029
..	..	..	..	..	*82,520
..	..	..	..	..	*83,380
..	..	..	..	..	987
..	..	..	..	..	996
..	..	..	..	..	1,005
..	..	..	..	..	1,005
..	..	..	..	..	87,136
..	..	..	..	..	89,041
..	..	..	..	..	72,518
..	..	..	..	..	74,870
..	..	..	..	..	29
..	..	..	..	..	29
..	..	..	..	..	29
..	..	..	..	..	29
..	..	..	..	..	51,530
..	..	..	..	..	51,905
..	..	..	..	..	52,298
..	..	..	..	..	52,757
..	..	..	..	..	128,074
..	..	..	..	..	128,074
..	..	..	..	..	128,074
..	..	..	..	..	128,820
..	..	*2,814	..	..	680,514
..	..	*2,894	..	..	693,628
..	..	*2,948	..	..	713,447
..	..	*3,028	..	..	717,887
..	..	..	..	..	..
..	..	..	..	..	8,064
..	..	..	..	..	7,826
..	..	..	..	..	8,138
..	..	..	..	..	5
..	..	..	..	..	5
..	..	..	..	..	6
..	..	..	..	..	*5
..	..	..	..	..	..
..	..	..	..	..	378,972
..	..	..	..	..	387,333
..	..	..	..	..	396,179
..	..	..	..	..	486,645
..	..	..	..	..	..
..	..	..	..	..	..
..	..	..	..	..	..
..	..	..	..	..	29,476
..	..	..	..	..	29,327
..	..	..	..	..	31,421
..	..	..	..	..	31,613
..	..	..	..	..	101,435
..	..	..	..	..	104,487
..	..	..	..	..	107,321
..	..	..	..	..	110,202
..	..	..	..	..	39,338
..	..	..	..	..	39,588
..	..	..	..	..	39,759
..	..	..	..	..	39,902

Table 12

Production of biofuels and wastes

Terajoules (TJ)

Country or area	Year	Solid biofuels					
		Fuelwood	Animal waste	Vegetal waste	Bagasse	Black liquor	Charcoal
Uganda	2011	380,016	..	..	*10,270	..	27,478
	2012	380,016	..	..	*11,506	..	27,478
	2013	380,016	..	..	*14,278	..	29,500
	2014	382,360	..	..	*17,375	..	30,604
United Rep. of Tanzania	2011	645,439	..	122,578	7,151	..	51,537
	2012	665,915	..	126,790	8,056	..	53,130
	2013	708,344	..	130,691	7,739	..	54,752
	2014	742,758	..	137,635	7,297	..	57,142
Zambia	2011	248,383	..	42,407	..	..	29,795
	2012	256,417	..	43,779	..	..	30,503
	2013	264,870	..	45,222	..	..	31,211
	2014	273,675	..	46,218	..	..	31,949
Zimbabwe	2011	174,648	..	106,871	3,665	..	266
	2012	178,927	..	109,489	2,405	..	266
	2013	183,503	..	112,289	2,390	..	266
	2014	188,213	..	115,171	2,074	..	266
America, North	**2011**	**1,488,073**	**669,197**	**663,052**	**264,173**	**904,054**	**76,281**
	2012	**1,481,716**	**685,025**	**599,747**	**303,586**	**818,798**	**77,113**
	2013	**1,507,365**	**755,800**	**707,195**	**317,643**	**952,267**	**79,328**
	2014	**1,536,668**	**769,128**	**735,033**	**336,118**	**1,004,349**	**80,770**
Anguilla	2011	*2	..	..	..	..	..
	2012	*2	..	..	..	..	..
	2013	*2	..	..	..	..	..
	2014	*2	..	..	..	..	..
Aruba	2011	17	..	..	..	..	5
	2012	17	..	..	..	..	5
	2013	17	..	..	..	..	5
	2014	18	..	..	..	..	6
Bahamas	2011	303	..	..	..	..	32
	2012	304	..	..	..	..	32
	2013	305	..	..	..	..	33
	2014	31	..	..	..	..	33
Barbados	2011	45	..	..	857	..	9
	2012	45	..	..	769	..	9
	2013	45	..	..	575	..	9
	2014	45	..	..	518	..	9
Belize	2011	1,151	..	..	2,803	..	21
	2012	1,151	..	..	3,158	..	22
	2013	1,151	..	..	3,073	..	22
	2014	1,151	..	..	2,973	..	22
Bermuda	2011	..	..	..	..	..	..
	2012	..	..	..	..	..	..
	2013	..	..	..	..	..	..
	2014	..	..	..	..	..	..
British Virgin Islands	2011	13	..	..	..	..	2
	2012	13	..	..	..	..	2
	2013	13	..	..	..	..	2
	2014	14	..	..	..	..	2
Canada	2011	355,957	..	..	..	170,418	..
	2012	352,340	..	..	..	175,026	..
	2013	360,069	..	..	..	191,861	..
	2014	379,138	..	..	..	182,966	..
Costa Rica	2011	17,163	..	4,965	8,613	..	32
	2012	12,350	..	4,966	9,285	..	36
	2013	12,213	..	4,966	11,061	..	30
	2014	12,082	..	4,793	10,676	..	59

Production of biofuels and wastes

Terajoules (TJ)

Non-biomass wastes		Liquid biofuels		Biogases	Total
Industrial waste	Municipal waste	Biogasoline	Biodiesel		
..	..	..	..	..	390,286
..	..	..	..	..	391,522
..	..	..	..	..	394,294
..	..	..	..	..	399,735
..	..	..	..	..	775,168
..	..	..	..	..	800,761
..	..	..	..	..	846,774
..	..	..	..	..	887,690
..	..	..	..	..	290,790
..	..	..	..	..	300,196
..	..	..	..	..	310,092
..	..	..	..	..	319,893
..	..	..	..	..	285,184
..	..	..	..	..	290,821
..	..	..	..	..	298,182
..	..	1,152	..	..	306,610
89,517	**305,060**	**1,118,637**	**106,941**	**141,503**	**5,750,206**
89,331	**305,729**	**1,093,625**	**130,456**	**154,237**	**5,662,249**
66,915	**304,888**	**1,098,228**	**170,605**	**178,575**	**6,059,481**
64,987	**303,446**	**1,180,986**	**161,184**	**202,439**	**6,294,337**
..	..	..	..	..	*2
..	..	..	..	..	*2
..	..	..	..	..	*2
..	..	..	..	..	*2
..	..	..	..	..	17
..	..	..	..	..	17
..	..	..	..	..	17
..	..	..	..	..	18
..	..	..	..	..	303
..	..	..	..	..	304
..	..	..	..	..	305
..	..	..	..	..	31
..	..	..	..	..	902
..	..	..	..	..	814
..	..	..	..	..	620
..	..	..	..	..	562
..	..	..	..	..	3,954
..	..	..	..	..	4,309
..	..	..	..	..	4,224
..	..	..	..	..	4,124
..	*560	..	..	..	*560
..	*565	..	..	..	*565
..	*565	..	..	..	*565
..	*605	..	..	..	*605
..	..	..	..	..	13
..	..	..	..	..	13
..	..	..	..	..	13
..	..	..	..	..	14
5,080	5,264	35,108	3,938	15,923	591,688
5,252	5,815	36,475	3,754	14,969	593,630
5,429	7,147	36,073	4,048	16,664	621,291
5,606	7,147	37,413	9,826	16,977	639,072
..	..	525	..	..	31,266
..	..	670	..	..	27,271
..	..	742	..	4	28,986
..	..	..	..	4	27,555

Table 12

Production of biofuels and wastes

Terajoules (TJ)

Country or area	Year	Solid biofuels					
		Fuelwood	Animal waste	Vegetal waste	Bagasse	Black liquor	Charcoal
Cuba	2011	11,921	..	13	30,502	..	1,564
	2012	11,062	..	88	30,580	..	974
	2013	12,641	..	*88	28,085	..	3,339
	2014	10,931	..	*88	35,554	..	3,685
Dominica	2011	69	..	..	..	..	7
	2012	69	..	..	..	..	7
	2013	69	..	..	..	..	7
	2014	69	..	..	..	..	7
Dominican Republic	2011	9,842	..	..	11,600	..	1,421
	2012	9,743	..	..	12,771	..	1,435
	2013	9,773	..	..	11,981	..	1,471
	2014	9,963	..	..	12,519	..	1,532
El Salvador	2011	21,450	..	..	10,226	..	236
	2012	16,618	..	..	12,347	..	236
	2013	15,712	..	..	7,495	..	236
	2014	16,109	..	..	7,640	..	1,092
Greenland	2011	..	..	..	..	..	..
	2012	..	..	..	..	..	..
	2013	..	..	..	..	..	..
	2014	..	..	..	..	..	..
Grenada	2011	240	..	..	22	..	57
	2012	235	..	..	22	..	58
	2013	229	..	..	22	..	57
	2014	248	..	..	20	..	60
Guadeloupe	2011	137	..	..	*1,197	..	62
	2012	137	..	..	*1,236	..	62
	2013	137	..	..	*1,236	..	62
	2014	137	..	..	*1,236	..	62
Guatemala	2011	180,951	..	..	72,189	..	1,918
	2012	183,663	..	..	75,799	..	1,977
	2013	183,663	..	..	80,657	..	2,036
	2014	199,158	..	..	80,657	..	2,095
Haiti	2011	133,606	..	..	970	..	35,430
	2012	137,987	..	..	967	..	36,787
	2013	138,054	..	..	963	..	38,173
	2014	134,364	..	..	931	..	38,232
Honduras	2011	79,543	..	..	11,138	..	538
	2012	81,085	..	..	11,502	..	620
	2013	82,677	..	..	11,864	..	620
	2014	83,904	..	..	10,692	..	620
Jamaica	2011	2,773	..	..	3,409	..	1,133
	2012	2,773	..	..	3,357	..	1,133
	2013	2,773	..	..	3,687	..	1,133
	2014	2,773	..	..	3,926	..	1,133
Martinique	2011	91	..	..	*175	..	53
	2012	91	..	..	*175	..	53
	2013	91	..	..	*175	..	53
	2014	91	..	..	*175	..	53
Mexico	2011	253,496	..	41,671	31,790	..	2,508
	2012	252,826	..	40,993	41,468	..	2,360
	2013	255,520	..	61,991	53,758	..	2,419
	2014	254,274	..	37,708	71,125	..	2,419
Neth. Antilles (former)	2011	*31	..	..	..	..	*8
	2012	..	..	..	..	..	..
	2013	..	..	..	..	..	..
	2014	..	..	..	..	..	..

Production of biofuels and wastes

Terajoules (TJ)

Non-biomass wastes		Liquid biofuels		Biogases	Total
Industrial waste	Municipal waste	Biogasoline	Biodiesel		
..	..	399	..	..	42,835
..	..	375	..	..	42,104
..	..	374	..	..	41,188
..	..	372	..	..	46,945
..	..	..	..	..	69
..	..	..	..	..	69
..	..	..	..	..	69
..	..	..	..	..	69
..	..	..	..	..	21,442
..	..	..	..	..	22,514
..	..	..	..	..	21,754
..	..	..	..	..	22,482
..	..	..	..	..	31,676
..	..	..	..	..	28,965
..	..	..	..	526	23,733
..	..	..	..	409	24,158
..	91	..	..	..	91
..	103	..	..	..	103
..	117	..	..	..	117
..	92	..	..	..	92
..	..	..	..	..	262
..	..	..	..	..	257
..	..	..	..	..	251
..	..	..	..	..	268
..	..	..	..	..	*1,334
..	..	..	..	..	*1,373
..	..	..	..	..	*1,373
..	..	..	..	..	*1,373
..	..	..	..	..	253,140
..	..	..	..	..	259,462
..	..	..	..	..	264,320
..	..	..	..	..	279,815
..	..	..	..	..	134,576
..	..	..	..	..	138,954
..	..	..	..	..	139,017
..	..	..	..	..	135,295
..	..	..	..	..	90,681
..	..	..	..	..	92,587
..	..	..	..	..	94,541
..	..	..	..	..	94,596
..	..	..	..	..	6,183
..	..	..	..	..	6,130
..	..	..	..	..	6,460
..	..	..	..	..	6,699
..	*432	..	..	..	*698
..	*432	..	..	..	*698
..	*432	..	..	..	*698
..	*432	..	..	..	*698
914	..	..	..	1,471	329,342
8,922	..	..	..	1,822	346,031
1,690	..	..	..	1,970	374,929
861	..	..	..	1,939	365,907
..	..	..	..	..	*31
..	..	..	..	..	..
..	..	..	..	..	..
..	..	..	..	..	..

Table 12

Production of biofuels and wastes

Terajoules (TJ)

Country or area	Year	Solid biofuels					
		Fuelwood	Animal waste	Vegetal waste	Bagasse	Black liquor	Charcoal
Nicaragua	2011	42,087	..	2,196	9,523	..	502
	2012	42,078	..	2,755	11,670	..	531
	2013	41,933	..	3,531	15,269	..	531
	2014	41,935	..	4,775	15,695	..	561
Panama	2011	7,327	..	..	4,433	..	41
	2012	7,282	..	..	4,666	..	41
	2013	7,236	..	..	5,395	..	38
	2014	7,191	..	..	4,901	..	38
St. Lucia	2011	90	..	..	..	..	10
	2012	90	..	..	..	..	10
	2013	90	..	..	..	..	10
	2014	90	..	..	..	..	10
St. Vincent-Grenadines	2011	68	..	..	..	..	11
	2012	68	..	..	..	..	11
	2013	67	..	..	..	..	11
	2014	67	..	..	..	..	11
Trinidad and Tobago	2011	297	..	..	..	..	59
	2012	294	..	..	..	..	59
	2013	291	..	..	..	..	59
	2014	289	..	..	..	..	59
Turks and Caicos Islands	2011	9	..	..	..	..	4
	2012	9	..	..	..	..	4
	2013	10	..	..	..	..	4
	2014	10	..	..	..	..	4
United States	2011	369,392	*669,197	*614,207	64,727	*733,636	30,621
	2012	369,383	*685,025	*550,945	83,816	*643,772	30,651
	2013	382,583	*755,800	*636,619	82,348	*760,406	28,969
	2014	382,583	*769,128	*687,669	76,881	*821,383	28,969
America, South	**2011**	**1,634,474**	**..**	**610,399**	**1,331,916**	**35,246**	**270,308**
	2012	**1,752,930**	**..**	**640,842**	**1,387,015**	**34,508**	**260,872**
	2013	**1,749,411**	**..**	**646,244**	**1,467,322**	**34,293**	**238,307**
	2014	**1,608,323**	**..**	**659,838**	**1,473,087**	**112,779**	**229,498**
Argentina	2011	26,987	..	27,586	40,303	..	11,210
	2012	28,230	..	29,978	40,220	..	11,977
	2013	29,917	..	16,915	47,313	..	10,679
	2014	48,190	..	17,083	48,345	..	11,092
Bolivia (Plur. State of)	2011	21,630	..	4,262	7,980	..	1,062
	2012	21,798	..	4,431	11,001	..	1,062
	2013	21,970	..	4,606	10,195	..	1,062
	2014	22,145	..	4,815	11,832	..	1,062
Brazil	2011	1,049,520	..	511,647	1,134,709	..	225,262
	2012	1,039,325	..	534,041	1,189,968	..	217,917
	2013	987,722	..	554,035	1,259,722	..	195,143
	2014	1,005,206	..	574,738	1,244,602	..	191,957
Chile	2011	247,124	..	..	..	0	7,316
	2012	392,646	..	..	..	0	7,287
	2013	428,535	..	..	..	0	7,287
	2014	240,937	..	..	..	65,771	2,124
Colombia	2011	61,771	..	10,158	80,835	10,158	9,352
	2012	59,414	..	11,023	75,870	10,157	7,139
	2013	57,706	..	10,796	78,565	7,630	8,142
	2014	67,046	..	..	94,375	12,095	7,552
Ecuador	2011	10,676	..	..	19,552	..	..
	2012	10,171	..	..	21,457	..	..
	2013	9,784	..	..	20,435	..	..
	2014	9,804	..	..	22,765	..	..

Production of biofuels and wastes

Terajoules (TJ)

Non-biomass wastes		Liquid biofuels		Biogases	Total
Industrial waste	Municipal waste	Biogasoline	Biodiesel		
..	..	..	..	..	53,806
..	..	..	..	..	56,503
..	..	..	..	..	60,733
..	..	..	..	..	62,405
..	..	..	..	..	11,760
..	..	..	..	..	11,948
..	..	..	..	..	12,630
..	..	..	..	..	12,092
..	..	..	..	..	90
..	..	..	..	..	90
..	..	..	..	..	90
..	..	..	..	..	90
..	..	..	..	..	68
..	..	..	..	..	68
..	..	..	..	..	67
..	..	..	..	..	67
..	..	..	..	..	297
..	..	..	..	..	294
..	..	..	..	..	291
..	..	..	..	..	289
..	..	..	..	..	9
..	..	..	..	..	9
..	..	..	..	..	10
..	..	..	..	..	10
83,523	298,713	1,082,605	103,003	124,109	4,143,111
75,157	298,814	1,056,105	126,702	137,446	4,027,165
59,796	296,627	1,061,039	166,557	159,411	4,361,186
58,520	295,170	1,143,201	151,358	183,110	4,569,004
..	**..**	**485,085**	**115,866**	**418**	**4,213,403**
..	**..**	**498,090**	**118,588**	**1,150**	**4,433,122**
..	**..**	**588,714**	**93,848**	**1,394**	**4,581,226**
..	**..**	**606,121**	**134,404**	**3,160**	**4,597,712**
..	..	3,779	96,600	..	195,255
..	..	5,628	97,704	..	201,760
..	..	10,023	73,526	..	177,695
..	..	10,559	113,123	..	237,300
..	..	..	..	..	33,872
..	..	..	..	..	37,230
..	..	..	..	..	36,772
..	..	..	..	..	38,792
..	..	469,590	..	..	3,165,465
..	..	478,916	..	..	3,242,250
..	..	564,837	..	..	3,366,315
..	..	588,126	..	..	3,412,672
..	..	..	..	418	247,542
..	..	..	..	273	392,919
..	..	..	..	559	429,094
..	..	..	..	2,324	309,032
..	..	7,141	18,202	..	188,265
..	..	7,818	19,633	..	183,915
..	..	*7,818	19,151	..	181,665
..	..	456	19,713	..	193,685
..	..	134	..	..	30,362
..	..	161	..	..	31,789
..	..	188	..	..	30,407
..	..	*188	..	..	32,756

Table 12

Production of biofuels and wastes

Terajoules (TJ)

Country or area	Year	Solid biofuels					
		Fuelwood	Animal waste	Vegetal waste	Bagasse	Black liquor	Charcoal
Falkland Is. (Malvinas)	2011	6	..	..	..	..	3
	2012	7	..	..	..	..	3
	2013	7	..	..	..	..	3
	2014	7	..	..	..	..	3
French Guiana	2011	1,211	..	..	..	..	222
	2012	1,255	..	..	..	..	232
	2013	1,198	..	..	..	..	242
	2014	1,244	..	..	..	..	242
Guyana	2011	1,329	..	..	8,320	..	49
	2012	1,213	..	..	6,960	..	48
	2013	1,096	..	..	6,014	..	46
	2014	1,277	..	..	5,540	..	50
Paraguay	2011	99,660	..	23,362	3,617	..	13,561
	2012	96,611	..	23,412	3,605	..	12,966
	2013	90,272	..	23,627	5,236	..	13,610
	2014	92,418	..	24,319	5,035	..	13,381
Peru	2011	73,103	..	8,585	22,688	..	2,036
	2012	61,904	..	8,285	23,908	..	2,006
	2013	80,895	..	8,242	25,105	..	*1,859
	2014	81,302	..	7,701	26,479	..	1,829
Suriname	2011	963	..	*603	151	..	..
	2012	1,028	..	*784	0	..	..
	2013	1,066	..	*816	0	..	..
	2014	1,028	..	*823	0	..	..
Uruguay	2011	27,072	..	5,833	797	25,088	..
	2012	26,530	..	10,258	1,062	24,351	..
	2013	27,038	..	8,316	1,421	26,663	..
	2014	26,080	..	11,207	1,149	34,913	..
Venezuela (Bolivar. Rep.)	2011	13,420	..	18,363	12,965	..	236
	2012	12,798	..	18,629	12,965	..	236
	2013	12,205	..	18,892	13,317	..	236
	2014	11,639	..	19,153	12,965	..	207
Asia	**2011**	**10,526,092**	**133,278**	**4,810,354**	**1,385,666**	**180,164**	**392,320**
	2012	**10,668,688**	**134,640**	**4,924,073**	**1,516,603**	**184,005**	**409,071**
	2013	**10,796,862**	**135,138**	**5,011,815**	**1,597,315**	**203,345**	**422,244**
	2014	**11,059,953**	**138,352**	**5,126,548**	**1,648,156**	**211,714**	**417,339**
Afghanistan	2011	15,121	..	..	..	..	3,412
	2012	15,474	..	..	..	..	3,510
	2013	15,765	..	..	..	..	3,610
	2014	16,063	..	..	..	..	3,714
Armenia	2011	18,946	..	..	..	..	..
	2012	18,946	..	..	..	..	..
	2013	14,123	..	..	..	..	..
	2014	14,123	..	..	..	..	..
Azerbaijan	2011	3,672	..	..	..	..	..
	2012	3,700	..	..	..	..	..
	2013	3,855	..	..	..	..	148
	2014	3,289	..	..	..	..	177
Bahrain	2011	61	..	..	..	..	32
	2012	62	..	..	..	..	32
	2013	63	..	..	..	..	33
	2014	63	..	..	..	..	33
Bangladesh	2011	247,812	112,856	214,806	2,769	..	9,638
	2012	246,378	114,209	217,382	2,392	..	9,708
	2013	244,961	115,612	220,052	2,769	..	9,779
	2014	243,561	118,558	225,660	2,769	..	3,629

Production of biofuels and wastes

Terajoules (TJ)

Non-biomass wastes		Liquid biofuels		Biogases	Total
Industrial waste	Municipal waste	Biogasoline	Biodiesel		
..	..	..	..	..	6
..	..	..	..	..	7
..	..	..	..	..	7
..	..	..	..	..	7
..	..	..	..	..	1,211
..	..	..	..	..	1,255
..	..	..	..	..	1,198
..	..	..	..	..	1,244
..	..	..	..	..	9,649
..	..	..	..	..	8,173
..	..	..	..	..	7,110
..	..	..	..	..	6,817
..	..	3,107	..	..	129,747
..	..	3,265	..	..	126,894
..	..	3,853	..	..	122,987
..	..	4,279	..	..	126,051
..	..	992	478	0	105,846
..	..	1,876	589	877	97,439
..	..	1,394	37	835	116,507
..	..	1,608	110	836	118,037
..	..	..	..	..	1,717
..	..	..	..	..	1,812
..	..	..	..	..	1,881
..	..	..	..	..	1,850
..	..	343	585	..	59,717
..	..	426	662	..	63,289
..	..	603	1,133	..	65,174
..	..	906	1,457	..	75,712
..	..	..	..	..	44,748
..	..	..	..	..	44,392
..	..	..	..	..	44,414
..	..	..	..	..	43,757
322,634	**1,408,815**	**80,441**	**222,005**	**40,718**	**19,110,167**
340,192	**1,509,751**	**94,337**	**259,877**	**43,809**	**19,675,975**
385,308	**1,644,213**	**112,720**	**300,355**	**54,812**	**20,241,883**
381,303	**1,743,860**	**124,174**	**320,892**	**55,930**	**20,810,881**
..	..	..	..	..	15,121
..	..	..	..	..	15,474
..	..	..	..	..	15,765
..	..	..	..	..	16,063
..	..	..	..	..	18,946
..	..	..	..	..	18,946
..	..	..	..	..	14,123
..	..	..	..	..	14,123
..	..	..	..	..	3,672
..	..	..	..	..	3,700
..	2,104	..	..	..	5,959
..	2,628	..	..	..	5,917
..	..	..	..	..	61
..	..	..	..	..	62
..	..	..	..	..	63
..	..	..	..	..	63
..	..	..	..	..	578,244
..	..	..	..	..	580,361
..	..	..	..	..	583,394
..	..	..	..	..	590,548

Table 12

Production of biofuels and wastes

Terajoules (TJ)

Country or area	Year	Solid biofuels					
		Fuelwood	Animal waste	Vegetal waste	Bagasse	Black liquor	Charcoal
Bhutan	2011	44,734	..	..	..	..	826
	2012	45,218	..	..	..	..	826
	2013	45,712	..	..	..	..	826
	2014	46,214	..	..	..	..	826
Brunei Darussalam	2011	107	..	..	..	..	9
	2012	107	..	..	..	..	9
	2013	107	..	..	..	..	9
	2014	107	..	..	..	..	9
Cambodia	2011	158,385	..	288	..	..	25,134
	2012	162,658	..	288	..	..	25,577
	2013	167,311	..	158	..	..	26,049
	2014	171,447	..	201	..	..	26,403
China	2011	1,693,905	..	..	262,953	..	50,607
	2012	1,663,483	..	..	299,316	..	50,328
	2013	1,633,621	..	..	330,599	..	50,050
	2014	1,604,309	..	..	315,409	..	49,773
China, Macao SAR	2011	..	..	..	..	..	..
	2012	..	..	..	..	..	..
	2013	..	..	..	..	..	..
	2014	..	..	..	..	..	..
Cyprus	2011	181	..	30	..	..	30
	2012	208	..	28	..	..	30
	2013	181	..	28	..	..	30
	2014	105	..	250	..	..	30
Georgia	2011	13,188	..	..	..	..	..
	2012	12,840	..	..	..	..	..
	2013	23,231	..	..	..	..	..
	2014	22,607	..	1	..	..	..
India	2011	3,826,013	..	3,540,340	706,534	..	84,960
	2012	3,874,672	..	3,614,237	734,928	..	84,960
	2013	3,923,089	..	3,682,785	*749,044	..	92,601
	2014	4,005,976	..	3,771,228	*749,044	..	94,017
Indonesia	2011	1,783,344	..	464,165	56,094	..	19,854
	2012	1,768,546	..	469,995	64,848	..	19,913
	2013	1,795,239	..	475,710	64,269	..	17,671
	2014	1,884,621	..	481,657	65,411	..	21,624
Iran (Islamic Rep. of)	2011	45,675	1,175	7,500	..	..	24
	2012	73,994	1,175	7,500	..	..	32
	2013	73,994	1,175	7,500	..	..	35
	2014	*73,994	*1,175	8,400	..	..	30
Iraq	2011	1,710	..	..	..	..	443
	2012	1,758	..	..	..	..	443
	2013	1,807	..	..	..	..	738
	2014	1,856	..	..	..	..	738
Israel	2011	*594	..	..	..	..	..
	2012	*183	..	..	..	..	..
	2013	*183	..	..	..	..	..
	2014	*183	..	..	..	..	..
Japan	2011	39,282	..	116,418	1,537	172,134	354
	2012	39,054	..	123,311	1,614	165,523	354
	2013	42,856	..	129,642	1,591	181,750	354
	2014	44,525	..	128,546	1,637	185,782	354
Jordan	2011	2,832	..	..	..	..	1,215
	2012	2,905	..	..	..	..	1,248
	2013	2,978	..	..	..	..	1,281
	2014	3,051	..	..	..	..	1,315

Production of biofuels and wastes

Terajoules (TJ)

Non-biomass wastes		Liquid biofuels		Biogases	Total
Industrial waste	Municipal waste	Biogasoline	Biodiesel		
..	..	..	..	..	44,734
..	..	..	..	..	45,218
..	..	..	..	..	45,712
..	..	..	..	..	46,214
..	..	..	..	..	107
..	..	..	..	..	107
..	..	..	..	..	107
..	..	..	..	..	107
..	..	..	..	..	158,673
..	..	..	..	..	162,946
..	..	..	..	..	167,469
..	..	..	..	..	171,648
188,599	1,254,137	65,446	179,735	..	3,644,774
198,687	1,355,330	74,555	204,750	..	3,796,122
220,786	1,484,547	84,294	231,494	..	3,985,341
233,878	1,572,324	87,920	241,445	..	4,055,285
..	*1,541	..	..	..	*1,541
..	*1,627	..	..	..	*1,627
..	*1,842	..	..	..	*1,842
..	*2,058	..	..	..	*2,058
4	..	..	221	448	884
0	..	..	258	476	970
0	..	..	74	466	749
273	13	..	0	475	1,116
..	..	..	..	3	13,191
..	..	..	..	3	12,843
..	..	..	..	0	23,231
..	..	..	..	0	22,608
..	..	..	..	..	8,072,887
..	..	..	..	..	8,223,837
..	..	..	..	..	8,354,918
..	..	..	..	..	8,526,248
..	..	..	..	..	2,303,603
..	..	..	..	..	2,303,389
..	..	..	..	..	2,335,218
..	..	..	..	..	2,431,689
..	..	..	..	127	54,477
..	..	..	..	127	82,796
..	..	..	..	127	82,796
..	..	..	..	553	*84,122
..	..	..	..	..	1,710
..	..	..	..	..	1,758
..	..	..	..	..	1,807
..	..	..	..	..	1,856
..	..	..	..	391	*985
..	..	..	..	777	960
..	..	..	..	442	625
..	..	..	..	638	821
48,436	32,192	..	..	..	409,999
51,858	28,672	..	..	..	410,032
67,381	31,368	..	..	..	454,588
67,303	40,954	..	..	..	468,747
..	..	..	..	96	2,928
..	..	..	..	72	2,977
..	..	..	..	72	3,050
..	..	..	..	72	3,123

Production of biofuels and wastes

Terajoules (TJ)

Country or area	Year	Solid biofuels					
		Fuelwood	Animal waste	Vegetal waste	Bagasse	Black liquor	Charcoal
Kazakhstan	2011	3,243	..	47	..	..	14
	2012	2,441	..	49	..	..	14
	2013	2,762	..	..	..	..	14
	2014	920	..	..	..	..	14
Korea, Dem.Ppl's.Rep.	2011	45,202	..	..	..	..	4,661
	2012	45,444	..	..	..	..	4,691
	2013	45,686	..	..	..	..	4,750
	2014	45,928	..	..	..	..	4,779
Korea, Republic of	2011	7,050	..	10,196	..	..	..
	2012	21,178	..	718	..	9,560	..
	2013	28,710	..	1,573	..	9,598	..
	2014	74,353	..	..	..	13,494	..
Kuwait	2011	171	..	..	..	..	84
	2012	173	..	..	..	..	85
	2013	176	..	..	..	..	86
	2014	176	..	..	..	..	87
Kyrgyzstan	2011	158	..	..	..	..	..
	2012	160	..	..	..	..	..
	2013	140	..	..	..	..	..
	2014	130	..	..	..	..	..
Lao People's Dem. Rep.	2011	54,215	..	..	..	..	636
	2012	54,101	..	..	..	..	649
	2013	53,988	..	..	..	..	662
	2014	53,891	..	..	..	..	675
Lebanon	2011	5,006	..	..	..	..	325
	2012	4,997	..	..	..	..	325
	2013	4,988	..	..	..	..	325
	2014	4,988	..	..	..	..	295
Malaysia	2011	25,213	..	15,048	579	..	853
	2012	24,768	..	11,405	579	..	856
	2013	24,339	..	15,955	226	..	858
	2014	24,339	..	9,346	226	..	858
Maldives	2011	142	..	..	..	..	23
	2012	144	..	..	..	..	24
	2013	145	..	..	..	..	24
	2014	147	..	..	..	..	25
Mongolia	2011	*6,997	..	..	..	..	..
	2012	*5,481	..	..	..	..	..
	2013	*5,481	..	..	..	..	..
	2014	*5,481	..	..	..	..	..
Myanmar	2011	421,626	..	25,890	*2,386	..	797
	2012	425,015	..	26,098	*2,355	..	826
	2013	428,574	..	26,317	*2,355	..	826
	2014	432,238	..	26,542	*2,355	..	826
Nepal	2011	262,680	18,757	15,458	12,979	..	2,331
	2012	299,482	18,757	14,811	14,779	..	2,390
	2013	341,392	17,836	*14,811	16,873	..	2,419
	2014	381,166	18,087	14,811	17,082	..	2,419
Oman	2011	367	..	..	..	..	177
	2012	378	..	..	..	..	182
	2013	390	..	..	..	..	199
	2014	402	..	..	..	..	194
Other Asia	2011	12,898	..	..	..	..	..
	2012	10,274	..	..	..	..	..
	2013	8,839	..	..	..	..	..
	2014	10,383	..	..	..	..	..

Production of biofuels and wastes

Terajoules (TJ)

Non-biomass wastes		Liquid biofuels		Biogases	Total
Industrial waste	Municipal waste	Biogasoline	Biodiesel		
..	..	..	..	..	3,290
..	..	..	..	..	2,490
..	..	..	..	..	2,762
..	..	..	..	..	920
..	..	..	..	..	45,202
..	..	..	..	..	45,444
..	..	..	..	..	45,686
..	..	..	..	..	45,928
84,096	39,262	..	11,408	9,019	161,031
88,080	37,499	..	12,218	9,358	178,611
95,926	36,341	..	12,806	9,917	194,871
78,446	36,953	4,822	13,174	9,330	230,573
..	..	..	..	..	171
..	..	..	..	..	173
..	..	..	..	..	176
..	..	..	..	..	176
..	..	..	..	..	158
..	..	..	..	..	160
..	..	..	..	..	140
..	..	..	..	..	130
..	..	..	..	..	54,215
..	..	..	..	..	54,101
..	..	..	..	..	53,988
..	..	..	..	..	53,891
..	..	..	..	..	5,006
..	..	..	..	..	4,997
..	..	..	..	..	4,988
..	..	..	..	..	4,988
..	..	..	6,366	..	47,206
..	..	..	9,163	..	45,915
..	..	..	17,370	..	57,890
..	..	..	22,522	..	56,433
..	..	..	..	..	142
..	..	..	..	..	144
..	..	..	..	..	145
..	..	..	..	..	147
..	..	..	..	..	*6,997
..	..	..	..	..	*5,481
..	..	..	..	..	*5,481
..	..	..	..	..	*5,481
..	..	..	..	..	449,902
..	..	..	..	..	453,468
..	..	..	..	..	457,246
..	..	..	..	..	461,135
..	..	..	..	2,765	312,639
..	..	..	..	4,296	352,125
..	..	..	..	6,711	397,623
..	..	..	..	6,711	437,857
..	..	..	..	..	367
..	..	..	..	..	378
..	..	..	..	..	390
..	..	..	..	..	402
..	55,114	3,823	..	323	72,158
..	55,210	4,372	..	283	70,139
..	55,412	4,750	..	254	69,255
..	55,156	3,091	..	224	68,854

Table 12

Production of biofuels and wastes

Terajoules (TJ)

Country or area	Year	Solid biofuels					
		Fuelwood	Animal waste	Vegetal waste	Bagasse	Black liquor	Charcoal
Pakistan	2011	270,944	..	*160,697	108,951	..	2,187
	2012	270,944	..	*163,428	118,920	..	2,243
	2013	270,944	..	*165,000	130,959	..	2,300
	2014	269,784	..	*168,500	155,724	..	2,358
Philippines	2011	115,808	490	24,038	17,015	..	35,371
	2012	172,840	499	38,488	18,372	..	49,666
	2013	176,505	515	39,763	19,004	..	52,259
	2014	178,497	532	40,313	19,243	..	54,705
Qatar	2011	46	..	..	..	..	22
	2012	46	..	..	..	..	23
	2013	47	..	..	..	..	23
	2014	47	..	..	..	..	23
Singapore	2011	294	..	..	..	..	156
	2012	290	..	..	..	..	153
	2013	285	..	..	..	..	150
	2014	280	..	..	..	..	*148
Sri Lanka	2011	162,788	..	..	842	..	..
	2012	159,638	..	..	1,004	..	..
	2013	156,526	..	..	1,475	..	..
	2014	159,976	..	..	1,148	..	..
State of Palestine	2011	3,089	..	627	..	..	..
	2012	2,762	..	702	..	..	..
	2013	3,013	..	436	..	..	..
	2014	3,332	..	726	..	..	..
Syrian Arab Republic	2011	271	..	..	..	..	30
	2012	275	..	..	..	..	30
	2013	280	..	..	..	..	30
	2014	285	..	..	..	..	30
Tajikistan	2011	822	..	..	..	..	..
	2012	822	..	..	..	..	..
	2013	822	..	..	..	..	..
	2014	822	..	..	..	..	..
Thailand	2011	501,003	..	169,115	186,482	8,030	128,650
	2012	503,666	..	188,938	221,809	8,922	130,243
	2013	518,371	..	187,957	238,965	11,997	134,166
	2014	537,420	..	208,908	278,054	12,438	127,027
Timor-Leste	2011	1,052	..	..	..	..	..
	2012	1,009	..	..	..	..	..
	2013	1,009	..	..	..	..	..
	2014	1,009	..	..	..	..	..
Turkey	2011	102,418	..	45,692	..	..	..
	2012	98,396	..	46,696	..	..	..
	2013	93,848	..	44,128	..	..	..
	2014	90,519	..	41,459	..	..	..
Turkmenistan	2011	91	..	..	..	..	..
	2012	91	..	..	..	..	..
	2013	*91	..	..	..	..	..
	2014	*91	..	..	..	..	..
Uzbekistan	2011	210	..	..	..	..	..
	2012	238	..	..	..	..	..
	2013	238	..	..	..	..	..
	2014	238	..	..	..	..	..
Viet Nam	2011	622,311	..	..	26,545	..	17,376
	2012	628,921	..	..	35,687	..	17,553
	2013	635,552	..	..	39,186	..	17,730
	2014	642,253	..	..	40,053	..	17,936

Production of biofuels and wastes

Terajoules (TJ)

Non-biomass wastes		Liquid biofuels		Biogases	Total
Industrial waste	Municipal waste	Biogasoline	Biodiesel		
..	..	..	..	..	540,592
..	..	..	..	..	553,292
..	..	..	..	..	566,903
..	..	..	..	..	594,008
..	1,095	105	4,035	..	162,585
..	926	1,286	4,453	..	236,862
..	1,499	2,316	4,939	..	244,542
..	1,639	4,355	5,295	..	249,874
..	..	..	..	..	46
..	..	..	..	..	46
..	..	..	..	..	47
..	..	..	..	..	47
..	25,263	..	..	..	25,557
..	26,008	..	..	..	26,298
..	26,917	..	..	..	27,202
..	27,063	..	..	..	27,344
..	..	..	..	..	163,629
..	..	..	..	..	160,642
..	..	..	..	..	158,001
..	..	..	..	..	161,123
..	..	..	..	..	3,716
..	..	..	..	..	3,463
..	..	..	..	..	3,448
..	..	..	..	..	4,058
..	..	..	..	..	271
..	..	..	..	..	275
..	..	..	..	..	280
..	..	..	..	..	285
..	..	..	..	..	822
..	..	..	..	..	822
..	..	..	..	..	822
..	..	..	..	..	822
..	211	10,854	19,798	24,509	920,002
..	4,479	13,909	28,373	23,030	993,126
..	4,183	20,395	32,862	28,312	1,043,042
..	5,071	22,512	37,131	28,186	1,129,720
..	..	..	..	..	1,052
..	..	..	..	..	1,009
..	..	..	..	..	1,009
..	..	..	..	..	1,009
1,499	..	214	442	3,037	153,302
1,567	..	214	662	5,387	152,923
1,215	..	965	810	8,511	149,476
1,403	..	1,474	1,325	9,741	145,921
..	..	..	..	..	91
..	..	..	..	..	91
..	..	..	..	..	*91
..	..	..	..	..	*91
..	..	..	..	..	210
..	..	..	..	..	238
..	..	..	..	..	238
..	..	..	..	..	238
..	..	..	..	..	648,856
..	..	..	..	..	664,608
..	..	..	..	..	674,738
..	..	..	..	..	682,306

Table 12

Production of biofuels and wastes

Terajoules (TJ)

Country or area	Year	Solid biofuels					
		Fuelwood	Animal waste	Vegetal waste	Bagasse	Black liquor	Charcoal
Yemen	2011	4,413	..	..	..	..	2,095
	2012	4,531	..	..	..	..	2,154
	2013	4,649	..	..	..	..	2,213
	2014	4,768	..	..	..	..	2,272
Europe	**2011**	**2,979,537**	**7,398**	**317,062**	**13,478**	**492,743**	**8,948**
	2012	**3,286,216**	**7,183**	**306,457**	**13,218**	**478,454**	**8,596**
	2013	**3,312,759**	**9,391**	**299,798**	**13,645**	**494,971**	**9,116**
	2014	**3,264,644**	**11,090**	**288,419**	**12,591**	**496,756**	**13,445**
Albania	2011	8,706	..	..	..	..	..
	2012	8,642	..	..	..	..	..
	2013	8,433	..	..	..	..	..
	2014	8,161	..	..	..	..	..
Andorra	2011	..	..	..	..	..	..
	2012	..	..	..	..	..	..
	2013	..	..	..	..	..	..
	2014	..	..	..	..	..	..
Austria	2011	149,023	..	9,592	..	27,320	30
	2012	159,802	..	9,742	..	27,689	30
	2013	157,289	..	8,326	..	30,088	30
	2014	144,124	..	8,806	..	30,329	30
Belarus	2011	60,145	..	3,240	..	..	118
	2012	61,644	..	3,256	..	..	118
	2013	59,946	..	3,071	..	..	118
	2014	55,795	..	3,790	..	..	89
Belgium	2011	35,654	..	18,483	..	1,470	..
	2012	51,678	263	5,838	..	1,400	..
	2013	53,947	133	3,150	..	1,735	..
	2014	38,570	97	5,826	..	1,733	..
Bosnia and Herzegovina	2011	6,614	..	..	..	..	561
	2012	6,458	..	..	..	..	561
	2013	*6,458	..	..	..	..	679
	2014	72,578	..	..	..	..	2,596
Bulgaria	2011	37,269	..	234	..	5,653	472
	2012	41,388	..	772	..	4,265	118
	2013	40,906	..	850	..	5,235	148
	2014	40,501	..	0	..	4,987	177
Croatia	2011	58,983	..	34	..	..	118
	2012	60,238	..	21	..	..	148
	2013	61,312	..	12	..	..	177
	2014	57,543	..	12	..	..	236
Czechia	2011	69,603	..	5,076	..	12,395	..
	2012	72,181	..	5,208	..	12,749	..
	2013	77,009	..	7,015	..	*11,961	..
	2014	*77,291	..	6,717	..	*12,331	..
Denmark	2011	42,104	..	21,004	..	..	..
	2012	40,830	..	19,241	..	..	..
	2013	38,795	..	21,125	..	..	..
	2014	35,481	..	19,134	..	..	..
Estonia	2011	29,553	..	8,701	..	1,059	..
	2012	32,115	..	9,220	..	1,056	..
	2013	35,807	..	7,471	..	1,408	..
	2014	38,208	..	7,398	..	1,374	..
Finland	2011	183,839	292	737	..	135,109	..
	2012	194,673	271	702	..	135,768	..
	2013	196,762	256	535	..	140,701	..
	2014	194,944	197	680	..	141,905	..

Production of biofuels and wastes

Terajoules (TJ)

Non-biomass wastes		Liquid biofuels		Biogases	Total
Industrial waste	Municipal waste	Biogasoline	Biodiesel		
..	..	..	..	..	4,413
..	..	..	..	..	4,531
..	..	..	..	..	4,649
..	..	..	..	..	4,768
338,369	**728,497**	**87,063**	**356,316**	**441,381**	**5,761,845**
344,785	**750,282**	**100,331**	**376,828**	**516,080**	**6,179,833**
338,104	**783,442**	**120,342**	**402,518**	**589,584**	**6,364,555**
349,068	**804,581**	**108,188**	**463,698**	**630,928**	**6,429,964**
..	..	..	..	..	8,706
..	..	..	..	..	8,642
..	..	..	..	..	8,433
..	..	..	..	..	8,161
..	*211	..	..	..	*211
..	*211	..	..	..	*211
..	*232	..	..	..	*232
..	*241	..	..	..	*241
22,405	15,468	2,173	8,795	7,087	241,863
19,728	15,554	2,975	7,654	8,640	251,784
18,048	15,722	2,680	6,514	8,233	246,900
18,773	18,090	5,575	9,642	12,235	247,574
1,288	..	..	1,141	128	65,942
1,135	..	..	1,214	182	67,431
1,037	..	..	957	381	65,392
1,156	..	..	1,030	373	62,144
13,505	38,110	6,534	10,414	5,372	129,542
14,575	29,194	7,252	10,893	6,601	127,694
13,794	28,700	7,171	10,746	7,913	127,288
13,526	28,860	6,942	13,763	8,639	117,957
..	..	..	..	..	6,614
..	..	..	..	..	6,458
..	..	..	..	..	*6,458
..	..	..	..	..	72,578
345	414	0	589	139	44,643
276	871	0	294	5	47,871
322	644	322	1,582	93	49,954
951	272	563	2,208	435	49,917
179	..	..	294	290	59,780
341	..	..	1,435	479	62,514
366	..	..	1,214	693	63,597
425	..	..	1,288	1,096	60,364
6,961	5,574	1,447	7,728	10,460	119,244
7,073	5,840	2,734	6,366	15,698	127,849
6,721	5,786	2,787	6,698	23,910	141,887
8,170	5,755	2,787	8,059	25,457	*146,567
..	38,427	575	2,907	4,107	109,125
..	37,344	685	..	4,399	102,499
..	37,466	603	..	4,604	102,593
..	37,898	521	..	5,143	98,177
..	..	..	..	137	39,450
..	..	..	..	122	42,513
..	4,411	..	..	302	49,399
..	2,851	..	..	403	50,234
1,805	10,088	597	7,176	2,221	341,864
2,741	13,039	786	9,090	2,425	359,495
2,120	15,933	1,384	11,261	3,725	372,677
2,219	17,949	1,413	12,586	4,173	376,066

Table 12

Production of biofuels and wastes

Terajoules (TJ)

Country or area	Year	Solid biofuels					
		Fuelwood	Animal waste	Vegetal waste	Bagasse	Black liquor	Charcoal
France	2011	299,575	87	8,058	73	54,070	*1,475
	2012	346,541	97	10,980	0	33,188	*1,328
	2013	386,587	97	13,223	0	34,821	1,328
	2014	334,586	82	10,415	0	34,821	*1,180
Germany	2011	411,791	..	..	..	33,230	59
	2012	425,016	..	..	..	32,642	59
	2013	417,169	..	5,614	..	33,674	59
	2014	436,947	..	5,905	..	35,478	59
Greece	2011	28,014	..	11,330	..	..	59
	2012	32,894	..	8,985	..	..	30
	2013	28,171	..	5,985	..	..	59
	2014	29,663	..	6,633	..	..	30
Hungary	2011	46,022	..	13,525	..	299	0
	2012	44,060	106	13,525	..	290	0
	2013	45,497	230	13,796	..	82	0
	2014	45,842	87	12,474	..	330	0
Iceland	2011	..	..	..	..	..	..
	2012	..	..	..	..	..	..
	2013	..	..	..	..	..	..
	2014	..	..	..	..	..	..
Ireland	2011	6,241	1,397	..	..	..	..
	2012	6,851	623	..	..	..	..
	2013	7,046	628	..	..	..	..
	2014	7,386	1,423	..	..	..	..
Isle of Man	2011	..	..	..	..	..	..
	2012	..	..	..	..	..	..
	2013	..	..	..	..	..	..
	2014	..	..	..	..	..	..
Italy	2011	190,913	..	9,087	..	..	295
	2012	294,361	..	9,136	..	..	295
	2013	303,913	..	7,918	..	..	295
	2014	264,871	..	8,921	..	..	295
Jersey	2011	..	..	..	..	..	..
	2012	..	..	..	..	..	..
	2013	..	..	..	..	..	..
	2014	..	..	..	..	..	..
Latvia	2011	86,207	..	43	..	..	295
	2012	89,779	..	38	..	..	325
	2013	81,950	..	58	..	..	295
	2014	94,319	..	99	..	..	384
Liechtenstein	2011	*199	..	..	..	..	..
	2012	*203	..	..	..	..	..
	2013	172	..	..	..	..	..
	2014	*166	..	..	..	..	..
Lithuania	2011	40,955	..	212	..	..	30
	2012	41,291	..	242	..	..	30
	2013	43,355	..	238	..	..	30
	2014	46,749	..	457	..	..	30
Luxembourg	2011	1,533	..	275	..	..	..
	2012	1,561	..	269	..	..	..
	2013	1,752	..	265	..	..	..
	2014	2,513	..	243	..	..	..
Malta	2011	..	..	..	..	..	..
	2012	..	..	..	..	..	..
	2013	..	..	..	..	..	..
	2014	..	..	..	..	..	..

Production of biofuels and wastes

Terajoules (TJ)

Non-biomass wastes		Liquid biofuels		Biogases	Total
Industrial waste	Municipal waste	Biogasoline	Biodiesel		
..	94,960	17,902	67,786	14,878	557,389
..	104,926	17,554	81,438	16,511	611,235
2,149	98,982	21,065	80,187	18,279	655,390
3,058	97,804	20,636	86,848	18,442	606,692
61,520	201,342	23,390	113,418	216,897	1,061,588
59,464	217,346	25,109	100,611	268,856	1,129,045
49,437	245,060	21,481	107,125	287,845	1,167,405
52,968	254,310	23,878	123,354	311,259	1,244,099
1,175	..	..	3,974	3,048	47,541
625	..	..	5,042	3,710	51,256
883	..	..	5,594	3,704	44,337
868	..	..	5,741	3,640	46,545
2,187	3,484	643	5,851	2,540	74,551
1,804	3,766	6,432	5,299	2,212	77,494
1,680	3,475	7,584	5,152	3,215	80,711
2,250	3,533	7,906	4,894	3,181	80,497
..	0	..	..	48	48
..	0	..	..	73	73
..	0	..	..	71	71
..	0	..	..	101	101
..	1,039	..	994	2,411	12,082
13	3,680	..	994	2,341	14,502
4	4,455	..	920	2,018	15,071
..	4,799	..	994	2,187	16,789
..	*429	..	..	..	*429
..	*436	..	..	..	*436
..	*440	..	..	..	*440
..	*440	..	..	..	*440
12,232	70,588	3,298	21,749	46,218	354,085
13,626	67,560	3,686	10,562	49,352	448,282
12,994	69,302	4,485	16,891	76,013	491,516
12,532	71,882	3,178	21,344	82,105	464,833
..	679	..	..	..	679
..	*686	..	..	..	*686
..	*950	..	..	..	*950
..	*700	..	..	..	*700
84	69	54	2,208	920	89,585
84	51	54	3,312	2,175	95,492
84	85	80	2,429	2,695	87,381
168	171	0	2,760	3,136	100,653
..	315	..	..	..	513
..	328	..	..	..	531
..	356	..	..	..	529
..	378	..	..	..	544
..	..	482	2,944	463	45,056
..	..	563	3,938	484	46,517
155	943	616	4,306	649	50,262
258	970	268	4,416	876	53,994
576	1,225	..	..	511	4,120
603	1,203	..	..	658	4,294
641	1,185	..	..	653	4,496
640	1,146	..	..	701	5,243
..	..	..	37	59	96
..	..	..	37	93	130
..	..	..	37	59	96
..	..	..	37	65	102

Table 12

Production of biofuels and wastes

Terajoules (TJ)

Country or area	Year	Solid biofuels					
		Fuelwood	Animal waste	Vegetal waste	Bagasse	Black liquor	Charcoal
Montenegro	2011	7,637	..	..	..	..	*15
	2012	7,701	..	..	..	..	*15
	2013	7,281	..	..	..	..	*15
	2014	7,281	..	..	..	..	*20
Netherlands	2011	30,730	..	10,947	..	..	177
	2012	34,388	..	11,448	..	..	177
	2013	34,153	..	12,466	..	..	177
	2014	40,875	..	13,139	..	..	..
Norway	2011	49,408	..	..	..	..	..
	2012	48,183	..	..	..	..	..
	2013	44,138	..	..	..	..	..
	2014	35,191	..	..	..	..	..
Poland	2011	192,687	..	73,201	..	..	..
	2012	201,106	..	91,456	..	..	..
	2013	211,876	..	74,367	..	..	..
	2014	201,369	..	57,354	..	..	..
Portugal	2011	41,217	..	41,472	..	36,761	614
	2012	44,191	..	23,808	..	38,563	561
	2013	43,687	..	27,424	..	41,266	590
	2014	44,604	..	26,779	..	40,995	590
Republic of Moldova	2011	7,829	..	858	..	..	0
	2012	9,007	..	455	..	..	0
	2013	10,231	..	490	..	..	30
	2014	12,953	..	888	..	..	30
Romania	2011	145,527	..	..	..	..	472
	2012	158,895	..	..	..	..	528
	2013	153,097	..	..	..	..	628
	2014	152,640	..	..	..	..	487
Russian Federation	2011	126,651	..	..	..	..	1,505
	2012	144,541	..	..	..	..	1,564
	2013	126,851	..	..	..	..	1,510
	2014	115,921	..	..	..	..	1,552
Serbia	2011	44,354	..	..	..	..	236
	2012	44,187	..	..	..	..	266
	2013	46,823	..	..	..	..	325
	2014	46,518	..	..	..	..	354
Slovakia	2011	14,255	..	3,842	..	14,714	118
	2012	17,202	..	861	..	15,465	118
	2013	15,140	..	1,425	..	15,615	0
	2014	15,618	..	1,737	..	14,443	0
Slovenia	2011	25,422	..	328	..	..	..
	2012	24,799	..	290	..	..	..
	2013	26,033	..	269	..	..	..
	2014	23,178	..	263	..	..	..
Spain	2011	144,456	365	31,642	13,405	17,358	1,062
	2012	146,908	549	32,756	13,218	19,890	1,062
	2013	148,981	567	34,189	13,645	20,532	1,062
	2014	156,406	560	26,234	12,591	20,275	1,062
Sweden	2011	220,726	..	..	..	*153,305	..
	2012	244,757	153	..	..	155,489	..
	2013	227,662	149	..	..	157,853	..
	2014	215,728	110	..	..	157,755	..
Switzerland	2011	34,960	..	..	..	..	..
	2012	39,500	..	..	..	..	..
	2013	42,460	..	..	..	..	..
	2014	37,020	..	..	..	..	..

Production of biofuels and wastes

Terajoules (TJ)

Non-biomass wastes		Liquid biofuels		Biogases	Total
Industrial waste	Municipal waste	Biogasoline	Biodiesel		
..	..	..	..	..	7,637
..	..	..	..	..	7,701
..	..	..	..	..	7,281
..	..	..	..	..	7,281
..	67,939	110	18,069	11,968	139,762
..	63,855	0	43,314	12,165	165,170
..	60,809	11,095	50,600	12,777	181,900
..	61,576	..	63,296	13,094	191,980
115	14,328	0	3,349	1,235	68,435
295	15,062	0	2,613	1,187	67,340
112	16,907	192	1,214	1,105	63,668
103	18,586	110	0	969	54,959
12,992	6,793	3,699	13,358	5,732	308,462
13,250	5,672	4,636	22,264	7,033	345,417
14,761	5,612	5,119	23,626	7,593	342,953
16,993	6,099	3,832	27,195	8,671	321,514
2,101	8,245	137	13,469	1,883	145,285
2,424	7,197	137	11,187	2,383	129,890
1,963	8,097	301	11,003	2,763	136,505
2,680	6,845	466	11,923	3,432	137,724
180	..	0	..	0	8,867
374	..	27	..	0	9,864
130	..	274	..	36	11,161
207	..	0	..	178	14,226
1,200	0	1,447	3,938	547	152,659
1,131	4	1,769	3,717	1,143	166,659
1,695	14	1,099	5,042	822	161,768
2,738	189	322	4,048	810	160,747
170,277	..	..	..	..	296,928
177,809	..	..	..	..	322,350
177,615	..	..	..	..	304,466
179,858	..	..	..	..	295,779
288	..	..	..	..	44,642
50	..	..	..	39	44,276
38	..	..	..	184	47,045
34	..	..	..	215	46,767
750	1,416	2,814	4,600	1,918	44,309
1,175	1,407	2,734	4,048	2,594	45,486
5,165	1,099	2,787	3,864	2,300	47,395
5,072	1,720	2,841	3,717	4,025	49,173
1,228	..	..	0	1,506	28,484
1,333	..	..	37	1,597	28,056
1,453	..	..	74	1,454	29,283
1,809	..	..	..	1,290	26,540
..	16,332	9,862	24,987	11,520	269,928
..	14,710	8,120	18,253	12,177	266,581
..	16,724	9,702	26,496	20,072	290,908
..	17,098	10,291	44,602	14,791	302,848
1,583	49,786	11,281	9,568	4,993	451,242
1,283	53,698	11,809	13,763	5,303	486,255
1,580	57,233	8,366	8,979	6,070	467,892
1,360	59,853	5,671	4,379	6,422	451,278
10,080	45,410	0	221	3,421	94,092
10,260	46,060	0	258	3,745	99,823
10,500	44,220	0	184	4,087	101,451
11,830	45,050	..	184	4,354	98,438

Table 12

Production of biofuels and wastes

Terajoules (TJ)

Country or area	Year	Solid biofuels					
		Fuelwood	Animal waste	Vegetal waste	Bagasse	Black liquor	Charcoal
T.F.Yug.Rep. Macedonia	2011	7,786	..	..	..	..	..
	2012	7,445	..	..	..	..	..
	2013	6,606	..	93	..	..	..
	2014	6,502	..	108	..	..	..
Ukraine	2011	54,381	..	..	..	..	1,239
	2012	52,234	..	..	..	..	1,269
	2013	58,254	..	..	..	..	1,564
	2014	71,938	..	..	..	..	4,248
United Kingdom	2011	38,570	5,257	45,141	..	..	..
	2012	48,966	5,121	48,208	..	..	..
	2013	57,209	7,331	50,423	..	..	..
	2014	54,665	8,534	64,407	..	..	..
Oceania	**2011**	**132,943**	**..**	**8,315**	**124,304**	**22,449**	**358**
	2012	**130,481**	**..**	**8,379**	**126,500**	**21,737**	**360**
	2013	**129,346**	**..**	**8,765**	**138,335**	**20,946**	**374**
	2014	**133,256**	**..**	**8,910**	**132,164**	**21,991**	**380**
Australia	2011	31,848	..	380	118,678	22,449	..
	2012	29,677	..	417	121,740	21,737	..
	2013	30,797	..	708	132,171	20,946	..
	2014	34,467	..	828	125,219	21,991	..
Fiji	2011	338	..	..	4,629	..	71
	2012	338	..	..	3,762	..	71
	2013	338	..	..	5,297	..	71
	2014	338	..	..	6,001	..	71
French Polynesia	2011	41	..	..	..	..	15
	2012	41	..	..	..	..	15
	2013	41	..	..	..	..	15
	2014	42	..	..	..	..	15
Kiribati	2011	27	..	..	..	..	10
	2012	28	..	..	..	..	11
	2013	28	..	..	..	..	11
	2014	29	..	..	..	..	11
Micronesia (Fed. States of)	2011	23	..	..	..	..	10
	2012	23	..	..	..	..	10
	2013	23	..	..	..	..	11
	2014	23	..	..	..	..	11
New Zealand	2011	47,450	..	..	..	..	..
	2012	47,149	..	..	..	..	..
	2013	44,884	..	..	..	..	..
	2014	45,113	..	..	..	..	..
Niue	2011	1	..	..	..	..	0
	2012	1	..	..	..	..	0
	2013	1	..	..	..	..	0
	2014	1	..	..	..	..	0
Papua New Guinea	2011	50,544	..	*5,263	934	..	177
	2012	50,544	..	*5,263	934	..	177
	2013	50,544	..	*5,263	803	..	188
	2014	50,544	..	*5,263	880	..	192
Samoa	2011	639	..	*652	*63	..	14
	2012	639	..	*679	*64	..	14
	2013	639	..	*774	*64	..	14
	2014	*639	..	*796	*64	..	15
Solomon Islands	2011	1,182	..	*2,020	..	..	38
	2012	1,191	..	*2,020	..	..	39
	2013	1,200	..	*2,020	..	..	40
	2014	1,209	..	*2,023	..	..	41

Production of biofuels and wastes

Terajoules (TJ)

Non-biomass wastes		Liquid biofuels		Biogases	Total
Industrial waste	Municipal waste	Biogasoline	Biodiesel		
..	..	..	166	..	7,952
..	..	..	33	..	7,478
..	..	..	0	..	6,699
..	..	..	128	..	6,739
..	..	..	..	..	54,381
..	..	..	..	..	52,234
..	..	..	..	..	58,254
..	..	..	..	..	71,938
13,313	35,827	616	6,587	78,724	224,036
13,313	40,582	3,270	9,163	81,698	250,321
12,657	38,600	11,149	9,826	85,266	272,460
8,422	39,516	10,988	5,262	89,030	280,824
4,193	**..**	**6,325**	**3,275**	**16,410**	**318,214**
4,029	**..**	**6,112**	**2,870**	**17,757**	**317,864**
4,022	**..**	**6,675**	**3,054**	**16,879**	**328,023**
4,298	**..**	**5,978**	**3,974**	**19,268**	**329,839**
4,193	..	6,191	3,202	13,438	200,378
4,029	..	5,950	2,834	14,851	201,234
4,022	..	6,486	3,054	13,992	212,176
4,298	..	5,816	3,938	16,366	212,922
..	..	..	..	..	4,967
..	..	..	..	..	4,100
..	..	..	..	..	5,635
..	..	..	..	..	6,339
..	..	..	..	..	41
..	..	..	..	..	41
..	..	..	..	..	41
..	..	..	..	..	42
..	..	..	..	..	27
..	..	..	..	..	28
..	..	..	..	..	28
..	..	..	..	..	29
..	..	..	..	..	23
..	..	..	..	..	23
..	..	..	..	..	23
..	..	..	..	..	23
..	..	107	74	2,972	50,603
..	..	107	37	2,906	50,199
..	..	107	0	2,887	47,878
..	..	80	37	2,902	48,132
..	..	..	..	..	1
..	..	..	..	..	1
..	..	..	..	..	1
..	..	..	..	..	1
..	..	..	..	..	56,741
..	..	..	..	..	56,741
..	..	..	..	..	56,610
..	..	..	..	..	56,687
..	..	..	..	..	*1,354
..	..	..	..	..	*1,382
..	..	..	..	..	*1,477
..	..	..	..	..	*1,499
..	..	..	..	..	*3,202
..	..	..	..	..	*3,211
..	..	..	..	..	*3,220
..	..	..	..	..	*3,232

Table 12

Production of biofuels and wastes

Terajoules (TJ)

Country or area	Year	Solid biofuels					
		Fuelwood	Animal waste	Vegetal waste	Bagasse	Black liquor	Charcoal
Tonga	2011	19	..	..	..	..	7
	2012	19	..	..	..	..	7
	2013	19	..	..	..	..	7
	2014	19	..	..	..	..	7
Vanuatu	2011	831	..	..	..	..	16
	2012	831	..	..	..	..	16
	2013	831	..	..	..	..	17
	2014	831	..	..	..	..	17

Production of biofuels and wastes

Terajoules (TJ)

Non-biomass wastes		Liquid biofuels		Biogases	Total
Industrial waste	Municipal waste	Biogasoline	Biodiesel		
..	..	..	..	..	19
..	..	..	..	..	19
..	..	..	..	..	19
..	..	..	..	..	19
..	..	*27	..	..	859
..	..	*55	..	..	886
..	..	*82	..	..	913
..	..	*82	..	..	913

Table 13

Production, trade and supply of conventional crude oil

Thousand metric tons; kilograms per capita; reserves production ratio (R/P) in years

Table Notes

The **Reserves / Production** ratio refers to crude oil and natural gas liquids.

Data on reserves were taken from the Survey of Energy Resources, World Energy Council.

Please refer to the Definitions Section on pages ix to xv for the appropriate product description/classification.

Figure 36: Production of conventional crude oil by region, 1990-2014

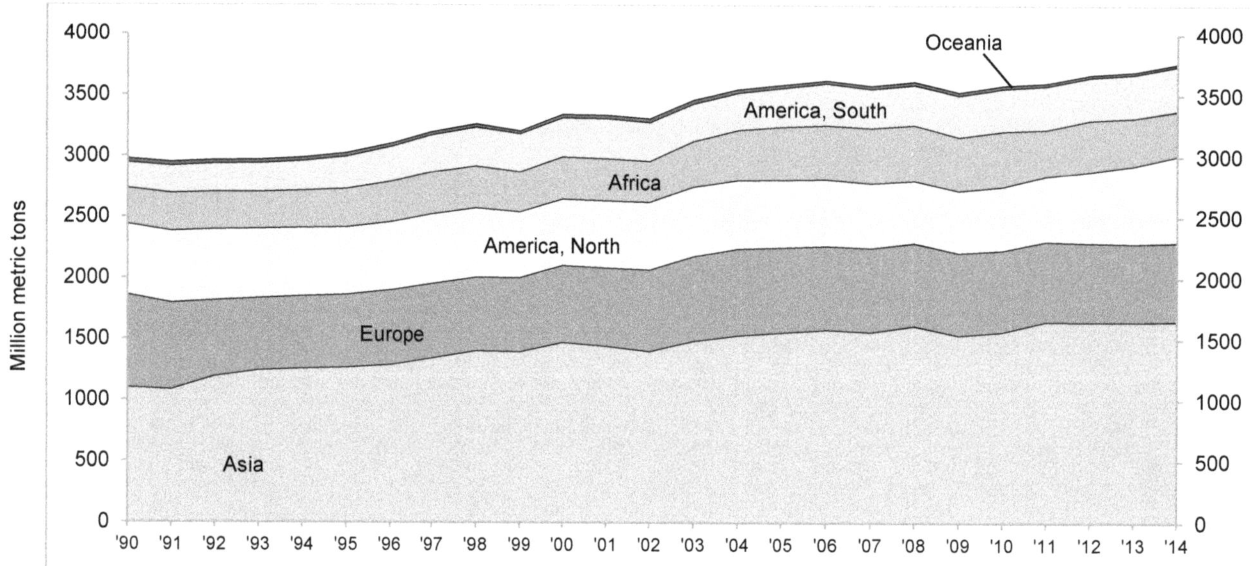

Figure 37: Exports of conventional crude oil by region, 1990-2014

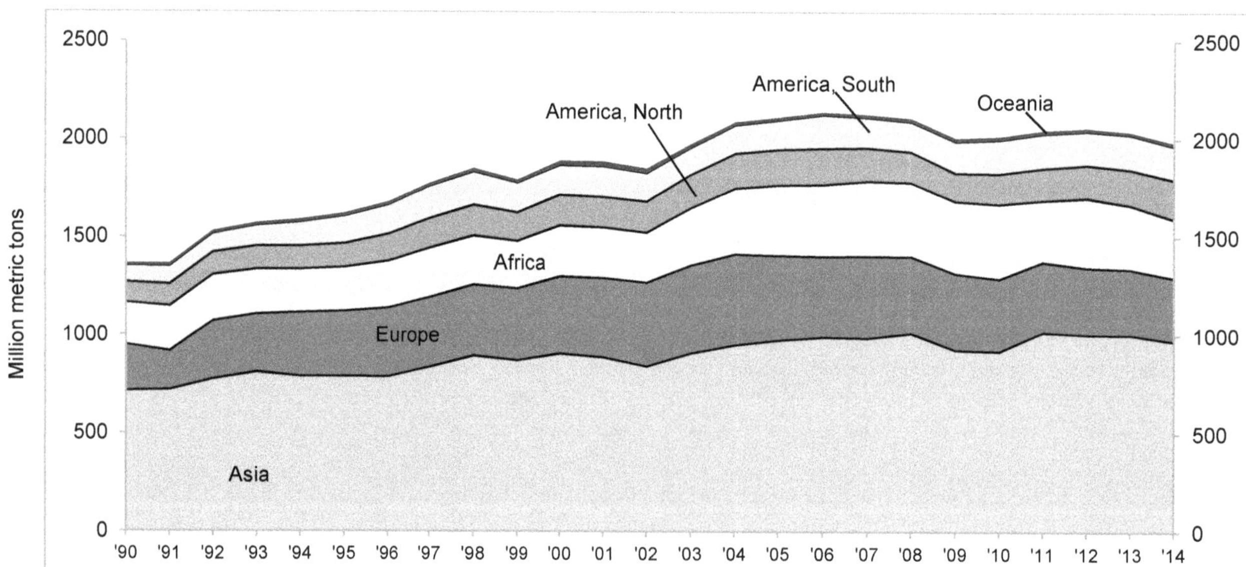

Figure 38: Production, trade and supply of conventional crude oil by region, 2014

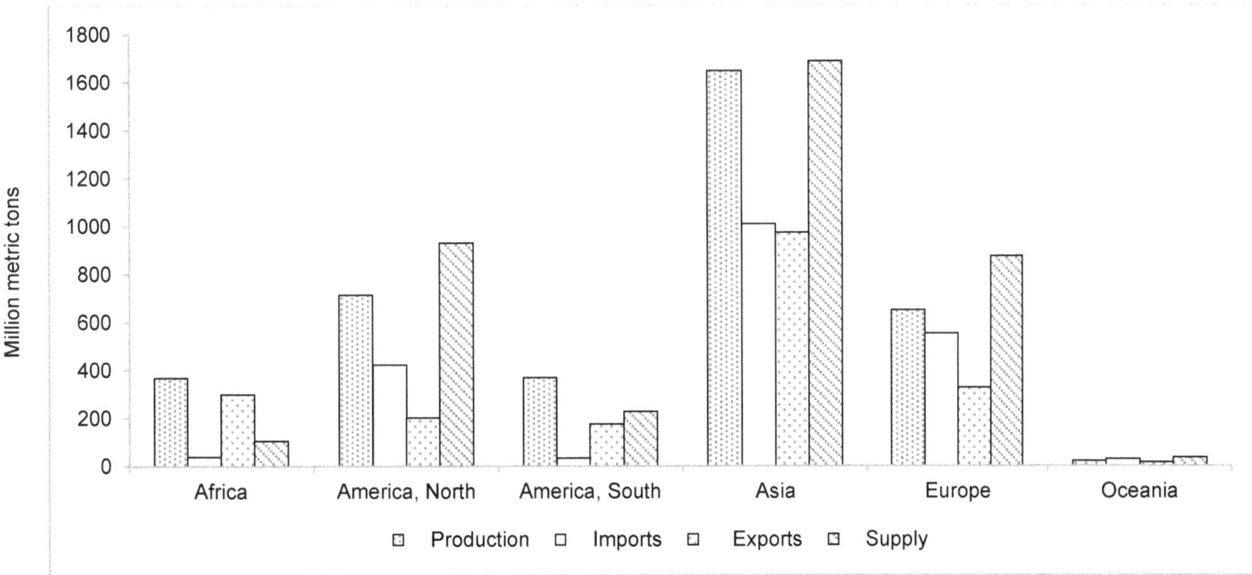

Figure 39: Major crude oil producing countries, 2014

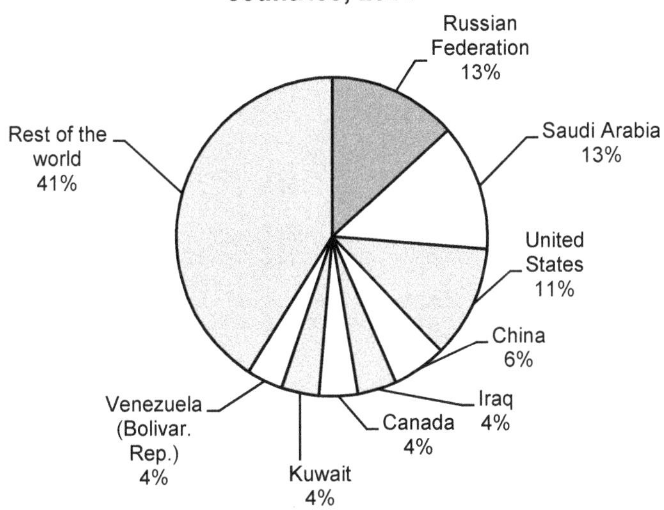

Figure 40: Major crude oil consuming countries, 2014

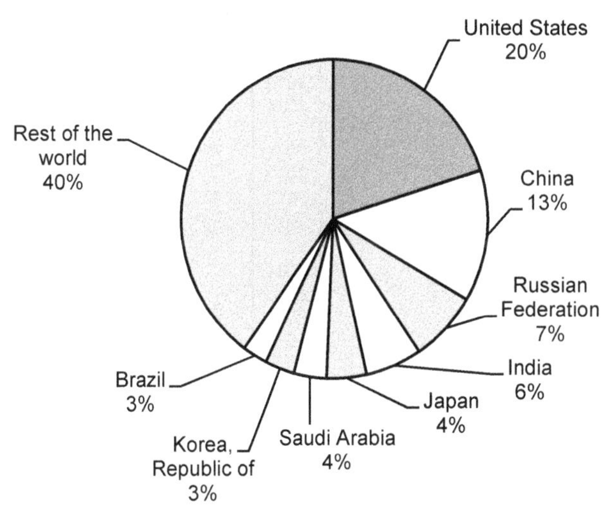

Table 13

Production, trade and supply of conventional crude oil

Thousand metric tons; kilograms per capita; reserves/production ratio (R/P) in years

Country or area	Year	Production		Imports	Exports	Changes in stocks	Supply	
		Total	Reserves/ Production				Total	Per Capita
World	**2011**	**3,602,614**	**46**	**2,148,345**	**2,045,360**	**7,217**	**3,698,382**	**527**
	2012	**3,671,192**	**45**	**2,162,451**	**2,057,204**	**1,848**	**3,774,592**	**532**
	2013	**3,693,537**	**45**	**2,098,583**	**2,037,825**	**-2,398**	**3,756,692**	**523**
	2014	**3,758,941**	**53**	**2,085,587**	**1,981,320**	**10,888**	**3,852,320**	**530**
Africa	**2011**	**383,314**	**45**	**35,579**	**313,105**	**4,598**	**101,189**	**95**
	2012	**418,694**	**40**	**39,919**	**355,673**	**-2,623**	**105,562**	**96**
	2013	**391,005**	**43**	**40,025**	**326,609**	**-4,341**	**108,762**	**96**
	2014	**365,234**	**43**	**39,005**	**298,058**	**2,657**	**103,524**	**90**
Algeria	2011	54,538	44	206	32,424	229	22,091	602
	2012	51,063	47	295	31,881	-398	19,875	531
	2013	49,574	49	146	28,130	109	21,481	563
	2014	50,805	22	294	23,550	747	26,802	688
Angola	2011	80,900	16	..	78,947	..	1,953	89
	2012	85,433	15	..	83,371	..	2,062	91
	2013	85,778	15	..	83,510	..	2,268	97
	2014	83,584	20	..	81,274	..	2,310	95
Cameroon	2011	2,997	56	1,710	2,559	0	2,148	102
	2012	3,103	54	1,794	2,650	0	2,247	104
	2013	3,367	50	1,880	2,875	0	2,372	107
	2014	3,814	5	1,956	3,257	..	2,513	110
Chad	2011	5,734	36	..	5,644	..	90	7
	2012	5,211	39	..	5,119	..	92	7
	2013	4,888	42	..	4,775	..	112	9
	2014	5,158	42	..	*5,040	..	119	9
Congo	2011	14,995	14	..	13,586	823	586	140
	2012	13,793	15	..	12,884	0	909	212
	2013	12,153	17	..	11,551	-287	889	202
	2014	12,809	14	..	11,637	327	845	188
Côte d'Ivoire	2011	1,678	8	2,313	1,623	-1	2,369	115
	2012	1,459	10	3,413	1,433	0	3,439	163
	2013	1,254	11	3,546	1,234	0	3,566	165
	2014	948	15	3,277	888	0	3,337	151
Dem. Rep. of the Congo	2011	1,205	21	..	1,205	..	0	0
	2012	1,197	21	..	1,197	..	0	0
	2013	1,129	22	..	1,129	..	0	0
	2014	1,061	23	..	1,061	..	0	0
Egypt	2011	27,634	18	2,429	4,678	..	25,385	303
	2012	32,936	17	2,919	9,530	..	26,325	307
	2013	33,007	17	2,757	11,752	..	24,012	274
	2014	29,364	15	3,047	9,945	..	22,466	251
Equatorial Guinea	2011	11,965	12	..	*11,965	..	0	0
	2012	13,072	11	..	*13,072	..	0	0
	2013	12,823	11	..	*12,823	..	0	0
	2014	13,511	10	..	*13,511	..	0	0
Gabon	2011	12,512	40	..	11,482	..	1,030	653
	2012	12,655	40	..	11,884	..	771	478
	2013	11,608	44	..	10,770	..	838	508
	2014	11,607	24	..	10,792	..	815	483
Ghana	2011	3,405	6	1,532	3,887	..	1,049	42
	2012	4,134	5	1,210	4,202	..	1,141	45
	2013	5,267	4	1,302	5,109	157	1,303	50
	2014	5,328	17	693	5,386	147	488	18

Production, trade and supply of conventional crude oil

Thousand metric tons; kilograms per capita; reserves/production ratio (R/P) in years

Country or area	Year	Production		Imports	Exports	Changes in stocks	Supply	
		Total	Reserves/ Production				Total	Per Capita
Kenya	2011	..	..	1,772	89	-24	1,707	41
	2012	..	..	997	27	1	969	23
	2013	..	..	567	18	0	549	13
	2014	..	..	0	0	0	0	0
Libya	2011	23,562	253	..	14,420	3,148	5,994	953
	2012	70,013	86	..	61,807	-406	8,612	1,371
	2013	48,100	124	..	41,689	-4,383	10,794	1,723
	2014	23,113	249	..	16,425	1,748	4,940	789
Mauritania	2011	386	36	..	131	*255	0	0
	2012	330	42	..	354	*-24	0	0
	2013	300	47	..	*300	0	0	0
	2014	306	10	..	*306	..	0	0
Morocco	2011	8	50	5,147	..	205	4,950	150
	2012	7	57	6,305	..	40	6,272	187
	2013	8	50	5,487	..	-53	5,548	163
	2014	5	50	5,189	..	-13	5,207	151
Mozambique	2011	31	..	..	31	..	0	0
	2012	35	..	..	35	..	0	0
	2013	49	..	..	49	..	0	0
	2014	59	34	..	59	..	0	0
Niger	2011	93	..	..	..	11	82	5
	2012	636	..	..	..	15	621	35
	2013	890	..	..	..	4	886	48
	2014	857	23	..	..	*26	831	43
Nigeria	2011	115,499	41	..	109,611	0	5,888	36
	2012	113,704	42	..	110,770	-1,792	4,726	28
	2013	106,732	45	..	101,606	38	5,088	29
	2014	106,472	44	..	103,178	-219	3,513	20
Senegal	2011	..	..	738	..	-66	804	60
	2012	..	..	756	..	-22	778	56
	2013	..	..	862	..	0	862	61
	2014	..	..	903	..	..	903	62
South Africa	2011	304	4	18,838	..	..	19,142	366
	2012	53	8	20,481	..	..	20,534	389
	2013	66	9	21,726	..	..	21,792	408
	2014	85	9	21,906	..	..	21,991	407
South Sudan	2011	..	..	..	..	..	..	..
	2012	1,550	..	..	1,550	..	0	0
	2013	4,954	..	..	4,938	..	16	1
	2014	7,775	61	..	7,760	..	15	1
Sudan	2011	..	..	..	..	..	..	..
	2012	5,117	..	..	1,315	..	3,802	101
	2013	6,178	..	..	2,169	..	4,009	104
	2014	6,000	34	..	1,788	..	4,212	107
Sudan (former)	2011	22,632	30	..	18,145	..	4,487	96
	2012	..	..	..	..	..	..	..
	2013	..	..	..	..	..	..	..
	2014	..	..	..	..	..	..	..
Tunisia	2011	3,236	16	317	2,678	18	857	80
	2012	3,193	16	1,106	2,592	-25	1,732	159
	2013	2,881	17	1,146	2,182	95	1,750	159
	2014	2,572	20	1,180	2,201	35	1,516	136

Table 13

Production, trade and supply of conventional crude oil

Thousand metric tons; kilograms per capita; reserves/production ratio (R/P) in years

Country or area	Year	Production		Imports	Exports	Changes in stocks	Supply	
		Total	Reserves/ Production				Total	Per Capita
Zambia	2011	..	..	577	..	0	577	40
	2012	..	..	643	..	-12	655	44
	2013	..	..	606	..	-21	627	41
	2014	..	..	560	..	-141	701	45
America, North	**2011**	**534,473**	**10**	**541,988**	**164,967**	**-5,681**	**917,175**	**1,660**
	2012	**585,403**	**9**	**487,143**	**167,542**	**3,725**	**901,278**	**1,615**
	2013	**641,246**	**8**	**442,591**	**181,680**	**-425**	**902,583**	**1,602**
	2014	**712,204**	**44**	**422,138**	**200,358**	**4,876**	**929,109**	**1,634**
Aruba	2011	*100	..	*10,450	..	..	*10,550	*103,496
	2012	*80	..	*2,500	..	..	*2,580	*25,197
	2013	0	..	0	..	..	0	0
	2014	0	..	0	..	..	0	0
Barbados	2011	40	14	..	40	0	0	0
	2012	37	15	..	38	0	0	0
	2013	35	16	..	34	1	0	0
	2014	32	8	..	29	3	0	0
Belize	2011	212	5	..	192	8	12	36
	2012	151	7	..	132	5	15	44
	2013	120	8	..	101	4	14	42
	2014	97	3	..	89	-4	12	35
Canada	2011	114,482	5	40,770	89,540	-800	66,512	1,928
	2012	124,862	5	41,783	95,715	-159	71,089	2,039
	2013	135,352	5	39,527	109,803	316	64,760	1,838
	2014	149,375	170	38,614	120,996	954	66,039	1,856
Costa Rica	2011	..	..	176	0	24	152	33
	2012	..	..	0	64	-59	-5	-1
	2013	..	..	..	0	-3	3	1
	2014	..	..	..	..	-2	2	0
Cuba	2011	3,012	6	4,917	..	..	7,929	700
	2012	2,999	6	4,976	..	..	7,975	703
	2013	2,897	7	4,778	..	..	7,675	675
	2014	2,905	1	4,625	..	..	7,531	662
Curaçao	2011	..	..	..	..	..	..	..
	2012	..	..	8,558	..	..	8,558	56,074
	2013	..	..	8,817	..	..	8,817	57,140
	2014	..	..	9,803	..	..	9,803	62,935
Dominican Republic	2011	..	..	1,320	..	0	1,321	132
	2012	..	..	1,272	..	39	1,233	121
	2013	..	..	1,332	..	-35	1,367	133
	2014	..	..	1,321	..	25	1,296	125
El Salvador	2011	..	..	732	..	-40	772	127
	2012	..	..	497	..	-85	582	96
	2013	..	..	0	11	-11	0	0
	2014	..	..	0	0	0	0	0
Guatemala	2011	596	17	..	527	-16	85	6
	2012	579	16	..	488	19	72	5
	2013	498	18	..	440	-3	60	4
	2014	501	22	..	411	17	73	5
Jamaica	2011	..	..	1,211	..	..	1,211	440
	2012	..	..	1,204	..	..	1,204	436
	2013	..	..	1,167	..	..	1,167	421
	2014	..	..	942	..	..	942	339

Production, trade and supply of conventional crude oil

Thousand metric tons; kilograms per capita; reserves/production ratio (R/P) in years

Country or area	Year	Production		Imports	Exports	Changes in stocks	Supply	
		Total	Reserves/ Production				Total	Per Capita
Martinique	2011	..	..	908	..	..	908	2,300
	2012	..	..	908	..	..	908	2,298
	2013	..	..	*908	..	..	*908	*2,295
	2014	..	..	*953	..	..	*953	*2,406
Mexico	2011	132,041	10	..	70,361	-463	62,143	516
	2012	132,183	10	..	66,181	267	65,735	538
	2013	130,488	10	..	62,912	410	67,166	543
	2014	125,472	11	..	59,823	706	64,943	518
Neth. Antilles (former)	2011	..	..	8,344	..	..	8,344	39,143
	2012	..	..	..	..	..	..	..
	2013	..	..	..	..	..	..	..
	2014	..	..	..	..	..	..	..
Nicaragua	2011	..	..	838	..	29	809	139
	2012	..	..	542	..	1	541	92
	2013	..	..	672	..	-45	717	121
	2014	..	..	709	..	-2	711	118
Trinidad and Tobago	2011	4,739	7	4,284	1,989	-42	7,076	5,301
	2012	4,225	8	2,959	1,587	80	5,517	4,112
	2013	4,184	9	4,109	1,761	-285	6,817	5,056
	2014	4,189	21	2,923	1,681	34	5,397	3,985
United States	2011	279,251	12	468,038	2,318	-4,381	749,352	2,399
	2012	320,286	11	421,945	3,337	3,619	735,275	2,336
	2013	367,673	9	381,281	6,617	-775	743,112	2,343
	2014	429,633	13	362,247	17,329	3,145	771,406	2,415
America, South	**2011**	**356,046**	**87**	**31,552**	**175,903**	**872**	**210,822**	**525**
	2012	**350,796**	**88**	**32,706**	**172,466**	**-1,118**	**212,154**	**523**
	2013	**356,779**	**86**	**35,732**	**177,464**	**333**	**214,713**	**524**
	2014	**366,616**	**80**	**33,635**	**173,796**	**317**	**226,137**	**546**
Argentina	2011	29,861	11	0	3,057	342	26,462	635
	2012	28,324	11	219	3,080	-331	25,794	613
	2013	27,730	12	373	1,932	116	26,055	613
	2014	28,664	10	509	1,979	-184	27,378	637
Bolivia (Plur. State of)	2011	2,177	22	..	0	0	2,177	216
	2012	2,534	19	..	0	-1	2,535	248
	2013	2,906	17	..	353	-3	2,556	246
	2014	3,096	6	..	453	-3	2,646	251
Brazil	2011	97,550	20	16,831	30,660	745	82,976	414
	2012	95,652	20	17,495	27,051	-532	86,628	428
	2013	93,777	21	20,006	20,142	5	93,636	458
	2014	111,816	16	17,757	26,318	-25	103,280	501
Chile	2011	266	7	8,931	..	-64	9,261	538
	2012	290	8	8,470	..	-116	8,876	510
	2013	340	8	9,259	..	193	9,406	535
	2014	317	43	9,383	..	-23	9,723	547
Colombia	2011	46,102	6	..	30,118	-62	16,046	346
	2012	47,686	5	..	31,445	-77	16,318	348
	2013	56,312	5	..	38,365	108	17,839	377
	2014	51,064	7	..	33,073	559	17,432	365
Ecuador	2011	23,919	41	..	15,809	9	8,101	534
	2012	24,268	40	..	16,858	-7	7,416	481
	2013	25,338	38	..	18,311	0	7,027	449
	2014	26,625	44	..	20,283	-9	6,351	399

Table 13

Production, trade and supply of conventional crude oil

Thousand metric tons; kilograms per capita; reserves/production ratio (R/P) in years

Country or area	Year	Production		Imports	Exports	Changes in stocks	Supply	
		Total	Reserves/ Production				Total	Per Capita
Peru	2011	3,422	12	4,629	802	71	7,178	241
	2012	3,289	12	4,466	768	-176	7,163	238
	2013	3,095	11	4,178	718	84	6,471	212
	2014	3,410	23	4,105	763	76	6,676	216
Suriname	2011	911	13	..	..	0	911	1,740
	2012	929	13	..	..	0	929	1,758
	2013	913	13	..	..	0	913	1,712
	2014	936	13	..	..	..	936	1,738
Uruguay	2011	..	..	1,161	..	-168	1,329	393
	2012	..	..	2,056	..	121	1,935	570
	2013	..	..	1,916	..	-169	2,085	612
	2014	..	..	1,881	..	-74	1,955	572
Venezuela (Bolivar. Rep.)	2011	151,838	180	..	95,457	..	56,381	1,916
	2012	147,824	185	..	93,264	..	54,560	1,828
	2013	146,368	188	..	97,643	..	48,725	1,609
	2014	140,688	182	..	90,927	..	49,761	1,621
Asia	**2011**	**1,645,225**	**63**	**930,392**	**1,017,260**	**11,614**	**1,546,743**	**367**
	2012	**1,642,917**	**63**	**987,721**	**1,004,437**	**1,153**	**1,625,049**	**381**
	2013	**1,642,042**	**64**	**993,337**	**1,002,446**	**-1,392**	**1,634,324**	**380**
	2014	**1,647,587**	**65**	**1,010,096**	**971,272**	**-989**	**1,687,401**	**388**
Azerbaijan	2011	43,445	21	..	37,268	59	6,118	663
	2012	41,015	22	..	34,728	38	6,249	668
	2013	40,726	22	..	34,260	-45	6,511	686
	2014	39,443	23	..	32,903	14	6,526	678
Bahrain	2011	9,488	16	10,828	7,475	..	12,841	9,832
	2012	8,648	18	10,951	6,649	..	12,950	9,711
	2013	9,853	16	10,735	7,120	..	13,468	9,981
	2014	10,093	1	10,385	7,693	..	12,785	9,387
Bangladesh	2011	..	13	1,211	..	-95	1,306	9
	2012	..	13	1,266	..	116	1,150	7
	2013	..	12	1,183	..	-118	1,301	8
	2014	..	16	1,308	..	155	1,153	7
Brunei Darussalam	2011	7,288	19	..	6,740	56	492	1,232
	2012	7,280	19	..	6,920	36	324	799
	2013	6,061	23	..	5,822	-35	274	666
	2014	5,680	24	..	5,362	87	231	553
China	2011	202,876	14	253,779	2,514	14,530	439,610	326
	2012	207,478	13	271,027	2,432	9,226	466,847	344
	2013	209,919	13	281,742	1,617	3,334	486,709	357
	2014	211,429	12	308,374	600	3,757	515,446	376
Georgia	2011	49	102	3	49	0	3	1
	2012	44	114	0	37	3	4	1
	2013	48	104	0	57	-9	0	0
	2014	43	117	10	51	-2	3	1
India	2011	38,090	18	167,727	..	0	205,817	165
	2012	37,862	18	184,173	..	0	222,035	176
	2013	37,788	18	189,238	..	703	226,323	177
	2014	37,461	19	189,432	..	0	226,893	175
Indonesia	2011	44,981	11	13,232	18,520	-5,036	44,729	183
	2012	42,987	11	13,110	15,757	-1,638	41,978	169
	2013	42,110	11	16,015	10,450	991	46,684	186
	2014	40,746	11	16,186	11,091	183	45,658	179

Production, trade and supply of conventional crude oil

Thousand metric tons; kilograms per capita; reserves/production ratio (R/P) in years

Country or area	Year	Production		Imports	Exports	Changes in stocks	Supply	
		Total	Reserves/ Production				Total	Per Capita
Iran (Islamic Rep. of)	2011	194,613	97	..	110,700	-1,814	85,728	1,140
	2012	142,127	129	..	56,476	-2,521	88,173	1,158
	2013	140,613	129	..	50,916	-701	90,398	1,172
	2014	139,732	132	..	53,415	-751	87,068	1,114
Iraq	2011	130,575	119	..	106,931	0	23,644	742
	2012	144,902	107	..	119,328	0	25,574	776
	2013	146,340	106	..	117,466	557	28,317	830
	2014	152,906	126	..	123,677	2,179	27,050	767
Israel	2011	20	14	10,484	..	-488	10,992	1,453
	2012	12	23	11,127	..	385	10,754	1,398
	2013	64	4	10,879	..	-308	11,251	1,439
	2014	64	31	11,396	..	..	11,460	1,443
Japan	2011	243	13	168,301	..	435	168,109	1,321
	2012	240	14	170,309	..	-559	171,108	1,346
	2013	226	16	168,028	..	652	167,602	1,320
	2014	224	8	158,705	..	-2,575	161,504	1,274
Jordan	2011	1	152	3,193	..	-89	3,283	486
	2012	1	125	3,626	..	-22	3,650	522
	2013	1	166	3,141	..	-41	3,183	441
	2014	1	357	3,225	..	46	3,179	429
Kazakhstan	2011	67,765	51	4,526	*57,621	-1,108	15,778	953
	2012	66,475	52	5,919	*57,943	87	14,364	854
	2013	69,483	50	7,494	58,078	-65	18,964	1,109
	2014	67,908	49	969	53,879	-411	15,409	887
Korea, Dem.Ppl's.Rep.	2011	..	..	526	..	..	526	21
	2012	..	..	525	..	..	525	21
	2013	..	..	580	..	..	580	23
	2014	..	..	532	..	..	532	21
Korea, Republic of	2011	39	..	125,197	..	-86	125,322	2,539
	2012	36	..	128,275	..	938	127,373	2,568
	2013	39	..	122,580	..	-1,072	123,691	2,481
	2014	27	..	125,058	..	575	124,510	2,487
Kuwait	2011	132,571	100	..	87,897	..	44,674	13,792
	2012	148,833	89	..	101,296	..	47,537	13,901
	2013	148,110	90	..	102,646	..	45,464	12,651
	2014	145,319	92	..	101,123	..	44,196	11,776
Kyrgyzstan	2011	77	65	14	2	-6	95	17
	2012	79	63	6	0	6	79	14
	2013	84	60	0	1	-2	85	15
	2014	82	61	5	..	-2	89	15
Malaysia	2011	27,370	18	8,797	11,020	1,636	23,511	823
	2012	28,133	18	9,658	11,584	-934	27,141	935
	2013	27,612	19	8,794	10,421	-509	26,494	899
	2014	28,549	16	9,450	11,432	480	26,087	872
Mongolia	2011	302	7	..	302	-1	0	0
	2012	430	5	..	422	8	0	0
	2013	607	3	..	621	-14	0	0
	2014	876	40	..	815	62	0	0
Myanmar	2011	907	8	..	0	-25	932	18
	2012	849	8	..	140	28	682	13
	2013	838	8	..	236	3	599	11
	2014	801	4	..	241	-14	575	11

Table 13

Production, trade and supply of conventional crude oil

Thousand metric tons; kilograms per capita; reserves/production ratio (R/P) in years

Country or area	Year	Production		Imports	Exports	Changes in stocks	Supply	
		Total	Reserves/ Production				Total	Per Capita
Oman	2011	43,240	17	..	36,064	..	7,176	2,236
	2012	45,067	17	..	37,507	..	7,560	2,132
	2013	46,148	16	..	40,497	..	5,651	1,446
	2014	46,228	15	..	39,221	..	7,007	1,654
Other Asia	2011	10	27	41,508	..	-730	42,248	1,816
	2012	10	27	46,215	..	-461	46,686	2,003
	2013	9	30	43,716	..	-479	44,204	1,894
	2014	8	4	45,066	..	821	44,253	1,894
Pakistan	2011	3,297	11	6,113	..	54	9,356	54
	2012	3,731	10	7,402	..	0	11,133	63
	2013	4,232	9	8,037	25	0	12,244	68
	2014	4,040	12	8,298	24	0	12,314	67
Philippines	2011	318	22	9,489	334	114	9,357	99
	2012	224	26	8,857	191	-110	8,999	94
	2013	257	27	7,676	190	51	7,692	79
	2014	420	17	8,861	387	548	8,346	84
Qatar	2011	35,692	47	..	31,863	..	3,829	2,010
	2012	35,797	46	..	31,767	..	4,030	1,999
	2013	35,226	46	..	29,133	..	6,093	2,900
	2014	34,508	36	..	30,920	..	3,588	1,652
Saudi Arabia	2011	464,088	70	..	351,154	4,069	108,865	3,782
	2012	487,969	66	..	370,327	-4,924	122,566	4,155
	2013	480,352	68	..	375,945	-1,540	105,947	3,508
	2014	484,110	68	..	353,920	-2,844	133,034	4,307
Singapore	2011	..	..	45,216	0	..	45,216	8,711
	2012	..	..	50,430	0	..	50,430	9,516
	2013	..	..	46,921	0	..	46,920	8,681
	2014	..	..	47,148	0	..	47,148	8,562
Sri Lanka	2011	..	..	1,932	..	-72	2,004	99
	2012	..	..	1,626	..	0	1,626	80
	2013	..	..	1,743	..	-188	1,931	94
	2014	..	..	1,824	..	-153	1,977	96
Syrian Arab Republic	2011	18,033	18	0	5,332	1,073	11,628	567
	2012	8,253	39	0	0	0	8,253	413
	2013	2,544	125	2,913	0	0	5,457	282
	2014	1,300	244	4,157	0	0	5,457	291
Tajikistan	2011	28	71	..	0	..	28	4
	2012	30	67	..	0	..	30	4
	2013	27	74	..	4	..	23	3
	2014	25	80	..	..	..	25	3
Thailand	2011	6,982	5	39,596	1,620	*-1,531	46,489	695
	2012	7,430	6	43,003	2,052	*-139	48,520	722
	2013	7,455	6	42,054	1,359	-4,160	52,310	776
	2014	11,663	3	41,524	334	-4,231	57,084	843
Timor-Leste	2011	4,177	..	..	*4,177	..	0	0
	2012	3,965	..	..	*3,965	..	0	0
	2013	3,958	..	..	*3,958	..	0	0
	2014	3,347	18	..	*3,347	..	0	0
Turkey	2011	2,370	18	18,092	..	-318	20,780	283
	2012	2,338	18	19,488	..	277	21,549	288
	2013	2,399	18	18,556	..	84	20,871	274
	2014	2,465	19	17,480	..	80	19,865	256

Production, trade and supply of conventional crude oil

Thousand metric tons; kilograms per capita; reserves/production ratio (R/P) in years

Country or area	Year	Production		Imports	Exports	Changes in stocks	Supply	
		Total	Reserves/ Production				Total	Per Capita
Turkmenistan	2011	10,465	7	..	2,300	..	8,165	1,599
	2012	10,903	7	..	3,100	..	7,803	1,508
	2013	11,651	7	..	3,600	..	8,051	1,536
	2014	11,800	6	..	3,200	..	8,600	1,620
United Arab Emirates	2011	127,675	90	..	122,347	..	5,328	610
	2012	132,469	87	..	126,477	..	5,992	669
	2013	139,278	84	..	134,497	..	4,781	529
	2014	139,352	82	..	124,524	..	14,828	1,632
Uzbekistan	2011	3,550	23	..	..	..	3,550	126
	2012	3,165	26	..	..	..	3,165	111
	2013	2,997	27	..	..	..	2,997	103
	2014	2,821	29	..	..	..	2,821	96
Viet Nam	2011	15,185	39	629	8,240	*987	6,587	74
	2012	16,739	35	728	9,458	*1,314	6,695	74
	2013	16,705	35	1,312	8,568	*1,517	7,932	87
	2014	17,392	34	704	9,306	*1,009	7,781	84
Yemen	2011	9,416	41	..	6,789	..	2,627	108
	2012	7,396	51	..	5,881	..	1,515	61
	2013	8,283	45	..	4,960	..	3,323	130
	2014	6,725	56	..	3,807	..	2,918	111
Europe	**2011**	**660,283**	**14**	**577,685**	**357,263**	**-3,864**	**884,569**	**1,201**
	2012	**651,051**	**15**	**585,249**	**341,554**	**951**	**893,795**	**1,213**
	2013	**642,948**	**15**	**557,177**	**335,654**	**3,469**	**861,001**	**1,167**
	2014	**648,615**	**23**	**552,665**	**324,254**	**4,134**	**872,892**	**1,183**
Albania	2011	895	34	..	598	..	297	103
	2012	1,031	29	..	920	..	111	38
	2013	1,207	25	..	1,167	..	40	14
	2014	1,368	19	..	1,007	..	361	125
Austria	2011	838	7	7,293	..	-167	8,298	985
	2012	839	8	7,473	..	-37	8,349	987
	2013	843	8	7,778	..	55	8,566	1,009
	2014	890	8	7,510	..	28	8,372	983
Belarus	2011	1,682	16	20,436	1,676	-45	20,487	2,159
	2012	1,660	16	21,669	1,645	10	21,674	2,284
	2013	1,645	16	21,261	1,619	105	21,182	2,230
	2014	1,645	16	22,508	1,617	220	22,316	2,349
Belgium	2011	..	..	29,853	..	79	29,774	2,705
	2012	..	..	31,569	..	-80	31,649	2,857
	2013	..	..	27,618	..	-9	27,627	2,477
	2014	..	..	32,188	..	67	32,121	2,861
Bosnia and Herzegovina	2011	..	..	1,161	..	0	1,161	303
	2012	..	..	1,006	..	0	1,006	263
	2013	..	..	1,017	..	59	958	251
	2014	..	..	934	..	-56	990	259
Bulgaria	2011	22	7,273	4,989	..	-72	5,083	691
	2012	24	6,667	5,885	..	15	5,894	807
	2013	28	5,714	5,641	..	46	5,623	775
	2014	26	77	5,103	..	-25	5,154	716
Croatia	2011	580	15	2,838	..	145	3,273	761
	2012	549	17	2,325	..	-184	3,058	713
	2013	536	17	2,462	..	30	2,968	695
	2014	541	12	1,851	..	3	2,389	561

Table 13

Production, trade and supply of conventional crude oil

Thousand metric tons; kilograms per capita; reserves/production ratio (R/P) in years

Country or area	Year	Production		Imports	Exports	Changes in stocks	Supply	
		Total	Reserves/ Production				Total	Per Capita
Czechia	2011	165	10	6,925	19	-27	7,098	674
	2012	156	11	7,074	21	-38	7,247	687
	2013	154	11	6,552	25	17	6,664	632
	2014	150	13	7,371	27	18	7,476	709
Denmark	2011	10,941	10	3,020	7,275	-34	6,720	1,205
	2012	9,980	11	3,717	6,059	135	7,503	1,340
	2013	8,683	13	4,670	6,165	32	7,156	1,272
	2014	8,131	9	3,458	4,799	-127	6,917	1,225
Finland	2011	..	..	11,036	..	-127	11,163	2,069
	2012	..	..	10,821	..	98	10,723	1,977
	2013	..	..	11,154	..	83	11,071	2,030
	2014	..	..	11,267	..	-82	11,349	2,071
France	2011	895	15	64,183	461	-456	65,073	1,028
	2012	807	17	56,568	183	405	56,787	893
	2013	793	17	55,484	68	594	55,615	871
	2014	766	14	53,583	20	-318	54,647	852
Germany	2011	2,627	8	90,519	375	-792	93,563	1,163
	2012	2,602	8	93,421	194	761	95,068	1,181
	2013	2,636	8	90,567	34	555	92,614	1,150
	2014	2,435	13	89,397	30	368	91,434	1,134
Greece	2011	98	10	16,514	684	-481	16,409	1,471
	2012	94	11	20,953	258	242	20,547	1,849
	2013	70	14	19,434	145	-497	19,856	1,796
	2014	64	16	20,826	59	140	20,691	1,881
Hungary	2011	659	5	5,887	0	-48	6,594	660
	2012	649	5	5,449	0	-16	6,114	614
	2013	599	6	5,377	61	-53	5,968	601
	2014	584	5	6,058	10	102	6,530	660
Ireland	2011	..	..	3,010	39	22	2,949	634
	2012	..	..	2,939	0	-129	3,068	657
	2013	..	..	2,933	0	95	2,838	608
	2014	..	..	2,701	..	-51	2,752	589
Italy	2011	5,284	14	72,216	440	-1,098	78,158	1,309
	2012	5,397	14	68,826	881	-297	73,639	1,232
	2013	5,502	14	58,357	522	652	62,685	1,048
	2014	5,765	14	53,843	358	269	58,981	986
Lithuania	2011	114	18	8,930	80	-43	9,007	2,933
	2012	102	20	8,512	79	2	8,533	2,829
	2013	86	23	8,962	67	-31	9,012	3,041
	2014	82	12	7,467	63	-11	7,497	2,570
Netherlands	2011	1,081	27	49,316	378	-244	50,263	3,012
	2012	1,111	27	50,040	506	298	50,347	3,006
	2013	1,122	26	47,184	544	55	47,707	2,838
	2014	1,525	16	47,480	614	-1,388	49,779	2,951
Norway	2011	85,616	8	1,238	72,045	458	14,351	2,897
	2012	78,453	8	1,369	65,256	102	14,464	2,882
	2013	74,380	9	1,816	60,702	56	15,438	3,037
	2014	75,691	12	1,221	64,072	-301	13,141	2,553
Poland	2011	617	27	23,792	292	-52	24,169	626
	2012	681	24	24,633	211	25	25,078	650
	2013	962	17	23,347	403	-297	24,203	627
	2014	951	16	23,713	420	112	24,132	625

Production, trade and supply of conventional crude oil

Thousand metric tons; kilograms per capita; reserves/production ratio (R/P) in years

Country or area	Year	Production		Imports	Exports	Changes in stocks	Supply	
		Total	Reserves/ Production				Total	Per Capita
Portugal	2011	..	..	10,344	..	69	10,275	973
	2012	..	..	11,058	..	4	11,054	1,051
	2013	..	..	12,209	..	297	11,912	1,139
	2014	..	..	10,543	..	-249	10,792	1,037
Republic of Moldova	2011	13	..	0	..	1	12	3
	2012	11	..	0	..	0	11	3
	2013	10	..	1	..	0	11	3
	2014	8	..	0	..	-1	9	2
Romania	2011	4,075	13	5,452	77	-141	9,591	477
	2012	3,860	14	5,128	49	115	8,824	442
	2013	4,043	13	5,299	60	-125	9,407	475
	2014	3,963	20	6,727	51	275	10,364	527
Russian Federation	2011	491,937	16	..	244,490	-503	247,950	1,731
	2012	497,425	16	..	236,716	-494	261,203	1,823
	2013	497,483	16	..	233,686	1,442	262,355	1,830
	2014	503,092	27	756	222,846	4,656	276,346	1,927
Serbia	2011	1,020	10	1,389	..	50	2,359	261
	2012	1,125	9	1,017	..	-1	2,143	239
	2013	1,160	9	1,608	..	20	2,748	307
	2014	1,108	6	1,435	..	-71	2,614	294
Slovakia	2011	15	56	6,022	15	31	5,991	1,107
	2012	11	67	5,376	12	-24	5,399	997
	2013	10	77	5,861	10	70	5,791	1,069
	2014	9	83	5,288	9	68	5,220	963
Spain	2011	100	200	52,147	..	-81	52,328	1,120
	2012	142	141	58,807	..	-296	59,245	1,270
	2013	368	54	57,871	..	85	58,154	1,252
	2014	305	62	59,054	..	330	59,029	1,276
Sweden	2011	..	..	18,799	33	91	18,675	1,974
	2012	..	..	20,673	29	-147	20,791	2,179
	2013	..	..	16,770	0	276	16,494	1,714
	2014	..	..	18,733	0	-106	18,839	1,942
Switzerland	2011	..	..	4,359	..	13	4,346	546
	2012	..	..	3,397	..	2	3,395	421
	2013	..	..	4,850	..	4	4,846	594
	2014	..	..	4,900	..	11	4,889	593
T.F.Yug.Rep. Macedonia	2011	..	..	689	..	-16	705	341
	2012	..	..	259	..	0	260	125
	2013	..	..	52	..	-8	60	29
	2014	..	..	0	7	-7	0	0
Ukraine	2011	2,438	16	5,679	0	82	8,035	177
	2012	2,290	16	1,522	0	-78	3,890	86
	2013	2,172	18	729	0	1	2,900	64
	2014	2,042	20	179	41	92	2,088	46
United Kingdom	2011	48,571	7	49,649	28,286	-478	70,412	1,115
	2012	42,052	9	53,763	28,535	558	66,722	1,050
	2013	38,456	9	50,313	30,376	-140	58,533	915
	2014	37,474	9	46,571	28,204	168	55,673	865
Oceania	**2011**	**23,273**	**19**	**31,150**	**16,862**	**-323**	**37,884**	**1,024**
	2012	**22,332**	**20**	**29,713**	**15,531**	**-239**	**36,753**	**977**
	2013	**19,517**	**22**	**29,722**	**13,973**	**-42**	**35,308**	**925**
	2014	**18,685**	**23**	**28,047**	**13,582**	**-107**	**33,257**	**858**

Table 13

Production, trade and supply of conventional crude oil

Thousand metric tons; kilograms per capita; reserves/production ratio (R/P) in years

Country or area	Year	Production		Imports	Exports	Changes in stocks	Supply	
		Total	Reserves/ Production				Total	Per Capita
Australia	2011	21,162	19	25,045	14,826	-281	31,662	1,405
	2012	19,731	21	23,340	13,021	-136	30,186	1,318
	2013	17,114	24	23,348	11,715	-50	28,797	1,237
	2014	16,111	24	21,835	11,099	-60	26,907	1,139
New Zealand	2011	2,111	9	5,135	2,036	-55	5,265	1,195
	2012	1,853	10	5,273	1,763	-147	5,510	1,242
	2013	1,592	11	5,172	1,464	69	5,231	1,171
	2014	1,770	9	4,959	1,679	24	5,026	1,118
Papua New Guinea	2011	0	..	970	0	13	957	137
	2012	748	16	1,100	747	44	1,057	148
	2013	811	15	1,202	794	-61	1,280	175
	2014	*804	30	*1,253	*804	*-71	*1,324	*177

Production of additives and other hydrocarbons

Thousand metric tons

Table Notes

Additives production refers to any quantities of additives (as per the definition in page x) which come from non-energy sources, such as Tetraethyllead (TEL) or detergents.

Additives from other sources refers to quantities of additives (as per the definition in page x) already accounted for in energy statistics, such as oxygenates from a biomass or hydrocarbon origin.

Other hydrocarbons production refers to quantities of other hydrocarbons which are oils obtained by non-conventional production techniques and are primary for the purposes of energy statistics.

Other hydrocarbons from other sources refers to the oils extracted from extra heavy oils, oil sands, coal and oil shale whose production has been already accounted for in energy statistics, together with the output of GTL plants.

Please refer to the Definitions Section on pages ix to xv for the appropriate product description/classification.

Figure 41: Major additives producing countries, 2014

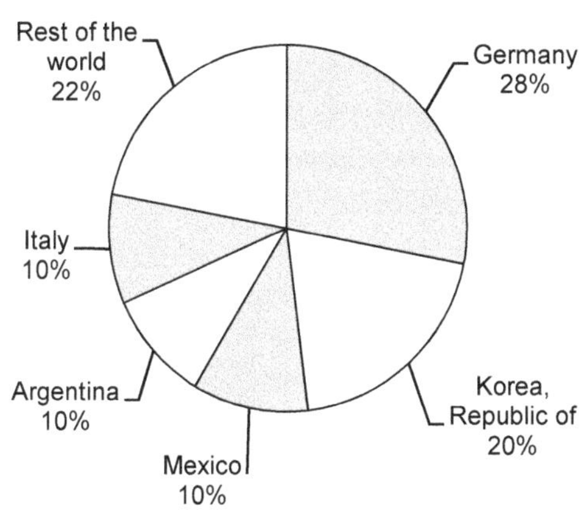

Figure 42: Major other hydrocarbons producing countries, 2014

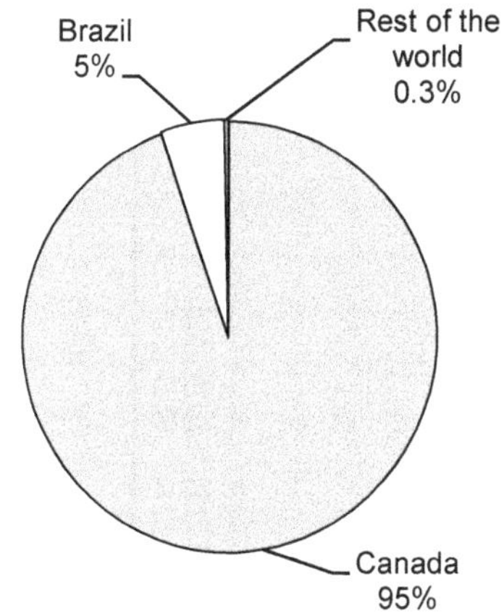

Table 14

Production of additives and other hydrocarbons

Thousand metric tons

Country or area	Year	Additives Production (kt)	Additives From other sources (kt)	Other hydrocarbons Production (kt)	Other hydrocarbons From other sources (kt)
World	**2011**	**4,022**	**3,122**	**59,840**	**11,766**
	2012	**3,884**	**2,983**	**59,249**	**18,333**
	2013	**3,689**	**2,959**	**60,384**	**20,691**
	2014	**3,778**	**2,888**	**54,773**	**22,561**
Africa	**2011**	**..**	**..**	**..**	**4,820**
	2012	**..**	**..**	**..**	**4,846**
	2013	**..**	**..**	**..**	**5,055**
	2014	**..**	**..**	**..**	**5,009**
South Africa	2011	..	..	..	4,820
	2012	..	..	..	4,846
	2013	..	..	..	5,055
	2014	..	..	..	5,009
America, North	**2011**	**371**	**2,861**	**50,191**	**2,046**
	2012	**407**	**2,788**	**49,582**	**5,945**
	2013	**364**	**2,747**	**51,108**	**6,224**
	2014	**397**	**2,667**	**51,799**	**6,473**
Canada	2011	..	..	46,892	1,989
	2012	..	..	49,582	2,359
	2013	..	..	51,108	2,237
	2014	..	..	51,799	2,296
Mexico	2011	371	..	..	57
	2012	407	..	..	63
	2013	364	..	..	163
	2014	397	..	..	251
United States	2011	..	2,861	3,299	..
	2012	..	2,788	..	3,523
	2013	..	2,747	..	3,824
	2014	..	2,667	..	3,926
America, South	**2011**	**391**	**..**	**9,467**	**..**
	2012	**431**	**..**	**9,442**	**..**
	2013	**414**	**..**	**9,102**	**..**
	2014	**404**	**..**	**2,791**	**..**
Argentina	2011	355	..	..	..
	2012	393	..	..	..
	2013	379	..	..	..
	2014	369	..	..	..
Brazil	2011	..	..	9,467	..
	2012	..	..	9,442	..
	2013	..	..	9,102	..
	2014	..	..	2,791	..
Colombia	2011	36	..	..	..
	2012	38	..	..	..
	2013	35	..	..	..
	2014	*35	..	..	..
Asia	**2011**	**667**	**1**	**176**	**3,584**
	2012	**681**	**1**	**157**	**5,914**
	2013	**565**	**1**	**149**	**7,464**
	2014	**755**	**1**	**140**	**9,100**
China	2011	..	..	..	*1,348
	2012	..	..	..	*1,474
	2013	..	..	..	*1,767
	2014	..	..	..	2,536

Production of additives and other hydrocarbons

Thousand metric tons

Country or area	Year	Additives Production (kt)	Additives From other sources (kt)	Other hydrocarbons Production (kt)	Other hydrocarbons From other sources (kt)
China, Macao SAR	2011	..	1	..	..
	2012	..	1	..	..
	2013	..	1	..	..
	2014	..	1	..	..
Korea, Republic of	2011	667	..	..	..
	2012	681	..	..	..
	2013	565	..	..	..
	2014	755	..	..	..
Malaysia	2011	..	..	..	344
	2012	..	..	..	467
	2013	..	..	..	459
	2014	..	..	..	460
Qatar	2011	..	..	..	1,892
	2012	..	..	..	3,973
	2013	..	..	..	5,060
	2014	..	..	..	5,902
Turkey	2011	..	..	..	0
	2012	..	..	..	0
	2013	..	..	..	178
	2014	..	..	..	202
Uzbekistan	2011	..	..	176	..
	2012	..	..	157	..
	2013	..	..	149	..
	2014	..	..	140	..
Europe	**2011**	**2,593**	**260**	**6**	**1,312**
	2012	**2,365**	**194**	**68**	**1,613**
	2013	**2,346**	**211**	**25**	**1,855**
	2014	**2,222**	**220**	**43**	**1,885**
Bulgaria	2011	..	..	..	29
	2012	..	..	..	27
	2013	..	..	..	28
	2014	..	..	..	28
Croatia	2011	..	..	..	53
	2012	..	..	..	145
	2013	..	..	..	58
	2014	..	..	..	61
Czechia	2011	177	7	..	..
	2012	165	6	..	..
	2013	106	2	..	..
	2014	115	3	..	..
Estonia	2011	..	..	..	561
	2012	..	..	..	599
	2013	..	..	..	624
	2014	..	..	..	697
Finland	2011	60	..	..	192
	2012	70	..	..	196
	2013	70	..	..	235
	2014	65	..	..	179
France	2011	201	..	..	..
	2012	188	..	..	..
	2013	267	..	..	..
	2014	282	..	..	..

Table 14

Production of additives and other hydrocarbons

Thousand metric tons

Country or area	Year	Additives Production (kt)	Additives From other sources (kt)	Other hydrocarbons Production (kt)	Other hydrocarbons From other sources (kt)
Germany	2011	1,286	..	..	..
	2012	1,193	..	..	..
	2013	1,137	..	..	..
	2014	1,061	..	..	..
Hungary	2011	..	10	6	..
	2012	..	0	68	..
	2013	..	4	0	..
	2014	..	0	4	..
Ireland	2011	..	..	..	45
	2012	..	..	..	53
	2013	..	..	..	50
	2014	..	..	..	49
Italy	2011	496	45	..	..
	2012	389	..	..	..
	2013	411	..	..	..
	2014	365	..	..	..
Netherlands	2011	130	1	..	..
	2012	175	..	..	..
	2013	165	..	..	..
	2014	138	..	..	..
Norway	2011	142	..	..	..
	2012	113	..	..	..
	2013	90	..	..	..
	2014	59	..	..	..
Poland	2011	57	..	..	230
	2012	10	..	..	401
	2013	1	..	..	538
	2014	..	..	..	542
Portugal	2011	..	..	..	..
	2012	..	..	..	16
	2013	..	..	..	76
	2014	..	..	..	72
Romania	2011	..	..	..	49
	2012	..	21	..	49
	2013	..	..	..	65
	2014	10	..	..	79
Serbia	2011	13	..	..	..
	2012	5	..	..	..
	2013	15	..	..	28
	2014	8	..	..	27
Slovakia	2011	..	197	..	153
	2012	..	167	..	127
	2013	..	205	..	153
	2014	..	217	..	151
Ukraine	2011	31	..	..	..
	2012	57	..	..	..
	2013	84	..	..	..
	2014	67	..	..	..
United Kingdom	2011	..	..	..	..
	2012	..	..	..	..
	2013	..	..	25	..
	2014	52	..	39	..

Production of additives and other hydrocarbons

Thousand metric tons

Country or area	Year	Additives Production (kt)	Additives From other sources (kt)	Other hydrocarbons Production (kt)	Other hydrocarbons From other sources (kt)
Oceania	**2011**	..	..	..	**4**
	2012	..	..	..	**15**
	2013	..	..	..	**93**
	2014	..	..	..	**94**
Australia	2011	..	..	..	4
	2012	..	..	..	15
	2013	..	..	..	93
	2014	..	..	..	94

Table 15

International trade of conventional crude oil (Principal importers/exporters)

Thousand metric tons

2013

Importers	Exporters							
	World	Angola	Azerbaijan	Canada	Colombia	Iran (Islamic Rep. of)	Iraq	Kazakhstan
Australia	23,348	..	95	..	..	..	..	..
Bahrain	10,735	..	..	..	..	..	..	..
Belarus	21,261	..	..	..	..	..	..	..
Belgium	27,618	402	81	..	133	..	302	..
Brazil	20,006	762	..	..	71	..	959	..
Canada	39,527	1,929	589	..	141	..	4,072	3,058
Chile	9,259	..	..	908	721	..	..	..
China	281,742	40,010	190	155	3,941	21,412	23,513	11,981
Finland	11,154	..	..	..	..	..	..	101
France	55,484	1,583	2,832	302	207	..	1,248	7,220
Germany	90,567	796	3,692	93	961	..	799	7,128
Greece	19,434	..	228	..	..	..	6,202	1,615
India	189,238	8,527	1,168	49	5,895	10,663	26,670	439
Indonesia	16,015	255	2,153	..	..	..	..	..
Italy	58,357	901	12,996	375	274	..	5,094	4,179
Japan	168,028	170	..	..	..	6,276	2,902	142
Korea, Republic of	122,580	28	..	..	45	6,279	12,044	..
Netherlands	47,184	1,420	..	..	450	..	1,219	758
Philippines	7,676	..	..	..	..	..	..	..
Poland	23,347	..	..	..	..	..	..	5
Portugal	12,209	4,352	962	..	..	..	407	859
Singapore	46,921	171	..	38	193	..	3,596	33
South Africa	21,726	2,444	..	..	..	..	..	..
Spain	57,871	3,595	174	135	3,091	..	2,008	1,072
Sweden	16,770	..	..	..	..	..	142	..
Thailand	42,054	224	2,227	..	..	..	..	..
Turkey	18,556	..	121	..	..	5,256	5,999	1,544
United Kingdom	50,313	1,481	..	1,733	..	..	..	80
United States	381,281	9,925	1,407	*101,756	18,103	..	16,811	..

International trade of conventional crude oil
(Principal importers/exporters)

Thousand metric tons

2013

Exporters									
Kuwait	Mexico	Nigeria	Norway	Qatar	Russian Federation	Saudi Arabia	United Arab Emirates	United Kingdom	Venezuela (Bolivar. Rep.)
..	..	2,545	42	..	779	..	3,482	..	..
..	..	..	..	..	..	10,147	..	..	..
..	..	..	..	..	21,261	..	..	..	..
275	63	2,889	2,626	..	12,443	5,451	..	1,452	437
..	..	10,359	..	..	..	3,668	..	..	125
145	1,962	1,351	3,706	..	195	3,465	..	1,370	415
..	..	..	344	..	..	..	..	622	87
9,343	1,029	1,052	..	131	24,348	53,903	10,276	..	15,552
..	..	..	415	..	9,688	..	..	235	..
219	..	4,852	5,009	..	6,414	10,157	141	1,877	..
563	198	7,306	10,953	..	31,480	2,433	31	9,445	325
..	..	..	..	..	7,035	1,618	..	..	..
20,468	5,214	15,263	..	*3,312	113	39,231	14,823	193	22,516
80	..	3,684	..	..	233	4,983	30	..	..
79	..	2,998	625	..	10,391	6,756	30	330	..
14,415	75	268	113	*9,084	12,867	56,695	40,833	..	666
18,782	..	..	298	*8,049	5,441	38,304	14,993	2,794	..
3,453	104	4,364	6,448	..	14,160	5,152	..	5,460	537
..	..	..	..	1,154	1,480	3,361	1,141	..	..
..	..	..	757	..	*21,157	..	..	246	..
..	..	772	..	..	783	1,052	..	..	..
3,484	..	..	..	*5,452	1,481	8,516	13,438	..	..
..	..	4,336	..	..	140	9,723	307	2,477	..
..	8,941	7,611	870	..	8,127	8,140	..	399	2,371
..	..	1,896	4,707	..	6,575	..	..	262	1,013
432	..	221	66	*1,588	2,283	7,740	18,532	76	..
..	..	479	..	..	1,462	2,755	..	..	..
..	..	6,034	20,371	..	3,751	1,903	..	..	541
16,057	41,946	11,811	823	..	2,079	65,347	121	1,040	37,224

Table 15

International trade of conventional crude oil
(Principal importers/exporters)

Thousand metric tons

2014

Importers	Exporters							
	World	Angola	Azerbaijan	Canada	Colombia	Iran (Islamic Rep. of)	Iraq	Kazakhstan
Australia	21,835	35	292	..	..	..	..	..
Bahrain	10,385	..	..	..	..	..	..	..
Belarus	22,508	..	..	..	..	..	..	..
Belgium	32,188	792	83	53	326	..	371	..
Brazil	17,757	1,302	0	..	424	..	1,551	..
Canada	38,614	1,569	558	..	286	..	2,373	293
Chile	9,383	538	..	423	140	..	..	..
China	308,374	40,648	222	155	*8,073	*24,167	28,578	5,686
Finland	11,267	..	..	..	..	..	..	244
France	53,583	3,166	2,595	95	124	..	1,209	7,119
Germany	89,397	251	4,132	0	1,275	..	919	6,777
Greece	20,826	..	336	..	79	..	7,857	2,879
India	189,432	7,244	499	49	*4,214	*11,922	22,794	943
Indonesia	16,186	255	3,033	..	..	..	..	..
Italy	53,843	1,832	9,158	1,994	1,242	446	6,371	4,464
Japan	158,705	380	..	..	767	5,515	2,185	433
Korea, Republic of	125,058	..	..	52	684	6,095	9,919	1,043
Netherlands	47,480	1,703	..	..	379	..	2,781	76
Philippines	8,861	..	..	..	..	..	..	..
Poland	23,713	..	..	..	..	..	256	567
Portugal	10,543	2,788	885	..	..	..	274	1,088
Singapore	47,148	132	..	38	203	..	3,196	..
South Africa	21,906	2,614	..	..	274	..	585	..
Spain	59,054	5,274	1,235	246	*3,146	..	1,867	161
Sweden	18,733	75	..	..	..	..	75	..
Thailand	41,524	413	1,525	..	64	..	36	..
Turkey	17,480	..	..	..	555	5,194	5,483	1,526
United Kingdom	46,571	605	..	1,061	..	..	..	..
United States	362,247	6,876	1,135	*115,146	*10,458	..	18,195	..

International trade of conventional crude oil
(Principal importers/exporters)
Thousand metric tons

2014

Exporters									
Kuwait	Mexico	Nigeria	Norway	Qatar	Russian Federation	Saudi Arabia	United Arab Emirates	United Kingdom	Venezuela (Bolivar. Rep.)
..	..	2,104	0	..	1,363	..	2,967	..	..
..	..	..	..	..	..	10,147	..	..	..
..	..	..	..	..	22,508	..	..	..	..
229	584	4,348	2,727	..	11,653	7,607	..	1,777	653
145	..	*9,786	..	..	..	3,461	..	..	0
0	1,352	505	1,653	..	0	3,650	..	286	0
..	..	..	0	..	..	..	..	291	0
10,619	1,029	1,996	..	361	33,108	49,667	11,652	..	13,786
..	..	..	1,428	..	8,584	..	..	237	..
138	244	6,120	4,291	..	5,179	11,079	339	1,350	..
234	432	7,119	15,183	..	30,028	1,414	0	9,727	8
..	..	..	..	..	5,146	2,108	..	..	..
18,328	5,214	18,343	..	*3,383	202	35,431	15,012	193	21,599
..	..	3,856	..	..	81	5,378	..	..	..
211	515	1,382	901	..	8,582	5,879	..	165	..
12,473	481	..	..	*7,926	13,950	55,320	41,630	..	530
18,432	..	113	..	*9,438	4,577	39,056	14,702	3,341	..
3,505	395	5,876	8,353	..	11,082	5,018	..	4,436	929
..	..	..	..	750	1,165	4,275	895	..	..
..	..	..	518	..	22,140	..	..	158	..
..	..	1,013	82	..	262	1,321	..	..	..
2,906	..	..	..	*6,275	1,278	9,350	12,549	..	..
153	..	6,658	..	202	..	*7,714	924	2,477	..
..	8,559	9,900	1,176	..	7,074	7,242	..	1,357	2,917
..	..	2,250	4,757	..	8,532	..	..	241	1,183
179	..	1,019	..	2,244	3,092	7,276	13,350	..	..
..	..	1,721	..	..	608	2,014	..	..	..
..	..	5,587	21,411	..	1,386	1,689	..	..	493
15,259	38,509	2,871	427	..	902	57,171	658	470	36,157

Table 16

Refinery distillation capacity, throughput and output

Thousand metric tons

Table Notes

Output is equal to the sum of "production from refineries" for aviation gasoline, bitumen, gas/diesel oil, ethane, feedstocks, gasoline-type jet fuel, kerosene-type jet fuel, other kerosene, liquefied petroleum gas, lubricants, motor gasoline, naphtha, petroleum coke, other oil products, paraffin waxes, fuel oil, refinery gas and white spirits.

Throughput is equal to the sum of "transformation in refineries" for conventional crude oil, feedstocks, other hydrocarbons, additives and oxygenates, and natural gas liquids.

Please refer to the Definitions Section on pages ix to xv for the appropriate product description/ classification.

Figure 43: Refinery distillation capacity, throughput and output by region, 2014

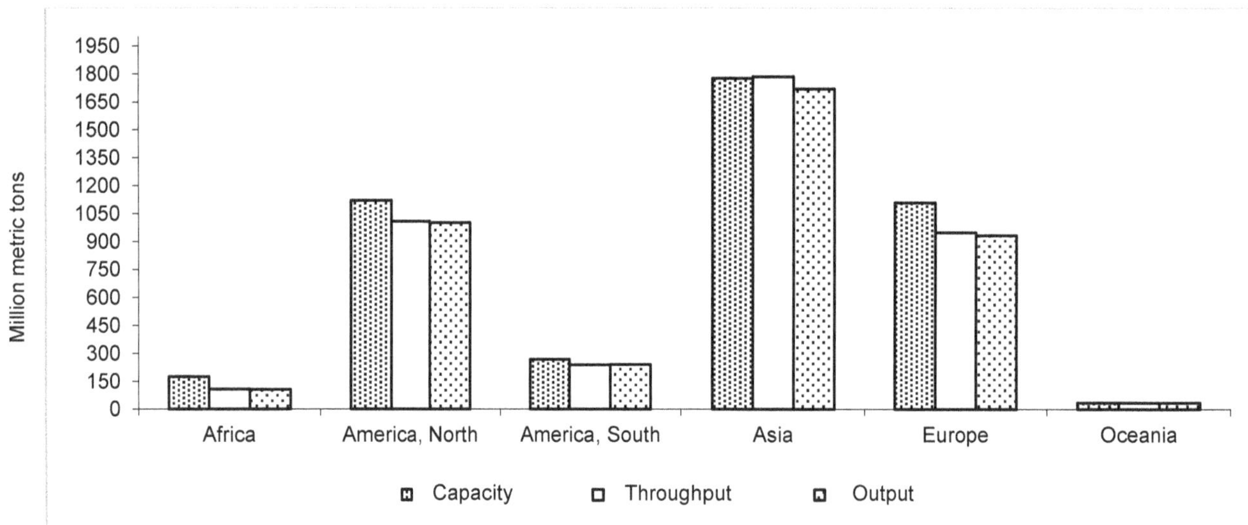

Figure 44: Total refinery distillation capacity by region, 1990-2014

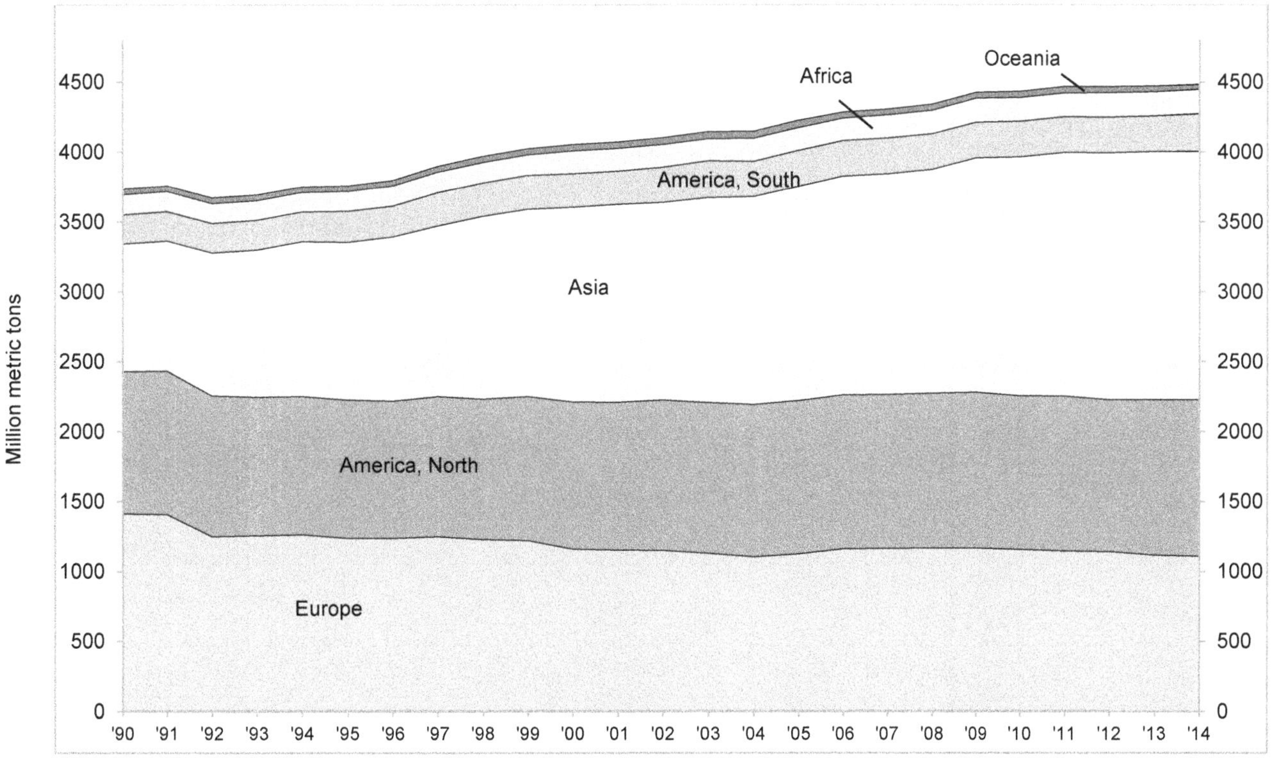

Table 16

Refinery distillation capacity, throughput and output

Thousand metric tons

Country or area	Year	Capacity	Throughput	Output
World	**2011**	**4,464,885**	**3,988,298**	**3,934,121**
	2012	**4,466,072**	**4,045,571**	**3,955,474**
	2013	**4,469,428**	**4,054,240**	**3,975,553**
	2014	**4,481,229**	**4,129,064**	**4,032,370**
Africa	**2011**	**170,301**	**112,297**	**107,498**
	2012	**173,228**	**111,072**	**108,356**
	2013	**172,502**	**113,527**	**108,825**
	2014	**174,199**	**108,683**	**105,591**
Algeria	2011	29,526	25,995	25,320
	2012	32,459	23,710	23,007
	2013	32,459	24,326	23,975
	2014	32,459	31,108	31,037
Angola	2011	2,750	1,953	1,906
	2012	2,750	2,062	1,963
	2013	2,750	2,268	2,187
	2014	2,750	2,310	2,235
Cameroon	2011	*2,244	2,148	2,142
	2012	*2,244	2,247	2,240
	2013	*2,244	2,372	2,365
	2014	*2,993	2,513	2,506
Congo	2011	1,047	603	573
	2012	1,047	870	711
	2013	1,047	915	775
	2014	1,047	870	791
Côte d'Ivoire	2011	3,192	2,425	2,249
	2012	3,192	3,448	3,224
	2013	3,192	3,336	3,288
	2014	3,192	3,202	3,112
Egypt	2011	36,228	30,387	28,502
	2012	36,222	27,230	27,790
	2013	36,222	25,598	24,010
	2014	36,222	22,466	22,778
Gabon	2011	1,198	1,030	1,028
	2012	1,198	771	770
	2013	1,198	838	837
	2014	1,198	815	814
Ghana	2011	2,244	1,091	1,000
	2012	2,244	524	474
	2013	2,244	438	424
	2014	2,244	70	129
Kenya	2011	4,488	1,707	1,689
	2012	4,488	969	968
	2013	4,488	627	597
	2014	4,488	0	0
Liberia	2011	749	..	..
	2012	749	..	..
	2013	..	..	..
	2014	..	..	..
Libya	2011	18,852	5,994	5,820
	2012	18,852	8,612	8,362
	2013	18,852	10,794	10,479
	2014	18,853	4,940	4,797

Refinery distillation capacity, throughput and output

Thousand metric tons

Country or area	Year	Capacity	Throughput	Output
Mauritania	2011	948	..	..
	2012	948	..	..
	2013	948	..	..
	2014	998	..	..
Morocco	2011	7,725	7,072	6,979
	2012	7,725	7,488	7,049
	2013	7,725	7,238	7,080
	2014	7,725	6,729	6,581
Niger	2011	*1,200	80	57
	2012	*1,200	611	560
	2013	*1,200	886	784
	2014	*1,200	809	744
Nigeria	2011	22,195	5,887	5,868
	2012	22,195	4,615	4,673
	2013	22,195	4,794	4,767
	2014	22,195	3,246	3,253
Senegal	2011	1,248	803	714
	2012	1,248	778	768
	2013	1,248	862	764
	2014	1,248	903	802
Sierra Leone	2011	498	..	..
	2012	498	..	..
	2013	498	..	..
	2014	498	..	..
South Africa	2011	24,167	19,338	18,070
	2012	24,167	20,731	19,547
	2013	24,190	21,939	20,531
	2014	25,087	22,138	19,709
Sudan	2011	..	..	..
	2012	6,074	4,008	3,905
	2013	6,074	3,978	3,734
	2014	6,074	4,212	4,003
Sudan (former)	2011	6,074	4,487	4,326
	2012	..	..	..
	2013	..	..	..
	2014	..	..	..
Tunisia	2011	1,800	719	712
	2012	1,800	1,743	1,728
	2013	1,800	1,692	1,640
	2014	1,800	1,651	1,641
United Rep. of Tanzania	2011	743	..	..
	2012	743	..	..
	2013	743	..	..
	2014	743	..	..
Zambia	2011	1,185	577	543
	2012	1,185	655	616
	2013	1,185	627	589
	2014	1,185	701	659
America, North	**2011**	**1,107,347**	**993,334**	**1,016,030**
	2012	**1,086,663**	**978,558**	**975,061**
	2013	**1,113,447**	**990,502**	**987,625**
	2014	**1,118,721**	**1,008,569**	**999,773**

Table 16

Refinery distillation capacity, throughput and output

Thousand metric tons

Country or area	Year	Capacity	Throughput	Output
Aruba	2011	11,463	*10,550	*10,050
	2012	11,463	*2,580	*2,440
	2013	*11,463	0	0
	2014	*11,463	0	0
Canada	2011	95,683	89,907	91,300
	2012	95,249	93,313	92,633
	2013	97,594	91,227	91,037
	2014	100,125	91,500	89,357
Costa Rica	2011	1,247	151	150
	2012	1,197	0	0
	2013	1,197	0	0
	2014	1,197	..	..
Cuba	2011	5,985	*4,962	4,871
	2012	6,135	5,145	5,107
	2013	6,135	5,056	5,138
	2014	6,135	5,001	4,877
Curaçao	2011	..	..	..
	2012	15,971	9,105	8,619
	2013	15,971	9,314	9,044
	2014	15,971	9,952	9,574
Dominican Republic	2011	1,721	1,321	1,243
	2012	1,721	1,234	1,163
	2013	1,721	1,337	1,295
	2014	1,721	1,351	1,278
El Salvador	2011	1,097	772	743
	2012	1,097	582	575
	2013	1,097	0	0
	2014	1,097	0	0
Guatemala	2011	748	86	81
	2012	748	72	71
	2013	748	60	59
	2014	748	73	72
Haiti	2011	..	..	..
	2012	..	..	8
	2013	..	..	..
	2014	..	..	..
Jamaica	2011	1,795	1,173	1,259
	2012	1,795	1,145	1,218
	2013	1,795	1,096	1,171
	2014	1,795	903	963
Martinique	2011	864	*908	*845
	2012	864	*908	*858
	2013	864	*908	*867
	2014	864	*953	*904
Mexico	2011	76,808	66,050	65,508
	2012	76,808	66,620	65,987
	2013	76,808	70,403	69,691
	2014	76,808	69,604	66,719
Neth. Antilles (former)	2011	15,971	8,818	8,718
	2012	..	..	..
	2013	..	..	..
	2014	..	..	..

Refinery distillation capacity, throughput and output

Thousand metric tons

Country or area	Year	Capacity	Throughput	Output
Nicaragua	2011	997	803	780
	2012	997	538	518
	2013	997	717	682
	2014	997	713	680
Trinidad and Tobago	2011	8,379	7,076	6,860
	2012	8,379	5,517	5,415
	2013	8,379	6,817	6,613
	2014	8,379	5,397	5,275
United States	2011	884,589	800,757	823,622
	2012	864,239	791,799	790,449
	2013	888,678	803,566	802,029
	2014	891,421	823,122	820,074
America, South	**2011**	**253,917**	**231,753**	**227,808**
	2012	**254,531**	**234,422**	**233,893**
	2013	**254,717**	**232,484**	**230,965**
	2014	**267,901**	**239,068**	**239,249**
Argentina	2011	31,472	29,993	29,192
	2012	31,472	31,223	31,128
	2013	31,472	30,377	28,044
	2014	30,483	31,405	31,297
Bolivia (Plur. State of)	2011	3,476	2,180	2,115
	2012	3,708	2,527	2,451
	2013	3,584	2,510	2,435
	2014	*3,584	2,718	2,636
Brazil	2011	106,286	94,555	96,038
	2012	106,812	98,498	101,717
	2013	104,993	104,992	108,480
	2014	119,300	107,150	111,078
Chile	2011	11,312	10,583	8,892
	2012	11,312	10,398	9,022
	2013	11,312	10,373	9,378
	2014	11,628	11,341	9,618
Colombia	2011	15,472	15,473	14,898
	2012	15,328	15,328	15,574
	2013	17,456	14,533	14,910
	2014	*17,456	16,937	16,596
Ecuador	2011	8,778	*8,402	7,516
	2012	8,778	*8,264	7,352
	2013	8,778	*7,622	6,709
	2014	8,778	*6,854	6,556
Paraguay	2011	370	..	..
	2012	370	..	..
	2013	370	..	..
	2014	370	..	..
Peru	2011	9,923	9,424	9,480
	2012	9,923	8,947	8,818
	2013	9,923	8,796	8,945
	2014	9,923	9,056	9,065
Suriname	2011	390	908	911
	2012	390	926	931
	2013	390	910	1,000
	2014	390	932	1,035

Table 16

Refinery distillation capacity, throughput and output

Thousand metric tons

Country or area	Year	Capacity	Throughput	Output
Uruguay	2011	2,493	1,329	1,297
	2012	2,493	1,934	1,951
	2013	2,494	2,083	2,021
	2014	2,494	1,947	1,876
Venezuela (Bolivar. Rep.)	2011	63,945	58,906	57,469
	2012	63,945	56,378	54,949
	2013	63,945	50,288	49,043
	2014	63,495	50,728	49,491
Asia	**2011**	**1,742,176**	**1,650,905**	**1,602,648**
	2012	**1,766,315**	**1,716,944**	**1,663,277**
	2013	**1,773,663**	**1,741,120**	**1,691,425**
	2014	**1,776,779**	**1,786,692**	**1,720,585**
Azerbaijan	2011	16,000	6,339	6,200
	2012	16,000	6,205	6,056
	2013	16,000	6,571	6,375
	2014	16,000	6,723	6,536
Bahrain	2011	12,964	12,841	12,998
	2012	12,964	12,950	13,918
	2013	12,964	13,468	13,480
	2014	12,965	12,785	13,526
Bangladesh	2011	1,646	1,402	1,344
	2012	1,646	1,261	1,193
	2013	1,646	1,383	1,347
	2014	1,646	1,229	1,216
Brunei Darussalam	2011	*1,000	627	573
	2012	*1,000	471	418
	2013	*1,000	564	485
	2014	*1,000	401	381
China	2011	342,443	484,550	468,311
	2012	352,419	515,819	496,858
	2013	352,419	544,255	526,665
	2014	352,419	582,289	564,104
Georgia	2011	*5,440	0	0
	2012	*5,440	9	7
	2013	*5,440	0	0
	2014	*5,440	0	0
India	2011	213,066	215,006	214,526
	2012	215,066	231,799	229,362
	2013	216,597	232,407	227,580
	2014	215,066	232,765	231,555
Indonesia	2011	57,711	49,853	45,130
	2012	57,711	47,635	44,942
	2013	*57,711	48,011	47,125
	2014	*57,711	50,677	44,593
Iran (Islamic Rep. of)	2011	85,596	87,873	88,666
	2012	85,596	89,911	91,584
	2013	85,596	92,478	93,062
	2014	88,890	90,118	87,684
Iraq	2011	40,898	27,796	27,213
	2012	41,397	29,176	28,654
	2013	41,397	29,747	29,343
	2014	43,690	24,384	23,850

Refinery distillation capacity, throughput and output

Thousand metric tons

Country or area	Year	Capacity	Throughput	Output
Israel	2011	12,500	12,408	12,032
	2012	12,500	11,865	11,720
	2013	*14,000	13,420	13,923
	2014	*14,300	13,940	14,394
Japan	2011	235,905	170,682	169,743
	2012	237,201	170,470	169,173
	2013	220,593	173,552	172,618
	2014	222,775	164,513	162,815
Jordan	2011	*5,977	3,414	3,218
	2012	*5,977	3,757	3,537
	2013	*5,977	3,284	3,146
	2014	5,977	3,282	3,075
Kazakhstan	2011	17,212	*14,429	13,627
	2012	17,212	14,458	13,528
	2013	17,212	14,760	13,762
	2014	16,969	15,537	14,544
Korea, Dem.Ppl's.Rep.	2011	3,541	526	508
	2012	3,541	525	508
	2013	3,541	580	561
	2014	3,541	532	515
Korea, Republic of	2011	137,631	130,662	127,495
	2012	147,556	133,852	131,006
	2013	147,556	128,952	126,256
	2014	148,055	131,507	128,633
Kuwait	2011	46,683	42,014	40,974
	2012	46,683	45,273	42,128
	2013	46,683	43,767	41,381
	2014	46,683	42,546	41,639
Kyrgyzstan	2011	499	92	90
	2012	499	84	81
	2013	499	83	77
	2014	499	89	88
Malaysia	2011	26,862	23,943	23,040
	2012	26,862	26,720	25,114
	2013	26,862	25,784	24,086
	2014	26,862	25,526	23,519
Myanmar	2011	2,843	827	789
	2012	2,843	690	648
	2013	2,843	552	510
	2014	2,843	555	521
Oman	2011	11,072	10,116	9,929
	2012	11,072	10,033	9,876
	2013	11,072	9,274	9,160
	2014	11,072	10,733	10,598
Other Asia	2011	65,337	44,084	41,883
	2012	65,337	48,871	45,973
	2013	65,382	45,856	43,382
	2014	65,382	45,724	42,658
Pakistan	2011	13,080	9,191	8,964
	2012	*13,080	10,746	10,468
	2013	*13,080	11,758	11,496
	2014	*13,080	12,314	11,976

Table 16

Refinery distillation capacity, throughput and output

Thousand metric tons

Country or area	Year	Capacity	Throughput	Output
Philippines	2011	13,616	9,204	8,909
	2012	13,616	8,237	7,954
	2013	13,616	7,657	7,384
	2014	13,766	8,101	7,843
Qatar	2011	16,893	12,937	12,912
	2012	16,893	13,618	13,478
	2013	16,893	14,920	12,640
	2014	16,893	12,122	12,038
Saudi Arabia	2011	105,337	93,813	91,974
	2012	105,337	97,530	95,618
	2013	125,287	93,401	91,570
	2014	124,489	109,580	107,431
Singapore	2011	67,681	*56,740	51,769
	2012	67,681	*58,041	52,148
	2013	67,057	*54,138	49,221
	2014	67,057	*54,394	45,520
Sri Lanka	2011	2,350	2,004	1,740
	2012	2,350	1,626	1,494
	2013	2,350	1,743	1,569
	2014	2,350	1,824	1,686
Syrian Arab Republic	2011	11,963	11,628	11,546
	2012	11,963	8,253	8,152
	2013	11,963	5,457	5,392
	2014	11,963	5,457	5,391
Tajikistan	2011	*100	28	25
	2012	*100	30	27
	2013	150	23	22
	2014	*150	25	23
Thailand	2011	61,321	46,679	39,148
	2012	61,670	48,715	41,293
	2013	62,419	*52,115	45,347
	2014	62,419	56,902	43,005
Turkey	2011	35,625	22,350	22,170
	2012	35,625	23,332	23,139
	2013	35,625	24,207	23,995
	2014	33,067	22,032	21,876
Turkmenistan	2011	11,819	8,548	8,292
	2012	11,819	8,157	7,912
	2013	11,819	8,241	7,911
	2014	11,819	8,903	8,547
United Arab Emirates	2011	34,414	26,157	25,758
	2012	34,508	26,078	25,437
	2013	35,262	26,490	28,068
	2014	35,262	*26,930	26,663
Uzbekistan	2011	11,186	3,636	3,583
	2012	11,186	3,245	3,204
	2013	11,186	3,075	3,058
	2014	11,186	2,897	2,864
Viet Nam	2011	6,983	6,150	5,325
	2012	6,983	6,200	5,429
	2013	6,983	6,063	6,493
	2014	6,510	*6,632	6,682

Refinery distillation capacity, throughput and output

Thousand metric tons

Country or area	Year	Capacity	Throughput	Output
Yemen	2011	6,983	2,356	2,244
	2012	6,983	1,302	1,240
	2013	6,983	3,084	2,937
	2014	6,983	2,724	2,594
Europe	**2011**	**1,146,273**	**961,041**	**941,886**
	2012	**1,140,439**	**966,140**	**937,950**
	2013	**1,114,520**	**939,617**	**920,105**
	2014	**1,109,657**	**951,027**	**933,276**
Albania	2011	1,500	297	297
	2012	1,500	111	111
	2013	1,500	40	40
	2014	1,500	361	336
Austria	2011	10,404	8,997	8,816
	2012	9,645	8,868	8,756
	2013	10,169	8,999	8,917
	2014	10,169	8,914	8,904
Belarus	2011	24,605	20,474	20,187
	2012	24,605	21,667	21,375
	2013	24,605	21,156	20,880
	2014	24,605	22,289	21,973
Belgium	2011	36,899	33,339	32,782
	2012	36,293	35,155	34,828
	2013	36,293	32,361	32,145
	2014	41,060	35,680	35,250
Bosnia and Herzegovina	2011	*1,320	1,313	1,260
	2012	11,970	1,119	1,079
	2013	*11,970	1,036	965
	2014	*11,970	990	989
Bulgaria	2011	5,748	6,162	5,999
	2012	5,748	6,606	6,425
	2013	5,748	6,639	6,443
	2014	5,748	6,335	6,129
Croatia	2011	12,485	3,613	3,591
	2012	12,485	3,727	3,707
	2013	12,485	3,389	3,357
	2014	12,485	2,979	2,953
Czechia	2011	9,127	7,510	7,417
	2012	9,127	7,662	7,587
	2013	9,127	7,001	6,969
	2014	9,676	7,862	7,822
Denmark	2011	8,698	6,817	6,557
	2012	8,698	7,635	7,427
	2013	8,698	7,173	6,989
	2014	8,449	6,910	6,685
Finland	2011	12,996	15,054	14,794
	2012	12,996	14,415	14,293
	2013	12,996	14,837	14,758
	2014	12,996	14,327	14,141
France	2011	85,725	72,217	71,630
	2012	87,153	61,406	60,903
	2013	75,233	58,296	57,891
	2014	70,120	58,111	57,621

Table 16

Refinery distillation capacity, throughput and output

Thousand metric tons

Country or area	Year	Capacity	Throughput	Output
Germany	2011	120,557	101,977	99,740
	2012	112,082	102,874	100,742
	2013	112,082	100,062	97,921
	2014	109,122	98,953	96,755
Greece	2011	24,381	19,355	19,253
	2012	24,381	23,917	23,766
	2013	*24,381	24,251	24,033
	2014	*24,381	27,579	27,515
Hungary	2011	8,030	8,262	8,234
	2012	8,030	7,862	7,818
	2013	8,030	7,681	7,644
	2014	8,030	7,986	7,944
Ireland	2011	3,541	2,964	2,918
	2012	3,541	3,073	3,019
	2013	3,541	2,850	2,788
	2014	3,541	2,763	2,710
Italy	2011	116,570	86,065	85,642
	2012	109,488	82,020	81,598
	2013	105,523	70,111	69,628
	2014	102,060	66,817	66,337
Lithuania	2011	10,220	9,468	9,376
	2012	10,220	9,059	8,977
	2013	10,220	9,584	9,486
	2014	10,220	8,036	7,943
Netherlands	2011	59,679	57,710	56,954
	2012	59,585	57,525	56,639
	2013	59,560	55,837	55,183
	2014	60,503	56,906	56,515
Norway	2011	15,910	15,795	15,686
	2012	15,910	15,641	15,023
	2013	15,910	16,423	16,018
	2014	15,910	14,592	14,400
Poland	2011	24,586	25,738	24,944
	2012	24,586	26,767	25,768
	2013	24,586	25,707	25,049
	2014	24,586	25,636	24,821
Portugal	2011	15,171	10,820	10,820
	2012	15,171	11,673	11,673
	2013	15,171	13,941	13,941
	2014	15,171	12,662	12,662
Republic of Moldova	2011	*19	19	19
	2012	*19	17	18
	2013	*19	20	19
	2014	*19	18	18
Romania	2011	26,797	10,446	10,445
	2012	26,797	9,850	9,850
	2013	23,032	10,122	10,122
	2014	23,032	11,666	11,666
Russian Federation	2011	270,868	260,057	249,860
	2012	274,314	274,310	255,290
	2013	274,314	278,600	266,656
	2014	274,314	292,845	283,328

Refinery distillation capacity, throughput and output

Thousand metric tons

Country or area	Year	Capacity	Throughput	Output
Serbia	2011	10,715	2,462	2,462
	2012	10,715	2,260	2,260
	2013	10,715	3,111	3,111
	2014	10,715	3,132	3,110
Slovakia	2011	5,736	6,690	6,662
	2012	5,736	5,907	5,877
	2013	5,736	6,572	6,552
	2014	5,736	5,745	5,725
Slovenia	2011	674	..	..
	2012	674	..	..
	2013	674	..	..
	2014	674	..	..
Spain	2011	63,416	57,152	56,649
	2012	63,416	61,892	61,346
	2013	64,414	61,037	60,501
	2014	71,197	61,169	60,290
Sweden	2011	21,796	19,914	19,622
	2012	21,796	22,105	21,410
	2013	21,796	18,138	17,584
	2014	21,796	20,280	19,799
Switzerland	2011	3,591	4,441	4,407
	2012	3,591	3,454	3,424
	2013	3,591	4,933	4,914
	2014	3,591	4,964	4,940
T.F.Yug.Rep. Macedonia	2011	2,494	725	680
	2012	2,494	261	244
	2013	2,494	80	78
	2014	2,494	3	1
Ukraine	2011	43,878	9,641	9,366
	2012	43,878	5,073	4,968
	2013	43,878	3,944	4,018
	2014	43,878	3,206	3,321
United Kingdom	2011	88,138	75,548	74,817
	2012	83,795	72,229	71,750
	2013	76,029	65,687	65,505
	2014	69,910	61,311	60,673
Oceania	**2011**	**44,870**	**38,968**	**38,251**
	2012	**44,895**	**38,435**	**36,937**
	2013	**40,581**	**36,988**	**36,607**
	2014	**33,973**	**35,025**	**33,897**
Australia	2011	37,913	32,641	31,976
	2012	37,938	31,806	30,346
	2013	33,623	30,302	30,038
	2014	26,990	28,463	27,441
New Zealand	2011	5,337	5,372	5,370
	2012	5,337	5,577	5,577
	2013	5,337	5,363	5,342
	2014	5,337	5,238	5,208
Papua New Guinea	2011	1,621	955	905
	2012	1,621	1,052	1,014
	2013	1,621	1,323	1,227
	2014	*1,646	1,324	*1,248

Table 17

Production, trade and supply of liquefied petroleum gases (LPG)

Thousand metric tons and kilograms per capita

Table Notes

Production includes output from refineries and plants.

Please refer to the Definitions Section on pages ix to xv for the appropriate product description/ classification.

Figure 45: Production of LPG by region, 1990-2014

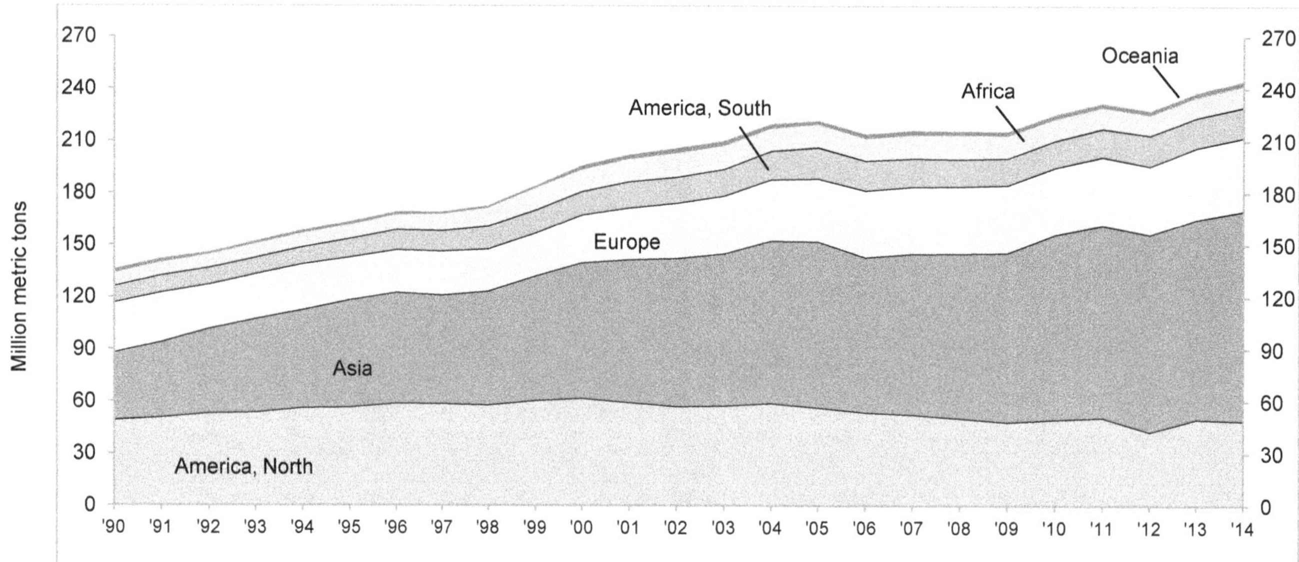

Figure 46: Major LPG producing countries, 2014

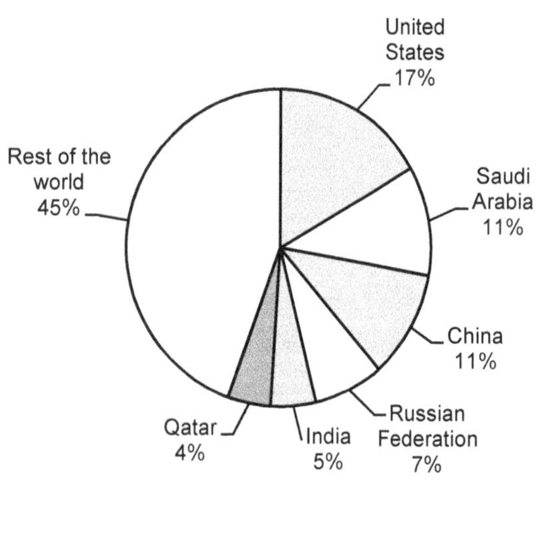

Figure 47: Major LPG consuming countries, 2014

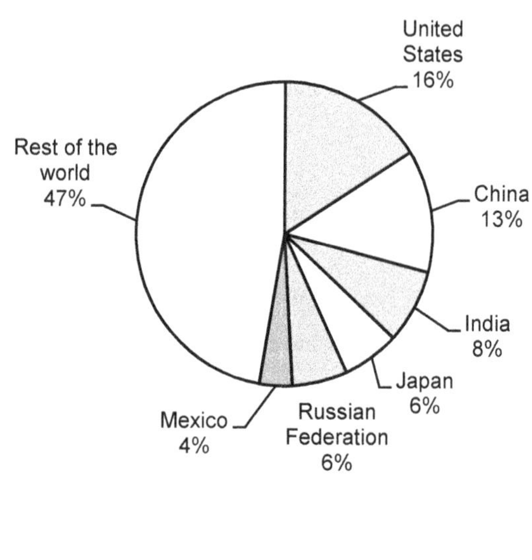

Production, trade and supply of liquefied petroleum gases (LPG)

Thousand metric tons and kilograms per capita

Country or area	Year	Production	Imports	Exports	Changes in stocks	Supply	
						Total	Per Capita
World	**2011**	**232,033**	**77,074**	**77,294**	**1,473**	**230,340**	**33**
	2012	**227,672**	**75,968**	**77,148**	**1,648**	**224,844**	**32**
	2013	**237,825**	**79,535**	**75,289**	**-753**	**242,824**	**34**
	2014	**244,950**	**86,661**	**78,976**	**1,020**	**251,614**	**35**
Africa	**2011**	**12,052**	**5,454**	**6,522**	**-82**	**11,066**	**10**
	2012	**11,668**	**5,652**	**6,095**	**-77**	**11,303**	**10**
	2013	**11,833**	**5,736**	**5,992**	**-70**	**11,648**	**10**
	2014	**12,821**	**6,232**	**7,660**	**19**	**11,373**	**10**
Algeria	2011	8,014	..	6,019	2	1,993	54
	2012	7,429	..	5,505	28	1,896	51
	2013	7,484	..	5,346	-4	2,142	56
	2014	9,176	..	7,191	-12	1,997	51
Angola	2011	596	0	445	..	151	7
	2012	592	64	402	..	254	11
	2013	712	91	537	..	266	11
	2014	551	52	356	..	247	10
Benin	2011	..	15	4	..	11	1
	2012	..	16	4	..	12	1
	2013	..	17	4	..	13	1
	2014	..	18	4	..	14	1
Botswana	2011	..	34	..	..	34	16
	2012	..	19	..	..	19	9
	2013	..	22	..	..	22	10
	2014	..	24	..	..	24	11
Burkina Faso	2011	..	37	..	..	37	2
	2012	..	46	..	..	46	3
	2013	..	46	..	..	46	3
	2014	..	61	..	..	61	3
Burundi	2011	..	*2	..	..	*2	0
	2012	..	*2	..	..	*2	0
	2013	..	*2	..	..	*2	0
	2014	..	1	..	..	1	0
Cabo Verde	2011	..	11	..	..	11	22
	2012	..	11	..	..	11	21
	2013	..	11	..	..	11	21
	2014	..	11	..	..	11	22
Cameroon	2011	21	35	..	..	56	3
	2012	22	35	..	..	57	3
	2013	23	35	..	..	58	3
	2014	24	35	..	..	59	3
Chad	2011	..	*1	..	..	*1	0
	2012	..	*1	..	..	*1	0
	2013	..	*1	..	..	*1	0
	2014	..	*1	..	..	*1	0
Congo	2011	7	4	..	..	11	3
	2012	7	..	..	..	7	2
	2013	8	..	..	..	8	2
	2014	8	..	..	..	8	2
Côte d'Ivoire	2011	12	85	2	..	95	5
	2012	13	104	2	..	115	5
	2013	3	166	0	..	169	8
	2014	17	188	5	..	200	9

Table 17

Production, trade and supply of liquefied petroleum gases (LPG)

Thousand metric tons and kilograms per capita

Country or area	Year	Production	Imports	Exports	Changes in stocks	Supply	
						Total	Per Capita
Djibouti	2011	..	*20	..	..	*20	*24
	2012	..	*20	..	..	*20	*23
	2013	..	*20	..	..	*20	*23
	2014	..	*20	..	..	*20	*23
Egypt	2011	1,980	2,184	..	..	4,164	50
	2012	1,996	2,077	..	..	4,073	48
	2013	2,102	2,013	..	..	4,115	47
	2014	1,754	2,168	..	..	3,922	44
Equatorial Guinea	2011	..	*20	..	..	*20	*27
	2012	..	*20	..	..	*20	*26
	2013	..	*20	..	..	*20	*25
	2014	..	*20	..	..	*20	*24
Eritrea	2011	..	2	..	..	2	0
	2012	..	2	..	..	2	0
	2013	..	2	..	..	2	0
	2014	..	2	..	..	2	0
Ethiopia	2011	..	5	..	..	5	0
	2012	..	5	..	..	5	0
	2013	..	7	..	..	7	0
	2014	..	7	..	..	7	0
Gabon	2011	12	19	..	..	31	20
	2012	14	19	..	..	33	20
	2013	14	21	..	..	35	21
	2014	13	23	..	..	36	21
Gambia	2011	..	2	..	..	2	1
	2012	..	2	..	..	2	1
	2013	..	2	..	..	2	1
	2014	..	2	..	..	2	1
Ghana	2011	45	178	..	..	222	9
	2012	27	242	..	..	268	11
	2013	26	204	..	-3	232	9
	2014	3	236	..	..	240	9
Guinea-Bissau	2011	..	*3	..	..	*3	*2
	2012	..	*3	..	..	*3	*2
	2013	..	*3	..	..	*3	*2
	2014	..	*3	..	..	*3	*2
Kenya	2011	28	63	0	..	91	2
	2012	17	77	0	..	94	2
	2013	12	81	4	4	85	2
	2014	..	159	9	..	150	3
Lesotho	2011	..	*7	..	..	*7	*3
	2012	..	*8	..	..	*8	*4
	2013	..	*9	..	..	*9	*4
	2014	..	*9	..	..	*9	*4
Liberia	2011	..	*3	..	..	*3	*1
	2012	..	*3	..	..	*3	*1
	2013	..	*3	..	..	*3	*1
	2014	..	*3	..	..	*3	*1
Libya	2011	331	..	35	..	296	47
	2012	446	..	153	..	293	47
	2013	374	..	53	..	321	51
	2014	343	..	40	..	303	48

Production, trade and supply of liquefied petroleum gases (LPG)

Thousand metric tons and kilograms per capita

Country or area	Year	Production	Imports	Exports	Changes in stocks	Supply	
						Total	Per Capita
Madagascar	2011	..	8	..	1	7	0
	2012	..	9	..	1	8	0
	2013	..	13	..	..	13	1
	2014	..	*13	..	..	*13	*1
Malawi	2011	..	*2	..	..	*2	0
	2012	..	*2	..	..	*2	0
	2013	..	*2	..	..	*2	0
	2014	..	*2	..	..	*2	0
Mauritania	2011	..	47	..	..	47	13
	2012	..	46	..	..	46	12
	2013	..	*46	..	..	*46	*12
	2014	..	*46	..	..	*46	*12
Mauritius	2011	..	66	..	0	66	53
	2012	..	68	..	1	67	53
	2013	..	68	..	..	68	54
	2014	..	76	..	..	76	60
Mayotte	2011	..	3	..	0	3	12
	2012	..	*3	..	..	*3	*11
	2013	..	*3	..	..	*3	*11
	2014	..	*3	..	..	*3	*12
Morocco	2011	38	1,946	..	-86	2,070	63
	2012	115	2,050	..	-110	2,275	68
	2013	96	2,039	..	-69	2,204	65
	2014	99	2,244	..	28	2,315	67
Mozambique	2011	..	15	..	..	15	1
	2012	..	17	..	..	17	1
	2013	..	13	..	..	13	0
	2014	..	23	..	..	23	1
Namibia	2011	..	8	..	..	8	4
	2012	..	6	..	..	6	3
	2013	..	7	..	..	7	3
	2014	..	12	..	..	12	5
Niger	2011	3	4	0	0	7	0
	2012	37	0	11	0	27	2
	2013	50	0	30	6	14	1
	2014	57	0	*21	0	36	2
Nigeria	2011	129	..	..	..	129	1
	2012	104	..	..	..	104	1
	2013	195	..	..	..	195	1
	2014	72	..	..	..	72	0
Réunion	2011	..	23	..	-2	24	29
	2012	..	23	..	1	23	27
	2013	..	21	..	-1	22	26
	2014	..	21	..	0	21	25
Sao Tome and Principe	2011	..	0	..	..	0	*1
	2012	..	0	..	..	0	*1
	2013	..	0	..	..	0	*1
	2014	..	0	..	..	0	*1
Senegal	2011	2	102	2	..	102	8
	2012	1	102	..	..	103	7
	2013	1	96	..	..	97	7
	2014	1	105	..	..	106	7

Table 17

Production, trade and supply of liquefied petroleum gases (LPG)

Thousand metric tons and kilograms per capita

Country or area	Year	Production	Imports	Exports	Changes in stocks	Supply	
						Total	Per Capita
Seychelles	2011	..	5	..	..	5	51
	2012	..	5	..	..	5	48
	2013	..	4	..	..	4	43
	2014	..	3	..	..	3	35
Somalia	2011	..	*3	..	..	*3	0
	2012	..	*6	..	..	*6	*1
	2013	..	*6	..	..	*6	*1
	2014	..	*6	..	..	*6	*1
South Africa	2011	375	19	3	..	391	7
	2012	354	10	14	..	350	7
	2013	262	40	16	..	286	5
	2014	215	25	33	..	207	4
South Sudan	2011	..	..	..	..	..	..
	2012	..	3	..	..	3	0
	2013	..	2	..	..	2	0
	2014	..	2	..	..	2	0
Sudan	2011	..	..	..	..	..	..
	2012	320	136	3	..	453	12
	2013	289	177	1	..	465	12
	2014	310	161	1	..	470	12
Sudan (former)	2011	312	88	9	..	391	8
	2012	..	..	..	..	..	..
	2013	..	..	..	..	..	..
	2014	..	..	..	..	..	..
Swaziland	2011	..	*7	..	..	*7	*5
	2012	..	*7	..	..	*7	*5
	2013	..	*7	..	..	*7	*5
	2014	..	6	..	..	6	5
Togo	2011	..	6	..	..	6	1
	2012	..	7	..	..	7	1
	2013	..	7	..	..	7	1
	2014	..	7	..	..	7	1
Tunisia	2011	138	337	..	2	473	44
	2012	164	319	..	2	481	44
	2013	173	325	..	-4	502	46
	2014	167	342	..	3	506	45
Uganda	2011	..	6	..	..	6	0
	2012	..	*7	..	..	*7	0
	2013	..	7	..	..	7	0
	2014	..	7	..	..	7	0
United Rep. of Tanzania	2011	..	26	..	..	26	1
	2012	..	44	..	..	44	1
	2013	..	69	..	..	69	1
	2014	..	69	..	..	69	1
Zambia	2011	9	..	3	..	6	0
	2012	10	..	1	..	9	1
	2013	9	..	1	..	8	1
	2014	11	..	..	..	11	1
Zimbabwe	2011	..	4	..	..	4	0
	2012	..	7	..	..	7	0
	2013	..	9	..	..	9	1
	2014	..	14	..	..	14	1

Production, trade and supply of liquefied petroleum gases (LPG)

Thousand metric tons and kilograms per capita

Country or area	Year	Production	Imports	Exports	Changes in stocks	Supply	
						Total	Per Capita
America, North	**2011**	**50,655**	**6,472**	**1,810**	**596**	**54,721**	**99**
	2012	**42,462**	**5,590**	**1,626**	**482**	**45,944**	**82**
	2013	**49,647**	**5,410**	**1,844**	**-281**	**53,495**	**95**
	2014	**48,828**	**5,781**	**1,724**	**592**	**52,293**	**92**
Anguilla	2011	..	5	..	..	5	380
	2012	..	6	..	..	6	389
	2013	..	*5	..	..	*5	*378
	2014	..	*6	..	..	*6	*380
Antigua and Barbuda	2011	..	*3	..	..	*3	*36
	2012	..	*3	..	..	*3	*37
	2013	..	*3	..	..	*3	*37
	2014	..	*3	..	..	*3	*37
Aruba	2011	..	*10	..	..	*10	*97
	2012	..	*10	..	..	*10	*94
	2013	..	*10	..	..	*10	*94
	2014	..	*10	..	..	*10	*100
Bahamas	2011	..	11	..	..	11	30
	2012	..	14	..	..	14	37
	2013	..	11	..	..	11	29
	2014	..	*11	..	..	*11	*29
Barbados	2011	..	8	..	..	8	30
	2012	..	9	..	..	9	34
	2013	..	9	..	0	8	30
	2014	..	7	..	0	7	24
Belize	2011	2	13	..	..	15	45
	2012	1	11	..	..	12	36
	2013	1	17	..	..	18	52
	2014	1	4	..	..	5	14
Bermuda	2011	..	0	..	..	0	1
	2012	..	1	..	..	1	21
	2013	..	4	..	..	4	65
	2014	..	7	..	..	7	108
Bonaire, St Eustatius, Saba	2011	..	..	..	..	..	..
	2012	..	*8	..	..	*8	*349
	2013	..	*8	..	..	*8	*346
	2014	..	*8	..	..	*8	*337
Canada	2011	1,727	379	419	30	1,657	48
	2012	1,534	188	483	108	1,131	32
	2013	1,437	85	571	37	914	26
	2014	1,295	203	548	63	887	25
Cayman Islands	2011	..	4	..	..	4	71
	2012	..	3	..	..	3	52
	2013	..	4	..	..	4	69
	2014	..	3	..	..	3	54
Costa Rica	2011	1	115	..	-1	118	26
	2012	..	124	..	3	121	26
	2013	..	127	..	1	126	27
	2014	..	134	..	..	134	28
Cuba	2011	50	93	..	..	143	13
	2012	42	107	..	..	149	13
	2013	74	65	..	..	139	12
	2014	67	80	..	..	147	13

Table 17

Production, trade and supply of liquefied petroleum gases (LPG)

Thousand metric tons and kilograms per capita

Country or area	Year	Production	Imports	Exports	Changes in stocks	Supply	
						Total	Per Capita
Curaçao	2011	..	..	..	..	..	..
	2012	51	30	43	..	38	249
	2013	47	6	39	..	14	91
	2014	55	..	46	..	9	58
Dominica	2011	..	*2	..	..	*2	*28
	2012	..	*2	..	..	*2	*28
	2013	..	*2	..	..	*2	*28
	2014	..	*2	..	..	*2	*28
Dominican Republic	2011	21	728	..	0	749	75
	2012	14	716	..	2	728	72
	2013	13	773	..	3	783	76
	2014	16	829	..	-5	849	82
El Salvador	2011	12	188	4	-15	211	35
	2012	6	231	0	19	218	36
	2013	..	237	0	0	237	39
	2014	..	270	6	-13	277	45
Greenland	2011	..	0	..	..	0	2
	2012	..	0	..	..	0	1
	2013	..	0	..	..	0	1
	2014	..	0	..	..	0	1
Grenada	2011	..	5	..	..	5	43
	2012	..	5	..	..	5	43
	2013	..	5	..	..	5	49
	2014	..	4	..	..	4	40
Guadeloupe	2011	..	*14	..	..	*14	*31
	2012	..	*14	..	..	*14	*30
	2013	..	*15	..	..	*15	*32
	2014	..	*15	..	..	*15	*32
Guatemala	2011	..	351	104	-12	259	17
	2012	..	371	103	-10	278	18
	2013	..	422	121	12	289	18
	2014	..	418	140	-23	301	19
Haiti	2011	..	10	..	..	10	1
	2012	..	13	..	..	13	1
	2013	..	21	..	..	21	2
	2014	..	8	..	..	8	1
Honduras	2011	..	366	297	-11	80	10
	2012	..	480	377	0	103	13
	2013	..	527	414	0	113	14
	2014	..	497	394	-17	120	15
Jamaica	2011	11	72	..	1	81	29
	2012	9	76	..	1	85	31
	2013	6	69	..	1	74	27
	2014	6	67	..	0	73	26
Martinique	2011	*21	*3	*12	..	*11	*28
	2012	*21	*3	*12	..	*12	*29
	2013	*21	*3	*12	..	*12	*31
	2014	*22	*3	*13	..	*12	*30
Mexico	2011	6,371	2,691	46	54	8,962	74
	2012	6,096	2,918	4	2	9,008	74
	2013	6,154	2,691	6	-7	8,846	71
	2014	6,006	2,893	18	15	8,866	71

Production, trade and supply of liquefied petroleum gases (LPG)

Thousand metric tons and kilograms per capita

Country or area	Year	Production	Imports	Exports	Changes in stocks	Supply	
						Total	Per Capita
Montserrat	2011	..	1	..	0	1	260
	2012	..	1	..	*-1	1	239
	2013	..	1	..	0	1	198
	2014	..	1	..	0	1	118
Neth. Antilles (former)	2011	49	72	41	..	80	375
	2012	..	..	..	..	..	..
	2013	..	..	..	..	..	..
	2014	..	..	..	..	..	..
Nicaragua	2011	17	57	..	1	73	13
	2012	9	73	..	-1	83	14
	2013	11	81	..	0	92	15
	2014	10	83	..	-5	98	16
Panama	2011	..	152	0	-1	153	42
	2012	..	158	0	-1	159	42
	2013	..	194	0	-16	210	55
	2014	..	199	0	29	170	44
Sint Maarten (Dutch part)	2011	..	..	..	..	..	..
	2012	..	*2	..	..	*2	*51
	2013	..	2	..	..	2	50
	2014	..	*2	..	..	*2	*50
St. Kitts-Nevis	2011	..	2	..	0	2	41
	2012	..	2	..	..	2	36
	2013	..	*2	..	..	*2	*35
	2014	..	*2	..	..	*2	*35
St. Lucia	2011	..	*6	..	..	*6	*33
	2012	..	*6	..	..	*6	*33
	2013	..	*6	..	..	*6	*33
	2014	..	*6	..	..	*6	*34
St. Vincent-Grenadines	2011	..	4	..	..	4	35
	2012	..	6	..	..	6	51
	2013	..	7	..	..	7	61
	2014	..	*7	..	..	*7	*61
Trinidad and Tobago	2011	932	..	887	16	29	22
	2012	708	..	603	90	15	11
	2013	753	..	681	-9	81	60
	2014	659	..	558	73	28	21
United States	2011	41,441	1,108	..	534	42,015	134
	2012	33,970	..	..	269	33,701	107
	2013	41,130	..	..	-303	41,433	131
	2014	40,691	..	..	475	40,216	126
America, South	**2011**	**16,156**	**4,129**	**3,544**	**-73**	**16,814**	**42**
	2012	**17,624**	**3,710**	**3,423**	**62**	**17,850**	**44**
	2013	**17,304**	**4,285**	**3,121**	**76**	**18,392**	**45**
	2014	**17,379**	**4,551**	**2,881**	**97**	**18,952**	**46**
Argentina	2011	2,305	..	1,186	-57	1,176	28
	2012	2,775	..	919	7	1,849	44
	2013	*2,624	..	748	7	*1,869	*44
	2014	2,847	..	867	5	1,975	46
Bolivia (Plur. State of)	2011	328	34	..	..	362	36
	2012	328	39	..	..	367	36
	2013	360	5	6	..	359	35
	2014	396	..	37	..	359	34

Table 17

Production, trade and supply of liquefied petroleum gases (LPG)

Thousand metric tons and kilograms per capita

Country or area	Year	Production	Imports	Exports	Changes in stocks	Supply	
						Total	Per Capita
Brazil	2011	5,101	1,829	23	6	6,901	34
	2012	5,437	1,540	17	53	6,907	34
	2013	5,422	1,819	49	-4	7,196	35
	2014	5,297	2,010	10	9	7,288	35
Chile	2011	350	762	0	-6	1,118	65
	2012	593	755	2	-17	1,363	78
	2013	257	1,024	67	13	1,201	68
	2014	316	945	24	-5	1,242	70
Colombia	2011	*630	18	15	..	*633	*14
	2012	*652	..	57	..	*595	*13
	2013	586	..	49	53	484	10
	2014	557	..	31	-66	592	12
Ecuador	2011	388	1,300	..	0	1,688	111
	2012	405	1,203	..	0	1,608	104
	2013	381	1,277	..	0	1,658	106
	2014	300	1,433	..	-1	1,735	109
French Guiana	2011	..	*4	..	..	*4	*17
	2012	..	*4	..	..	*4	*17
	2013	..	*4	..	..	*4	*17
	2014	..	*4	..	..	*4	*16
Guyana	2011	..	18	..	-1	19	25
	2012	..	16	..	-1	16	22
	2013	..	16	..	-1	17	22
	2014	..	17	..	0	17	22
Paraguay	2011	..	86	..	..	86	14
	2012	..	88	..	..	88	14
	2013	..	78	..	-5	83	13
	2014	..	82	..	1	81	12
Peru	2011	1,406	0	176	-13	1,243	42
	2012	1,517	1	128	20	1,370	45
	2013	1,795	0	238	11	1,546	51
	2014	1,774	0	18	155	1,601	52
Suriname	2011	..	15	..	..	15	29
	2012	..	16	..	..	16	30
	2013	..	17	..	..	17	31
	2014	..	14	..	..	14	27
Uruguay	2011	57	63	2	-2	120	35
	2012	76	49	0	0	125	37
	2013	91	45	0	2	135	40
	2014	81	44	0	-2	126	37
Venezuela (Bolivar. Rep.)	2011	5,591	..	2,142	..	3,449	117
	2012	5,841	..	2,300	..	3,541	119
	2013	5,787	..	1,964	..	3,823	126
	2014	5,811	..	1,894	..	3,917	128
Asia	**2011**	**110,669**	**43,539**	**49,184**	**587**	**104,436**	**25**
	2012	**113,579**	**44,248**	**49,332**	**1,111**	**107,384**	**25**
	2013	**115,102**	**44,428**	**47,137**	**-403**	**112,796**	**26**
	2014	**120,909**	**48,435**	**49,433**	**87**	**119,825**	**28**
Armenia	2011	..	7	..	..	7	3
	2012	..	7	..	..	7	2
	2013	..	7	..	..	7	2
	2014	..	7	..	..	7	2

Production, trade and supply of liquefied petroleum gases (LPG)

Thousand metric tons and kilograms per capita

Country or area	Year	Production	Imports	Exports	Changes in stocks	Supply	
						Total	Per Capita
Azerbaijan	2011	250	..	109	2	139	15
	2012	215	..	90	-2	127	14
	2013	263	..	93	0	170	18
	2014	202	..	35	-3	170	18
Bahrain	2011	122	..	76	-3	49	38
	2012	122	..	61	9	52	39
	2013	104	..	38	11	55	41
	2014	104	..	39	7	58	43
Bangladesh	2011	19	..	..	..	19	0
	2012	21	..	..	..	21	0
	2013	22	..	..	..	22	0
	2014	20	..	..	..	20	0
Bhutan	2011	..	7	..	..	7	10
	2012	..	8	..	..	8	10
	2013	..	7	..	..	7	9
	2014	..	7	..	..	7	9
Brunei Darussalam	2011	15	..	..	..	15	38
	2012	16	..	..	..	16	39
	2013	16	..	..	..	16	39
	2014	16	..	..	..	16	38
Cambodia	2011	..	72	..	..	72	5
	2012	..	85	..	..	85	6
	2013	..	99	..	..	99	7
	2014	..	112	..	..	112	7
China	2011	22,408	3,496	1,191	-30	24,742	18
	2012	22,687	3,585	1,282	30	24,960	18
	2013	25,133	4,517	1,269	20	28,361	21
	2014	27,058	7,394	1,444	80	32,927	24
China, Hong Kong SAR	2011	..	400	0	0	400	57
	2012	..	391	0	-2	393	55
	2013	..	378	2	-14	390	54
	2014	..	399	0	-5	404	56
China, Macao SAR	2011	..	42	..	..	42	77
	2012	..	43	..	..	43	77
	2013	..	45	..	..	45	79
	2014	..	44	..	..	44	76
Cyprus	2011	..	54	..	-1	55	49
	2012	..	53	..	1	52	46
	2013	..	50	..	-2	52	46
	2014	..	48	..	0	48	42
Georgia	2011	..	19	..	-2	21	5
	2012	..	18	..	0	18	4
	2013	..	17	..	0	17	4
	2014	..	18	..	0	18	4
India	2011	9,547	8,534	174	452	17,455	14
	2012	9,824	8,176	200	563	17,237	14
	2013	10,030	6,607	227	-289	16,699	13
	2014	11,727	8,318	254	0	19,791	15
Indonesia	2011	2,286	1,992	0	-69	4,347	18
	2012	2,493	2,574	0	36	5,031	20
	2013	2,388	3,426	10	0	5,804	23
	2014	3,140	3,590	4	0	6,726	26

Table 17

Production, trade and supply of liquefied petroleum gases (LPG)

Thousand metric tons and kilograms per capita

Country or area	Year	Production	Imports	Exports	Changes in stocks	Supply	
						Total	Per Capita
Iran (Islamic Rep. of)	2011	6,491	184	3,576	..	3,099	41
	2012	5,577	154	2,604	..	3,128	41
	2013	5,418	124	1,733	..	3,809	49
	2014	7,216	38	2,928	176	4,150	53
Iraq	2011	1,394	177	..	-34	1,605	50
	2012	1,538	391	..	225	1,704	52
	2013	1,310	440	..	-75	1,825	54
	2014	715	328	..	-625	1,668	47
Israel	2011	569	225	123	..	671	89
	2012	498	275	129	..	644	84
	2013	582	231	132	-2	683	87
	2014	605	231	132	..	704	89
Japan	2011	3,934	12,675	85	-46	16,570	130
	2012	4,014	13,240	33	385	16,836	132
	2013	4,509	11,996	69	212	16,224	128
	2014	4,242	11,645	46	216	15,625	123
Jordan	2011	84	295	..	-6	385	57
	2012	102	295	..	12	385	55
	2013	78	275	..	-5	359	50
	2014	91	282	..	5	367	49
Kazakhstan	2011	2,028	3	1,660	-68	439	27
	2012	2,136	7	1,682	32	429	26
	2013	2,202	3	1,650	15	540	32
	2014	2,388	1	1,917	16	456	26
Korea, Republic of	2011	1,579	6,410	78	51	7,860	159
	2012	1,487	6,234	153	-261	7,829	158
	2013	1,728	5,623	76	162	7,113	143
	2014	2,099	5,344	114	174	7,155	143
Kuwait	2011	4,428	..	4,081	..	347	107
	2012	4,896	..	4,492	..	404	118
	2013	4,620	..	4,376	..	244	68
	2014	4,539	..	4,398	..	141	38
Kyrgyzstan	2011	..	11	..	..	11	2
	2012	..	14	..	..	14	2
	2013	..	13	..	..	13	2
	2014	..	..	..	..	..	..
Lao People's Dem. Rep.	2011	..	*20	..	..	*20	*3
	2012	..	*21	..	..	*21	*3
	2013	..	*21	..	..	*21	*3
	2014	..	*21	..	..	*21	*3
Lebanon	2011	..	197	..	..	197	43
	2012	..	243	..	..	243	49
	2013	..	270	..	..	270	51
	2014	..	247	..	..	247	44
Malaysia	2011	3,045	494	911	1	2,627	92
	2012	2,725	411	399	-9	2,746	95
	2013	2,387	388	313	-204	2,666	90
	2014	2,249	382	318	-61	2,374	79
Maldives	2011	..	12	..	..	12	34
	2012	..	10	..	..	10	29
	2013	..	12	..	..	12	34
	2014	..	14	..	..	14	38

Production, trade and supply of liquefied petroleum gases (LPG)

Thousand metric tons and kilograms per capita

Country or area	Year	Production	Imports	Exports	Changes in stocks	Supply	
						Total	Per Capita
Myanmar	2011	10	4	..	0	14	0
	2012	16	0	..	0	16	0
	2013	6	16	..	-2	23	0
	2014	4	23	..	3	23	0
Nepal	2011	..	181	..	..	181	7
	2012	..	207	..	..	207	8
	2013	..	233	..	..	233	8
	2014	..	258	..	..	258	9
Oman	2011	626	..	531	..	95	30
	2012	586	..	488	..	98	28
	2013	557	..	462	..	95	24
	2014	573	..	455	..	118	28
Other Asia	2011	1,047	1,119	31	35	2,100	90
	2012	1,176	742	91	11	1,816	78
	2013	1,329	1,148	89	-97	2,485	106
	2014	976	1,700	44	72	2,560	110
Pakistan	2011	386	80	..	..	466	3
	2012	423	68	..	..	491	3
	2013	471	71	..	..	542	3
	2014	469	93	..	..	562	3
Philippines	2011	386	720	6	-19	1,120	12
	2012	351	705	0	-2	1,058	11
	2013	314	768	0	8	1,074	11
	2014	294	798	0	2	1,090	11
Qatar	2011	10,360	..	9,689	..	671	352
	2012	10,820	..	9,999	..	821	407
	2013	10,886	..	10,013	..	873	415
	2014	10,967	..	10,041	..	926	426
Saudi Arabia	2011	26,006	..	22,501	13	3,492	121
	2012	27,798	..	23,097	1	4,700	159
	2013	26,273	..	22,224	2	4,047	134
	2014	27,325	..	22,871	4	4,450	144
Singapore	2011	302	12	295	..	19	4
	2012	394	10	446	..	-43	-8
	2013	285	945	190	..	1,040	192
	2014	317	115	240	..	191	35
Sri Lanka	2011	24	181	..	0	205	10
	2012	17	193	..	0	210	10
	2013	22	197	..	-12	231	11
	2014	28	198	..	-5	232	11
State of Palestine	2011	..	123	..	..	123	29
	2012	..	126	..	..	126	29
	2013	..	133	..	..	133	30
	2014	..	148	..	..	148	33
Syrian Arab Republic	2011	444	414	..	..	858	42
	2012	324	362	..	..	686	34
	2013	190	427	..	..	617	32
	2014	151	320	..	..	471	25
Tajikistan	2011	..	68	..	..	68	9
	2012	..	175	..	..	175	22
	2013	..	*177	..	..	*177	*22
	2014	..	295	..	..	295	36

Table 17

Production, trade and supply of liquefied petroleum gases (LPG)

Thousand metric tons and kilograms per capita

Country or area	Year	Production	Imports	Exports	Changes in stocks	Supply	
						Total	Per Capita
Thailand	2011	5,421	1,437	16	314	6,528	98
	2012	5,687	1,730	10	29	7,378	110
	2013	5,447	1,949	6	-125	7,515	111
	2014	5,506	2,099	10	78	7,517	111
Turkey	2011	739	3,048	105	-3	3,685	50
	2012	775	2,993	88	53	3,627	48
	2013	733	3,077	82	-6	3,734	49
	2014	699	3,118	113	-47	3,751	48
Turkmenistan	2011	712	..	..	..	712	139
	2012	734	..	..	..	734	142
	2013	1,038	..	..	..	1,038	198
	2014	994	..	..	..	994	187
United Arab Emirates	2011	4,814	81	3,918	..	977	112
	2012	4,782	44	3,961	..	865	97
	2013	*5,257	42	*4,004	..	*1,295	*143
	2014	4,827	..	3,952	..	875	96
Uzbekistan	2011	21	..	..	..	21	1
	2012	19	..	..	..	19	1
	2013	18	..	..	..	18	1
	2014	17	..	..	..	17	1
Viet Nam	2011	588	745	13	..	1,320	15
	2012	656	659	9	..	1,306	14
	2013	707	697	57	..	1,347	15
	2014	661	801	62	..	1,400	15
Yemen	2011	585	..	16	..	569	23
	2012	670	..	18	..	652	26
	2013	778	..	21	..	757	30
	2014	691	..	15	..	676	26
Europe	**2011**	**39,583**	**16,966**	**14,870**	**523**	**41,155**	**56**
	2012	**39,557**	**16,183**	**15,490**	**18**	**40,231**	**55**
	2013	**41,298**	**19,137**	**15,863**	**7**	**44,564**	**60**
	2014	**42,229**	**21,235**	**15,885**	**184**	**47,395**	**64**
Albania	2011	..	101	..	..	101	35
	2012	..	88	..	..	88	31
	2013	..	141	..	29	112	39
	2014	..	157	0	36	121	42
Andorra	2011	..	3	..	..	3	31
	2012	..	2	..	..	2	29
	2013	..	2	..	..	2	30
	2014	..	2	..	..	2	29
Austria	2011	101	91	29	0	163	19
	2012	67	80	22	0	125	15
	2013	66	62	24	-1	105	12
	2014	65	64	21	3	105	12
Belarus	2011	462	1	233	2	228	24
	2012	532	6	297	25	216	23
	2013	568	0	407	-31	192	20
	2014	573	40	432	11	170	18
Belgium	2011	570	1,788	919	-4	1,443	131
	2012	580	1,611	736	6	1,449	131
	2013	576	1,743	732	-1	1,588	142
	2014	707	1,830	669	6	1,862	166

Production, trade and supply of liquefied petroleum gases (LPG)

Thousand metric tons and kilograms per capita

Country or area	Year	Production	Imports	Exports	Changes in stocks	Supply	
						Total	Per Capita
Bosnia and Herzegovina	2011	25	55	..	0	80	21
	2012	23	56	..	0	79	21
	2013	23	40	..	1	62	16
	2014	24	43	..	1	66	17
Bulgaria	2011	118	268	13	7	366	50
	2012	119	287	20	-5	391	54
	2013	93	331	19	-1	406	56
	2014	98	348	16	-6	436	61
Croatia	2011	266	14	127	3	150	35
	2012	281	14	146	2	147	34
	2013	250	40	145	-1	146	34
	2014	225	44	132	-2	139	33
Czechia	2011	195	74	109	4	156	15
	2012	200	78	112	-2	168	16
	2013	180	107	90	0	197	19
	2014	208	89	82	3	212	20
Denmark	2011	132	8	94	0	46	8
	2012	168	5	117	1	55	10
	2013	103	18	74	3	44	8
	2014	118	18	98	-4	42	7
Estonia	2011	..	9	0	2	7	5
	2012	..	8	0	-2	10	8
	2013	..	13	4	0	9	7
	2014	..	16	6	1	9	7
Finland	2011	387	307	1	42	651	121
	2012	256	290	6	-34	574	106
	2013	310	293	10	-17	610	112
	2014	293	316	15	35	559	102
France	2011	1,877	2,694	1,098	11	3,462	55
	2012	1,466	2,915	1,302	-34	3,113	49
	2013	1,472	3,221	1,238	-20	3,475	54
	2014	1,474	3,479	1,061	-3	3,895	61
Germany	2011	2,612	870	272	38	3,172	39
	2012	2,624	1,088	303	-31	3,440	43
	2013	2,645	1,001	290	-7	3,363	42
	2014	2,518	672	276	-14	2,928	36
Greece	2011	559	54	240	-4	377	34
	2012	631	53	216	15	453	41
	2013	671	40	177	-1	535	48
	2014	735	39	244	-2	532	48
Hungary	2011	300	144	72	9	363	36
	2012	272	165	62	3	372	37
	2013	372	176	74	-7	481	48
	2014	385	243	76	10	542	55
Iceland	2011	..	3	..	..	3	9
	2012	..	3	..	..	3	9
	2013	..	3	..	..	3	9
	2014	..	2	..	..	2	6
Ireland	2011	67	89	30	2	124	27
	2012	73	80	35	0	118	25
	2013	62	92	26	1	127	27
	2014	60	98	30	1	127	27

Table 17

Production, trade and supply of liquefied petroleum gases (LPG)

Thousand metric tons and kilograms per capita

Country or area	Year	Production	Imports	Exports	Changes in stocks	Supply	
						Total	Per Capita
Italy	2011	2,072	2,018	394	190	3,506	59
	2012	1,541	2,135	391	-161	3,446	58
	2013	1,681	2,322	423	30	3,550	59
	2014	1,327	2,204	215	-35	3,351	56
Latvia	2011	..	139	85	1	53	26
	2012	..	153	81	0	72	35
	2013	..	187	104	-1	84	42
	2014	..	197	102	2	93	47
Lithuania	2011	323	112	162	1	272	89
	2012	299	112	143	-2	270	90
	2013	322	106	179	0	249	84
	2014	275	112	156	1	230	79
Luxembourg	2011	..	10	1	..	9	17
	2012	..	12	1	..	11	21
	2013	..	11	1	..	10	18
	2014	..	9	1	..	8	14
Malta	2011	..	21	..	0	21	51
	2012	..	23	..	1	22	53
	2013	..	22	..	1	21	50
	2014	..	22	..	1	21	50
Montenegro	2011	..	19	..	..	19	30
	2012	..	17	..	..	17	27
	2013	..	14	..	..	14	22
	2014	..	18	2	..	16	26
Netherlands	2011	1,454	2,308	1,483	22	2,257	135
	2012	1,477	1,732	1,574	7	1,628	97
	2013	1,424	3,570	1,428	-2	3,568	212
	2014	1,632	4,823	1,629	38	4,788	284
Norway	2011	5,685	173	5,055	-14	817	165
	2012	5,729	183	4,727	45	1,140	227
	2013	5,560	138	4,689	-27	1,036	204
	2014	5,775	167	4,923	-18	1,037	201
Poland	2011	446	2,000	67	11	2,368	61
	2012	539	1,945	114	6	2,364	61
	2013	548	2,053	301	0	2,300	60
	2014	560	2,032	225	8	2,359	61
Portugal	2011	318	475	75	11	707	67
	2012	221	431	77	4	571	54
	2013	261	604	68	-17	814	78
	2014	198	824	39	8	975	94
Republic of Moldova	2011	..	77	1	8	68	17
	2012	..	78	7	-4	75	18
	2013	..	90	12	3	75	18
	2014	..	78	1	-1	78	19
Romania	2011	582	103	198	1	486	24
	2012	527	84	180	3	428	21
	2013	523	112	203	-6	438	22
	2014	560	132	204	2	486	25
Russian Federation	2011	14,167	206	2,573	2	11,798	82
	2012	15,314	147	2,681	298	12,482	87
	2013	16,899	79	3,135	214	13,629	95
	2014	18,227	138	3,391	-173	15,147	106

Production, trade and supply of liquefied petroleum gases (LPG)

Thousand metric tons and kilograms per capita

Country or area	Year	Production	Imports	Exports	Changes in stocks	Supply	
						Total	Per Capita
Serbia	2011	105	315	25	3	392	43
	2012	113	269	9	4	369	41
	2013	193	164	14	-11	354	40
	2014	158	161	16	4	299	34
Slovakia	2011	131	42	9	-1	165	30
	2012	122	51	19	1	153	28
	2013	172	51	67	0	156	29
	2014	155	49	71	..	133	25
Slovenia	2011	..	90	7	3	80	39
	2012	..	81	1	-4	84	41
	2013	..	76	1	-2	77	37
	2014	..	74	..	..	74	36
Spain	2011	1,439	569	249	-33	1,792	38
	2012	1,702	355	300	66	1,691	36
	2013	1,713	326	398	-98	1,739	37
	2014	1,575	643	425	78	1,715	37
Sweden	2011	271	1,188	255	191	1,013	107
	2012	391	819	571	-181	820	86
	2013	308	1,025	307	-13	1,039	108
	2014	423	1,100	387	175	961	99
Switzerland	2011	184	41	45	..	180	23
	2012	119	75	20	..	174	22
	2013	193	34	55	..	172	21
	2014	173	35	39	1	168	20
T.F.Yug.Rep. Macedonia	2011	20	40	4	-1	56	27
	2012	10	51	3	2	55	26
	2013	3	57	3	-1	57	28
	2014	0	63	1	-1	62	30
Ukraine	2011	665	252	125	4	788	17
	2012	563	307	70	-2	802	18
	2013	500	361	..	2	859	19
	2014	440	431	..	-12	883	20
United Kingdom	2011	4,050	195	820	12	3,413	54
	2012	3,598	293	1,147	-9	2,753	43
	2013	3,607	413	1,165	-11	2,866	45
	2014	3,268	423	899	30	2,762	43
Oceania	**2011**	**2,919**	**514**	**1,362**	**-77**	**2,148**	**58**
	2012	**2,782**	**585**	**1,182**	**52**	**2,132**	**57**
	2013	**2,643**	**537**	**1,332**	**-81**	**1,929**	**51**
	2014	**2,784**	**427**	**1,394**	**41**	**1,776**	**46**
Australia	2011	2,765	454	1,346	-72	1,945	86
	2012	2,606	524	1,156	50	1,924	84
	2013	2,454	472	1,298	-80	1,708	73
	2014	2,553	372	1,338	42	1,545	65
Fiji	2011	..	18	0	..	18	21
	2012	..	19	0	..	19	22
	2013	..	21	0	..	21	24
	2014	..	22	0	..	22	25
French Polynesia	2011	..	*8	..	..	*8	*30
	2012	..	*8	..	..	*8	*29
	2013	..	*8	..	..	*8	*29
	2014	..	*8	..	..	*8	*27

Table 17

Production, trade and supply of liquefied petroleum gases (LPG)

Thousand metric tons and kilograms per capita

Country or area	Year	Production	Imports	Exports	Changes in stocks	Supply	
						Total	Per Capita
Kiribati	2011	..	0	..	..	0	*1
	2012	..	0	..	..	0	*1
	2013	..	0	..	..	0	*1
	2014	..	0	..	..	0	*1
Nauru	2011	..	0	..	..	0	*1
	2012	..	0	..	..	0	*1
	2013	..	0	..	..	0	*1
	2014	..	0	..	..	0	*1
New Caledonia	2011	..	11	..	2	9	36
	2012	..	12	..	-1	12	49
	2013	..	19	..	0	19	73
	2014	..	*18	..	..	*18	*68
New Zealand	2011	148	7	16	-7	146	33
	2012	166	14	26	3	151	34
	2013	177	2	34	-2	147	33
	2014	210	1	56	-1	156	35
Niue	2011	..	0	..	..	0	17
	2012	..	0	..	..	0	19
	2013	..	0	..	..	0	20
	2014	..	0	..	..	0	21
Papua New Guinea	2011	6	10	..	0	16	2
	2012	10	2	..	0	12	2
	2013	12	9	..	1	19	3
	2014	*21	0	..	0	*21	*3
Samoa	2011	..	2	..	0	1	7
	2012	..	*1	..	..	*1	*7
	2013	..	1	..	0	1	7
	2014	..	1	..	..	1	7
Solomon Islands	2011	..	*2	..	..	*2	*3
	2012	..	*2	..	..	*2	*3
	2013	..	*2	..	..	*2	*3
	2014	..	*2	..	..	*2	*3
Tonga	2011	..	1	..	..	1	13
	2012	..	1	..	..	1	13
	2013	..	1	..	..	1	14
	2014	..	2	..	..	2	15
Tuvalu	2011	..	0	..	..	0	2
	2012	..	0	..	..	0	*2
	2013	..	0	..	..	0	*2
	2014	..	0	..	..	0	*2
Vanuatu	2011	..	*1	..	..	*1	*5
	2012	..	*1	..	..	*1	*5
	2013	..	*1	..	..	*1	*5
	2014	..	*1	..	..	*1	*5
Wallis and Futuna Is.	2011	..	0	..	..	0	*17
	2012	..	0	..	..	0	*19
	2013	..	0	..	..	0	*18
	2014	..	0	..	..	0	*21

Production, trade and supply of motor gasoline

Thousand metric tons and kilograms per capita

Table Notes

Production includes output from refineries and plants, together with quantities from other sources which include any blended biogasoline.

Please refer to the Definitions Section on pages ix to xv for the appropriate product description/ classification.

Figure 48: Production of motor gasoline by region, 1990-2014

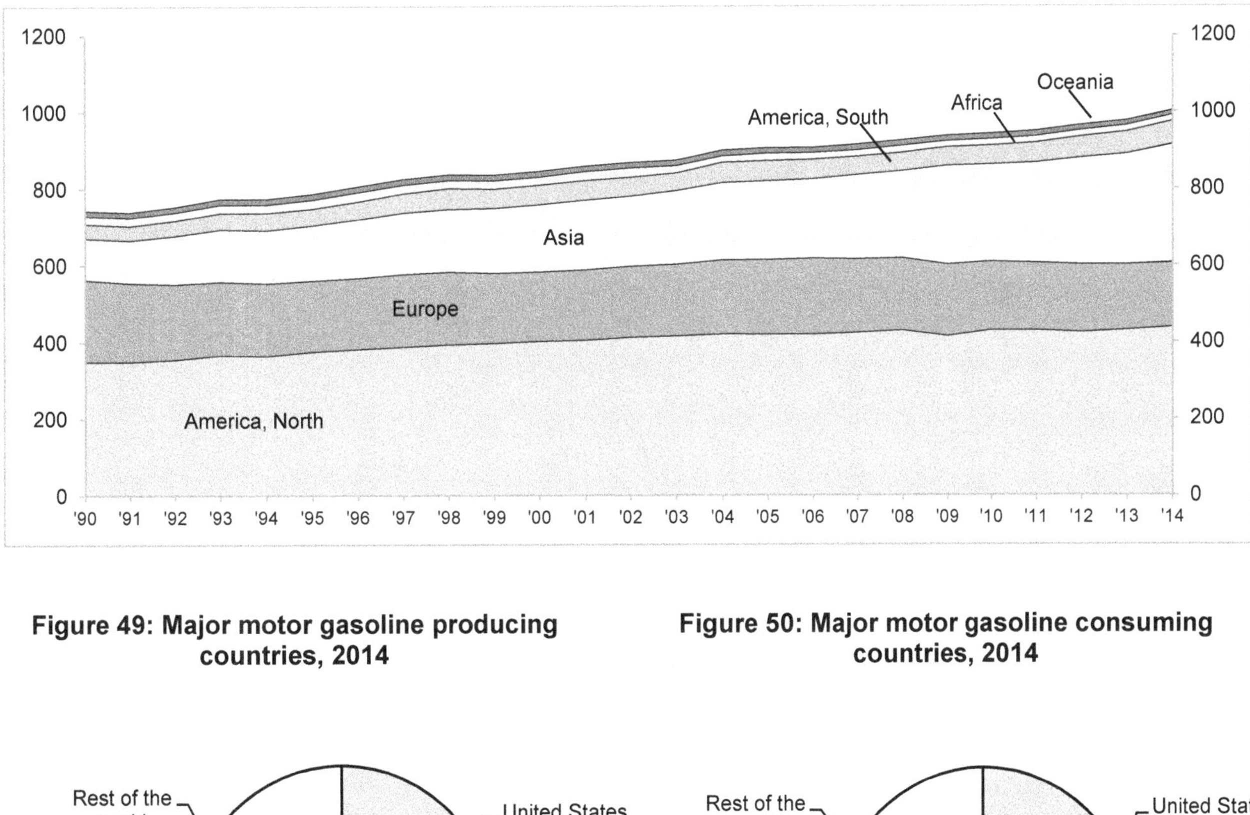

Figure 49: Major motor gasoline producing countries, 2014

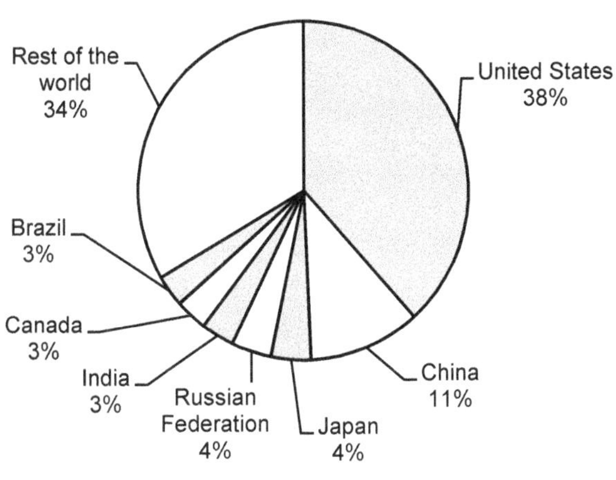

Figure 50: Major motor gasoline consuming countries, 2014

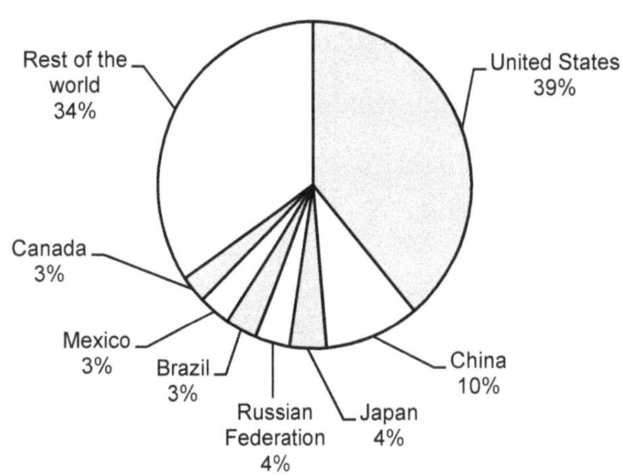

Table 18

Production, trade and supply of motor gasoline

Thousand metric tons and kilograms per capita

Country or area	Year	Production	Imports	Exports	Changes in stocks	Supply	
						Total	Per Capita
World	**2011**	**949,239**	**176,915**	**172,984**	**-1,114**	**954,284**	**136**
	2012	**964,630**	**183,063**	**184,778**	**-1,847**	**964,762**	**136**
	2013	**975,893**	**184,672**	**178,997**	**-1,833**	**983,400**	**137**
	2014	**1,002,243**	**190,008**	**187,790**	**7,267**	**997,195**	**137**
Africa	**2011**	**17,203**	**18,971**	**1,109**	**-62**	**35,126**	**33**
	2012	**17,096**	**21,473**	**1,200**	**19**	**37,350**	**34**
	2013	**16,565**	**23,033**	**2,061**	**181**	**37,356**	**33**
	2014	**15,670**	**25,514**	**4,553**	**-109**	**36,740**	**32**
Algeria	2011	2,568	395	..	-36	2,999	82
	2012	2,273	1,181	..	24	3,430	92
	2013	2,300	1,592	..	107	3,785	99
	2014	3,007	1,033	..	-18	4,058	104
Angola	2011	63	937	0	..	1,000	46
	2012	73	1,313	6	..	1,380	61
	2013	69	1,005	7	..	1,067	46
	2014	27	1,400	7	..	1,420	59
Benin	2011	..	616	13	..	603	62
	2012	..	645	14	..	631	63
	2013	..	689	15	..	674	65
	2014	..	734	16	..	718	68
Botswana	2011	..	353	..	..	353	169
	2012	..	348	..	..	348	163
	2013	..	343	..	..	343	158
	2014	..	355	..	..	355	160
Burkina Faso	2011	..	167	..	..	167	10
	2012	..	190	..	..	190	11
	2013	..	255	..	..	255	15
	2014	..	207	..	..	207	12
Burundi	2011	..	27	..	0	27	3
	2012	..	31	..	-1	32	3
	2013	..	30	..	1	29	3
	2014	..	50	..	-1	51	5
Cabo Verde	2011	..	7	..	..	7	15
	2012	..	7	..	..	7	14
	2013	..	7	..	..	7	14
	2014	..	7	..	..	7	14
Cameroon	2011	399	52	95	..	356	17
	2012	417	54	99	..	372	17
	2013	440	57	104	..	393	18
	2014	466	60	110	..	416	18
Central African Rep.	2011	..	*25	..	..	*25	*6
	2012	..	*27	..	..	*27	*6
	2013	..	*28	..	..	*28	*6
	2014	..	*28	..	..	*28	*6
Chad	2011	..	*10	..	..	*10	*1
	2012	..	*10	..	..	*10	*1
	2013	..	*10	..	..	*10	*1
	2014	..	*10	..	..	*10	*1
Comoros	2011	..	*10	..	..	*10	*15
	2012	..	*10	..	..	*10	*14
	2013	..	*11	..	..	*11	*15
	2014	..	*11	..	..	*11	*15

Production, trade and supply of motor gasoline

Thousand metric tons and kilograms per capita

Country or area	Year	Production	Imports	Exports	Changes in stocks	Supply	
						Total	Per Capita
Congo	2011	94	65	..	-2	161	39
	2012	80	54	..	..	134	31
	2013	70	76	..	..	146	33
	2014	78	86	..	..	164	36
Côte d'Ivoire	2011	390	..	251	0	139	7
	2012	559	..	417	0	142	7
	2013	647	..	361	0	286	13
	2014	548	..	252	-12	308	14
Dem. Rep. of the Congo	2011	..	251	..	..	251	4
	2012	..	239	..	..	239	3
	2013	..	376	..	..	376	5
	2014	..	513	..	..	513	7
Djibouti	2011	..	4	..	..	4	5
	2012	..	*4	..	..	*4	*5
	2013	..	*4	..	..	*4	*5
	2014	..	*5	..	..	*5	*5
Egypt	2011	4,476	1,189	..	..	5,665	68
	2012	4,201	1,193	..	..	5,394	63
	2013	2,839	1,610	..	..	4,449	51
	2014	2,830	1,005	..	..	3,835	43
Equatorial Guinea	2011	..	*23	..	..	*23	*31
	2012	..	*23	..	..	*23	*30
	2013	..	*23	..	..	*23	*29
	2014	..	*23	..	..	*23	*28
Eritrea	2011	..	9	..	..	9	2
	2012	..	10	..	..	10	2
	2013	..	10	..	..	10	2
	2014	..	10	..	..	10	2
Ethiopia	2011	..	156	..	2	154	2
	2012	..	193	..	7	186	2
	2013	..	212	..	1	211	2
	2014	..	233	..	..	233	2
Gabon	2011	79	..	7	..	72	46
	2012	76	..	8	..	68	42
	2013	62	16	..	..	78	47
	2014	68	12	..	..	80	47
Gambia	2011	..	22	..	..	22	13
	2012	..	21	..	..	21	12
	2013	..	22	..	..	22	12
	2014	..	25	..	..	25	13
Ghana	2011	344	713	258	0	799	32
	2012	158	812	54	0	915	36
	2013	167	1,017	36	6	1,143	44
	2014	40	1,254	10	11	1,273	48
Guinea	2011	..	227	..	..	227	20
	2012	..	241	..	..	241	21
	2013	..	251	..	..	251	21
	2014	..	*255	..	..	*255	*21
Guinea-Bissau	2011	..	*2	..	..	*2	*1
	2012	..	*2	..	..	*2	*1
	2013	..	*2	..	..	*2	*1
	2014	..	*2	..	..	*2	*1

Table 18

Production, trade and supply of motor gasoline

Thousand metric tons and kilograms per capita

Country or area	Year	Production	Imports	Exports	Changes in stocks	Supply	
						Total	Per Capita
Kenya	2011	178	384	0	..	562	14
	2012	117	502	0	..	619	15
	2013	76	1,207	424	..	859	20
	2014	0	3,850	2,956	..	895	20
Lesotho	2011	..	75	..	..	75	37
	2012	..	75	..	..	75	37
	2013	..	77	..	..	77	37
	2014	..	*78	..	..	*78	*37
Liberia	2011	..	95	..	..	95	23
	2012	..	92	..	..	92	22
	2013	..	113	..	..	113	26
	2014	..	60	..	..	60	14
Libya	2011	569	1,997	..	..	2,566	408
	2012	615	2,819	..	..	3,434	547
	2013	640	3,160	..	..	3,800	606
	2014	516	3,306	..	..	3,822	611
Madagascar	2011	..	86	..	-1	87	4
	2012	..	91	..	6	85	4
	2013	..	94	..	0	94	4
	2014	..	*95	..	0	*95	*4
Malawi	2011	..	72	..	..	72	5
	2012	..	76	..	..	76	5
	2013	..	82	..	..	82	5
	2014	..	96	..	..	96	6
Mali	2011	..	*73	..	..	*73	*5
	2012	..	*74	..	..	*74	*5
	2013	..	*76	..	..	*76	*5
	2014	..	*76	..	..	*76	*4
Mauritania	2011	..	28	..	..	28	8
	2012	..	29	..	..	29	8
	2013	..	26	..	..	26	7
	2014	..	*26	..	..	*26	*6
Mauritius	2011	..	117	..	-3	120	96
	2012	..	128	..	2	126	100
	2013	..	138	..	6	132	104
	2014	..	138	..	-3	141	111
Mayotte	2011	..	11	..	-1	12	57
	2012	..	*12	..	..	*12	*55
	2013	..	*12	..	..	*12	*53
	2014	..	*12	..	..	*12	*51
Morocco	2011	415	142	0	11	546	17
	2012	407	159	0	3	563	17
	2013	369	173	2	-13	553	16
	2014	331	181	3	-31	540	16
Mozambique	2011	..	168	..	0	168	7
	2012	..	177	..	0	177	7
	2013	..	277	..	98	179	7
	2014	..	404	..	-30	434	16
Namibia	2011	..	294	..	..	294	131
	2012	..	315	..	..	315	137
	2013	..	177	..	..	177	75
	2014	..	301	..	..	301	125

Production, trade and supply of motor gasoline

Thousand metric tons and kilograms per capita

Country or area	Year	Production	Imports	Exports	Changes in stocks	Supply	
						Total	Per Capita
Niger	2011	18	63	0	*13	68	4
	2012	195	0	47	0	148	8
	2013	253	0	135	0	118	6
	2014	235	3	*98	0	140	7
Nigeria	2011	1,277	6,078	..	..	7,355	45
	2012	1,135	6,539	..	..	7,674	46
	2013	1,237	5,917	..	..	7,154	41
	2014	542	5,781	..	..	6,323	36
Réunion	2011	..	100	..	-9	109	130
	2012	..	100	..	2	98	116
	2013	..	89	..	-7	96	112
	2014	..	93	..	0	93	108
Rwanda	2011	..	*65	..	..	*65	*6
	2012	..	74	..	..	74	7
	2013	..	77	..	..	77	7
	2014	..	78	..	..	78	7
Sao Tome and Principe	2011	..	*8	..	..	*8	*43
	2012	..	*8	..	..	*8	*45
	2013	..	*8	..	..	*8	*44
	2014	..	*8	..	..	*8	*43
Senegal	2011	99	23	2	-2	122	9
	2012	75	52	3	..	124	9
	2013	29	68	40	..	57	4
	2014	30	58	41	..	47	3
Seychelles	2011	..	13	..	..	13	140
	2012	..	15	..	..	15	155
	2013	..	16	..	..	16	170
	2014	..	20	..	..	20	210
Sierra Leone	2011	..	77	..	..	77	13
	2012	..	68	..	..	68	11
	2013	..	86	..	..	86	14
	2014	..	92	..	..	92	15
Somalia	2011	..	*34	..	..	*34	*3
	2012	..	*34	..	..	*34	*3
	2013	..	*34	..	..	*34	*3
	2014	..	*34	..	..	*34	*3
South Africa	2011	5,005	1,815	176	..	6,644	127
	2012	5,527	1,352	328	..	6,551	124
	2013	6,273	1,116	708	..	6,681	125
	2014	5,771	851	758	..	5,864	109
South Sudan	2011	..	..	..	..	..	..
	2012	..	50	..	..	50	5
	2013	..	47	..	..	47	4
	2014	..	49	..	..	49	4
St. Helena and Depend.	2011	..	1	..	..	1	143
	2012	..	1	..	..	1	159
	2013	..	1	..	..	1	149
	2014	..	*1	..	..	*1	*150
Sudan	2011	..	..	..	..	..	..
	2012	1,099	32	224	..	907	24
	2013	1,009	41	230	..	820	21
	2014	1,085	40	302	..	823	21

Table 18

Production, trade and supply of motor gasoline

Thousand metric tons and kilograms per capita

Country or area	Year	Production	Imports	Exports	Changes in stocks	Supply	
						Total	Per Capita
Sudan (former)	2011	1,150	..	307	..	843	18
	2012	..	..	..	..	..	..
	2013	..	..	..	..	..	..
	2014	..	..	..	..	..	..
Swaziland	2011	..	*90	..	..	*90	*74
	2012	..	*92	..	..	*92	*75
	2013	..	*93	..	..	*93	*74
	2014	..	95	..	..	95	75
Togo	2011	..	116	..	-53	169	26
	2012	..	125	..	-20	145	21
	2013	..	128	..	-21	149	22
	2014	..	131	..	-22	153	22
Tunisia	2011	..	527	..	18	509	47
	2012	..	489	..	-4	493	45
	2013	..	483	..	3	480	44
	2014	..	535	..	-4	539	48
Uganda	2011	..	354	..	..	354	10
	2012	..	367	..	..	367	10
	2013	..	420	..	..	420	11
	2014	..	477	..	..	477	13
United Rep. of Tanzania	2011	..	392	..	..	392	8
	2012	..	415	..	..	415	9
	2013	..	564	..	..	564	11
	2014	..	687	..	..	687	13
Zambia	2011	78	99	..	..	177	12
	2012	89	142	..	..	231	16
	2013	85	185	..	..	270	18
	2014	95	203	..	..	298	19
Zimbabwe	2011	..	312	..	..	312	22
	2012	..	362	..	..	362	25
	2013	..	401	..	..	401	27
	2014	..	337	..	..	337	22
America, North	**2011**	**429,288**	**60,056**	**30,970**	**499**	**457,876**	**829**
	2012	**424,597**	**56,391**	**30,312**	**-3,853**	**454,529**	**815**
	2013	**430,362**	**53,018**	**30,938**	**-416**	**452,858**	**804**
	2014	**437,683**	**51,870**	**32,600**	**1,001**	**455,951**	**802**
Anguilla	2011	..	6	..	..	6	437
	2012	..	8	..	..	8	545
	2013	..	*8	..	..	*8	*524
	2014	..	*8	..	..	*8	*526
Antigua and Barbuda	2011	..	*40	*3	..	*37	*420
	2012	..	*41	*3	..	*38	*427
	2013	..	*41	*3	..	*38	*422
	2014	..	*42	*3	..	*39	*429
Aruba	2011	..	*60	..	..	*60	*589
	2012	..	*60	..	..	*60	*589
	2013	..	*61	..	..	*61	*589
	2014	..	*61	..	..	*61	*593
Bahamas	2011	..	195	..	..	195	532
	2012	..	210	..	..	210	564
	2013	..	269	*1	..	268	710
	2014	..	247	*1	..	246	643

Production, trade and supply of motor gasoline

Thousand metric tons and kilograms per capita

Country or area	Year	Production	Imports	Exports	Changes in stocks	Supply	
						Total	Per Capita
Barbados	2011	..	90	..	0	90	322
	2012	..	90	..	0	90	320
	2013	..	88	..	0	88	312
	2014	..	82	..	-5	87	307
Belize	2011	..	89	..	..	89	270
	2012	..	47	..	..	47	139
	2013	..	49	..	..	49	143
	2014	..	52	..	..	52	149
Bermuda	2011	..	33	..	..	33	524
	2012	..	66	..	..	66	1,050
	2013	..	23	..	..	23	364
	2014	..	23	..	..	23	376
Bonaire, St Eustatius, Saba	2011	..	..	..	..	..	..
	2012	..	*10	..	..	*10	*414
	2013	..	*10	..	..	*10	*409
	2014	..	*10	..	..	*10	*399
British Virgin Islands	2011	..	*14	..	..	*14	*502
	2012	..	*14	..	..	*14	*491
	2013	..	*14	..	..	*14	*482
	2014	..	*15	..	..	*15	*490
Canada	2011	30,648	4,092	5,580	-21	29,181	846
	2012	31,555	2,932	5,933	-96	28,650	822
	2013	30,993	2,395	6,158	-194	27,424	778
	2014	30,948	2,784	6,121	277	27,334	768
Cayman Islands	2011	..	43	..	..	43	760
	2012	..	37	..	..	37	643
	2013	..	38	..	..	38	651
	2014	..	38	..	..	38	635
Costa Rica	2011	31	671	0	-6	707	154
	2012	..	753	0	7	746	160
	2013	..	759	..	-9	768	163
	2014	..	794	..	22	772	162
Cuba	2011	503	394	..	..	897	79
	2012	415	560	..	..	975	86
	2013	672	274	..	..	946	83
	2014	554	311	..	..	866	76
Curaçao	2011	..	..	..	..	..	..
	2012	1,574	64	1,574	..	64	419
	2013	1,712	64	1,712	..	64	415
	2014	1,890	65	1,890	..	65	417
Dominica	2011	..	*13	..	..	*13	*182
	2012	..	*13	..	..	*13	*181
	2013	..	*13	..	..	*13	*181
	2014	..	*13	..	..	*13	*180
Dominican Republic	2011	242	969	..	-1	1,212	121
	2012	231	1,022	..	-11	1,263	124
	2013	265	743	..	7	1,001	97
	2014	267	572	..	13	825	79
El Salvador	2011	92	345	4	-28	461	76
	2012	72	365	0	-3	440	72
	2013	0	453	0	0	453	74
	2014	0	476	0	0	476	78

Table 18

Production, trade and supply of motor gasoline

Thousand metric tons and kilograms per capita

Country or area	Year	Production	Imports	Exports	Changes in stocks	Supply	
						Total	Per Capita
Greenland	2011	..	15	..	1	14	240
	2012	..	12	..	-2	14	240
	2013	..	14	..	0	14	240
	2014	..	13	..	-1	14	241
Grenada	2011	..	25	..	0	25	241
	2012	..	24	..	0	24	230
	2013	..	24	..	0	25	232
	2014	..	24	..	0	24	221
Guadeloupe	2011	..	*108	..	..	*108	*235
	2012	..	*110	..	..	*110	*239
	2013	..	*111	..	..	*111	*239
	2014	..	*112	..	..	*112	*240
Guatemala	2011	..	938	20	-14	932	62
	2012	..	929	0	0	929	60
	2013	..	997	33	-49	1,014	65
	2014	..	1,195	60	32	1,103	69
Haiti	2011	..	130	..	..	130	13
	2012	8	154	..	..	162	16
	2013	..	149	..	..	149	14
	2014	..	178	..	..	178	17
Honduras	2011	..	508	..	4	504	66
	2012	..	509	..	0	509	66
	2013	..	518	..	0	518	66
	2014	..	497	..	-48	545	68
Jamaica	2011	144	347	..	-7	498	181
	2012	133	354	..	3	484	175
	2013	124	372	..	-14	510	184
	2014	103	316	..	0	419	151
Martinique	2011	*147	*17	*47	..	*117	*296
	2012	*149	*17	*47	..	*119	*301
	2013	*149	*16	*44	..	*121	*306
	2014	*156	*17	*50	..	*123	*311
Mexico	2011	16,725	16,437	..	233	32,929	274
	2012	17,518	16,126	..	0	33,644	276
	2013	18,262	14,522	..	-46	32,830	265
	2014	17,581	14,891	..	27	32,445	259
Montserrat	2011	..	4	..	..	4	821
	2012	..	4	..	..	4	855
	2013	..	5	..	..	5	968
	2014	..	5	..	..	5	1,060
Neth. Antilles (former)	2011	1,511	96	1,511	..	96	450
	2012	..	..	..	..	..	..
	2013	..	..	..	..	..	..
	2014	..	..	..	..	..	..
Nicaragua	2011	97	124	..	-3	224	39
	2012	68	181	..	7	242	41
	2013	97	160	..	13	244	41
	2014	96	163	..	-9	268	45
Panama	2011	..	591	..	-29	620	168
	2012	..	630	..	-49	679	181
	2013	..	652	..	-56	708	186
	2014	..	687	..	-50	738	191

Production, trade and supply of motor gasoline

Thousand metric tons and kilograms per capita

Country or area	Year	Production	Imports	Exports	Changes in stocks	Supply	
						Total	Per Capita
Sint Maarten (Dutch part)	2011	..	..	..	..	..	..
	2012	..	*13	..	..	*13	*370
	2013	..	*13	..	..	*13	*365
	2014	..	*14	..	..	*14	*361
St. Kitts-Nevis	2011	..	16	..	..	16	296
	2012	..	17	..	1	17	309
	2013	..	*16	..	..	*16	*295
	2014	..	*17	..	..	*17	*300
St. Lucia	2011	..	*48	..	..	*48	*268
	2012	..	*48	..	..	*48	*265
	2013	..	*48	..	..	*48	*263
	2014	..	*49	..	..	*49	*267
St. Pierre-Miquelon	2011	..	*4	..	..	*4	*638
	2012	..	*4	..	..	*4	*637
	2013	..	*4	..	..	*4	*637
	2014	..	*4	..	..	*4	*669
St. Vincent-Grenadines	2011	..	22	..	..	22	205
	2012	..	21	..	..	21	194
	2013	..	17	..	..	17	152
	2014	..	*17	..	..	*17	*153
Trinidad and Tobago	2011	1,674	122	1,256	38	502	376
	2012	1,132	337	951	0	518	386
	2013	1,640	..	1,281	-151	510	378
	2014	1,259	..	823	-83	519	383
United States	2011	377,475	33,449	22,549	331	388,044	1,242
	2012	371,742	30,609	21,804	-3,710	384,257	1,221
	2013	376,449	30,078	21,706	84	384,737	1,213
	2014	384,828	28,079	23,653	826	388,428	1,216
America, South	**2011**	**51,214**	**5,128**	**3,831**	**190**	**52,321**	**130**
	2012	**54,771**	**7,025**	**3,186**	**-1,324**	**59,935**	**148**
	2013	**57,335**	**6,155**	**3,271**	**-881**	**61,100**	**149**
	2014	**61,065**	**6,419**	**2,475**	**-385**	**65,394**	**158**
Argentina	2011	5,335	105	1	396	5,043	121
	2012	5,539	39	..	-141	5,719	136
	2013	6,031	39	..	77	5,993	141
	2014	5,244	330	..	14	5,560	129
Bolivia (Plur. State of)	2011	629	137	..	5	761	76
	2012	686	159	..	0	845	83
	2013	783	153	..	-1	937	90
	2014	794	211	..	..	1,005	95
Brazil	2011	24,204	1,619	229	-110	25,704	128
	2012	25,353	2,798	91	-972	29,032	143
	2013	28,844	1,677	246	-990	31,265	153
	2014	30,678	1,563	258	-480	32,463	158
Chile	2011	2,194	609	125	-21	2,699	157
	2012	2,246	572	243	-28	2,603	150
	2013	2,619	542	197	3	2,961	168
	2014	2,789	440	77	12	3,140	177
Colombia	2011	*2,312	125	574	-2	*1,865	*40
	2012	2,930	477	352	23	3,032	65
	2013	3,244	449	383	111	3,199	68
	2014	3,764	609	0	119	4,254	89

Table 18

Production, trade and supply of motor gasoline

Thousand metric tons and kilograms per capita

Country or area	Year	Production	Imports	Exports	Changes in stocks	Supply	
						Total	Per Capita
Ecuador	2011	1,467	1,507	90	-6	2,891	190
	2012	1,448	1,700	175	-20	2,993	194
	2013	1,300	1,912	165	23	3,026	193
	2014	1,593	2,376	151	-31	3,850	242
Falkland Is. (Malvinas)	2011	..	*1	..	..	*1	*350
	2012	..	*1	..	..	*1	*349
	2013	..	*1	..	..	*1	*348
	2014	..	*1	..	..	*1	*346
French Guiana	2011	..	*31	..	..	*31	*129
	2012	..	*32	..	..	*32	*129
	2013	..	*33	..	..	*33	*130
	2014	..	*33	..	..	*33	*126
Guyana	2011	..	117	..	-4	121	160
	2012	..	136	..	3	133	175
	2013	..	132	..	-5	137	180
	2014	..	139	..	-8	147	193
Paraguay	2011	110	350	..	-35	495	79
	2012	118	347	..	-52	518	81
	2013	138	370	..	-4	512	79
	2014	149	409	..	3	555	85
Peru	2011	3,199	221	1,571	-64	1,913	64
	2012	3,311	185	1,872	-174	1,798	60
	2013	3,720	409	1,904	43	2,182	71
	2014	3,724	108	1,562	8	2,262	73
Suriname	2011	..	131	..	..	131	250
	2012	..	141	..	..	141	268
	2013	..	149	..	..	149	280
	2014	..	149	..	..	149	277
Uruguay	2011	349	165	48	31	436	129
	2012	473	82	81	-6	479	141
	2013	537	15	22	14	515	151
	2014	496	51	23	-22	546	160
Venezuela (Bolivar. Rep.)	2011	11,414	11	1,194	0	10,231	348
	2012	12,668	356	371	44	12,609	422
	2013	10,119	273	354	-152	10,190	337
	2014	11,835	..	405	..	11,430	372
Asia	**2011**	**262,183**	**57,153**	**60,442**	**-1,327**	**260,221**	**62**
	2012	**279,192**	**61,509**	**64,682**	**3,639**	**272,380**	**64**
	2013	**288,791**	**68,129**	**61,466**	**-1,691**	**297,144**	**69**
	2014	**307,959**	**71,617**	**67,706**	**7,149**	**304,721**	**70**
Afghanistan	2011	..	1,590	..	..	1,590	55
	2012	..	*1,261	..	..	*1,261	*42
	2013	..	932	..	..	932	30
	2014	..	713	..	..	713	23
Armenia	2011	..	164	..	..	164	55
	2012	..	136	..	..	136	45
	2013	..	132	..	..	132	44
	2014	..	129	..	..	129	43
Azerbaijan	2011	1,296	..	96	26	1,174	127
	2012	1,297	..	60	-25	1,262	135
	2013	1,407	..	25	66	1,316	139
	2014	1,238	194	13	30	1,389	144

Production, trade and supply of motor gasoline

Thousand metric tons and kilograms per capita

Country or area	Year	Production	Imports	Exports	Changes in stocks	Supply	
						Total	Per Capita
Bahrain	2011	735	..	89	46	600	459
	2012	664	..	0	-5	669	502
	2013	596	..	1	-108	703	521
	2014	753	..	0	4	749	550
Bangladesh	2011	99	86	..	..	185	1
	2012	114	116	..	..	230	1
	2013	137	98	..	..	235	1
	2014	130	36	..	..	166	1
Bhutan	2011	..	20	..	..	20	28
	2012	..	22	..	..	22	30
	2013	..	22	..	..	22	30
	2014	..	23	..	..	23	30
Brunei Darussalam	2011	168	74	..	-4	246	616
	2012	146	116	..	1	261	644
	2013	167	99	..	1	265	644
	2014	118	163	..	9	272	652
Cambodia	2011	..	399	..	..	399	27
	2012	..	388	..	..	388	26
	2013	..	392	..	..	392	26
	2014	..	427	..	..	427	28
China	2011	80,596	29	4,060	1,170	75,396	56
	2012	89,135	5	2,917	5,203	81,020	60
	2013	97,665	0	4,687	-42	93,020	68
	2014	109,574	34	5,075	7,550	96,982	71
China, Hong Kong SAR	2011	..	544	8	13	523	74
	2012	..	557	11	-1	547	77
	2013	..	560	14	-19	565	79
	2014	..	515	18	-91	588	81
China, Macao SAR	2011	..	62	..	..	62	113
	2012	..	65	..	..	65	117
	2013	..	71	..	..	71	125
	2014	..	72	..	-3	75	130
Cyprus	2011	..	376	..	0	376	337
	2012	..	364	..	0	364	322
	2013	..	343	..	-2	345	302
	2014	..	342	..	2	340	295
Georgia	2011	0	402	0	3	399	95
	2012	4	384	0	0	388	94
	2013	0	376	2	-8	383	94
	2014	0	374	1	-8	381	95
India	2011	27,186	654	14,524	0	13,316	11
	2012	30,118	147	16,657	0	13,608	11
	2013	30,275	235	13,514	-132	17,128	13
	2014	32,325	328	15,942	-1,876	18,587	14
Indonesia	2011	9,326	11,504	10	..	20,820	85
	2012	9,506	13,176	15	..	22,667	91
	2013	8,418	14,648	9	..	23,057	92
	2014	8,891	14,932	19	..	23,804	94
Iran (Islamic Rep. of)	2011	12,786	1,335	19	..	14,101	188
	2012	14,695	411	..	..	15,106	198
	2013	16,358	992	..	..	17,349	225
	2014	16,356	1,230	..	..	17,586	225

Table 18

Production, trade and supply of motor gasoline

Thousand metric tons and kilograms per capita

Country or area	Year	Production	Imports	Exports	Changes in stocks	Supply	
						Total	Per Capita
Iraq	2011	2,739	2,269	..	-99	5,107	160
	2012	3,094	2,332	..	16	5,410	164
	2013	3,089	2,024	..	-150	5,263	154
	2014	2,413	1,851	..	-40	4,304	122
Israel	2011	2,702	929	825	-49	2,855	377
	2012	2,809	1,007	762	11	3,043	395
	2013	3,110	566	883	16	2,777	355
	2014	3,149	566	883	..	2,832	357
Japan	2011	40,139	2,140	923	121	41,235	324
	2012	39,142	2,125	847	71	40,349	317
	2013	40,183	1,222	1,266	0	40,139	316
	2014	39,365	1,107	2,211	-48	38,309	302
Jordan	2011	681	405	..	-11	1,097	162
	2012	716	423	..	34	1,105	158
	2013	663	507	..	-11	1,182	164
	2014	634	520	..	2	1,153	155
Kazakhstan	2011	2,775	643	13	-218	3,623	219
	2012	2,877	1,264	41	298	3,802	226
	2013	2,738	1,140	38	-6	3,846	225
	2014	3,023	1,143	22	319	3,825	220
Korea, Dem.Ppl's.Rep.	2011	176	12	..	..	188	8
	2012	176	11	..	..	187	8
	2013	194	11	..	..	205	8
	2014	178	11	..	..	189	8
Korea, Republic of	2011	13,673	0	6,255	-75	7,493	152
	2012	15,306	12	7,422	121	7,775	157
	2013	15,162	222	7,486	188	7,710	155
	2014	16,359	..	8,575	245	7,539	151
Kuwait	2011	2,206	..	56	-379	2,529	781
	2012	2,202	..	67	-491	2,626	768
	2013	2,420	..	69	-462	2,813	783
	2014	2,455	..	70	-486	2,871	765
Kyrgyzstan	2011	16	559	10	-27	592	107
	2012	11	908	5	29	885	157
	2013	9	802	17	-11	805	140
	2014	7	676	25	217	441	75
Lao People's Dem. Rep.	2011	..	*26	..	..	*26	*4
	2012	..	*26	..	..	*26	*4
	2013	..	*28	..	..	*28	*4
	2014	..	*29	..	..	*29	*4
Lebanon	2011	..	1,598	..	..	1,598	348
	2012	..	1,685	..	..	1,685	342
	2013	..	1,596	..	..	1,596	302
	2014	..	1,720	..	..	1,720	306
Malaysia	2011	3,425	3,968	0	-13	7,406	259
	2012	4,395	5,633	1,978	-166	8,216	283
	2013	4,475	8,386	82	700	12,079	410
	2014	4,681	7,741	34	33	12,355	413
Maldives	2011	..	31	..	..	31	92
	2012	..	38	..	..	38	110
	2013	..	43	..	..	43	122
	2014	..	44	..	..	44	124

Production, trade and supply of motor gasoline

Thousand metric tons and kilograms per capita

Country or area	Year	Production	Imports	Exports	Changes in stocks	Supply	
						Total	Per Capita
Mongolia	2011	..	325	..	..	325	118
	2012	..	389	..	..	389	139
	2013	..	380	..	..	380	133
	2014	..	416	..	..	416	143
Myanmar	2011	432	91	..	-25	549	11
	2012	334	216	..	12	538	10
	2013	230	323	..	2	551	10
	2014	234	894	..	-209	1,337	25
Nepal	2011	..	203	..	..	203	7
	2012	..	223	..	..	223	8
	2013	..	253	..	..	253	9
	2014	..	288	..	..	288	10
Oman	2011	2,727	139	765	-37	2,138	666
	2012	2,752	77	404	12	2,413	681
	2013	2,548	186	250	-92	2,576	659
	2014	2,973	..	232	-18	2,759	651
Other Asia	2011	6,110	57	2,676	-114	3,605	155
	2012	6,740	..	3,220	44	3,476	149
	2013	7,274	..	4,068	30	3,176	136
	2014	7,460	..	4,032	-143	3,571	153
Pakistan	2011	1,227	1,554	1	-4	2,784	16
	2012	1,514	1,818	0	-21	3,353	19
	2013	1,595	2,296	0	13	3,878	21
	2014	1,670	2,392	0	0	4,062	22
Philippines	2011	1,508	1,312	32	10	2,779	29
	2012	1,412	1,456	14	18	2,837	30
	2013	1,255	1,709	10	-78	3,033	31
	2014	1,303	1,744	7	28	3,013	30
Qatar	2011	1,859	..	714	-9	1,154	606
	2012	1,956	..	598	11	1,347	668
	2013	1,600	..	335	..	1,265	602
	2014	1,976	..	300	..	1,676	772
Saudi Arabia	2011	16,715	2,220	1,850	-1,961	19,046	662
	2012	17,103	3,945	1,774	-1,350	20,624	699
	2013	15,790	5,669	1,629	-1,757	21,587	715
	2014	18,867	6,766	2,027	1,249	22,357	724
Singapore	2011	*10,219	13,974	23,155	193	845	163
	2012	*10,224	13,213	22,948	-271	760	143
	2013	10,118	13,132	22,209	180	862	159
	2014	*10,736	13,963	23,847	1	852	155
Sri Lanka	2011	206	508	..	5	709	35
	2012	152	575	..	-24	751	37
	2013	144	539	..	-104	787	38
	2014	152	585	..	-99	836	41
State of Palestine	2011	..	134	..	0	134	32
	2012	..	144	..	0	144	33
	2013	..	176	..	0	176	40
	2014	..	198	..	-11	209	46
Syrian Arab Republic	2011	1,014	711	..	..	1,725	84
	2012	677	620	..	..	1,297	65
	2013	448	825	..	..	1,273	66
	2014	448	523	..	..	971	52

Table 18

Production, trade and supply of motor gasoline

Thousand metric tons and kilograms per capita

Country or area	Year	Production	Imports	Exports	Changes in stocks	Supply	
						Total	Per Capita
Tajikistan	2011	0	85	..	..	85	11
	2012	1	149	..	..	150	19
	2013	1	138	..	..	139	17
	2014	1	246	..	..	247	30
Thailand	2011	5,797	24	676	*41	5,104	76
	2012	6,354	217	1,092	*128	5,351	80
	2013	6,474	191	1,211	146	5,308	79
	2014	6,307	644	976	518	5,457	81
Timor-Leste	2011	..	*18	..	..	*18	*17
	2012	..	*20	..	..	*20	*18
	2013	..	*23	..	..	*23	*20
	2014	..	*23	..	..	*23	*20
Turkey	2011	4,280	136	2,420	57	1,939	26
	2012	4,379	48	2,631	2	1,794	24
	2013	4,314	20	2,506	-81	1,909	25
	2014	4,008	..	2,088	-25	1,945	25
Turkmenistan	2011	2,130	..	902	..	1,228	240
	2012	2,010	..	851	..	1,159	224
	2013	2,220	..	940	..	1,280	244
	2014	2,610	..	1,105	..	1,505	284
United Arab Emirates	2011	3,535	1,984	..	..	5,519	632
	2012	3,750	1,961	..	..	5,711	638
	2013	3,912	3,104	..	..	7,016	776
	2014	3,744	4,364	..	..	8,108	892
Uzbekistan	2011	1,308	..	28	..	1,280	45
	2012	1,226	..	26	..	1,200	42
	2013	1,105	..	23	..	1,082	37
	2014	1,041	..	22	..	1,019	35
Viet Nam	2011	1,953	2,915	335	14	4,519	51
	2012	2,072	2,805	342	-18	4,553	50
	2013	2,515	2,323	193	31	4,614	50
	2014	2,618	2,421	182	0	4,857	53
Yemen	2011	438	943	..	..	1,381	57
	2012	128	1,022	..	..	1,150	46
	2013	182	1,393	..	..	1,575	62
	2014	161	1,230	..	..	1,391	53
Europe	**2011**	**175,567**	**32,338**	**76,505**	**-454**	**131,854**	**179**
	2012	**175,884**	**32,557**	**85,270**	**-325**	**123,496**	**168**
	2013	**170,318**	**30,170**	**81,187**	**869**	**118,432**	**161**
	2014	**168,564**	**30,472**	**80,359**	**-261**	**118,938**	**161**
Albania	2011	..	119	..	..	119	41
	2012	..	100	..	..	100	35
	2013	..	147	..	36	111	39
	2014	..	177	..	57	120	42
Andorra	2011	..	21	..	..	21	252
	2012	..	20	..	..	20	254
	2013	..	19	..	..	19	248
	2014	..	19	..	..	19	255
Austria	2011	1,614	779	569	63	1,761	209
	2012	1,635	884	776	-48	1,791	212
	2013	1,631	814	829	-102	1,718	202
	2014	1,868	771	871	64	1,704	200

Production, trade and supply of motor gasoline

Thousand metric tons and kilograms per capita

Country or area	Year	Production	Imports	Exports	Changes in stocks	Supply	
						Total	Per Capita
Belarus	2011	3,135	2	1,975	-47	1,209	127
	2012	3,729	1	2,606	37	1,087	115
	2013	3,683	3	2,517	48	1,121	118
	2014	3,945	1	2,752	58	1,136	120
Belgium	2011	3,857	1,350	4,176	-71	1,102	100
	2012	4,577	931	4,446	-27	1,089	98
	2013	4,267	631	3,874	12	1,012	91
	2014	4,622	828	4,220	93	1,137	101
Bosnia and Herzegovina	2011	74	159	5	0	228	59
	2012	109	129	46	0	192	50
	2013	146	118	63	1	200	52
	2014	152	118	78	3	189	50
Bulgaria	2011	1,531	236	1,214	15	538	73
	2012	1,658	254	1,386	13	513	70
	2013	1,757	228	1,526	37	422	58
	2014	1,600	238	1,358	-2	482	67
Croatia	2011	871	216	501	-49	635	148
	2012	990	169	549	20	590	138
	2013	928	99	472	-21	576	135
	2014	805	142	428	-14	533	125
Czechia	2011	1,415	622	299	-56	1,794	170
	2012	1,534	460	298	27	1,669	158
	2013	1,341	559	332	-2	1,570	149
	2014	1,576	449	459	-27	1,593	151
Denmark	2011	1,931	718	1,065	38	1,546	277
	2012	2,011	758	1,364	-13	1,418	253
	2013	2,033	696	1,348	2	1,379	245
	2014	1,854	715	1,215	-6	1,360	241
Estonia	2011	..	261	0	0	261	197
	2012	..	511	261	-2	252	190
	2013	..	358	118	5	235	178
	2014	..	434	194	-2	242	184
Faeroe Islands	2011	..	*12	..	..	*12	*239
	2012	..	*11	..	..	*11	*230
	2013	..	*14	..	..	*14	*295
	2014	..	*11	..	..	*11	*220
Finland	2011	4,212	28	2,443	5	1,792	332
	2012	4,313	62	2,491	9	1,875	346
	2013	4,303	614	3,102	64	1,751	321
	2014	4,436	493	3,123	-72	1,878	343
France	2011	13,418	449	4,783	-15	9,099	144
	2012	12,331	373	4,513	-143	8,334	131
	2013	11,253	203	3,701	-110	7,865	123
	2014	11,627	284	3,813	59	8,039	125
Germany	2011	21,720	1,902	4,745	-95	18,972	236
	2012	20,734	1,558	4,992	-37	17,337	215
	2013	20,448	1,832	5,036	-48	17,292	215
	2014	20,094	1,790	4,820	-29	17,093	212
Gibraltar	2011	..	27	..	..	27	869
	2012	..	27	..	..	27	860
	2013	..	29	..	..	29	914
	2014	..	32	..	..	32	1,000

Table 18

Production, trade and supply of motor gasoline

Thousand metric tons and kilograms per capita

Country or area	Year	Production	Imports	Exports	Changes in stocks	Supply	
						Total	Per Capita
Greece	2011	3,950	563	1,385	-101	3,229	290
	2012	4,592	380	2,382	-11	2,601	234
	2013	4,205	334	2,018	71	2,450	222
	2014	4,649	351	2,571	-21	2,450	223
Hungary	2011	1,206	319	274	-54	1,305	131
	2012	1,211	318	235	11	1,283	129
	2013	1,115	337	250	9	1,193	120
	2014	1,166	442	294	22	1,292	131
Iceland	2011	..	145	..	2	143	445
	2012	..	138	..	1	137	424
	2013	..	133	..	-2	135	415
	2014	..	133	..	1	132	403
Ireland	2011	543	1,141	250	30	1,404	302
	2012	546	1,013	406	-15	1,168	250
	2013	546	833	166	-2	1,215	260
	2014	528	882	262	-3	1,151	246
Italy	2011	18,875	127	8,422	37	10,543	177
	2012	18,333	44	9,229	-219	9,367	157
	2013	15,431	357	7,897	188	7,703	129
	2014	15,431	385	7,023	-55	8,848	148
Latvia	2011	..	256	4	1	251	122
	2012	..	235	7	9	219	108
	2013	..	225	20	1	204	101
	2014	..	215	20	-18	213	107
Lithuania	2011	2,694	25	2,554	-3	168	55
	2012	2,504	22	2,353	33	140	46
	2013	2,535	15	2,466	-45	129	44
	2014	2,103	22	1,998	-8	135	46
Luxembourg	2011	..	368	..	-1	369	710
	2012	..	350	..	-1	351	659
	2013	..	317	..	-2	319	585
	2014	..	310	..	..	310	557
Malta	2011	..	82	..	9	73	176
	2012	..	64	..	-9	73	176
	2013	..	73	..	1	72	173
	2014	..	78	..	5	73	175
Montenegro	2011	..	55	12	..	43	69
	2012	..	48	11	..	37	59
	2013	..	45	13	..	32	51
	2014	..	43	11	-3	35	56
Netherlands	2011	6,546	10,406	14,877	174	1,901	114
	2012	7,097	11,972	19,233	189	-353	-21
	2013	6,527	9,679	17,370	116	-1,280	-76
	2014	6,528	9,500	18,879	116	-2,967	-176
Norway	2011	3,820	164	2,804	53	1,127	227
	2012	3,697	177	2,433	22	1,419	283
	2013	3,873	160	2,186	27	1,820	358
	2014	3,314	166	2,138	-65	1,407	273
Poland	2011	3,969	530	518	-30	4,011	104
	2012	4,020	437	678	-99	3,878	100
	2013	4,052	414	872	-95	3,689	96
	2014	3,833	371	672	-56	3,588	93

Production, trade and supply of motor gasoline

Thousand metric tons and kilograms per capita

Country or area	Year	Production	Imports	Exports	Changes in stocks	Supply	
						Total	Per Capita
Portugal	2011	1,875	191	799	18	1,249	118
	2012	1,864	160	898	-26	1,152	110
	2013	2,338	102	1,293	67	1,080	103
	2014	1,924	154	976	21	1,081	104
Republic of Moldova	2011	..	212	..	4	208	51
	2012	..	161	..	-8	169	41
	2013	..	162	..	0	162	40
	2014	..	151	..	-8	159	39
Romania	2011	2,886	68	1,714	-49	1,289	64
	2012	2,751	113	1,549	-45	1,360	68
	2013	2,863	129	1,692	31	1,269	64
	2014	2,998	80	1,729	-35	1,384	70
Russian Federation	2011	36,686	1,013	3,864	70	33,765	236
	2012	38,232	219	3,205	185	35,061	245
	2013	39,379	615	4,334	-29	35,689	249
	2014	38,935	1,059	3,657	-166	36,503	255
Serbia	2011	430	25	0	1	454	50
	2012	372	55	0	12	415	46
	2013	401	52	68	-11	396	44
	2014	464	53	120	2	395	44
Slovakia	2011	1,473	182	1,045	-15	625	115
	2012	1,407	220	1,008	28	591	109
	2013	1,449	215	1,068	-16	612	113
	2014	1,349	209	971	10	577	106
Slovenia	2011	..	714	144	4	566	275
	2012	..	686	150	22	514	249
	2013	..	579	113	-4	470	228
	2014	..	494	68	-15	441	213
Spain	2011	7,823	113	3,341	-87	4,682	100
	2012	7,543	81	3,378	-90	4,336	93
	2013	7,735	77	3,407	-25	4,430	95
	2014	7,569	77	3,262	57	4,327	94
Sweden	2011	4,335	1,487	2,327	-61	3,556	376
	2012	4,715	1,177	2,829	-55	3,118	327
	2013	3,706	1,244	1,952	178	2,820	293
	2014	5,081	1,815	3,520	13	3,363	347
Switzerland	2011	1,266	1,716	5	-61	3,038	382
	2012	1,032	1,910	18	-8	2,932	364
	2013	1,391	1,403	20	-27	2,801	343
	2014	1,428	1,248	14	-20	2,682	325
T.F.Yug.Rep. Macedonia	2011	110	63	55	-13	130	63
	2012	46	106	38	8	106	51
	2013	22	101	18	-16	122	59
	2014	1	104	4	-1	102	49
Ukraine	2011	2,837	2,214	973	-59	4,137	91
	2012	1,636	2,039	460	-31	3,246	72
	2013	1,015	1,761	237	73	2,466	55
	2014	712	2,147	157	-135	2,837	63
United Kingdom	2011	20,465	3,259	9,363	-111	14,472	229
	2012	20,665	4,184	11,042	-64	13,871	218
	2013	19,945	4,443	10,809	459	13,120	205
	2014	17,972	3,482	8,682	-81	12,853	200

Table 18

Production, trade and supply of motor gasoline

Thousand metric tons and kilograms per capita

Country or area	Year	Production	Imports	Exports	Changes in stocks	Supply	
						Total	Per Capita
Oceania	**2011**	**13,785**	**3,269**	**128**	**40**	**16,886**	**456**
	2012	**13,091**	**4,107**	**129**	**-3**	**17,072**	**454**
	2013	**12,521**	**4,168**	**74**	**105**	**16,510**	**432**
	2014	**11,303**	**4,116**	**97**	**-129**	**15,451**	**399**
Australia	2011	12,453	1,939	127	58	14,207	630
	2012	11,719	2,683	128	-9	14,283	623
	2013	11,153	2,671	74	77	13,673	588
	2014	10,007	2,630	97	-108	12,648	535
Cook Islands	2011	..	*5	..	..	*5	*245
	2012	..	*5	..	..	*5	*244
	2013	..	*5	..	..	*5	*242
	2014	..	*5	..	..	*5	*241
Fiji	2011	..	79	..	..	79	91
	2012	..	69	..	..	69	79
	2013	..	*70	..	..	*70	*80
	2014	..	*72	..	..	*72	*81
French Polynesia	2011	..	48	..	..	48	176
	2012	..	46	..	..	46	169
	2013	..	46	..	..	46	167
	2014	..	46	..	..	46	164
Kiribati	2011	..	*4	..	..	*4	*39
	2012	..	*4	..	..	*4	*39
	2013	..	*4	..	..	*4	*38
	2014	..	*4	..	..	*4	*38
Micronesia (Fed. States of)	2011	..	*12	..	..	*12	*116
	2012	..	*14	..	..	*14	*135
	2013	..	*15	..	..	*15	*145
	2014	..	*18	..	..	*18	*172
Nauru	2011	..	*1	..	..	*1	*110
	2012	..	*1	..	..	*1	*109
	2013	..	*1	..	..	*1	*119
	2014	..	*1	..	..	*1	*124
New Caledonia	2011	..	71	*1	*2	68	271
	2012	..	74	*1	*4	69	271
	2013	..	65	..	-3	68	267
	2014	..	70	..	..	70	271
New Zealand	2011	1,332	951	..	-22	2,305	523
	2012	1,359	1,100	..	7	2,452	553
	2013	1,341	1,118	..	25	2,434	545
	2014	1,273	1,090	..	-21	2,384	530
Niue	2011	..	1	..	..	1	322
	2012	..	1	..	..	1	351
	2013	..	1	..	..	1	386
	2014	..	1	..	..	1	421
Palau	2011	..	*26	..	..	*26	*1,262
	2012	..	*26	..	..	*26	*1,253
	2013	..	*26	..	..	*26	*1,243
	2014	..	*27	..	..	*27	*1,294
Papua New Guinea	2011	..	78	..	1	77	11
	2012	13	28	..	-5	46	6
	2013	27	89	..	7	110	15
	2014	*23	*90	..	..	*113	*15

Production, trade and supply of motor gasoline

Thousand metric tons and kilograms per capita

Country or area	Year	Production	Imports	Exports	Changes in stocks	Supply	
						Total	Per Capita
Samoa	2011	..	23	..	0	22	118
	2012	..	*23	..	..	*23	*119
	2013	..	22	..	0	22	117
	2014	..	22	..	..	22	115
Solomon Islands	2011	..	*16	..	..	*16	*29
	2012	..	*16	..	..	*16	*29
	2013	..	*17	..	..	*17	*30
	2014	..	*17	..	..	*17	*30
Tonga	2011	..	8	..	..	8	77
	2012	..	9	..	..	9	82
	2013	..	10	..	..	10	94
	2014	..	11	..	..	11	101
Tuvalu	2011	..	1	..	..	1	79
	2012	..	*1	..	..	*1	*112
	2013	..	*1	..	..	*1	*111
	2014	..	*1	..	..	*1	*121
Vanuatu	2011	..	6	..	..	6	26
	2012	..	6	..	..	6	25
	2013	..	6	..	..	6	23
	2014	..	9	..	..	9	36
Wallis and Futuna Is.	2011	..	*1	..	..	*1	*108
	2012	..	1	..	..	1	102
	2013	..	*1	..	..	*1	*101
	2014	..	*1	..	..	*1	*96

Table 19

Production, trade and supply of other kerosene

Thousand metric tons and kilograms per capita

Table Notes

Production includes output from refineries and plants.

For China, data up to 2013 include kerosene-type jet fuel.

Please refer to the Definitions Section on pages ix to xv for the appropriate product description/ classification.

Figure 51: Production of other kerosene by region, 1990-2014

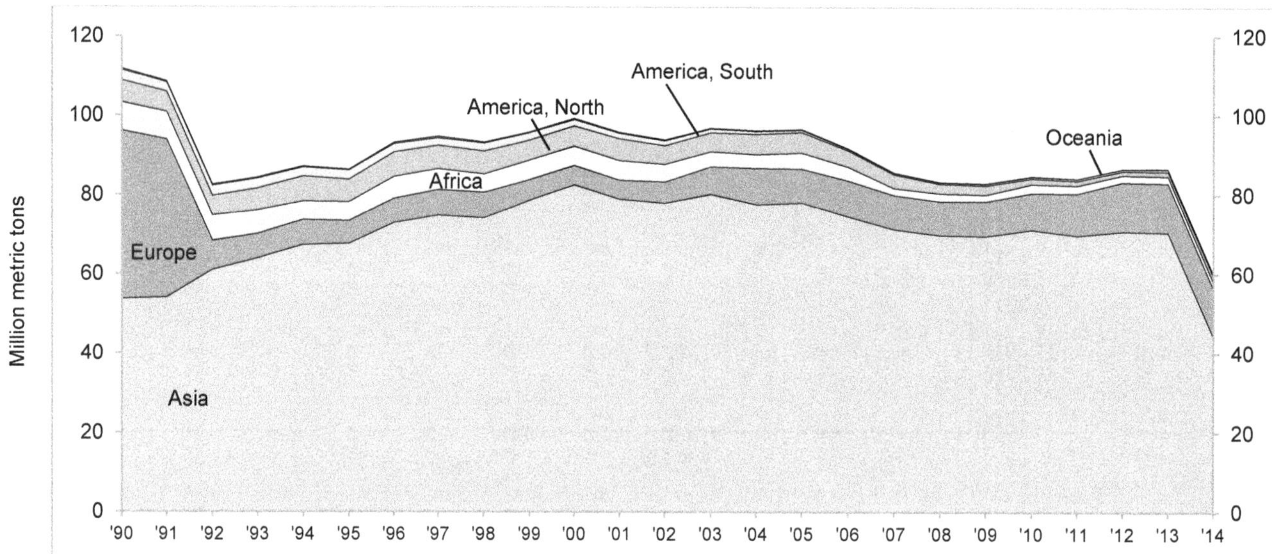

Figure 52: Major other kerosene producing countries, 2014

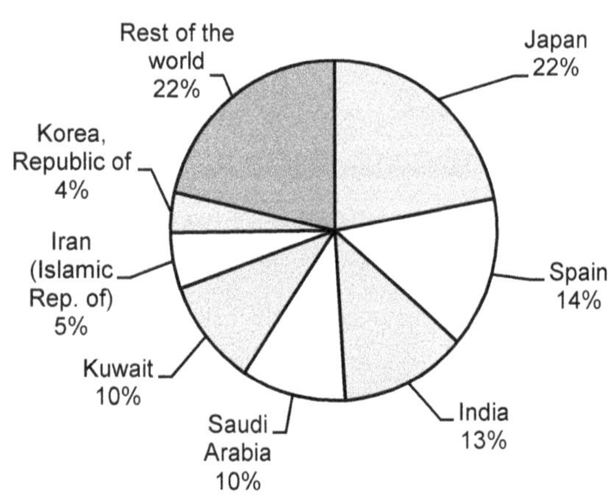

Figure 53: Major other kerosene consuming countries, 2014

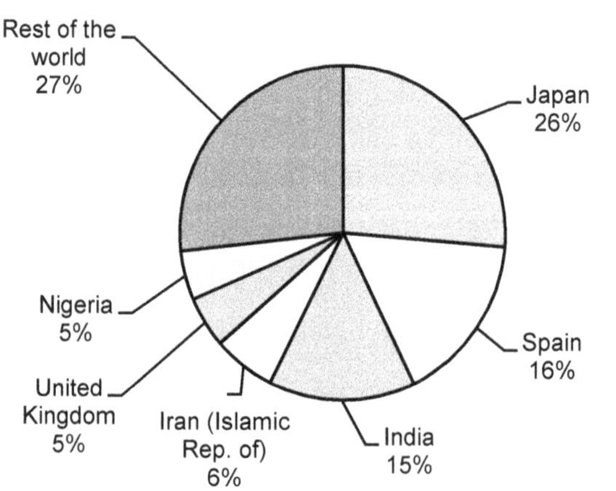

Production, trade and supply of other kerosene

Thousand metric tons and kilograms per capita

Country or area	Year	Production	Imports	Exports	Changes in stocks	Supply	
						Total	Per Capita
World	**2011**	**84,102**	**19,071**	**22,723**	**1,096**	**79,354**	**11**
	2012	**86,547**	**17,409**	**23,367**	**-534**	**81,124**	**11**
	2013	**86,623**	**18,572**	**24,873**	**849**	**79,473**	**11**
	2014	**60,353**	**8,852**	**15,907**	**968**	**52,330**	**7**
Africa	**2011**	**2,006**	**2,815**	**383**	**-33**	**4,471**	**4**
	2012	**1,814**	**3,631**	**284**	**-1**	**5,162**	**5**
	2013	**1,805**	**3,509**	**152**	**30**	**5,132**	**5**
	2014	**1,707**	**3,597**	**336**	**40**	**4,927**	**4**
Angola	2011	73	0	..	0	73	3
	2012	76	0	..	6	70	3
	2013	89	0	..	23	66	3
	2014	69	6	..	39	36	1
Benin	2011	..	405	48	..	357	37
	2012	..	424	50	..	374	37
	2013	..	453	53	..	400	39
	2014	..	483	56	..	427	40
Botswana	2011	..	16	..	..	16	8
	2012	..	22	..	..	22	10
	2013	..	17	..	..	17	8
	2014	..	20	..	..	20	9
Burkina Faso	2011	..	4	..	..	4	0
	2012	..	1	..	..	1	0
	2013	..	2	..	..	2	0
	2014	..	4	..	..	4	0
Burundi	2011	..	3	..	0	3	0
	2012	..	3	..	0	2	0
	2013	..	3	..	0	3	0
	2014	..	0	..	0	0	0
Cabo Verde	2011	..	1	..	..	1	1
	2012	..	1	..	..	1	1
	2013	..	0	..	..	0	1
	2014	..	0	..	..	0	1
Central African Rep.	2011	..	*27	..	..	*27	*6
	2012	..	*28	..	..	*28	*6
	2013	..	*28	..	..	*28	*6
	2014	..	*28	..	..	*28	*6
Chad	2011	..	*1	..	..	*1	0
	2012	..	*1	..	..	*1	0
	2013	..	*1	..	..	*1	0
	2014	..	*1	..	..	*1	0
Comoros	2011	..	*9	..	..	*9	*13
	2012	..	*9	..	..	*9	*12
	2013	..	*10	..	..	*10	*13
	2014	..	*10	..	..	*10	*13
Congo	2011	22	1	..	..	23	6
	2012	20	..	..	..	20	5
	2013	21	..	..	..	21	5
	2014	20	..	..	..	20	4
Côte d'Ivoire	2011	45	..	..	..	45	2
	2012	59	..	..	..	59	3
	2013	85	..	..	..	85	4
	2014	54	..	..	..	54	2

Table 19

Production, trade and supply of other kerosene

Thousand metric tons and kilograms per capita

Country or area	Year	Production	Imports	Exports	Changes in stocks	Supply	
						Total	Per Capita
Dem. Rep. of the Congo	2011	..	9	..	..	9	0
	2012	..	9	..	..	9	0
	2013	..	10	..	..	10	0
	2014	..	3	..	..	3	0
Djibouti	2011	..	12	..	..	12	14
	2012	..	*12	..	..	*12	*14
	2013	..	*13	..	..	*13	*15
	2014	..	*13	..	..	*13	*15
Egypt	2011	285	..	*275	..	10	0
	2012	296	..	*215	..	81	1
	2013	86	..	0	..	86	1
	2014	249	..	181	..	68	1
Equatorial Guinea	2011	..	*10	..	..	*10	*13
	2012	..	*10	..	..	*10	*13
	2013	..	*10	..	..	*10	*13
	2014	..	*10	..	..	*10	*12
Eritrea	2011	..	14	..	..	14	3
	2012	..	14	..	..	14	3
	2013	..	14	..	..	14	3
	2014	..	14	..	..	14	3
Ethiopia	2011	..	215	..	-22	237	3
	2012	..	267	..	-2	269	3
	2013	..	260	..	3	257	3
	2014	..	263	..	..	263	3
Gabon	2011	38	5	..	..	43	27
	2012	22	15	..	..	37	23
	2013	28	13	..	..	41	25
	2014	27	10	..	..	37	22
Ghana	2011	53	..	..	..	53	2
	2012	21	..	..	..	21	1
	2013	15	..	..	..	15	1
	2014	5	..	..	..	5	0
Guinea	2011	..	16	..	..	16	1
	2012	..	12	..	..	12	1
	2013	..	11	..	..	11	1
	2014	..	*11	..	..	*11	*1
Guinea-Bissau	2011	..	*8	..	..	*8	*5
	2012	..	*8	..	..	*8	*5
	2013	..	*8	..	..	*8	*5
	2014	..	*9	..	..	*9	*5
Kenya	2011	118	150	..	..	268	6
	2012	67	242	..	..	309	7
	2013	33	339	77	..	295	7
	2014	0	373	77	..	296	7
Lesotho	2011	..	30	..	..	30	15
	2012	..	28	..	..	28	14
	2013	..	27	..	..	27	13
	2014	..	*28	..	..	*28	*13
Libya	2011	139	0	45	..	94	15
	2012	200	190	0	..	390	62
	2013	251	40	0	..	291	46
	2014	115	108	0	..	223	36

Production, trade and supply of other kerosene

Thousand metric tons and kilograms per capita

Country or area	Year	Production	Imports	Exports	Changes in stocks	Supply	
						Total	Per Capita
Madagascar	2011	..	42	..	-2	44	2
	2012	..	44	..	0	43	2
	2013	..	25	..	0	25	1
	2014	..	*32	..	0	*32	*1
Malawi	2011	..	6	..	..	6	0
	2012	..	5	..	..	5	0
	2013	..	1	..	..	1	0
	2014	..	1	..	..	1	0
Mali	2011	..	*23	..	..	*23	*1
	2012	..	*23	..	..	*23	*1
	2013	..	*24	..	..	*24	*1
	2014	..	*24	..	..	*24	*1
Mauritius	2011	..	4	..	-1	5	4
	2012	..	7	..	3	4	3
	2013	..	3	..	2	1	1
	2014	..	2	..	1	1	1
Mayotte	2011	..	3	..	0	3	13
	2012	..	*3	..	..	*3	*11
	2013	..	*2	..	..	*2	*10
	2014	..	*2	..	..	*2	*9
Mozambique	2011	..	18	..	..	18	1
	2012	..	0	..	..	0	0
	2013	..	0	..	..	0	0
	2014	..	0	..	..	0	0
Namibia	2011	..	2	..	..	2	1
	2012	..	3	..	..	3	1
	2013	..	2	..	..	2	1
	2014	..	2	..	..	2	1
Niger	2011	..	3	..	0	3	0
	2012	..	2	..	..	2	0
	2013	..	2	..	..	2	0
	2014	..	3	..	..	3	0
Nigeria	2011	674	1,416	..	..	2,090	13
	2012	544	1,948	..	..	2,492	15
	2013	671	1,912	..	..	2,583	15
	2014	629	1,862	..	..	2,491	14
Rwanda	2011	..	*14	..	..	*14	*1
	2012	..	*16	..	..	*16	*1
	2013	..	*15	..	..	*15	*1
	2014	..	*15	..	..	*15	*1
Sao Tome and Principe	2011	..	*3	..	..	*3	*19
	2012	..	*4	..	..	*4	*20
	2013	..	*4	..	..	*4	*19
	2014	..	*4	..	..	*4	*19
Senegal	2011	3	..	..	..	3	0
	2012	1	..	..	..	1	0
	2013	0	..	..	..	0	0
	2014	2	..	..	..	2	0
Seychelles	2011	..	*1	..	..	*1	*11
	2012	..	*1	..	..	*1	*11
	2013	..	*1	..	..	*1	*11
	2014	..	*1	..	..	*1	*10

Table 19

Production, trade and supply of other kerosene

Thousand metric tons and kilograms per capita

Country or area	Year	Production	Imports	Exports	Changes in stocks	Supply	
						Total	Per Capita
Sierra Leone	2011	..	13	..	..	13	2
	2012	..	14	..	..	14	2
	2013	..	13	..	..	13	2
	2014	..	14	..	..	14	2
Somalia	2011	..	*38	..	..	*38	*4
	2012	..	*38	..	..	*38	*4
	2013	..	*38	..	..	*38	*4
	2014	..	*38	..	..	*38	*4
South Africa	2011	463	..	2	..	461	9
	2012	430	..	1	..	429	8
	2013	424	..	1	..	423	8
	2014	446	..	1	..	445	8
South Sudan	2011	..	..	..	..	..	..
	2012	..	2	..	..	2	0
	2013	..	2	..	..	2	0
	2014	..	2	..	..	2	0
Sudan	2011	..	..	..	..	..	..
	2012	8	..	..	..	8	0
	2013	19	..	..	..	19	0
	2014	3	..	..	..	3	0
Sudan (former)	2011	24	..	..	..	24	1
	2012	..	..	..	..	..	..
	2013	..	..	..	..	..	..
	2014	..	..	..	..	..	..
Swaziland	2011	..	*8	..	..	*8	*7
	2012	..	*8	..	..	*8	*6
	2013	..	*9	..	..	*9	*7
	2014	..	9	..	..	9	7
Togo	2011	..	53	..	..	53	8
	2012	..	57	..	..	57	8
	2013	..	59	..	..	59	9
	2014	..	61	..	..	61	9
Tunisia	2011	40	14	..	-8	62	6
	2012	37	0	..	-8	45	4
	2013	52	0	..	2	50	5
	2014	53	0	..	0	53	5
Uganda	2011	..	60	..	..	60	2
	2012	..	60	..	..	60	2
	2013	..	59	..	..	59	2
	2014	..	54	..	..	54	1
United Rep. of Tanzania	2011	..	124	..	..	124	3
	2012	..	65	..	..	65	1
	2013	..	37	..	..	37	1
	2014	..	31	..	..	31	1
Zambia	2011	29	..	13	..	16	1
	2012	33	..	18	..	15	1
	2013	31	..	21	..	10	1
	2014	35	..	21	..	14	1
Zimbabwe	2011	..	35	..	..	35	2
	2012	..	38	..	..	38	3
	2013	..	41	..	..	41	3
	2014	..	45	..	..	45	3

Production, trade and supply of other kerosene

Thousand metric tons and kilograms per capita

Country or area	Year	Production	Imports	Exports	Changes in stocks	Supply	
						Total	Per Capita
America, North	**2011**	**1,273**	**558**	**506**	**-8**	**1,333**	**2**
	2012	**992**	**342**	**370**	**-55**	**1,018**	**2**
	2013	**1,172**	**379**	**365**	**17**	**1,168**	**2**
	2014	**1,577**	**280**	**470**	**99**	**1,288**	**2**
Bahamas	2011	..	*3	..	..	*3	*9
	2012	..	*3	..	..	*3	*9
	2013	..	*3	..	..	*3	*9
	2014	..	*3	..	..	*3	*9
Barbados	2011	..	9	..	0	9	32
	2012	..	8	..	1	7	24
	2013	..	*9	..	*1	*8	*29
	2014	..	*7	..	0	*7	*24
Bermuda	2011	..	2	..	..	2	36
	2012	..	4	..	..	4	66
	2013	..	3	..	..	3	41
	2014	..	*4	..	..	*4	*71
British Virgin Islands	2011	..	*4	..	..	*4	*143
	2012	..	*4	..	..	*4	*140
	2013	..	*4	..	..	*4	*138
	2014	..	*4	..	..	*4	*142
Canada	2011	489	339	6	-23	845	24
	2012	495	108	76	33	494	14
	2013	604	171	73	-19	721	20
	2014	588	94	13	-3	672	19
Costa Rica	2011	2	..	..	0	2	1
	2012	..	..	..	0	0	0
	2013	..	..	..	..	..	..
	2014	..	..	..	..	..	..
Cuba	2011	4	..	..	..	4	0
	2012	31	..	..	..	31	3
	2013	29	..	..	..	29	3
	2014	25	..	..	..	25	2
Dominica	2011	..	*1	..	..	*1	*14
	2012	..	*1	..	..	*1	*14
	2013	..	*1	..	..	*1	*14
	2014	..	*1	..	..	*1	*14
Dominican Republic	2011	10	..	..	..	10	1
	2012	9	..	..	..	9	1
	2013	10	..	..	..	10	1
	2014	10	..	..	..	10	1
El Salvador	2011	2	1	..	..	3	0
	2012	0	0	..	..	0	0
	2013	0	0	..	..	0	0
	2014	0	1	..	..	1	0
Grenada	2011	..	0	..	..	0	1
	2012	..	0	..	..	0	1
	2013	..	0	..	..	0	1
	2014	..	0	..	..	0	2
Guadeloupe	2011	..	*1	..	..	*1	*2
	2012	..	*1	..	..	*1	*2
	2013	..	*1	..	..	*1	*2
	2014	..	*1	..	..	*1	*2

Table 19

Production, trade and supply of other kerosene

Thousand metric tons and kilograms per capita

Country or area	Year	Production	Imports	Exports	Changes in stocks	Supply	
						Total	Per Capita
Guatemala	2011	1	21	..	3	19	1
	2012	1	39	..	0	40	3
	2013	0	61	..	20	42	3
	2014	1	113	..	92	22	1
Haiti	2011	..	56	..	..	56	6
	2012	..	13	..	..	13	1
	2013	..	23	..	..	23	2
	2014	..	29	..	..	29	3
Honduras	2011	..	37	..	-7	44	6
	2012	..	41	..	0	41	5
	2013	..	35	..	0	35	4
	2014	..	3	..	-21	24	3
Jamaica	2011	2	..	..	..	2	1
	2012	1	..	..	..	1	1
	2013	1	..	..	..	1	0
	2014	2	..	..	..	2	1
Martinique	2011	..	*2	..	..	*2	*4
	2012	..	*2	..	..	*2	*4
	2013	..	*2	..	..	*2	*4
	2014	..	*2	..	..	*2	*4
Montserrat	2011	..	*1	..	..	*1	*100
	2012	..	*1	..	..	*1	*99
	2013	..	*1	..	..	*1	*99
	2014	..	*1	..	..	*1	*98
Nicaragua	2011	5	..	..	..	5	1
	2012	5	..	..	..	5	1
	2013	5	..	..	..	5	1
	2014	6	..	..	..	6	1
Panama	2011	..	30	..	..	30	8
	2012	..	65	..	..	65	17
	2013	..	16	..	..	16	4
	2014	..	10	..	..	10	3
St. Kitts-Nevis	2011	..	5	..	..	5	92
	2012	..	5	..	..	5	91
	2013	..	*5	..	..	*5	*92
	2014	..	*5	..	..	*5	*91
St. Lucia	2011	..	*1	..	..	*1	*6
	2012	..	*1	..	..	*1	*6
	2013	..	*1	..	..	*1	*6
	2014	..	*1	..	..	*1	*7
Trinidad and Tobago	2011	30	..	..	..	30	22
	2012	29	..	..	..	29	22
	2013	32	..	..	..	32	24
	2014	33	..	..	..	33	24
United States	2011	728	45	500	19	254	1
	2012	420	47	294	-89	262	1
	2013	491	43	292	16	226	1
	2014	913	1	457	31	426	1
America, South	**2011**	**444**	**18**	**..**	**34**	**429**	**1**
	2012	**501**	**22**	**..**	**-23**	**545**	**1**
	2013	**549**	**25**	**..**	**39**	**535**	**1**
	2014	**511**	**17**	**..**	**33**	**495**	**1**

Production, trade and supply of other kerosene

Thousand metric tons and kilograms per capita

Country or area	Year	Production	Imports	Exports	Changes in stocks	Supply	
						Total	Per Capita
Argentina	2011	36	..	..	-5	41	1
	2012	20	..	..	6	14	0
	2013	21	..	..	1	20	0
	2014	16	..	..	-7	23	1
Bolivia (Plur. State of)	2011	16	..	..	0	16	2
	2012	15	..	..	0	15	1
	2013	14	..	..	-1	15	1
	2014	15	..	..	..	15	1
Brazil	2011	25	..	..	1	24	0
	2012	19	..	..	0	19	0
	2013	12	..	..	0	12	0
	2014	10	..	..	0	10	0
Chile	2011	78	0	..	29	49	3
	2012	100	0	..	-30	130	7
	2013	141	8	..	39	110	6
	2014	143	0	..	38	105	6
Colombia	2011	196	..	..	..	196	4
	2012	216	..	..	..	216	5
	2013	236	..	..	..	236	5
	2014	249	..	..	..	249	5
Ecuador	2011	5	..	..	..	5	0
	2012	5	..	..	..	5	0
	2013	5	..	..	..	5	0
	2014	3	..	..	..	3	0
Falkland Is. (Malvinas)	2011	..	*4	..	..	*4	*1,331
	2012	..	*4	..	..	*4	*1,397
	2013	..	*4	..	..	*4	*1,390
	2014	..	*4	..	..	*4	*1,384
French Guiana	2011	..	*1	..	..	*1	*4
	2012	..	*1	..	..	*1	*5
	2013	..	*1	..	..	*1	*5
	2014	..	*1	..	..	*1	*5
Guyana	2011	..	13	..	3	10	13
	2012	..	16	..	1	15	20
	2013	..	11	..	0	11	14
	2014	..	11	..	2	9	12
Peru	2011	66	..	..	6	60	2
	2012	49	..	..	0	49	2
	2013	46	..	..	0	46	2
	2014	0	..	..	0	0	0
Suriname	2011	..	0	..	..	0	1
	2012	..	0	..	..	0	1
	2013	..	0	..	..	0	1
	2014	..	0	..	..	0	1
Uruguay	2011	8	..	..	0	8	2
	2012	9	..	..	1	8	2
	2013	7	..	..	0	8	2
	2014	7	..	..	0	7	2
Venezuela (Bolivar. Rep.)	2011	14	..	..	..	14	0
	2012	68	..	..	..	68	2
	2013	67	..	..	..	67	2
	2014	68	..	..	..	68	2

Table 19

Production, trade and supply of other kerosene

Thousand metric tons and kilograms per capita

Country or area	Year	Production	Imports	Exports	Changes in stocks	Supply	
						Total	Per Capita
Asia	**2011**	**69,695**	**11,922**	**20,227**	**1,122**	**60,268**	**14**
	2012	**70,748**	**10,951**	**21,340**	**-345**	**60,706**	**14**
	2013	**70,537**	**11,522**	**22,728**	**518**	**58,814**	**14**
	2014	**44,477**	**2,208**	**13,956**	**921**	**31,808**	**7**
Afghanistan	2011	..	36	..	..	36	1
	2012	..	26	..	..	26	1
	2013	..	*26	..	..	*26	*1
	2014	..	*26	..	..	*26	*1
Azerbaijan	2011	0	0	..	0	1	0
	2012	..	20	..	0	19	2
	2013	..	28	..	3	26	3
	2014	..	5	..	0	5	0
Bahrain	2011	0	..	0	-25	25	19
	2012	0	..	0	-24	24	18
	2013	0	..	0	-28	28	21
	2014	0	..	0	-24	24	18
Bangladesh	2011	283	154	..	53	384	3
	2012	226	20	..	-94	340	2
	2013	286	28	..	20	294	2
	2014	231	..	..	-34	265	2
Bhutan	2011	..	5	..	..	5	6
	2012	..	4	..	..	4	6
	2013	..	4	..	..	4	5
	2014	..	0	..	..	0	0
Brunei Darussalam	2011	3	..	..	..	3	8
	2012	7	..	..	..	7	17
	2013	7	..	..	..	7	17
	2014	6	..	..	..	6	14
Cambodia	2011	..	12	..	..	12	1
	2012	..	6	..	..	6	0
	2013	..	3	..	..	3	0
	2014	..	0	..	..	0	0
China	2011	19,224	8,751	6,563	90	21,322	16
	2012	21,640	8,773	7,451	-37	22,999	17
	2013	25,239	9,452	9,175	-6	25,522	19
	2014	1,003	227	28	14	1,188	1
Cyprus	2011	..	16	..	1	15	13
	2012	..	13	..	-2	15	13
	2013	..	12	..	-1	13	11
	2014	..	13	..	0	13	11
Georgia	2011	0	0	..	1	-1	0
	2012	3	0	..	-1	4	1
	2013	0	0	..	0	0	0
	2014	0	0	..	0	0	0
India	2011	7,789	564	34	0	8,319	7
	2012	7,868	0	23	0	7,845	6
	2013	7,338	0	15	-179	7,502	6
	2014	7,634	30	15	0	7,649	6
Indonesia	2011	1,857	0	349	..	1,508	6
	2012	1,396	0	248	..	1,148	5
	2013	1,233	5	206	..	1,032	4
	2014	957	0	51	..	906	4

Production, trade and supply of other kerosene

Thousand metric tons and kilograms per capita

Country or area	Year	Production	Imports	Exports	Changes in stocks	Supply	
						Total	Per Capita
Iran (Islamic Rep. of)	2011	4,378	..	128	..	4,250	57
	2012	4,606	..	..	..	4,606	60
	2013	3,738	..	..	..	3,738	48
	2014	3,207	..	..	..	3,207	41
Iraq	2011	2,258	403	..	..	2,661	84
	2012	2,205	394	..	..	2,599	79
	2013	2,049	366	..	..	2,415	71
	2014	1,398	250	..	..	1,648	47
Israel	2011	527	393	470	40	410	54
	2012	367	430	377	0	420	55
	2013	139	237	486	11	-121	-15
	2014	121	173	264	..	30	4
Japan	2011	15,615	1,209	488	55	16,281	128
	2012	14,779	987	117	118	15,531	122
	2013	14,404	747	596	-294	14,849	117
	2014	13,237	1,115	579	13	13,760	109
Jordan	2011	58	..	..	-18	75	11
	2012	96	..	..	15	81	12
	2013	34	..	..	-28	62	9
	2014	63	..	..	14	49	7
Kazakhstan	2011	177	1	19	-2	161	10
	2012	167	48	6	32	177	11
	2013	168	77	16	5	224	13
	2014	114	71	12	3	170	10
Korea, Dem.Ppl's.Rep.	2011	36	1	..	..	37	2
	2012	36	1	..	..	37	1
	2013	40	1	..	..	41	2
	2014	37	1	..	..	38	2
Korea, Republic of	2011	4,632	59	1,159	-65	3,597	73
	2012	4,200	19	981	-370	3,608	73
	2013	2,918	23	863	370	1,708	34
	2014	2,330	13	570	269	1,504	30
Kuwait	2011	5,912	..	5,869	..	43	13
	2012	6,401	..	6,353	..	48	14
	2013	6,287	..	6,245	..	42	12
	2014	6,146	..	6,106	..	40	11
Lao People's Dem. Rep.	2011	..	*22	..	..	*22	*4
	2012	..	*23	..	..	*23	*4
	2013	..	*24	..	..	*24	*4
	2014	..	*24	..	..	*24	*4
Malaysia	2011	406	144	680	-5	-125	-4
	2012	633	5	680	-1	-41	-1
	2013	375	280	521	-6	139	5
	2014	97	140	204	-5	38	1
Maldives	2011	..	0	..	..	0	1
	2012	..	0	..	..	0	0
	2013	..	7	..	..	7	19
	2014	..	7	..	..	7	20
Mongolia	2011	..	20	..	..	20	7
	2012	..	22	..	..	22	8
	2013	..	39	..	..	39	14
	2014	..	1	..	..	1	0

Table 19

Production, trade and supply of other kerosene

Thousand metric tons and kilograms per capita

Country or area	Year	Production	Imports	Exports	Changes in stocks	Supply	
						Total	Per Capita
Myanmar	2011	1	..	..	..	1	0
	2012	1	..	..	..	1	0
	2013	0	..	..	..	0	0
	2014	0	..	..	..	0	0
Nepal	2011	..	42	..	..	42	2
	2012	..	24	..	..	24	1
	2013	..	18	..	..	18	1
	2014	..	20	..	..	20	1
Oman	2011	4	..	..	..	4	1
	2012	4	..	..	..	4	1
	2013	4	..	..	..	4	1
	2014	4	..	..	..	4	1
Pakistan	2011	132	..	..	3	129	1
	2012	161	..	..	-4	165	1
	2013	174	..	..	-2	176	1
	2014	182	..	..	0	182	1
Philippines	2011	100	32	7	-2	127	1
	2012	92	29	0	-3	125	1
	2013	87	46	0	21	112	1
	2014	59	55	0	-22	137	1
Saudi Arabia	2011	4,878	..	3,404	1,139	335	12
	2012	5,124	..	4,712	65	347	12
	2013	4,775	..	4,251	166	358	12
	2014	6,209	..	5,913	-86	382	12
Singapore	2011	881	39	961	-38	-2	0
	2012	325	21	362	-21	6	1
	2013	305	16	336	-16	1	0
	2014	172	8	187	-7	0	0
Sri Lanka	2011	93	0	..	-54	147	7
	2012	75	55	..	4	126	6
	2013	57	56	..	-12	126	6
	2014	65	..	..	-57	122	6
State of Palestine	2011	..	1	..	..	1	0
	2012	..	2	..	..	2	0
	2013	..	1	..	..	1	0
	2014	..	1	..	..	1	0
Syrian Arab Republic	2011	4	..	..	..	4	0
	2012	3	..	..	..	3	0
	2013	2	..	..	..	2	0
	2014	2	..	..	..	2	0
Thailand	2011	123	..	86	*26	11	0
	2012	61	..	25	*25	11	0
	2013	543	..	15	519	9	0
	2014	868	..	24	835	9	0
Turkey	2011	60	..	..	-5	65	1
	2012	45	..	..	-3	48	1
	2013	53	..	..	-3	56	1
	2014	56	..	..	22	34	0
Turkmenistan	2011	150	..	..	..	150	29
	2012	140	..	..	..	140	27
	2013	140	..	..	..	140	27
	2014	150	..	..	..	150	28

Production, trade and supply of other kerosene

Thousand metric tons and kilograms per capita

Country or area	Year	Production	Imports	Exports	Changes in stocks	Supply	
						Total	Per Capita
Uzbekistan	2011	56	..	..	..	56	2
	2012	50	..	..	..	50	2
	2013	47	..	..	..	47	2
	2014	44	..	..	..	44	1
Viet Nam	2011	..	17	11	-72	78	1
	2012	..	29	5	-46	70	1
	2013	..	25	3	-18	40	0
	2014	..	29	3	-16	42	0
Yemen	2011	58	..	..	..	58	2
	2012	37	..	..	..	37	1
	2013	96	..	..	..	96	4
	2014	84	..	..	..	84	3
Europe	**2011**	**10,682**	**3,705**	**1,607**	**-18**	**12,798**	**17**
	2012	**12,477**	**2,411**	**1,373**	**-111**	**13,626**	**18**
	2013	**12,446**	**3,112**	**1,628**	**239**	**13,691**	**19**
	2014	**12,031**	**2,724**	**1,145**	**-126**	**13,736**	**19**
Austria	2011	0	1	..	..	1	0
	2012	16	6	..	..	22	3
	2013	18	4	..	..	22	3
	2014	18	1	..	..	19	2
Belarus	2011	274	..	241	3	30	3
	2012	146	..	162	-24	8	1
	2013	102	..	99	-2	5	1
	2014	12	..	8	1	3	0
Belgium	2011	36	113	112	19	18	2
	2012	35	100	79	2	54	5
	2013	26	83	44	-5	70	6
	2014	20	71	43	5	43	4
Croatia	2011	..	1	..	..	1	0
	2012	..	1	..	..	1	0
	2013	..	1	..	..	1	0
	2014	..	2	..	..	2	0
Czechia	2011	..	3	..	..	3	0
	2012	..	3	..	..	3	0
	2013	..	2	..	..	2	0
	2014	..	2	..	..	2	0
Finland	2011	..	888	71	..	817	151
	2012	..	94	88	..	6	1
	2013	..	312	125	..	187	34
	2014	..	254	81	..	173	32
France	2011	1	293	32	1	261	4
	2012	0	461	67	-4	398	6
	2013	0	480	67	0	413	6
	2014	0	376	66	0	310	5
Germany	2011	5	18	0	0	23	0
	2012	3	11	0	-1	15	0
	2013	4	9	0	1	12	0
	2014	1	9	0	-2	12	0
Greece	2011	10	..	0	2	8	1
	2012	6	..	0	-2	8	1
	2013	0	..	2	1	-3	0
	2014	1	..	6	-1	-4	0

Table 19

Production, trade and supply of other kerosene

Thousand metric tons and kilograms per capita

Country or area	Year	Production	Imports	Exports	Changes in stocks	Supply	
						Total	Per Capita
Hungary	2011	..	2	..	..	2	0
	2012	..	0	..	..	0	0
	2013	..	0	..	..	0	0
	2014	..	0	..	..	0	0
Ireland	2011	117	596	0	-17	730	157
	2012	131	503	6	27	601	129
	2013	106	471	5	28	544	116
	2014	137	371	8	-36	536	115
Italy	2011	563	553	351	2	763	13
	2012	672	193	144	-30	751	13
	2013	396	565	305	116	540	9
	2014	639	537	256	-73	993	17
Latvia	2011	..	6	2	4	0	0
	2012	..	0	4	-4	0	0
	2013	..	1	1	0	0	0
	2014	..	1	1	0	0	0
Luxembourg	2011	..	1	..	..	1	2
	2012	..	1	..	..	1	2
	2013	..	1	..	..	1	2
	2014	..	0	..	..	0	0
Netherlands	2011	383	382	474	-15	306	18
	2012	580	215	434	-38	399	24
	2013	407	230	299	19	319	19
	2014	382	189	371	-14	214	13
Norway	2011	434	177	104	-43	550	111
	2012	223	10	248	0	-15	-3
	2013	196	142	275	19	44	9
	2014	40	150	115	-9	84	16
Poland	2011	1	1	..	..	2	0
	2012	1	1	..	..	2	0
	2013	1	1	..	..	2	0
	2014	..	1	..	..	1	0
Portugal	2011	1	1	..	0	2	0
	2012	1	1	..	-2	4	0
	2013	1	1	..	0	2	0
	2014	..	1	..	..	1	0
Romania	2011	26	1	..	..	27	1
	2012	23	1	..	..	24	1
	2013	5	1	..	..	6	0
	2014	8	3	..	..	11	1
Slovakia	2011	2	..	..	-1	3	1
	2012	1	..	..	..	1	0
	2013	1	..	..	..	1	0
	2014	1	..	..	..	1	0
Spain	2011	6,452	15	..	0	6,467	138
	2012	8,371	0	..	2	8,369	179
	2013	8,478	0	..	-15	8,493	183
	2014	8,679	..	..	48	8,631	187
Sweden	2011	..	1	..	..	1	0
	2012	..	1	..	..	1	0
	2013	..	1	..	..	1	0
	2014	..	1	..	..	1	0

Production, trade and supply of other kerosene

Thousand metric tons and kilograms per capita

Country or area	Year	Production	Imports	Exports	Changes in stocks	Supply	
						Total	Per Capita
Switzerland	2011	..	3	..	..	3	0
	2012	..	3	..	..	3	0
	2013	..	2	..	..	2	0
	2014	..	2	..	..	2	0
Ukraine	2011	..	31	47	-2	-14	0
	2012	..	104	29	3	72	2
	2013	..	168	28	-4	144	3
	2014	..	136	26	-1	111	2
United Kingdom	2011	2,377	618	173	29	2,793	44
	2012	2,268	702	112	-40	2,898	46
	2013	2,705	637	378	81	2,883	45
	2014	2,093	617	164	-44	2,590	40
Oceania	**2011**	**2**	**53**	**0**	**-1**	**55**	**1**
	2012	**15**	**52**	**0**	**1**	**66**	**2**
	2013	**114**	**25**	**0**	**5**	**134**	**4**
	2014	**51**	**25**	**0**	**1**	**75**	**2**
Australia	2011	0	..	..	-1	1	0
	2012	0	1	..	1	0	0
	2013	2	0	..	0	2	0
	2014	5	0	..	1	4	0
Fiji	2011	..	0	0	..	0	0
	2012	..	1	0	..	1	1
	2013	..	1	0	..	1	1
	2014	..	1	0	..	1	1
French Polynesia	2011	..	*1	..	..	*1	*4
	2012	..	*1	..	..	*1	*4
	2013	..	*1	..	..	*1	*4
	2014	..	*1	..	..	*1	*4
Kiribati	2011	..	*3	..	..	*3	*28
	2012	..	*3	..	..	*3	*27
	2013	..	*3	..	..	*3	*27
	2014	..	*3	..	..	*3	*27
New Caledonia	2011	..	*4	..	..	*4	*16
	2012	..	*4	..	..	*4	*14
	2013	..	3	..	..	3	13
	2014	..	*3	..	..	*3	*13
New Zealand	2011	2	..	..	..	2	0
	2012	0	..	..	..	0	0
	2013	0	..	..	..	0	0
	2014	0	..	..	..	0	0
Palau	2011	..	*2	..	..	*2	*97
	2012	..	*2	..	..	*2	*96
	2013	..	*2	..	..	*2	*96
	2014	..	*2	..	..	*2	*95
Papua New Guinea	2011	0	27	..	0	27	4
	2012	15	26	..	0	41	6
	2013	112	0	..	5	107	15
	2014	*46	0	..	..	*46	*6
Samoa	2011	..	12	..	0	12	63
	2012	..	*12	..	..	*12	*62
	2013	..	12	..	1	11	59
	2014	..	*12	..	..	*12	*63

Table 19

Production, trade and supply of other kerosene

Thousand metric tons and kilograms per capita

Country or area	Year	Production	Imports	Exports	Changes in stocks	Supply	
						Total	Per Capita
Solomon Islands	2011	..	*3	..	..	*3	*5
	2012	..	*3	..	..	*3	*5
	2013	..	*3	..	..	*3	*5
	2014	..	*3	..	..	*3	*5
Tonga	2011	..	1	..	..	1	7
	2012	..	0	..	..	0	3
	2013	..	0	..	..	0	1
	2014	..	0	..	..	0	1
Vanuatu	2011	..	1	..	0	0	1
	2012	..	0	..	0	0	*1
	2013	..	0	..	0	0	1
	2014	..	0	..	0	0	0

Production, trade and supply of jet fuels

Thousand metric tons and kilograms per capita

Table Notes

Data refer to aviation gasoline, gasoline-type jet fuel and kerosene-type jet fuel.

Production includes output from refineries and plants.

For China, data on kerosene-type jet fuel up to 2013 are included with other kerosene (Table 19).

Please refer to the Definitions Section on pages ix to xv for the appropriate product description/ classification.

Figure 54: Production of jet fuels by region, 1990-2014

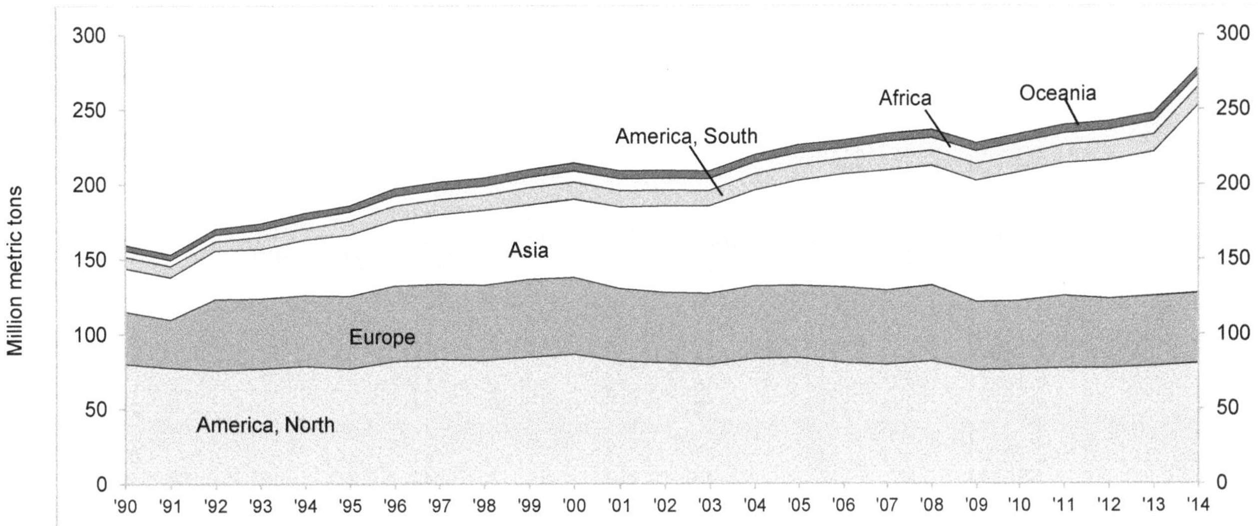

Figure 55: Major jet fuels producing countries, 2014

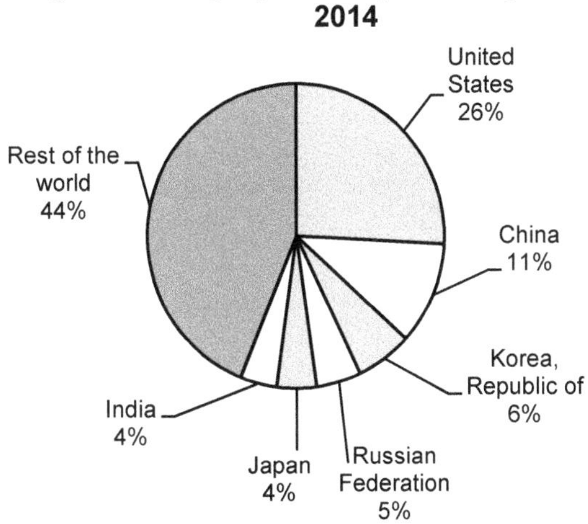

Figure 56: Major jet fuels consuming[1] countries, 2014

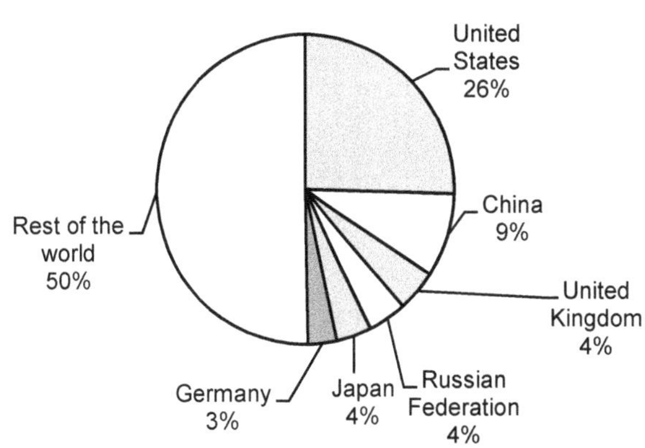

[1] Consumption includes bunkers

Table 20

Production, trade and supply of jet fuels

Thousand metric tons and kilograms per capita

Country or area	Year	Production	Imports	Exports	Bunkers	Changes in stocks	Supply	
							Total	Per capita
World	**2011**	**239,575**	**61,952**	**63,092**	**149,274**	**-2**	**89,163**	**13**
	2012	**241,938**	**61,137**	**65,366**	**151,510**	**-1,434**	**87,634**	**12**
	2013	**247,461**	**66,303**	**69,820**	**153,918**	**43**	**89,983**	**13**
	2014	**278,366**	**70,290**	**78,043**	**161,837**	**-931**	**109,707**	**15**
Africa	**2011**	**7,712**	**3,676**	**2,402**	**6,682**	**-14**	**2,318**	**2**
	2012	**7,944**	**4,069**	**2,541**	**7,257**	**-23**	**2,237**	**2**
	2013	**8,994**	**4,154**	**3,960**	**6,798**	**23**	**2,367**	**2**
	2014	**7,940**	**4,445**	**3,259**	**6,701**	**38**	**2,388**	**2**
Algeria	2011	1,209	..	750	454	2	3	0
	2012	1,204	..	683	506	0	15	0
	2013	1,713	..	1,124	523	8	58	2
	2014	2,001	..	1,481	415	-14	119	3
Angola	2011	268	176	7	208	0	229	10
	2012	303	68	8	173	-1	191	8
	2013	339	109	8	201	17	222	9
	2014	393	137	8	225	49	248	10
Benin	2011	..	153	..	153	..	0	0
	2012	..	160	..	160	..	0	0
	2013	..	171	..	171	..	0	0
	2014	..	182	..	182	..	0	0
Botswana	2011	..	19	..	17	..	2	1
	2012	..	14	..	12	..	2	1
	2013	..	15	..	13	..	2	1
	2014	..	13	..	11	..	2	1
Burkina Faso	2011	..	41	..	*41	..	0	0
	2012	..	43	..	*43	..	0	0
	2013	..	55	..	*55	..	0	0
	2014	..	34	..	*34	..	0	0
Burundi	2011	..	5	..	4	1	0	0
	2012	..	4	..	4	0	0	0
	2013	..	4	..	4	0	0	0
	2014	..	4	..	4	0	0	0
Cabo Verde	2011	..	66	..	46	..	19	39
	2012	..	68	..	54	..	14	28
	2013	..	67	..	50	..	17	33
	2014	..	69	..	54	..	15	29
Cameroon	2011	346	3	184	68	..	97	5
	2012	362	3	193	71	..	101	5
	2013	382	3	205	75	..	105	5
	2014	405	3	220	79	..	109	5
Central African Rep.	2011	..	*31	..	*31	..	0	0
	2012	..	*33	..	*33	..	0	0
	2013	..	*33	..	*33	..	0	0
	2014	..	*34	..	*34	..	0	0
Chad	2011	..	*25	..	*18	..	*7	*1
	2012	..	*25	..	*18	..	*7	*1
	2013	..	*25	..	*18	..	*7	*1
	2014	..	*25	..	*18	..	*7	*1
Comoros	2011	..	*4	..	*3	..	*1	*1
	2012	..	*4	..	*3	..	*1	*1
	2013	..	*4	..	*3	..	*1	*1
	2014	..	*4	..	*3	..	*1	*1

Production, trade and supply of jet fuels

Thousand metric tons and kilograms per capita

Country or area	Year	Production	Imports	Exports	Bunkers	Changes in stocks	Supply	
							Total	Per capita
Congo	2011	39	52	..	49	-4	46	11
	2012	67	30	..	47	0	50	12
	2013	56	29	..	43	0	42	10
	2014	60	36	..	40	0	56	12
Côte d'Ivoire	2011	579	..	525	54	..	0	0
	2012	792	..	732	60	..	0	0
	2013	831	..	775	56	..	0	0
	2014	683	..	617	66	..	0	0
Dem. Rep. of the Congo	2011	..	145	..	145	..	0	0
	2012	..	139	..	139	..	0	0
	2013	..	151	..	151	..	0	0
	2014	..	226	..	226	..	0	0
Djibouti	2011	..	82	..	82	..	0	0
	2012	..	*83	..	*83	..	0	0
	2013	..	*87	..	*87	..	0	0
	2014	..	*89	..	*89	..	0	0
Egypt	2011	1,519	0	183	771	..	565	7
	2012	1,505	129	311	815	..	508	6
	2013	1,629	73	457	564	..	681	8
	2014	1,269	178	262	687	..	498	6
Equatorial Guinea	2011	..	*35	..	*35	..	0	0
	2012	..	*35	..	*35	..	0	0
	2013	..	*35	..	*35	..	0	0
	2014	..	*35	..	*35	..	0	0
Eritrea	2011	..	1	..	1	..	0	0
	2012	..	1	..	1	..	0	0
	2013	..	1	..	1	..	0	0
	2014	..	1	..	1	..	0	0
Ethiopia	2011	..	320	..	272	..	48	1
	2012	..	347	..	318	-27	56	1
	2013	..	441	..	350	29	62	1
	2014	..	486	..	386	32	68	1
Gabon	2011	85	27	29	66	17	0	0
	2012	49	30	17	74	-12	0	0
	2013	63	30	17	76	0	0	0
	2014	60	28	25	63	..	0	0
Gambia	2011	..	22	*1	*21	..	0	0
	2012	..	16	*1	*15	..	0	0
	2013	..	22	*1	*21	..	0	0
	2014	..	15	*1	*14	..	0	0
Ghana	2011	116	*5	..	127	0	-6	0
	2012	48	101	..	125	0	23	1
	2013	60	46	..	127	-4	-18	-1
	2014	9	117	..	111	..	16	1
Guinea	2011	..	*30	..	*25	..	*5	0
	2012	..	*30	..	*25	..	*5	0
	2013	..	*30	..	*25	..	*5	0
	2014	..	*30	..	*25	..	*5	0
Guinea-Bissau	2011	..	*11	..	*11	..	0	0
	2012	..	*11	..	*11	..	0	0
	2013	..	*11	..	*11	..	0	0
	2014	..	*12	..	*12	..	0	0

Table 20

Production, trade and supply of jet fuels

Thousand metric tons and kilograms per capita

Country or area	Year	Production	Imports	Exports	Bunkers	Changes in stocks	Supply	
							Total	Per capita
Kenya	2011	275	399	0	671	0	3	0
	2012	157	516	0	671	0	2	0
	2013	106	447	44	551	0	-43	-1
	2014	0	212	38	210	..	-36	-1
Liberia	2011	..	19	..	*5	..	15	4
	2012	..	29	..	*7	..	22	5
	2013	..	18	..	*7	..	12	3
	2014	..	28	..	*7	..	21	5
Libya	2011	648	..	595	50	0	3	0
	2012	736	..	524	208	0	4	1
	2013	908	..	747	156	0	5	1
	2014	371	..	250	119	0	2	0
Madagascar	2011	..	63	..	62	2	0	0
	2012	..	68	..	63	5	0	0
	2013	..	53	..	53	1	0	0
	2014	..	*56	..	*55	*1	0	0
Malawi	2011	..	*8	..	*7	..	*1	0
	2012	..	*8	..	*7	..	*1	0
	2013	..	*8	..	*7	..	*1	0
	2014	..	*8	..	*7	..	*1	0
Mali	2011	..	*21	..	*21	..	0	0
	2012	..	*22	..	*22	..	0	0
	2013	..	*23	..	*23	..	0	0
	2014	..	*23	..	*23	..	0	0
Mauritania	2011	..	18	..	17	..	1	0
	2012	..	24	..	23	..	1	0
	2013	..	30	..	29	..	1	0
	2014	..	30	..	29	..	1	0
Mauritius	2011	..	226	..	119	-22	129	103
	2012	..	213	..	110	-8	111	88
	2013	..	241	..	116	9	116	92
	2014	..	232	..	122	-12	122	96
Mayotte	2011	..	4	..	4	0	0	0
	2012	..	*4	..	*4	..	0	0
	2013	..	*4	..	*4	..	0	0
	2014	..	*4	..	*4	..	0	0
Morocco	2011	553	88	0	629	12	0	0
	2012	877	50	0	923	4	0	0
	2013	1,060	15	514	576	-15	0	0
	2014	906	30	291	630	15	0	0
Mozambique	2011	..	59	..	59	..	0	0
	2012	..	70	..	65	..	5	0
	2013	..	77	..	70	..	7	0
	2014	..	156	..	120	..	36	1
Namibia	2011	..	15	..	*13	..	3	1
	2012	..	46	..	*39	..	7	3
	2013	..	22	..	*20	..	2	1
	2014	..	58	..	47	..	11	5
Niger	2011	..	15	..	15	..	0	0
	2012	..	12	..	12	..	0	0
	2013	..	22	..	22	..	0	0
	2014	..	32	..	29	..	3	0

Production, trade and supply of jet fuels

Thousand metric tons and kilograms per capita

Country or area	Year	Production	Imports	Exports	Bunkers	Changes in stocks	Supply	
							Total	Per capita
Nigeria	2011	77	105	..	182	..	0	0
	2012	62	110	..	172	..	0	0
	2013	76	263	..	339	..	0	0
	2014	47	255	..	302	..	0	0
Réunion	2011	..	199	..	*61	-3	141	169
	2012	..	185	..	*54	6	125	149
	2013	..	147	..	*44	-1	103	122
	2014	..	153	..	*48	-8	112	132
Rwanda	2011	..	*31	..	*17	..	*14	*1
	2012	..	*35	..	*20	..	*16	*1
	2013	..	*34	..	*19	..	*15	*1
	2014	..	*35	..	*20	..	*15	*1
Sao Tome and Principe	2011	..	*11	..	*10	..	*1	*6
	2012	..	*11	..	*10	..	*1	*6
	2013	..	*11	..	*10	..	*1	*5
	2014	..	*11	..	*10	..	*1	*5
Senegal	2011	63	154	..	207	-2	12	1
	2012	8	194	..	199	-1	4	0
	2013	12	251	..	250	-3	16	1
	2014	13	248	..	262	-3	2	0
Seychelles	2011	..	40	..	41	..	-1	-12
	2012	..	32	..	28	..	4	42
	2013	..	30	..	28	..	2	17
	2014	..	32	..	*27	..	5	48
Sierra Leone	2011	..	*8	..	*8	..	0	0
	2012	..	*8	..	*8	..	0	0
	2013	..	*8	..	*8	..	0	0
	2014	..	*8	..	*8	..	0	0
Somalia	2011	..	*42	..	42	..	0	0
	2012	..	*42	..	42	..	0	0
	2013	..	*42	..	42	..	0	0
	2014	..	*42	..	42	..	0	0
South Africa	2011	1,772	225	128	888	..	981	19
	2012	1,623	271	72	871	..	951	18
	2013	1,649	219	68	863	..	937	18
	2014	1,568	259	66	841	..	920	17
South Sudan	2011	..	..	..	..	..	..	..
	2012	..	39	..	31	..	8	1
	2013	..	46	..	36	..	10	1
	2014	..	48	..	37	..	11	1
Sudan	2011	..	..	..	..	..	..	..
	2012	118	101	..	219	*4	-4	0
	2013	79	78	..	*192	*-19	*-16	0
	2014	120	124	..	244	..	0	0
Sudan (former)	2011	134	158	..	307	..	-15	0
	2012	..	..	..	..	..	..	..
	2013	..	..	..	..	..	..	..
	2014	..	..	..	..	..	..	..
Swaziland	2011	..	*1	..	*1	..	0	0
	2012	..	*1	..	*1	..	0	0
	2013	..	*1	..	*1	..	0	0
	2014	..	*1	..	*1	..	0	0

Table 20

Production, trade and supply of jet fuels

Thousand metric tons and kilograms per capita

Country or area	Year	Production	Imports	Exports	Bunkers	Changes in stocks	Supply	
							Total	Per capita
Togo	2011	..	59	..	65	-7	1	0
	2012	..	65	..	70	-7	2	0
	2013	..	68	..	74	-7	1	0
	2014	..	72	..	78	-7	1	0
Tunisia	2011	..	238	..	234	-9	13	1
	2012	..	300	..	282	14	4	0
	2013	..	304	..	283	8	13	1
	2014	..	258	..	264	-15	9	1
Uganda	2011	..	90	..	*90	..	0	0
	2012	..	*92	..	*92	..	0	0
	2013	..	92	..	*92	..	0	0
	2014	..	120	..	115	..	6	0
United Rep. of Tanzania	2011	..	114	..	114	..	0	0
	2012	..	122	..	122	..	0	0
	2013	..	129	..	129	..	0	0
	2014	..	138	..	138	..	0	0
Zambia	2011	29	4	..	33	..	0	0
	2012	33	17	..	50	..	0	0
	2013	31	19	..	50	..	0	0
	2014	35	4	..	39	..	0	0
Zimbabwe	2011	..	9	..	9	..	0	0
	2012	..	10	..	10	..	0	0
	2013	..	10	..	10	..	0	0
	2014	..	11	..	11	..	0	0
America, North	**2011**	**77,050**	**8,413**	**6,196**	**26,348**	**-266**	**53,185**	**96**
	2012	**77,057**	**7,525**	**7,385**	**26,398**	**-1,271**	**52,070**	**93**
	2013	**78,646**	**9,062**	**8,684**	**26,810**	**-158**	**52,373**	**93**
	2014	**80,481**	**10,067**	**8,926**	**27,515**	**-118**	**54,226**	**95**
Antigua and Barbuda	2011	..	*77	..	45	..	*32	*366
	2012	..	*78	..	46	..	*32	*362
	2013	..	*78	..	*46	..	*32	*358
	2014	..	*79	..	*47	..	*32	*355
Aruba	2011	..	*78	..	*78	..	0	0
	2012	..	*81	..	*81	..	0	0
	2013	..	*82	..	*82	..	0	0
	2014	..	*83	..	*83	..	0	0
Bahamas	2011	..	86	..	*86	..	0	1
	2012	..	89	..	*88	..	0	1
	2013	..	60	..	*59	..	1	4
	2014	..	56	..	*54	..	1	4
Barbados	2011	..	*141	..	*141	..	0	0
	2012	..	*114	..	*114	..	0	0
	2013	..	*128	..	*128	..	0	0
	2014	..	*32	..	*32	..	0	0
Belize	2011	..	12	..	11	..	0	1
	2012	..	11	..	11	..	0	1
	2013	..	14	..	14	..	0	1
	2014	..	13	..	12	..	0	1
Bermuda	2011	..	12	..	*12	..	0	0
	2012	..	*12	..	*12	..	0	0
	2013	..	*12	..	*12	..	0	0
	2014	..	*18	..	*18	..	0	0

Production, trade and supply of jet fuels

Thousand metric tons and kilograms per capita

Country or area	Year	Production	Imports	Exports	Bunkers	Changes in stocks	Supply	
							Total	Per capita
Bonaire, St Eustatius, Saba	2011	..	..	..	..	..	..	..
	2012	..	*7	..	*7	..	0	0
	2013	..	*8	..	*8	..	0	0
	2014	..	*8	..	*8	..	0	0
Canada	2011	3,286	3,110	253	911	12	5,220	151
	2012	3,805	2,364	381	944	73	4,771	137
	2013	3,995	2,234	480	775	-25	4,999	142
	2014	4,168	2,407	394	689	25	5,467	154
Cayman Islands	2011	..	12	..	*11	..	1	18
	2012	..	13	..	*12	..	1	17
	2013	..	11	..	10	..	1	17
	2014	..	11	..	10	..	1	9
Costa Rica	2011	0	184	0	0	5	179	39
	2012	..	175	0	0	1	174	37
	2013	..	162	0	151	-1	12	3
	2014	..	174	0	153	-7	28	6
Cuba	2011	354	578	..	*134	..	798	70
	2012	284	264	..	*134	..	414	36
	2013	293	553	..	*207	..	638	56
	2014	276	576	..	*209	..	643	56
Curaçao	2011	..	..	..	..	..	..	..
	2012	668	57	666	59	..	0	0
	2013	609	58	607	60	..	0	0
	2014	707	59	705	61	..	0	0
Dominica	2011	..	*1	..	*1	..	0	0
	2012	..	*1	..	*1	..	0	0
	2013	..	*1	..	*1	..	0	0
	2014	..	*1	..	*1	..	0	0
Dominican Republic	2011	234	192	..	426	-1	1	0
	2012	219	260	..	478	0	1	0
	2013	228	205	..	434	-2	1	0
	2014	232	246	..	484	-6	1	0
El Salvador	2011	54	56	..	110	..	0	0
	2012	41	42	..	83	..	0	0
	2013	0	144	..	144	..	0	0
	2014	0	152	..	152	..	0	0
Greenland	2011	..	21	..	..	-2	23	412
	2012	..	17	..	..	-2	19	332
	2013	..	16	..	..	-1	17	299
	2014	..	17	..	..	2	15	271
Grenada	2011	..	5	..	5	..	0	0
	2012	..	4	..	4	..	0	0
	2013	..	4	..	4	..	0	0
	2014	..	6	..	6	..	0	0
Guadeloupe	2011	..	*107	..	107	..	0	0
	2012	..	*108	..	*108	..	0	0
	2013	..	*108	..	*108	..	0	0
	2014	..	*109	..	*109	..	0	0
Guatemala	2011	..	57	..	57	0	0	0
	2012	..	39	..	41	-2	0	0
	2013	..	40	..	40	0	0	0
	2014	..	24	..	20	4	0	0

Table 20

Production, trade and supply of jet fuels

Thousand metric tons and kilograms per capita

Country or area	Year	Production	Imports	Exports	Bunkers	Changes in stocks	Supply	
							Total	Per capita
Haiti	2011	..	17	..	17	..	0	0
	2012	..	54	..	52	..	2	0
	2013	..	94	..	89	..	5	0
	2014	..	34	..	34	..	0	0
Honduras	2011	..	52	..	52	..	0	0
	2012	..	49	..	49	..	0	0
	2013	..	41	..	41	..	0	0
	2014	..	67	..	67	..	0	0
Jamaica	2011	93	98	..	185	4	2	1
	2012	101	93	..	181	6	6	2
	2013	98	86	..	184	-2	1	1
	2014	82	112	..	196	3	-6	-2
Martinique	2011	*116	0	*29	*87	..	0	0
	2012	*114	0	*29	*85	..	0	0
	2013	*112	0	*27	*85	..	0	0
	2014	*110	0	*25	*85	..	0	0
Mexico	2011	2,568	63	80	2,573	34	-56	0
	2012	2,590	168	0	2,723	-4	39	0
	2013	2,774	166	56	2,850	14	20	0
	2014	2,436	563	..	3,046	-39	-8	0
Neth. Antilles (former)	2011	641	85	638	88	..	0	0
	2012	..	..	..	..	..	..	..
	2013	..	..	..	..	..	..	..
	2014	..	..	..	..	..	..	..
Nicaragua	2011	30	1	0	20	2	9	2
	2012	20	7	4	21	-1	3	1
	2013	28	0	2	20	1	5	1
	2014	29	1	..	22	1	7	1
Panama	2011	..	428	..	389	-39	78	21
	2012	..	468	..	434	0	34	9
	2013	..	579	..	509	31	39	10
	2014	..	558	..	585	48	-76	-20
Sint Maarten (Dutch part)	2011	..	..	..	..	..	..	..
	2012	..	*48	..	*48	..	0	0
	2013	..	49	..	49	..	0	0
	2014	..	*49	..	*49	..	0	0
Trinidad and Tobago	2011	655	0	246	301	6	102	76
	2012	397	0	138	169	-9	99	74
	2013	605	0	221	270	6	108	80
	2014	476	..	174	212	-27	117	86
United States	2011	69,019	2,940	4,950	20,501	-287	46,795	150
	2012	68,819	2,902	6,167	20,413	-1,333	46,474	148
	2013	69,903	4,130	7,291	20,430	-180	46,492	147
	2014	71,966	4,614	7,628	21,071	-122	48,003	150
America, South	**2011**	**12,146**	**2,042**	**3,973**	**5,785**	**-505**	**4,935**	**12**
	2012	**12,321**	**2,293**	**3,390**	**6,568**	**-525**	**5,181**	**13**
	2013	**11,484**	**2,281**	**2,864**	**6,602**	**-638**	**4,937**	**12**
	2014	**12,331**	**2,067**	**3,315**	**6,764**	**-563**	**4,882**	**12**
Argentina	2011	1,276	90	..	694	14	658	16
	2012	1,344	40	..	693	8	683	16
	2013	1,337	55	..	863	13	516	12
	2014	1,413	74	..	860	-8	635	15

Production, trade and supply of jet fuels

Thousand metric tons and kilograms per capita

Country or area	Year	Production	Imports	Exports	Bunkers	Changes in stocks	Supply	
							Total	Per capita
Bolivia (Plur. State of)	2011	151	..	..	48	1	102	10
	2012	136	..	..	42	1	93	9
	2013	167	..	..	52	3	112	11
	2014	181	..	..	59	..	122	12
Brazil	2011	4,413	1,465	94	2,054	77	3,653	18
	2012	4,449	1,520	109	2,135	-20	3,745	19
	2013	4,567	1,446	131	2,228	-35	3,689	18
	2014	4,992	1,218	255	2,228	55	3,672	18
Chile	2011	570	202	0	415	-30	387	22
	2012	617	466	0	668	23	392	23
	2013	613	456	6	591	25	447	25
	2014	629	402	0	615	14	402	23
Colombia	2011	1,313	17	92	*1,144	8	87	2
	2012	1,339	35	31	*1,233	16	95	2
	2013	1,108	41	29	995	25	99	2
	2014	1,084	40	26	1,032	55	11	0
Ecuador	2011	360	3	..	342	-13	34	2
	2012	360	3	..	330	-2	35	2
	2013	384	14	..	373	-5	30	2
	2014	359	38	..	375	19	1	0
French Guiana	2011	..	*39	..	*39	..	0	0
	2012	..	*39	..	*39	..	0	0
	2013	..	*39	..	*39	..	0	0
	2014	..	*39	..	*39	..	0	0
Guyana	2011	..	11	..	9	0	2	2
	2012	..	17	..	15	0	3	4
	2013	..	11	..	10	0	1	1
	2014	..	10	..	9	..	1	1
Paraguay	2011	..	30	..	26	0	3	1
	2012	..	31	..	27	0	3	0
	2013	..	33	..	32	-1	3	0
	2014	..	40	..	37	0	3	0
Peru	2011	597	92	500	753	-565	1	0
	2012	749	72	532	661	-496	124	4
	2013	692	135	574	722	-497	28	1
	2014	748	155	784	796	-701	24	1
Suriname	2011	..	44	..	*39	..	4	8
	2012	..	45	..	*41	..	5	9
	2013	..	47	..	*43	..	5	9
	2014	..	48	..	*43	..	5	9
Uruguay	2011	50	50	0	92	4	4	1
	2012	72	24	0	92	0	5	2
	2013	72	3	0	70	-3	7	2
	2014	84	3	0	79	3	5	2
Venezuela (Bolivar. Rep.)	2011	3,415	..	3,286	129	0	0	0
	2012	3,255	..	2,718	593	-55	-1	0
	2013	2,545	..	2,124	584	-163	0	0
	2014	2,841	..	2,250	591	..	0	0
Asia	**2011**	**88,712**	**16,281**	**36,638**	**55,003**	**816**	**12,536**	**3**
	2012	**92,489**	**15,070**	**37,455**	**56,199**	**917**	**12,988**	**3**
	2013	**96,065**	**16,092**	**37,530**	**58,827**	**274**	**15,527**	**4**
	2014	**125,663**	**19,912**	**46,283**	**65,595**	**-35**	**33,732**	**8**

Table 20

Production, trade and supply of jet fuels

Thousand metric tons and kilograms per capita

Country or area	Year	Production	Imports	Exports	Bunkers	Changes in stocks	Supply	
							Total	Per capita
Afghanistan	2011	..	60	..	10	..	50	2
	2012	..	51	..	10	..	41	1
	2013	..	*51	..	*10	..	*41	*1
	2014	..	*52	..	*10	..	*41	*1
Armenia	2011	..	44	..	*44	..	0	0
	2012	..	47	..	*47	..	0	0
	2013	..	46	..	*46	..	0	0
	2014	..	40	..	*40	..	0	0
Azerbaijan	2011	621	..	79	425	13	104	11
	2012	626	..	146	371	-7	116	12
	2013	703	..	131	380	28	164	17
	2014	709	..	196	347	-6	172	18
Bahrain	2011	3,116	..	2,534	593	-35	24	18
	2012	3,176	..	2,512	563	71	30	22
	2013	3,424	..	2,930	402	46	46	34
	2014	2,993	..	2,802	395	-246	42	31
Bangladesh	2011	2	318	..	336	-16	0	0
	2012	4	340	..	312	32	0	0
	2013	0	311	..	318	-7	0	0
	2014	1	334	..	323	12	0	0
Bhutan	2011	..	*1	..	*1	..	0	0
	2012	..	2	..	2	..	0	0
	2013	..	3	..	*3	..	0	0
	2014	..	3	..	*3	..	0	0
Brunei Darussalam	2011	102	17	11	107	1	0	0
	2012	47	32	7	80	-3	-5	-12
	2013	65	16	4	83	-1	-5	-12
	2014	58	18	6	71	0	-1	-2
Cambodia	2011	..	55	..	44	..	11	1
	2012	..	69	..	55	..	14	1
	2013	..	79	..	63	..	16	1
	2014	..	92	..	74	..	18	1
China	2011	583	..	..	..	..	583	0
	2012	626	..	..	..	..	626	0
	2013	676	..	..	..	..	676	0
	2014	30,532	3,914	10,637	6,962	82	16,766	12
China, Hong Kong SAR	2011	..	7,299	308	6,946	28	17	2
	2012	..	6,927	253	6,857	-199	16	2
	2013	..	7,052	1	6,939	96	16	2
	2014	..	6,960	1	7,205	-260	14	2
Cyprus	2011	..	284	..	294	-5	-5	-4
	2012	..	265	..	264	-4	5	4
	2013	..	238	..	235	8	-5	-4
	2014	..	221	..	231	-4	-6	-5
Georgia	2011	..	57	..	35	4	18	4
	2012	..	51	..	37	0	14	3
	2013	..	89	..	85	3	1	0
	2014	..	78	..	80	-3	1	0
India	2011	10,051	0	4,561	4,719	0	771	1
	2012	10,077	0	4,664	4,719	0	694	1
	2013	11,210	0	5,745	3,853	39	1,573	1
	2014	11,103	0	*5,645	*3,798	*-40	1,700	1

Production, trade and supply of jet fuels

Thousand metric tons and kilograms per capita

Country or area	Year	Production	Imports	Exports	Bunkers	Changes in stocks	Supply	
							Total	Per capita
Indonesia	2011	2,153	649	1	680	..	2,121	9
	2012	2,403	564	2	722	..	2,243	9
	2013	2,538	781	2	893	..	2,424	10
	2014	2,274	768	2	1,298	..	1,742	7
Iran (Islamic Rep. of)	2011	1,140	1	..	1,145	..	-4	0
	2012	1,165	0	..	1,137	..	28	0
	2013	1,192	0	0	1,173	..	19	0
	2014	1,252	0	..	1,248	..	4	0
Iraq	2011	308	482	..	790	..	0	0
	2012	133	183	..	316	..	0	0
	2013	144	251	..	395	..	0	0
	2014	215	312	..	527	..	0	0
Israel	2011	813	..	0	813	..	0	0
	2012	786	..	0	786	..	0	0
	2013	1,072	..	..	1,072	..	0	0
	2014	932	129	..	1,061	..	0	0
Japan	2011	10,037	998	2,007	5,802	-76	3,302	26
	2012	10,404	1,235	2,171	6,086	196	3,186	25
	2013	12,063	783	2,749	6,166	-60	3,991	31
	2014	12,054	467	2,144	6,026	87	4,264	34
Jordan	2011	338	1	..	113	6	220	33
	2012	366	1	..	129	5	233	33
	2013	328	26	..	*134	-4	225	31
	2014	326	26	..	127	16	209	28
Kazakhstan	2011	210	113	..	158	-1	166	10
	2012	200	95	..	197	9	89	5
	2013	200	37	..	164	-1	74	4
	2014	302	27	1	226	12	90	5
Korea, Republic of	2011	14,902	3	10,244	3,808	50	803	16
	2012	16,484	2	11,545	3,810	70	1,061	21
	2013	16,397	2	11,290	4,023	92	994	20
	2014	17,053	1	11,831	4,033	-120	1,310	26
Kuwait	2011	2,538	..	1,855	683	..	0	0
	2012	2,748	..	1,947	801	..	0	0
	2013	2,495	..	1,768	727	..	0	0
	2014	2,438	..	1,730	708	..	0	0
Kyrgyzstan	2011	..	118	67	51	0	0	0
	2012	..	102	73	30	0	-1	0
	2013	..	126	111	48	-33	0	0
	2014	..	96	89	4	1	2	0
Lao People's Dem. Rep.	2011	..	*22	..	*10	..	*12	*2
	2012	..	*22	..	*10	..	*12	*2
	2013	..	*23	..	*11	..	*12	*2
	2014	..	*24	..	*11	..	*12	*2
Lebanon	2011	..	224	..	224	..	0	0
	2012	..	207	..	207	..	0	0
	2013	..	259	..	259	..	0	0
	2014	..	229	..	229	..	0	0
Malaysia	2011	3,348	480	661	1,730	799	638	22
	2012	3,794	336	976	1,709	743	702	24
	2013	2,663	831	554	2,323	19	598	20
	2014	2,824	369	517	2,447	-374	603	20

Table 20

Production, trade and supply of jet fuels

Thousand metric tons and kilograms per capita

Country or area	Year	Production	Imports	Exports	Bunkers	Changes in stocks	Supply	
							Total	Per capita
Maldives	2011	..	31	..	16	..	15	45
	2012	..	94	..	71	..	23	67
	2013	..	133	..	*103	..	30	86
	2014	..	223	..	*187	..	36	101
Myanmar	2011	42	68	..	18	-13	105	2
	2012	33	60	..	34	9	50	1
	2013	21	104	..	36	-11	100	2
	2014	17	114	..	40	1	89	2
Nepal	2011	..	*89	..	*89	..	0	0
	2012	..	*94	..	*94	..	0	0
	2013	..	*94	..	*94	..	0	0
	2014	..	139	..	*139	..	0	0
Oman	2011	632	..	257	375	..	0	0
	2012	712	..	309	403	..	0	0
	2013	564	..	192	372	..	0	0
	2014	617	..	144	507	-34	0	0
Other Asia	2011	3,373	40	1,119	1,972	-47	369	16
	2012	3,863	..	1,317	2,073	32	441	19
	2013	3,866	..	1,262	2,198	48	358	15
	2014	3,738	..	1,011	2,398	54	275	12
Pakistan	2011	703	271	118	200	-13	669	4
	2012	742	132	2	184	0	688	4
	2013	785	130	61	199	0	655	4
	2014	758	130	0	208	0	680	4
Philippines	2011	783	622	..	1,085	-12	332	4
	2012	693	758	..	1,096	0	355	4
	2013	644	972	..	1,112	0	503	5
	2014	768	847	..	1,054	-6	567	6
Qatar	2011	3,581	..	2,105	1,454	22	0	0
	2012	3,387	..	1,650	1,797	-60	0	0
	2013	3,534	..	1,033	1,988	252	261	124
	2014	2,881	..	827	1,254	800	0	0
Saudi Arabia	2011	2,898	..	219	2,009	..	670	23
	2012	3,045	..	270	2,081	..	694	24
	2013	2,837	..	252	1,869	..	716	24
	2014	3,689	..	630	2,294	..	765	25
Singapore	2011	9,441	2,116	5,837	*5,720	..	0	0
	2012	9,751	1,507	4,859	*6,399	..	0	0
	2013	9,866	1,589	4,762	*6,692	..	0	0
	2014	9,596	1,938	4,795	*6,739	..	0	0
Sri Lanka	2011	155	244	..	308	84	7	0
	2012	93	288	..	388	-37	30	1
	2013	125	296	..	348	70	2	0
	2014	168	235	..	458	-58	3	0
Syrian Arab Republic	2011	347	..	259	29	..	59	3
	2012	246	..	176	23	..	47	2
	2013	163	..	99	22	..	42	2
	2014	163	..	115	16	..	32	2
Tajikistan	2011	..	31	..	31	..	0	0
	2012	..	33	..	33	..	0	0
	2013	..	35	..	35	..	0	0
	2014	..	61	..	61	..	0	0

Production, trade and supply of jet fuels

Thousand metric tons and kilograms per capita

Country or area	Year	Production	Imports	Exports	Bunkers	Changes in stocks	Supply	
							Total	Per capita
Thailand	2011	4,816	8	919	*3,342	*19	544	8
	2012	4,471	2	574	*3,342	*13	544	8
	2013	5,007	12	1,152	3,648	-301	520	8
	2014	5,224	3	972	3,637	122	496	7
Timor-Leste	2011	..	*3	..	*3	..	0	0
	2012	..	*3	..	*3	..	0	0
	2013	..	*3	..	*3	..	0	0
	2014	..	*3	..	*3	..	0	0
Turkey	2011	2,879	568	1,997	1,133	3	314	4
	2012	3,277	589	2,437	1,003	45	381	5
	2013	3,584	522	2,022	1,308	-7	783	10
	2014	3,548	850	959	2,540	-69	968	12
Turkmenistan	2011	480	..	..	480	..	0	0
	2012	415	..	..	415	..	0	0
	2013	455	..	..	455	..	0	0
	2014	470	..	..	470	..	0	0
United Arab Emirates	2011	7,894	..	989	*6,581	..	324	37
	2012	8,411	..	1,262	*6,813	..	336	38
	2013	8,919	..	879	*7,663	..	377	42
	2014	8,478	..	723	5,233	..	2,522	278
Uzbekistan	2011	148	..	..	..	..	148	5
	2012	132	..	..	..	..	132	5
	2013	125	..	..	..	..	125	4
	2014	118	..	..	..	..	118	4
Viet Nam	2011	83	964	358	536	4	149	2
	2012	54	979	218	650	2	163	2
	2013	75	1,200	309	774	-2	194	2
	2014	78	1,209	312	781	*-2	196	2
Yemen	2011	195	..	133	62	..	0	0
	2012	125	..	85	40	..	0	0
	2013	326	..	222	104	..	0	0
	2014	285	..	194	91	..	0	0
Europe	**2011**	**48,398**	**29,754**	**13,850**	**51,456**	**-70**	**12,916**	**18**
	2012	**46,542**	**30,256**	**14,562**	**50,926**	**-534**	**11,844**	**16**
	2013	**46,830**	**31,963**	**16,756**	**50,522**	**538**	**10,977**	**15**
	2014	**46,871**	**30,739**	**16,246**	**50,647**	**-258**	**10,975**	**15**
Albania	2011	..	20	..	*8	..	12	4
	2012	..	20	..	*8	..	12	4
	2013	..	28	..	*8	..	20	7
	2014	..	8	..	7	..	1	0
Austria	2011	615	125	9	688	6	37	4
	2012	618	96	24	659	-4	35	4
	2013	654	103	23	627	73	34	4
	2014	581	85	32	627	-23	30	4
Belarus	2011	0	0	0	0	0	0	0
	2012	0	0	0	0	0	0	0
	2013	361	16	255	100	9	13	1
	2014	337	17	226	110	-4	22	2
Belgium	2011	1,498	1,813	1,548	1,434	-76	405	37
	2012	1,819	2,104	2,675	1,316	-116	48	4
	2013	1,497	1,544	1,653	1,240	44	104	9
	2014	1,666	945	1,318	1,295	-73	71	6

Table 20

Production, trade and supply of jet fuels

Thousand metric tons and kilograms per capita

Country or area	Year	Production	Imports	Exports	Bunkers	Changes in stocks	Supply	
							Total	Per capita
Bosnia and Herzegovina	2011	..	5	..	5	..	0	0
	2012	..	5	..	5	..	0	0
	2013	..	4	..	4	..	0	0
	2014	..	5	..	4	..	1	0
Bulgaria	2011	186	8	11	165	-2	20	3
	2012	171	8	10	159	-2	12	2
	2013	166	5	4	155	0	12	2
	2014	173	7	8	165	-4	11	2
Croatia	2011	117	2	6	99	3	11	3
	2012	97	20	4	105	-2	10	2
	2013	109	16	0	115	..	10	2
	2014	106	28	4	115	5	10	2
Czechia	2011	134	182	0	300	-56	72	7
	2012	143	198	1	284	22	34	3
	2013	111	173	0	274	-9	19	2
	2014	160	128	..	283	-13	18	2
Denmark	2011	275	834	440	808	-204	65	12
	2012	224	683	151	808	-108	56	10
	2013	145	796	68	800	7	66	12
	2014	137	749	64	868	-100	54	10
Estonia	2011	..	36	0	34	2	0	0
	2012	..	58	21	36	0	1	1
	2013	..	50	21	28	1	0	0
	2014	..	53	13	40	0	0	0
Faeroe Islands	2011	..	*4	..	4	..	0	0
	2012	..	0	..	0	..	0	0
	2013	..	0	..	0	..	0	0
	2014	..	0	..	0	..	0	0
Finland	2011	795	126	78	617	14	212	39
	2012	618	213	78	596	-13	170	31
	2013	671	130	11	615	15	160	29
	2014	542	243	15	606	-8	172	31
France	2011	4,761	3,610	680	5,648	-31	2,074	33
	2012	3,851	4,106	852	5,491	-52	1,666	26
	2013	4,288	3,923	1,033	5,436	52	1,690	26
	2014	3,712	4,032	1,155	5,483	47	1,059	17
Germany	2011	4,967	4,197	766	7,489	-9	918	11
	2012	5,216	4,845	1,134	7,990	77	860	11
	2013	4,757	5,273	1,145	8,150	130	605	8
	2014	4,862	4,825	893	7,826	-58	1,026	13
Gibraltar	2011	..	7	..	7	..	0	0
	2012	..	6	..	6	..	0	0
	2013	..	7	..	7	..	0	0
	2014	..	7	..	7	..	0	0
Greece	2011	1,845	404	1,094	720	1	434	39
	2012	1,784	120	831	641	-10	442	40
	2013	2,371	39	1,248	656	-10	516	47
	2014	2,237	164	1,065	757	67	512	47
Hungary	2011	239	2	1	230	10	0	0
	2012	169	0	7	166	-4	0	0
	2013	181	3	19	162	-1	4	0
	2014	191	0	16	171	4	0	0

Production, trade and supply of jet fuels

Thousand metric tons and kilograms per capita

Country or area	Year	Production	Imports	Exports	Bunkers	Changes in stocks	Supply	
							Total	Per capita
Iceland	2011	..	142	..	134	-1	9	28
	2012	..	148	..	140	2	6	19
	2013	..	167	..	158	3	6	18
	2014	..	188	..	177	-2	13	40
Ireland	2011	..	870	..	658	9	203	44
	2012	..	723	..	551	-10	182	39
	2013	..	873	60	609	25	179	38
	2014	..	898	..	704	4	190	41
Italy	2011	3,068	1,283	205	3,164	155	827	14
	2012	2,605	1,038	51	3,019	-228	801	13
	2013	2,435	1,434	3	2,922	169	775	13
	2014	2,269	1,574	19	2,995	91	738	12
Latvia	2011	..	310	167	114	29	0	0
	2012	..	271	184	115	-29	1	0
	2013	..	138	19	119	-1	1	0
	2014	..	112	5	106	0	1	1
Lithuania	2011	829	7	229	54	-5	558	182
	2012	815	168	336	61	20	566	188
	2013	1,019	48	275	68	-19	743	251
	2014	948	29	153	75	5	744	255
Luxembourg	2011	..	388	..	392	-4	0	0
	2012	..	361	..	361	0	0	0
	2013	..	363	..	363	0	0	0
	2014	..	394	..	394	..	0	0
Malta	2011	..	107	0	102	4	1	2
	2012	..	101	1	97	1	2	5
	2013	..	113	10	102	0	1	2
	2014	..	108	..	108	-1	1	2
Montenegro	2011	..	12	2	10	..	0	0
	2012	..	12	0	12	..	0	0
	2013	..	13	0	13	..	0	0
	2014	..	18	1	17	..	0	0
Netherlands	2011	7,062	2,684	6,342	3,416	-73	61	4
	2012	6,589	2,647	5,845	3,283	-2	110	7
	2013	6,467	3,847	6,830	3,354	81	49	3
	2014	7,003	2,614	6,103	3,481	-38	71	4
Norway	2011	415	302	128	413	46	130	26
	2012	501	346	98	392	-19	376	75
	2013	598	434	85	478	9	460	90
	2014	577	385	12	491	-48	507	98
Poland	2011	873	2	390	485	33	-33	-1
	2012	937	34	446	524	-53	54	1
	2013	877	53	380	512	26	12	0
	2014	1,156	27	554	569	33	27	1
Portugal	2011	791	249	0	891	7	142	13
	2012	910	181	43	899	-10	159	15
	2013	1,132	17	43	909	49	148	14
	2014	976	109	1	961	-44	167	16
Republic of Moldova	2011	..	19	0	13	0	6	1
	2012	..	22	0	22	0	0	0
	2013	..	21	0	21	0	0	0
	2014	..	27	0	25	2	0	0

Production, trade and supply of jet fuels

Thousand metric tons and kilograms per capita

Country or area	Year	Production	Imports	Exports	Bunkers	Changes in stocks	Supply	
							Total	Per capita
Romania	2011	274	57	149	126	24	32	2
	2012	254	41	165	115	-11	26	1
	2013	297	12	168	141	2	-2	0
	2014	372	17	224	178	-1	-12	-1
Russian Federation	2011	12,555	0	0	6,257	..	6,298	44
	2012	12,863	0	0	6,431	..	6,432	45
	2013	13,615	73	1,849	5,913	..	5,926	41
	2014	13,583	38	2,479	5,513	..	5,629	39
Serbia	2011	75	8	36	46	1	0	0
	2012	73	1	37	38	-1	0	0
	2013	73	1	32	42	-1	1	0
	2014	109	1	40	67	1	2	0
Slovakia	2011	58	..	9	42	-1	8	1
	2012	47	..	6	36	3	2	0
	2013	118	..	72	40	-3	9	2
	2014	78	..	43	34	1	0	0
Slovenia	2011	..	24	0	23	0	1	0
	2012	..	24	1	23	-1	1	0
	2013	..	26	0	25	0	1	0
	2014	..	317	317	25	-26	1	0
Spain	2011	257	2,736	275	3,548	50	-880	-19
	2012	163	2,303	177	3,505	15	-1,231	-26
	2013	149	1,840	422	3,494	-106	-1,821	-39
	2014	197	1,955	339	3,556	26	-1,769	-38
Sweden	2011	190	823	52	721	-46	286	30
	2012	245	656	44	686	-26	197	21
	2013	169	559	48	774	76	-170	-18
	2014	248	873	58	703	116	244	25
Switzerland	2011	81	1,455	0	1,469	16	51	6
	2012	38	1,541	3	1,493	10	73	9
	2013	38	1,572	2	1,500	7	101	12
	2014	16	1,558	..	1,530	-19	63	8
T.F.Yug.Rep. Macedonia	2011	27	0	23	4	1	0	0
	2012	17	10	17	8	2	0	0
	2013	5	13	9	10	-1	0	0
	2014	0	27	16	12	-1	0	0
Ukraine	2011	..	0	0	240	0	-240	-5
	2012	..	0	0	298	0	-298	-7
	2013	..	0	0	*123	0	*-123	*-3
	2014	..	0	0	*128	0	*-128	*-3
United Kingdom	2011	6,411	6,901	1,210	10,879	27	1,196	19
	2012	5,775	7,146	1,320	10,547	15	1,039	16
	2013	4,527	8,236	969	10,455	-89	1,428	22
	2014	4,635	8,174	1,073	10,434	-197	1,499	23
Oceania	**2011**	**5,557**	**1,786**	**34**	**3,999**	**37**	**3,272**	**88**
	2012	**5,585**	**1,925**	**33**	**4,161**	**2**	**3,314**	**88**
	2013	**5,443**	**2,750**	**26**	**4,359**	**5**	**3,803**	**100**
	2014	**5,080**	**3,059**	**15**	**4,614**	**5**	**3,505**	**90**
Australia	2011	4,412	1,652	22	3,125	22	2,895	128
	2012	4,415	1,787	17	3,243	-12	2,954	129
	2013	4,484	2,536	26	3,414	25	3,555	153
	2014	4,044	2,757	15	3,683	12	3,091	131

Production, trade and supply of jet fuels

Thousand metric tons and kilograms per capita

Country or area	Year	Production	Imports	Exports	Bunkers	Changes in stocks	Supply	
							Total	Per capita
Cook Islands	2011	..	*2	..	*2	..	0	0
	2012	..	*2	..	*2	..	0	0
	2013	..	*2	..	*2	..	0	0
	2014	..	*2	..	*2	..	0	0
Fiji	2011	..	18	0	*10	..	7	9
	2012	..	20	0	*12	..	8	9
	2013	..	*24	0	*15	..	*8	*10
	2014	..	*31	0	*22	..	*8	*10
French Polynesia	2011	..	*18	..	*14	..	*4	*15
	2012	..	*18	..	*14	..	*4	*15
	2013	..	*18	..	*14	..	*4	*14
	2014	..	*18	..	*14	..	*4	*14
Kiribati	2011	..	*4	..	*4	..	0	0
	2012	..	4	..	*4	..	0	0
	2013	..	*4	..	*4	..	0	0
	2014	..	*4	..	*4	..	0	0
Micronesia (Fed. States of)	2011	..	*4	..	*1	..	*3	*32
	2012	..	*4	..	*1	..	*3	*32
	2013	..	*4	..	*1	..	*3	*32
	2014	..	*4	..	*1	..	*3	*32
Nauru	2011	..	*1	..	*1	..	0	0
	2012	..	*1	..	*1	..	0	0
	2013	..	*1	..	*1	..	0	0
	2014	..	*1	..	*1	..	0	0
New Caledonia	2011	..	*30	..	*25	0	*5	*21
	2012	..	*30	..	*25	0	*5	*21
	2013	..	6	..	*7	-2	0	1
	2014	..	10	..	*5	0	5	20
New Zealand	2011	1,048	24	12	739	14	307	70
	2012	1,051	26	0	775	16	286	64
	2013	959	81	0	816	-18	242	54
	2014	968	150	0	791	-7	334	74
Niue	2011	..	0	..	0	..	0	15
	2012	..	0	..	0	..	0	16
	2013	..	1	..	0	..	0	15
	2014	..	1	..	0	..	0	15
Palau	2011	..	*17	..	*17	..	0	0
	2012	..	*17	..	*17	..	0	0
	2013	..	*17	..	*17	..	0	0
	2014	..	*18	..	*18	..	0	0
Papua New Guinea	2011	97	0	0	*47	1	49	7
	2012	119	0	16	*52	-2	52	7
	2013	0	44	0	*55	0	*-11	*-2
	2014	*68	*45	0	*57	0	*57	*8
Solomon Islands	2011	..	*5	..	*3	..	*1	*2
	2012	..	*5	..	*3	..	*1	*2
	2013	..	*5	..	*4	..	*1	*2
	2014	..	*5	..	*4	..	*1	*2
Tonga	2011	..	4	..	*4	..	0	0
	2012	..	4	..	*4	..	0	0
	2013	..	1	..	*1	..	0	0
	2014	..	5	..	*4	..	1	7

Table 20

Production, trade and supply of jet fuels

Thousand metric tons and kilograms per capita

Country or area	Year	Production	Imports	Exports	Bunkers	Changes in stocks	Supply	
							Total	Per capita
Vanuatu	2011	..	*6	..	*6	..	0	0
	2012	..	*6	..	*6	..	0	0
	2013	..	*6	..	*6	..	0	0
	2014	..	*7	..	*7	..	0	0
Wallis and Futuna Is.	2011	..	*1	..	1	..	0	0
	2012	..	*1	..	1	..	0	0
	2013	..	*1	..	1	..	0	0
	2014	..	*1	..	1	..	0	0

Production, trade and supply of gas oil/diesel oil

Thousand metric tons and kilograms per capita

Table Notes

Production includes output from refineries and plants, together with quantities from other sources which include any blended biodiesel.

Please refer to the Definitions Section on pages ix to xv for the appropriate product description/ classification.

Figure 57: Production of gas oil/diesel oil by region, 1990-2014

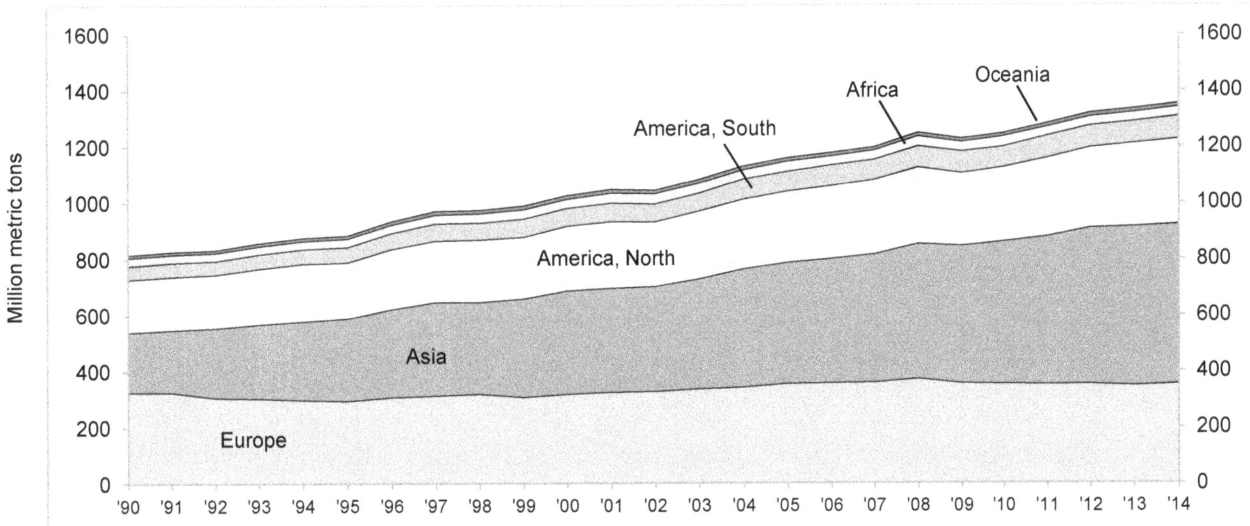

Figure 58: Major gas oil/diesel oil producing countries, 2014

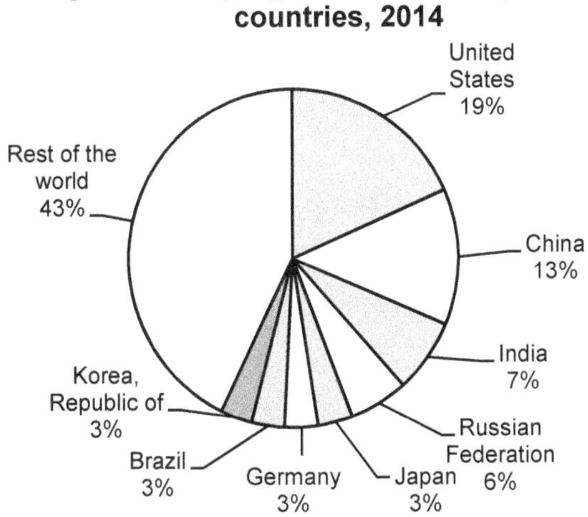

Figure 59: Major gas oil/diesel oil consuming countries, 2014

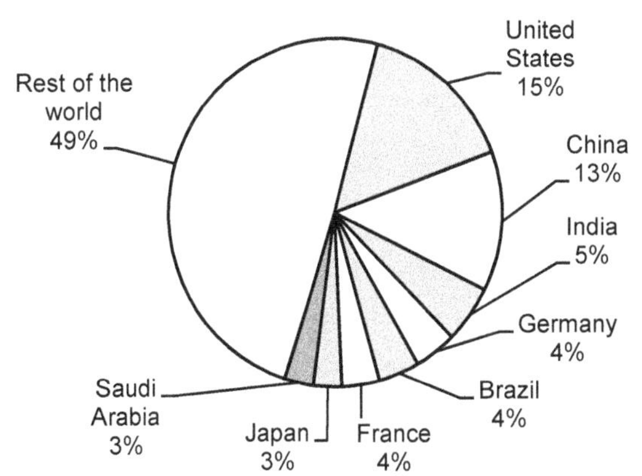

Table 21

Production, trade and supply of gas oil/diesel oil

Thousand metric tons and kilograms per capita

Country or area	Year	Production	Imports	Exports	Bunkers	Changes in stocks	Supply	
							Total	Per capita
World	**2011**	**1,281,361**	**313,561**	**318,711**	**29,549**	**-3,222**	**1,249,885**	**178**
	2012	**1,318,271**	**328,369**	**342,743**	**20,381**	**-4,461**	**1,287,977**	**181**
	2013	**1,333,679**	**344,835**	**348,461**	**21,598**	**1,830**	**1,306,625**	**182**
	2014	**1,353,304**	**357,574**	**356,784**	**31,178**	**1,253**	**1,321,663**	**182**
Africa	**2011**	**31,654**	**31,366**	**1,166**	**2,323**	**292**	**59,240**	**55**
	2012	**32,581**	**34,700**	**1,526**	**2,370**	**693**	**62,692**	**57**
	2013	**32,371**	**40,752**	**2,952**	**2,392**	**-290**	**68,070**	**60**
	2014	**32,980**	**47,682**	**8,706**	**2,410**	**-250**	**69,796**	**60**
Algeria	2011	7,573	1,230	..	45	-34	8,792	239
	2012	6,867	2,762	..	53	203	9,373	250
	2013	7,042	2,728	..	64	-113	9,819	257
	2014	9,027	1,088	..	56	162	9,897	254
Angola	2011	518	2,234	13	..	..	2,739	125
	2012	527	2,862	21	..	..	3,368	148
	2013	564	3,227	25	..	..	3,766	161
	2014	532	3,579	32	..	..	4,079	168
Benin	2011	..	557	83	..	4	470	48
	2012	..	583	87	..	5	491	49
	2013	..	623	93	..	6	524	51
	2014	..	664	99	..	7	558	53
Botswana	2011	..	430	..	..	..	430	206
	2012	..	493	..	..	..	493	231
	2013	..	545	..	..	..	545	250
	2014	..	566	..	..	..	566	255
Burkina Faso	2011	..	370	..	..	..	370	23
	2012	..	465	..	..	..	465	28
	2013	..	537	..	..	..	537	31
	2014	..	454	..	..	..	454	26
Burundi	2011	..	31	..	..	1	31	3
	2012	..	30	..	..	-1	31	3
	2013	..	29	..	..	1	28	3
	2014	..	64	..	..	0	64	6
Cabo Verde	2011	..	124	..	20	..	104	210
	2012	..	108	..	30	..	78	156
	2013	..	102	..	33	..	69	136
	2014	..	99	..	31	..	68	132
Cameroon	2011	770	51	0	41	..	780	37
	2012	805	53	0	43	..	815	38
	2013	850	40	0	45	..	845	38
	2014	901	9	100	48	..	762	33
Central African Rep.	2011	..	*34	..	..	..	*34	*8
	2012	..	*34	..	..	..	*34	*7
	2013	..	*35	..	..	..	*35	*8
	2014	..	*36	..	..	..	*36	*7
Chad	2011	..	*66	..	..	..	*66	*5
	2012	..	*66	..	..	..	*66	*5
	2013	..	*68	..	..	..	*68	*5
	2014	..	*68	..	..	..	*68	*5
Comoros	2011	..	*24	..	..	..	*24	*34
	2012	..	*27	..	..	..	*27	*37
	2013	..	*35	..	..	..	*35	*47
	2014	..	*28	..	..	..	*28	*36

Production, trade and supply of gas oil/diesel oil

Thousand metric tons and kilograms per capita

Country or area	Year	Production	Imports	Exports	Bunkers	Changes in stocks	Supply	
							Total	Per capita
Congo	2011	195	138	..	..	-16	349	84
	2012	243	118	..	..	..	361	84
	2013	276	185	..	..	..	461	105
	2014	270	189	..	..	..	459	102
Côte d'Ivoire	2011	712	..	160	0	..	552	27
	2012	1,057	..	252	0	..	805	38
	2013	1,072	..	221	0	..	851	39
	2014	1,192	..	319	32	..	841	38
Dem. Rep. of the Congo	2011	..	420	..	..	..	420	6
	2012	..	427	..	..	..	427	6
	2013	..	669	..	..	..	669	9
	2014	..	910	..	..	..	910	12
Djibouti	2011	..	44	..	*8	..	36	42
	2012	..	*60	..	*9	..	*51	*60
	2013	..	*63	..	*10	..	*54	*62
	2014	..	*64	..	*10	..	*55	*62
Egypt	2011	7,497	4,442	..	91	..	11,848	141
	2012	7,116	4,902	..	65	..	11,953	140
	2013	5,842	5,622	..	52	..	11,412	130
	2014	5,996	4,442	..	54	-355	10,739	120
Equatorial Guinea	2011	..	*145	..	..	..	*145	*193
	2012	..	*145	..	..	..	*145	*187
	2013	..	*145	..	..	..	*145	*182
	2014	..	*145	..	..	..	*145	*177
Eritrea	2011	..	71	..	..	..	71	15
	2012	..	76	..	..	..	76	16
	2013	..	77	..	..	..	77	15
	2014	..	79	..	..	..	79	15
Ethiopia	2011	..	1,206	..	..	-26	1,232	14
	2012	..	1,323	..	..	34	1,289	14
	2013	..	1,558	..	..	-9	1,567	17
	2014	..	1,727	..	..	..	1,727	18
Gabon	2011	264	343	7	80	..	520	330
	2012	212	454	6	83	..	577	358
	2013	229	522	6	75	..	670	406
	2014	223	541	6	75	..	683	405
Gambia	2011	..	70	*1	..	..	69	39
	2012	..	68	*1	..	..	67	37
	2013	..	65	*1	..	..	64	34
	2014	..	73	*1	..	..	72	37
Ghana	2011	310	1,201	357	81	0	1,073	43
	2012	122	1,309	184	81	0	1,166	46
	2013	113	1,639	0	52	-89	1,789	68
	2014	28	1,742	0	11	-65	1,824	68
Guinea	2011	..	283	..	..	..	283	25
	2012	..	344	..	..	..	344	30
	2013	..	321	..	..	..	321	27
	2014	..	*325	..	..	..	*325	*26
Guinea-Bissau	2011	..	*39	..	..	..	*39	*23
	2012	..	*40	..	..	..	*40	*24
	2013	..	*42	..	..	..	*42	*24
	2014	..	*44	..	..	..	*44	*25

Table 21

Production, trade and supply of gas oil/diesel oil

Thousand metric tons and kilograms per capita

Country or area	Year	Production	Imports	Exports	Bunkers	Changes in stocks	Supply	
							Total	Per capita
Kenya	2011	429	1,117	0	3	..	1,543	37
	2012	239	1,288	0	1	..	1,526	36
	2013	155	2,224	688	16	..	1,675	38
	2014	0	8,289	6,569	12	..	1,708	38
Lesotho	2011	..	68	..	..	..	68	33
	2012	..	84	..	..	..	84	41
	2013	..	83	..	..	..	83	40
	2014	..	*84	..	..	..	*84	*40
Liberia	2011	..	139	..	*4	..	135	33
	2012	..	171	..	*4	..	167	40
	2013	..	128	..	*4	..	123	29
	2014	..	148	..	*5	..	143	33
Libya	2011	1,522	1,738	..	..	..	3,260	518
	2012	2,251	1,661	..	..	..	3,912	623
	2013	2,775	2,149	..	..	..	4,924	786
	2014	1,278	4,218	..	..	..	5,496	878
Madagascar	2011	..	364	..	*3	-12	373	17
	2012	..	408	..	*3	12	394	18
	2013	..	390	..	..	..	390	17
	2014	..	*400	..	..	..	*400	*17
Malawi	2011	..	183	..	..	..	183	12
	2012	..	168	..	..	..	168	11
	2013	..	181	..	..	..	181	11
	2014	..	170	..	..	..	170	10
Mali	2011	..	*208	..	..	..	*208	*13
	2012	..	*190	..	..	..	*190	*12
	2013	..	*197	..	..	..	*197	*12
	2014	..	*323	..	..	..	*323	*19
Mauritania	2011	..	427	..	..	..	427	116
	2012	..	477	..	..	..	477	126
	2013	..	*477	..	..	..	*477	*123
	2014	..	*477	..	..	..	*477	*120
Mauritius	2011	..	310	..	100	2	208	166
	2012	..	314	..	103	0	211	168
	2013	..	336	..	114	*17	205	162
	2014	..	304	..	117	*-19	206	162
Mayotte	2011	..	71	..	3	-1	69	322
	2012	..	*74	..	*3	..	*71	*321
	2013	..	*77	..	*3	..	*75	*328
	2014	..	*81	..	*3	..	*79	*336
Morocco	2011	2,403	2,226	0	*13	-33	4,649	141
	2012	2,481	2,309	32	*13	131	4,614	138
	2013	2,240	2,788	208	23	-278	5,075	149
	2014	2,275	2,990	9	23	59	5,174	150
Mozambique	2011	..	589	..	..	0	589	24
	2012	..	556	..	..	0	556	22
	2013	..	970	..	..	*200	770	29
	2014	..	1,327	..	..	20	1,307	48
Namibia	2011	..	513	..	..	..	513	229
	2012	..	567	..	..	..	567	247
	2013	..	479	..	..	..	479	204
	2014	..	726	..	..	..	726	302

Production, trade and supply of gas oil/diesel oil

Thousand metric tons and kilograms per capita

Country or area	Year	Production	Imports	Exports	Bunkers	Changes in stocks	Supply	
							Total	Per capita
Niger	2011	36	168	0	..	13	190	11
	2012	328	0	102	..	0	226	13
	2013	481	0	204	..	0	276	15
	2014	452	10	*158	..	0	304	16
Nigeria	2011	1,085	121	..	124	429	653	4
	2012	1,030	0	..	125	368	537	3
	2013	664	86	..	115	37	598	3
	2014	647	..	..	117	-23	553	3
Réunion	2011	..	399	..	10	-5	395	471
	2012	..	370	..	9	-30	391	464
	2013	..	383	..	11	0	372	438
	2014	..	376	..	9	-8	375	439
Rwanda	2011	..	*104	..	..	..	*104	*10
	2012	..	114	..	..	..	114	11
	2013	..	132	..	..	..	132	12
	2014	..	139	..	..	..	139	12
Sao Tome and Principe	2011	..	*22	..	..	..	*22	*123
	2012	..	*24	..	..	..	*24	*132
	2013	..	*24	..	..	..	*24	*129
	2014	..	*24	..	..	..	*24	*126
Senegal	2011	329	594	77	82	-23	787	59
	2012	412	403	0	69	-12	758	55
	2013	451	403	81	67	-23	729	51
	2014	472	476	85	70	..	793	54
Seychelles	2011	..	150	..	124	..	26	276
	2012	..	203	..	151	..	52	546
	2013	..	188	..	150	..	39	409
	2014	..	209	..	*150	..	59	612
Sierra Leone	2011	..	134	..	..	..	134	23
	2012	..	179	..	..	..	179	30
	2013	..	207	..	..	..	207	34
	2014	..	224	..	..	..	224	35
Somalia	2011	..	*106	..	*11	..	*95	*10
	2012	..	*106	..	*13	..	*93	*9
	2013	..	*106	..	*13	..	*93	*9
	2014	..	*106	..	*13	..	*93	*9
South Africa	2011	5,704	3,265	468	1,436	..	7,065	135
	2012	6,222	3,250	809	1,468	..	7,195	136
	2013	7,050	4,307	1,409	1,500	..	8,448	158
	2014	7,000	4,347	1,313	1,523	..	8,511	158
South Sudan	2011	..	..	..	..	..	..	..
	2012	..	371	..	..	..	371	34
	2013	..	395	..	..	..	395	34
	2014	..	408	..	..	..	408	34
St. Helena and Depend.	2011	..	3	..	..	..	3	664
	2012	..	3	..	..	..	3	746
	2013	..	3	..	..	..	3	731
	2014	..	*3	..	..	..	*3	*740
Sudan	2011	..	..	..	..	..	..	..
	2012	1,666	616	32	19	..	2,231	59
	2013	1,594	600	15	19	..	2,160	56
	2014	1,686	476	15	20	..	2,127	54

Table 21

Production, trade and supply of gas oil/diesel oil

Thousand metric tons and kilograms per capita

Country or area	Year	Production	Imports	Exports	Bunkers	Changes in stocks	Supply	
							Total	Per capita
Sudan (former)	2011	1,752	669	..	21	..	2,400	51
	2012	..	..	..	..	..	..	..
	2013	..	..	..	..	..	..	..
	2014	..	..	..	..	..	..	..
Swaziland	2011	..	*110	..	..	..	*110	*91
	2012	..	*112	..	..	..	*112	*91
	2013	..	*114	..	..	..	*114	*91
	2014	..	120	..	..	..	120	95
Togo	2011	..	308	..	12	-8	304	46
	2012	..	263	..	15	-10	258	38
	2013	..	270	..	16	-11	265	38
	2014	..	277	..	17	-12	272	38
Tunisia	2011	254	1,390	..	12	1	1,631	152
	2012	661	1,014	..	11	-7	1,671	154
	2013	646	1,013	..	11	-29	1,677	152
	2014	634	1,151	..	15	-16	1,786	160
Uganda	2011	..	652	..	..	..	652	19
	2012	..	561	..	..	..	561	16
	2013	..	585	..	..	..	585	16
	2014	..	633	..	..	..	633	17
United Rep. of Tanzania	2011	..	883	..	..	..	883	19
	2012	..	1,185	..	..	..	1,185	24
	2013	..	1,564	..	..	..	1,564	31
	2014	..	1,337	..	..	..	1,337	26
Zambia	2011	302	74	..	..	..	376	26
	2012	343	114	..	..	..	457	31
	2013	328	128	..	..	..	456	30
	2014	367	91	..	..	..	458	29
Zimbabwe	2011	..	710	..	..	..	710	50
	2012	..	795	..	..	..	795	55
	2013	..	886	..	..	..	886	59
	2014	..	804	..	..	..	804	53
America, North	**2011**	**279,274**	**24,543**	**51,707**	**10,319**	**-1,785**	**243,576**	**441**
	2012	**285,837**	**25,769**	**59,565**	**2,877**	**-2,431**	**251,595**	**451**
	2013	**296,620**	**25,804**	**66,024**	**3,480**	**-756**	**253,677**	**450**
	2014	**303,665**	**28,894**	**63,127**	**5,910**	**683**	**262,839**	**462**
Anguilla	2011	..	*35	..	..	..	*35	*2,479
	2012	..	*32	..	..	..	*32	*2,271
	2013	..	*31	..	..	..	*31	*2,196
	2014	..	*32	..	..	..	*32	*2,192
Antigua and Barbuda	2011	..	*62	*1	..	..	*61	*692
	2012	..	*64	*1	..	..	*63	*707
	2013	..	*64	*1	..	..	*63	*700
	2014	..	*65	*1	..	..	*64	*704
Aruba	2011	..	*24	..	..	..	*24	*231
	2012	..	*22	..	..	..	*22	*215
	2013	..	*22	..	..	..	*22	*214
	2014	..	*22	..	..	..	*22	*215
Bahamas	2011	..	383	..	..	..	383	1,044
	2012	..	401	..	..	..	401	1,077
	2013	..	757	*44	..	*100	613	1,623
	2014	..	409	..	..	*-100	509	1,330

Production, trade and supply of gas oil/diesel oil

Thousand metric tons and kilograms per capita

Country or area	Year	Production	Imports	Exports	Bunkers	Changes in stocks	Supply	
							Total	Per capita
Barbados	2011	..	125	..	..	0	125	444
	2012	..	109	..	..	0	109	387
	2013	..	103	..	..	0	103	366
	2014	..	76	..	..	-2	78	277
Belize	2011	..	55	..	..	..	55	167
	2012	..	59	..	..	..	59	176
	2013	..	61	..	..	..	61	178
	2014	..	69	..	..	..	69	196
Bermuda	2011	..	*35	..	..	..	*35	*551
	2012	..	*34	..	..	..	*34	*538
	2013	..	32	..	..	..	32	507
	2014	..	59	..	..	..	59	940
Bonaire, St Eustatius, Saba	2011	..	..	..	..	..	..	..
	2012	..	*65	..	*26	..	*39	*1,704
	2013	..	*66	..	*26	..	*40	*1,697
	2014	..	*66	..	*26	..	*40	*1,653
British Virgin Islands	2011	..	*39	..	..	..	*39	*1,394
	2012	..	*39	..	..	..	*39	*1,357
	2013	..	*39	..	..	..	*39	*1,332
	2014	..	*39	..	..	..	*39	*1,315
Canada	2011	30,505	1,678	6,334	28	190	25,631	743
	2012	31,800	726	7,879	20	83	24,544	704
	2013	30,962	1,154	7,940	31	-325	24,470	695
	2014	30,511	1,232	6,756	33	247	24,707	694
Cayman Islands	2011	..	143	..	..	..	143	2,527
	2012	..	133	..	..	..	133	2,312
	2013	..	131	..	..	..	131	2,244
	2014	..	135	..	..	..	135	2,278
Costa Rica	2011	52	930	17	..	-4	968	210
	2012	..	910	0	..	-2	911	196
	2013	..	930	..	..	-5	935	199
	2014	..	965	..	..	-3	968	203
Cuba	2011	1,221	702	..	*30	..	1,893	167
	2012	1,319	716	..	*30	..	2,005	177
	2013	1,256	829	..	*30	..	2,055	181
	2014	1,124	950	..	*32	..	2,041	179
Curaçao	2011	..	..	..	..	..	..	..
	2012	1,886	465	1,872	195	..	284	1,861
	2013	1,712	469	1,699	197	..	285	1,847
	2014	1,991	477	1,976	200	..	292	1,875
Dominica	2011	..	*25	..	..	..	*25	*347
	2012	..	*27	..	..	..	*27	*379
	2013	..	*26	..	..	..	*26	*354
	2014	..	*27	..	..	..	*27	*372
Dominican Republic	2011	350	987	..	..	-11	1,349	135
	2012	345	1,040	..	..	5	1,380	136
	2013	373	872	..	..	-6	1,250	122
	2014	372	778	..	..	28	1,123	108
El Salvador	2011	156	433	9	..	-21	601	99
	2012	107	463	0	..	-7	577	95
	2013	0	569	10	..	-1	560	92
	2014	0	561	11	..	0	550	90

Table 21

Production, trade and supply of gas oil/diesel oil

Thousand metric tons and kilograms per capita

Country or area	Year	Production	Imports	Exports	Bunkers	Changes in stocks	Supply	
							Total	Per capita
Greenland	2011	..	213	..	40	-19	192	3,399
	2012	..	165	..	14	1	150	2,670
	2013	..	155	..	5	5	146	2,589
	2014	..	131	..	16	-9	125	2,220
Grenada	2011	..	52	..	..	0	52	493
	2012	..	53	..	..	-6	59	555
	2013	..	70	..	..	0	70	659
	2014	..	51	..	..	0	51	482
Guadeloupe	2011	..	*237	..	..	..	*237	*517
	2012	..	*272	..	..	..	*272	*590
	2013	..	*272	..	..	..	*272	*587
	2014	..	*274	..	..	..	*274	*588
Guatemala	2011	32	1,236	23	292	13	940	62
	2012	31	1,280	2	301	24	984	64
	2013	31	1,350	36	322	-49	1,073	68
	2014	38	1,519	85	344	-24	1,152	72
Haiti	2011	..	447	..	..	..	447	44
	2012	..	457	..	..	..	457	44
	2013	..	463	..	..	..	463	44
	2014	..	530	..	..	..	530	50
Honduras	2011	..	690	..	..	-32	722	95
	2012	..	756	..	..	0	756	98
	2013	..	773	..	..	0	773	98
	2014	..	761	..	..	-80	841	106
Jamaica	2011	216	356	..	64	14	494	180
	2012	225	296	..	63	-11	469	170
	2013	193	288	..	44	-23	460	166
	2014	147	301	..	17	12	419	151
Martinique	2011	*173	*37	*31	*34	..	*145	*368
	2012	*177	*37	*31	*34	..	*149	*376
	2013	*179	*36	*29	*34	..	*151	*382
	2014	*185	*36	*28	*38	..	*155	*392
Mexico	2011	17,768	6,693	236	815	152	23,258	193
	2012	18,854	6,570	130	816	-47	24,525	201
	2013	20,906	5,284	456	702	-196	25,228	204
	2014	19,489	6,588	107	720	-38	25,288	202
Montserrat	2011	..	8	..	*1	..	7	1,402
	2012	..	8	..	*1	..	7	1,471
	2013	..	11	..	*1	..	10	2,054
	2014	..	10	..	*1	..	9	1,747
Neth. Antilles (former)	2011	1,810	689	1,797	289	..	413	1,937
	2012	..	..	..	..	..	..	..
	2013	..	..	..	..	..	..	..
	2014	..	..	..	..	..	..	..
Nicaragua	2011	233	236	1	..	5	463	80
	2012	156	293	0	..	6	443	75
	2013	209	276	1	..	20	464	78
	2014	213	285	1	..	-18	515	86
Panama	2011	..	1,071	..	153	-238	1,156	314
	2012	..	1,079	..	*150	-94	1,023	273
	2013	..	1,090	..	0	-165	1,255	330
	2014	..	1,302	..	0	141	1,161	300

Production, trade and supply of gas oil/diesel oil

Thousand metric tons and kilograms per capita

Country or area	Year	Production	Imports	Exports	Bunkers	Changes in stocks	Supply	
							Total	Per capita
Sint Maarten (Dutch part)	2011	..	..	..	..	..	..	..
	2012	..	*153	..	*61	..	*93	*2,644
	2013	..	*157	..	*62	..	*95	*2,601
	2014	..	*160	..	*63	..	*97	*2,565
St. Kitts-Nevis	2011	..	57	..	..	..	57	1,066
	2012	..	54	..	..	..	54	1,012
	2013	..	*55	..	..	..	*55	*1,013
	2014	..	*56	..	..	..	*56	*1,010
St. Lucia	2011	..	78	..	..	..	78	435
	2012	..	77	..	..	..	77	427
	2013	..	77	..	..	..	77	422
	2014	..	77	..	..	..	77	418
St. Pierre-Miquelon	2011	..	*25	..	*6	..	*19	*3,060
	2012	..	*25	..	*6	..	*19	*2,996
	2013	..	*26	..	*6	..	*20	*3,108
	2014	..	*27	..	*6	..	*21	*3,344
St. Vincent-Grenadines	2011	..	38	..	..	..	38	351
	2012	..	55	..	..	..	55	504
	2013	..	44	..	..	..	44	399
	2014	..	*44	..	..	..	*44	*399
Trinidad and Tobago	2011	1,381	..	750	196	-146	581	435
	2012	921	..	379	126	-110	526	392
	2013	1,269	..	517	173	0	579	429
	2014	831	..	247	83	-34	535	395
Turks and Caicos Islands	2011	..	*62	..	..	..	*62	*1,945
	2012	..	*65	..	..	..	*65	*2,004
	2013	..	*65	..	..	..	*65	*1,964
	2014	..	*67	..	..	..	*67	*1,986
United States	2011	225,377	6,664	42,508	8,372	-1,689	182,850	585
	2012	230,017	8,769	49,271	1,035	-2,273	190,753	606
	2013	239,530	9,159	55,291	1,847	-111	191,662	604
	2014	248,764	10,713	53,915	4,330	564	200,668	628
America, South	**2011**	**78,619**	**23,535**	**8,273**	**1,174**	**170**	**92,537**	**231**
	2012	**77,783**	**25,603**	**3,775**	**970**	**338**	**98,303**	**242**
	2013	**77,711**	**30,197**	**2,090**	**1,530**	**1,962**	**102,326**	**250**
	2014	**80,899**	**30,708**	**2,923**	**1,554**	**1,776**	**105,354**	**254**
Argentina	2011	10,024	1,752	0	189	-39	11,626	279
	2012	9,931	1,185	65	177	-50	10,924	260
	2013	8,803	3,652	92	179	-10	12,194	287
	2014	9,311	3,390	302	157	33	12,209	284
Bolivia (Plur. State of)	2011	574	658	..	..	5	1,227	122
	2012	616	754	..	..	1	1,369	134
	2013	697	777	..	..	-1	1,475	142
	2014	758	826	..	..	-9	1,593	151
Brazil	2011	37,383	8,111	528	437	237	44,292	221
	2012	39,881	8,446	287	401	-328	47,967	237
	2013	43,267	8,712	327	568	133	50,951	249
	2014	43,570	9,799	246	568	184	52,371	254
Chile	2011	3,168	4,464	273	17	-51	7,393	430
	2012	2,863	4,815	401	9	102	7,166	412
	2013	3,076	4,971	428	8	55	7,556	430
	2014	3,176	4,909	379	0	30	7,676	432

Table 21

Production, trade and supply of gas oil/diesel oil

Thousand metric tons and kilograms per capita

Country or area	Year	Production	Imports	Exports	Bunkers	Changes in stocks	Supply	
							Total	Per capita
Colombia	2011	4,867	2,380	805	*308	11	6,123	132
	2012	4,806	2,615	709	*156	2	6,554	140
	2013	4,578	3,162	686	588	193	6,273	133
	2014	5,422	3,937	734	627	1,490	6,508	136
Ecuador	2011	1,968	2,111	..	..	-43	4,122	272
	2012	1,846	2,382	..	..	-22	4,249	276
	2013	1,653	2,916	..	..	-25	4,593	293
	2014	1,534	3,493	..	..	54	4,973	313
Falkland Is. (Malvinas)	2011	..	*9	..	..	..	*9	*2,977
	2012	..	*9	..	..	..	*9	*3,004
	2013	..	*9	..	..	..	*9	*2,989
	2014	..	*9	..	..	..	*9	*2,975
French Guiana	2011	..	*112	..	..	..	*112	*466
	2012	..	*109	..	..	..	*109	*441
	2013	..	*114	..	..	..	*114	*448
	2014	..	*146	..	..	..	*146	*558
Guyana	2011	..	262	..	..	-8	270	357
	2012	..	288	..	..	-14	302	398
	2013	..	270	..	..	-3	272	358
	2014	..	296	..	..	6	289	379
Paraguay	2011	..	1,041	..	..	-71	1,112	177
	2012	..	1,130	..	..	108	1,022	160
	2013	..	1,021	..	..	-97	1,118	173
	2014	..	1,170	..	..	8	1,161	177
Peru	2011	4,631	1,572	75	..	82	6,046	203
	2012	4,465	1,628	21	..	38	6,034	200
	2013	4,268	2,330	52	..	436	6,110	200
	2014	4,185	2,316	1	..	668	5,832	188
Suriname	2011	55	182	40	*27	..	170	324
	2012	53	240	21	*39	..	233	440
	2013	58	226	79	*27	..	178	334
	2014	61	219	51	*34	..	196	365
Uruguay	2011	450	881	..	119	46	1,166	344
	2012	795	667	..	126	41	1,296	382
	2013	843	319	..	108	1	1,052	309
	2014	800	198	..	109	29	861	252
Venezuela (Bolivar. Rep.)	2011	15,499	0	6,552	77	0	8,870	301
	2012	12,527	1,337	2,271	63	460	11,070	371
	2013	10,468	1,720	426	52	1,279	10,431	345
	2014	12,082	..	1,210	60	-717	11,529	376
Asia	**2011**	**526,417**	**94,764**	**117,303**	**7,270**	**-850**	**497,459**	**118**
	2012	**556,085**	**100,367**	**126,687**	**5,649**	**-1,776**	**525,892**	**123**
	2013	**565,955**	**100,160**	**128,361**	**5,546**	**-272**	**532,479**	**124**
	2014	**568,570**	**107,255**	**136,343**	**5,307**	**-1,110**	**535,285**	**123**
Afghanistan	2011	..	801	..	..	..	801	28
	2012	..	873	..	..	..	873	29
	2013	..	*873	..	..	..	*873	*28
	2014	..	*882	..	..	..	*882	*28
Armenia	2011	..	135	0	..	..	135	46
	2012	..	149	..	..	..	149	50
	2013	..	138	..	..	..	138	46
	2014	..	143	..	..	..	143	47

Production, trade and supply of gas oil/diesel oil

Thousand metric tons and kilograms per capita

Country or area	Year	Production	Imports	Exports	Bunkers	Changes in stocks	Supply	
							Total	Per capita
Azerbaijan	2011	2,483	0	1,359	78	6	1,040	113
	2012	2,369	3	1,139	86	11	1,136	121
	2013	2,495	2	1,085	80	3	1,329	140
	2014	2,946	1	1,523	74	32	1,318	137
Bahrain	2011	3,927	..	3,640	..	-46	333	255
	2012	4,223	..	3,806	..	56	361	271
	2013	3,942	..	3,589	..	38	315	233
	2014	4,083	..	3,900	..	-123	306	225
Bangladesh	2011	417	2,956	..	..	134	3,239	21
	2012	410	2,619	..	..	-211	3,240	21
	2013	456	2,609	..	..	100	2,965	19
	2014	362	2,904	..	..	88	3,178	20
Bhutan	2011	..	87	..	..	..	87	119
	2012	..	102	..	..	..	102	137
	2013	..	110	..	..	..	110	146
	2014	..	102	..	..	..	102	134
Brunei Darussalam	2011	164	145	14	..	-1	296	741
	2012	114	197	5	..	-4	310	764
	2013	138	154	5	..	7	280	680
	2014	113	179	9	..	6	277	664
Cambodia	2011	..	815	..	..	..	815	56
	2012	..	898	..	..	..	898	61
	2013	..	918	..	..	..	918	61
	2014	..	1,005	..	..	..	1,005	66
China	2011	156,897	2,433	2,012	383	780	156,156	116
	2012	170,638	999	1,854	292	-89	169,580	125
	2013	172,757	353	2,782	247	-894	170,975	125
	2014	176,353	474	4,105	211	937	171,575	125
China, Hong Kong SAR	2011	..	4,279	77	2,413	-100	1,889	268
	2012	..	3,557	112	1,364	172	1,909	269
	2013	..	3,429	209	1,309	22	1,889	264
	2014	..	3,279	94	1,437	-160	1,908	264
China, Macao SAR	2011	..	134	..	..	1	133	243
	2012	..	149	..	..	1	148	265
	2013	..	170	..	..	9	161	283
	2014	..	171	..	66	0	105	182
Cyprus	2011	..	620	..	58	-7	569	510
	2012	..	704	..	69	2	633	561
	2013	..	673	..	83	14	576	505
	2014	..	549	..	80	6	463	401
Georgia	2011	..	428	0	..	-9	437	104
	2012	..	459	0	..	0	459	111
	2013	..	479	0	..	14	465	114
	2014	..	530	0	..	10	519	129
India	2011	82,864	1,059	20,491	..	-869	64,301	52
	2012	91,085	626	22,464	..	-458	69,705	55
	2013	94,165	84	26,499	9	-17	67,758	53
	2014	95,914	191	*25,565	..	..	70,540	54
Indonesia	2011	15,842	11,954	15	84	..	27,697	113
	2012	16,774	14,200	12	89	..	30,873	124
	2013	19,532	10,349	1	160	..	29,720	118
	2014	16,246	9,739	1	164	..	25,820	101

Table 21

Production, trade and supply of gas oil/diesel oil

Thousand metric tons and kilograms per capita

Country or area	Year	Production	Imports	Exports	Bunkers	Changes in stocks	Supply	
							Total	Per capita
Iran (Islamic Rep. of)	2011	30,076	0	640	108	..	29,328	390
	2012	29,814	38	558	153	..	29,141	383
	2013	31,032	213	225	144	..	30,876	400
	2014	29,548	259	8	127	..	29,672	380
Iraq	2011	6,364	5,519	..	..	..	11,883	373
	2012	6,262	8,133	..	..	..	14,395	437
	2013	6,477	9,073	..	..	..	15,550	456
	2014	5,099	11,229	..	..	..	16,328	463
Israel	2011	3,562	758	1,628	50	-50	2,692	356
	2012	3,915	1,934	1,199	64	93	4,493	584
	2013	4,673	499	2,451	37	45	2,639	338
	2014	5,191	499	2,969	25	..	2,696	340
Japan	2011	46,080	817	6,646	67	44	40,140	315
	2012	45,381	571	5,992	77	275	39,608	312
	2013	48,557	262	9,011	186	-339	39,961	315
	2014	45,670	570	7,475	91	233	38,441	303
Jordan	2011	1,030	1,366	..	6	-55	2,445	362
	2012	1,109	2,097	..	6	62	3,139	449
	2013	980	1,657	..	3	-80	2,715	376
	2014	931	2,373	..	2	26	3,276	442
Kazakhstan	2011	4,774	242	954	19	-118	4,161	251
	2012	4,890	495	1,796	0	34	3,555	211
	2013	5,299	617	1,766	0	31	4,119	241
	2014	5,282	618	1,765	..	324	3,811	219
Korea, Dem.Ppl's.Rep.	2011	188	46	..	..	..	234	10
	2012	188	41	..	..	..	229	9
	2013	208	43	..	..	..	251	10
	2014	191	44	..	..	..	235	9
Korea, Republic of	2011	39,914	349	21,337	863	85	17,978	364
	2012	42,206	899	23,648	815	-309	18,951	382
	2013	40,587	1,395	21,898	823	650	18,611	373
	2014	42,480	831	23,345	885	-505	19,586	391
Kuwait	2011	10,872	..	7,037	95	..	3,740	1,155
	2012	11,563	..	7,485	100	..	3,978	1,163
	2013	10,532	..	6,985	103	..	3,444	958
	2014	10,806	..	7,112	106	..	3,588	956
Kyrgyzstan	2011	28	422	17	..	-12	445	80
	2012	31	566	6	..	45	546	97
	2013	27	615	11	..	-52	683	119
	2014	30	509	8	..	28	503	86
Lao People's Dem. Rep.	2011	..	*24	..	..	..	*24	*4
	2012	..	*24	..	..	..	*24	*4
	2013	..	*25	..	..	..	*25	*4
	2014	..	*25	..	..	..	*25	*4
Lebanon	2011	..	2,448	..	..	..	2,448	533
	2012	..	3,269	..	..	..	3,269	664
	2013	..	3,076	..	..	..	3,076	582
	2014	..	3,317	..	..	..	3,317	591
Malaysia	2011	8,788	3,708	4,381	14	-1,518	9,619	337
	2012	9,878	3,449	5,225	2	-1,482	9,582	330
	2013	10,892	5,027	5,236	2	837	9,844	334
	2014	9,575	5,834	5,158	2	-258	10,507	351

Production, trade and supply of gas oil/diesel oil

Thousand metric tons and kilograms per capita

Country or area	Year	Production	Imports	Exports	Bunkers	Changes in stocks	Supply	
							Total	Per capita
Maldives	2011	..	304	..	*35	..	269	794
	2012	..	338	..	*39	..	298	865
	2013	..	308	..	*35	..	273	778
	2014	..	386	..	*44	..	342	957
Mongolia	2011	..	647	..	..	..	647	235
	2012	..	716	..	..	..	716	255
	2013	..	774	..	..	..	774	271
	2014	..	686	..	..	..	686	236
Myanmar	2011	195	1,253	..	1	-122	1,570	30
	2012	165	1,017	..	2	3	1,177	22
	2013	150	1,224	..	3	28	1,343	25
	2014	161	2,226	..	0	-495	2,882	54
Nepal	2011	..	654	..	..	..	654	24
	2012	..	721	..	..	..	721	26
	2013	..	809	..	..	..	809	29
	2014	..	922	..	..	..	922	33
Oman	2011	2,147	0	244	..	..	1,903	593
	2012	2,074	188	179	..	..	2,083	588
	2013	1,891	477	137	..	..	2,231	571
	2014	2,633	..	153	..	..	2,480	585
Other Asia	2011	12,458	0	7,646	137	63	4,612	198
	2012	13,904	0	9,306	123	123	4,352	187
	2013	14,162	37	9,818	118	-69	4,332	186
	2014	14,140	..	9,811	109	-187	4,407	189
Pakistan	2011	3,252	3,263	26	9	93	6,387	37
	2012	3,902	2,957	0	9	-6	6,856	39
	2013	4,361	2,568	2	7	-20	6,940	38
	2014	4,567	2,709	0	7	0	7,269	39
Philippines	2011	3,486	2,541	36	25	-177	6,143	65
	2012	3,130	3,450	51	17	101	6,410	67
	2013	2,951	3,633	0	22	-125	6,686	69
	2014	3,058	4,197	0	23	37	7,195	73
Qatar	2011	2,616	89	2,090	..	0	615	323
	2012	2,205	0	3,326	..	-278	-843	-418
	2013	2,190	125	3,513	..	2	-1,200	-571
	2014	2,201	12	3,665	..	46	-1,498	-690
Saudi Arabia	2011	30,750	5,895	2,622	..	2,169	31,854	1,106
	2012	31,383	8,188	4,637	..	355	34,579	1,172
	2013	29,460	12,052	5,543	..	269	35,700	1,182
	2014	36,842	11,474	11,758	..	-297	36,855	1,193
Singapore	2011	13,439	17,116	27,360	1,576	-450	2,068	398
	2012	14,627	14,200	27,331	1,456	271	-231	-44
	2013	12,332	12,590	21,388	1,376	-455	2,613	483
	2014	10,605	13,946	22,083	1,313	18	1,137	206
Sri Lanka	2011	501	1,402	..	65	-56	1,894	93
	2012	394	1,652	..	54	56	1,936	95
	2013	390	1,199	..	0	-186	1,774	86
	2014	496	1,394	..	0	-129	2,020	98
State of Palestine	2011	..	469	..	..	0	469	112
	2012	..	445	..	..	0	445	103
	2013	..	484	..	..	0	484	110
	2014	..	520	..	..	-44	564	124

Table 21

Production, trade and supply of gas oil/diesel oil

Thousand metric tons and kilograms per capita

Country or area	Year	Production	Imports	Exports	Bunkers	Changes in stocks	Supply	
							Total	Per capita
Syrian Arab Republic	2011	3,624	1,587	..	861	..	4,350	212
	2012	2,572	1,305	..	448	..	3,429	172
	2013	1,701	1,780	..	412	..	3,069	159
	2014	1,701	909	..	220	..	2,390	127
Tajikistan	2011	3	143	..	..	..	146	19
	2012	3	122	..	..	..	125	16
	2013	4	148	..	..	..	152	19
	2014	4	244	..	..	..	248	30
Thailand	2011	19,208	48	3,920	..	*-626	15,962	239
	2012	20,836	13	4,463	..	*-670	17,056	254
	2013	21,955	169	4,193	..	-51	17,982	267
	2014	20,709	300	3,714	..	-853	18,148	268
Timor-Leste	2011	..	*62	..	..	..	*62	*58
	2012	..	*75	..	..	..	*75	*68
	2013	..	*120	..	..	..	*120	*106
	2014	..	*130	..	..	..	*130	*112
Turkey	2011	7,056	10,154	124	87	-38	17,037	232
	2012	7,874	11,094	227	153	93	18,495	247
	2013	7,690	12,011	91	126	-28	19,512	256
	2014	6,266	12,959	55	102	179	18,889	244
Turkmenistan	2011	2,520	..	770	..	..	1,750	343
	2012	2,470	..	320	..	..	2,150	416
	2013	2,540	..	560	..	..	1,980	378
	2014	2,770	..	690	..	..	2,080	392
United Arab Emirates	2011	6,205	826	988	*171	..	5,872	672
	2012	5,478	691	571	*164	..	5,434	607
	2013	6,029	1,308	*768	*192	..	6,377	705
	2014	6,219	1,802	718	148	..	7,155	787
Uzbekistan	2011	1,058	..	144	..	..	914	32
	2012	1,021	..	139	..	..	882	31
	2013	1,082	..	147	..	..	935	32
	2014	1,019	..	138	..	..	881	30
Viet Nam	2011	2,867	5,311	1,085	..	29	7,064	79
	2012	2,854	4,565	835	..	-21	6,605	73
	2013	3,456	3,157	448	..	-25	6,190	68
	2014	3,598	4,112	522	..	-29	7,217	78
Yemen	2011	762	1,455	..	65	..	2,152	89
	2012	343	1,580	..	67	..	1,856	75
	2013	862	2,345	..	70	..	3,137	123
	2014	761	2,070	..	71	..	2,760	105
Europe	**2011**	**352,085**	**129,985**	**140,159**	**8,278**	**-1,257**	**334,889**	**455**
	2012	**352,916**	**131,024**	**151,078**	**8,351**	**-1,346**	**325,857**	**442**
	2013	**347,514**	**135,214**	**148,958**	**8,492**	**1,084**	**324,194**	**440**
	2014	**354,129**	**129,448**	**145,625**	**15,826**	**516**	**321,610**	**436**
Albania	2011	85	630	..	..	..	715	248
	2012	20	668	..	..	..	688	239
	2013	0	935	..	20	171	745	258
	2014	89	980	53	20	215	782	271
Andorra	2011	..	136	..	..	..	136	1,655
	2012	..	137	..	..	..	137	1,724
	2013	..	134	..	..	..	134	1,772
	2014	..	130	..	..	..	130	1,786

Production, trade and supply of gas oil/diesel oil

Thousand metric tons and kilograms per capita

Country or area	Year	Production	Imports	Exports	Bunkers	Changes in stocks	Supply	
							Total	Per capita
Austria	2011	4,105	4,261	899	20	121	7,326	870
	2012	4,006	4,448	1,011	20	20	7,403	876
	2013	4,071	4,639	962	22	-53	7,779	917
	2014	3,981	4,490	1,012	20	-53	7,492	880
Belarus	2011	7,440	3,293	5,750	..	-215	5,198	548
	2012	7,539	3,218	7,576	..	2	3,179	335
	2013	7,687	19	4,356	..	-2	3,352	353
	2014	8,257	22	5,160	..	-9	3,128	329
Belgium	2011	11,806	8,255	7,863	433	402	11,363	1,033
	2012	12,892	7,884	9,579	500	-90	10,787	974
	2013	11,439	12,047	11,011	505	438	11,532	1,034
	2014	13,766	9,782	11,768	600	184	10,996	979
Bosnia and Herzegovina	2011	503	491	103	..	0	891	232
	2012	434	478	56	..	0	856	224
	2013	440	504	38	..	11	895	234
	2014	431	541	13	..	24	935	245
Bulgaria	2011	1,992	715	652	47	22	1,986	270
	2012	2,205	663	965	45	-101	1,959	268
	2013	2,087	859	1,039	60	144	1,703	235
	2014	1,970	919	842	56	-47	2,038	283
Croatia	2011	1,131	945	378	1	17	1,680	391
	2012	1,287	694	388	..	4	1,589	371
	2013	1,242	787	444	..	-17	1,602	375
	2014	1,063	1,020	528	..	-32	1,587	373
Czechia	2011	3,256	1,646	651	..	75	4,176	396
	2012	3,357	1,488	759	..	-38	4,124	391
	2013	3,124	1,710	690	..	-22	4,166	395
	2014	3,568	1,786	879	..	103	4,372	415
Denmark	2011	2,951	2,276	1,246	244	-210	3,947	708
	2012	3,668	1,968	1,480	244	-36	3,948	705
	2013	3,451	2,524	1,679	263	163	3,870	688
	2014	3,294	2,762	1,673	299	273	3,811	675
Estonia	2011	..	597	0	24	-8	581	437
	2012	..	713	13	86	4	610	461
	2013	..	713	17	40	50	606	459
	2014	..	691	9	50	-9	641	487
Faeroe Islands	2011	..	*125	..	7	..	*118	*2,425
	2012	..	*133	..	8	..	*125	*2,584
	2013	..	*152	..	8	..	*144	*2,975
	2014	..	*130	..	6	..	*124	*2,562
Finland	2011	6,611	2,443	2,839	61	86	6,068	1,125
	2012	6,502	2,360	3,123	42	-110	5,807	1,070
	2013	6,691	3,094	4,005	47	538	5,195	953
	2014	6,396	2,159	3,638	33	-49	4,933	900
France	2011	31,788	20,397	3,930	100	695	47,460	750
	2012	27,514	23,462	2,781	91	-169	48,273	759
	2013	27,030	23,160	1,976	73	180	47,961	751
	2014	27,371	22,482	1,805	84	287	47,677	743
Germany	2011	44,809	13,657	6,488	490	-342	51,830	644
	2012	46,015	13,591	6,364	433	-580	53,389	663
	2013	45,093	17,792	7,076	427	-82	55,464	688
	2014	44,270	16,871	7,694	490	-650	53,607	665

Table 21

Production, trade and supply of gas oil/diesel oil

Thousand metric tons and kilograms per capita

Country or area	Year	Production	Imports	Exports	Bunkers	Changes in stocks	Supply	
							Total	Per capita
Gibraltar	2011	..	547	..	469	..	78	2,511
	2012	..	553	..	473	..	80	2,548
	2013	..	518	..	433	..	85	2,680
	2014	..	414	..	322	..	92	2,875
Greece	2011	5,776	2,892	3,075	306	-328	5,615	503
	2012	7,835	1,391	4,740	264	-220	4,442	400
	2013	8,241	1,338	5,706	281	211	3,381	306
	2014	9,694	936	6,592	264	-320	4,094	372
Hungary	2011	3,711	1,297	1,277	..	-17	3,748	375
	2012	3,685	1,195	1,438	..	-13	3,455	347
	2013	3,599	1,248	1,384	..	-20	3,483	351
	2014	3,647	1,421	1,576	..	3	3,489	353
Iceland	2011	..	300	..	31	2	267	832
	2012	..	279	..	32	-1	248	767
	2013	2	273	..	12	17	246	756
	2014	2	288	..	15	-16	291	889
Ireland	2011	1,124	2,165	185	72	91	2,941	632
	2012	1,310	2,211	309	72	99	3,041	651
	2013	1,144	2,121	0	113	-68	3,220	689
	2014	1,087	2,227	17	107	-8	3,198	684
Italy	2011	37,087	2,577	9,341	216	336	29,771	499
	2012	36,528	2,102	10,746	147	-209	27,946	468
	2013	32,691	3,038	8,484	126	103	27,016	452
	2014	30,684	3,168	5,498	114	448	27,792	465
Latvia	2011	..	1,063	161	75	-3	830	402
	2012	..	1,177	211	87	51	828	406
	2013	..	1,445	503	80	10	852	423
	2014	..	1,449	589	69	-97	888	446
Lithuania	2011	3,107	381	2,981	20	-2	489	159
	2012	3,025	461	2,986	20	-42	522	173
	2013	3,086	744	3,401	13	47	369	125
	2014	2,531	1,010	2,997	8	-3	539	185
Luxembourg	2011	..	2,121	1	..	-6	2,126	4,089
	2012	..	2,081	0	..	14	2,067	3,882
	2013	..	2,076	2	..	11	2,063	3,786
	2014	..	1,969	0	..	14	1,955	3,513
Malta	2011	..	421	3	222	8	188	454
	2012	..	415	7	221	-4	191	460
	2013	..	449	64	174	20	191	458
	2014	..	406	25	185	1	195	467
Montenegro	2011	..	167	0	..	..	167	268
	2012	..	170	0	..	..	170	272
	2013	..	171	1	..	..	170	272
	2014	..	177	7	..	..	170	272
Netherlands	2011	20,502	18,362	30,918	1,461	-994	7,479	448
	2012	20,327	18,198	29,277	1,475	368	7,405	442
	2013	20,331	17,877	29,535	1,336	-102	7,439	443
	2014	20,974	15,403	27,798	1,342	-145	7,382	438
Norway	2011	6,643	1,206	2,989	187	238	4,435	895
	2012	6,298	1,471	2,524	148	88	5,009	998
	2013	6,866	1,287	2,658	175	25	5,295	1,042
	2014	6,366	1,113	2,761	68	38	4,612	896

Production, trade and supply of gas oil/diesel oil

Thousand metric tons and kilograms per capita

Country or area	Year	Production	Imports	Exports	Bunkers	Changes in stocks	Supply	
							Total	Per capita
Poland	2011	11,513	1,946	112	67	-82	13,362	346
	2012	11,982	1,422	337	66	40	12,961	336
	2013	11,677	943	608	76	-50	11,986	310
	2014	11,237	1,452	968	75	-99	11,745	304
Portugal	2011	3,919	1,349	120	53	-41	5,136	486
	2012	4,182	890	350	49	34	4,639	441
	2013	5,701	578	1,693	59	0	4,527	433
	2014	5,166	699	1,295	58	-90	4,602	442
Republic of Moldova	2011	4	445	0	..	3	446	109
	2012	3	398	1	..	-3	403	99
	2013	1	455	6	..	7	443	109
	2014	1	457	4	..	-11	465	114
Romania	2011	3,743	1,269	722	9	-231	4,512	224
	2012	3,806	1,546	709	14	2	4,627	232
	2013	4,209	1,117	865	41	64	4,356	220
	2014	5,170	994	1,635	79	33	4,417	225
Russian Federation	2011	70,293	795	39,526	1,148	246	30,168	211
	2012	69,436	252	41,144	1,277	110	27,157	190
	2013	71,599	174	42,424	1,388	244	27,717	193
	2014	77,001	89	40,308	8,585	19	28,178	196
Serbia	2011	776	791	3	0	35	1,529	169
	2012	691	712	5	3	-45	1,440	160
	2013	1,122	397	29	9	-13	1,494	167
	2014	1,172	434	59	15	-6	1,538	173
Slovakia	2011	3,209	608	2,204	..	9	1,604	296
	2012	2,863	586	1,899	..	-4	1,554	287
	2013	3,041	645	2,126	..	-1	1,561	288
	2014	2,787	707	1,941	..	9	1,544	285
Slovenia	2011	36	2,236	541	..	-11	1,742	846
	2012	52	2,420	643	..	87	1,742	844
	2013	60	2,335	806	..	-58	1,647	798
	2014	41	2,505	951	..	-12	1,607	778
Spain	2011	25,312	8,541	2,373	1,314	-602	30,768	659
	2012	28,639	6,696	6,192	1,166	-451	28,428	610
	2013	27,608	4,658	5,400	1,216	-426	26,076	561
	2014	27,985	3,812	5,071	1,264	528	24,934	539
Sweden	2011	7,335	2,198	3,939	203	-522	5,913	625
	2012	8,340	2,080	5,099	237	-66	5,150	540
	2013	6,811	2,165	3,577	241	-262	5,420	563
	2014	8,034	2,400	4,829	313	-32	5,324	549
Switzerland	2011	2,220	3,301	38	8	-106	5,581	701
	2012	1,714	4,241	56	8	89	5,802	720
	2013	2,604	3,970	66	6	-67	6,569	805
	2014	2,637	2,936	64	5	47	5,457	662
T.F.Yug.Rep. Macedonia	2011	278	335	184	..	11	417	202
	2012	74	502	154	..	3	419	202
	2013	28	516	100	..	-9	452	218
	2014	0	560	86	..	6	469	226
Ukraine	2011	2,713	3,757	873	..	44	5,553	122
	2012	1,411	4,765	509	..	-59	5,726	126
	2013	1,120	4,284	78	..	148	5,178	115
	2014	832	4,775	76	..	-182	5,713	127

Table 21

Production, trade and supply of gas oil/diesel oil

Thousand metric tons and kilograms per capita

Country or area	Year	Production	Imports	Exports	Bunkers	Changes in stocks	Supply	
							Total	Per capita
United Kingdom	2011	26,307	9,048	7,794	990	22	26,549	420
	2012	25,276	10,906	7,647	1,123	-120	27,532	433
	2013	23,618	11,323	6,149	1,248	-266	27,810	435
	2014	22,625	12,882	5,404	1,280	155	28,668	446
Oceania	**2011**	**13,312**	**9,368**	**103**	**185**	**208**	**22,184**	**599**
	2012	**13,067**	**10,906**	**112**	**164**	**60**	**23,637**	**629**
	2013	**13,508**	**12,707**	**76**	**159**	**103**	**25,878**	**678**
	2014	**13,060**	**13,586**	**60**	**171**	**-363**	**26,778**	**691**
Australia	2011	10,932	7,445	99	29	158	18,091	803
	2012	10,465	9,464	110	37	77	19,705	860
	2013	10,932	10,510	76	29	98	21,239	913
	2014	10,528	11,260	54	34	-358	22,058	934
Cook Islands	2011	..	*18	..	..	..	*18	*882
	2012	..	*18	..	..	..	*18	*877
	2013	..	*18	..	..	..	*18	*873
	2014	..	*18	..	..	..	*18	*869
Fiji	2011	..	240	0	*39	..	201	232
	2012	..	230	0	*38	..	191	219
	2013	..	*239	0	*39	..	*200	*227
	2014	..	*247	0	*42	..	*205	*232
French Polynesia	2011	..	137	..	*39	..	98	363
	2012	..	131	..	*37	..	94	342
	2013	..	134	..	*38	..	96	347
	2014	..	131	..	*37	..	94	337
Kiribati	2011	..	*13	..	..	..	*13	*121
	2012	..	*13	..	..	..	*13	*120
	2013	..	*13	..	..	..	*13	*120
	2014	..	13	..	..	..	13	119
Marshall Islands	2011	..	*34	..	..	..	*34	*644
	2012	..	33	..	..	..	33	631
	2013	..	*34	..	..	..	*34	*641
	2014	..	*33	..	..	..	*33	*628
Micronesia (Fed. States of)	2011	..	*24	..	..	..	*24	*232
	2012	..	*26	..	..	..	*26	*251
	2013	..	*27	..	..	..	*27	*260
	2014	..	*27	..	..	..	*27	*260
Nauru	2011	..	*8	..	..	0	*8	*788
	2012	..	*8	..	..	0	*8	*801
	2013	..	*8	..	..	0	*8	*820
	2014	..	*9	..	..	0	*9	*855
New Caledonia	2011	..	209	*4	*7	*-19	217	870
	2012	..	214	*2	*6	*-11	216	853
	2013	..	338	..	6	2	331	1,289
	2014	..	343	..	*6	*2	335	1,291
New Zealand	2011	1,925	694	..	45	83	2,491	566
	2012	2,140	466	..	45	-14	2,575	580
	2013	2,052	653	..	46	7	2,652	594
	2014	1,997	765	6	52	-7	2,711	603
Niue	2011	..	1	..	..	..	1	687
	2012	..	1	..	..	..	1	712
	2013	..	1	..	..	..	1	743
	2014	..	1	..	..	..	1	732

Production, trade and supply of gas oil/diesel oil

Thousand metric tons and kilograms per capita

Country or area	Year	Production	Imports	Exports	Bunkers	Changes in stocks	Supply	
							Total	Per capita
Palau	2011	..	*23	..	..	..	*23	*1,116
	2012	..	*22	..	..	..	*22	*1,060
	2013	..	*24	..	..	..	*24	*1,123
	2014	..	*24	..	..	..	*24	*1,123
Papua New Guinea	2011	455	373	..	*26	-14	816	117
	2012	462	148	..	0	8	603	84
	2013	524	588	..	0	-3	1,115	153
	2014	*535	*581	..	0	0	*1,116	*150
Samoa	2011	..	39	..	..	-1	40	214
	2012	..	*39	..	..	..	*39	*204
	2013	..	39	..	..	-2	41	214
	2014	..	*40	..	..	..	*40	*211
Solomon Islands	2011	..	*44	..	*1	..	*43	*79
	2012	..	*44	..	*1	..	*43	*78
	2013	..	*44	..	*1	..	*43	*77
	2014	..	*44	..	*1	..	*43	*76
Tonga	2011	..	23	..	..	..	23	223
	2012	..	12	..	..	..	12	113
	2013	..	0	..	..	..	0	2
	2014	..	1	..	..	..	1	9
Tuvalu	2011	..	1	..	..	..	1	138
	2012	..	*2	..	..	..	*2	*183
	2013	..	*2	..	..	..	*2	*182
	2014	..	*2	..	..	..	*2	*182
Vanuatu	2011	..	35	..	..	..	35	144
	2012	..	30	..	..	..	30	123
	2013	..	28	..	..	..	28	110
	2014	..	40	..	..	..	40	153
Wallis and Futuna Is.	2011	..	7	..	..	..	7	510
	2012	..	7	..	..	..	7	489
	2013	..	6	..	..	..	6	482
	2014	..	6	..	..	..	6	478

Table 22

Production, trade and supply of fuel oil

Thousand metric tons and kilograms per capita

Table Notes

Production includes output from refineries and plants.

Please refer to the Definitions Section on pages ix to xv for the appropriate product description /classification.

Figure 60: Production of fuel oil by region, 1990-2014

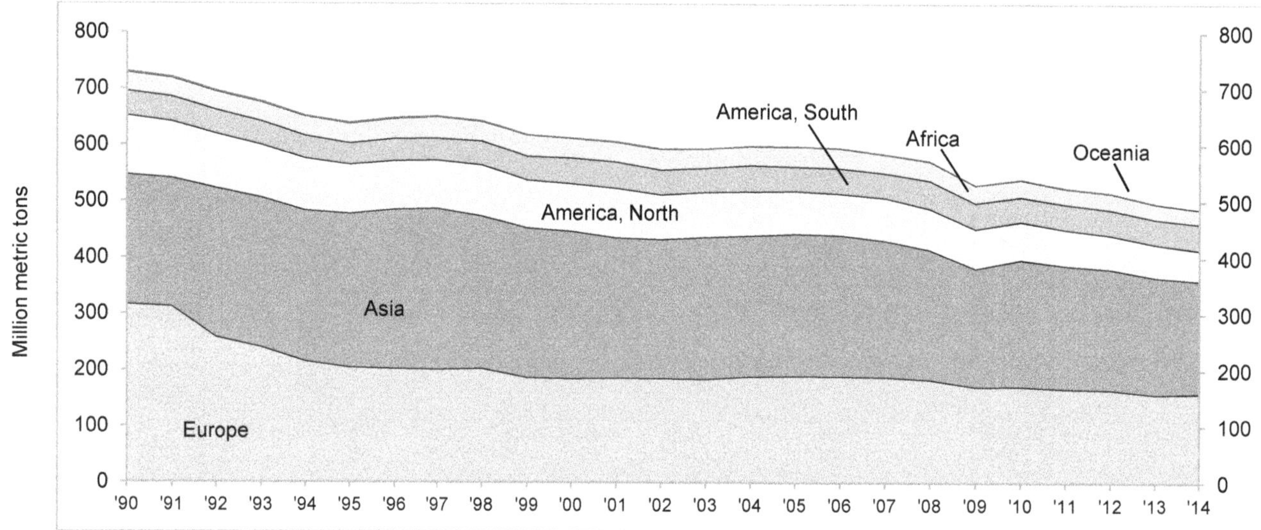

Figure 61: Major fuel oil producing countries, 2014

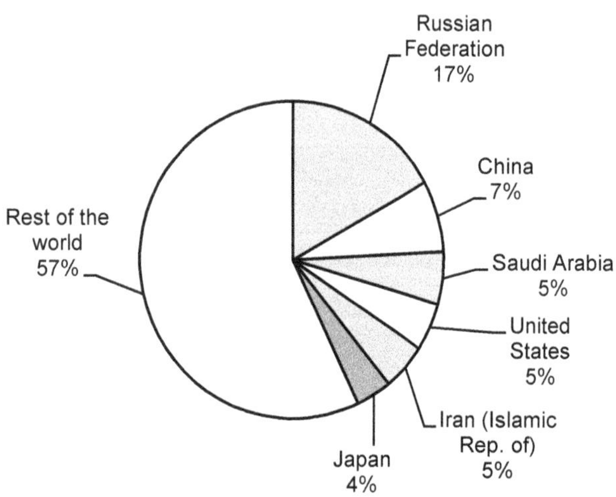

Figure 62: Major fuel oil consuming countries, 2014

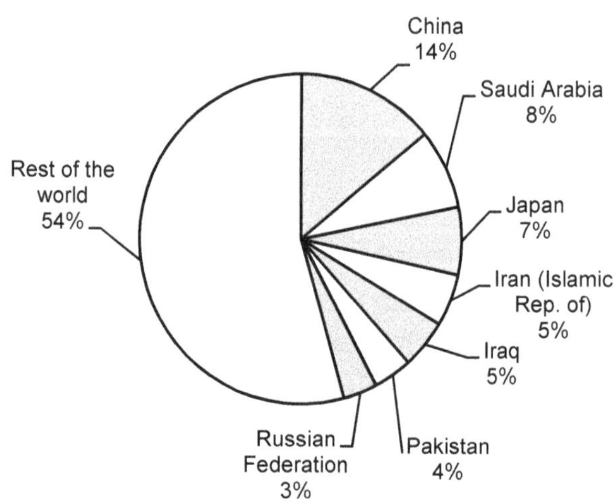

Production, trade and supply of fuel oil

Thousand metric tons and kilograms per capita

Country or area	Year	Production	Imports	Exports	Bunkers	Changes in stocks	Supply	
							Total	Per capita
World	**2011**	**525,293**	**263,438**	**301,950**	**180,736**	**3,341**	**302,704**	**43**
	2012	**516,578**	**258,545**	**289,744**	**170,022**	**3,788**	**311,570**	**44**
	2013	**498,464**	**263,715**	**307,946**	**175,482**	**18**	**278,733**	**39**
	2014	**487,609**	**240,304**	**298,062**	**167,585**	**-1,589**	**263,855**	**36**
Africa	**2011**	**27,867**	**4,146**	**9,287**	**3,239**	**135**	**19,352**	**18**
	2012	**29,187**	**5,099**	**8,660**	**3,997**	**-252**	**21,881**	**20**
	2013	**27,644**	**8,652**	**11,065**	**3,566**	**1,567**	**20,098**	**18**
	2014	**25,463**	**6,646**	**10,450**	**4,116**	**156**	**17,387**	**15**
Algeria	2011	5,848	193	5,694	213	1	133	4
	2012	5,115	201	5,127	211	-68	46	1
	2013	5,358	204	5,308	216	-3	41	1
	2014	6,594	180	6,464	209	79	22	1
Angola	2011	727	..	190	159	..	378	17
	2012	737	..	181	377	..	179	8
	2013	815	..	455	237	..	123	5
	2014	921	25	408	370	..	168	7
Benin	2011	..	114	66	..	2	46	5
	2012	..	119	69	..	2	48	5
	2013	..	127	74	..	2	51	5
	2014	..	135	79	..	2	54	5
Botswana	2011	..	10	..	..	..	10	5
	2012	..	10	..	..	..	10	5
	2013	..	11	..	..	..	11	5
	2014	..	11	..	..	..	11	5
Burkina Faso	2011	..	47	..	..	..	47	3
	2012	..	47	..	..	..	47	3
	2013	..	62	..	..	..	62	4
	2014	..	136	..	..	..	136	8
Burundi	2011	..	*6	..	..	..	*6	*1
	2012	..	*2	..	..	..	*2	0
	2013	..	*7	..	..	..	*7	*1
	2014	..	*11	..	..	..	*11	*1
Cabo Verde	2011	..	102	..	44	..	59	118
	2012	..	72	..	20	..	53	105
	2013	..	74	..	17	..	56	111
	2014	..	96	..	38	..	58	113
Cameroon	2011	343	..	163	4	..	176	8
	2012	359	..	171	4	..	184	8
	2013	379	..	182	4	..	193	9
	2014	402	..	193	4	..	205	9
Central African Rep.	2011	..	*5	..	..	..	*5	*1
	2012	..	*6	..	..	..	*6	*1
	2013	..	*6	..	..	..	*6	*1
	2014	..	*6	..	..	..	*6	*1
Congo	2011	197	8	197	..	-4	12	3
	2012	272	0	256	..	0	16	4
	2013	321	0	311	..	0	10	2
	2014	332	0	332	..	0	0	0
Côte d'Ivoire	2011	328	..	255	12	..	61	3
	2012	543	..	312	20	..	211	10
	2013	467	..	341	17	..	109	5
	2014	396	..	168	14	..	214	10

Table 22

Production, trade and supply of fuel oil

Thousand metric tons and kilograms per capita

Country or area	Year	Production	Imports	Exports	Bunkers	Changes in stocks	Supply	
							Total	Per capita
Dem. Rep. of the Congo	2011	..	44	..	..	..	44	1
	2012	..	41	..	..	..	41	1
	2013	..	41	..	..	..	41	1
	2014	..	41	..	..	..	41	1
Djibouti	2011	..	28	..	*9	*-51	*70	*83
	2012	..	*80	..	*10	0	*70	*82
	2013	..	*84	..	*11	0	*74	*85
	2014	..	*86	..	*11	0	*75	*86
Egypt	2011	8,902	349	..	296	0	8,955	107
	2012	10,155	1,005	..	290	0	10,870	127
	2013	*7,858	2,958	..	196	0	*10,620	*121
	2014	7,736	847	..	1,530	-822	7,875	88
Eritrea	2011	..	67	..	..	..	67	14
	2012	..	71	..	..	..	71	15
	2013	..	72	..	..	..	72	14
	2014	..	73	..	..	..	73	14
Ethiopia	2011	..	144	..	..	-4	148	2
	2012	..	160	..	..	0	160	2
	2013	..	152	..	..	0	152	2
	2014	..	168	..	..	..	168	2
Gabon	2011	536	..	366	116	..	54	34
	2012	386	..	211	120	..	55	34
	2013	429	..	264	109	..	56	34
	2014	411	..	244	109	..	58	34
Gambia	2011	..	51	..	..	..	51	29
	2012	..	56	..	..	..	56	31
	2013	..	51	..	..	..	51	27
	2014	..	66	..	..	..	66	34
Ghana	2011	91	..	44	..	..	47	2
	2012	79	..	43	..	..	36	1
	2013	44	44	4	..	45	39	2
	2014	44	49	..	..	..	92	3
Guinea	2011	..	315	..	..	..	315	28
	2012	..	187	..	..	..	187	16
	2013	..	100	..	..	..	100	8
	2014	..	*120	..	..	..	*120	*10
Guinea-Bissau	2011	..	*28	..	..	..	*28	*17
	2012	..	*29	..	..	..	*29	*17
	2013	..	*29	..	..	..	*29	*16
	2014	..	*30	..	..	..	*30	*17
Kenya	2011	520	276	0	24	0	772	19
	2012	300	149	0	12	0	437	10
	2013	184	187	42	12	-21	339	8
	2014	0	386	53	5	0	328	7
Liberia	2011	..	*40	..	*10	..	*30	*7
	2012	..	*40	..	*10	..	*30	*7
	2013	..	*40	..	*10	..	*30	*7
	2014	..	*40	..	*10	..	*30	*7
Libya	2011	2,056	..	254	83	..	1,719	273
	2012	2,895	..	108	358	..	2,429	387
	2013	3,973	..	1,965	241	..	1,767	282
	2014	1,655	159	..	95	..	1,719	275

Production, trade and supply of fuel oil

Thousand metric tons and kilograms per capita

Country or area	Year	Production	Imports	Exports	Bunkers	Changes in stocks	Supply	
							Total	Per capita
Madagascar	2011	..	100	..	*15	0	85	4
	2012	..	108	..	*16	5	87	4
	2013	..	106	..	*18	0	88	4
	2014	..	*110	..	*20	0	*90	*4
Malawi	2011	..	*2	..	..	..	*2	0
	2012	..	*2	..	..	..	*2	0
	2013	..	*2	..	..	..	*2	0
	2014	..	*2	..	..	..	*2	0
Mali	2011	..	*36	..	..	..	*36	*2
	2012	..	*37	..	..	..	*37	*2
	2013	..	*37	..	..	..	*37	*2
	2014	..	*37	..	..	..	*37	*2
Mauritania	2011	..	185	..	..	..	185	50
	2012	..	207	..	..	..	207	55
	2013	..	*207	..	..	..	*207	*53
	2014	..	*207	..	..	..	*207	*52
Mauritius	2011	..	435	..	185	-8	258	206
	2012	..	401	..	163	-18	256	203
	2013	..	429	..	156	14	259	205
	2014	..	406	..	171	*-31	266	210
Morocco	2011	2,441	218	76	0	102	2,481	75
	2012	2,118	660	47	0	-82	2,813	84
	2013	2,185	2,126	71	114	1,700	2,426	71
	2014	1,656	1,548	188	114	1,241	1,661	48
Mozambique	2011	..	1	..	..	..	1	0
	2012	..	0	..	..	..	0	0
	2013	..	0	..	..	..	0	0
	2014	..	0	..	..	..	0	0
Namibia	2011	..	15	..	..	0	15	7
	2012	..	26	..	..	0	26	11
	2013	..	51	..	..	*12	39	17
	2014	..	17	..	..	..	17	7
Niger	2011	..	5	..	..	0	4	0
	2012	..	7	..	..	..	7	0
	2013	..	3	..	..	..	3	0
	2014	..	0	..	..	..	0	0
Nigeria	2011	1,691	..	755	276	60	600	4
	2012	1,287	..	365	279	-66	709	4
	2013	1,222	..	342	256	-115	739	4
	2014	959	..	333	260	-299	665	4
Réunion	2011	..	103	..	..	12	91	109
	2012	..	77	..	..	-1	77	92
	2013	..	88	..	..	-1	89	105
	2014	..	127	..	..	-12	139	162
Rwanda	2011	..	*3	..	..	..	*3	0
	2012	..	*4	..	..	..	*4	0
	2013	..	*4	..	..	..	*4	0
	2014	..	*4	..	..	..	*4	0
Senegal	2011	175	336	15	..	-22	518	39
	2012	205	302	15	..	-26	518	38
	2013	171	328	4	..	-8	503	35
	2014	179	436	4	..	-8	619	42

Table 22

Production, trade and supply of fuel oil

Thousand metric tons and kilograms per capita

Country or area	Year	Production	Imports	Exports	Bunkers	Changes in stocks	Supply	
							Total	Per capita
Seychelles	2011	..	79	..	12	..	67	711
	2012	..	76	..	10	..	66	699
	2013	..	77	..	7	..	70	738
	2014	..	77	..	*4	..	73	760
Sierra Leone	2011	..	18	..	..	..	18	3
	2012	..	20	..	..	..	20	3
	2013	..	31	..	..	..	31	5
	2014	..	41	..	..	..	41	6
Somalia	2011	..	*39	..	*12	..	*27	*3
	2012	..	*39	..	*12	..	*27	*3
	2013	..	*39	..	*12	..	*27	*3
	2014	..	*39	..	*12	..	*27	*3
South Africa	2011	3,060	98	964	1,720	..	474	9
	2012	3,644	86	1,160	2,032	..	538	10
	2013	3,355	89	1,114	1,876	..	454	8
	2014	3,124	58	1,495	1,080	..	607	11
Sudan	2011	..	..	..	..	..	..	..
	2012	398	21	..	..	..	419	11
	2013	253	27	..	..	..	280	7
	2014	403	26	..	..	..	429	11
Sudan (former)	2011	629	..	37	..	..	592	13
	2012	..	..	..	..	..	..	..
	2013	..	..	..	..	..	..	..
	2014	..	..	..	..	..	..	..
Swaziland	2011	..	*5	..	..	..	*5	*4
	2012	..	*5	..	..	..	*5	*4
	2013	..	*5	..	..	..	*5	*4
	2014	..	*5	..	..	..	*5	*4
Togo	2011	..	*88	..	..	12	*76	*12
	2012	..	*67	..	..	-3	*70	*10
	2013	..	*78	..	..	-3	*81	*12
	2014	..	*62	..	..	-3	*65	*9
Tunisia	2011	251	239	189	0	34	267	25
	2012	612	279	573	1	4	313	29
	2013	551	241	560	0	-54	286	26
	2014	563	258	489	0	9	323	29
Uganda	2011	..	46	..	..	..	46	1
	2012	..	*42	..	..	..	*42	*1
	2013	..	*38	..	..	..	*38	*1
	2014	..	31	..	..	..	31	1
United Rep. of Tanzania	2011	..	267	..	49	..	218	5
	2012	..	357	..	52	..	305	6
	2013	..	399	..	57	..	342	7
	2014	..	465	..	61	..	404	8
Zambia	2011	72	..	22	..	..	50	3
	2012	82	..	22	..	..	60	4
	2013	79	..	28	..	..	51	3
	2014	88	27	..	..	..	115	7
America, North	**2011**	**63,687**	**34,760**	**38,184**	**24,881**	**-2,051**	**37,433**	**68**
	2012	**60,460**	**32,785**	**34,769**	**19,546**	**-637**	**39,567**	**71**
	2013	**58,937**	**30,123**	**36,686**	**20,815**	**342**	**31,217**	**55**
	2014	**54,800**	**25,866**	**37,826**	**16,310**	**-1,094**	**27,624**	**49**

Production, trade and supply of fuel oil

Thousand metric tons and kilograms per capita

Country or area	Year	Production	Imports	Exports	Bunkers	Changes in stocks	Supply	
							Total	Per capita
Antigua and Barbuda	2011	..	*37	..	*3	..	*34	*386
	2012	..	*38	..	*3	..	*35	*393
	2013	..	*38	..	*3	..	*35	*389
	2014	..	*39	..	*4	..	*35	*385
Aruba	2011	..	*230	..	*10	..	*220	*2,158
	2012	..	*218	..	*11	..	*207	*2,022
	2013	..	*199	..	*11	..	*188	*1,827
	2014	..	*202	..	*11	..	*191	*1,846
Bahamas	2011	..	116	..	*102	..	14	38
	2012	..	148	..	*136	..	12	32
	2013	..	151	..	*136	..	15	40
	2014	..	*151	..	*136	..	*15	*39
Barbados	2011	..	219	..	..	..	219	779
	2012	..	224	..	..	..	224	796
	2013	..	223	..	..	1	223	789
	2014	..	199	..	..	-1	199	702
Belize	2011	..	*21	..	..	..	*21	*62
	2012	..	*21	..	..	..	*21	*62
	2013	..	*22	..	..	..	*22	*62
	2014	..	*22	..	..	..	*22	*61
Bermuda	2011	..	78	..	*9	..	69	1,078
	2012	..	58	..	*9	..	49	777
	2013	..	95	..	*9	..	86	1,363
	2014	..	103	..	*10	..	93	1,487
Bonaire, St Eustatius, Saba	2011	..	..	..	..	..	..	..
	2012	..	*228	..	*183	..	*46	*1,992
	2013	..	*235	..	*188	..	*47	*1,980
	2014	..	*235	..	*188	..	*47	*1,929
Canada	2011	6,630	459	2,933	491	-64	3,729	108
	2012	7,292	771	3,895	406	-19	3,781	108
	2013	6,751	1,158	4,578	420	-60	2,971	84
	2014	6,175	957	3,613	347	97	3,075	86
Costa Rica	2011	61	219	26	0	8	247	54
	2012	0	266	8	0	-26	283	61
	2013	0	335	0	4	9	322	68
	2014	..	318	..	1	9	308	65
Cuba	2011	2,322	1,371	..	..	..	3,693	326
	2012	2,520	1,490	..	..	..	4,010	354
	2013	2,352	1,357	..	..	..	3,709	326
	2014	2,315	1,238	..	..	..	3,553	312
Curaçao	2011	..	..	..	..	..	..	..
	2012	3,101	1,723	3,020	1,393	..	411	2,693
	2013	3,792	1,732	3,692	1,405	..	427	2,767
	2014	3,505	1,757	3,413	1,430	..	419	2,690
Dominica	2011	..	*2	..	*1	..	*1	*14
	2012	..	*2	..	*1	..	*1	*14
	2013	..	*2	..	*1	..	*1	*14
	2014	..	*2	..	*1	..	*1	*14
Dominican Republic	2011	373	1,023	..	..	9	1,387	138
	2012	333	997	..	..	4	1,325	131
	2013	392	1,150	..	..	-11	1,553	151
	2014	367	1,168	..	..	-12	1,547	149

Table 22

Production, trade and supply of fuel oil

Thousand metric tons and kilograms per capita

Country or area	Year	Production	Imports	Exports	Bunkers	Changes in stocks	Supply	
							Total	Per capita
El Salvador	2011	379	297	57	..	-13	632	104
	2012	302	393	0	..	3	692	114
	2013	0	585	0	..	0	585	96
	2014	0	580	0	..	0	580	95
Greenland	2011	..	9	..	7	0	2	36
	2012	..	11	..	8	1	1	24
	2013	..	14	..	11	-1	3	58
	2014	..	14	..	1	1	11	202
Guadeloupe	2011	..	*210	..	..	..	*210	*458
	2012	..	*210	..	..	..	*210	*456
	2013	..	*211	..	..	..	*211	*455
	2014	..	*212	..	..	..	*212	*455
Guatemala	2011	..	628	9	..	2	617	41
	2012	..	582	1	..	-18	599	39
	2013	..	479	2	..	-480	957	61
	2014	..	466	17	..	-496	944	59
Haiti	2011	..	31	..	..	..	31	3
	2012	..	66	..	..	..	66	6
	2013	..	53	..	..	-21	74	7
	2014	..	128	..	..	-10	138	13
Honduras	2011	..	1,024	..	1	-64	1,087	143
	2012	..	1,020	..	1	0	1,019	132
	2013	..	1,016	..	1	0	1,015	129
	2014	..	1,057	..	..	-49	1,106	139
Jamaica	2011	729	794	219	..	2	1,301	473
	2012	695	721	233	..	0	1,182	428
	2013	702	855	196	..	5	1,356	489
	2014	566	890	180	..	3	1,274	458
Martinique	2011	*389	*3	*12	*19	..	*360	*913
	2012	*397	*3	*14	*20	..	*366	*927
	2013	*406	*3	*21	*20	..	*368	*931
	2014	*430	*4	*29	*22	..	*383	*967
Mexico	2011	17,486	1,423	5,736	99	-146	13,220	110
	2012	15,594	2,544	3,978	103	230	13,827	113
	2013	15,287	1,781	5,413	101	-23	11,577	94
	2014	14,742	740	7,325	93	-134	8,198	65
Neth. Antilles (former)	2011	2,974	2,555	2,900	2,063	..	566	2,655
	2012	..	..	..	..	..	..	..
	2013	..	..	..	..	..	..	..
	2014	..	..	..	..	..	..	..
Nicaragua	2011	355	279	..	..	40	594	102
	2012	234	293	..	..	-9	536	91
	2013	301	169	..	..	2	468	79
	2014	292	197	..	..	15	474	79
Panama	2011	..	3,613	..	3,279	-298	632	172
	2012	..	3,683	..	*3,333	-147	497	133
	2013	..	3,879	..	3,313	139	427	112
	2014	..	3,234	..	3,024	175	35	9
Sint Maarten (Dutch part)	2011	..	..	..	..	..	..	..
	2012	..	*246	..	*126	..	*120	*3,418
	2013	..	*250	..	*127	..	*123	*3,372
	2014	..	*254	..	*128	..	*126	*3,337

Production, trade and supply of fuel oil

Thousand metric tons and kilograms per capita

Country or area	Year	Production	Imports	Exports	Bunkers	Changes in stocks	Supply	
							Total	Per capita
St. Lucia	2011	..	*6	..	*6	..	0	0
	2012	..	*6	..	*6	..	0	0
	2013	..	*6	..	*6	..	0	0
	2014	..	*6	..	*6	..	0	0
Trinidad and Tobago	2011	2,461	..	2,287	178	-5	1	1
	2012	2,297	..	2,039	332	-74	0	0
	2013	2,897	..	2,374	386	136	1	1
	2014	2,588	..	2,317	377	-105	-1	-1
United States	2011	29,529	20,114	24,005	18,612	-1,521	8,547	27
	2012	27,696	16,823	21,581	13,475	-582	10,045	32
	2013	26,057	14,126	20,410	14,673	647	4,453	14
	2014	23,820	11,694	20,932	10,531	-588	4,639	15
America, South	**2011**	**45,359**	**3,153**	**23,141**	**7,346**	**185**	**17,839**	**44**
	2012	**44,830**	**2,446**	**23,062**	**6,828**	**-735**	**18,121**	**45**
	2013	**44,239**	**1,315**	**21,449**	**6,295**	**-349**	**18,159**	**44**
	2014	**46,107**	**1,611**	**21,281**	**6,122**	**1,274**	**19,042**	**46**
Argentina	2011	3,351	696	34	1,183	-33	2,863	69
	2012	3,953	696	50	1,541	-53	3,111	74
	2013	4,309	335	133	1,591	13	2,907	68
	2014	4,617	592	189	1,408	17	3,595	84
Bolivia (Plur. State of)	2011	416	..	454	..	-59	21	2
	2012	332	..	317	..	15	0	0
	2013	376	..	388	..	-12	0	0
	2014	411	..	408	..	3	0	0
Brazil	2011	13,289	674	4,877	3,960	26	5,100	25
	2012	13,747	216	5,140	3,509	-72	5,386	27
	2013	14,618	91	5,493	2,929	-34	6,321	31
	2014	16,010	379	5,123	2,929	54	8,283	40
Chile	2011	1,238	598	66	492	20	1,258	73
	2012	1,192	373	146	163	2	1,254	72
	2013	1,252	112	215	207	-86	1,028	58
	2014	1,185	66	67	206	-13	991	56
Colombia	2011	3,923	8	3,760	0	-16	187	4
	2012	4,093	8	3,844	0	33	224	5
	2013	4,333	13	3,987	226	75	58	1
	2014	4,521	12	3,692	254	773	-186	-4
Ecuador	2011	2,845	475	1,443	485	1	1,391	92
	2012	2,689	364	1,144	400	-4	1,512	98
	2013	2,479	280	775	368	0	1,616	103
	2014	2,211	140	281	402	4	1,664	105
French Guiana	2011	..	*47	..	..	..	*47	*196
	2012	..	*30	..	..	..	*30	*121
	2013	..	*38	..	..	..	*38	*149
	2014	..	*40	..	..	..	*40	*153
Guyana	2011	..	159	..	..	1	158	209
	2012	..	184	..	..	4	180	237
	2013	..	200	..	..	8	193	253
	2014	..	190	..	..	0	190	249
Paraguay	2011	..	38	..	..	-2	41	6
	2012	..	68	..	..	-8	76	12
	2013	..	36	..	..	0	36	5
	2014	..	50	..	..	7	43	7

Table 22

Production, trade and supply of fuel oil

Thousand metric tons and kilograms per capita

Country or area	Year	Production	Imports	Exports	Bunkers	Changes in stocks	Supply	
							Total	Per capita
Peru	2011	2,414	49	1,113	172	281	897	30
	2012	1,716	..	1,465	216	-679	714	24
	2013	1,694	..	1,244	..	-301	751	25
	2014	1,735	..	792	..	419	524	17
Suriname	2011	843	31	569	..	0	306	584
	2012	865	32	576	..	0	321	608
	2013	929	33	623	..	0	340	637
	2014	961	37	638	..	0	360	669
Uruguay	2011	307	378	..	258	-34	461	136
	2012	414	414	..	158	27	644	190
	2013	372	178	..	123	-11	439	129
	2014	356	106	..	105	10	347	101
Venezuela (Bolivar. Rep.)	2011	16,733	..	10,826	796	..	5,111	174
	2012	15,828	61	10,379	841	..	4,669	156
	2013	13,877	..	8,592	852	..	4,433	146
	2014	14,100	..	10,091	818	..	3,191	104
Asia	**2011**	**219,475**	**147,135**	**96,211**	**94,448**	**5,619**	**170,332**	**40**
	2012	**214,906**	**147,393**	**86,778**	**93,738**	**6,110**	**175,672**	**41**
	2013	**208,119**	**153,762**	**93,251**	**98,649**	**412**	**169,569**	**39**
	2014	**200,737**	**137,182**	**81,036**	**94,990**	**-1,335**	**163,229**	**38**
Afghanistan	2011	..	1	..	..	..	1	0
	2012	..	1	..	..	..	1	0
	2013	..	*1	..	..	..	*1	0
	2014	..	*1	..	..	..	*1	0
Armenia	2011	..	1	..	..	..	1	0
	2012	..	0	..	..	..	0	0
	2013	..	0	..	..	..	0	0
	2014	..	0	0	..	..	0	0
Azerbaijan	2011	254	0	213	..	-93	134	15
	2012	286	0	107	..	39	140	15
	2013	285	9	228	..	31	35	4
	2014	207	21	237	..	-29	20	2
Bahrain	2011	2,458	..	2,652	..	-194	0	0
	2012	2,443	..	2,702	..	-259	0	0
	2013	2,519	..	2,504	..	15	0	0
	2014	2,158	..	2,174	..	-16	0	0
Bangladesh	2011	367	231	..	84	..	514	3
	2012	275	681	..	89	..	867	6
	2013	323	804	..	94	..	1,033	7
	2014	302	1,016	..	100	..	1,218	8
Brunei Darussalam	2011	98	..	..	98	..	0	0
	2012	66	..	..	66	..	0	0
	2013	80	..	..	80	..	0	0
	2014	72	..	..	72	..	0	0
Cambodia	2011	..	229	..	..	..	229	16
	2012	..	214	..	..	..	214	14
	2013	..	166	..	..	..	166	11
	2014	..	89	..	..	..	89	6
China	2011	22,818	30,974	12,271	8,723	260	32,538	24
	2012	22,532	31,024	11,617	8,590	631	32,718	24
	2013	27,759	27,343	11,348	7,462	904	35,388	26
	2014	35,417	17,846	9,477	6,939	62	36,784	27

Production, trade and supply of fuel oil

Thousand metric tons and kilograms per capita

Country or area	Year	Production	Imports	Exports	Bunkers	Changes in stocks	Supply	
							Total	Per capita
China, Hong Kong SAR	2011	..	7,715	272	6,926	494	23	3
	2012	..	7,493	486	7,127	-213	93	13
	2013	..	7,625	397	7,539	-341	30	4
	2014	..	6,482	395	6,292	-207	2	0
China, Macao SAR	2011	..	93	..	..	-2	95	174
	2012	..	86	..	..	1	85	152
	2013	..	52	..	..	-1	53	93
	2014	..	55	..	..	-1	56	97
Cyprus	2011	..	1,153	..	141	-85	1,097	982
	2012	..	1,072	..	128	24	920	815
	2013	..	830	..	157	-4	677	593
	2014	..	942	..	153	-39	828	718
Georgia	2011	..	6	..	..	..	6	2
	2012	..	3	1	..	..	2	1
	2013	..	5	5	..	-1	1	0
	2014	..	14	12	..	..	2	0
India	2011	19,456	1,128	7,895	0	5,216	7,473	6
	2012	15,054	1,068	5,922	0	*4,609	5,591	4
	2013	13,405	1,283	6,159	1,290	16	7,223	6
	2014	11,919	945	4,762	*1,343	..	6,759	5
Indonesia	2011	8,833	942	4,507	106	..	5,162	21
	2012	8,472	397	3,950	113	..	4,806	19
	2013	8,501	402	4,073	196	..	4,634	18
	2014	8,369	406	4,529	124	..	4,122	16
Iran (Islamic Rep. of)	2011	26,689	..	10,372	2,180	..	14,136	188
	2012	27,701	..	5,798	1,801	..	20,103	264
	2013	26,879	..	8,531	2,969	..	15,380	199
	2014	23,441	..	6,365	3,558	..	13,518	173
Iraq	2011	13,985	..	57	163	..	13,765	432
	2012	15,187	..	118	182	..	14,887	452
	2013	15,656	..	561	179	..	14,916	437
	2014	13,268	..	388	188	..	12,692	360
Israel	2011	2,481	128	1,112	265	-30	1,262	167
	2012	1,813	687	630	261	-20	1,629	212
	2013	2,984	142	1,669	213	-3	1,247	160
	2014	3,040	142	2,242	97	..	843	106
Japan	2011	23,126	6,856	2,589	3,949	-26	23,470	184
	2012	24,969	8,579	2,691	3,886	65	26,906	212
	2013	20,393	6,277	1,889	3,731	-115	21,165	167
	2014	18,119	4,241	1,685	3,371	-159	17,463	138
Jordan	2011	868	658	..	3	-103	1,626	241
	2012	999	783	..	12	49	1,721	246
	2013	900	687	..	5	-74	1,656	230
	2014	812	1,232	..	0	2	2,041	275
Kazakhstan	2011	4,277	2	2,843	..	21	1,415	85
	2012	3,936	26	2,600	..	-103	1,465	87
	2013	3,734	231	2,500	..	165	1,300	76
	2014	3,877	3	2,769	..	-79	1,190	69
Korea, Dem.Ppl's.Rep.	2011	108	276	..	..	..	384	16
	2012	108	247	..	..	..	355	14
	2013	119	256	..	..	..	375	15
	2014	109	263	..	..	..	372	15

Table 22

Production, trade and supply of fuel oil

Thousand metric tons and kilograms per capita

Country or area	Year	Production	Imports	Exports	Bunkers	Changes in stocks	Supply	
							Total	Per capita
Korea, Republic of	2011	18,496	2,310	5,173	7,743	176	7,714	156
	2012	15,282	4,094	4,267	7,562	-81	7,628	154
	2013	11,801	6,932	3,368	7,367	-14	8,012	161
	2014	8,133	7,839	2,024	7,444	9	6,495	130
Kuwait	2011	9,766	..	3,072	962	..	5,732	1,770
	2012	10,825	..	3,167	1,004	..	6,654	1,946
	2013	10,207	..	2,118	1,040	..	7,049	1,961
	2014	10,204	..	1,634	1,040	..	7,530	2,006
Kyrgyzstan	2011	46	77	9	..	-19	133	24
	2012	39	50	2	..	6	81	14
	2013	41	9	2	..	-3	51	9
	2014	51	148	..	..	-11	210	36
Lao People's Dem. Rep.	2011	..	*33	..	..	..	*33	*5
	2012	..	*33	..	..	..	*33	*5
	2013	..	*34	..	..	..	*34	*5
	2014	..	*34	..	..	..	*34	*5
Lebanon	2011	..	1,347	..	29	..	1,318	287
	2012	..	1,138	..	30	..	1,108	225
	2013	..	1,332	..	30	..	1,302	246
	2014	..	1,487	..	31	..	1,456	259
Malaysia	2011	576	1,482	343	189	0	1,526	53
	2012	1,621	800	1,026	54	0	1,341	46
	2013	1,297	2,979	3,064	349	172	691	23
	2014	2,359	411	2,626	207	-639	576	19
Mongolia	2011	..	7	..	..	..	7	3
	2012	..	7	..	..	..	7	2
	2013	..	4	..	..	..	4	1
	2014	..	3	..	..	..	3	1
Myanmar	2011	71	..	..	..	-3	74	1
	2012	65	..	..	..	0	65	1
	2013	55	..	..	..	0	55	1
	2014	61	..	..	..	-11	72	1
Nepal	2011	..	0	..	..	..	0	0
	2012	..	3	..	..	..	3	0
	2013	..	2	..	..	..	2	0
	2014	..	1	..	..	*-1	*2	0
Oman	2011	2,933	..	511	580	..	1,842	574
	2012	2,867	..	492	801	..	1,574	444
	2013	3,065	..	326	885	..	1,854	475
	2014	3,039	..	389	1,180	..	1,470	347
Other Asia	2011	8,287	1,347	916	1,478	-235	7,475	321
	2012	8,828	1,030	1,321	1,049	539	6,949	298
	2013	5,812	1,443	597	1,096	-432	5,994	257
	2014	5,880	1,800	715	1,055	117	5,793	248
Pakistan	2011	2,174	6,419	1	91	-153	8,654	50
	2012	2,653	5,582	..	84	*-153	8,304	47
	2013	2,926	6,529	..	89	*-153	9,519	53
	2014	3,064	7,119	..	93	*-153	10,243	55
Philippines	2011	1,994	346	531	197	17	1,595	17
	2012	1,734	283	202	172	-20	1,663	17
	2013	1,605	408	238	148	-48	1,674	17
	2014	1,717	740	243	69	103	2,042	21

Production, trade and supply of fuel oil

Thousand metric tons and kilograms per capita

Country or area	Year	Production	Imports	Exports	Bunkers	Changes in stocks	Supply	
							Total	Per capita
Qatar	2011	224	..	236	..	-12	0	0
	2012	321	..	329	..	-8	0	0
	2013	384	..	329	..	46	9	4
	2014	245	..	202	..	43	0	0
Saudi Arabia	2011	22,849	1,300	6,084	3,198	699	14,168	492
	2012	25,282	3,309	10,223	3,382	181	14,805	502
	2013	24,955	5,574	10,002	3,424	240	16,863	558
	2014	26,378	3,305	5,800	3,224	-140	20,799	673
Singapore	2011	7,017	62,640	26,578	41,577	-272	1,774	342
	2012	4,925	59,444	22,490	41,229	391	259	49
	2013	5,445	63,692	26,314	41,306	35	1,482	274
	2014	2,528	62,280	24,688	41,104	86	-1,070	-194
Sri Lanka	2011	615	368	26	223	-216	951	47
	2012	648	564	0	217	-13	1,008	49
	2013	724	176	105	265	-144	674	33
	2014	641	348	..	338	-119	771	37
State of Palestine	2011	..	0	..	..	..	0	0
	2012	..	0	..	..	..	0	0
	2013	..	0	..	..	..	0	0
	2014	..	3	..	..	..	3	1
Syrian Arab Republic	2011	4,741	1,052	554	77	..	5,162	252
	2012	3,365	449	..	40	..	3,774	189
	2013	2,225	354	..	37	..	2,542	132
	2014	2,225	..	..	20	..	2,205	117
Tajikistan	2011	10	111	..	..	..	121	16
	2012	11	22	..	..	..	33	4
	2013	13	15	..	..	..	28	3
	2014	15	16	..	..	..	31	4
Thailand	2011	6,074	344	3,619	0	*216	2,583	39
	2012	6,393	376	3,993	0	*298	2,478	37
	2013	5,557	252	3,548	764	228	1,269	19
	2014	5,328	349	3,606	1,018	-259	1,312	19
Turkey	2011	2,492	676	2,148	65	-40	995	14
	2012	1,567	913	1,204	34	153	1,089	15
	2013	3,013	740	1,908	798	-98	1,145	15
	2014	3,837	1,130	2,621	930	114	1,302	17
Turkmenistan	2011	2,230	..	1,001	..	..	1,229	241
	2012	2,160	..	970	..	..	1,190	230
	2013	1,760	..	790	..	..	970	185
	2014	1,690	..	759	..	..	931	175
United Arab Emirates	2011	2,037	14,438	..	*15,135	..	1,340	153
	2012	1,935	15,092	..	*15,643	..	1,384	155
	2013	2,793	*15,724	..	*17,011	..	*1,506	*167
	2014	1,433	15,066	..	14,873	..	1,626	179
Uzbekistan	2011	456	..	..	..	..	456	16
	2012	278	..	..	..	..	278	10
	2013	228	..	..	..	..	228	8
	2014	198	..	..	..	..	198	7
Viet Nam	2011	79	1,470	323	236	3	987	11
	2012	59	823	280	160	-6	448	5
	2013	32	659	251	106	-8	342	4
	2014	33	676	258	109	-8	350	4

Table 22

Production, trade and supply of fuel oil

Thousand metric tons and kilograms per capita

Country or area	Year	Production	Imports	Exports	Bunkers	Changes in stocks	Supply	
							Total	Per capita
Yemen	2011	490	976	303	30	..	1,133	47
	2012	207	1,020	190	23	..	1,014	41
	2013	645	791	427	20	..	989	39
	2014	569	728	435	17	..	845	32
Europe	**2011**	**167,238**	**71,914**	**134,684**	**49,953**	**-552**	**55,066**	**75**
	2012	**165,549**	**68,314**	**135,744**	**44,905**	**-662**	**53,876**	**73**
	2013	**157,907**	**67,206**	**145,062**	**45,296**	**-2,043**	**36,797**	**50**
	2014	**159,172**	**66,744**	**147,224**	**45,129**	**-615**	**34,178**	**46**
Albania	2011	9	3	..	..	..	12	4
	2012	..	1	..	..	..	1	0
	2013	..	1	..	..	..	1	0
	2014	2	1	..	..	..	3	1
Austria	2011	822	86	266	..	-29	671	80
	2012	953	59	220	..	-4	796	94
	2013	1,011	58	325	..	-3	747	88
	2014	981	41	428	..	-50	644	76
Belarus	2011	6,407	848	6,538	..	3	714	75
	2012	7,284	3,321	6,566	..	22	4,017	423
	2013	6,615	18	6,146	..	22	465	49
	2014	5,865	5	5,420	..	-37	487	51
Belgium	2011	5,689	4,914	4,329	6,604	-107	-223	-20
	2012	5,179	4,420	3,953	5,794	17	-165	-15
	2013	4,916	5,270	4,780	5,906	40	-540	-48
	2014	5,512	3,700	4,551	4,993	79	-411	-37
Bosnia and Herzegovina	2011	218	31	55	..	0	194	51
	2012	176	30	8	..	0	198	52
	2013	116	33	0	..	1	148	39
	2014	143	10	30	..	-10	133	35
Bulgaria	2011	1,424	29	1,338	29	-12	98	13
	2012	1,488	59	1,394	19	6	128	18
	2013	1,640	38	1,561	31	-20	106	15
	2014	1,617	72	1,536	26	38	89	12
Croatia	2011	731	13	372	23	-100	449	104
	2012	563	63	326	..	-37	337	79
	2013	514	..	376	..	-60	198	46
	2014	423	30	335	..	-46	164	39
Czechia	2011	195	29	98	..	-21	147	14
	2012	145	29	73	..	0	101	10
	2013	148	13	120	..	-57	98	9
	2014	125	13	106	..	-13	45	4
Denmark	2011	1,147	2,040	2,884	465	-334	172	31
	2012	1,312	1,724	2,645	277	-75	189	34
	2013	1,213	2,953	3,446	374	127	219	39
	2014	1,242	3,317	3,860	449	203	47	8
Estonia	2011	..	169	0	169	9	-9	-7
	2012	..	906	581	325	1	-1	-1
	2013	..	448	54	392	-24	26	20
	2014	..	602	317	281	27	-23	-17
Faeroe Islands	2011	..	*78	..	23	..	*55	*1,129
	2012	..	*79	..	23	..	*57	*1,175
	2013	..	*78	..	15	..	*63	*1,305
	2014	..	*77	..	17	..	*60	*1,253

Production, trade and supply of fuel oil

Thousand metric tons and kilograms per capita

Country or area	Year	Production	Imports	Exports	Bunkers	Changes in stocks	Supply	
							Total	Per capita
Finland	2011	1,294	695	927	142	-1	921	171
	2012	1,287	809	1,097	82	-118	1,035	191
	2013	1,330	793	1,095	85	9	934	171
	2014	1,224	1,255	844	65	21	1,549	283
France	2011	9,024	7,402	6,497	2,447	-7	7,489	118
	2012	8,237	5,346	5,905	2,317	-222	5,583	88
	2013	6,735	5,045	5,633	2,124	-127	4,150	65
	2014	6,481	5,023	5,845	1,785	29	3,845	60
Germany	2011	7,754	2,791	1,967	2,296	-243	6,525	81
	2012	8,319	2,388	2,057	2,173	139	6,338	79
	2013	7,776	2,500	2,729	1,927	-48	5,668	70
	2014	7,696	2,485	3,142	1,866	-157	5,330	66
Gibraltar	2011	..	3,838	..	3,796	..	42	1,352
	2012	..	3,871	..	3,828	..	43	1,369
	2013	..	3,548	..	3,502	..	46	1,450
	2014	..	2,653	..	2,605	..	48	1,500
Greece	2011	5,053	1,852	2,165	2,535	42	2,163	194
	2012	5,111	2,196	2,731	2,064	77	2,435	219
	2013	5,447	1,222	3,162	1,912	-376	1,971	178
	2014	6,043	1,222	3,205	1,653	149	2,258	205
Hungary	2011	150	13	58	..	-9	114	11
	2012	81	10	24	..	-11	78	8
	2013	84	18	29	..	-5	78	8
	2014	67	20	23	..	1	63	6
Iceland	2011	..	89	..	32	1	56	174
	2012	..	106	..	27	3	76	235
	2013	..	109	..	13	-6	102	313
	2014	..	90	..	8	10	72	220
Ireland	2011	1,071	236	1,088	23	-5	201	43
	2012	967	202	984	24	-31	192	41
	2013	939	187	923	16	2	185	40
	2014	937	103	938	20	-29	111	24
Italy	2011	9,180	850	2,804	2,338	340	4,548	76
	2012	9,216	600	3,145	2,388	584	3,699	62
	2013	5,127	538	2,213	2,109	-1,204	2,547	43
	2014	5,731	329	2,819	1,845	-861	2,257	38
Latvia	2011	..	159	0	143	-2	18	9
	2012	..	169	1	157	-3	14	7
	2013	..	169	0	164	0	5	2
	2014	..	168	12	167	-12	1	1
Lithuania	2011	1,760	159	1,628	126	-23	188	61
	2012	1,709	155	1,590	105	-113	282	93
	2013	1,833	98	1,730	77	-30	154	52
	2014	1,475	12	1,423	3	-53	114	39
Luxembourg	2011	..	2	..	..	..	2	4
	2012	..	2	..	..	..	2	4
	2013	..	0	..	..	..	0	0
	2014	..	0	..	..	..	0	0
Malta	2011	..	1,718	0	1,158	15	545	1,316
	2012	..	1,631	0	1,003	43	585	1,408
	2013	..	1,674	77	1,045	75	477	1,145
	2014	..	1,683	181	1,071	-44	475	1,137

Table 22

Production, trade and supply of fuel oil

Thousand metric tons and kilograms per capita

Country or area	Year	Production	Imports	Exports	Bunkers	Changes in stocks	Supply	
							Total	Per capita
Montenegro	2011	..	10	1	..	..	9	14
	2012	..	8	1	..	..	7	11
	2013	..	12	0	..	..	12	19
	2014	..	4	0	..	..	4	6
Netherlands	2011	8,512	34,240	27,578	13,628	-378	1,924	115
	2012	8,726	33,209	28,240	12,399	244	1,052	63
	2013	8,443	33,938	28,251	11,970	15	2,145	128
	2014	8,501	35,081	29,470	11,777	118	2,217	131
Norway	2011	1,747	1,372	1,874	194	-62	1,113	225
	2012	1,591	1,202	1,659	182	-15	967	193
	2013	1,668	1,990	1,922	186	-1	1,551	305
	2014	1,688	1,186	1,573	78	-14	1,237	240
Poland	2011	3,391	71	2,004	106	2	1,350	35
	2012	3,430	61	2,310	80	-23	1,124	29
	2013	3,408	62	2,592	65	1	812	21
	2014	3,659	72	2,893	73	-18	783	20
Portugal	2011	2,534	381	1,258	527	37	1,093	104
	2012	2,740	233	1,439	579	44	911	87
	2013	2,461	177	1,594	609	-249	684	65
	2014	2,498	239	1,575	571	121	470	45
Republic of Moldova	2011	10	17	7	..	-4	24	6
	2012	12	7	8	..	-7	18	4
	2013	16	14	13	..	-1	18	4
	2014	16	18	13	..	5	16	4
Romania	2011	722	15	330	..	-77	484	24
	2012	562	2	304	..	-219	479	24
	2013	364	1	256	..	-96	205	10
	2014	321	..	203	..	-24	142	7
Russian Federation	2011	73,640	1,156	55,183	1,907	205	17,501	122
	2012	75,072	189	57,080	519	-299	17,961	125
	2013	77,993	11	64,467	3,838	157	9,542	67
	2014	82,109	5	66,245	6,469	350	9,050	63
Serbia	2011	542	4	83	..	-8	471	52
	2012	472	74	49	..	9	488	54
	2013	395	2	136	..	25	236	26
	2014	366	3	140	..	-12	241	27
Slovakia	2011	638	101	355	..	16	368	68
	2012	616	117	352	..	-1	382	71
	2013	677	181	430	..	-1	429	79
	2014	489	4	237	..	-4	260	48
Slovenia	2011	..	41	..	33	..	8	4
	2012	..	59	..	51	..	8	4
	2013	..	71	..	62	..	9	4
	2014	..	103	36	58	..	9	4
Spain	2011	7,927	3,785	2,195	7,506	-45	2,056	44
	2012	6,536	3,233	1,438	7,451	-186	1,066	23
	2013	5,296	4,391	2,992	6,183	-23	535	12
	2014	4,482	5,555	3,026	6,742	-229	498	11
Sweden	2011	4,970	1,223	3,369	1,564	190	1,070	113
	2012	5,099	783	3,221	1,498	-197	1,360	143
	2013	4,124	761	2,346	1,399	-176	1,316	137
	2014	3,161	389	1,813	1,448	25	264	27

Production, trade and supply of fuel oil

Thousand metric tons and kilograms per capita

Country or area	Year	Production	Imports	Exports	Bunkers	Changes in stocks	Supply	
							Total	Per capita
Switzerland	2011	344	15	306	..	-7	60	8
	2012	275	14	169	..	58	62	8
	2013	365	8	360	..	-34	47	6
	2014	394	..	340	..	8	46	6
T.F.Yug.Rep. Macedonia	2011	246	8	70	..	-36	219	106
	2012	97	133	24	..	-40	245	119
	2013	21	126	10	..	-13	150	72
	2014	0	127	6	..	-2	123	59
Ukraine	2011	2,181	623	1,917	..	-1	888	20
	2012	790	155	509	..	-90	526	12
	2013	658	25	617	..	14	52	1
	2014	516	44	489	..	-37	108	2
United Kingdom	2011	7,907	808	5,140	2,139	99	1,337	21
	2012	7,504	659	5,641	1,540	-218	1,200	19
	2013	6,574	625	4,677	1,292	23	1,207	19
	2014	5,408	1,003	4,150	1,059	-147	1,349	21
Oceania	**2011**	**1,668**	**2,330**	**444**	**868**	**4**	**2,682**	**72**
	2012	**1,646**	**2,508**	**730**	**1,008**	**-36**	**2,452**	**65**
	2013	**1,618**	**2,657**	**434**	**860**	**89**	**2,892**	**76**
	2014	**1,330**	**2,255**	**245**	**918**	**26**	**2,396**	**62**
Australia	2011	954	1,474	184	562	-8	1,690	75
	2012	945	1,533	460	735	-18	1,301	57
	2013	888	1,588	209	586	39	1,642	71
	2014	645	1,264	43	674	11	1,181	50
Fiji	2011	..	23	0	*2	..	21	25
	2012	..	34	0	*2	..	32	36
	2013	..	46	0	*2	..	44	50
	2014	..	43	0	*2	..	41	47
French Polynesia	2011	..	*114	..	*2	..	*111	*411
	2012	..	*114	..	*2	..	*111	*407
	2013	..	*115	..	*2	..	*113	*407
	2014	..	*111	..	*2	..	*108	*387
Nauru	2011	..	*8	..	*4	..	*4	*398
	2012	..	*8	..	*4	..	*4	*397
	2013	..	*9	..	*4	..	*5	*494
	2014	..	*10	..	4	..	*5	*521
New Caledonia	2011	..	452	..	..	3	449	1,796
	2012	..	454	..	..	12	442	1,745
	2013	..	517	..	..	11	506	1,972
	2014	..	440	..	..	..	440	1,692
New Zealand	2011	594	..	260	287	12	35	8
	2012	562	..	270	253	-19	58	13
	2013	553	..	225	253	41	34	8
	2014	512	6	202	231	15	70	16
Palau	2011	..	*32	..	..	..	*32	*1,529
	2012	..	*32	..	..	..	*32	*1,518
	2013	..	*32	..	..	..	*32	*1,506
	2014	..	*32	..	..	..	*32	*1,503
Papua New Guinea	2011	120	223	..	6	-3	340	49
	2012	139	317	..	7	-11	460	64
	2013	177	320	..	8	-2	491	67
	2014	*173	*322	..	0	0	*495	*66

Production, trade and supply of fuel oil

Thousand metric tons and kilograms per capita

Country or area	Year	Production	Imports	Exports	Bunkers	Changes in stocks	Supply	
							Total	Per capita
Tonga	2011	..	1	..	0	..	0	4
	2012	..	13	..	0	..	13	124
	2013	..	26	..	0	..	26	246
	2014	..	24	..	0	..	24	224
Vanuatu	2011	..	*4	..	*4	..	0	0
	2012	..	*4	..	*4	..	0	0
	2013	..	*4	..	*4	..	0	0
	2014	..	*5	..	*5	..	0	0

Production of minor oil products from refineries - by type

Thousand metric tons

Table Notes

Data refer to production from refineries of refinery gas, ethane, naphtha, lubricants, bitumen, paraffin waxes, petroleum coke, white spirit, and other oil products (which for this table includes refinery output of feedstocks).

Please refer to the Definitions Section on pages ix to xv for the appropriate product description /classification.

Figure 63: Production of minor oil products from refineries, 2014*

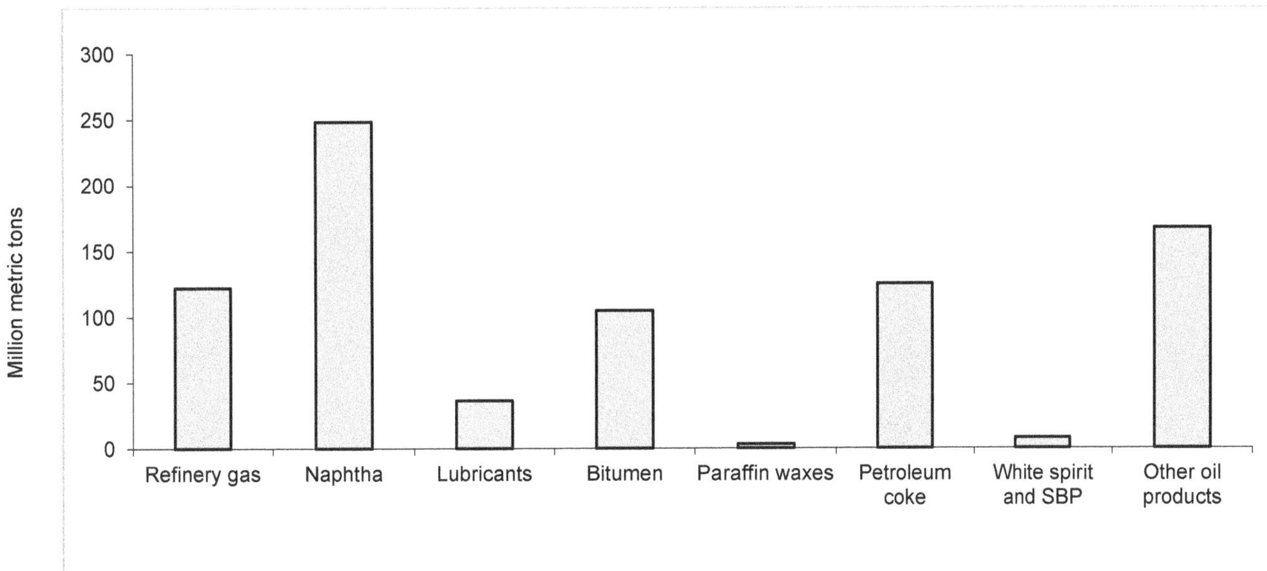

*Ethane is included with other oil products

Table 23

Production of minor oil products from refineries - by type

Thousand metric tons

Country or area	Year	Refinery gas	Ethane	Naphtha	Lubricants	Bitumen	Paraffin waxes	Petroleum coke	White spirit and SBP	Other oil products
World	**2011**	**144,337**	**499**	**231,821**	**37,584**	**104,147**	**3,278**	**114,962**	**8,008**	**164,703**
	2012	**124,917**	**349**	**232,565**	**36,402**	**102,506**	**3,229**	**121,299**	**8,430**	**166,346**
	2013	**122,979**	**212**	**239,020**	**35,543**	**102,813**	**3,267**	**124,868**	**7,449**	**164,326**
	2014	**122,217**	**144**	**248,368**	**36,485**	**104,883**	**3,269**	**124,992**	**7,987**	**167,074**
Africa	**2011**	**2,198**	**..**	**11,250**	**432**	**1,965**	**12**	**661**	**118**	**2,283**
	2012	**2,296**	**..**	**11,578**	**358**	**1,708**	**17**	**521**	**125**	**1,944**
	2013	**2,091**	**..**	**11,650**	**347**	**1,539**	**18**	**510**	**121**	**2,843**
	2014	**1,734**	**..**	**12,829**	**659**	**1,543**	**11**	**580**	**87**	**2,214**
Algeria	2011	304	..	6,939	118	175	..	..	..	69
	2012	273	..	6,432	100	209	..	..	..	64
	2013	293	..	6,336	113	159	..	..	..	64
	2014	363	..	8,721	146	102	..	..	..	272
Angola	2011	59	..	166	..	2	..	..	..	..
	2012	62	..	156	..	5	..	..	..	..
	2013	68	..	205	..	3	..	..	..	..
	2014	69	..	190	..	3	..	..	..	..
Cameroon	2011	79	..	..	..	..	..	..	..	184
	2012	83	..	..	..	..	..	..	..	192
	2013	88	..	..	..	..	..	..	..	203
	2014	93	..	..	..	..	..	..	..	215
Congo	2011	..	..	7	..	..	..	..	..	12
	2012	..	..	10	..	..	..	..	..	12
	2013	..	..	11	..	..	..	..	..	12
	2014	..	..	10	..	..	..	..	..	13
Côte d'Ivoire	2011	19	..	..	13	151	..	..	..	..
	2012	27	..	..	14	172	..	..	..	..
	2013	26	..	..	17	140	..	..	..	..
	2014	25	..	..	22	177	..	..	..	..
Egypt	2011	121	..	2,843	0	802	..	375	..	1,177
	2012	133	..	2,706	0	646	..	252	..	1,240
	2013	*133	..	2,634	0	*646	..	*252	..	1,531
	2014	112	..	2,280	0	545	..	212	..	*1,115
Gabon	2011	..	..	..	12	2	..	..	..	..
	2012	..	..	..	9	2	..	..	..	..
	2013	..	..	..	10	2	..	..	..	..
	2014	..	..	..	10	2	..	..	..	..
Ghana	2011	42	..	..	..	..	..	..	..	..
	2012	20	..	..	..	..	..	..	..	..
	2013	0	..	..	..	..	..	..	..	..
	2014	0	..	..	..	..	..	..	..	..
Kenya	2011	68	..	..	54	15	..	..	..	4
	2012	38	..	..	4	27	..	..	..	2
	2013	31	..	..	0	0	..	..	..	0
	2014	0	..	..	0	0	..	..	..	0
Libya	2011	120	..	649	..	31	..	..	..	5
	2012	172	..	1,348	..	45	..	..	..	7
	2013	216	..	1,533	..	56	..	..	..	9
	2014	99	..	682	..	26	..	..	..	4
Morocco	2011	*262	..	448	103	315	1	..	..	..
	2012	*280	..	493	107	171	1	..	..	..
	2013	185	..	482	63	181	1	..	..	218
	2014	174	..	494	70	281	..	..	..	295

Production of minor oil products from refineries - by type

Thousand metric tons

Country or area	Year	Refinery gas	Ethane	Naphtha	Lubricants	Bitumen	Paraffin waxes	Petroleum coke	White spirit and SBP	Other oil products
Nigeria	2011	111	..	..	10	33	..	1	..	780
	2012	125	..	..	10	0	..	1	..	375
	2013	125	..	..	11	5	..	2	..	559
	2014	84	..	..	12	3	..	10	..	248
Senegal	2011	12	..	31	..	..	..	..	..	..
	2012	13	..	53	..	..	..	..	..	..
	2013	15	..	85	..	..	..	..	..	..
	2014	16	..	89	..	..	..	..	..	..
South Africa	2011	967	..	..	122	439	11	..	112	40
	2012	1,037	..	..	114	431	16	..	110	39
	2013	878	..	..	133	347	17	..	106	37
	2014	664	..	..	399	404	11	..	69	38
Sudan	2011	..	..	..	..	..	..	..	..	..
	2012	20	..	8	..	..	..	268	..	0
	2013	20	..	18	..	..	..	256	..	*197
	2014	21	..	17	..	..	..	358	..	0
Sudan (former)	2011	22	..	18	..	..	..	285	..	..
	2012	..	..	..	..	..	..	..	..	..
	2013	..	..	..	..	..	..	..	..	..
	2014	..	..	..	..	..	..	..	..	..
Tunisia	2011	0	..	149	..	..	..	..	6	..
	2012	0	..	372	..	..	..	..	15	..
	2013	0	..	346	..	..	..	..	15	..
	2014	0	..	346	..	..	..	..	18	..
Zambia	2011	12	..	..	..	..	..	..	..	12
	2012	13	..	..	..	..	..	..	..	13
	2013	13	..	..	..	..	..	..	..	13
	2014	14	..	..	..	..	..	..	..	14
America, North	**2011**	**42,381**	**424**	**16,997**	**10,380**	**28,944**	**421**	**60,286**	**1,695**	**24,548**
	2012	**23,018**	**287**	**15,923**	**9,778**	**27,549**	**413**	**61,376**	**2,002**	**19,096**
	2013	**23,553**	**145**	**17,449**	**10,123**	**25,517**	**420**	**62,673**	**1,865**	**16,570**
	2014	**22,538**	**105**	**16,475**	**10,168**	**25,402**	**359**	**62,829**	**1,874**	**17,365**
Aruba	2011	..	..	..	..	..	..	..	..	*10,050
	2012	..	..	..	..	..	..	..	..	*2,440
	2013	..	..	..	..	..	..	..	..	0
	2014	..	..	..	..	..	..	..	..	0
Canada	2011	4,693	..	3,366	1,029	4,357	..	1,794	54	5,112
	2012	4,029	..	3,505	1,128	4,510	..	1,717	71	3,753
	2013	3,932	..	3,062	1,245	4,488	..	1,725	59	4,466
	2014	3,447	..	3,308	1,229	4,000	..	1,576	59	4,806
Costa Rica	2011	..	..	..	..	2	..	..	..	..
	2012	..	..	..	..	0	..	..	..	..
	2013	..	..	..	..	0	..	..	..	..
	2014	..	..	..	..	..	..	..	..	..
Cuba	2011	34	..	247	49	80	..	7	..	..
	2012	18	..	349	47	81	..	1	..	..
	2013	30	..	284	48	83	..	18	..	..
	2014	27	..	367	49	61	..	12	..	..
Curaçao	2011	..	..	..	..	..	..	..	..	..
	2012	194	..	310	253	451	..	75	..	56
	2013	177	..	337	131	408	..	68	..	51
	2014	208	..	373	103	601	..	81	..	60

Table 23

Production of minor oil products from refineries - by type

Thousand metric tons

Country or area	Year	Refinery gas	Ethane	Naphtha	Lubricants	Bitumen	Paraffin waxes	Petroleum coke	White spirit and SBP	Other oil products
Dominican Republic	2011	14	..	..	..	..	..	..	..	..
	2012	13	..	..	..	..	..	..	..	..
	2013	14	..	..	..	..	..	..	..	..
	2014	14	..	..	..	..	..	..	..	..
El Salvador	2011	13	..	..	..	..	..	..	..	35
	2012	12	..	..	..	..	..	..	..	35
	2013	0	..	..	..	..	..	..	..	0
	2014	0	..	..	..	..	..	..	..	0
Guatemala	2011	..	..	..	6	42	..	..	..	..
	2012	..	..	..	..	39	..	..	..	..
	2013	..	..	..	..	27	..	..	..	..
	2014	..	..	..	..	34	..	..	..	..
Jamaica	2011	..	..	42	..	23	..	..	..	..
	2012	..	..	31	..	22	..	..	..	..
	2013	..	..	30	..	16	..	..	..	..
	2014	..	..	41	..	17	..	..	..	..
Mexico	2011	3,213	..	3,640	195	1,513	35	1,628	..	66
	2012	3,494	..	2,910	203	1,345	37	2,579	..	72
	2013	3,632	..	3,427	229	1,086	34	3,182	..	84
	2014	3,282	..	3,646	192	1,388	30	3,018	..	87
Neth. Antilles (former)	2011	188	..	298	243	877	..	73	..	54
	2012	..	..	..	..	..	..	..	..	..
	2013	..	..	..	..	..	..	..	..	..
	2014	..	..	..	..	..	..	..	..	..
Nicaragua	2011	10	..	..	..	12	..	..	21	..
	2012	8	..	..	..	6	..	..	12	..
	2013	12	..	..	..	6	..	..	13	..
	2014	12	..	..	..	7	..	..	15	..
Trinidad and Tobago	2011	234	..	40	..	40	..	..	..	972
	2012	221	..	0	..	31	..	..	..	940
	2013	180	..	68	..	38	..	..	..	363
	2014	179	..	0	..	42	..	..	..	381
United States	2011	33,982	424	9,364	8,858	21,998	386	56,784	1,620	8,259
	2012	15,029	287	8,818	8,147	21,064	376	57,004	1,919	11,800
	2013	15,575	145	10,241	8,470	19,365	386	57,680	1,793	11,606
	2014	15,369	105	8,740	8,595	19,251	329	58,142	1,800	12,031
America, South	**2011**	**7,646**	**..**	**7,477**	**1,151**	**4,148**	**28**	**13,316**	**571**	**8,307**
	2012	**8,372**	**..**	**7,677**	**1,174**	**4,197**	**29**	**14,102**	**542**	**8,874**
	2013	**9,566**	**..**	**6,050**	**1,248**	**4,378**	**28**	**14,577**	**391**	**8,203**
	2014	**8,718**	**..**	**7,034**	**1,234**	**4,844**	**27**	**12,105**	**640**	**8,664**
Argentina	2011	762	..	2,510	369	675	12	1,393	263	2,694
	2012	1,218	..	2,643	372	532	11	1,474	301	2,970
	2013	1,226	..	1,773	358	607	10	1,383	30	2,472
	2014	1,039	..	2,954	360	548	10	1,411	329	2,731
Bolivia (Plur. State of)	2011	85	..	..	21	2	0	..	15	120
	2012	100	..	..	22	2	1	..	13	443
	2013	101	..	..	24	2	1	..	13	167
	2014	77	..	..	24	2	1	..	13	276
Brazil	2011	4,151	..	4,594	534	2,465	..	4,281	266	2,960
	2012	4,121	..	4,660	538	2,572	..	5,075	205	2,867
	2013	4,779	..	3,872	610	2,653	..	5,483	321	2,748
	2014	4,953	..	3,674	604	3,249	..	5,413	272	2,850

Production of minor oil products from refineries - by type

Thousand metric tons

Country or area	Year	Refinery gas	Ethane	Naphtha	Lubricants	Bitumen	Paraffin waxes	Petroleum coke	White spirit and SBP	Other oil products
Chile	2011	252	..	59	..	..	..	331	..	652
	2012	226	..	60	..	..	..	418	..	707
	2013	227	..	111	..	..	..	359	..	723
	2014	211	..	79	..	..	..	298	..	792
Colombia	2011	843	..	314	..	..	..	..	..	*601
	2012	835	..	*314	..	..	..	..	..	*504
	2013	285	..	294	..	..	..	..	..	366
	2014	375	..	327	..	..	..	..	..	424
Ecuador	2011	..	..	..	0	370	..	..	18	256
	2012	..	..	..	3	395	..	..	21	348
	2013	..	..	..	0	391	..	..	24	312
	2014	..	..	..	0	365	..	..	23	376
Peru	2011	177	..	..	..	..	..	..	..	337
	2012	169	..	..	..	..	..	..	..	309
	2013	155	..	..	..	..	..	..	..	679
	2014	162	..	..	..	..	..	..	..	508
Suriname	2011	..	..	..	13	..	..	..	..	..
	2012	..	..	..	13	..	..	..	..	..
	2013	..	..	..	13	..	..	..	..	..
	2014	..	..	..	13	..	..	..	..	..
Uruguay	2011	57	..	..	..	19	..	20	9	..
	2012	74	..	..	..	44	..	26	2	..
	2013	58	..	..	..	64	..	28	3	..
	2014	52	..	..	..	45	..	28	3	..
Venezuela (Bolivar. Rep.)	2011	1,319	..	..	214	617	16	7,291	..	687
	2012	1,629	..	..	226	652	17	7,109	..	726
	2013	2,735	..	..	243	661	17	7,324	..	736
	2014	1,849	..	..	233	635	16	4,955	..	707
Asia	**2011**	**54,694**	**34**	**137,477**	**14,330**	**41,675**	**1,607**	**28,028**	**2,904**	**105,589**
	2012	**54,457**	**39**	**138,734**	**14,447**	**44,234**	**1,758**	**31,673**	**3,064**	**111,857**
	2013	**51,869**	**32**	**144,704**	**13,322**	**46,641**	**1,679**	**32,567**	**2,491**	**114,978**
	2014	**52,206**	**5**	**148,774**	**13,035**	**48,926**	**1,674**	**34,397**	**2,193**	**116,483**
Azerbaijan	2011	245	..	219	91	266	..	269	..	206
	2012	263	..	149	63	288	..	231	..	269
	2013	276	..	125	47	313	..	223	..	238
	2014	246	..	235	54	242	..	252	..	205
Bahrain	2011	157	..	1,677	*64	355	..	..	..	387
	2012	165	..	1,816	*287	321	..	..	..	701
	2013	129	..	1,869	*39	427	..	..	..	431
	2014	125	..	1,913	*41	609	..	..	..	747
Bangladesh	2011	14	..	133	13	55	..	..	8	22
	2012	13	..	113	13	65	..	..	7	26
	2013	14	..	139	11	70	..	..	10	25
	2014	12	..	132	18	57	..	..	7	24
Brunei Darussalam	2011	38	..	..	..	..	..	..	..	..
	2012	38	..	..	..	..	..	..	..	..
	2013	28	..	..	..	..	..	..	..	..
	2014	14	..	..	..	..	..	..	..	..
China	2011	15,482	..	34,266	2,479	14,114	1,333	16,880	1,730	79,500
	2012	15,674	..	35,661	2,224	16,187	1,468	17,675	1,614	79,099
	2013	16,253	..	40,492	1,137	17,938	1,410	17,393	1,228	81,587
	2014	17,580	..	43,892	1,124	19,532	1,432	18,318	824	81,467

Table 23

Production of minor oil products from refineries - by type

Thousand metric tons

Country or area	Year	Refinery gas	Ethane	Naphtha	Lubricants	Bitumen	Paraffin waxes	Petroleum coke	White spirit and SBP	Other oil products
India	2011	15,422	..	17,135	1,028	4,638	47	7,837	..	13,740
	2012	16,560	..	17,354	896	4,658	52	10,943	..	17,003
	2013	11,125	..	17,038	941	4,785	56	12,068	..	17,284
	2014	*10,774	..	17,391	946	4,632	53	12,448	..	18,137
Indonesia	2011	661	..	3,423	432	323	*15	*576	*20	*964
	2012	636	..	2,795	421	337	*18	*584	*22	*916
	2013	644	..	3,174	377	242	15	301	62	1,524
	2014	759	..	1,840	354	182	0	634	14	3,314
Iran (Islamic Rep. of)	2011	1,439	34	3,290	1,595	3,630	..	..	45	208
	2012	1,663	39	3,067	1,538	4,116	..	..	57	477
	2013	1,806	32	3,532	1,581	3,896	..	..	110	418
	2014	1,592	5	3,413	1,613	4,713	..	..	249	248
Iraq	2011	205	..	413	118	514	27	..	..	..
	2012	205	..	468	80	681	27	..	..	..
	2013	196	..	471	91	852	53	..	..	..
	2014	160	..	403	35	708	6	..	..	..
Israel	2011	..	..	695	..	262	..	..	..	421
	2012	..	..	484	..	226	..	..	..	822
	2013	..	..	795	..	221	..	..	..	347
	2014	..	..	819	..	230	..	..	..	307
Japan	2011	7,264	..	13,822	2,210	3,939	70	1,203	76	2,228
	2012	7,239	..	13,916	2,084	3,960	73	919	79	2,214
	2013	8,267	..	15,023	2,189	3,659	66	1,021	73	2,211
	2014	7,796	..	13,404	2,159	3,254	67	1,177	72	2,199
Jordan	2011	33	..	..	*16	107	..	..	..	4
	2012	32	..	..	*16	96	..	..	..	4
	2013	33	..	..	25	101	..	..	..	4
	2014	32	..	..	*25	160	..	..	..	3
Kazakhstan	2011	134	..	..	..	257	..	212	..	..
	2012	150	..	..	..	287	..	198	..	..
	2013	148	..	..	..	332	..	262	..	..
	2014	111	..	..	..	602	..	278	..	..
Korea, Republic of	2011	2,686	..	22,953	2,431	3,022	15	277	496	2,729
	2012	2,689	..	24,315	2,637	3,109	13	290	489	2,831
	2013	3,803	..	24,429	2,424	3,614	15	305	410	3,014
	2014	4,077	..	25,658	2,867	4,027	16	327	366	3,210
Kuwait	2011	393	..	8,123	..	139	..	..	..	886
	2012	423	..	6,043	..	209	..	..	..	1,575
	2013	402	..	8,219	..	100	..	..	..	590
	2014	391	..	8,393	..	91	..	..	..	574
Malaysia	2011	1,633	..	3,898	181	119	39	16	..	..
	2012	194	..	3,617	172	113	37	15	..	..
	2013	192	..	1,747	947	154	13	181	..	..
	2014	189	..	1,787	633	128	61	173	..	..
Myanmar	2011	30	..	..	..	..	1	13	..	5
	2012	27	..	..	..	..	1	13	..	5
	2013	28	..	..	..	..	1	14	..	5
	2014	30	..	..	..	..	1	12	..	5
Oman	2011	232	..	542	52	..	..	..	..	285
	2012	198	..	585	53	..	..	..	..	296
	2013	174	..	366	47	..	..	..	..	212
	2014	224	..	478	61	..	..	..	..	242

Table 23

Production of minor oil products from refineries - by type

Thousand metric tons

Country or area	Year	Refinery gas	Ethane	Naphtha	Lubricants	Bitumen	Paraffin waxes	Petroleum coke	White spirit and SBP	Other oil products
Other Asia	2011	1,120	..	7,085	219	649	..	598	75	862
	2012	1,111	..	8,063	220	648	..	641	75	704
	2013	1,204	..	7,639	165	621	..	649	70	591
	2014	1,199	..	7,304	103	636	..	637	79	506
Pakistan	2011	132	..	724	200	214	11	..	..	26
	2012	159	..	736	204	155	*11	..	..	*26
	2013	168	..	878	202	160	*12	..	..	*29
	2014	176	..	920	212	168	*12	..	..	*31
Philippines	2011	..	..	158	..	..	..	..	..	493
	2012	..	..	135	..	..	..	..	..	407
	2013	..	..	238	..	..	..	..	..	290
	2014	..	..	243	..	..	..	..	..	400
Qatar	2011	62	..	4,006	..	..	..	..	..	221
	2012	65	..	3,923	..	..	..	..	..	1,264
	2013	60	..	3,726	..	..	..	..	..	781
	2014	58	..	3,556	..	..	..	..	..	759
Saudi Arabia	2011	2,520	..	7,308	..	2,666	..	..	..	358
	2012	2,351	..	7,551	..	2,146	..	..	..	663
	2013	2,536	..	6,900	..	2,448	..	..	..	674
	2014	2,570	..	8,267	..	2,657	..	..	..	558
Singapore	2011	3,338	..	1,604	2,649	1,891	30	123	236	599
	2012	3,178	..	2,164	3,137	2,141	32	143	237	870
	2013	3,005	..	1,853	2,829	2,250	25	131	218	559
	2014	2,790	..	2,168	2,561	2,882	25	122	205	813
Sri Lanka	2011	16	..	80	..	46	..	..	4	..
	2012	13	..	71	..	27	..	..	4	..
	2013	13	..	87	..	5	..	..	3	..
	2014	15	..	117	..	0	..	..	3	..
Syrian Arab Republic	2011	153	..	948	42	526	..	..	..	26
	2012	109	..	673	30	373	..	..	..	18
	2013	71	..	445	20	247	..	..	..	13
	2014	71	..	445	20	247	..	..	..	13
Tajikistan	2011	..	..	*9	..	4	..	..	..	..
	2012	..	..	*10	..	2	..	..	..	..
	2013	..	..	..	..	4	..	..	..	..
	2014	..	..	..	..	3	..	..	..	..
Thailand	2011	..	..	..	..	1,136	..	..	..	0
	2012	..	..	..	..	1,207	..	..	..	0
	2013	..	..	..	..	1,198	..	..	..	2,690
	2014	..	..	..	..	1,177	..	..	..	1,537
Turkey	2011	847	..	302	392	2,714	18	..	207	204
	2012	872	..	308	266	2,810	26	..	474	492
	2013	854	..	148	152	2,925	12	..	301	274
	2014	791	..	304	116	1,917	..	..	368	57
Turkmenistan	2011	112	..	..	..	..	..	..	..	350
	2012	117	..	..	..	..	..	..	..	320
	2013	126	..	..	..	..	..	..	..	360
	2014	127	..	..	..	..	..	..	..	370
United Arab Emirates	2011	213	..	*4,439	..	..	..	..	..	658
	2012	211	..	*4,391	..	..	..	..	..	670
	2013	217	..	*4,640	..	..	..	..	..	652
	2014	206	..	5,041	..	..	..	..	..	593

Table 23

Production of minor oil products from refineries - by type

Thousand metric tons

Country or area	Year	Refinery gas	Ethane	Naphtha	Lubricants	Bitumen	Paraffin waxes	Petroleum coke	White spirit and SBP	Other oil products
Uzbekistan	2011	114	..	..	118	65	1	24	7	207
	2012	102	..	..	105	58	1	21	6	185
	2013	97	..	..	99	55	1	20	6	175
	2014	91	..	..	93	52	1	19	6	165
Yemen	2011	..	..	225	..	24	..	..	..	..
	2012	..	..	327	..	13	..	..	..	..
	2013	..	..	732	..	24	..	..	..	..
	2014	..	..	651	..	21	..	..	..	..
Europe	**2011**	**35,905**	**41**	**58,236**	**11,226**	**26,703**	**1,209**	**12,084**	**2,649**	**23,826**
	2012	**35,366**	**23**	**58,237**	**10,570**	**24,227**	**1,012**	**13,108**	**2,627**	**24,461**
	2013	**34,570**	**35**	**58,692**	**10,424**	**24,389**	**1,122**	**13,996**	**2,504**	**21,497**
	2014	**35,830**	**34**	**62,750**	**11,310**	**23,898**	**1,198**	**14,548**	**3,128**	**22,176**
Albania	2011	10	..	19	..	26	..	62	..	86
	2012	5	..	4	..	25	..	15	..	42
	2013	0	..	1	..	29	..	0	..	10
	2014	12	..	17	..	62	..	76	..	78
Austria	2011	381	..	844	72	376	..	62	65	23
	2012	311	..	1,009	2	366	..	70	0	32
	2013	311	..	1,011	2	314	..	73	0	57
	2014	342	..	976	0	314	..	55	0	43
Belarus	2011	614	..	..	..	479	..	..	..	1,376
	2012	693	..	..	..	563	..	..	..	889
	2013	725	..	..	..	548	..	..	..	591
	2014	769	..	..	..	576	..	..	..	1,639
Belgium	2011	762	..	1,840	0	1,187	..	242	..	5,679
	2012	820	..	1,890	8	1,100	..	330	..	5,991
	2013	733	..	2,314	4	949	..	311	..	5,508
	2014	819	..	2,569	..	1,099	..	330	..	4,594
Bosnia and Herzegovina	2011	..	..	24	174	153	..	..	..	89
	2012	..	..	19	136	145	..	..	..	37
	2013	..	..	5	78	136	..	..	..	21
	2014	..	..	7	57	153	..	..	..	*22
Bulgaria	2011	187	..	276	..	169	..	81	2	33
	2012	198	..	303	..	188	..	78	3	14
	2013	222	..	275	..	107	..	89	6	1
	2014	209	..	249	..	125	..	80	4	4
Croatia	2011	267	..	90	9	50	..	65	..	46
	2012	294	..	59	12	26	..	85	..	55
	2013	175	..	30	14	36	..	60	..	40
	2014	276	..	31	13	3	..	35	..	9
Czechia	2011	145	..	773	159	459	13	..	1	1,002
	2012	158	..	794	90	477	11	..	0	1,015
	2013	134	..	737	100	477	9	..	0	925
	2014	151	..	847	88	521	10	..	..	990
Denmark	2011	288	..	0	..	..	..	..	..	..
	2012	301	..	0	..	..	..	..	..	..
	2013	297	..	0	..	..	..	..	..	..
	2014	299	..	0	..	..	..	..	..	..
Finland	2011	442	..	281	295	261	..	137	193	134
	2012	428	..	250	284	190	..	126	152	135
	2013	410	..	284	316	261	..	139	161	142
	2014	373	..	294	327	339	..	139	180	124

Production of minor oil products from refineries - by type

Thousand metric tons

Country or area	Year	Refinery gas	Ethane	Naphtha	Lubricants	Bitumen	Paraffin waxes	Petroleum coke	White spirit and SBP	Other oil products
France	2011	2,116	24	5,259	1,526	2,516	206	717	106	1,210
	2012	1,538	7	4,362	983	1,891	130	635	99	1,077
	2013	1,358	18	4,532	1,067	1,930	160	503	29	820
	2014	1,319	11	4,694	1,036	1,763	154	526	17	915
Germany	2011	3,884	..	7,750	2,431	3,869	127	1,763	26	1,626
	2012	3,795	..	8,104	2,469	3,595	139	1,743	22	1,646
	2013	3,763	..	7,812	2,419	3,410	137	1,858	10	1,286
	2014	3,646	..	7,851	2,746	3,461	154	1,764	9	1,271
Greece	2011	506	..	597	232	321	..	188	..	333
	2012	590	..	1,266	221	253	..	264	2	1,354
	2013	532	..	913	237	436	..	435	..	702
	2014	620	..	1,327	234	540	..	510	..	1,100
Hungary	2011	199	..	1,080	118	591	34	355	21	680
	2012	179	..	1,004	130	410	33	340	85	643
	2013	177	..	904	110	402	37	353	91	641
	2014	185	..	902	97	467	42	407	90	716
Ireland	2011	54	..	17	..	..	..	..	..	..
	2012	60	..	12	..	..	..	..	..	..
	2013	58	..	27	..	..	..	..	..	..
	2014	57	..	16	..	..	..	..	..	..
Italy	2011	3,077	..	3,803	1,283	3,305	86	1,733	5	3,094
	2012	2,729	..	3,122	1,269	2,604	81	1,445	5	3,000
	2013	2,659	..	2,729	1,175	3,104	95	1,342	4	2,179
	2014	2,262	..	2,705	1,216	2,514	86	758	4	1,919
Lithuania	2011	295	..	..	40	132	..	119	..	77
	2012	294	..	..	43	110	..	105	..	73
	2013	327	..	..	43	126	..	114	..	81
	2014	290	..	..	45	91	..	108	..	77
Netherlands	2011	1,949	..	8,216	634	247	345	370	111	914
	2012	2,066	..	7,720	583	315	276	395	103	797
	2013	1,885	..	7,644	489	452	329	343	51	868
	2014	2,052	..	7,144	554	516	422	414	29	880
Norway	2011	365	..	1,500	..	..	..	483	..	..
	2012	363	..	1,914	..	..	..	155	..	..
	2013	438	..	1,837	..	..	..	212	..	..
	2014	431	..	1,572	..	..	..	191	..	..
Poland	2011	955	..	1,332	408	1,788	76	..	176	852
	2012	1,032	..	1,542	455	1,550	78	..	168	944
	2013	922	..	1,491	465	1,314	69	..	97	1,008
	2014	603	..	1,557	541	1,229	75	..	123	1,075
Portugal	2011	15	..	961	135	202	15	..	14	383
	2012	19	..	1,218	135	163	14	..	17	499
	2013	24	..	960	102	115	8	..	13	1,127
	2014	19	..	1,180	66	117	9	..	17	793
Republic of Moldova	2011	..	..	..	5	..	..	..	..	0
	2012	..	..	..	2	..	..	..	..	1
	2013	..	..	..	1	..	..	..	..	1
	2014	..	..	..	0	..	..	..	..	1
Romania	2011	783	..	269	..	54	..	749	86	271
	2012	700	..	263	..	45	..	695	106	118
	2013	712	..	221	..	46	..	672	79	131
	2014	749	..	280	..	61	..	750	105	292

Table 23

Production of minor oil products from refineries - by type

Thousand metric tons

Country or area	Year	Refinery gas	Ethane	Naphtha	Lubricants	Bitumen	Paraffin waxes	Petroleum coke	White spirit and SBP	Other oil products
Russian Federation	2011	12,224	..	19,366	2,630	5,422	253	1,052	1,572	..
	2012	12,661	..	19,978	2,582	6,054	216	1,268	1,614	..
	2013	12,849	..	21,477	2,628	6,193	242	2,105	1,677	..
	2014	14,323	..	25,202	3,089	5,751	191	2,736	2,181	..
Serbia	2011	48	..	150	7	179	..	28	..	122
	2012	37	..	127	8	172	..	25	..	170
	2013	47	..	439	11	240	..	18	..	172
	2014	50	..	351	7	262	..	13	..	158
Slovakia	2011	422	17	531	..	11	..	57	..	328
	2012	321	16	371	..	5	..	55	..	243
	2013	452	17	470	..	0	..	56	..	281
	2014	341	23	385	..	..	..	40	..	262
Spain	2011	1,926	..	464	194	2,434	41	1,543	206	2,651
	2012	2,059	..	318	165	1,878	34	3,086	179	3,134
	2013	1,879	..	345	191	2,074	36	3,435	180	2,471
	2014	1,956	..	304	337	1,917	55	3,744	204	2,455
Sweden	2011	530	..	248	342	672	..	45	..	1,276
	2012	623	..	240	435	702	..	65	..	1,250
	2013	499	..	201	436	790	..	52	..	1,282
	2014	748	..	..	403	895	..	67	..	1,706
Switzerland	2011	217	..	20	..	..	..	53	..	36
	2012	168	..	20	..	..	..	56	..	17
	2013	227	..	20	..	..	..	53	..	38
	2014	225	..	..	..	..	..	60	..	34
Ukraine	2011	438	..	..	102	324	13	..	..	93
	2012	269	..	..	101	178	..	..	..	20
	2013	288	..	..	159	123	..	..	..	155
	2014	388	..	..	81	116	..	..	..	236
United Kingdom	2011	2,806	..	2,526	430	1,476	..	2,180	65	1,412
	2012	2,655	..	2,328	457	1,222	..	2,072	72	1,265
	2013	2,467	..	2,013	377	777	..	1,773	106	959
	2014	2,316	..	2,290	373	1,006	..	1,745	165	783
Oceania	**2011**	**1,513**	**..**	**384**	**64**	**713**	**1**	**587**	**72**	**150**
	2012	**1,408**	**..**	**416**	**76**	**592**	**0**	**520**	**70**	**115**
	2013	**1,331**	**..**	**474**	**80**	**350**	**0**	**545**	**77**	**235**
	2014	**1,191**	**..**	**506**	**79**	**270**	**0**	**533**	**65**	**171**
Australia	2011	1,219	..	157	64	597	1	587	72	83
	2012	1,120	..	160	76	475	0	520	70	47
	2013	1,040	..	99	80	253	0	545	77	181
	2014	912	..	124	79	156	0	533	65	101
New Zealand	2011	294	..	..	..	116	..	..	..	67
	2012	288	..	..	..	117	..	..	..	68
	2013	291	..	..	..	97	..	..	..	54
	2014	279	..	..	..	114	..	..	..	70
Papua New Guinea	2011	..	..	227	..	..	..	..	..	..
	2012	..	..	256	..	..	..	..	..	..
	2013	..	..	375	..	..	..	..	..	..
	2014	..	..	*382	..	..	..	..	..	..

Production of oil products from refineries – by type

Thousand metric tons

Table Notes

Production comprises all refinery output of all oil products.

The **Jet fuels** column comprises aviation gasoline, gasoline-type jet fuel and kerosene-type jet fuel.

The **Other products** column includes bitumen, ethane, feedstocks, lubricants, petroleum coke, paraffin waxes, white spirit, and other oil products.

Aruba produces only feedstocks to be further refined abroad (but the refinery ceased operation in 2012).

Please refer to the Definitions Section on pages ix to xv for the appropriate product description /classification.

Figure 64: Production of selected oil products from refineries, 1990-2014

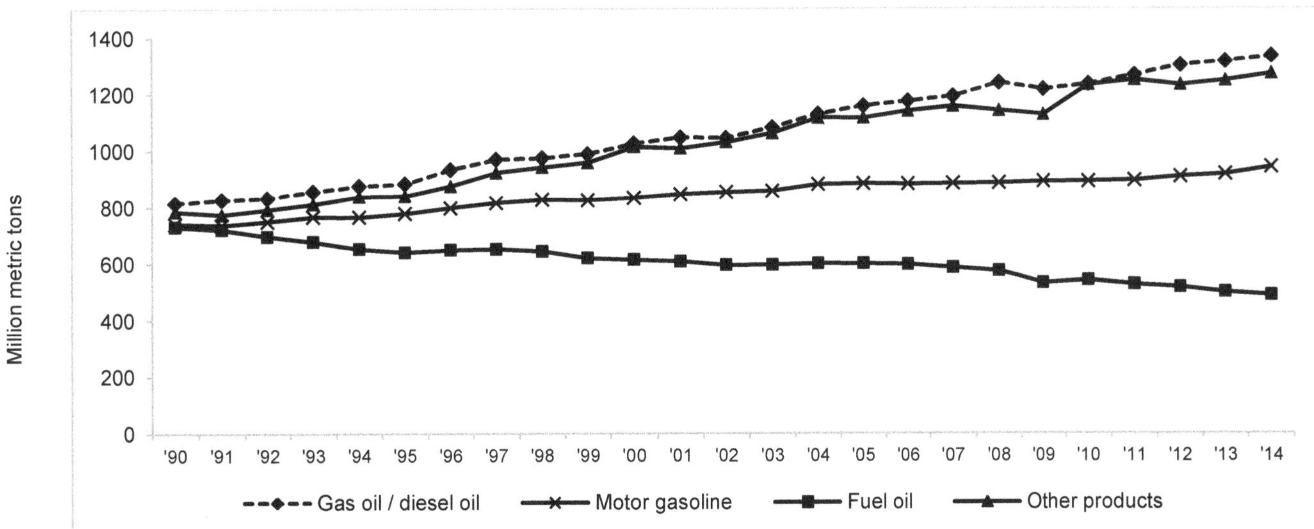

Table 24

Production of oil products from refineries - by type

Thousand metric tons

Country or area	Year	Total	Gas/ diesel oil	Motor gasoline	Fuel oil	Naphtha	Jet fuels	Refinery gas	LPG	Other kerosene	Other products
World	**2011**	**3,934,121**	**1,265,746**	**894,130**	**525,293**	**231,821**	**239,575**	**144,337**	**115,935**	**84,102**	**433,181**
	2012	**3,955,474**	**1,300,943**	**907,943**	**515,603**	**232,565**	**241,938**	**124,917**	**106,456**	**86,547**	**438,562**
	2013	**3,975,553**	**1,314,344**	**916,296**	**498,464**	**239,020**	**247,461**	**122,979**	**111,888**	**86,623**	**438,478**
	2014	**4,032,370**	**1,332,904**	**941,234**	**487,609**	**248,368**	**278,366**	**122,217**	**116,562**	**60,278**	**444,833**
Africa	**2011**	**107,498**	**31,654**	**17,203**	**27,867**	**11,250**	**7,712**	**2,198**	**2,138**	**2,006**	**5,471**
	2012	**108,356**	**32,581**	**17,096**	**28,212**	**11,578**	**7,944**	**2,296**	**2,162**	**1,814**	**4,673**
	2013	**108,825**	**32,371**	**16,565**	**27,644**	**11,650**	**8,994**	**2,091**	**2,328**	**1,805**	**5,378**
	2014	**105,591**	**32,980**	**15,670**	**25,463**	**12,829**	**7,940**	**1,734**	**2,175**	**1,707**	**5,094**
Algeria	2011	25,320	7,573	2,568	5,848	6,939	1,209	304	517	0	362
	2012	23,007	6,867	2,273	5,115	6,432	1,204	273	470	0	373
	2013	23,975	7,042	2,300	5,358	6,336	1,713	293	597	0	336
	2014	31,037	9,027	3,007	6,594	8,721	2,001	363	804	0	520
Angola	2011	1,906	518	63	727	166	268	59	30	73	2
	2012	1,963	527	73	737	156	303	62	24	76	5
	2013	2,187	564	69	815	205	339	68	35	89	3
	2014	2,235	532	27	921	190	393	69	31	69	3
Cameroon	2011	2,142	770	399	343	..	346	79	21	..	184
	2012	2,240	805	417	359	..	362	83	22	..	192
	2013	2,365	850	440	379	..	382	88	23	..	203
	2014	2,506	901	466	402	..	405	93	24	..	215
Congo	2011	573	195	94	197	7	39	..	7	22	12
	2012	711	243	80	272	10	67	..	7	20	12
	2013	775	276	70	321	11	56	..	8	21	12
	2014	791	270	78	332	10	60	..	8	20	13
Côte d'Ivoire	2011	2,249	712	390	328	..	579	19	12	45	164
	2012	3,224	1,057	559	543	..	792	27	1	59	186
	2013	3,288	1,072	647	467	..	831	26	3	85	157
	2014	3,112	1,192	548	396	..	683	25	15	54	199
Egypt	2011	28,502	7,497	4,476	8,902	2,843	1,519	121	505	285	2,354
	2012	27,790	7,116	4,201	9,180	2,706	1,505	133	515	296	2,138
	2013	24,010	5,842	2,839	*7,858	2,634	1,629	*133	560	86	2,429
	2014	22,778	5,996	2,830	7,736	2,280	1,269	112	434	249	*1,872
Gabon	2011	1,028	264	79	536	..	85	..	12	38	14
	2012	770	212	76	386	..	49	..	14	22	11
	2013	837	229	62	429	..	63	..	14	28	12
	2014	814	223	68	411	..	60	..	13	27	12
Ghana	2011	1,000	310	344	91	..	116	42	45	53	..
	2012	474	122	158	79	..	48	20	27	21	..
	2013	424	113	167	44	..	60	0	26	15	..
	2014	129	28	40	44	..	9	0	3	5	..
Kenya	2011	1,689	429	178	520	..	275	68	28	118	73
	2012	968	239	117	300	..	157	38	17	67	33
	2013	597	155	76	184	..	106	31	12	33	0
	2014	0	0	0	0	..	0	0	..	0	0
Libya	2011	5,820	1,522	569	2,056	649	648	120	81	139	36
	2012	8,362	2,251	615	2,895	1,348	736	172	93	200	52
	2013	10,479	2,775	640	3,973	1,533	908	216	118	251	65
	2014	4,797	1,278	516	1,655	682	371	99	51	115	30
Morocco	2011	6,979	2,403	415	2,441	448	553	*262	38	..	419
	2012	7,049	2,481	407	2,118	493	877	*280	115	..	278
	2013	7,080	2,240	369	2,185	482	1,060	185	96	..	463
	2014	6,581	2,275	331	1,656	494	906	174	99	..	646

Production of oil products from refineries - by type

Thousand metric tons

Country or area	Year	Total	Gas/ diesel oil	Motor gasoline	Fuel oil	Naphtha	Jet fuels	Refinery gas	LPG	Other kerosene	Other products
Niger	2011	57	36	18	..	..	..	..	3	..	..
	2012	560	328	195	..	..	..	..	37	..	..
	2013	784	481	253	..	..	..	..	50	..	..
	2014	744	452	235	..	..	..	..	57	..	..
Nigeria	2011	5,868	1,085	1,277	1,691	..	77	111	129	674	824
	2012	4,673	1,030	1,135	1,287	..	62	125	104	544	386
	2013	4,767	664	1,237	1,222	..	76	125	195	671	577
	2014	3,253	647	542	959	..	47	84	72	629	273
Senegal	2011	714	329	99	175	31	63	12	2	3	..
	2012	768	412	75	205	53	8	13	1	1	..
	2013	764	451	29	171	85	12	15	1	0	..
	2014	802	472	30	179	89	13	16	1	2	..
South Africa	2011	18,070	5,704	5,005	3,060	..	1,772	967	375	463	724
	2012	19,547	6,222	5,527	3,644	..	1,623	1,037	354	430	710
	2013	20,531	7,050	6,273	3,355	..	1,649	878	262	424	640
	2014	19,709	7,000	5,771	3,124	..	1,568	664	215	446	921
Sudan	2011	..	..	..	..	..	..	..	..	..	..
	2012	3,905	1,666	1,099	398	8	118	20	320	8	268
	2013	3,734	1,594	1,009	253	18	79	20	289	19	453
	2014	4,003	1,686	1,085	403	17	120	21	310	3	358
Sudan (former)	2011	4,326	1,752	1,150	629	18	134	22	312	24	285
	2012	..	..	..	..	..	..	..	..	..	..
	2013	..	..	..	..	..	..	..	..	..	..
	2014	..	..	..	..	..	..	..	..	..	..
Tunisia	2011	712	254	..	251	149	..	0	12	40	6
	2012	1,728	661	..	612	372	..	0	31	37	15
	2013	1,640	646	..	551	346	..	0	30	52	15
	2014	1,641	634	..	563	346	..	0	27	53	18
Zambia	2011	543	302	78	72	..	29	12	9	29	12
	2012	616	343	89	82	..	33	13	10	33	13
	2013	589	328	85	79	..	31	13	9	31	13
	2014	659	367	95	88	..	35	14	11	35	14
America, North	**2011**	**1,016,030**	**276,414**	**388,502**	**63,687**	**16,997**	**77,050**	**42,381**	**23,028**	**1,273**	**126,698**
	2012	**975,061**	**282,230**	**382,032**	**60,460**	**15,923**	**77,057**	**23,018**	**12,847**	**992**	**120,501**
	2013	**987,625**	**290,715**	**387,018**	**58,937**	**17,449**	**78,646**	**23,553**	**12,823**	**1,172**	**117,312**
	2014	**999,773**	**298,173**	**394,028**	**54,800**	**16,475**	**80,481**	**22,538**	**13,599**	**1,577**	**118,102**
Aruba	2011	*10,050	..	..	..	..	..	..	..	..	*10,050
	2012	*2,440	..	..	..	..	..	..	..	..	*2,440
	2013	0	..	..	..	..	..	..	..	..	0
	2014	0	..	..	..	..	..	..	..	..	0
Canada	2011	91,300	30,176	28,587	6,630	3,366	3,286	4,693	1,727	489	12,346
	2012	92,633	31,404	29,390	7,292	3,505	3,805	4,029	1,534	495	11,179
	2013	91,037	30,478	28,795	6,751	3,062	3,995	3,932	1,437	604	11,983
	2014	89,357	30,052	28,654	6,175	3,308	4,168	3,447	1,295	588	11,670
Costa Rica	2011	150	52	31	61	..	0	..	1	2	2
	2012	0	..	..	0	..	..	..	..	..	0
	2013	0	..	..	0	..	..	..	..	..	0
	2014	..	..	..	..	..	..	..	..	..	..
Cuba	2011	4,871	1,221	503	2,322	247	354	34	50	4	136
	2012	5,107	1,319	415	2,520	349	284	18	42	31	129
	2013	5,138	1,256	672	2,352	284	293	30	74	29	148
	2014	4,877	1,124	554	2,315	367	276	27	67	25	123

Table 24

Production of oil products from refineries - by type

Thousand metric tons

Country or area	Year	Total	Gas/ diesel oil	Motor gasoline	Fuel oil	Naphtha	Jet fuels	Refinery gas	LPG	Other kerosene	Other products
Curaçao	2011	..	..	..	..	..	..	..	..	..	..
	2012	8,619	1,886	1,574	3,101	310	668	194	51	..	835
	2013	9,044	1,712	1,712	3,792	337	609	177	47	..	658
	2014	9,574	1,991	1,890	3,505	373	707	208	55	..	845
Dominican Republic	2011	1,243	350	242	373	..	234	14	21	10	..
	2012	1,163	345	231	333	..	219	13	14	9	..
	2013	1,295	373	265	392	..	228	14	13	10	..
	2014	1,278	372	267	367	..	232	14	16	10	..
El Salvador	2011	743	156	92	379	..	54	13	12	2	35
	2012	575	107	72	302	..	41	12	6	0	35
	2013	0	0	0	0	..	0	0	..	0	0
	2014	0	0	0	0	..	0	0	..	0	0
Guatemala	2011	81	32	..	..	..	..	..	..	1	48
	2012	71	31	..	..	..	..	..	..	1	39
	2013	59	31	..	..	..	..	..	..	0	27
	2014	72	38	..	..	..	..	..	..	1	34
Haiti	2011	..	..	..	..	..	..	..	..	..	..
	2012	8	..	8	..	..	..	..	..	..	..
	2013	..	..	..	..	..	..	..	..	..	..
	2014	..	..	..	..	..	..	..	..	..	..
Jamaica	2011	1,259	216	144	729	42	93	..	11	2	23
	2012	1,218	225	133	695	31	101	..	9	1	22
	2013	1,171	193	124	702	30	98	..	6	1	16
	2014	963	147	103	566	41	82	..	6	2	17
Martinique	2011	*845	*173	*147	*389	..	*116	..	*21	..	..
	2012	*858	*177	*149	*397	..	*114	..	*21	..	..
	2013	*867	*179	*149	*406	..	*112	..	*21	..	..
	2014	*904	*185	*156	*430	..	*110	..	*22	..	..
Mexico	2011	65,508	17,768	16,725	17,486	3,640	2,568	3,213	671	..	3,437
	2012	65,987	18,854	17,517	15,594	2,910	2,590	3,494	792	..	4,236
	2013	69,691	20,906	18,262	15,287	3,427	2,774	3,632	788	..	4,615
	2014	66,719	19,489	17,581	14,742	3,646	2,436	3,282	828	..	4,715
Neth. Antilles (former)	2011	8,718	1,810	1,511	2,974	298	641	188	49	..	1,247
	2012	..	..	..	..	..	..	..	..	..	..
	2013	..	..	..	..	..	..	..	..	..	..
	2014	..	..	..	..	..	..	..	..	..	..
Nicaragua	2011	780	233	97	355	..	30	10	17	5	33
	2012	518	156	68	234	..	20	8	9	5	18
	2013	682	209	97	301	..	28	12	11	5	19
	2014	680	213	96	292	..	29	12	10	6	22
Trinidad and Tobago	2011	6,860	1,381	1,007	2,461	40	655	234	40	30	1,012
	2012	5,415	921	567	2,297	0	397	221	12	29	971
	2013	6,613	1,269	1,079	2,897	68	605	180	82	32	401
	2014	5,275	831	733	2,588	0	476	179	12	33	423
United States	2011	823,622	222,846	339,417	29,529	9,364	69,019	33,982	20,408	728	98,329
	2012	790,449	226,806	331,908	27,696	8,818	68,819	15,029	10,356	420	100,597
	2013	802,029	234,109	335,864	26,057	10,241	69,903	15,575	10,344	491	99,445
	2014	820,074	243,731	343,993	23,820	8,740	71,966	15,369	11,289	913	100,253
America, South	**2011**	**227,808**	**77,911**	**42,449**	**45,359**	**7,477**	**12,146**	**7,646**	**6,857**	**444**	**27,520**
	2012	**233,893**	**77,135**	**46,561**	**44,830**	**7,677**	**12,321**	**8,372**	**7,578**	**501**	**28,917**
	2013	**230,965**	**76,134**	**47,170**	**44,239**	**6,050**	**11,484**	**9,566**	**6,949**	**549**	**28,824**
	2014	**239,249**	**80,358**	**49,742**	**46,107**	**7,034**	**12,331**	**8,718**	**6,935**	**511**	**27,513**

Production of oil products from refineries - by type

Thousand metric tons

Country or area	Year	Total	Gas/ diesel oil	Motor gasoline	Fuel oil	Naphtha	Jet fuels	Refinery gas	LPG	Other kerosene	Other products
Argentina	2011	29,192	10,024	5,043	3,351	2,510	1,276	762	784	36	5,406
	2012	31,128	9,931	5,362	3,953	2,643	1,344	1,218	997	20	5,660
	2013	28,044	7,918	5,657	4,309	1,773	1,337	1,226	*943	21	4,860
	2014	31,297	9,311	5,244	4,617	2,954	1,413	1,039	1,314	16	5,389
Bolivia (Plur. State of)	2011	2,115	574	629	416	..	151	85	86	16	158
	2012	2,451	616	686	332	..	136	100	85	15	481
	2013	2,435	697	783	376	..	167	101	90	14	207
	2014	2,636	758	794	411	..	181	77	84	15	316
Brazil	2011	96,038	37,383	17,397	13,289	4,594	4,413	4,151	4,280	25	10,506
	2012	101,717	39,881	19,080	13,747	4,660	4,449	4,121	4,503	19	11,257
	2013	108,480	43,267	21,182	14,618	3,872	4,567	4,779	4,368	12	11,815
	2014	111,078	43,570	21,373	16,010	3,674	4,992	4,953	4,108	10	12,388
Chile	2011	8,892	3,168	2,194	1,238	59	570	252	350	78	983
	2012	9,022	2,863	2,246	1,192	60	617	226	593	100	1,125
	2013	9,378	3,076	2,619	1,252	111	613	227	257	141	1,082
	2014	9,618	3,176	2,789	1,185	79	629	211	316	143	1,090
Colombia	2011	14,898	4,867	*2,250	3,923	314	1,313	843	*591	196	*601
	2012	15,574	4,806	2,854	4,093	*314	1,339	835	*613	216	*504
	2013	14,910	4,578	3,158	4,333	294	1,108	285	552	236	366
	2014	16,596	5,422	3,671	4,521	327	1,084	375	523	249	424
Ecuador	2011	7,516	1,968	1,422	2,845	..	360	..	272	5	644
	2012	7,352	1,846	1,402	2,689	..	360	..	282	5	767
	2013	6,709	1,653	1,239	2,479	..	384	..	224	5	727
	2014	6,556	1,534	1,533	2,211	..	359	..	151	3	764
Peru	2011	9,480	3,939	1,763	2,414	..	597	177	187	66	337
	2012	8,818	3,835	1,806	1,716	..	749	169	185	49	309
	2013	8,945	3,607	1,899	1,694	..	692	155	173	46	679
	2014	9,065	3,683	2,041	1,735	..	748	162	188	0	508
Suriname	2011	911	55	..	843	..	..	..	..	..	13
	2012	931	53	..	865	..	..	..	..	..	13
	2013	1,000	58	..	929	..	..	..	..	..	13
	2014	1,035	61	..	961	..	..	..	..	..	13
Uruguay	2011	1,297	434	336	307	..	50	57	57	8	47
	2012	1,951	777	457	414	..	72	74	76	9	72
	2013	2,021	812	514	372	..	72	58	91	7	94
	2014	1,876	760	462	356	..	84	52	81	7	76
Venezuela (Bolivar. Rep.)	2011	57,469	15,499	11,414	16,733	..	3,415	1,319	250	14	8,825
	2012	54,949	12,527	12,668	15,828	..	3,255	1,629	244	68	8,730
	2013	49,043	10,468	10,119	13,877	..	2,545	2,735	251	67	8,981
	2014	49,491	12,082	11,835	14,100	..	2,841	1,849	170	68	6,546
Asia	**2011**	**1,602,648**	**526,056**	**262,128**	**219,475**	**137,477**	**88,712**	**54,694**	**50,243**	**69,695**	**194,168**
	2012	**1,663,277**	**555,700**	**279,129**	**214,906**	**138,734**	**92,489**	**54,457**	**50,044**	**70,748**	**207,070**
	2013	**1,691,425**	**565,512**	**288,687**	**208,119**	**144,704**	**96,065**	**51,869**	**54,223**	**70,537**	**211,710**
	2014	**1,720,585**	**566,679**	**307,833**	**200,737**	**148,774**	**125,663**	**52,206**	**57,578**	**44,402**	**216,714**
Azerbaijan	2011	6,200	2,483	1,296	254	219	621	245	250	0	832
	2012	6,056	2,369	1,297	286	149	626	263	215	..	851
	2013	6,375	2,495	1,407	285	125	703	276	263	..	821
	2014	6,536	2,946	1,238	207	235	709	246	202	..	753
Bahrain	2011	12,998	3,927	735	2,458	1,677	3,116	157	122	0	806
	2012	13,918	4,223	664	2,443	1,816	3,176	165	122	0	1,309
	2013	13,480	3,942	596	2,519	1,869	3,424	129	104	0	897
	2014	13,526	4,083	753	2,158	1,913	2,993	125	104	0	1,397

Table 24

Production of oil products from refineries - by type

Thousand metric tons

Country or area	Year	Total	Gas/ diesel oil	Motor gasoline	Fuel oil	Naphtha	Jet fuels	Refinery gas	LPG	Other kerosene	Other products
Bangladesh	2011	1,344	378	56	367	133	2	14	13	283	98
	2012	1,193	375	63	275	113	4	13	13	226	111
	2013	1,347	386	69	323	139	0	14	14	286	116
	2014	1,216	362	59	302	132	1	12	11	231	106
Brunei Darussalam	2011	573	164	168	98	..	102	38	..	3	..
	2012	418	114	146	66	..	47	38	..	7	..
	2013	485	138	167	80	..	65	28	..	7	..
	2014	381	113	118	72	..	58	14	..	6	..
China	2011	468,311	156,897	80,596	22,818	34,266	583	15,482	22,408	19,224	116,037
	2012	496,858	170,638	89,135	22,532	35,661	626	15,674	22,687	21,640	118,266
	2013	526,665	172,757	97,665	27,759	40,492	676	16,253	25,133	25,239	120,692
	2014	564,104	176,353	109,574	35,417	43,892	30,532	17,580	27,058	1,003	122,696
Georgia	2011	0	..	0	..	..	..	..	..	0	0
	2012	7	..	4	..	..	..	..	..	3	0
	2013	0	..	0	..	..	..	..	..	0	0
	2014	0	..	0	..	..	..	..	..	0	0
India	2011	214,526	82,864	27,186	19,456	17,135	10,051	15,422	7,333	7,789	27,290
	2012	229,362	91,085	30,118	15,054	17,354	10,077	16,560	7,694	7,868	33,552
	2013	227,580	94,165	30,275	13,405	17,038	11,210	11,125	7,890	7,338	35,134
	2014	231,555	94,428	32,325	11,919	17,391	11,103	*10,774	9,840	7,559	36,216
Indonesia	2011	45,130	15,842	9,326	8,833	3,423	2,153	661	705	1,857	*2,330
	2012	44,942	16,774	9,506	8,472	2,795	2,403	636	662	1,396	*2,298
	2013	47,125	19,532	8,418	8,501	3,174	2,538	644	564	1,233	2,521
	2014	44,593	16,246	8,891	8,369	1,840	2,274	759	759	957	4,498
Iran (Islamic Rep. of)	2011	88,666	30,076	12,786	26,689	3,290	1,140	1,439	3,356	4,378	5,512
	2012	91,584	29,814	14,695	27,701	3,067	1,165	1,663	2,646	4,606	6,227
	2013	93,062	31,032	16,358	26,879	3,532	1,192	1,806	2,490	3,738	6,036
	2014	87,684	29,548	16,356	23,441	3,413	1,252	1,592	2,047	3,207	6,828
Iraq	2011	27,213	6,364	2,739	13,985	413	308	205	282	2,258	659
	2012	28,654	6,262	3,094	15,187	468	133	205	312	2,205	788
	2013	29,343	6,477	3,089	15,656	471	144	196	265	2,049	996
	2014	23,850	5,099	2,413	13,268	403	215	160	145	1,398	749
Israel	2011	12,032	3,562	2,702	2,481	695	813	..	569	527	683
	2012	11,720	3,915	2,809	1,813	484	786	..	498	367	1,048
	2013	13,923	4,673	3,110	2,984	795	1,072	..	582	139	568
	2014	14,394	5,191	3,149	3,040	819	932	..	605	121	537
Japan	2011	169,743	46,080	40,139	23,126	13,822	10,037	7,264	3,934	15,615	9,726
	2012	169,173	45,381	39,142	24,969	13,916	10,404	7,239	4,014	14,779	9,329
	2013	172,618	48,557	40,183	20,393	15,023	12,063	8,267	4,509	14,404	9,219
	2014	162,815	45,670	39,365	18,119	13,404	12,054	7,796	4,242	13,237	8,928
Jordan	2011	3,218	1,030	681	868	..	338	33	84	58	127
	2012	3,537	1,109	716	999	..	366	32	102	96	116
	2013	3,146	980	663	900	..	328	33	78	34	130
	2014	3,075	931	634	812	..	326	32	91	63	187
Kazakhstan	2011	13,627	4,774	2,775	4,277	..	210	134	811	177	469
	2012	13,528	4,890	2,877	3,936	..	200	150	823	167	485
	2013	13,762	5,299	2,738	3,734	..	200	148	881	168	594
	2014	14,544	5,282	3,023	3,877	..	302	111	955	114	880
Korea, Dem.Ppl's.Rep.	2011	508	188	176	108	..	..	..	..	36	..
	2012	508	188	176	108	..	..	..	..	36	..
	2013	561	208	194	119	..	..	..	..	40	..
	2014	515	191	178	109	..	..	..	..	37	..

Production of oil products from refineries - by type

Thousand metric tons

Country or area	Year	Total	Gas/ diesel oil	Motor gasoline	Fuel oil	Naphtha	Jet fuels	Refinery gas	LPG	Other kerosene	Other products
Korea, Republic of	2011	127,495	39,604	13,673	18,496	22,953	14,902	2,686	1,579	4,632	8,970
	2012	131,006	41,874	15,306	15,282	24,315	16,484	2,689	1,487	4,200	9,369
	2013	126,256	40,236	15,162	11,801	24,429	16,397	3,803	1,728	2,918	9,782
	2014	128,633	42,111	16,359	8,133	25,658	17,053	4,077	2,099	2,330	10,813
Kuwait	2011	40,974	10,872	2,206	9,766	8,123	2,538	393	139	5,912	1,025
	2012	42,128	11,563	2,202	10,825	6,043	2,748	423	139	6,401	1,784
	2013	41,381	10,532	2,420	10,207	8,219	2,495	402	129	6,287	690
	2014	41,639	10,806	2,455	10,204	8,393	2,438	391	141	6,146	665
Kyrgyzstan	2011	90	28	16	46	..	..	..	..	..	..
	2012	81	31	11	39	..	..	..	..	..	..
	2013	77	27	9	41	..	..	..	..	..	..
	2014	88	30	7	51	..	..	..	..	..	..
Malaysia	2011	23,040	8,788	3,425	576	3,898	3,348	1,633	611	406	355
	2012	25,114	9,878	4,395	1,621	3,617	3,794	194	645	633	337
	2013	24,086	10,892	4,475	1,297	1,747	2,663	192	1,150	375	1,295
	2014	23,519	9,575	4,681	2,359	1,787	2,824	189	1,012	97	995
Myanmar	2011	789	195	428	71	..	42	30	3	1	19
	2012	648	165	330	65	..	33	27	8	1	19
	2013	510	150	230	55	..	21	28	6	0	19
	2014	521	161	234	61	..	17	30	..	0	18
Oman	2011	9,929	2,147	2,727	2,933	542	632	232	375	4	337
	2012	9,876	2,074	2,752	2,867	585	712	198	335	4	349
	2013	9,160	1,891	2,548	3,065	366	564	174	289	4	259
	2014	10,598	2,633	2,973	3,039	478	617	224	327	4	303
Other Asia	2011	41,883	12,458	6,110	8,287	7,085	3,373	1,120	1,047	..	2,403
	2012	45,973	13,904	6,740	8,828	8,063	3,863	1,111	1,176	..	2,288
	2013	43,382	14,162	7,274	5,812	7,639	3,866	1,204	1,329	..	2,096
	2014	42,658	14,140	7,460	5,880	7,304	3,738	1,199	976	..	1,961
Pakistan	2011	8,964	3,252	1,227	2,174	724	703	132	169	132	451
	2012	10,468	3,902	1,514	2,653	736	742	159	205	161	396
	2013	11,496	4,361	1,595	2,926	878	785	168	206	174	403
	2014	11,976	4,567	1,670	3,064	920	758	176	216	182	423
Philippines	2011	8,909	3,486	1,508	1,994	158	783	..	386	100	493
	2012	7,954	3,130	1,412	1,734	135	693	..	351	92	407
	2013	7,384	2,951	1,255	1,605	238	644	..	314	87	290
	2014	7,843	3,058	1,303	1,717	243	768	..	294	59	400
Qatar	2011	12,912	2,616	1,859	224	4,006	3,581	62	343	..	221
	2012	13,478	2,205	1,956	321	3,923	3,387	65	357	..	1,264
	2013	12,640	2,190	1,600	384	3,726	3,534	60	365	..	781
	2014	12,038	2,201	1,976	245	3,556	2,881	58	362	..	759
Saudi Arabia	2011	91,974	30,750	16,715	22,849	7,308	2,898	2,520	1,032	4,878	3,024
	2012	95,618	31,383	17,103	25,282	7,551	3,045	2,351	970	5,124	2,809
	2013	91,570	29,460	15,790	24,955	6,900	2,837	2,536	1,195	4,775	3,122
	2014	107,431	36,842	18,867	26,378	8,267	3,689	2,570	1,394	6,209	3,215
Singapore	2011	51,769	13,439	*10,219	7,017	1,604	9,441	3,338	302	881	5,528
	2012	52,148	14,627	*10,224	4,925	2,164	9,751	3,178	394	325	6,560
	2013	49,221	12,332	10,118	5,445	1,853	9,866	3,005	285	305	6,012
	2014	45,520	10,605	*10,736	2,528	2,168	9,596	2,790	317	172	6,608
Sri Lanka	2011	1,740	501	206	615	80	155	16	24	93	50
	2012	1,494	394	152	648	71	93	13	17	75	31
	2013	1,569	390	144	724	87	125	13	22	57	8
	2014	1,686	496	152	641	117	168	15	28	65	3

Production of oil products from refineries - by type

Thousand metric tons

Country or area	Year	Total	Gas/ diesel oil	Motor gasoline	Fuel oil	Naphtha	Jet fuels	Refinery gas	LPG	Other kerosene	Other products
Syrian Arab Republic	2011	11,546	3,624	1,014	4,741	948	347	153	121	4	594
	2012	8,152	2,572	677	3,365	673	246	109	86	3	421
	2013	5,392	1,701	448	2,225	445	163	71	57	2	280
	2014	5,391	1,701	448	2,225	445	163	71	56	2	280
Tajikistan	2011	25	3	0	10	*9	..	..	..	..	4
	2012	27	3	1	11	*10	..	..	..	..	2
	2013	22	4	1	13	..	..	..	..	..	4
	2014	23	4	1	15	..	..	..	..	..	3
Thailand	2011	39,148	19,208	5,797	6,074	..	4,816	..	1,994	123	1,136
	2012	41,293	20,836	6,354	6,393	..	4,471	..	1,971	61	1,207
	2013	45,347	21,955	6,474	5,557	..	5,007	..	1,923	543	3,888
	2014	43,005	20,709	6,307	5,328	..	5,224	..	1,855	868	2,714
Turkey	2011	22,170	7,044	4,272	2,492	302	2,879	847	739	60	3,535
	2012	23,139	7,856	4,371	1,567	308	3,277	872	775	45	4,068
	2013	23,995	7,668	4,278	3,013	148	3,584	854	733	53	3,664
	2014	21,876	6,230	3,953	3,837	304	3,548	791	699	56	2,458
Turkmenistan	2011	8,292	2,520	2,130	2,230	..	480	112	320	150	350
	2012	7,912	2,470	2,010	2,160	..	415	117	280	140	320
	2013	7,911	2,540	2,220	1,760	..	455	126	310	140	360
	2014	8,547	2,770	2,610	1,690	..	470	127	360	150	370
United Arab Emirates	2011	25,758	6,205	3,535	2,037	*4,439	7,894	213	777	..	658
	2012	25,437	5,478	3,750	1,935	*4,391	8,411	211	591	..	670
	2013	28,068	6,029	3,912	2,793	*4,640	8,919	217	906	..	652
	2014	26,663	6,219	3,744	1,433	5,041	8,478	206	949	..	593
Uzbekistan	2011	3,583	1,058	1,308	456	..	148	114	21	56	422
	2012	3,204	1,021	1,226	278	..	132	102	19	50	376
	2013	3,058	1,082	1,105	228	..	125	97	18	47	356
	2014	2,864	1,019	1,041	198	..	118	91	17	44	336
Viet Nam	2011	5,325	2,867	1,953	79	..	83	..	343	..	..
	2012	5,429	2,854	2,072	59	..	54	..	390	..	..
	2013	6,493	3,456	2,515	32	..	75	..	415	..	..
	2014	6,682	3,598	2,618	33	..	78	..	355	..	..
Yemen	2011	2,244	762	438	490	225	195	..	52	58	24
	2012	1,240	343	128	207	327	125	..	60	37	13
	2013	2,937	862	182	645	732	326	..	70	96	24
	2014	2,594	761	161	569	651	285	..	62	84	21
Europe	**2011**	**941,886**	**340,487**	**170,301**	**167,238**	**58,236**	**48,398**	**35,905**	**32,901**	**10,682**	**77,737**
	2012	**937,950**	**340,306**	**170,264**	**165,549**	**58,237**	**46,542**	**35,366**	**33,182**	**12,477**	**76,028**
	2013	**920,105**	**336,187**	**164,582**	**157,907**	**58,692**	**46,830**	**34,570**	**34,925**	**12,446**	**73,967**
	2014	**933,276**	**341,762**	**162,880**	**159,172**	**62,750**	**46,871**	**35,830**	**35,688**	**12,031**	**76,292**
Albania	2011	297	85	..	9	19	..	10	..	..	173
	2012	111	20	..	..	4	..	5	..	..	82
	2013	40	0	..	..	1	..	0	..	..	39
	2014	336	89	..	2	17	..	12	..	..	216
Austria	2011	8,816	3,924	1,531	822	844	615	381	101	0	598
	2012	8,756	3,759	1,553	953	1,009	618	311	67	16	470
	2013	8,917	3,851	1,549	1,011	1,011	654	311	66	18	446
	2014	8,904	3,756	1,773	981	976	581	342	65	18	412
Belarus	2011	20,187	7,440	3,135	6,407	..	0	614	462	274	1,855
	2012	21,375	7,539	3,729	7,284	..	0	693	532	146	1,452
	2013	20,880	7,687	3,683	6,615	..	361	725	568	102	1,139
	2014	21,973	8,257	3,945	5,865	..	337	769	573	12	2,215

Production of oil products from refineries - by type

Thousand metric tons

Country or area	Year	Total	Gas/ diesel oil	Motor gasoline	Fuel oil	Naphtha	Jet fuels	Refinery gas	LPG	Other kerosene	Other products
Belgium	2011	32,782	11,497	3,782	5,689	1,840	1,498	762	570	36	7,108
	2012	34,828	12,574	4,502	5,179	1,890	1,819	820	580	35	7,429
	2013	32,145	11,119	4,192	4,916	2,314	1,497	733	576	26	6,772
	2014	35,250	13,369	4,565	5,512	2,569	1,666	819	707	20	6,023
Bosnia and Herzegovina	2011	1,260	503	74	218	24	..	..	25	..	416
	2012	1,079	434	109	176	19	..	..	23	..	318
	2013	965	440	146	116	5	..	..	23	..	235
	2014	989	431	152	143	7	..	..	24	..	232
Bulgaria	2011	5,999	1,992	1,531	1,424	276	186	187	118	..	285
	2012	6,425	2,205	1,658	1,488	303	171	198	119	..	283
	2013	6,443	2,087	1,757	1,640	275	166	222	93	..	203
	2014	6,129	1,970	1,600	1,617	249	173	209	98	..	213
Croatia	2011	3,591	1,131	871	731	90	117	267	214	..	170
	2012	3,707	1,287	990	563	59	97	294	239	..	178
	2013	3,357	1,242	928	514	30	109	175	209	..	150
	2014	2,953	1,063	805	423	31	106	276	189	..	60
Czechia	2011	7,417	3,011	1,330	195	773	134	145	195	..	1,634
	2012	7,587	3,115	1,439	145	794	143	158	200	..	1,593
	2013	6,969	2,896	1,252	148	737	111	134	180	..	1,511
	2014	7,822	3,268	1,454	125	847	160	151	208	..	1,609
Denmark	2011	6,557	2,858	1,857	1,147	0	275	288	132	..	..
	2012	7,427	3,411	2,011	1,312	0	224	301	168	..	..
	2013	6,989	3,198	2,033	1,213	0	145	297	103	..	..
	2014	6,685	3,035	1,854	1,242	0	137	299	118	..	..
Finland	2011	14,794	6,505	4,070	1,294	281	795	442	387	..	1,020
	2012	14,293	6,401	4,166	1,287	250	618	428	256	..	887
	2013	14,758	6,539	4,195	1,330	284	671	410	310	..	1,019
	2014	14,141	5,983	4,323	1,224	294	542	373	293	..	1,109
France	2011	71,630	29,481	12,806	9,024	5,259	4,761	2,116	1,877	1	6,305
	2012	60,903	24,948	11,679	8,237	4,362	3,851	1,538	1,466	0	4,822
	2013	57,891	24,415	10,564	6,735	4,532	4,288	1,358	1,472	0	4,527
	2014	57,621	24,527	10,992	6,481	4,694	3,712	1,319	1,474	0	4,422
Germany	2011	99,740	42,429	20,497	7,754	7,750	4,967	3,884	2,612	5	9,842
	2012	100,742	43,580	19,487	8,319	8,104	5,216	3,795	2,624	3	9,614
	2013	97,921	42,808	19,236	7,776	7,812	4,757	3,763	2,645	4	9,120
	2014	96,755	41,920	18,856	7,696	7,851	4,862	3,646	2,518	1	9,405
Greece	2011	19,253	5,659	3,950	5,053	597	1,845	506	559	10	1,074
	2012	23,766	7,692	4,592	5,111	1,266	1,784	590	631	6	2,094
	2013	24,033	8,084	4,205	5,447	913	2,371	532	671	0	1,810
	2014	27,515	9,519	4,649	6,043	1,327	2,237	620	735	1	2,384
Hungary	2011	8,234	3,563	1,130	150	1,080	239	199	74	..	1,799
	2012	7,818	3,542	1,138	81	1,004	169	179	64	..	1,641
	2013	7,644	3,468	1,067	84	904	181	177	129	..	1,634
	2014	7,944	3,509	1,089	67	902	191	185	182	..	1,819
Ireland	2011	2,918	1,086	506	1,071	17	..	54	67	117	..
	2012	3,019	1,275	501	967	12	..	60	73	131	..
	2013	2,788	1,094	502	939	27	..	58	62	106	..
	2014	2,710	1,014	489	937	16	..	57	60	137	..
Italy	2011	85,642	35,631	18,742	9,180	3,803	3,068	3,077	2,072	563	9,506
	2012	81,598	35,099	18,210	9,216	3,122	2,605	2,729	1,541	672	8,404
	2013	69,628	31,358	15,344	5,127	2,729	2,435	2,659	1,681	396	7,899
	2014	66,337	29,488	15,419	5,731	2,705	2,269	2,262	1,327	639	6,497

Table 24

Production of oil products from refineries - by type

Thousand metric tons

Country or area	Year	Total	Gas/ diesel oil	Motor gasoline	Fuel oil	Naphtha	Jet fuels	Refinery gas	LPG	Other kerosene	Other products
Lithuania	2011	9,376	3,107	2,694	1,760	..	829	295	323	..	368
	2012	8,977	3,025	2,504	1,709	..	815	294	299	..	331
	2013	9,486	3,086	2,535	1,833	..	1,019	327	322	..	364
	2014	7,943	2,531	2,103	1,475	..	948	290	275	..	321
Netherlands	2011	56,954	20,282	6,475	8,512	8,216	7,062	1,949	1,454	383	2,621
	2012	56,639	20,054	6,958	8,726	7,720	6,589	2,066	1,477	580	2,469
	2013	55,183	20,057	6,324	8,443	7,644	6,467	1,885	1,424	407	2,532
	2014	56,515	20,652	6,334	8,501	7,144	7,003	2,052	1,632	382	2,815
Norway	2011	15,686	6,523	3,808	1,747	1,500	415	365	411	434	483
	2012	15,023	6,161	3,683	1,591	1,914	501	363	432	223	155
	2013	16,018	6,735	3,856	1,668	1,837	598	438	478	196	212
	2014	14,400	6,236	3,299	1,688	1,572	577	431	366	40	191
Poland	2011	24,944	10,929	3,717	3,391	1,332	873	955	446	1	3,300
	2012	25,768	11,289	3,803	3,430	1,542	937	1,032	539	1	3,195
	2013	25,049	11,023	3,826	3,408	1,491	877	922	548	1	2,953
	2014	24,821	10,616	3,627	3,659	1,557	1,156	603	560	..	3,043
Portugal	2011	10,820	3,576	1,875	2,534	961	791	15	318	1	749
	2012	11,673	3,872	1,864	2,740	1,218	910	19	221	1	828
	2013	13,941	5,399	2,338	2,461	960	1,132	24	261	1	1,365
	2014	12,662	4,865	1,924	2,498	1,180	976	19	198	..	1,002
Republic of Moldova	2011	19	4	..	10	..	..	..	..	..	5
	2012	18	3	..	12	..	..	..	..	..	3
	2013	19	1	..	16	..	..	..	..	..	2
	2014	18	1	..	16	..	..	..	..	..	1
Romania	2011	10,445	3,743	2,886	722	269	274	783	582	26	1,160
	2012	9,850	3,806	2,751	562	263	254	700	527	23	964
	2013	10,122	4,209	2,863	364	221	297	712	523	5	928
	2014	11,666	5,170	2,998	321	280	372	749	560	8	1,208
Russian Federation	2011	249,860	70,293	36,686	73,640	19,366	12,555	12,224	14,167	..	10,929
	2012	255,290	69,436	38,232	75,072	19,978	12,863	12,661	15,314	..	11,734
	2013	266,656	71,599	39,379	77,993	21,477	13,615	12,849	16,899	..	12,845
	2014	283,328	77,001	38,935	82,109	25,202	13,583	14,323	18,227	..	13,948
Serbia	2011	2,462	776	430	542	150	75	48	105	..	336
	2012	2,260	691	372	472	127	73	37	113	..	375
	2013	3,111	1,122	401	395	439	73	47	193	..	441
	2014	3,110	1,172	464	366	351	109	50	158	..	440
Slovakia	2011	6,662	3,056	1,411	638	531	58	422	131	2	413
	2012	5,877	2,732	1,348	616	371	47	321	122	1	319
	2013	6,552	2,913	1,395	677	470	118	452	172	1	354
	2014	5,725	2,660	1,291	489	385	78	341	155	1	325
Spain	2011	56,649	23,644	7,471	7,927	464	257	1,926	1,439	6,452	7,069
	2012	61,346	26,490	7,231	6,536	318	163	2,059	1,702	8,371	8,476
	2013	60,501	26,783	7,471	5,296	345	149	1,879	1,713	8,478	8,387
	2014	60,290	27,110	7,275	4,482	304	197	1,956	1,575	8,679	8,712
Sweden	2011	19,622	7,075	4,003	4,970	248	190	530	271	..	2,335
	2012	21,410	7,966	4,394	5,099	240	245	623	391	..	2,452
	2013	17,584	6,298	3,425	4,124	201	169	499	308	..	2,560
	2014	19,799	7,325	4,823	3,161	..	248	748	423	..	3,071
Switzerland	2011	4,407	2,209	1,263	344	20	81	217	184	..	89
	2012	3,424	1,703	1,028	275	20	38	168	119	..	73
	2013	4,914	2,592	1,388	365	20	38	227	193	..	91
	2014	4,940	2,618	1,420	394	..	16	225	173	..	94

Production of oil products from refineries - by type

Thousand metric tons

Country or area	Year	Total	Gas/ diesel oil	Motor gasoline	Fuel oil	Naphtha	Jet fuels	Refinery gas	LPG	Other kerosene	Other products
T.F.Yug.Rep. Macedonia	2011	680	278	110	246	..	27	0	20	..	..
	2012	244	74	46	97	..	17	0	10	..	..
	2013	78	28	22	21	..	5	0	3	..	..
	2014	1	0	1	0	..	0	0	0	..	..
Ukraine	2011	9,366	2,713	2,837	2,181	..	..	438	665	..	532
	2012	4,968	1,411	1,636	790	..	..	269	563	..	299
	2013	4,018	1,120	1,015	658	..	..	288	500	..	437
	2014	3,321	832	712	516	..	..	388	440	..	433
United Kingdom	2011	74,817	25,484	18,823	7,907	2,526	6,411	2,806	2,920	2,377	5,563
	2012	71,750	24,712	18,650	7,504	2,328	5,775	2,655	2,770	2,268	5,088
	2013	65,505	22,936	17,691	6,574	2,013	4,527	2,467	2,600	2,705	3,992
	2014	60,673	21,775	15,709	5,408	2,290	4,635	2,316	2,375	2,093	4,072
Oceania	**2011**	**38,251**	**13,223**	**13,548**	**1,668**	**384**	**5,557**	**1,513**	**769**	**2**	**1,587**
	2012	**36,937**	**12,989**	**12,862**	**1,646**	**416**	**5,585**	**1,408**	**643**	**15**	**1,373**
	2013	**36,607**	**13,425**	**12,274**	**1,618**	**474**	**5,443**	**1,331**	**641**	**114**	**1,287**
	2014	**33,897**	**12,952**	**11,082**	**1,330**	**506**	**5,080**	**1,191**	**587**	**51**	**1,118**
Australia	2011	31,976	10,845	12,222	954	157	4,412	1,219	763	0	1,404
	2012	30,346	10,388	11,497	945	160	4,415	1,120	633	0	1,188
	2013	30,038	10,849	10,911	888	99	4,484	1,040	629	2	1,136
	2014	27,441	10,421	9,790	645	124	4,044	912	566	5	934
New Zealand	2011	5,370	1,923	1,326	594	..	1,048	294	..	2	183
	2012	5,577	2,139	1,352	562	..	1,051	288	..	0	185
	2013	5,342	2,052	1,336	553	..	959	291	..	0	151
	2014	5,208	1,996	1,269	512	..	968	279	..	0	184
Papua New Guinea	2011	905	455	..	120	227	97	..	6	0	..
	2012	1,014	462	13	139	256	119	..	10	15	..
	2013	1,227	524	27	177	375	0	..	12	112	..
	2014	*1,248	*535	*23	*173	*382	*68	..	*21	*46	..

Table 25

Production, trade and supply of oil products

Thousand metric tons

Table Notes

Production refers to the production from refineries and plants of all oil products, together with quantities from other sources which include any blended biofuels.

Imports, Exports, Bunkers and Changes in stocks refer to all oil products.

Please refer to the Definitions Section on pages ix to xv for the appropriate product description /classification.

Figure 65: Production of oil products by region, 1990-2014

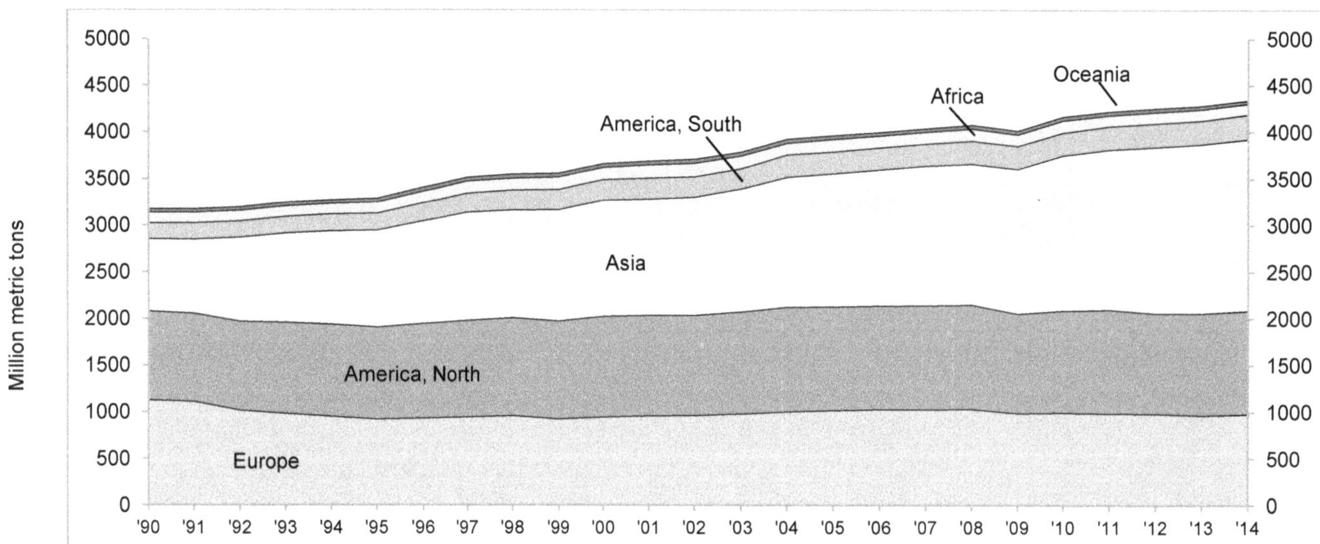

Production, trade and supply of oil products

Thousand metric tons and kilograms per capita

Country or area	Year	Production	Imports	Exports	Bunkers	Changes in stocks	Supply	
							Total	Per capita
World	**2011**	**4,219,266**	**1,188,027**	**1,188,256**	**365,452**	**3,961**	**3,849,624**	**549**
	2012	**4,252,154**	**1,192,042**	**1,209,361**	**348,056**	**4,623**	**3,882,156**	**547**
	2013	**4,282,793**	**1,247,732**	**1,238,652**	**357,542**	**4,306**	**3,930,024**	**547**
	2014	**4,340,734**	**1,234,832**	**1,245,469**	**360,744**	**8,308**	**3,961,046**	**545**
Africa	**2011**	**118,264**	**74,966**	**33,077**	**12,243**	**78**	**147,831**	**138**
	2012	**119,783**	**82,427**	**31,737**	**13,626**	**434**	**156,413**	**142**
	2013	**119,645**	**94,033**	**37,557**	**12,755**	**1,357**	**162,009**	**144**
	2014	**116,854**	**102,496**	**47,694**	**13,227**	**-113**	**158,542**	**137**
Algeria	2011	33,013	2,115	19,522	712	-2	14,896	406
	2012	30,159	4,490	17,864	770	181	15,834	423
	2013	31,033	5,253	18,181	803	65	17,237	451
	2014	39,577	3,073	23,893	680	121	17,956	461
Angola	2011	2,472	3,459	783	367	19	4,762	217
	2012	2,531	4,559	720	550	62	5,758	254
	2013	2,864	4,693	1,235	438	4	5,880	251
	2014	2,755	5,571	999	595	198	6,534	270
Benin	2011	..	1,860	214	153	6	1,487	152
	2012	..	1,947	224	160	7	1,556	155
	2013	..	2,080	239	171	8	1,662	161
	2014	..	2,216	254	182	9	1,771	167
Botswana	2011	..	879	..	17	..	862	412
	2012	..	924	..	12	..	912	428
	2013	..	972	..	13	..	959	441
	2014	..	1,009	..	11	..	998	450
Burkina Faso	2011	..	667	..	*41	..	626	39
	2012	..	792	..	*43	..	749	45
	2013	..	958	..	*55	..	904	53
	2014	..	897	..	*34	..	863	49
Burundi	2011	..	92	..	4	1	87	9
	2012	..	80	..	4	-2	78	8
	2013	..	95	..	4	2	89	9
	2014	..	139	..	4	-1	136	13
Cabo Verde	2011	..	313	..	110	..	203	410
	2012	..	269	..	104	..	165	329
	2013	..	263	..	100	..	163	320
	2014	..	284	..	123	..	161	314
Cameroon	2011	2,142	167	622	113	0	1,574	75
	2012	2,240	172	651	118	0	1,643	76
	2013	2,365	164	690	124	0	1,715	77
	2014	2,506	138	834	131	..	1,679	74
Central African Rep.	2011	..	*123	..	*31	..	*92	*20
	2012	..	*128	..	*33	..	*96	*21
	2013	..	*132	..	*33	..	*98	*21
	2014	..	*133	..	*34	..	*100	*21
Chad	2011	..	*105	..	*18	..	*87	*7
	2012	..	*105	..	*18	..	*87	*7
	2013	..	*107	..	*18	..	*89	*7
	2014	..	*107	..	*18	..	*89	*7
Comoros	2011	..	*48	..	*3	..	*45	*63
	2012	..	*50	..	*3	..	*47	*65
	2013	..	*61	..	*3	..	*57	*76
	2014	..	*54	..	*3	..	*51	*66

Table 25

Production, trade and supply of oil products

Thousand metric tons and kilograms per capita

Country or area	Year	Production	Imports	Exports	Bunkers	Changes in stocks	Supply	
							Total	Per capita
Congo	2011	573	277	197	49	-26	630	151
	2012	711	211	256	47	0	619	144
	2013	775	300	311	43	0	721	164
	2014	791	321	332	40	0	740	164
Côte d'Ivoire	2011	2,249	109	1,335	66	-33	990	48
	2012	3,236	138	1,877	80	-48	1,465	69
	2013	3,288	201	1,795	73	-49	1,670	77
	2014	3,114	221	1,512	112	-59	1,770	80
Dem. Rep. of the Congo	2011	..	873	..	145	..	728	11
	2012	..	860	..	139	..	721	10
	2013	..	1,252	..	151	..	1,101	15
	2014	..	1,698	..	226	..	1,472	20
Djibouti	2011	..	232	*22	99	*-51	161	191
	2012	..	*300	*21	*102	0	*177	*208
	2013	..	*310	*21	*107	0	*182	*211
	2014	..	*316	*21	*109	0	*185	*211
Egypt	2011	30,199	8,233	2,803	1,158	-123	34,594	413
	2012	30,464	9,343	2,491	1,171	-102	36,247	423
	2013	25,552	12,334	2,025	812	0	35,049	400
	2014	24,098	8,682	1,940	2,271	-1,177	29,746	332
Equatorial Guinea	2011	..	*238	..	*35	..	*203	*270
	2012	..	*238	..	*35	..	*203	*262
	2013	..	*238	..	*35	..	*203	*255
	2014	..	*238	..	*35	..	*203	*247
Eritrea	2011	..	166	..	1	0	165	34
	2012	..	176	..	1	0	175	36
	2013	..	178	..	1	0	177	35
	2014	..	181	..	1	0	180	35
Ethiopia	2011	..	2,326	..	272	-50	2,104	23
	2012	..	2,433	..	318	12	2,103	23
	2013	..	2,855	..	350	24	2,481	26
	2014	..	3,130	..	386	32	2,712	28
Gabon	2011	1,028	394	409	262	17	734	465
	2012	770	518	242	277	-12	781	484
	2013	837	602	287	260	0	892	540
	2014	814	614	275	247	..	906	537
Gambia	2011	..	167	*2	*21	..	144	82
	2012	..	164	*2	*15	..	147	81
	2013	..	162	*2	*21	..	139	75
	2014	..	180	*2	*14	..	164	85
Ghana	2011	1,000	2,159	658	207	0	2,294	92
	2012	474	2,526	282	206	0	2,513	98
	2013	424	3,014	40	179	-44	3,263	125
	2014	129	3,462	10	121	-53	3,513	131
Guinea	2011	..	898	..	*25	..	873	77
	2012	..	851	..	*25	..	826	71
	2013	..	741	..	*25	..	717	60
	2014	..	*772	..	*25	..	*747	*61
Guinea-Bissau	2011	..	*112	..	*11	..	*101	*60
	2012	..	*114	..	*11	..	*103	*60
	2013	..	*117	..	*11	..	*106	*60
	2014	..	*123	..	*12	..	*111	*62

Production, trade and supply of oil products

Thousand metric tons and kilograms per capita

Country or area	Year	Production	Imports	Exports	Bunkers	Changes in stocks	Supply	
							Total	Per capita
Kenya	2011	1,689	2,646	93	698	0	3,544	86
	2012	968	3,087	44	684	8	3,319	78
	2013	597	4,536	1,288	579	-17	3,283	75
	2014	0	13,326	9,715	227	0	3,385	75
Lesotho	2011	..	180	..	0	..	180	88
	2012	..	196	..	0	..	195	95
	2013	..	197	..	0	..	196	94
	2014	..	*199	..	0	..	*198	*94
Liberia	2011	..	303	..	*19	..	285	70
	2012	..	342	..	*21	..	321	77
	2013	..	309	..	*21	..	288	67
	2014	..	286	..	*22	..	264	60
Libya	2011	6,504	3,766	1,517	133	0	8,620	1,371
	2012	9,250	4,710	1,640	566	0	11,754	1,871
	2013	11,760	5,394	4,154	397	0	12,603	2,011
	2014	5,360	7,828	723	214	0	12,251	1,957
Madagascar	2011	..	673	..	79	-12	605	28
	2012	..	747	..	82	29	636	29
	2013	..	728	..	71	1	657	29
	2014	..	*745	..	*75	*1	*669	*28
Malawi	2011	..	287	..	*7	..	280	18
	2012	..	275	..	*7	..	268	17
	2013	..	291	..	*7	..	284	18
	2014	..	294	..	*7	..	287	17
Mali	2011	..	*361	..	*21	..	*340	*22
	2012	..	*346	..	*22	..	*324	*20
	2013	..	*357	..	*23	..	*334	*20
	2014	..	*483	..	*23	..	*460	*27
Mauritania	2011	..	712	..	17	..	695	189
	2012	..	791	..	23	..	768	203
	2013	..	*793	..	29	..	*765	*197
	2014	..	*793	..	29	..	*765	*193
Mauritius	2011	..	1,175	..	404	-32	803	640
	2012	..	1,148	..	376	-20	792	630
	2013	..	1,238	..	386	48	804	637
	2014	..	1,178	..	410	*-64	832	656
Mayotte	2011	..	92	..	7	-2	87	403
	2012	..	*95	..	*7	..	*88	*399
	2013	..	*98	..	*7	..	*92	*403
	2014	..	*102	..	*7	..	*95	*408
Morocco	2011	6,979	8,084	581	642	-76	13,916	421
	2012	7,049	7,894	527	936	101	13,379	399
	2013	7,199	10,124	1,318	713	1,134	14,158	416
	2014	6,749	9,708	1,154	767	1,344	13,192	383
Mozambique	2011	..	850	..	59	0	791	32
	2012	..	820	..	65	0	755	29
	2013	..	1,337	..	70	*298	969	37
	2014	..	1,910	..	120	-10	1,800	66
Namibia	2011	..	878	..	*13	0	866	386
	2012	..	995	..	*39	0	956	417
	2013	..	776	..	*20	*12	744	317
	2014	..	1,157	4	47	..	1,106	460

Table 25

Production, trade and supply of oil products

Thousand metric tons and kilograms per capita

Country or area	Year	Production	Imports	Exports	Bunkers	Changes in stocks	Supply	
							Total	Per capita
Niger	2011	57	269	0	15	27	284	17
	2012	560	29	159	12	0	419	24
	2013	784	35	369	22	6	422	23
	2014	744	56	*277	29	0	494	26
Nigeria	2011	5,868	7,720	1,369	582	477	11,160	68
	2012	4,673	8,597	703	576	241	11,750	70
	2013	4,767	8,178	743	710	34	11,458	66
	2014	3,253	7,898	557	679	-401	10,316	58
Réunion	2011	..	860	..	*70	-6	795	950
	2012	..	790	..	*63	-22	749	888
	2013	..	762	..	*56	-10	716	844
	2014	..	804	..	*57	-28	775	906
Rwanda	2011	..	*229	0	*17	..	*211	*20
	2012	..	258	0	*20	..	238	22
	2013	..	278	0	*19	..	259	23
	2014	..	287	0	*20	..	267	24
Sao Tome and Principe	2011	..	*44	..	*10	..	*34	*192
	2012	..	*46	..	*10	..	*36	*203
	2013	..	*46	..	*10	..	*36	*198
	2014	..	*46	..	*10	..	*36	*194
Senegal	2011	714	1,287	160	289	-50	1,602	120
	2012	768	1,131	104	268	-40	1,567	114
	2013	764	1,248	174	317	-42	1,563	110
	2014	802	1,430	181	332	-19	1,738	118
Seychelles	2011	..	289	..	177	..	111	1,186
	2012	..	332	..	189	..	143	1,510
	2013	..	318	..	185	..	133	1,397
	2014	..	343	..	*181	..	162	1,686
Sierra Leone	2011	..	254	..	*8	..	246	42
	2012	..	294	..	*8	..	286	47
	2013	..	350	..	*8	..	343	55
	2014	..	385	..	*8	..	377	60
Somalia	2011	..	*262	..	65	..	*197	*20
	2012	..	*265	..	67	..	*198	*20
	2013	..	*265	..	67	..	*198	*19
	2014	..	*265	..	67	..	*198	*19
South Africa	2011	18,070	8,090	1,981	4,044	..	20,135	385
	2012	19,547	7,423	2,690	4,371	..	19,909	377
	2013	20,531	7,952	3,490	4,239	..	20,754	389
	2014	19,709	7,739	3,837	3,444	..	20,167	374
South Sudan	2011	..	..	..	..	..	..	..
	2012	..	465	..	31	..	434	40
	2013	..	492	..	36	..	456	40
	2014	..	509	..	37	..	472	40
St. Helena and Depend.	2011	..	3	..	..	..	3	807
	2012	..	4	..	..	..	4	905
	2013	..	4	..	..	..	4	880
	2014	..	*4	..	..	..	*4	*890
Sudan	2011	..	..	..	..	..	..	..
	2012	3,905	1,119	259	238	*4	4,523	120
	2013	3,734	1,144	256	*211	*-19	4,430	115
	2014	4,003	1,055	318	264	..	4,476	114

Production, trade and supply of oil products

Thousand metric tons and kilograms per capita

Country or area	Year	Production	Imports	Exports	Bunkers	Changes in stocks	Supply	
							Total	Per capita
Sudan (former)	2011	4,326	1,150	443	328	..	4,705	101
	2012	..	..	..	..	..	..	..
	2013	..	..	..	..	..	..	..
	2014	..	..	..	..	..	..	..
Swaziland	2011	..	*220	..	*1	..	*220	*181
	2012	..	*224	..	*1	..	*224	*182
	2013	..	*229	..	*1	..	*228	*182
	2014	..	238	..	*1	..	237	187
Togo	2011	..	635	..	77	-56	614	94
	2012	..	586	..	85	-41	542	80
	2013	..	612	..	90	-42	564	81
	2014	..	612	..	95	-44	561	79
Tunisia	2011	838	3,264	328	246	49	3,479	323
	2012	1,861	3,109	941	294	76	3,660	336
	2013	1,783	3,090	890	294	-56	3,745	340
	2014	1,791	3,539	835	279	39	4,177	375
Uganda	2011	..	1,219	..	*90	..	1,129	33
	2012	..	1,140	..	*92	..	1,048	30
	2013	..	1,213	..	*92	..	1,121	31
	2014	..	1,335	..	115	..	1,220	32
United Rep. of Tanzania	2011	..	1,843	..	163	..	1,680	36
	2012	..	2,228	..	174	..	2,054	42
	2013	..	2,804	..	186	..	2,618	52
	2014	..	2,772	..	199	..	2,573	50
Zambia	2011	543	224	38	33	..	696	49
	2012	616	323	41	50	..	848	57
	2013	589	386	50	50	..	875	57
	2014	659	382	21	39	..	981	62
Zimbabwe	2011	..	1,086	..	9	..	1,077	76
	2012	..	1,230	..	10	..	1,220	84
	2013	..	1,366	..	10	..	1,356	91
	2014	..	1,231	..	11	..	1,220	80
America, North	**2011**	**1,109,720**	**194,777**	**188,748**	**61,553**	**-3,443**	**1,057,639**	**1,914**
	2012	**1,074,729**	**176,867**	**187,152**	**48,827**	**-7,319**	**1,022,935**	**1,833**
	2013	**1,097,179**	**174,405**	**198,400**	**51,110**	**-1,562**	**1,023,636**	**1,817**
	2014	**1,109,539**	**167,117**	**202,251**	**49,740**	**1,551**	**1,023,114**	**1,800**
Anguilla	2011	..	*46	..	..	..	*46	*3,296
	2012	..	*45	..	..	..	*45	*3,205
	2013	..	*44	..	..	..	*44	*3,098
	2014	..	*45	..	..	..	*45	*3,098
Antigua and Barbuda	2011	..	*223	*4	48	..	*172	*1,946
	2012	..	*230	*4	49	..	*177	*1,982
	2013	..	*230	*4	*49	..	*177	*1,961
	2014	..	*234	*4	*51	..	*179	*1,966
Aruba	2011	*10,050	*402	*10,050	*88	..	*314	*3,084
	2012	*2,440	*392	*2,440	*92	..	*300	*2,929
	2013	0	*374	0	*93	..	*281	*2,733
	2014	0	*380	0	*94	..	*286	*2,763
Bahamas	2011	..	805	..	*188	..	617	1,682
	2012	..	880	..	*224	..	655	1,760
	2013	..	1,260	*45	*195	*100	921	2,437
	2014	..	887	*1	*190	*-100	795	2,076

Table 25

Production, trade and supply of oil products

Thousand metric tons and kilograms per capita

Country or area	Year	Production	Imports	Exports	Bunkers	Changes in stocks	Supply	
							Total	Per capita
Barbados	2011	..	592	..	*141	0	451	1,606
	2012	..	555	..	*114	1	439	1,561
	2013	..	560	..	*128	1	431	1,526
	2014	..	402	..	*32	-8	378	1,335
Belize	2011	2	191	..	11	..	182	553
	2012	1	152	..	11	..	143	423
	2013	1	166	..	14	..	153	444
	2014	1	162	..	12	..	151	429
Bermuda	2011	..	166	..	*21	..	145	2,280
	2012	..	178	..	*21	..	157	2,484
	2013	..	170	..	*21	..	149	2,376
	2014	..	216	..	*28	..	188	3,013
Bonaire, St Eustatius, Saba	2011	..	..	..	..	..	..	..
	2012	..	*333	..	*216	..	*117	*5,103
	2013	..	*342	..	*222	..	*120	*5,074
	2014	..	*342	..	*222	..	*120	*4,944
British Virgin Islands	2011	..	*57	..	..	..	*57	*2,039
	2012	..	*57	..	..	..	*57	*1,989
	2013	..	*57	..	..	..	*57	*1,951
	2014	..	*58	..	..	..	*58	*1,947
Canada	2011	93,690	15,450	18,694	1,430	203	88,813	2,574
	2012	95,194	12,376	22,643	1,370	-49	83,606	2,398
	2013	93,719	12,288	24,222	1,226	-531	81,090	2,302
	2014	92,110	12,278	22,456	1,069	895	79,968	2,247
Cayman Islands	2011	..	205	..	*11	..	194	3,429
	2012	..	190	..	*12	..	178	3,094
	2013	..	187	..	10	..	177	3,032
	2014	..	188	..	10	..	178	3,008
Costa Rica	2011	150	2,155	43	0	0	2,262	492
	2012	0	2,283	9	0	-17	2,291	492
	2013	0	2,366	0	155	-7	2,218	471
	2014	..	2,446	0	154	21	2,271	477
Cuba	2011	4,871	3,154	..	*164	..	7,861	694
	2012	5,107	3,142	..	*164	..	8,085	713
	2013	5,138	3,080	..	*237	..	7,981	702
	2014	4,877	3,158	..	*241	..	7,794	685
Curaçao	2011	..	..	..	..	..	..	..
	2012	8,619	2,982	7,851	1,647	..	2,103	13,779
	2013	9,044	2,924	8,315	1,662	..	1,991	12,903
	2014	9,574	2,606	8,742	1,691	..	1,747	11,216
Dominica	2011	..	*43	..	*2	..	*42	*585
	2012	..	*46	..	*2	..	*44	*617
	2013	..	*44	..	*2	..	*43	*590
	2014	..	*46	..	*2	..	*44	*607
Dominican Republic	2011	1,243	4,349	..	426	-5	5,171	516
	2012	1,163	4,528	..	478	0	5,213	513
	2013	1,295	4,213	..	434	-9	5,082	494
	2014	1,278	4,052	..	484	18	4,829	464
El Salvador	2011	743	1,324	93	110	-81	1,945	321
	2012	575	1,506	0	83	12	1,986	327
	2013	0	2,011	11	144	3	1,853	304
	2014	0	2,061	17	152	-13	1,905	312

Production, trade and supply of oil products

Thousand metric tons and kilograms per capita

Country or area	Year	Production	Imports	Exports	Bunkers	Changes in stocks	Supply	
							Total	Per capita
Greenland	2011	..	259	..	47	-19	232	4,106
	2012	..	205	..	22	-2	185	3,285
	2013	..	199	..	16	3	180	3,206
	2014	..	176	..	17	-7	166	2,953
Grenada	2011	..	86	..	5	0	82	778
	2012	..	86	..	4	-6	87	829
	2013	..	104	..	4	0	100	940
	2014	..	85	..	6	0	79	745
Guadeloupe	2011	..	*706	..	107	..	*599	*1,306
	2012	..	*744	..	*108	..	*636	*1,380
	2013	..	*747	..	*108	..	*639	*1,379
	2014	..	*752	..	*109	..	*643	*1,380
Guatemala	2011	81	3,460	181	349	10	3,001	199
	2012	71	3,457	118	342	30	3,038	198
	2013	59	3,512	205	362	-561	3,565	227
	2014	72	5,026	313	365	-391	4,813	300
Haiti	2011	..	696	..	17	..	679	67
	2012	8	761	..	52	..	717	70
	2013	..	807	..	89	-21	739	71
	2014	..	911	..	34	-10	887	84
Honduras	2011	..	2,677	297	53	-110	2,437	320
	2012	..	2,855	377	50	0	2,428	314
	2013	..	2,910	414	42	0	2,454	313
	2014	..	2,882	394	67	-215	2,636	331
Jamaica	2011	1,259	1,672	225	248	16	2,442	887
	2012	1,218	1,542	233	244	-1	2,283	826
	2013	1,171	1,673	198	228	-31	2,448	883
	2014	963	1,688	185	214	19	2,233	802
Martinique	2011	*845	*66	*131	*140	..	*640	*1,620
	2012	*858	*66	*133	*139	..	*652	*1,650
	2013	*867	*64	*133	*139	..	*658	*1,664
	2014	*904	*66	*145	*145	..	*680	*1,716
Mexico	2011	73,763	31,761	9,010	3,488	449	92,577	769
	2012	73,735	31,618	7,508	3,643	138	94,064	771
	2013	77,379	28,019	9,217	3,654	-231	92,758	750
	2014	74,353	28,944	10,067	3,860	-148	89,518	714
Montserrat	2011	..	14	..	*1	0	13	2,604
	2012	..	14	..	*1	*-1	13	2,684
	2013	..	18	..	*1	0	17	3,338
	2014	..	17	..	*1	0	16	3,043
Neth. Antilles (former)	2011	8,718	4,117	7,974	2,440	..	2,421	11,357
	2012	..	..	..	..	..	..	..
	2013	..	..	..	..	..	..	..
	2014	..	..	..	..	..	..	..
Nicaragua	2011	780	794	25	20	52	1,477	254
	2012	518	925	19	21	-4	1,407	239
	2013	682	752	13	20	31	1,370	230
	2014	680	808	12	22	-22	1,476	245
Panama	2011	..	6,010	1	3,821	-548	2,736	743
	2012	..	6,196	1	*3,917	-234	2,513	671
	2013	..	6,502	3	3,822	-10	2,687	706
	2014	..	6,078	3	3,609	378	2,088	540

Table 25

Production, trade and supply of oil products

Thousand metric tons and kilograms per capita

Country or area	Year	Production	Imports	Exports	Bunkers	Changes in stocks	Supply	
							Total	Per capita
Sint Maarten (Dutch part)	2011	..	..	..	..	..	..	..
	2012	..	*498	..	*235	..	*263	*7,486
	2013	..	*506	..	*237	..	*269	*7,374
	2014	..	*515	..	*240	..	*275	*7,285
St. Kitts-Nevis	2011	..	80	..	5	0	75	1,416
	2012	..	79	..	5	1	73	1,369
	2013	..	*78	..	*5	..	*74	*1,355
	2014	..	*79	..	*5	..	*75	*1,358
St. Lucia	2011	..	141	..	*6	..	135	754
	2012	..	141	..	*6	..	134	743
	2013	..	140	..	*6	..	134	735
	2014	..	142	..	*6	..	135	737
St. Pierre-Miquelon	2011	..	*29	..	*6	..	*23	*3,698
	2012	..	*29	..	*6	..	*23	*3,633
	2013	..	*30	..	*6	..	*24	*3,745
	2014	..	*31	..	*6	..	*25	*4,013
St. Vincent-Grenadines	2011	..	66	..	0	..	66	602
	2012	..	83	..	0	..	83	756
	2013	..	68	..	0	..	68	620
	2014	..	*68	..	0	..	*68	*622
Trinidad and Tobago	2011	8,419	253	6,569	675	-91	1,519	1,138
	2012	6,676	380	5,093	627	-103	1,439	1,073
	2013	7,845	77	5,583	829	-18	1,528	1,133
	2014	6,448	68	4,568	672	-176	1,452	1,072
Turks and Caicos Islands	2011	..	*62	..	..	..	*62	*1,945
	2012	..	*65	..	..	..	*65	*2,004
	2013	..	*65	..	..	..	*65	*1,964
	2014	..	*67	..	..	..	*67	*1,986
United States	2011	905,106	112,664	135,451	47,485	-3,319	838,153	2,683
	2012	878,546	97,251	140,723	34,923	-7,084	807,235	2,564
	2013	899,979	97,819	150,037	36,950	-281	811,092	2,558
	2014	918,279	89,155	155,344	35,932	1,311	814,847	2,551
America, South	**2011**	**248,323**	**52,491**	**52,031**	**14,311**	**188**	**234,283**	**584**
	2012	**255,256**	**55,685**	**45,774**	**14,375**	**-2,096**	**252,889**	**623**
	2013	**255,690**	**57,582**	**42,238**	**14,429**	**-987**	**257,592**	**628**
	2014	**264,202**	**58,732**	**39,641**	**14,447**	**2,081**	**266,765**	**644**
Argentina	2011	32,024	5,136	2,449	2,072	273	32,366	777
	2012	34,580	4,751	2,076	2,419	-135	34,971	831
	2013	32,569	6,527	1,359	2,635	26	35,076	825
	2014	34,449	6,909	2,041	2,432	58	36,827	857
Bolivia (Plur. State of)	2011	2,357	845	454	48	-33	2,733	271
	2012	2,694	975	318	42	16	3,293	322
	2013	2,705	952	395	52	-12	3,222	310
	2014	2,948	1,056	446	59	-6	3,505	332
Brazil	2011	104,359	24,953	6,466	6,451	255	116,140	579
	2012	109,876	24,842	6,248	6,045	-1,330	123,755	611
	2013	118,174	24,064	7,050	5,725	-1,116	130,579	639
	2014	122,557	25,386	6,908	5,725	-117	135,427	657
Chile	2011	8,923	7,096	649	924	-63	14,509	843
	2012	9,032	7,483	852	840	-171	14,994	862
	2013	9,415	7,386	913	806	16	15,066	857
	2014	9,625	6,999	547	821	26	15,230	857

Production, trade and supply of oil products

Thousand metric tons and kilograms per capita

Country or area	Year	Production	Imports	Exports	Bunkers	Changes in stocks	Supply	
							Total	Per capita
Colombia	2011	14,999	2,548	5,246	*1,452	1	10,849	234
	2012	15,689	3,135	4,993	*1,389	74	12,369	264
	2013	15,030	3,665	5,134	1,809	457	11,294	239
	2014	16,723	4,598	4,483	1,913	2,371	12,554	263
Ecuador	2011	7,677	5,396	1,593	827	-62	10,714	706
	2012	7,520	5,652	1,503	730	-48	10,987	713
	2013	6,929	6,400	1,085	741	-7	11,510	735
	2014	6,765	7,480	608	777	45	12,815	806
Falkland Is. (Malvinas)	2011	..	*13	..	..	..	*13	*4,658
	2012	..	*14	..	..	..	*14	*4,750
	2013	..	*14	..	..	..	*14	*4,727
	2014	..	*14	..	..	..	*14	*4,704
French Guiana	2011	..	*239	..	*39	..	*200	*830
	2012	..	*220	..	*39	..	*181	*733
	2013	..	*234	..	*39	..	*195	*767
	2014	..	*269	..	*39	..	*229	*877
Guyana	2011	..	581	..	9	-9	580	768
	2012	..	657	..	15	-8	650	857
	2013	..	641	..	10	-1	632	830
	2014	..	665	..	9	0	656	858
Paraguay	2011	110	1,582	..	26	-108	1,774	282
	2012	118	1,692	..	27	49	1,734	272
	2013	138	1,572	..	32	-107	1,785	276
	2014	149	1,787	..	37	20	1,878	287
Peru	2011	12,827	2,093	3,462	925	-45	10,578	355
	2012	12,285	2,736	4,171	877	-997	10,970	364
	2013	13,078	3,047	5,003	722	-1,140	11,540	378
	2014	12,869	2,579	3,157	796	395	11,100	358
Suriname	2011	911	403	608	*66	0	640	1,222
	2012	931	475	597	*80	0	729	1,379
	2013	1,000	472	702	*69	0	701	1,315
	2014	1,035	468	688	*77	0	738	1,371
Uruguay	2011	1,325	1,595	50	469	30	2,372	701
	2012	1,985	1,299	82	375	58	2,768	815
	2013	2,075	616	23	301	-12	2,379	698
	2014	1,949	523	23	293	58	2,099	614
Venezuela (Bolivar. Rep.)	2011	62,810	11	31,053	1,002	-51	30,817	1,047
	2012	60,546	1,754	24,934	1,497	395	35,474	1,188
	2013	54,579	1,993	20,575	1,488	909	33,600	1,110
	2014	55,132	..	20,740	1,469	-770	33,693	1,098
Asia	**2011**	**1,716,528**	**474,334**	**464,030**	**162,417**	**9,584**	**1,554,832**	**369**
	2012	**1,780,396**	**487,403**	**471,931**	**161,576**	**17,140**	**1,617,151**	**380**
	2013	**1,808,790**	**517,796**	**480,550**	**169,435**	**6,392**	**1,670,209**	**388**
	2014	**1,836,270**	**506,822**	**478,263**	**165,912**	**5,252**	**1,693,665**	**389**
Afghanistan	2011	..	2,498	..	10	..	2,488	86
	2012	..	*2,222	..	10	..	*2,212	*74
	2013	..	*1,893	..	*10	..	*1,883	*61
	2014	..	*1,684	..	*10	..	*1,674	*53
Armenia	2011	..	392	0	*44	..	349	117
	2012	..	384	..	*47	..	337	113
	2013	..	364	..	*46	..	318	106
	2014	..	356	0	*40	..	315	105

Table 25

Production, trade and supply of oil products

Thousand metric tons and kilograms per capita

Country or area	Year	Production	Imports	Exports	Bunkers	Changes in stocks	Supply	
							Total	Per capita
Azerbaijan	2011	6,200	26	2,211	503	-21	3,534	383
	2012	6,056	44	1,795	457	32	3,815	408
	2013	6,375	120	1,844	460	124	4,068	428
	2014	6,536	277	2,356	421	3	4,033	419
Bahrain	2011	13,422	*597	11,089	593	9	2,328	1,783
	2012	14,365	*630	11,643	563	68	2,721	2,040
	2013	13,926	*628	11,727	402	-21	2,446	1,813
	2014	13,913	*546	11,828	395	-91	2,327	1,709
Bangladesh	2011	1,432	3,754	133	420	185	4,448	29
	2012	1,287	3,785	114	401	-289	4,846	31
	2013	1,493	3,855	101	412	163	4,672	30
	2014	1,296	4,290	80	423	113	4,970	31
Bhutan	2011	..	121	..	*1	..	120	164
	2012	..	138	..	2	..	137	184
	2013	..	146	..	*3	..	143	190
	2014	..	136	..	*3	..	133	174
Brunei Darussalam	2011	588	255	25	205	-3	616	1,542
	2012	434	360	12	146	-6	642	1,583
	2013	501	286	9	163	8	607	1,475
	2014	397	373	15	143	15	597	1,430
Cambodia	2011	..	1,600	..	44	..	1,556	107
	2012	..	1,678	..	55	..	1,623	109
	2013	..	1,676	..	63	..	1,613	107
	2014	..	1,737	..	74	..	1,663	108
China	2011	468,311	62,158	30,690	14,779	6,520	478,481	355
	2012	496,858	59,861	28,402	14,855	11,651	501,812	370
	2013	526,665	60,906	32,767	14,102	7,527	533,176	391
	2014	564,104	46,660	34,190	14,112	8,711	553,751	404
China, Hong Kong SAR	2011	..	20,524	671	16,285	389	3,179	451
	2012	..	19,256	867	15,348	-270	3,311	466
	2013	..	19,419	627	15,787	-261	3,266	456
	2014	..	17,898	513	14,934	-757	3,208	444
China, Macao SAR	2011	..	430	..	..	-1	431	788
	2012	..	436	..	..	2	434	778
	2013	..	397	..	..	8	389	685
	2014	..	405	..	66	-4	343	594
Cyprus	2011	..	2,651	..	493	-124	2,282	2,044
	2012	..	2,618	..	461	30	2,127	1,883
	2013	..	2,324	..	475	23	1,826	1,599
	2014	..	2,292	..	464	-49	1,877	1,627
Georgia	2011	0	999	1	35	-3	966	230
	2012	7	1,020	2	37	-1	989	239
	2013	0	1,071	12	85	5	968	237
	2014	0	1,129	16	80	4	1,029	255
India	2011	216,740	15,689	60,810	4,719	4,799	162,101	130
	2012	231,492	14,413	64,881	4,719	*4,714	171,591	136
	2013	231,188	16,717	69,960	5,152	-160	172,953	135
	2014	236,361	13,928	*59,292	*5,141	-1,916	187,772	145
Indonesia	2011	46,711	33,156	5,722	870	-69	73,344	300
	2012	46,773	37,531	4,927	924	36	78,417	316
	2013	48,949	38,328	5,043	1,249	0	80,985	322
	2014	46,974	39,458	6,161	1,586	0	78,686	309

Production, trade and supply of oil products

Thousand metric tons and kilograms per capita

Country or area	Year	Production	Imports	Exports	Bunkers	Changes in stocks	Supply	
							Total	Per capita
Iran (Islamic Rep. of)	2011	101,680	1,809	18,538	3,434	..	81,518	1,084
	2012	102,918	742	12,828	3,091	..	87,742	1,152
	2013	107,872	1,697	13,946	4,285	..	91,338	1,184
	2014	97,737	2,163	12,855	4,933	176	81,936	1,049
Iraq	2011	28,325	8,850	57	953	-133	36,298	1,139
	2012	29,880	11,433	118	498	241	40,456	1,228
	2013	30,388	12,154	561	574	-225	41,632	1,221
	2014	24,420	13,970	388	715	-665	37,952	1,076
Israel	2011	12,032	4,881	4,492	1,128	-240	11,533	1,525
	2012	11,720	6,542	3,446	1,111	88	13,617	1,770
	2013	13,923	6,106	6,320	1,322	180	12,207	1,561
	2014	14,394	5,857	7,189	1,183	..	11,879	1,496
Japan	2011	175,024	47,831	13,548	9,832	-343	199,818	1,570
	2012	173,985	50,572	12,685	10,058	1,946	199,868	1,572
	2013	177,317	45,575	16,489	10,095	-1,296	197,604	1,556
	2014	167,376	44,190	15,000	9,500	-338	187,404	1,478
Jordan	2011	3,218	2,725	..	121	-187	6,008	889
	2012	3,537	3,598	..	147	176	6,811	974
	2013	3,146	3,153	..	*141	-204	6,362	882
	2014	3,075	4,433	..	129	65	7,314	986
Kazakhstan	2011	14,844	1,279	5,856	177	-406	10,496	634
	2012	14,841	2,428	6,387	197	291	10,394	618
	2013	15,083	2,573	6,328	164	210	10,954	641
	2014	15,977	1,912	6,759	226	630	10,274	591
Korea, Dem.Ppl's.Rep.	2011	508	335	..	..	..	843	34
	2012	508	300	..	..	..	808	33
	2013	561	311	..	..	..	872	35
	2014	515	319	..	..	..	834	33
Korea, Republic of	2011	135,451	31,381	53,274	12,414	248	100,896	2,044
	2012	138,924	35,812	57,053	12,187	-314	105,810	2,133
	2013	133,935	39,088	55,186	12,213	1,537	104,087	2,088
	2014	137,411	39,033	57,319	12,362	380	106,383	2,124
Kuwait	2011	47,198	..	30,101	1,740	-408	15,765	4,867
	2012	48,820	..	29,564	1,905	-468	17,819	5,211
	2013	47,807	..	29,790	1,870	-484	16,631	4,628
	2014	47,972	..	29,452	1,854	-486	17,152	4,570
Kyrgyzstan	2011	90	1,267	103	51	-58	1,261	227
	2012	81	1,720	86	30	80	1,605	284
	2013	77	1,627	141	48	-96	1,611	280
	2014	88	1,485	123	4	234	1,212	207
Lao People's Dem. Rep.	2011	..	*149	..	*10	..	*139	*22
	2012	..	*150	..	*10	..	*140	*22
	2013	..	*156	..	*11	..	*145	*22
	2014	..	*159	..	*11	..	*147	*22
Lebanon	2011	..	5,873	..	253	..	5,620	1,224
	2012	..	6,621	..	237	..	6,384	1,296
	2013	..	6,590	..	289	..	6,301	1,192
	2014	..	7,054	..	260	..	6,794	1,211
Malaysia	2011	25,713	12,473	8,967	1,933	-736	28,022	981
	2012	27,501	12,922	13,104	1,765	-915	26,470	912
	2013	25,613	18,856	11,837	2,674	1,665	28,292	960
	2014	25,023	15,494	10,329	2,656	-1,391	28,923	967

Table 25

Production, trade and supply of oil products

Thousand metric tons and kilograms per capita

Country or area	Year	Production	Imports	Exports	Bunkers	Changes in stocks	Supply	
							Total	Per capita
Maldives	2011	..	381	..	*57	..	324	956
	2012	..	482	..	118	..	365	1,057
	2013	..	505	..	*146	..	360	1,024
	2014	..	677	..	*239	..	438	1,225
Mongolia	2011	..	1,000	..	..	..	1,000	362
	2012	..	1,134	..	..	..	1,134	404
	2013	..	1,197	..	..	..	1,197	419
	2014	..	1,106	..	..	..	1,106	380
Myanmar	2011	806	1,494	..	19	-169	2,450	47
	2012	664	1,402	..	36	15	2,016	38
	2013	513	1,859	..	39	19	2,314	44
	2014	529	3,571	..	40	-706	4,765	89
Nepal	2011	..	1,186	..	*89	..	1,097	40
	2012	..	1,289	..	*94	..	1,195	43
	2013	..	1,431	..	*94	..	1,337	48
	2014	..	1,650	..	*139	*-1	1,512	54
Oman	2011	10,180	315	2,308	955	-37	7,269	2,264
	2012	10,127	309	1,872	1,204	42	7,318	2,064
	2013	9,428	706	1,367	1,257	-97	7,607	1,947
	2014	10,844	43	1,373	1,687	-45	7,872	1,858
Other Asia	2011	41,883	13,998	13,127	3,587	-400	39,567	1,701
	2012	45,973	12,758	16,103	3,245	420	38,963	1,672
	2013	43,382	15,527	16,712	3,412	-782	39,567	1,695
	2014	42,658	17,130	16,632	3,562	15	39,579	1,694
Pakistan	2011	9,181	11,587	872	300	-65	19,661	113
	2012	10,686	10,557	708	277	*-154	20,412	115
	2013	11,761	11,594	919	295	*-140	22,281	123
	2014	12,229	12,443	920	308	*-153	23,597	128
Philippines	2011	8,909	5,696	1,147	1,307	-170	12,322	130
	2012	7,954	6,877	650	1,285	75	12,821	134
	2013	7,384	7,806	634	1,282	-222	13,496	138
	2014	7,843	8,804	700	1,147	182	14,618	147
Qatar	2011	23,952	89	19,474	1,454	-162	3,275	1,719
	2012	24,966	0	20,909	1,797	-301	2,561	1,271
	2013	24,177	125	20,402	1,988	352	1,560	742
	2014	24,040	12	20,421	1,254	671	1,706	785
Saudi Arabia	2011	137,095	10,127	51,031	5,207	2,059	88,925	3,089
	2012	144,697	16,579	60,335	5,463	-748	96,226	3,262
	2013	136,824	24,291	57,869	5,293	-1,080	99,033	3,279
	2014	154,500	23,591	65,041	5,518	730	106,802	3,458
Singapore	2011	51,769	104,442	91,123	48,873	-567	16,781	3,233
	2012	52,148	96,624	86,229	49,084	370	13,089	2,470
	2013	49,221	101,269	82,049	49,374	-256	19,322	3,575
	2014	45,520	102,522	83,267	49,156	98	15,521	2,819
Sri Lanka	2011	1,740	2,826	63	598	-221	4,126	203
	2012	1,494	3,433	0	659	34	4,233	207
	2013	1,569	2,606	139	614	-441	3,864	188
	2014	1,686	2,817	21	797	-503	4,189	203
State of Palestine	2011	..	755	0	..	0	755	181
	2012	..	723	0	..	0	723	168
	2013	..	814	0	..	0	813	184
	2014	..	890	0	..	-54	944	208

Production, trade and supply of oil products

Thousand metric tons and kilograms per capita

Country or area	Year	Production	Imports	Exports	Bunkers	Changes in stocks	Supply	
							Total	Per capita
Syrian Arab Republic	2011	11,869	3,776	1,789	967	0	12,889	629
	2012	8,390	2,749	847	511	0	9,781	490
	2013	5,525	3,408	543	471	0	7,919	410
	2014	5,486	1,760	560	256	0	6,430	343
Tajikistan	2011	25	492	*9	31	..	478	62
	2012	27	625	*10	33	..	609	77
	2013	22	624	..	35	..	611	75
	2014	23	875	..	61	..	837	101
Thailand	2011	44,373	1,875	9,710	*3,342	*-18	33,214	496
	2012	47,118	2,356	10,551	*3,342	*-183	35,764	532
	2013	50,945	2,834	10,598	4,412	413	38,356	569
	2014	48,762	3,454	9,775	4,655	438	37,348	551
Timor-Leste	2011	..	*83	..	*3	..	*80	*74
	2012	..	*98	..	*3	..	*95	*86
	2013	..	*146	..	*3	..	*143	*127
	2014	..	*156	..	*3	..	*153	*132
Turkey	2011	22,991	17,989	7,365	1,285	-61	32,391	441
	2012	23,913	20,197	7,626	1,190	567	34,727	464
	2013	24,675	21,413	7,361	2,232	-54	36,549	479
	2014	22,665	22,358	6,951	3,572	2	34,498	445
Turkmenistan	2011	8,684	..	2,673	480	..	5,531	1,083
	2012	8,366	..	2,141	415	..	5,810	1,123
	2013	8,639	..	2,290	455	..	5,894	1,125
	2014	9,181	..	2,554	470	..	6,157	1,160
United Arab Emirates	2011	33,605	17,337	*14,025	*21,887	..	15,030	1,721
	2012	33,096	17,798	*13,517	*22,620	..	14,757	1,648
	2013	36,387	*20,188	*14,103	*24,866	..	*17,606	*1,948
	2014	33,624	21,242	13,348	20,254	..	21,264	2,340
Uzbekistan	2011	3,583	..	224	..	..	3,359	119
	2012	3,204	..	211	..	..	2,993	105
	2013	3,058	..	213	..	..	2,845	98
	2014	2,864	..	200	..	..	2,664	90
Viet Nam	2011	5,618	11,809	2,125	772	-22	14,552	163
	2012	5,735	10,575	1,689	810	-89	13,900	154
	2013	6,818	8,910	1,261	880	-22	13,609	149
	2014	7,023	10,455	1,339	890	-55	15,304	166
Yemen	2011	2,777	3,374	677	157	..	5,317	219
	2012	1,850	3,622	620	130	..	4,722	190
	2013	3,645	4,529	1,402	194	..	6,578	258
	2014	3,223	4,028	1,295	179	..	5,777	221
Europe	**2011**	**985,704**	**371,670**	**447,227**	**109,860**	**-2,318**	**802,604**	**1,090**
	2012	**982,607**	**367,009**	**469,396**	**104,307**	**-3,649**	**779,563**	**1,058**
	2013	**962,549**	**378,408**	**476,308**	**104,423**	**-934**	**761,159**	**1,032**
	2014	**977,446**	**373,285**	**474,353**	**111,701**	**-74**	**764,751**	**1,036**
Albania	2011	297	1,050	71	*8	0	1,268	439
	2012	111	977	20	*8	0	1,060	368
	2013	40	1,496	34	28	235	1,239	430
	2014	336	1,465	152	27	308	1,314	455
Andorra	2011	..	160	..	0	..	160	1,941
	2012	..	159	..	0	..	159	2,009
	2013	..	156	..	0	..	156	2,052
	2014	..	151	..	0	..	151	2,073

Table 25

Production, trade and supply of oil products

Thousand metric tons and kilograms per capita

Country or area	Year	Production	Imports	Exports	Bunkers	Changes in stocks	Supply	
							Total	Per capita
Austria	2011	9,080	6,286	2,266	708	137	12,255	1,455
	2012	9,085	6,207	2,499	679	-26	12,140	1,436
	2013	9,219	6,166	2,575	649	-109	12,270	1,446
	2014	9,224	6,013	2,555	647	-65	12,100	1,421
Belarus	2011	20,187	4,144	15,599	0	-267	8,999	948
	2012	21,375	6,561	17,652	0	62	10,222	1,077
	2013	20,880	68	13,982	100	44	6,822	718
	2014	21,973	96	14,170	110	20	7,769	818
Belgium	2011	33,963	23,831	23,935	8,471	102	25,286	2,298
	2012	35,949	22,515	26,343	7,610	-115	24,626	2,223
	2013	33,442	28,408	26,946	7,651	576	26,677	2,392
	2014	36,687	25,375	29,108	6,888	270	25,796	2,298
Bosnia and Herzegovina	2011	1,260	759	270	5	0	1,744	455
	2012	1,079	714	203	5	0	1,585	414
	2013	965	727	289	4	23	1,376	360
	2014	989	744	321	4	5	1,403	368
Bulgaria	2011	5,999	2,321	3,539	241	34	4,506	613
	2012	6,425	2,204	4,129	223	-60	4,337	594
	2013	6,443	2,687	4,429	246	154	4,301	593
	2014	6,129	2,981	4,020	247	10	4,833	671
Croatia	2011	3,672	1,633	1,572	123	-132	3,742	870
	2012	3,749	1,641	1,558	105	4	3,723	868
	2013	3,398	1,514	1,540	115	-90	3,347	784
	2014	2,989	2,023	1,549	115	-153	3,501	823
Czechia	2011	7,810	3,178	1,573	300	-41	9,156	869
	2012	7,990	2,967	1,685	284	-36	9,024	856
	2013	7,359	3,321	1,759	274	-63	8,710	826
	2014	8,330	3,303	2,012	283	41	9,297	882
Denmark	2011	6,724	6,444	5,904	1,519	-682	6,427	1,152
	2012	7,684	6,262	6,494	1,331	-219	6,340	1,132
	2013	7,242	7,813	7,197	1,439	211	6,208	1,104
	2014	6,944	8,638	7,731	1,618	441	5,792	1,026
Estonia	2011	..	1,152	10	227	4	911	686
	2012	..	2,272	888	447	2	935	706
	2013	..	1,680	228	460	31	961	728
	2014	..	1,903	552	371	17	963	732
Faeroe Islands	2011	..	*218	..	34	..	*184	*3,794
	2012	..	*223	..	30	..	*193	*3,990
	2013	..	*245	..	24	..	*221	*4,576
	2014	..	*218	..	23	..	*195	*4,036
Finland	2011	15,196	5,518	7,183	820	191	12,520	2,320
	2012	14,705	4,779	7,597	720	-414	11,581	2,135
	2013	15,184	6,068	9,051	747	689	10,765	1,974
	2014	14,836	5,744	8,502	704	-70	11,444	2,088
France	2011	76,073	40,566	22,673	8,214	134	85,618	1,352
	2012	65,225	43,273	19,928	7,917	-867	81,520	1,282
	2013	62,013	41,411	18,822	7,650	-21	76,973	1,205
	2014	62,336	41,287	18,775	7,369	351	77,128	1,202
Germany	2011	109,259	32,981	18,035	10,280	-748	114,673	1,426
	2012	110,194	32,241	18,688	10,597	-310	113,460	1,410
	2013	107,193	37,649	20,197	10,505	-176	114,316	1,419
	2014	106,304	37,039	21,236	10,183	-847	112,771	1,398

Production, trade and supply of oil products

Thousand metric tons and kilograms per capita

Country or area	Year	Production	Imports	Exports	Bunkers	Changes in stocks	Supply	
							Total	Per capita
Gibraltar	2011	..	4,440	..	4,272	..	168	5,408
	2012	..	4,479	..	4,307	..	172	5,477
	2013	..	4,126	..	3,942	..	184	5,801
	2014	..	3,130	..	2,934	..	196	6,126
Greece	2011	19,370	8,156	8,865	3,580	-407	15,488	1,389
	2012	23,909	7,105	12,127	2,977	114	15,796	1,422
	2013	24,190	7,469	14,032	2,857	-173	14,943	1,352
	2014	27,690	8,444	15,492	2,691	-60	18,011	1,637
Hungary	2011	9,088	2,223	2,968	232	-86	8,197	821
	2012	8,548	2,083	2,867	166	-40	7,638	767
	2013	8,382	2,276	2,900	162	52	7,544	760
	2014	8,703	2,703	3,222	171	-40	8,053	814
Iceland	2011	..	695	..	197	4	494	1,539
	2012	..	688	..	199	5	484	1,497
	2013	2	705	..	183	12	512	1,573
	2014	2	746	..	200	-7	555	1,696
Ireland	2011	2,993	5,551	1,590	753	93	6,108	1,313
	2012	3,099	5,211	1,772	647	87	5,804	1,243
	2013	2,882	5,049	1,241	738	-28	5,980	1,280
	2014	2,822	5,018	1,287	831	-47	5,769	1,234
Italy	2011	88,985	17,713	26,626	5,761	529	73,782	1,236
	2012	84,850	16,586	28,981	5,590	-73	66,938	1,120
	2013	72,125	19,385	23,902	5,193	-480	62,895	1,052
	2014	68,986	17,249	20,534	4,988	-817	61,530	1,029
Latvia	2011	..	2,067	454	332	30	1,251	606
	2012	..	2,131	511	359	28	1,233	605
	2013	..	2,297	673	363	7	1,254	623
	2014	..	2,326	816	342	-123	1,291	649
Lithuania	2011	9,376	1,087	7,736	200	-44	2,571	837
	2012	8,977	1,396	7,593	186	-86	2,680	888
	2013	9,486	1,520	8,229	158	-44	2,663	899
	2014	7,943	1,771	6,903	86	-74	2,799	960
Luxembourg	2011	..	2,932	7	392	-11	2,544	4,892
	2012	..	2,857	8	361	13	2,475	4,648
	2013	..	2,812	7	363	9	2,433	4,465
	2014	..	2,725	5	394	14	2,312	4,154
Malta	2011	..	2,358	3	1,482	36	837	2,021
	2012	..	2,245	8	1,321	32	884	2,127
	2013	..	2,337	151	1,321	97	768	1,843
	2014	..	2,301	206	1,364	-38	769	1,841
Montenegro	2011	..	316	15	10	..	291	467
	2012	..	294	12	12	..	270	433
	2013	..	286	14	13	..	259	415
	2014	..	294	22	17	-3	258	413
Netherlands	2011	60,640	85,277	98,190	18,572	-1,137	30,292	1,815
	2012	60,626	89,595	101,481	17,205	725	30,810	1,839
	2013	58,985	89,390	100,939	16,695	186	30,555	1,818
	2014	60,334	87,032	100,143	16,620	400	30,203	1,791
Norway	2011	24,343	4,341	15,673	794	257	11,960	2,414
	2012	24,299	4,320	14,777	722	181	12,939	2,578
	2013	25,202	4,962	14,444	839	99	14,782	2,908
	2014	23,109	4,294	15,342	637	-180	11,604	2,254

Table 25

Production, trade and supply of oil products

Thousand metric tons and kilograms per capita

Country or area	Year	Production	Imports	Exports	Bunkers	Changes in stocks	Supply	
							Total	Per capita
Poland	2011	26,633	5,852	4,580	658	-73	27,320	708
	2012	27,631	4,903	5,771	670	-116	26,209	679
	2013	26,615	4,428	6,571	653	-141	23,960	620
	2014	26,317	4,906	7,115	717	-110	23,501	609
Portugal	2011	11,163	4,315	2,856	1,472	9	11,141	1,055
	2012	11,983	3,135	3,703	1,527	-80	9,968	948
	2013	14,244	3,542	5,526	1,578	-600	11,282	1,079
	2014	13,155	4,313	4,734	1,590	-27	11,171	1,074
Republic of Moldova	2011	19	803	15	13	10	784	192
	2012	18	745	18	22	-10	733	180
	2013	19	789	33	21	1	753	185
	2014	18	781	18	25	-13	769	189
Romania	2011	10,924	2,384	3,631	135	-283	9,825	489
	2012	10,280	2,726	3,306	129	-249	9,820	492
	2013	10,435	2,434	3,788	182	-64	8,963	453
	2014	12,158	2,394	4,614	257	-60	9,741	496
Russian Federation	2011	249,860	4,345	105,229	9,312	523	139,141	972
	2012	255,290	1,856	107,978	8,227	294	140,647	982
	2013	266,656	1,927	119,099	11,139	566	137,779	961
	2014	283,328	2,270	117,993	20,567	39	146,999	1,025
Serbia	2011	2,509	1,535	283	46	0	3,715	412
	2012	2,298	1,377	292	41	41	3,301	367
	2013	3,258	858	564	51	-100	3,601	403
	2014	3,217	1,113	658	82	3	3,587	403
Slovakia	2011	7,082	1,296	3,923	42	19	4,394	812
	2012	6,201	1,262	3,506	36	20	3,901	720
	2013	6,871	1,380	4,112	40	-26	4,125	761
	2014	5,962	1,302	3,644	34	14	3,572	659
Slovenia	2011	36	3,270	710	56	-9	2,549	1,238
	2012	52	3,438	812	74	107	2,497	1,210
	2013	60	3,260	935	87	-65	2,363	1,144
	2014	41	3,707	1,393	83	-53	2,325	1,125
Spain	2011	58,749	27,196	13,054	12,380	98	60,413	1,293
	2012	63,890	20,040	19,539	12,134	-1,583	53,840	1,154
	2013	61,668	20,025	22,278	10,905	-1,297	49,807	1,072
	2014	61,532	20,439	21,666	11,570	992	47,743	1,032
Sweden	2011	20,214	8,314	11,474	2,490	-309	14,873	1,572
	2012	22,105	6,821	13,379	2,422	-501	13,626	1,428
	2013	18,378	7,175	9,916	2,414	-226	13,449	1,397
	2014	20,766	7,836	12,565	2,464	321	13,252	1,366
Switzerland	2011	4,421	7,056	441	1,477	-159	9,718	1,220
	2012	3,439	8,260	309	1,501	159	9,730	1,207
	2013	4,929	7,491	578	1,506	-126	10,462	1,283
	2014	4,967	6,273	523	1,535	19	9,163	1,111
T.F.Yug.Rep. Macedonia	2011	680	642	384	4	-34	968	469
	2012	244	991	275	8	-14	966	467
	2013	78	1,009	203	10	-50	924	446
	2014	1	1,097	179	12	8	900	433
Ukraine	2011	9,366	7,595	4,183	240	-32	12,570	276
	2012	4,968	8,180	1,663	298	-169	11,356	251
	2013	4,018	7,097	1,007	*123	239	9,746	216
	2014	3,321	7,881	786	*128	-398	10,686	237

Production, trade and supply of oil products

Thousand metric tons and kilograms per capita

Country or area	Year	Production	Imports	Exports	Bunkers	Changes in stocks	Supply	
							Total	Per capita
United Kingdom	2011	79,733	29,670	31,707	14,008	-74	63,762	1,009
	2012	76,330	31,289	31,024	13,210	-555	63,940	1,006
	2013	70,688	34,970	28,117	12,995	-286	64,832	1,014
	2014	65,997	33,960	23,809	12,773	-161	63,536	988
Oceania	**2011**	**40,727**	**19,788**	**3,142**	**5,067**	**-129**	**52,435**	**1,417**
	2012	**39,383**	**22,651**	**3,371**	**5,345**	**112**	**53,205**	**1,415**
	2013	**38,939**	**25,508**	**3,599**	**5,389**	**40**	**55,420**	**1,451**
	2014	**36,423**	**26,380**	**3,268**	**5,715**	**-390**	**54,209**	**1,399**
Australia	2011	34,296	15,068	2,849	3,716	-200	42,999	1,907
	2012	32,618	18,197	3,056	4,015	152	43,592	1,903
	2013	32,188	20,009	2,983	4,029	-63	45,248	1,944
	2014	29,752	20,723	2,644	4,391	-358	43,798	1,854
Cook Islands	2011	..	*25	..	*2	..	*23	*1,127
	2012	..	*25	..	*2	..	*23	*1,121
	2013	..	*25	..	*2	..	*23	*1,115
	2014	..	*25	..	*2	..	*23	*1,110
Fiji	2011	..	384	0	*51	..	332	383
	2012	..	377	0	*52	..	325	371
	2013	..	*407	0	*56	..	*351	*399
	2014	..	*422	0	*66	..	*356	*401
French Polynesia	2011	..	328	..	*55	..	273	1,009
	2012	..	321	..	*53	..	267	976
	2013	..	325	..	*54	..	270	977
	2014	..	317	..	*53	..	264	942
Kiribati	2011	..	*24	..	*4	..	*20	*189
	2012	..	*24	..	*4	..	*20	*187
	2013	..	*24	..	*4	..	*20	*186
	2014	..	25	..	*4	..	21	186
Marshall Islands	2011	..	*34	..	..	..	*34	*644
	2012	..	33	..	..	..	33	631
	2013	..	*34	..	..	..	*34	*641
	2014	..	*33	..	..	..	*33	*628
Micronesia (Fed. States of)	2011	..	*43	..	*1	..	*43	*411
	2012	..	*47	..	*1	..	*47	*450
	2013	..	*50	..	*1	..	*49	*469
	2014	..	*52	..	*1	..	*51	*495
Nauru	2011	..	*18	..	*5	0	*13	*1,296
	2012	..	*18	..	*5	0	*13	*1,308
	2013	..	*20	..	*5	0	*15	*1,433
	2014	..	*21	..	6	0	*15	*1,501
New Caledonia	2011	..	797	*5	*32	*-12	772	3,090
	2012	..	804	*3	*31	*4	766	3,026
	2013	..	964	..	*13	8	943	3,678
	2014	..	898	..	*10	*2	886	3,410
New Zealand	2011	5,526	2,002	288	1,074	98	6,068	1,378
	2012	5,751	1,932	296	1,074	-10	6,323	1,425
	2013	5,524	2,249	259	1,115	84	6,315	1,414
	2014	5,423	2,450	264	1,074	-34	6,569	1,461
Niue	2011	..	2	..	0	..	2	1,041
	2012	..	2	..	0	..	2	1,098
	2013	..	2	..	0	..	2	1,164
	2014	..	2	..	0	..	2	1,189

Table 25

Production, trade and supply of oil products

Thousand metric tons and kilograms per capita

Country or area	Year	Production	Imports	Exports	Bunkers	Changes in stocks	Supply	
							Total	Per capita
Palau	2011	..	*101	..	*17	..	*84	*4,052
	2012	..	*100	..	*17	..	*83	*3,975
	2013	..	*101	..	*17	..	*84	*4,015
	2014	..	*104	..	*18	..	*86	*4,062
Papua New Guinea	2011	905	711	0	*79	-16	1,553	222
	2012	1,014	521	16	*59	-34	1,493	209
	2013	1,227	1,050	357	*63	12	1,845	252
	2014	*1,248	*1,039	*360	*57	0	*1,870	*251
Samoa	2011	..	76	..	11	0	65	346
	2012	..	*76	..	*11	..	*65	*343
	2013	..	76	..	11	-1	67	352
	2014	..	*78	..	*11	..	*66	*346
Solomon Islands	2011	..	*70	..	*4	..	*66	*122
	2012	..	*71	..	*4	..	*66	*120
	2013	..	*72	..	*5	..	*67	*120
	2014	..	*73	..	*4	..	*68	*119
Tonga	2011	..	39	..	*4	..	35	335
	2012	..	40	..	*4	..	36	344
	2013	..	40	..	*1	..	38	362
	2014	..	42	..	*5	..	38	358
Tuvalu	2011	..	2	..	..	..	2	219
	2012	..	*3	..	..	..	*3	*296
	2013	..	*3	..	..	..	*3	*295
	2014	..	*3	..	..	..	*3	*305
Vanuatu	2011	..	55	..	*10	0	45	184
	2012	..	50	..	*10	0	40	162
	2013	..	48	..	*10	0	37	148
	2014	..	64	..	*12	0	53	203
Wallis and Futuna Is.	2011	..	9	..	1	..	9	635
	2012	..	9	..	1	..	8	610
	2013	..	9	..	1	..	8	601
	2014	..	9	..	1	..	8	595

Production of natural gas liquids – by end use

Thousand metric tons

Table Notes

Production refers to production of natural gas liquids.

End-use products refers to the output of natural gas liquids plants, arranged by product. **Net imports** refers to imports minus exports. **Inputs of NGLs to refineries** refers to quantities of natural gas liquids that are transformed in oil refineries. Data may not add up due to statistical differences or the presence of other natural gas liquid flows such as stock changes, transfers or energy industries own use.

Please refer to the Definitions Section on pages ix to xv for the appropriate product description /classification.

Figure 66: Production of natural gas liquids by region, 1990-2014

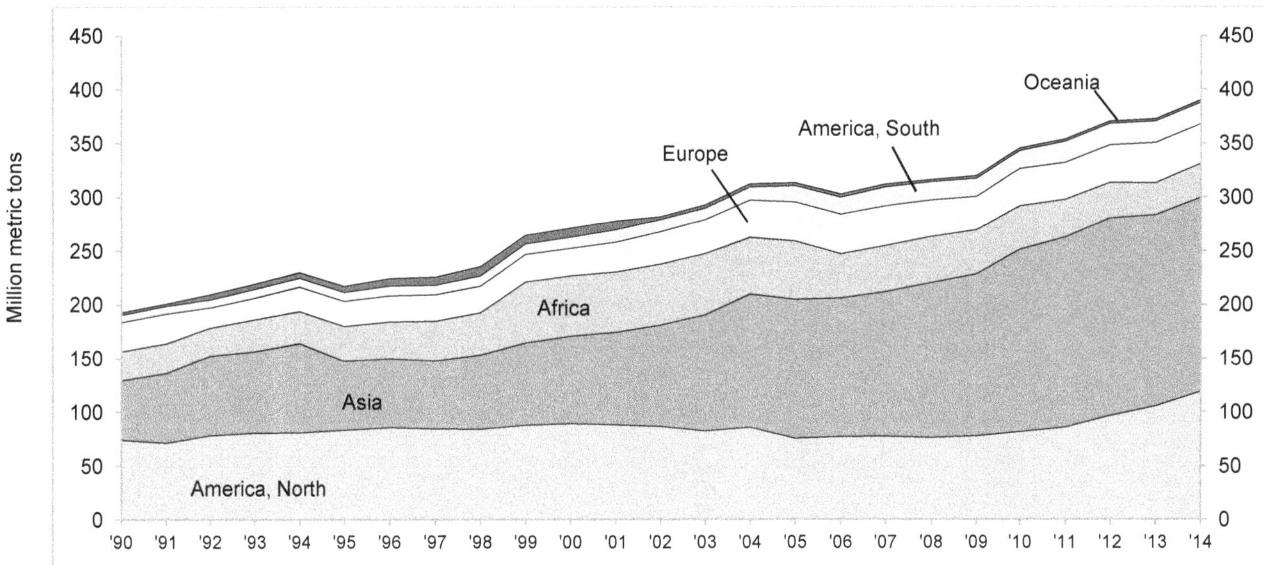

Figure 67: Production of natural gas liquids by end-use product, 2014

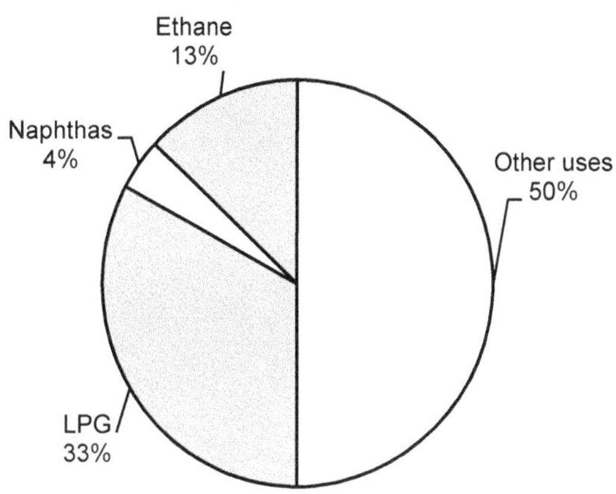

Table 26

Production of natural gas liquids - by end use

Thousand metric tons

Country or area	Year	Total production	End-use products				Net imports	Inputs of NGLs to refineries
			LPG	Naphthas	Ethane	Other products		
World	**2011**	**353,611**	**114,787**	**14,490**	**46,891**	**6,462**	**-43,618**	**105,150**
	2012	**370,259**	**120,079**	**16,213**	**48,848**	**7,456**	**-57,289**	**98,022**
	2013	**372,291**	**124,947**	**16,980**	**51,352**	**7,590**	**-56,490**	**100,953**
	2014	**390,022**	**128,388**	**15,859**	**50,991**	**7,425**	**-65,179**	**96,352**
Africa	**2011**	**35,075**	**9,914**	**630**	**0**	**..**	**-15,162**	**9,361**
	2012	**33,027**	**9,506**	**701**	**27**	**..**	**-17,720**	**4,911**
	2013	**30,013**	**9,505**	**1,196**	**0**	**..**	**-15,918**	**3,288**
	2014	**31,599**	**10,646**	**439**	**0**	**..**	**-15,342**	**5,239**
Algeria	2011	17,746	7,497	196	0	..	-5,946	4,146
	2012	16,474	6,959	166	27	..	-5,105	4,118
	2013	15,656	6,887	171	0	..	-5,416	3,115
	2014	18,318	8,372	168	0	..	-4,805	5,067
Angola	2011	566	566	..	..	..	..	..
	2012	568	568	..	..	..	..	..
	2013	677	677	..	..	..	..	..
	2014	520	520	..	..	..	..	..
Congo	2011	308	..	..	..	..	-265	17
	2012	517	..	..	..	..	-492	25
	2013	601	..	..	..	..	-575	26
	2014	730	..	..	..	..	-705	25
Côte d'Ivoire	2011	..	..	..	..	..	..	..
	2012	12	12	..	..	..	..	..
	2013	0	0	..	..	..	..	..
	2014	2	2	..	..	..	..	..
Egypt	2011	6,464	1,475	..	..	..	..	5,002
	2012	2,077	1,481	..	..	..	..	571
	2013	*1,542	1,542	..	..	..	..	0
	2014	1,320	1,320	..	..	..	..	0
Libya	2011	1,814	250	434	..	..	-1,130	..
	2012	4,624	353	535	..	..	-3,736	..
	2013	3,541	256	1,025	..	..	-2,260	..
	2014	2,154	292	271	..	..	-1,591	..
South Africa	2011	196	..	..	..	..	..	196
	2012	197	..	..	..	..	..	197
	2013	147	..	..	..	..	..	147
	2014	147	..	..	..	..	..	147
Tunisia	2011	200	126	..	..	..	-40	..
	2012	231	133	..	..	..	-60	..
	2013	272	143	..	..	..	-90	..
	2014	245	140	..	..	..	-78	..
America, North	**2011**	**85,890**	**27,627**	**..**	**22,367**	**2,054**	**-3,380**	**19,936**
	2012	**96,399**	**29,615**	**..**	**23,839**	**2,582**	**-8,772**	**21,688**
	2013	**105,485**	**36,824**	**..**	**23,422**	**2,295**	**-11,957**	**21,271**
	2014	**119,086**	**35,228**	**..**	**25,329**	**1,753**	**-20,007**	**20,490**
Belize	2011	*2	2	..	..	..	..	..
	2012	*1	1	..	..	..	..	..
	2013	*1	1	..	..	..	..	..
	2014	*1	1	..	..	..	..	..
Canada	2011	12,769	..	..	..	..	-2,177	1,435
	2012	13,022	..	..	..	..	-2,616	2,291
	2013	13,544	..	..	..	..	-1,736	2,599
	2014	13,663	..	..	..	..	-1,497	1,510

Production of natural gas liquids - by end use

Thousand metric tons

Country or area	Year	Total production	End-use products				Net imports	Inputs of NGLs to refineries
			LPG	Naphthas	Ethane	Other products		
Curaçao	2011	..	..	..	..	..	..	..
	2012	..	..	..	..	..	12	..
	2013	..	..	..	..	..	36	..
	2014	..	..	..	..	..	44	..
Mexico	2011	10,927	5,700	..	2,505	..	..	2,612
	2012	10,371	5,304	..	2,401	1	..	2,438
	2013	10,225	5,366	..	2,264	0	..	2,393
	2014	10,217	5,178	..	2,394	..	..	2,416
Trinidad and Tobago	2011	1,506	892	..	..	667	..	..
	2012	1,210	696	..	..	565	..	..
	2013	1,180	671	..	..	561	..	..
	2014	1,126	647	..	..	526	..	..
United States	2011	60,686	21,033	..	19,862	1,387	-1,203	15,889
	2012	71,795	23,614	..	21,438	2,016	-6,168	16,959
	2013	80,535	30,786	..	21,158	1,734	-10,257	16,279
	2014	94,079	29,402	..	22,935	1,227	-18,554	16,564
America, South	**2011**	**19,480**	**9,299**	**..**	**1,019**	**3,148**	**1**	**2,607**
	2012	**19,827**	**10,046**	**..**	**1,048**	**3,317**	**-9**	**1,887**
	2013	**19,869**	**10,354**	**..**	**1,046**	**3,443**	**-15**	**1,624**
	2014	**19,895**	**10,444**	**..**	**1,131**	**3,948**	**-15**	**1,027**
Argentina	2011	3,004	1,521	..	1,019	292	..	82
	2012	3,162	1,778	..	1,048	177	..	69
	2013	2,839	*1,681	..	1,046	0	..	61
	2014	2,934	1,533	..	1,131	..	..	60
Bolivia (Plur. State of)	2011	242	242	..	..	..	..	..
	2012	243	243	..	..	..	..	..
	2013	270	270	..	..	..	..	..
	2014	312	312	..	..	..	..	..
Brazil	2011	4,658	821	..	..	828	..	..
	2012	5,251	934	..	..	1,087	..	..
	2013	5,247	1,054	..	..	978	..	..
	2014	5,636	1,189	..	..	1,577	..	..
Chile	2011	316	..	..	..	..	..	..
	2012	199	..	..	..	..	..	..
	2013	160	..	..	..	..	..	..
	2014	153	..	..	..	..	..	..
Colombia	2011	96	*39	..	..	*62	1	..
	2012	*124	*39	..	..	76	-9	..
	2013	89	34	..	..	86	-15	..
	2014	*142	34	..	..	93	*-15	..
Ecuador	2011	*161	116	..	..	45	..	..
	2012	*168	123	..	..	46	..	..
	2013	*219	157	..	..	62	..	..
	2014	*209	149	..	..	60	..	..
Peru	2011	3,137	1,219	..	..	1,921	..	..
	2012	3,265	1,332	..	..	1,931	..	..
	2013	3,946	1,622	..	..	2,317	..	..
	2014	3,901	1,586	..	..	2,218	..	..
Venezuela (Bolivar. Rep.)	2011	7,866	5,341	..	..	..	..	2,525
	2012	7,415	5,597	..	..	..	..	1,818
	2013	7,099	5,536	..	..	..	..	1,563
	2014	6,608	5,641	..	..	..	..	967

Table 26

Production of natural gas liquids - by end use

Thousand metric tons

Country or area	Year	Total production	End-use products				Net imports	Inputs of NGLs to refineries
			LPG	Naphthas	Ethane	Other products		
Asia	**2011**	**176,643**	**59,114**	**11,775**	**21,524**	**134**	**-31,120**	**48,160**
	2012	**183,890**	**62,398**	**13,093**	**21,785**	**130**	**-37,479**	**43,550**
	2013	**177,730**	**59,888**	**13,406**	**24,519**	**171**	**-36,639**	**45,496**
	2014	**180,717**	**63,332**	**14,022**	**21,883**	**106**	**-39,272**	**39,692**
Azerbaijan	2011	2,245	..	..	..	..	-1,760	421
	2012	2,411	..	..	..	..	-2,028	332
	2013	2,781	..	..	..	..	-2,447	284
	2014	2,680	..	..	..	..	-2,324	262
Bahrain	2011	378	..	..	..	..	-394	..
	2012	368	..	..	..	..	-369	..
	2013	336	..	..	..	..	-320	..
	2014	357	..	..	..	..	-354	..
Bangladesh	2011	237	6	..	..	82	-39	96
	2012	234	8	..	..	86	-11	111
	2013	249	8	..	..	138	..	82
	2014	246	9	..	..	71	..	76
Brunei Darussalam	2011	544	15	..	..	..	-402	125
	2012	572	16	..	..	..	-420	139
	2013	583	16	..	..	..	-412	161
	2014	538	16	..	..	..	-408	117
China, Macao SAR	2011	..	..	..	..	..	..	..
	2012	..	..	..	..	..	..	..
	2013	..	..	..	..	..	..	..
	2014	..	..	..	..	..	44	..
India	2011	4,641	2,214	..	..	..	..	..
	2012	4,517	2,130	..	..	..	..	..
	2013	4,301	2,140	1,468	..	..	..	..
	2014	*3,245	1,887	1,358	..	..	..	..
Indonesia	2011	*4,500	1,581	..	..	..	*-2,693	*26
	2012	*4,666	1,831	..	..	..	*-2,909	..
	2013	4,320	1,824	..	..	..	-2,567	..
	2014	4,670	2,381	..	..	..	-1,968	*321
Iran (Islamic Rep. of)	2011	18,417	1,823	..	5,180	..	-8,889	1,019
	2012	18,234	1,794	..	4,220	..	-8,492	1,305
	2013	19,135	1,938	..	7,814	..	-9,038	1,640
	2014	22,272	5,169	..	4,884	..	-9,581	1,501
Iraq	2011	1,427	1,112	..	..	..	..	315
	2012	1,573	1,226	..	..	..	..	347
	2013	1,341	1,045	..	..	..	..	296
	2014	731	570	..	..	..	..	161
Japan	2011	386	..	..	..	..	8,817	8,854
	2012	342	..	..	..	..	8,443	8,412
	2013	289	..	..	..	..	9,882	9,886
	2014	261	..	..	..	..	6,665	6,703
Kazakhstan	2011	12,296	1,217	..	..	..	*-11,079	..
	2012	12,749	1,313	..	..	..	*-11,436	..
	2013	12,303	1,321	..	..	..	*-10,982	..
	2014	12,918	1,433	..	..	..	*-11,485	..
Kuwait	2011	6,224	4,289	..	1,935	..	..	..
	2012	6,692	4,757	..	1,935	..	..	..
	2013	6,555	4,491	..	1,935	..	..	..
	2014	6,333	4,398	..	1,935	..	..	..

Production of natural gas liquids - by end use

Thousand metric tons

Country or area	Year	Total production	End-use products				Net imports	Inputs of NGLs to refineries
			LPG	Naphthas	Ethane	Other products		
Malaysia	2011	2,434	2,434	..	..	..	..	..
	2012	2,080	2,080	..	..	..	..	..
	2013	1,237	1,237	..	..	..	..	..
	2014	1,649	1,237	..	..	..	..	..
Myanmar	2011	8	7	5	..	4	4	..
	2012	8	8	4	..	4	2	..
	2013	8	0	3	..	..	2	..
	2014	10	4	4	..	..	1	..
Oman	2011	251	251	..	..	..	..	..
	2012	251	251	..	..	..	..	..
	2013	268	268	..	..	..	..	..
	2014	246	246	..	..	..	..	..
Pakistan	2011	217	217	..	..	..	..	..
	2012	218	218	..	..	..	..	..
	2013	265	265	..	..	..	..	..
	2014	253	253	..	..	..	..	..
Qatar	2011	37,533	10,017	..	1,023	..	-17,385	9,108
	2012	39,701	10,463	..	1,025	..	-18,625	9,588
	2013	39,872	10,521	..	1,016	..	-19,508	8,827
	2014	39,487	10,605	..	1,397	..	-18,951	8,534
Saudi Arabia	2011	56,080	24,974	9,233	10,914	..	-1,524	9,435
	2012	58,716	26,828	10,455	11,796	..	-1,019	8,618
	2013	55,466	25,078	9,200	10,976	..	-671	9,541
	2014	56,893	25,931	10,224	10,914	..	-212	9,612
Singapore	2011	..	..	..	..	..	*4,819	*4,819
	2012	..	..	..	..	..	..	..
	2013	..	..	..	..	..	..	..
	2014	..	..	..	..	..	..	..
Syrian Arab Republic	2011	323	323	..	..	..	..	..
	2012	238	238	..	..	..	..	..
	2013	133	133	..	..	..	..	..
	2014	95	95	..	..	..	..	..
Thailand	2011	5,758	3,427	..	1,798	..	-107	..
	2012	6,431	3,716	..	2,109	..	-158	..
	2013	6,192	3,524	..	2,074	..	-170	..
	2014	6,375	3,651	..	2,106	..	-211	..
Turkmenistan	2011	775	392	..	..	..	..	383
	2012	808	454	..	..	..	..	354
	2013	918	728	..	..	..	..	190
	2014	937	634	..	..	..	..	303
United Arab Emirates	2011	20,655	4,037	2,537	674	..	..	13,559
	2012	21,708	4,191	2,634	700	..	..	14,344
	2013	19,737	*4,351	*2,735	704	..	..	14,589
	2014	19,063	3,878	2,436	647	..	..	12,102
Viet Nam	2011	293	245	..	..	48	..	..
	2012	306	266	..	..	40	..	..
	2013	325	292	..	..	33	..	..
	2014	341	306	..	..	35	..	..
Yemen	2011	533	533	..	..	..	..	..
	2012	610	610	..	..	..	..	..
	2013	708	708	..	..	..	..	..
	2014	629	629	..	..	..	..	..

Table 26

Production of natural gas liquids - by end use

Thousand metric tons

Country or area	Year	Total production	End-use products				Net imports	Inputs of NGLs to refineries
			LPG	Naphthas	Ethane	Other products		
Europe	**2011**	**34,357**	**6,682**	**2,085**	**1,981**	**1,126**	**6,057**	**25,086**
	2012	**34,964**	**6,375**	**2,419**	**2,149**	**1,427**	**6,698**	**25,986**
	2013	**37,182**	**6,373**	**2,378**	**2,365**	**1,681**	**8,048**	**29,274**
	2014	**36,519**	**6,541**	**1,398**	**2,648**	**1,618**	**9,464**	**29,904**
Austria	2011	120	..	..	..	..	113	194
	2012	80	..	..	..	..	44	124
	2013	30	..	..	..	..	51	81
	2014	23	..	..	..	..	152	175
Belgium	2011	..	..	..	..	..	115	114
	2012	..	..	..	..	..	0	1
	2013	..	..	..	..	..	360	333
	2014	..	..	..	..	..	544	242
Croatia	2011	104	52	..	29	..	..	23
	2012	65	42	..	0	..	..	23
	2013	61	41	..	0	..	..	20
	2014	54	36	..	0	..	..	18
Finland	2011	..	..	..	..	..	568	740
	2012	..	..	..	..	..	845	863
	2013	..	..	..	..	..	688	780
	2014	..	..	..	..	..	641	652
France	2011	28	..	..	..	2	0	0
	2012	27	..	..	..	27	0	0
	2013	19	..	..	..	77	321	105
	2014	0	..	..	..	0	407	9
Hungary	2011	287	226	..	..	..	93	146
	2012	309	208	..	..	..	110	185
	2013	272	243	..	..	..	116	128
	2014	244	203	..	..	..	106	121
Netherlands	2011	383	..	..	..	..	7,917	3,954
	2012	356	..	..	..	..	7,792	4,119
	2013	397	..	..	..	..	7,624	3,861
	2014	369	..	..	..	..	6,788	2,754
Norway	2011	8,477	5,274	2,085	1,118	..	..	0
	2012	9,082	5,297	2,419	1,366	..	..	0
	2013	8,982	5,082	2,378	1,522	..	..	0
	2014	9,517	5,409	1,398	1,705	..	..	1,005
Romania	2011	147	..	..	..	..	..	150
	2012	122	..	..	..	..	..	114
	2013	88	..	..	..	..	..	83
	2014	89	..	..	..	..	..	84
Russian Federation	2011	20,451	..	..	..	..	-2,631	17,825
	2012	21,322	..	..	..	..	-2,409	18,892
	2013	24,205	..	..	..	..	-2,075	22,125
	2014	23,034	..	..	..	..	129	23,155
Serbia	2011	31	..	..	..	..	8	39
	2012	34	..	..	..	..	5	39
	2013	36	..	..	..	..	66	103
	2014	36	..	..	..	..	60	96
Slovakia	2011	3	..	..	..	..	..	3
	2012	4	..	..	..	..	..	4
	2013	3	..	..	..	..	..	3
	2014	3	..	..	..	..	..	3

Production of natural gas liquids - by end use

Thousand metric tons

Country or area	Year	Total production	End-use products				Net imports	Inputs of NGLs to refineries
			LPG	Naphthas	Ethane	Other products		
Ukraine	2011	925	..	..	..	..	0	873
	2012	1,055	..	..	..	..	25	872
	2013	899	..	..	..	..	32	782
	2014	697	..	..	..	..	..	703
United Kingdom	2011	3,401	1,130	..	834	1,124	-126	1,025
	2012	2,508	828	..	783	1,400	286	750
	2013	2,190	1,007	..	843	1,604	865	870
	2014	2,453	893	..	943	1,618	637	887
Oceania	**2011**	**2,166**	**2,150**	**..**	**..**	**..**	**-14**	**..**
	2012	**2,152**	**2,139**	**..**	**..**	**..**	**-7**	**..**
	2013	**2,012**	**2,002**	**..**	**..**	**..**	**-9**	**..**
	2014	**2,206**	**2,197**	**..**	**..**	**..**	**-7**	**..**
Australia	2011	2,002	2,002	..	..	..	..	..
	2012	1,973	1,973	..	..	..	..	..
	2013	1,825	1,825	..	..	..	..	..
	2014	1,987	1,987	..	..	..	..	..
New Zealand	2011	164	148	..	..	..	-14	..
	2012	179	166	..	..	..	-7	..
	2013	187	177	..	..	..	-9	..
	2014	219	210	..	..	..	-7	..

Table 27

Production, trade and supply of natural gas

Terajoules

Table Notes

Production includes quantities from other sources, which would include any quantities of biogas, LPG or manufactured gas blended with natural gas.

Please refer to the Definitions Section on pages ix to xv for the appropriate product description /classification.

Figure 68: Production of natural gas by region, 1990-2014

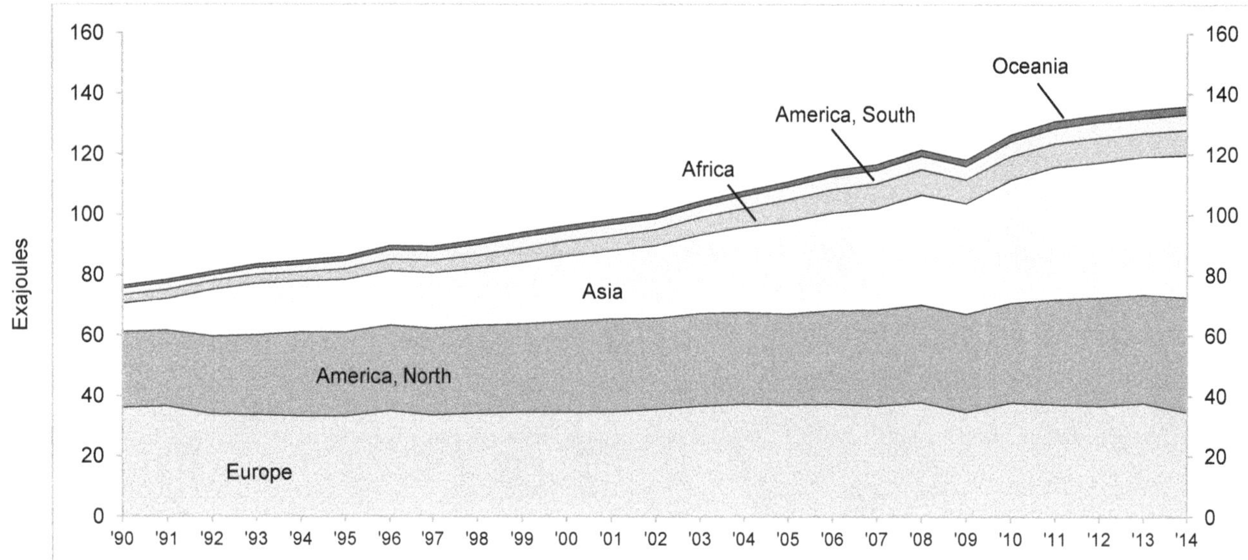

Figure 69: Major natural gas producing countries, 2014

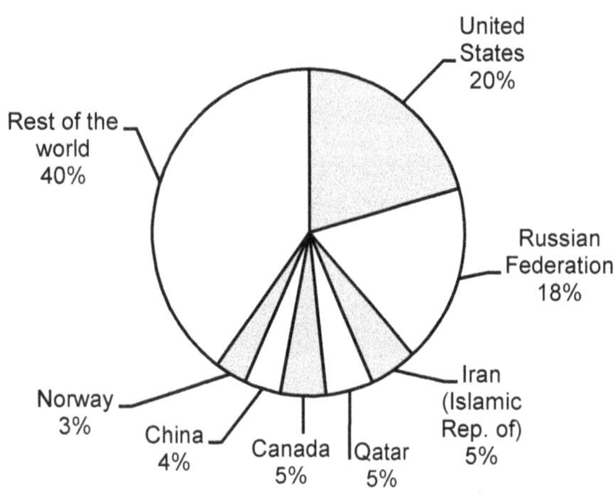

Figure 70: Major natural gas consuming countries, 2014

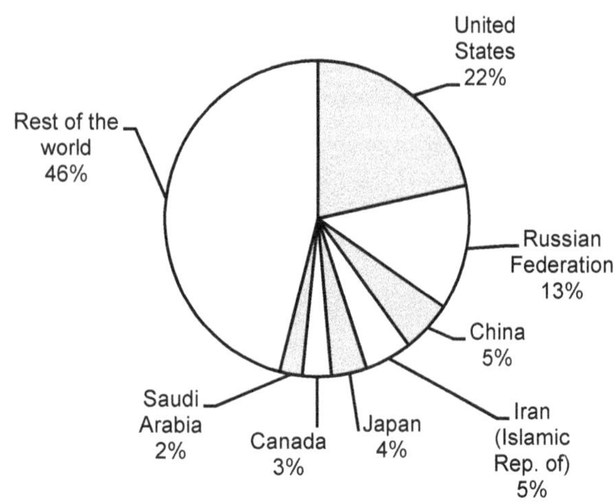

Production, trade and supply of natural gas

Terajoules and megajoules per capita

Country or area	Year	Production	Imports	Exports	Changes in stocks	Supply	
						Total	Per capita
World	**2011**	**130,991,068**	**39,957,667**	**39,932,666**	**809,251**	**130,206,817**	**18,567**
	2012	**133,006,038**	**39,608,324**	**40,347,677**	**-117,428**	**132,384,112**	**18,652**
	2013	**134,656,863**	**40,920,965**	**41,628,194**	**-677,465**	**134,627,099**	**18,746**
	2014	**135,952,475**	**39,288,389**	**40,024,671**	**407,654**	**134,808,539**	**18,554**
Africa	**2011**	**7,920,683**	**284,869**	**3,695,891**	**835**	**4,508,827**	**4,212**
	2012	**8,274,585**	**321,791**	**3,858,324**	**539**	**4,737,514**	**4,311**
	2013	**7,784,349**	**333,096**	**3,361,177**	**207**	**4,756,061**	**4,218**
	2014	**8,288,465**	**332,194**	**3,557,730**	**216**	**5,062,714**	**4,377**
Algeria	2011	3,237,585	..	2,084,557	..	1,153,028	31,403
	2012	3,374,117	..	2,070,772	..	1,303,345	34,812
	2013	3,205,532	..	1,875,080	..	1,330,452	34,841
	2014	3,266,332	..	1,765,448	..	1,500,884	38,549
Angola	2011	28,576	..	0	..	28,576	1,302
	2012	28,880	..	0	..	28,880	1,273
	2013	35,340	..	19,760	..	15,580	664
	2014	28,120	..	16,340	..	11,780	486
Cameroon	2011	8,621	..	..	..	8,621	408
	2012	10,345	..	..	..	10,345	478
	2013	17,816	..	..	..	17,816	802
	2014	25,862	..	..	..	25,862	1,136
Congo	2011	5,734	..	..	..	5,734	1,373
	2012	8,293	..	..	..	8,293	1,935
	2013	8,584	..	..	..	8,584	1,953
	2014	9,166	..	..	..	9,166	2,035
Côte d'Ivoire	2011	60,836	..	..	..	60,836	2,953
	2012	66,359	..	..	..	66,359	3,145
	2013	76,205	..	..	..	76,205	3,524
	2014	76,838	..	..	..	76,838	3,468
Dem. Rep. of the Congo	2011	..	914	..	..	914	13
	2012	..	0	..	..	0	0
	2013	..	0	..	..	0	0
	2014	..	80	..	..	80	1
Egypt	2011	2,210,762	0	326,726	..	1,884,036	22,486
	2012	2,124,400	0	235,503	..	1,888,897	22,051
	2013	1,898,480	17,965	78,780	..	1,837,665	20,975
	2014	2,145,567	0	245,918	..	1,899,649	21,206
Equatorial Guinea	2011	243,296	..	*177,759	..	65,537	87,276
	2012	253,637	..	184,179	..	69,458	89,770
	2013	245,454	..	184,179	..	61,275	76,874
	2014	257,730	..	193,388	..	64,342	78,381
Gabon	2011	14,003	..	..	..	14,003	8,878
	2012	14,343	..	..	..	14,343	8,889
	2013	11,755	..	..	..	11,755	7,123
	2014	12,864	..	..	..	12,864	7,622
Gambia	2011	..	*102	..	..	*102	*58
	2012	..	*102	..	..	*102	*56
	2013	..	*102	..	..	*102	*55
	2014	..	*102	..	..	*102	*53
Ghana	2011	..	32,203	..	..	32,203	1,292
	2012	..	16,344	..	..	16,344	640
	2013	..	12,211	..	..	12,211	467
	2014	2,200	23,781	..	..	25,981	970

Table 27

Production, trade and supply of natural gas

Terajoules and megajoules per capita

Country or area	Year	Production	Imports	Exports	Changes in stocks	Supply	
						Total	Per capita
Guinea	2011	3	..	..	..	3	0
	2012	*3	..	..	..	*3	0
	2013	*3	..	..	..	*3	0
	2014	*3	..	..	..	*3	0
Libya	2011	298,680	..	91,960	..	206,720	32,872
	2012	463,600	..	245,860	..	217,740	34,653
	2013	486,400	..	216,600	..	269,800	43,058
	2014	473,100	..	247,380	..	225,720	36,063
Morocco	2011	2,343	31,203	..	..	33,546	1,015
	2012	2,511	45,986	..	..	48,497	1,447
	2013	3,852	44,289	..	..	48,141	1,416
	2014	3,935	42,883	..	..	46,818	1,358
Mozambique	2011	130,866	..	127,161	..	3,705	148
	2012	146,793	..	143,301	..	3,492	136
	2013	163,740	..	156,478	..	7,262	274
	2014	177,257	..	150,276	..	26,981	991
Nigeria	2011	1,456,954	..	887,711	..	569,243	3,476
	2012	1,565,634	..	978,696	..	586,938	3,489
	2013	1,412,306	..	830,300	..	582,006	3,368
	2014	1,611,963	..	938,980	..	672,983	3,792
Rwanda	2011	*23	..	..	..	*23	*2
	2012	*23	..	..	..	*23	*2
	2013	*24	..	..	..	*24	*2
	2014	*24	..	..	..	*24	*2
Senegal	2011	1,588	..	..	835	753	56
	2012	1,622	..	..	539	1,083	79
	2013	1,726	..	..	207	1,519	107
	2014	1,807	..	..	216	1,591	108
South Africa	2011	51,745	127,161	17	..	178,889	3,425
	2012	44,468	143,301	13	..	187,756	3,553
	2013	47,702	142,687	0	..	190,389	3,564
	2014	40,425	138,651	0	..	179,076	3,318
Togo	2011	..	*311	..	..	*311	*47
	2012	..	*654	..	..	*654	*97
	2013	..	*844	..	..	*844	*122
	2014	..	*1,481	..	..	*1,481	*208
Tunisia	2011	136,022	92,975	..	..	228,997	21,284
	2012	131,759	115,405	..	..	247,164	22,714
	2013	131,467	114,998	..	..	246,465	22,394
	2014	119,881	125,217	..	..	245,098	22,021
United Rep. of Tanzania	2011	33,047	..	..	..	33,047	701
	2012	37,799	..	..	..	37,799	777
	2013	37,964	..	..	..	37,964	756
	2014	35,391	..	..	..	35,391	683
America, North	**2011**	**34,535,748**	**5,696,193**	**5,934,880**	**268,550**	**34,028,511**	**61,593**
	2012	**35,624,385**	**5,570,046**	**5,918,823**	**-74,111**	**35,349,719**	**63,360**
	2013	**35,906,088**	**5,250,624**	**5,651,528**	**-665,427**	**36,170,611**	**64,218**
	2014	**37,877,853**	**4,938,317**	**5,412,941**	**358,235**	**37,044,995**	**65,160**
Barbados	2011	644	..	..	..	644	2,294
	2012	737	..	..	..	737	2,618
	2013	864	..	..	..	864	3,057
	2014	554	..	..	..	554	1,955

Production, trade and supply of natural gas

Terajoules and megajoules per capita

Country or area	Year	Production	Imports	Exports	Changes in stocks	Supply	
						Total	Per capita
Belize	2011	*202	..	..	..	*202	*612
	2012	*129	..	..	..	*129	*383
	2013	*116	..	..	..	*116	*337
	2014	*80	..	..	..	*80	*227
Canada	2011	6,158,610	1,208,073	3,575,218	-91,700	3,883,165	112,556
	2012	6,033,139	1,213,081	3,420,362	-46,767	3,872,625	111,065
	2013	6,063,762	1,021,945	3,203,183	-127,665	4,010,189	113,827
	2014	6,400,017	837,699	3,040,885	67,860	4,128,971	116,022
Cuba	2011	39,801	..	..	..	39,801	3,515
	2012	40,387	..	..	..	40,387	3,561
	2013	41,596	..	..	..	41,596	3,661
	2014	46,821	..	..	..	46,821	4,115
Dominican Republic	2011	..	33,340	..	574	32,766	3,268
	2012	..	42,188	..	-1,375	43,563	4,290
	2013	..	40,429	..	980	39,448	3,837
	2014	..	39,662	..	1,115	38,547	3,704
Mexico	2011	1,990,452	675,183	3,230	-26,048	2,688,453	22,336
	2012	1,923,390	879,032	345	16,289	2,785,788	22,821
	2013	1,909,734	1,022,996	1,416	12,748	2,918,566	23,586
	2014	1,757,953	1,099,188	1,796	15,596	2,839,749	22,648
Puerto Rico	2011	..	29,257	..	..	29,257	7,903
	2012	..	53,059	..	..	53,059	14,357
	2013	..	60,856	..	..	60,856	16,489
	2014	..	64,855	..	..	64,855	17,592
Trinidad and Tobago	2011	1,588,473	..	745,851	..	842,622	631,277
	2012	1,597,441	..	772,572	..	824,869	614,849
	2013	1,604,456	..	769,610	..	834,846	619,212
	2014	1,595,661	..	756,857	..	838,804	619,280
United States	2011	24,757,567	3,750,340	1,610,581	385,724	26,511,602	84,867
	2012	26,029,162	3,382,686	1,725,544	-42,258	27,728,562	88,083
	2013	26,285,560	3,104,399	1,677,319	-551,490	28,264,130	89,123
	2014	28,076,767	2,896,914	1,613,403	273,664	29,086,614	91,053
America, South	**2011**	**5,032,193**	**923,440**	**793,502**	**8,143**	**5,153,988**	**12,839**
	2012	**5,275,736**	**1,101,199**	**856,259**	**1,671**	**5,519,005**	**13,602**
	2013	**5,051,267**	**1,284,803**	**972,580**	**893**	**5,362,597**	**13,078**
	2014	**5,241,283**	**1,374,411**	**950,037**	**2,149**	**5,663,508**	**13,671**
Argentina	2011	1,623,757	266,132	7,795	..	1,882,094	45,182
	2012	1,589,406	353,589	4,081	..	1,938,914	46,060
	2013	1,449,051	398,751	2,805	..	1,844,997	43,373
	2014	1,512,840	453,541	2,465	..	1,963,916	45,694
Bolivia (Plur. State of)	2011	623,529	..	490,116	7,908	125,505	12,453
	2012	711,747	..	568,814	0	142,933	13,960
	2013	810,573	..	666,515	0	144,058	13,852
	2014	856,197	..	696,662	..	159,535	15,105
Brazil	2011	620,321	405,917	..	..	1,026,238	5,118
	2012	705,582	510,593	..	..	1,216,175	6,009
	2013	787,981	656,887	..	..	1,444,868	7,074
	2014	838,045	748,204	..	..	1,586,249	7,697
Chile	2011	62,245	154,754	..	235	216,764	12,602
	2012	48,825	153,170	..	1,671	200,324	11,521
	2013	38,173	152,036	..	893	189,316	10,771
	2014	31,226	136,813	..	2,149	165,890	9,339

Table 27

Production, trade and supply of natural gas

Terajoules and megajoules per capita

Country or area	Year	Production	Imports	Exports	Changes in stocks	Supply	
						Total	Per capita
Colombia	2011	501,343	..	73,428	..	427,915	9,221
	2012	532,554	..	67,045	..	465,509	9,930
	2013	521,179	..	51,492	..	469,687	9,921
	2014	496,589	..	35,810	..	460,779	9,641
Ecuador	2011	27,155	..	..	..	27,155	1,789
	2012	34,572	..	..	..	34,572	2,242
	2013	40,150	..	..	..	40,150	2,564
	2014	40,781	..	..	..	40,781	2,564
Peru	2011	459,289	..	222,163	..	237,126	7,968
	2012	479,468	..	216,319	..	263,149	8,725
	2013	492,937	..	251,768	..	241,169	7,890
	2014	551,194	..	215,100	..	336,094	10,851
Suriname	2011	172	..	..	..	172	328
	2012	175	..	..	..	175	331
	2013	179	..	..	..	179	335
	2014	149	..	..	..	149	277
Uruguay	2011	..	2,995	..	..	2,995	885
	2012	..	2,188	..	..	2,188	644
	2013	..	2,043	..	..	2,043	599
	2014	..	1,885	..	..	1,885	551
Venezuela (Bolivar. Rep.)	2011	1,114,382	93,642	..	..	1,208,024	41,051
	2012	1,173,407	81,659	..	..	1,255,066	42,040
	2013	911,044	75,086	..	..	986,130	32,571
	2014	914,262	33,968	..	..	948,230	30,893
Asia	**2011**	**43,879,127**	**13,296,352**	**12,879,184**	**-6,338**	**44,302,633**	**10,511**
	2012	**44,735,244**	**13,731,179**	**12,795,198**	**-13,911**	**45,685,136**	**10,724**
	2013	**45,729,426**	**15,337,860**	**13,793,591**	**104,438**	**47,169,257**	**10,957**
	2014	**47,187,535**	**15,408,675**	**13,255,108**	**70,711**	**49,270,392**	**11,328**
Afghanistan	2011	6,282	..	..	..	6,282	218
	2012	6,243	..	..	..	6,243	210
	2013	6,029	..	..	..	6,029	196
	2014	5,537	..	..	..	5,537	175
Armenia	2011	..	80,469	4,143	0	76,325	25,716
	2012	..	95,532	2,710	0	92,822	31,166
	2013	..	91,953	2,079	0	89,875	30,036
	2014	..	91,295	707	..	90,587	30,134
Azerbaijan	2011	638,553	..	266,272	-14,452	386,733	41,911
	2012	673,082	..	258,460	11,484	403,138	43,064
	2013	698,627	..	285,450	8,281	404,896	42,632
	2014	735,031	..	316,112	-11,718	430,637	44,719
Bahrain	2011	493,642	..	..	-44	493,686	378,010
	2012	495,896	..	..	-1	495,897	371,855
	2013	532,149	..	..	-1	532,150	394,353
	2014	576,341	..	..	-1	576,342	423,180
Bangladesh	2011	779,158	..	..	-26,361	805,519	5,251
	2012	817,943	..	..	-30,030	847,973	5,462
	2013	882,129	..	..	306	881,823	5,611
	2014	904,663	..	..	-29,076	933,739	5,870
Brunei Darussalam	2011	497,813	..	350,476	-838	148,175	370,954
	2012	488,733	..	343,653	472	144,608	356,606
	2013	474,775	..	359,143	1,296	114,336	277,852
	2014	461,001	..	324,848	-1,514	137,667	329,825

Production, trade and supply of natural gas

Terajoules and megajoules per capita

Country or area	Year	Production	Imports	Exports	Changes in stocks	Supply	
						Total	Per capita
China	2011	3,957,609	1,161,703	118,785	..	5,000,528	3,709
	2012	4,182,496	1,568,254	107,681	..	5,643,069	4,163
	2013	4,650,688	1,959,202	102,279	..	6,507,611	4,776
	2014	5,080,315	2,204,773	97,211	..	7,187,877	5,249
China, Hong Kong SAR	2011	..	95,030	..	-12,096	107,126	15,208
	2012	..	99,313	..	0	99,313	13,984
	2013	..	96,693	..	0	96,693	13,497
	2014	..	94,480	..	0	94,480	13,073
China, Macao SAR	2011	..	2,870	..	..	2,870	5,250
	2012	..	0	..	..	0	0
	2013	..	14	..	..	14	25
	2014	..	2,278	..	..	2,278	3,942
Georgia	2011	202	70,135	..	208	70,129	16,712
	2012	202	76,462	..	0	76,664	18,523
	2013	183	64,371	..	0	64,554	15,811
	2014	399	84,913	..	0	85,312	21,144
India	2011	1,857,361	0	..	..	1,857,361	1,489
	2012	1,587,335	0	..	..	1,587,335	1,256
	2013	1,350,882	718,562	..	..	2,069,444	1,617
	2014	1,279,108	603,655	..	..	1,882,763	1,454
Indonesia	2011	3,305,194	..	1,647,351	..	1,657,843	6,772
	2012	3,130,119	..	1,502,600	..	1,627,519	6,562
	2013	2,819,216	..	1,502,882	..	1,316,334	5,239
	2014	2,689,941	..	1,421,118	..	1,268,823	4,986
Iran (Islamic Rep. of)	2011	5,870,358	460,877	369,724	..	5,961,511	79,292
	2012	6,122,785	182,072	363,442	..	5,941,416	78,015
	2013	6,147,993	209,660	362,193	..	5,995,460	77,709
	2014	6,872,781	296,185	381,588	..	6,787,378	86,858
Iraq	2011	232,893	..	..	..	232,893	7,308
	2012	229,187	..	..	..	229,187	6,954
	2013	265,434	..	..	..	265,434	7,782
	2014	256,790	..	..	..	256,790	7,280
Israel	2011	163,448	26,484	..	..	189,932	25,112
	2012	95,536	2,303	..	..	97,839	12,715
	2013	244,788	22,262	..	..	267,050	34,159
	2014	288,144	4,606	..	..	292,750	36,873
Japan	2011	220,074	4,546,523	..	39,539	4,727,058	37,147
	2012	214,309	4,748,508	..	-9,574	4,972,391	39,110
	2013	201,935	4,824,640	..	7,341	5,019,234	39,526
	2014	196,477	4,898,459	..	3,004	5,091,932	40,159
Jordan	2011	6,229	34,369	..	..	40,598	6,005
	2012	5,624	25,032	..	..	30,657	4,383
	2013	4,668	33,293	..	..	37,961	5,262
	2014	4,082	8,508	..	..	12,590	1,698
Kazakhstan	2011	1,542,539	176,296	237,129	-19,411	1,501,117	90,678
	2012	1,572,507	169,698	418,484	40,932	1,282,789	76,259
	2013	1,654,686	201,109	398,355	48,697	1,408,743	82,385
	2014	1,694,994	162,714	409,593	312	1,447,803	83,343
Korea, Republic of	2011	42,306	1,949,271	..	33,503	1,958,074	39,672
	2012	44,215	1,993,500	..	-80,728	2,118,443	42,703
	2013	21,477	2,217,444	..	21,622	2,217,299	44,482
	2014	13,431	2,047,587	..	54,274	2,006,744	40,075

Table 27

Production, trade and supply of natural gas

Terajoules and megajoules per capita

Country or area	Year	Production	Imports	Exports	Changes in stocks	Supply	
						Total	Per capita
Kuwait	2011	536,445	131,480	..	..	667,925	206,202
	2012	605,324	100,700	..	..	706,024	206,465
	2013	619,818	82,840	..	..	702,658	195,526
	2014	571,094	131,480	..	..	702,574	187,197
Kyrgyzstan	2011	1,038	11,905	..	..	12,943	2,330
	2012	1,132	15,452	..	..	16,584	2,936
	2013	1,288	11,433	..	..	12,721	2,214
	2014	1,249	9,248	..	..	10,497	1,796
Malaysia	2011	2,197,136	292,189	1,172,400	..	1,316,925	46,090
	2012	2,145,053	329,325	1,126,849	..	1,347,529	46,431
	2013	2,370,004	357,879	1,196,809	..	1,531,074	51,962
	2014	2,534,918	270,788	1,211,853	..	1,593,853	53,303
Myanmar	2011	513,425	..	401,710	..	111,715	2,143
	2012	471,228	..	398,510	..	72,718	1,384
	2013	483,794	..	398,390	..	85,404	1,612
	2014	594,533	..	496,360	..	98,173	1,837
Oman	2011	1,170,308	80,730	455,890	..	795,148	247,709
	2012	1,218,502	80,730	452,862	..	846,370	238,737
	2013	1,317,306	80,730	470,376	..	927,660	237,441
	2014	1,260,364	81,558	427,842	..	914,080	215,786
Other Asia	2011	12,302	662,617	..	22,420	652,499	28,054
	2012	14,806	629,062	..	27,671	616,197	26,442
	2013	12,764	633,001	..	34,972	610,793	26,173
	2014	12,706	666,548	..	44,223	635,031	27,183
Pakistan	2011	1,416,147	..	..	285	1,415,862	8,153
	2012	*1,421,196	..	..	-370	*1,421,566	*8,014
	2013	*1,332,755	..	..	561	*1,332,194	*7,352
	2014	*1,321,592	..	..	556	*1,321,036	*7,139
Philippines	2011	155,101	..	..	..	155,101	1,641
	2012	148,682	..	..	..	148,682	1,548
	2013	136,949	..	..	..	136,949	1,404
	2014	144,028	..	..	..	144,028	1,453
Qatar	2011	6,259,605	..	4,904,096	..	1,355,509	711,390
	2012	6,572,435	..	4,912,283	..	1,660,152	823,642
	2013	6,687,240	..	5,282,840	..	1,404,400	668,352
	2014	6,623,756	..	4,633,231	..	1,990,525	916,421
Saudi Arabia	2011	2,838,153	..	..	..	2,838,153	98,587
	2012	3,081,481	..	..	..	3,081,481	104,471
	2013	3,114,266	..	..	..	3,114,266	103,118
	2014	3,234,801	..	..	..	3,234,801	104,732
Singapore	2011	..	338,982	..	..	338,982	65,306
	2012	..	367,584	..	..	367,584	69,362
	2013	..	413,405	..	..	413,405	76,485
	2014	..	431,961	..	..	431,961	78,444
Syrian Arab Republic	2011	296,699	9,425	..	..	306,124	14,932
	2012	243,165	0	..	..	243,165	12,171
	2013	199,810	0	..	..	199,810	10,341
	2014	184,730	0	..	..	184,730	9,840
Tajikistan	2011	703	6,840	..	..	7,543	973
	2012	422	5,054	..	..	5,476	690
	2013	144	11,625	..	..	11,769	1,451
	2014	122	12,249	..	..	12,371	1,491

Production, trade and supply of natural gas

Terajoules and megajoules per capita

Country or area	Year	Production	Imports	Exports	Changes in stocks	Supply	
						Total	Per capita
Thailand	2011	*1,193,224	374,269	..	..	*1,567,493	*23,429
	2012	*1,332,995	383,142	..	..	*1,716,137	*25,551
	2013	1,320,549	440,315	..	..	1,760,864	26,106
	2014	1,348,418	411,810	..	..	1,760,228	25,990
Turkey	2011	29,096	1,680,569	27,353	-29,091	1,711,403	23,279
	2012	24,223	1,759,024	23,591	26,233	1,733,423	23,159
	2013	20,592	1,733,986	26,120	-18,637	1,747,095	22,921
	2014	18,357	1,886,840	24,231	10,651	1,870,315	24,126
Turkmenistan	2011	2,508,631	..	1,620,432	..	888,199	173,929
	2012	2,616,235	..	1,700,139	..	916,096	177,094
	2013	2,972,013	..	2,031,476	..	940,537	179,489
	2014	3,033,014	..	2,078,710	..	954,304	179,813
United Arab Emirates	2011	2,041,110	1,103,319	498,298	..	2,646,131	302,944
	2012	2,118,840	1,100,431	523,662	..	2,695,610	301,100
	2013	2,130,547	1,133,443	491,274	..	2,772,715	306,717
	2014	2,116,679	1,006,742	527,564	..	2,595,856	285,694
Uzbekistan	2011	2,388,523	..	454,668	..	1,933,855	68,678
	2012	2,383,597	..	386,468	..	1,997,129	69,848
	2013	2,259,321	..	511,502	..	1,747,819	60,200
	2014	2,339,267	..	556,968	..	1,782,299	60,479
Viet Nam	2011	330,898	..	..	..	330,898	3,705
	2012	365,041	..	..	..	365,041	4,041
	2013	380,494	..	..	..	380,494	4,164
	2014	398,404	..	..	..	398,404	4,311
Yemen	2011	376,922	..	350,457	..	26,465	1,092
	2012	304,674	..	273,805	..	30,869	1,241
	2013	414,115	..	372,423	..	41,692	1,633
	2014	390,467	..	347,171	..	43,296	1,654
Europe	**2011**	**37,238,783**	**19,495,766**	**15,543,552**	**534,780**	**40,656,218**	**55,213**
	2012	**36,791,240**	**18,651,660**	**15,892,763**	**-29,671**	**39,579,808**	**53,700**
	2013	**37,567,476**	**18,453,912**	**16,570,755**	**-119,614**	**39,570,248**	**53,647**
	2014	**34,686,483**	**16,957,269**	**15,584,273**	**-23,290**	**36,082,768**	**48,889**
Albania	2011	627	..	..	..	627	217
	2012	614	..	..	..	614	213
	2013	678	..	..	..	678	235
	2014	1,155	..	..	..	1,155	400
Austria	2011	64,042	510,953	138,898	79,449	356,648	42,339
	2012	68,829	541,128	243,589	25,532	340,836	40,309
	2013	55,364	396,671	151,273	-27,669	328,431	38,698
	2014	50,475	386,748	96,661	40,749	299,813	35,202
Belarus	2011	8,574	772,363	..	-18,500	799,437	84,261
	2012	8,420	782,173	..	2,626	787,967	83,023
	2013	8,806	782,482	..	-3,785	795,073	83,716
	2014	8,574	774,448	..	-5,909	788,931	83,042
Belgium	2011	..	829,807	157,038	3,357	669,412	60,827
	2012	..	709,209	50,174	-9,272	668,307	60,319
	2013	..	709,887	36,740	3,476	669,671	60,043
	2014	..	626,763	33,730	6,915	586,118	52,209
Bosnia and Herzegovina	2011	..	10,564	..	..	10,564	2,757
	2012	..	9,730	..	..	9,730	2,542
	2013	..	7,572	..	..	7,572	1,980
	2014	..	7,043	..	..	7,043	1,845

Table 27

Production, trade and supply of natural gas

Terajoules and megajoules per capita

Country or area	Year	Production	Imports	Exports	Changes in stocks	Supply	
						Total	Per capita
Bulgaria	2011	16,316	105,337	..	-707	122,360	16,636
	2012	14,341	95,029	..	-4,658	114,028	15,612
	2013	10,879	103,540	..	2,876	111,543	15,380
	2014	7,419	103,460	73	898	109,908	15,262
Croatia	2011	93,364	33,097	9,769	-2,875	119,567	27,793
	2012	76,050	51,291	9,701	5,376	112,264	26,187
	2013	70,119	47,993	14,208	-2,248	106,152	24,851
	2014	67,163	43,542	16,681	81	93,943	22,074
Czechia	2011	8,811	355,528	6,378	41,199	316,762	30,070
	2012	9,951	284,123	284	-25,156	318,946	30,246
	2013	9,579	324,154	310	10,276	323,147	30,644
	2014	9,859	276,915	45	-863	287,592	27,279
Denmark	2011	274,405	34,967	130,057	6,382	172,933	31,011
	2012	240,000	38,228	124,354	-8,221	162,095	28,941
	2013	199,194	55,984	91,820	9,099	154,259	27,427
	2014	192,944	26,024	87,155	936	130,877	23,177
Estonia	2011	..	23,413	..	..	23,413	17,629
	2012	..	25,372	..	..	25,372	19,163
	2013	..	25,814	..	..	25,814	19,555
	2014	..	20,262	..	..	20,262	15,394
Finland	2011	3	156,304	0	..	156,307	28,968
	2012	19	139,799	2	..	139,816	25,774
	2013	127	132,901	20	..	133,008	24,391
	2014	135	116,925	11	..	117,049	21,361
France	2011	23,523	1,937,315	156,859	80,792	1,723,187	27,220
	2012	21,021	1,970,780	253,152	-39,363	1,778,012	27,957
	2013	13,511	1,978,049	209,898	-32,994	1,814,656	28,406
	2014	702	1,867,235	295,551	55,968	1,516,418	23,635
Germany	2011	506,740	3,495,078	686,337	77,877	3,237,604	40,256
	2012	445,142	3,446,585	663,468	-19,736	3,247,995	40,359
	2013	412,433	3,812,138	856,197	-32,341	3,400,715	42,210
	2014	319,315	3,504,877	868,427	7,587	2,948,178	36,557
Greece	2011	299	184,847	..	357	184,789	16,568
	2012	321	170,908	..	868	170,361	15,334
	2013	270	150,477	..	195	150,552	13,618
	2014	253	114,747	..	-570	115,570	10,506
Hungary	2011	98,402	306,979	21,661	-51,425	435,145	43,563
	2012	82,252	313,776	32,128	-25,796	389,696	39,133
	2013	71,840	314,831	56,337	-23,247	353,581	35,627
	2014	66,863	345,937	28,733	59,282	324,785	32,841
Ireland	2011	8,662	183,928	..	711	191,879	41,240
	2012	7,896	178,557	..	-689	187,142	40,092
	2013	6,751	172,594	..	966	178,379	38,186
	2014	5,711	167,089	..	-372	173,172	37,041
Italy	2011	321,905	2,681,059	4,726	29,604	2,968,634	49,718
	2012	327,851	2,580,323	5,296	48,615	2,854,263	47,755
	2013	294,704	2,360,905	8,687	-22,708	2,669,630	44,641
	2014	272,377	2,124,342	9,030	28,842	2,358,847	39,432
Latvia	2011	..	65,579	..	5,642	59,937	29,044
	2012	..	64,115	..	7,772	56,343	27,659
	2013	..	64,778	..	8,736	56,042	27,856
	2014	..	36,291	..	-14,027	50,318	25,294

Production, trade and supply of natural gas

Terajoules and megajoules per capita

Country or area	Year	Production	Imports	Exports	Changes in stocks	Supply	
						Total	Per capita
Liechtenstein	2011	..	*956	..	..	*956	*26,176
	2012	..	*973	..	..	*973	*26,452
	2013	..	1,032	..	..	1,032	27,870
	2014	..	*859	..	..	*859	*23,025
Lithuania	2011	..	126,778	..	335	126,443	41,179
	2012	..	123,540	..	74	123,466	40,930
	2013	..	100,731	..	38	100,693	33,974
	2014	..	99,725	37	3,647	96,041	32,927
Luxembourg	2011	30	48,021	..	..	48,051	92,409
	2012	169	48,895	..	..	49,064	92,143
	2013	176	41,398	..	..	41,574	76,299
	2014	190	39,224	..	..	39,414	70,816
Netherlands	2011	2,691,504	767,094	1,862,051	52	1,596,495	95,657
	2012	2,681,728	869,744	2,013,973	3,963	1,533,536	91,558
	2013	2,881,835	899,944	2,230,449	-7,059	1,558,389	92,711
	2014	2,343,041	970,952	1,958,572	3,659	1,351,762	80,138
Norway	2011	4,174,449	..	3,948,118	1,250	225,081	45,435
	2012	4,687,823	..	4,474,726	-788	213,885	42,620
	2013	4,447,408	..	4,180,121	471	266,816	52,487
	2014	4,418,921	..	4,189,149	-8	229,780	44,635
Poland	2011	179,095	449,540	1,099	30,387	597,149	15,472
	2012	181,744	467,218	119	12,436	636,407	16,483
	2013	177,852	477,723	3,582	13,002	638,991	16,546
	2014	173,349	451,673	2,880	-1,432	623,574	16,146
Portugal	2011	..	210,887	..	3,238	207,649	19,666
	2012	..	182,460	..	-486	182,946	17,399
	2013	..	177,377	..	2,654	174,723	16,704
	2014	..	161,638	..	58	161,580	15,533
Republic of Moldova	2011	2	39,011	..	25	38,988	9,561
	2012	4	37,091	..	1	37,094	9,103
	2013	4	34,921	..	-66	34,991	8,589
	2014	3	35,655	..	44	35,614	8,745
Romania	2011	403,180	114,615	..	1,188	516,607	25,687
	2012	403,919	106,877	..	9,432	501,364	25,137
	2013	400,076	54,207	..	-3,425	457,708	23,123
	2014	407,772	21,643	..	-6,213	435,628	22,168
Russian Federation	2011	25,720,226	304,846	7,690,500	-86,894	18,421,466	128,631
	2012	25,157,696	307,331	7,388,394	67,870	18,008,763	125,683
	2013	26,203,721	313,371	8,099,677	34,545	18,382,870	128,222
	2014	24,082,912	328,549	7,247,337	-130,976	17,295,100	120,583
Serbia	2011	18,823	64,732	..	-4,954	88,509	9,808
	2012	19,762	66,297	..	8,005	78,054	8,689
	2013	19,676	69,916	..	2,760	86,832	9,715
	2014	20,639	51,661	..	-2,548	74,848	8,417
Slovakia	2011	4,781	226,185	115	15,135	215,716	39,863
	2012	5,917	184,190	1,803	-14,771	203,075	37,499
	2013	4,856	202,729	577	-5,032	212,040	39,127
	2014	3,900	184,049	116	12,344	175,489	32,361
Slovenia	2011	80	34,234	..	..	34,314	16,665
	2012	80	32,953	..	..	33,033	16,013
	2013	120	32,065	..	..	32,185	15,587
	2014	120	29,002	..	..	29,122	14,095

Table 27

Production, trade and supply of natural gas

Terajoules and megajoules per capita

Country or area	Year	Production	Imports	Exports	Changes in stocks	Supply	
						Total	Per capita
Spain	2011	2,115	1,436,483	68,666	23,830	1,346,102	28,819
	2012	2,409	1,419,084	113,118	-20,905	1,329,280	28,503
	2013	2,318	1,436,380	236,069	-14,468	1,217,097	26,199
	2014	968	1,472,594	332,704	39,902	1,100,956	23,799
Sweden	2011	..	53,900	..	..	53,900	5,696
	2012	..	46,827	..	..	46,827	4,907
	2013	..	44,440	..	..	44,440	4,618
	2014	..	36,939	..	..	36,939	3,807
Switzerland	2011	323	124,193	..	..	124,516	15,638
	2012	356	136,134	..	..	136,490	16,935
	2013	511	143,363	..	..	143,874	17,641
	2014	767	124,205	..	..	124,972	15,151
T.F.Yug.Rep. Macedonia	2011	..	5,131	..	0	5,131	2,484
	2012	..	5,325	..	0	5,325	2,573
	2013	..	6,032	..	4	6,028	2,909
	2014	..	5,166	..	-7	5,173	2,492
Ukraine	2011	722,386	1,683,533	0	226,398	2,179,521	47,925
	2012	716,824	1,237,329	0	-47,651	2,001,804	44,170
	2013	745,543	1,051,134	0	-38,787	1,835,464	40,639
	2014	699,017	731,493	..	-124,268	1,554,778	34,549
United Kingdom	2011	1,896,116	2,118,509	661,280	72,927	3,280,418	51,934
	2012	1,630,102	1,978,266	518,482	-4,749	3,094,635	48,678
	2013	1,529,125	1,926,379	394,790	5,117	3,055,597	47,777
	2014	1,531,939	1,669,594	417,381	2,991	2,781,161	43,232
Oceania	**2011**	**2,384,533**	**261,046**	**1,085,657**	**3,281**	**1,556,641**	**42,056**
	2012	**2,304,847**	**232,449**	**1,026,310**	**-1,945**	**1,512,931**	**40,233**
	2013	**2,618,256**	**260,670**	**1,278,563**	**2,038**	**1,598,325**	**41,859**
	2014	**2,670,856**	**277,522**	**1,264,582**	**-367**	**1,684,163**	**43,454**
Australia	2011	2,216,444	261,044	1,085,657	..	1,391,831	61,743
	2012	2,122,863	232,449	1,026,310	..	1,329,002	58,006
	2013	2,425,770	260,670	1,278,563	..	1,407,877	60,501
	2014	2,461,826	277,519	1,264,582	..	1,474,763	62,431
Fiji	2011	..	2	..	..	2	2
	2012	..	..	..	..	..	..
	2013	..	..	..	..	..	..
	2014	..	3	..	..	3	3
New Zealand	2011	161,939	..	..	3,281	158,658	36,022
	2012	176,667	..	..	-1,945	178,612	40,265
	2013	187,169	..	..	2,038	185,131	41,460
	2014	204,030	..	..	-367	204,397	45,467
Papua New Guinea	2011	6,150	..	..	..	6,150	878
	2012	5,317	..	..	..	5,317	743
	2013	5,317	..	..	..	5,317	728
	2014	*5,000	..	..	..	*5,000	*670

Table 28

International trade of natural gas (Principal importers/exporters)

Terajoules

2013

Importers	Exporters							
	World	Algeria	Australia	Bolivia (Plur. State of)	Canada	Germany	Indonesia	Malaysia
Argentina	398,751	..	..	203,391	..	..	..	..
Austria	396,671	..	..	..	..	..	..	..
Belarus	782,482	..	..	..	..	..	..	..
Belgium	709,887	..	..	..	..	47,935	..	..
Brazil	656,887	3,154	..	445,993	..	..	..	..
Canada	1,021,945	..	..	..	..	..	..	..
China	1,959,202	3,114	178,693	..	..	..	139,373	*124,385
Czechia	324,154	..	..	..	..	..	..	..
France	1,978,049	231,287	..	..	..	10,108	..	..
Germany	3,812,138	..	..	..	..	..	..	..
Hungary	314,831	..	..	..	..	..	..	..
India	718,562	6,624	..	..	..	..	..	..
Italy	2,360,905	476,936	..	..	..	80,086	..	..
Japan	4,824,640	36,987	1,010,611	..	..	..	*325,056	*660,158
Kazakhstan	201,109	..	..	..	..	..	..	..
Korea, Republic of	2,217,444	6,913	36,817	..	..	..	316,002	*206,348
Malaysia	357,879	12,000	..	..	..	..	*291,212	..
Mexico	1,022,996	..	..	..	..	..	13,150	..
Netherlands	899,944	1,788	..	..	..	28,610	..	..
Poland	477,723	..	..	..	..	86,939	..	..
Portugal	177,377	87,600	..	..	..	..	..	..
Russian Federation	313,371	..	..	..	..	..	..	..
Singapore	413,405	..	..	..	..	..	306,558	*54,784
Slovakia	202,729	..	..	..	..	..	..	..
Spain	1,436,380	781,826	..	..	..	..	..	..
Thailand	440,315	..	..	..	..	..	..	..
Turkey	1,733,986	150,051	..	..	..	..	..	..
Ukraine	1,051,134	..	..	..	..	33,489	..	..
United Arab Emirates	1,133,443	3,040	..	..	..	..	..	..
United Kingdom	1,926,379	16,170	..	..	..	..	..	..
United States	3,104,399	..	..	..	2,995,340	..	..	..

International trade of natural gas
(Principal importers/exporters)
Terajoules

2013

Exporters							
Netherlands	Nigeria	Norway	Qatar	Russian Federation	Trinidad and Tobago	Turkmenistan	United States
..	21,225	3,116	33,687	..	113,486	..	..
..	..	55,060	..	248,733	..	..	..
..	..	..	..	782,482	..	..	..
320,361	..	238,562	60,823	..	..	..	..
..	46,127	15,770	9,462	..	96,986	..	..
..	..	..	30,761	..	9,347	..	*932,745
..	22,191	..	369,845	..	5,840	977,593	..
..	..	152	..	324,002	..	..	..
296,868	48,334	773,021	68,564	382,402	..	..	..
*1,147,122	..	789,784	..	*1,511,872	..	..	..
..	..	..	..	299,089	..	..	..
..	44,713	3,312	609,822	..	..	..	..
105,956	..	76,352	202,768	1,069,581	..	..	..
2,541	*196,920	30,071	889,418	472,079	15,475	..	..
..	..	..	..	66,490	..	49,009	..
..	*137,140	3,370	741,170	107,862	34,731	..	..
..	16,000	3,040	6,000	..	..	..	..
..	40,632	12,921	60,071	..	27,139	..	743,453
..	2,948	538,356	..	143,048	3,544	..	..
..	..	..	..	368,358	..	..	..
..	43,040	9,939	12,998	..	3,118	..	..
..	..	..	..	..	..	..	..
..	..	..	6,000	..	10,640	..	..
..	..	..	..	202,729	..	..	..
..	*121,116	157,651	146,192	..	83,978	..	..
..	11,647	..	49,863	..	..	..	..
1,915	48,770	3,366	11,873	1,004,023	..	..	..
..	..	..	..	971,004	..	..	..
..	..	..	747,870	..	..	..	..
303,266	..	1,103,702	342,735	..	3,965	..	..
..	3,033	6,412	8,689	..	77,340	..	..

Table 28

International trade of natural gas (Principal importers/exporters)

Terajoules

2014

Importers	Exporters							
	World	Algeria	Australia	Bolivia (Plur. State of)	Canada	Germany	Indonesia	Malaysia
Argentina	453,541	..	..	224,933	..	..	..	..
Austria	386,748	..	..	..	..	..	..	..
Belarus	774,448	..	..	..	..	..	..	..
Belgium	626,763	106	..	..	..	13,676	..	..
Brazil	748,204	3,154	..	464,907	..	..	..	..
Canada	837,699	..	..	..	..	..	..	..
China	2,204,773	15,959	205,556	..	..	..	127,694	*124,650
Czechia	276,915	..	..	..	..	..	..	..
France	1,867,235	204,304	..	..	..	15,490	..	..
Germany	3,504,877	..	..	..	..	..	..	..
Hungary	345,937	..	..	..	..	..	..	..
India	603,655	6,624	..	..	..	..	..	..
Italy	2,124,342	258,204	..	..	..	64,427	..	..
Japan	4,898,459	30,820	*988,209	..	..	..	285,062	*673,902
Kazakhstan	162,714	..	..	..	..	..	..	..
Korea, Republic of	2,047,587	20,370	45,494	..	..	..	281,839	*186,451
Malaysia	270,788	*18,135	3,200	..	..	..	*228,316	..
Mexico	1,099,188	..	..	..	..	..	12,957	..
Netherlands	970,952	2,268	..	..	..	42,708	..	..
Poland	451,673	..	..	..	..	90,034	..	..
Portugal	161,638	94,776	..	..	..	..	..	..
Russian Federation	328,549	..	..	..	..	..	..	..
Singapore	431,961	..	..	..	..	..	293,858	55,286
Slovakia	184,049	..	..	..	..	..	..	..
Spain	1,472,594	853,019	..	..	..	..	..	..
Thailand	411,810	..	..	..	..	..	..	3,116
Turkey	1,886,840	160,046	..	..	..	..	..	..
Ukraine	731,493	..	..	..	..	104,636	..	..
United Arab Emirates	1,006,742	..	3,420	..	..	..	..	3,420
United Kingdom	1,669,594	20,480	..	..	..	..	..	..
United States	2,896,914	..	..	..	2,832,482	..	..	..

International trade of natural gas
(Principal importers/exporters)

Terajoules

2014

Exporters							
Netherlands	Nigeria	Norway	Qatar	Russian Federation	Trinidad and Tobago	Turkmenistan	United States
..	33,103	7,790	35,829	..	110,993	..	..
..	..	..	..	..	..	..	..
..	..	..	..	774,448	..	..	..
*258,629	..	216,103	51,251	..	..	..	..
..	*64,223	39,425	23,655	..	67,029	..	..
..	..	..	..	..	21,592	..	816,107
..	22,775	6,940	315,146	7,200	7,787	1,091,357	..
..	..	26,702	..	250,213	..	..	..
*214,220	45,063	825,171	39,500	260,183	..	..	..
*1,006,232	..	811,005	..	1,450,379	..	..	..
..	..	..	..	328,641	..	..	..
..	*60,437	..	*525,183	..	2,412	..	..
*228,890	..	101,575	167,145	915,773	2,096	..	..
..	*258,411	9,758	907,411	468,192	6,637	..	13,929
..	..	..	..	65,749	..	11,082	..
..	*164,864	3,963	714,206	104,849	6,906	..	..
..	12,700	..	3,600	..	..	..	..
..	*82,599	3,125	52,589	..	11,852	..	770,849
..	3,701	562,108	..	213,572	4,467	..	..
25	..	3,266	..	342,143	..	..	..
..	11,088	2,509	21,608	..	7,029	..	..
..	..	..	..	..	..	..	..
..	..	..	..	..	14,060	..	..
..	..	..	..	184,049	..	..	..
5,212	*105,541	169,237	126,142	..	81,207	..	..
..	5,460	..	46,745	3,384	2,912	..	..
..	54,146	10,373	38,478	1,033,217	3,262	..	..
..	..	33,528	..	543,019	..	..	..
..	6,080	..	731,430	..	..	..	..
*234,176	1,894	961,348	402,916	..	14,202	..	..
..	..	6,399	..	..	47,481	..	..

Table 29

Production of manufactured and recovered gases - by type

Table Notes

Production refers to gasworks gas, coke oven gas, blast furnace gas and other recovered gases.

Please refer to the Definitions Section on pages ix to xv for the appropriate product description /classification.

Figure 71: Production of manufactured and recovered gases by region, 1990-2014

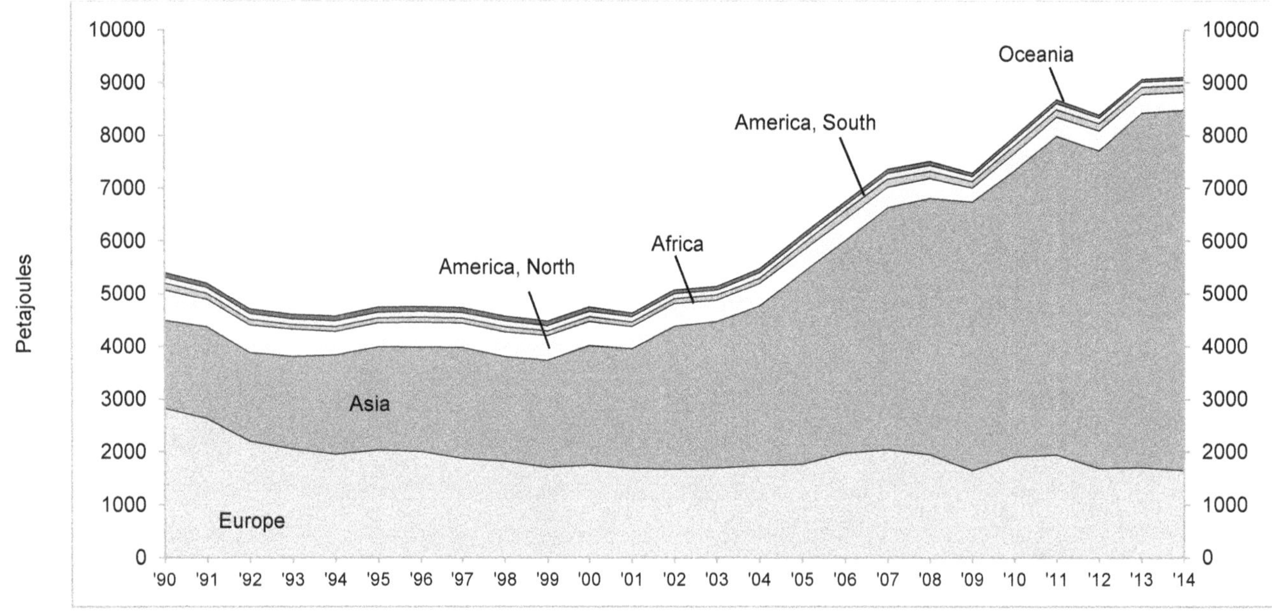

Figure 72: World production of manufactured and recovered gases by type, 2014

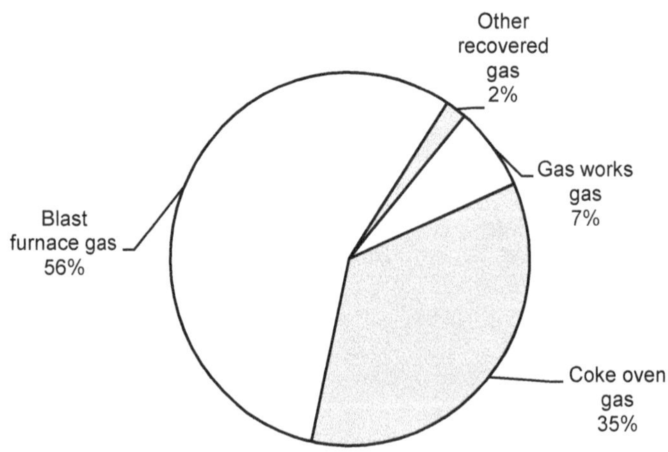

Production of manufactured and recovered gases - by type

Terajoules

Country or area	Year	Total	Gas works gas	Coke oven gas	Blast furnace gas	Other recovered gas
World	**2011**	**8,677,907**	**589,088**	**3,219,552**	**4,731,559**	**137,708**
	2012	**8,385,073**	**593,388**	**3,142,745**	**4,509,894**	**139,046**
	2013	**9,063,233**	**644,782**	**3,249,708**	**5,020,701**	**148,041**
	2014	**9,106,495**	**663,325**	**3,189,689**	**5,081,912**	**171,569**
Africa	**2011**	**141,664**	**82,628**	**23,484**	**35,552**	**..**
	2012	**130,962**	**76,066**	**23,601**	**31,295**	**..**
	2013	**134,178**	**78,519**	**25,259**	**30,400**	**..**
	2014	**125,325**	**72,002**	**24,007**	**29,316**	**..**
Algeria	2011	4,583	..	..	4,583	..
	2012	4,880	..	..	4,880	..
	2013	2,515	..	..	2,515	..
	2014	2,042	..	..	2,042	..
Egypt	2011	9,147	140	..	9,007	..
	2012	*4,559	84	..	*4,475	..
	2013	*4,441	65	..	*4,376	..
	2014	4,992	93	..	4,899	..
South Africa	2011	126,688	82,488	22,238	21,962	..
	2012	120,147	75,982	22,225	21,940	..
	2013	125,785	78,454	23,822	23,509	..
	2014	116,941	71,909	22,657	22,375	..
Zimbabwe	2011	1,246	..	1,246	..	..
	2012	1,376	..	1,376	..	..
	2013	1,437	..	1,437	..	..
	2014	1,350	..	1,350	..	..
America, North	**2011**	**364,216**	**56,572**	**126,346**	**181,298**	**..**
	2012	**383,126**	**57,614**	**131,815**	**193,697**	**..**
	2013	**352,802**	**51,244**	**127,794**	**173,764**	**..**
	2014	**349,767**	**53,629**	**121,688**	**174,450**	**..**
Canada	2011	46,940	..	24,062	22,878	..
	2012	50,698	..	26,125	24,573	..
	2013	45,834	..	24,595	21,239	..
	2014	46,260	..	22,932	23,328	..
Costa Rica	2011	782	..	..	782	..
	2012	*847	..	..	*847	..
	2013	848	..	..	848	..
	2014	848	..	..	848	..
Cuba	2011	3,886	3,852	..	34	..
	2012	4,086	4,063	..	23	..
	2013	4,159	4,148	..	11	..
	2014	3,807	3,796	..	*11	..
Mexico	2011	35,283	..	15,703	19,580	..
	2012	34,977	..	15,429	19,548	..
	2013	37,047	..	16,398	20,649	..
	2014	38,328	..	16,502	21,826	..
United States	2011	277,325	52,720	86,581	138,024	..
	2012	292,518	53,551	90,261	148,706	..
	2013	264,914	47,096	86,801	131,017	..
	2014	260,524	49,833	82,254	128,437	..
America, South	**2011**	**119,389**	**451**	**98,017**	**20,921**	**..**
	2012	**108,134**	**639**	**89,955**	**17,540**	**..**
	2013	**102,482**	**598**	**87,014**	**14,870**	**..**
	2014	**105,044**	**536**	**89,615**	**14,893**	**..**

Table 29

Production of manufactured and recovered gases - by type

Terajoules

Country or area	Year	Total	Gas works gas	Coke oven gas	Blast furnace gas	Other recovered gas
Argentina	2011	23,505	..	10,852	12,653	..
	2012	15,510	..	6,048	9,462	..
	2013	14,905	..	5,443	9,462	..
	2014	19,604	..	9,556	10,048	..
Brazil	2011	78,743	..	78,743	..	..
	2012	75,987	..	75,987	..	..
	2013	73,712	..	73,712	..	..
	2014	74,561	..	74,561	..	..
Chile	2011	11,152	451	4,466	6,235	..
	2012	10,949	639	3,964	6,346	..
	2013	8,624	598	4,160	3,866	..
	2014	8,196	536	3,990	3,670	..
Colombia	2011	5,989	..	3,956	2,033	..
	2012	5,688	..	3,956	1,732	..
	2013	5,241	..	3,699	1,542	..
	2014	2,683	..	1,508	1,175	..
Asia	**2011**	**6,042,155**	**421,261**	**2,152,254**	**3,371,899**	**96,741**
	2012	**6,022,048**	**432,404**	**2,106,386**	**3,384,405**	**98,853**
	2013	**6,724,394**	**488,669**	**2,221,942**	**3,906,638**	**107,144**
	2014	**6,828,702**	**510,581**	**2,228,227**	**3,980,716**	**109,178**
China	2011	4,207,374	388,989	1,485,202	2,333,184	..
	2012	4,136,293	399,882	1,437,323	2,299,088	..
	2013	4,738,852	455,941	1,536,283	2,746,628	..
	2014	4,977,099	477,518	1,543,031	2,956,550	..
China, Hong Kong SAR	2011	28,147	28,147	..	..	..
	2012	*28,360	*28,360	..	..	..
	2013	28,556	28,556	..	..	..
	2014	28,835	28,835	..	..	..
India	2011	337,634	825	101,585	235,224	..
	2012	378,369	862	116,299	261,208	..
	2013	430,675	872	127,385	302,418	..
	2014	*251,226	928	*108,998	*141,300	..
Indonesia	2011	*3,300	*3,300	..	..	..
	2012	*3,300	*3,300	..	..	..
	2013	*3,300	*3,300	..	..	..
	2014	*3,300	*3,300	..	..	..
Iran (Islamic Rep. of)	2011	24,205	..	4,257	19,948	..
	2012	19,867	..	3,131	16,736	..
	2013	24,013	..	3,113	20,900	..
	2014	24,739	..	7,504	17,235	..
Japan	2011	831,823	..	337,116	429,625	65,082
	2012	844,497	..	333,791	442,758	67,948
	2013	893,201	..	334,640	488,319	70,242
	2014	873,910	..	318,620	485,388	69,902
Kazakhstan	2011	37,796	..	16,565	21,231	..
	2012	34,176	..	15,540	18,636	..
	2013	22,569	..	8,690	13,879	..
	2014	36,051	..	15,983	20,068	..
Korea, Dem.Ppl's.Rep.	2011	1,849	..	..	1,849	..
	2012	1,844	..	..	1,844	..
	2013	1,826	..	..	1,826	..
	2014	1,841	..	..	1,841	..

Production of manufactured and recovered gases - by type

Terajoules

Country or area	Year	Total	Gas works gas	Coke oven gas	Blast furnace gas	Other recovered gas
Korea, Republic of	2011	381,648	..	125,092	233,841	22,715
	2012	382,844	..	126,679	234,178	21,987
	2013	366,032	..	126,773	212,335	26,924
	2014	405,263	..	146,697	230,682	27,884
Other Asia	2011	121,576	..	55,795	56,837	8,944
	2012	112,234	..	44,465	58,851	8,918
	2013	130,496	..	53,353	67,165	9,978
	2014	135,822	..	53,610	70,820	11,392
Pakistan	2011	1,697	..	..	1,697	..
	2012	1,669	..	..	1,669	..
	2013	699	..	..	699	..
	2014	733	..	..	733	..
Philippines	2011	2,346	..	..	2,346	..
	2012	2,414	..	..	2,414	..
	2013	2,436	..	..	2,436	..
	2014	3,294	..	..	3,294	..
Syrian Arab Republic	2011	45	..	..	45	..
	2012	6	..	..	6	..
	2013	6	..	..	6	..
	2014	6	..	..	6	..
Thailand	2011	384	..	..	384	..
	2012	305	..	..	305	..
	2013	327	..	..	327	..
	2014	372	..	..	372	..
Turkey	2011	62,331	..	26,643	35,688	..
	2012	75,870	..	29,158	46,712	..
	2013	81,405	..	31,705	49,700	..
	2014	86,211	..	33,784	52,427	..
Europe	**2011**	**1,943,256**	**27,669**	**791,107**	**1,083,790**	**40,690**
	2012	**1,686,596**	**26,500**	**765,163**	**855,023**	**39,910**
	2013	**1,697,224**	**25,644**	**761,731**	**869,228**	**40,621**
	2014	**1,648,109**	**26,464**	**701,021**	**858,502**	**62,122**
Austria	2011	45,142	..	10,595	34,547	..
	2012	43,647	..	10,696	32,951	..
	2013	45,656	..	10,323	35,333	..
	2014	44,905	..	10,087	34,818	..
Belgium	2011	43,716	..	16,791	23,953	2,972
	2012	39,536	..	15,171	21,190	3,175
	2013	39,523	..	14,613	21,684	3,226
	2014	38,180	..	12,529	22,171	3,480
Bosnia and Herzegovina	2011	11,753	..	7,212	4,541	..
	2012	10,642	..	5,778	4,864	..
	2013	*11,258	..	*6,265	*4,993	..
	2014	*11,300	..	*6,300	*5,000	..
Croatia	2011	144	144	..	..	..
	2012	109	109	..	..	..
	2013	76	76	..	..	..
	2014	42	42	..	..	..
Czechia	2011	61,617	17,962	20,602	21,181	1,872
	2012	58,546	16,917	19,843	20,013	1,773
	2013	59,272	16,396	20,550	20,513	1,813
	2014	64,275	17,246	20,560	20,823	5,646

Table 29

Production of manufactured and recovered gases - by type

Terajoules

Country or area	Year	Total	Gas works gas	Coke oven gas	Blast furnace gas	Other recovered gas
Denmark	2011	615	615	..	..	..
	2012	613	613	..	..	..
	2013	686	686	..	..	..
	2014	753	753	..	..	..
Estonia	2011	7,204	7,204	..	..	..
	2012	7,422	7,422	..	..	..
	2013	7,203	7,203	..	..	..
	2014	7,201	7,201	..	..	..
Finland	2011	23,921	..	9,321	12,717	1,883
	2012	21,663	..	8,927	10,970	1,766
	2013	21,833	..	8,316	10,249	3,268
	2014	23,146	..	8,781	10,881	3,484
France	2011	68,011	..	21,552	46,459	..
	2012	73,679	..	22,366	51,313	..
	2013	76,530	..	24,297	52,233	..
	2014	79,726	..	24,456	55,270	..
Germany	2011	247,932	..	66,551	163,289	18,092
	2012	244,869	..	66,735	158,930	19,204
	2013	251,697	..	68,582	164,517	18,598
	2014	255,120	..	72,709	163,611	18,800
Hungary	2011	17,260	..	9,575	7,685	..
	2012	16,665	..	9,375	7,290	..
	2013	11,949	..	8,014	3,935	..
	2014	13,791	..	8,444	5,347	..
Italy	2011	55,285	..	19,198	34,635	1,452
	2012	63,363	..	30,354	32,249	760
	2013	44,003	..	19,494	23,838	671
	2014	43,652	..	18,022	25,544	86
Netherlands	2011	55,393	..	18,315	37,078	..
	2012	54,693	..	17,691	37,002	..
	2013	53,207	..	17,570	35,637	..
	2014	54,284	..	18,096	36,188	..
Norway	2011	2,559	..	..	2,559	..
	2012	3,743	..	..	3,743	..
	2013	3,442	..	..	3,442	..
	2014	3,541	..	..	3,541	..
Poland	2011	106,176	103	77,156	22,271	6,646
	2012	101,621	75	72,579	22,684	6,283
	2013	105,601	73	76,493	22,530	6,505
	2014	109,793	44	77,504	25,802	6,443
Romania	2011	7,462	..	..	7,462	..
	2012	8,899	..	..	8,899	..
	2013	9,478	..	..	9,478	..
	2014	8,924	..	..	8,924	..
Russian Federation	2011	718,111	..	287,171	430,940	..
	2012	489,998	..	269,890	220,108	..
	2013	501,051	..	278,399	222,652	..
	2014	482,977	..	261,799	221,178	..
Serbia	2011	9,779	..	..	9,779	..
	2012	3,033	..	..	3,033	..
	2013	3,182	..	..	3,182	..
	2014	4,821	..	..	4,821	..

Production of manufactured and recovered gases - by type

Terajoules

Country or area	Year	Total	Gas works gas	Coke oven gas	Blast furnace gas	Other recovered gas
Slovakia	2011	37,347	..	13,263	20,718	3,366
	2012	36,529	..	12,976	21,165	2,388
	2013	36,294	..	12,695	21,056	2,543
	2014	38,153	..	12,964	21,904	3,285
Spain	2011	35,849	149	17,329	18,371	..
	2012	31,504	71	14,366	17,067	..
	2013	32,298	26	13,598	18,674	..
	2014	33,251	14	12,690	20,547	..
Sweden	2011	22,963	435	9,376	12,896	256
	2012	21,606	406	8,460	11,512	1,228
	2013	20,702	381	7,995	11,157	1,169
	2014	22,066	343	8,265	12,307	1,151
Ukraine	2011	295,139	1,057	155,034	134,897	4,151
	2012	282,177	887	150,014	127,943	3,333
	2013	275,454	803	143,771	128,052	2,828
	2014	221,787	821	96,785	104,434	19,747
United Kingdom	2011	69,878	..	32,066	37,812	..
	2012	72,039	..	29,942	42,097	..
	2013	86,829	..	30,756	56,073	..
	2014	86,421	..	31,030	55,391	..
Oceania	**2011**	**67,227**	**507**	**28,344**	**38,099**	**277**
	2012	**54,207**	**165**	**25,825**	**27,934**	**283**
	2013	**52,153**	**108**	**25,968**	**25,801**	**276**
	2014	**49,548**	**113**	**25,131**	**24,035**	**269**
Australia	2011	56,720	507	24,265	31,948	..
	2012	42,910	165	21,723	21,022	..
	2013	40,218	108	21,235	18,875	..
	2014	37,967	113	20,671	17,183	..
New Zealand	2011	10,507	..	4,079	6,151	277
	2012	11,297	..	4,102	6,912	283
	2013	11,935	..	4,733	6,926	276
	2014	11,581	..	4,460	6,852	269

Net installed capacity of electric plants - by type

Megawatts

Table Notes

Electricity capacity broken down by source, and main activity producers against autoproducers. For total producers, **Other** refers to geothermal, tide, wave, marine and other non-specified net installed capacity. For autoproducers, **Other** refers to nuclear (if any), wind, solar, geothermal, tide, wave, marine and other non-specified net installed capacity. For main activity producers, **Other** refers to wind, solar, geothermal, tide, wave, marine and other non-specified net installed capacity.

For Austria data are given in gross maximum capacity.

Please refer to the Definitions Section on pages ix to xv for the appropriate product description /classification.

Figure 73: Change in net installed capacity of electricity generating plants* by source, 2013-2014

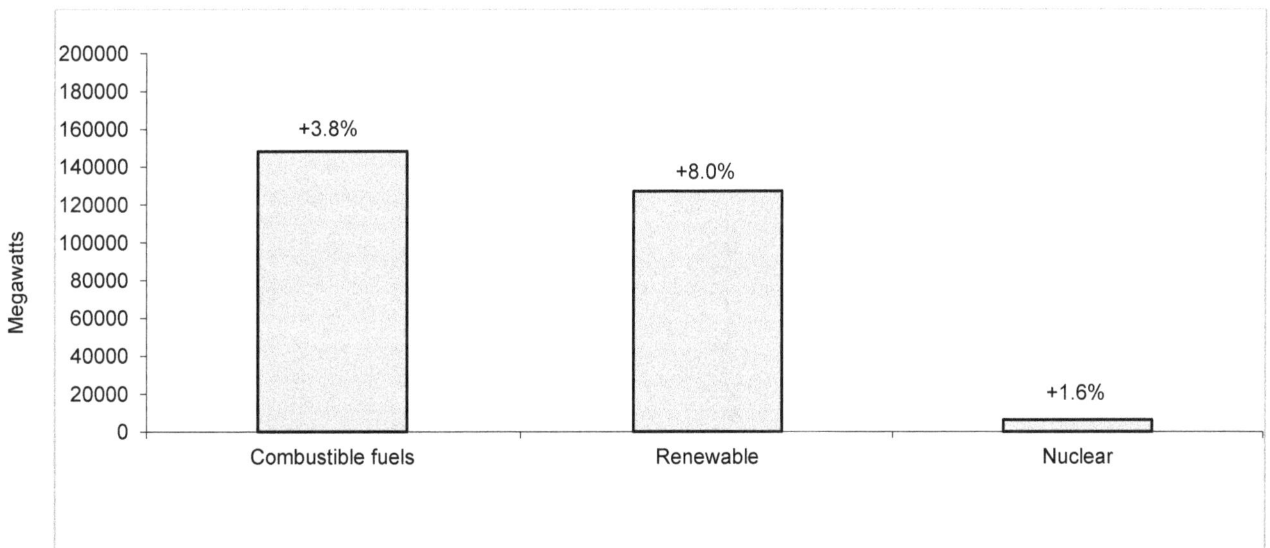

* Renewable includes capacity from other sources

Table 30

Net installed capacity of electricity plants by type

Megawatts

Country or area	Year	Total producers						
		Total	Combustible fuels	Hydro	Nuclear	Wind	Solar	Other
World	**2011**	**5,432,842**	**3,685,625**	**1,058,053**	**386,249**	**220,511**	**69,234**	**13,170**
	2012	**5,655,184**	**3,807,161**	**1,090,530**	**378,706**	**268,265**	**95,903**	**14,619**
	2013	**5,879,042**	**3,912,728**	**1,132,324**	**376,611**	**303,637**	**138,194**	**15,548**
	2014	**6,154,160**	**4,054,606**	**1,174,038**	**382,814**	**351,357**	**174,624**	**16,721**
Africa	**2011**	**147,742**	**118,449**	**25,841**	**1,890**	**1,063**	**302**	**198**
	2012	**152,281**	**122,807**	**25,831**	**1,890**	**1,205**	**342**	**207**
	2013	**159,037**	**128,482**	**26,840**	**1,800**	**1,238**	**433**	**244**
	2014	**168,611**	**135,449**	**27,988**	**1,880**	**1,956**	**773**	**565**
Algeria	2011	*11,545	*11,270	*275	..	..	..	..
	2012	*13,087	*12,787	*300	..	..	..	..
	2013	*15,097	*14,797	*300	..	..	..	..
	2014	*15,957	*15,657	*300	..	..	..	..
Angola	2011	1,530	*770	760	..	..	..	..
	2012	1,530	*770	760	..	..	..	..
	2013	*1,530	*770	*760	..	..	..	..
	2014	*2,070	*770	*1,300	..	..	..	..
Benin	2011	100	99	*1	..	..	..	..
	2012	62	61	*1	..	..	..	..
	2013	208	207	*1	..	..	..	..
	2014	*208	*207	*1	..	..	..	..
Botswana	2011	152	152	..	..	..	..	..
	2012	153	152	..	..	..	1	..
	2013	753	752	..	..	..	1	..
	2014	621	620	..	..	..	1	..
Burkina Faso	2011	261	229	32	..	..	..	..
	2012	298	266	32	..	..	..	..
	2013	*298	*266	*32	..	..	..	..
	2014	*298	*266	*32	..	..	..	..
Burundi	2011	37	*5	32	..	..	..	..
	2012	37	*5	32	..	..	..	..
	2013	39	*7	32	..	..	..	..
	2014	41	*9	32	..	..	..	..
Cabo Verde	2011	108	100	..	..	2	6	..
	2012	134	102	..	..	26	6	..
	2013	142	109	..	..	26	8	..
	2014	154	121	..	..	26	8	..
Cameroon	2011	1,297	*568	728	..	..	..	..
	2012	1,348	*620	728	..	..	..	..
	2013	1,571	839	732	..	..	..	..
	2014	*1,549	817	*732	..	..	..	..
Central African Rep.	2011	*44	*19	*25	..	..	..	..
	2012	*44	*19	*25	..	..	..	..
	2013	*44	*19	*25	..	..	..	..
	2014	*44	*19	*25	..	..	..	..
Chad	2011	*41	*41	..	..	..	..	..
	2012	*41	*41	..	..	..	..	..
	2013	*47	*47	..	..	..	..	..
	2014	*47	*47	..	..	..	..	..
Comoros	2011	*23	*23	..	..	..	..	..
	2012	*23	*23	..	..	..	..	..
	2013	*24	*24	..	..	..	..	..
	2014	*25	*25	..	..	..	..	..
Congo	2011	*369	*149	*220	..	..	..	..
	2012	*390	*170	*220	..	..	..	..
	2013	*390	*170	*220	..	..	..	..
	2014	*390	*170	*220	..	..	..	..

Table 30

Net installed capacity of electricity plants by type

Megawatts

Autoproducers				Main activity producers				
Total	Combustible fuels	Hydro	Other	Total	Combustible fuels	Hydro	Nuclear	Other
449,749	**396,252**	**29,488**	**24,009**	**4,983,093**	**3,289,373**	**1,028,565**	**386,249**	**278,906**
469,723	**406,249**	**29,607**	**33,868**	**5,185,460**	**3,400,912**	**1,060,923**	**378,706**	**344,919**
496,715	**416,637**	**30,885**	**49,193**	**5,382,326**	**3,496,091**	**1,101,439**	**376,611**	**408,186**
538,324	**441,100**	**31,686**	**65,538**	**5,615,836**	**3,613,507**	**1,142,352**	**382,814**	**477,164**
9,668	**8,922**	**739**	**7**	**138,074**	**109,527**	**25,102**	**1,890**	**1,555**
11,196	**10,443**	**739**	**14**	**141,086**	**112,364**	**25,092**	**1,890**	**1,740**
11,677	**10,915**	**739**	**23**	**147,360**	**117,567**	**26,101**	**1,800**	**1,892**
12,979	**12,140**	**778**	**61**	**155,632**	**123,309**	**27,211**	**1,880**	**3,232**
*1,154	*1,154	..	..	*10,391	*10,116	*275	..	..
*2,617	*2,617	..	..	*10,470	*10,170	*300	..	..
*3,019	*3,019	..	..	*12,078	*11,778	*300	..	..
*3,191	*3,191	..	..	*12,766	*12,466	*300	..	..
*130	*130	..	..	1,400	*640	760	..	..
*130	*130	..	..	1,400	*640	760	..	..
*130	*130	..	..	*1,400	*640	*760	..	..
*130	*130	..	..	*1,940	*640	*1,300	..	..
*8	*8	..	..	92	91	*1	..	..
*8	*8	..	..	54	53	*1	..	..
*8	*8	..	..	200	199	*1	..	..
*8	*8	..	..	*200	*199	*1	..	..
*20	*20	..	..	132	132	..	..	..
*20	*20	..	..	133	132	..	..	1
*20	*20	..	..	733	732	..	..	1
*20	*20	..	..	601	600	..	..	1
61	61	..	..	200	168	32	..	..
61	61	..	..	237	205	32	..	..
*61	*61	..	..	*237	*205	*32	..	..
*61	*61	..	..	*237	*205	*32	..	..
..	..	..	..	37	*5	32	..	..
..	..	..	..	37	*5	32	..	..
..	..	..	..	39	*7	32	..	..
..	..	..	..	41	*9	32	..	..
0	0	..	..	108	99	..	..	9
0	0	..	..	134	101	..	..	32
0	0	..	..	142	108	..	..	33
0	0	..	..	154	121	..	..	33
*300	*300	..	..	997	268	728	..	..
*300	*300	..	..	1,048	*320	728	..	..
*300	*300	..	..	1,271	539	732	..	..
*300	*300	..	..	*1,249	517	*732	..	..
..	..	..	..	*44	*19	*25	..	..
..	..	..	..	*44	*19	*25	..	..
..	..	..	..	*44	*19	*25	..	..
..	..	..	..	*44	*19	*25	..	..
..	..	..	..	*41	*41	..	..	..
..	..	..	..	*41	*41	..	..	..
..	..	..	..	*47	*47	..	..	..
..	..	..	..	*47	*47	..	..	..
..	..	..	..	*23	*23	..	..	..
..	..	..	..	*23	*23	..	..	..
..	..	..	..	*24	*24	..	..	..
..	..	..	..	*25	*25	..	..	..
*120	*120	..	..	*249	*29	*220	..	..
*160	*160	..	..	*230	*10	*220	..	..
*160	*160	..	..	*230	*10	*220	..	..
*160	*160	..	..	*230	*10	*220	..	..

Table 30

Net installed capacity of electricity plants by type

Megawatts

Country or area	Year	Total producers						
		Total	Combustible fuels	Hydro	Nuclear	Wind	Solar	Other
Côte d'Ivoire	2011	*1,414	*843	*571	..	..	..	..
	2012	*1,414	*1,043	*371	..	..	..	..
	2013	*1,525	*1,154	*371	..	..	..	..
	2014	*1,655	*1,223	*432	..	..	..	..
Dem. Rep. of the Congo	2011	*2,506	*34	*2,472	..	..	..	..
	2012	*2,506	*34	*2,472	..	..	..	..
	2013	*2,606	*34	*2,572	..	..	..	..
	2014	*2,606	*34	*2,572	..	..	..	..
Djibouti	2011	130	130	..	..	..	..	..
	2012	130	130	..	..	..	..	..
	2013	130	130	..	..	..	..	..
	2014	130	130	..	..	..	..	..
Egypt	2011	29,997	26,275	3,035	..	547	140	..
	2012	31,487	28,000	2,800	..	547	140	..
	2013	32,702	29,215	2,800	..	547	140	..
	2014	35,907	32,420	2,800	..	547	140	..
Equatorial Guinea	2011	*214	*184	*30	..	..	..	..
	2012	*334	*184	*150	..	..	..	..
	2013	*334	*184	*150	..	..	..	..
	2014	*334	*184	*150	..	..	..	..
Eritrea	2011	*140	*139	..	..	..	*1	..
	2012	*140	*139	..	..	..	*1	..
	2013	*140	*139	..	..	..	*1	..
	2014	*140	*139	..	..	..	*1	..
Ethiopia	2011	2,310	189	1,943	..	171	..	7
	2012	2,310	189	1,943	..	171	..	7
	2013	2,311	190	1,943	..	171	..	7
	2014	2,311	190	1,943	..	171	..	7
Gabon	2011	443	273	170	..	..	..	..
	2012	464	293	170	..	..	*1	..
	2013	514	343	170	..	..	*1	..
	2014	509	338	170	..	..	*1	..
Gambia	2011	*74	*74	..	..	..	..	..
	2012	*74	*74	..	..	..	..	..
	2013	*91	*91	..	..	..	..	..
	2014	116	114	..	..	1	1	..
Ghana	2011	2,800	1,620	1,180	..	..	0	..
	2012	2,910	1,730	1,180	..	..	0	..
	2013	3,041	1,458	1,580	..	..	3	..
	2014	3,041	1,458	1,580	..	..	3	..
Guinea	2011	385	*236	149	..	..	..	..
	2012	394	*246	149	..	..	..	..
	2013	518	370	149	..	..	..	..
	2014	*518	*370	*149	..	..	..	..
Guinea-Bissau	2011	*27	*27	..	..	..	..	..
	2012	*27	*27	..	..	..	..	..
	2013	*27	*27	..	..	..	..	..
	2014	*28	*28	..	..	..	..	..
Kenya	2011	1,540	609	735	..	*6	..	191
	2012	1,611	637	770	..	*5	..	200
	2013	1,723	715	767	..	*5	..	237
	2014	2,094	734	797	..	5	..	558
Lesotho	2011	80	..	80	..	..	..	..
	2012	80	..	80	..	..	..	..
	2013	80	..	80	..	..	..	..
	2014	80	..	80	..	..	..	..

Net installed capacity of electricity plants by type

Megawatts

Autoproducers				Main activity producers				
Total	Combustible fuels	Hydro	Other	Total	Combustible fuels	Hydro	Nuclear	Other
*23	*23	..	..	*1,391	*820	*571	..	..
*23	*23	..	..	*1,391	*1,020	*371	..	..
*23	*23	..	..	*1,502	*1,131	*371	..	..
*23	*23	..	..	*1,632	*1,200	*432	..	..
*577	*4	*573	..	*1,929	*30	*1,899	..	..
*577	*4	*573	..	*1,929	*30	*1,899	..	..
*577	*4	*573	..	*2,029	*30	*1,999	..	..
*577	*4	*573	..	*2,029	*30	*1,999	..	..
..	..	..	..	130	130	..	..	..
..	..	..	..	130	130	..	..	..
..	..	..	..	130	130	..	..	..
..	..	..	..	130	130	..	..	..
687	687	..	..	29,310	25,588	3,035	..	687
687	687	..	..	30,800	27,313	2,800	..	687
687	687	..	..	32,015	28,528	2,800	..	687
*687	*687	..	..	35,220	31,733	2,800	..	687
..	..	..	..	*214	*184	*30	..	..
..	..	..	..	*334	*184	*150	..	..
..	..	..	..	*334	*184	*150	..	..
..	..	..	..	*334	*184	*150	..	..
*6	*6	..	..	*134	*133	..	..	*1
*6	*6	..	..	*134	*133	..	..	*1
*6	*6	..	..	*134	*133	..	..	*1
*6	*6	..	..	*134	*133	..	..	*1
..	..	..	..	2,310	189	1,943	..	178
..	..	..	..	2,310	189	1,943	..	178
..	..	..	..	2,311	190	1,943	..	178
..	..	..	..	2,311	190	1,943	..	178
70	70	..	..	373	203	170	..	..
71	70	..	*1	393	223	170	..	..
71	70	..	*1	443	273	170	..	..
*71	*70	..	*1	438	268	170	..	..
*34	*34	..	..	*40	*40	..	..	..
*34	*34	..	..	*40	*40	..	..	..
*34	*34	..	..	*57	*57	..	..	..
*34	*34	..	..	82	80	..	..	2
*210	*210	..	..	2,590	1,410	1,180	..	0
*210	*210	..	..	2,700	1,520	1,180	..	0
*210	*210	..	..	2,831	1,248	1,580	..	3
*210	*210	..	..	2,831	1,248	1,580	..	3
*180	*159	*21	..	205	77	128	..	..
*180	*159	*21	..	214	87	128	..	..
*180	*159	*21	..	338	211	128	..	..
*180	*159	*21	..	*338	*211	*128	..	..
*2	*2	..	..	*25	*25	..	..	..
*2	*2	..	..	*25	*25	..	..	..
*2	*2	..	..	*25	*25	..	..	..
*2	*2	..	..	*26	*26	..	..	..
..	..	..	..	1,540	609	735	..	197
..	..	..	..	1,611	637	770	..	204
..	..	..	..	1,723	715	767	..	242
..	..	..	..	2,094	734	797	..	563
..	..	..	..	80	..	80	..	..
..	..	..	..	80	..	80	..	..
..	..	..	..	80	..	80	..	..
..	..	..	..	80	..	80	..	..

Table 30

Net installed capacity of electricity plants by type

Megawatts

Country or area	Year	Total producers						
		Total	Combustible fuels	Hydro	Nuclear	Wind	Solar	Other
Liberia	2011	*91	*91	..	..	..	..	..
	2012	*91	*91	..	..	..	..	..
	2013	*91	*91	..	..	..	..	..
	2014	*91	*91	..	..	..	..	..
Libya	2011	8,907	8,907	..	..	..	..	..
	2012	8,907	8,907	..	..	..	..	..
	2013	9,455	9,455	..	..	..	..	..
	2014	9,455	9,455	..	..	..	..	..
Madagascar	2011	*507	*379	*128	..	..	0	..
	2012	*507	*379	*128	..	..	0	..
	2013	*508	*379	*128	..	..	*1	..
	2014	*508	*379	*128	..	..	*1	..
Malawi	2011	502	*211	291	..	..	..	..
	2012	502	*211	291	..	..	..	..
	2013	502	*211	291	..	..	..	..
	2014	501	*211	290	..	..	..	..
Mali	2011	655	291	365	..	..	..	..
	2012	689	323	367	..	..	..	..
	2013	811	363	448	..	..	..	..
	2014	*890	*400	*490	..	..	..	..
Mauritania	2011	*196	*196	..	..	..	..	..
	2012	*196	*196	..	..	..	..	..
	2013	*263	*263	..	..	..	..	..
	2014	*315	*315	..	..	..	..	..
Mauritius	2011	738	677	60	..	1	0	..
	2012	781	718	60	..	1	1	..
	2013	778	714	61	..	1	2	..
	2014	783	703	61	..	1	18	..
Mayotte	2011	90	78	..	..	..	12	..
	2012	91	78	..	..	..	*13	..
	2013	91	78	..	..	..	13	..
	2014	*91	*78	..	..	..	13	..
Morocco	2011	6,713	4,693	1,770	..	250	..	..
	2012	7,013	4,993	1,770	..	250	..	..
	2013	*7,013	*4,993	*1,770	..	*250	..	..
	2014	*7,263	*4,993	*1,770	..	*500	..	..
Mozambique	2011	2,479	249	2,230	..	..	..	..
	2012	2,486	249	2,237	..	..	..	..
	2013	2,646	359	2,287	..	..	..	..
	2014	2,682	359	2,322	..	..	1	..
Namibia	2011	393	144	249	..	..	..	..
	2012	508	167	341	..	..	..	..
	2013	487	155	332	..	..	..	..
	2014	487	155	332	..	..	..	..
Niger	2011	*135	*134	..	..	..	*1	..
	2012	*135	*134	..	..	..	*1	..
	2013	176	172	..	..	..	4	..
	2014	176	172	..	..	..	4	..
Nigeria	2011	*8,902	*6,972	*1,930	..	..	..	..
	2012	*8,902	*6,972	*1,930	..	..	..	..
	2013	*9,822	*7,892	*1,930	..	..	..	..
	2014	*10,396	*8,418	*1,978	..	..	..	..
Réunion	2011	766	477	146	..	15	129	..
	2012	868	567	134	..	15	152	..
	2013	875	564	134	..	15	163	..
	2014	834	510	136	..	15	173	..

Net installed capacity of electricity plants by type

Megawatts

Autoproducers				Main activity producers				
Total	Combustible fuels	Hydro	Other	Total	Combustible fuels	Hydro	Nuclear	Other
*70	*70	..	..	*21	*21	..	..	..
*70	*70	..	..	*21	*21	..	..	..
*70	*70	..	..	*21	*21	..	..	..
*70	*70	..	..	*21	*21	..	..	..
..	..	..	..	8,907	8,907	..	..	..
..	..	..	..	8,907	8,907	..	..	..
..	..	..	..	9,455	9,455	..	..	..
..	..	..	..	9,455	9,455	..	..	..
*32	*32	..	..	*475	*347	*128	..	0
*32	*32	..	..	*475	*347	*128	..	0
*32	*32	..	..	*476	*347	*128	..	*1
*32	*32	..	..	*476	*347	*128	..	*1
*215	*210	*5	..	287	*1	286	..	..
*215	*210	*5	..	287	*1	286	..	..
*215	*210	*5	..	287	*1	286	..	..
*215	*210	5	..	287	*1	286	..	..
97	97	..	..	558	194	365	..	..
97	97	..	..	592	226	367	..	..
97	97	..	..	714	266	448	..	..
*110	*110	..	..	*780	*290	*490	..	..
*70	*70	..	..	*126	*126	..	..	..
*70	*70	..	..	*126	*126	..	..	..
*70	*70	..	..	*193	*193	..	..	..
*79	*79	..	..	*236	*236	..	..	..
288	288	..	0	450	389	60	..	1
274	273	..	1	506	445	60	..	1
277	275	..	2	501	439	61	..	1
293	275	..	18	490	428	61	..	1
..	..	..	..	90	78	..	..	12
..	..	..	..	91	78	..	..	*13
..	..	..	..	91	78	..	..	13
..	..	..	..	*91	*78	..	..	13
336	336	..	..	6,377	4,357	1,770	..	250
336	336	..	..	6,677	4,657	1,770	..	250
*336	*336	..	..	*6,677	*4,657	*1,770	..	*250
*336	*336	..	..	*6,927	*4,657	*1,770	..	*500
*50	..	*50	..	2,429	249	2,180	..	..
*50	..	*50	..	2,436	249	2,187	..	..
*50	..	*50	..	2,596	359	2,237	..	..
*86	..	*85	1	2,596	359	2,237	..	..
..	..	..	..	393	144	249	..	..
..	..	..	..	508	167	341	..	..
..	..	..	..	487	155	332	..	..
..	..	..	..	487	155	332	..	..
*30	*29	..	*1	*105	*105	..	..	..
*30	*29	..	*1	*105	*105	..	..	..
42	38	..	4	134	134	..	..	..
42	38	..	4	134	134	..	..	..
*2,132	*2,092	*40	..	*6,770	*4,880	*1,890	..	..
*2,132	*2,092	*40	..	*6,770	*4,880	*1,890	..	..
*2,132	*2,092	*40	..	*7,690	*5,800	*1,890	..	..
*3,040	*3,000	*40	..	*7,356	*5,418	*1,938	..	..
..	..	..	..	766	477	146	..	144
..	..	..	..	868	567	134	..	167
..	..	..	..	875	564	134	..	178
..	..	..	..	834	510	136	..	188

Table 30

Net installed capacity of electricity plants by type

Megawatts

Country or area	Year	Total producers						
		Total	Combustible fuels	Hydro	Nuclear	Wind	Solar	Other
Rwanda	2011	89	33	57	..	..	0	..
	2012	*99	*36	*63	..	..	0	..
	2013	*109	*40	*69	..	..	0	..
	2014	*119	*42	*76	..	..	*1	..
Sao Tome and Principe	2011	18	16	*2	..	..	..	..
	2012	*18	*16	*2	..	..	..	..
	2013	*19	*16	*3	..	..	..	..
	2014	19	16	*3	..	..	..	..
Senegal	2011	827	827	..	..	..	..	..
	2012	854	854	..	..	..	..	..
	2013	*854	*854	..	..	..	..	..
	2014	*875	*854	..	..	..	*21	..
Seychelles	2011	79	79	..	..	0	..	..
	2012	79	79	..	..	0	..	..
	2013	85	79	..	..	6	..	..
	2014	86	80	..	..	6	..	..
Sierra Leone	2011	77	27	50	..	..	..	..
	2012	77	27	50	..	..	..	..
	2013	77	27	50	..	..	..	..
	2014	77	27	50	..	..	..	..
Somalia	2011	80	80	..	..	..	..	..
	2012	80	80	..	..	..	..	..
	2013	80	80	..	..	..	..	..
	2014	80	80	..	..	..	..	..
South Africa	2011	*44,813	*42,186	*714	*1,890	*17	*6	..
	2012	*44,820	*42,186	*714	*1,890	*17	*13	..
	2013	*44,170	*41,559	*714	*1,800	*17	*80	..
	2014	*46,963	*43,538	*725	*1,880	*450	*370	..
South Sudan	2011	..	..	..	..	..	..	..
	2012	*82	*80	..	..	..	*2	..
	2013	*82	*80	..	..	..	*2	..
	2014	*82	*80	..	..	..	*2	..
St. Helena and Depend.	2011	5	5	..	..	0	0	..
	2012	8	8	..	..	0	0	..
	2013	8	8	..	..	0	0	..
	2014	9	8	..	..	1	0	..
Sudan	2011	..	..	..	..	..	..	..
	2012	2,877	1,287	1,590	..	..	..	..
	2013	3,117	1,287	1,830	..	..	..	..
	2014	*3,117	*1,287	*1,830	..	..	..	..
Sudan (former)	2011	2,723	1,133	1,590	..	..	..	..
	2012	..	..	..	..	..	..	..
	2013	..	..	..	..	..	..	..
	2014	..	..	..	..	..	..	..
Swaziland	2011	*159	*95	64	..	..	..	..
	2012	*164	*100	64	..	..	..	..
	2013	*164	*100	64	..	..	..	..
	2014	*173	*100	73	..	..	..	..
Togo	2011	201	134	67	..	..	..	..
	2012	227	160	67	..	..	..	..
	2013	227	160	67	..	..	..	..
	2014	227	160	67	..	..	..	..
Tunisia	2011	4,217	4,101	62	..	53	1	..
	2012	4,279	4,041	62	..	173	3	..
	2013	4,539	4,270	62	..	200	7	..
	2014	4,998	4,695	62	..	233	8	..

Net installed capacity of electricity plants by type

Megawatts

Autoproducers				Main activity producers				
Total	Combustible fuels	Hydro	Other	Total	Combustible fuels	Hydro	Nuclear	Other
..	..	..	..	89	33	57	..	0
..	..	..	..	*99	*36	*63	..	0
..	..	..	..	*109	*40	*69	..	0
..	..	..	..	*119	*42	*76	..	*1
..	..	..	..	18	16	*2	..	..
..	..	..	..	*18	*16	*2	..	..
..	..	..	..	*19	*16	*3	..	..
..	..	..	..	19	16	*3	..	..
344	344	..	..	483	483	..	..	..
365	365	..	..	489	489	..	..	..
*365	*365	..	..	*489	*489	..	..	..
*386	*365	..	*21	*489	*489	..	..	..
..	..	..	..	79	79	..	..	0
..	..	..	..	79	79	..	..	0
..	..	..	..	85	79	..	..	6
..	..	..	..	86	80	..	..	6
..	..	..	..	77	27	50	..	..
..	..	..	..	77	27	50	..	..
..	..	..	..	77	27	50	..	..
..	..	..	..	77	27	50	..	..
..	..	..	..	80	80	..	..	..
..	..	..	..	80	80	..	..	..
..	..	..	..	80	80	..	..	..
..	..	..	..	80	80	..	..	..
*2,036	*2,000	*36	..	*42,777	*40,186	*678	*1,890	*23
*2,036	*2,000	*36	..	*42,784	*40,186	*678	*1,890	*30
*2,036	*2,000	*36	..	*42,134	*39,559	*678	*1,800	*97
*2,140	*2,100	*40	..	*44,823	*41,438	*685	*1,880	*820
..	..	..	..	..	..	..	..	..
*41	*40	..	*1	*41	*40	..	..	*1
*41	*40	..	*1	*41	*40	..	..	*1
*41	*40	..	*1	*41	*40	..	..	*1
..	..	..	..	5	5	..	..	0
..	..	..	..	8	8	..	..	1
..	..	..	..	8	8	..	..	1
..	..	..	..	9	8	..	..	1
..	..	..	..	..	..	..	..	..
..	..	..	..	2,877	1,287	1,590	..	..
..	..	..	..	3,117	1,287	1,830	..	..
..	..	..	..	*3,117	*1,287	*1,830	..	..
..	..	..	..	2,723	1,133	1,590	..	..
..	..	..	..	..	..	..	..	..
..	..	..	..	..	..	..	..	..
..	..	..	..	..	..	..	..	..
*89	*85	*4	..	70	10	60	..	..
*94	*90	*4	..	70	10	60	..	..
*94	*90	*4	..	70	10	60	..	..
*94	*90	*4	..	79	10	69	..	..
*8	*8	..	..	193	126	67	..	..
*8	*8	..	..	219	152	67	..	..
*8	*8	..	..	219	152	67	..	..
*8	*8	..	..	219	152	67	..	..
*193	*192	..	1	4,024	3,909	62	..	53
*162	*159	..	3	4,117	3,882	62	..	173
*205	*198	..	7	4,334	4,072	62	..	200
*206	*198	..	8	4,792	4,497	62	..	233

Table 30

Net installed capacity of electricity plants by type

Megawatts

Country or area	Year	Total producers						
		Total	Combustible fuels	Hydro	Nuclear	Wind	Solar	Other
Uganda	2011	628	186	442	..	..	..	..
	2012	728	136	592	..	..	..	..
	2013	838	136	702	..	..	..	..
	2014	883	178	705	..	..	..	..
United Rep. of Tanzania	2011	1,113	*546	562	..	..	*5	..
	2012	1,114	*546	562	..	..	*6	..
	2013	1,115	*546	562	..	..	*7	..
	2014	1,115	*546	562	..	..	*7	..
Zambia	2011	1,959	91	1,868	..	..	..	..
	2012	1,959	91	1,868	..	..	..	..
	2013	2,038	141	1,897	..	..	..	..
	2014	2,452	195	2,257	..	..	..	..
Zimbabwe	2011	*2,112	*1,355	757	..	..	..	..
	2012	*2,112	*1,355	757	..	..	..	..
	2013	*2,112	*1,355	757	..	..	..	..
	2014	1,992	1,235	757	..	..	..	..
America, North	**2011**	**1,284,905**	**912,950**	**194,042**	**115,449**	**51,971**	**6,255**	**4,240**
	2012	**1,298,059**	**904,235**	**194,553**	**116,655**	**67,737**	**9,556**	**5,323**
	2013	**1,300,238**	**899,530**	**195,177**	**114,673**	**70,745**	**14,483**	**5,631**
	2014	**1,314,613**	**901,300**	**197,082**	**114,002**	**77,542**	**18,678**	**6,009**
Anguilla	2011	28	28	..	..	..	..	..
	2012	28	28	..	..	..	..	..
	2013	28	28	..	..	..	..	..
	2014	28	28	..	..	..	..	..
Antigua and Barbuda	2011	84	84	..	..	..	..	..
	2012	84	84	..	..	..	..	..
	2013	*84	*84	..	..	..	..	..
	2014	*84	*84	..	..	..	..	..
Aruba	2011	320	290	..	..	30	..	..
	2012	320	290	..	..	30	..	..
	2013	288	258	..	..	30	..	..
	2014	288	258	..	..	30	..	..
Bahamas	2011	591	591	..	..	..	..	..
	2012	591	591	..	..	..	..	..
	2013	591	591	..	..	..	..	..
	2014	576	576	..	..	..	..	..
Barbados	2011	242	242	..	..	..	..	..
	2012	241	241	..	..	..	..	..
	2013	241	241	..	..	..	..	..
	2014	249	249	..	..	..	..	..
Belize	2011	144	91	53	..	..	..	..
	2012	144	91	53	..	..	..	..
	2013	143	88	55	..	..	..	..
	2014	154	99	55	..	..	..	..
Bermuda	2011	169	169	..	..	..	..	..
	2012	171	171	..	..	..	..	..
	2013	*171	*171	..	..	..	..	..
	2014	*171	*171	..	..	..	..	..
Bonaire, St Eustatius, Saba	2011	..	..	..	..	..	..	..
	2012	*43	*14	..	..	*29	..	..
	2013	*43	*14	..	..	*29	..	..
	2014	*43	14	..	..	*29	..	..
British Virgin Islands	2011	45	44	..	..	1	0	..
	2012	45	44	..	..	1	0	..
	2013	45	44	..	..	1	0	..
	2014	45	44	..	..	1	0	..

Net installed capacity of electricity plants by type

Megawatts

Autoproducers				Main activity producers				
Total	Combustible fuels	Hydro	Other	Total	Combustible fuels	Hydro	Nuclear	Other
*26	*16	*10	..	602	170	432	..	..
*26	*16	*10	..	702	120	582	..	..
*46	*36	*10	..	792	100	692	..	..
*70	*60	*10	..	814	119	695	..	..
*25	*20	..	*5	1,088	*526	562	..	..
*26	*20	..	*6	1,088	*526	562	..	..
*27	*20	..	*7	1,088	*526	562	..	..
*27	*20	..	*7	1,088	*526	562	..	..
..	..	..	..	1,959	91	1,868	..	..
..	..	..	..	1,959	91	1,868	..	..
..	..	..	..	2,038	141	1,897	..	..
..	..	..	..	2,452	195	2,257	..	..
*45	*45	..	..	*2,067	*1,310	757	..	..
*45	*45	..	..	*2,067	*1,310	757	..	..
*45	*45	..	..	*2,067	*1,310	757	..	..
*45	*45	..	..	1,947	1,190	757	..	..
57,850	**46,154**	**6,415**	**5,281**	**1,227,055**	**866,796**	**187,627**	**115,449**	**57,184**
61,836	**47,651**	**6,211**	**7,974**	**1,236,224**	**856,585**	**188,342**	**116,655**	**74,642**
60,805	**44,987**	**6,389**	**9,429**	**1,239,433**	**854,542**	**188,788**	**114,673**	**81,430**
61,885	**45,891**	**6,081**	**9,913**	**1,252,727**	**855,409**	**191,001**	**114,002**	**92,316**
..	..	..	..	28	28	..	..	..
..	..	..	..	28	28	..	..	..
..	..	..	..	28	28	..	..	..
..	..	..	..	28	28	..	..	..
*1	*1	..	..	83	83	..	..	..
*1	*1	..	..	83	83	..	..	..
*1	*1	..	..	*83	*83	..	..	..
*1	*1	..	..	*83	*83	..	..	..
..	..	..	..	320	290	..	..	30
..	..	..	..	320	290	..	..	30
..	..	..	..	288	258	..	..	30
..	..	..	..	288	258	..	..	30
*55	*55	..	..	536	536	..	..	..
*55	*55	..	..	536	536	..	..	..
*55	*55	..	..	536	536	..	..	..
*39	*39	..	..	537	537	..	..	..
*3	*3	..	..	239	239	..	..	..
*2	*2	..	..	239	239	..	..	..
*2	*2	..	..	239	239	..	..	..
*10	*10	..	..	239	239	..	..	..
26	26	..	..	118	65	53	..	..
26	26	..	..	118	65	53	..	..
23	23	..	..	120	66	55	..	..
34	34	..	..	120	66	55	..	..
4	4	..	..	165	165	..	..	..
4	4	..	..	167	167	..	..	..
*4	*4	..	..	*167	*167	..	..	..
*4	*4	..	..	*167	*167	..	..	..
..	..	..	..	..	..	..	..	..
..	..	..	..	*43	*14	..	..	*29
..	..	..	..	*43	*14	..	..	*29
..	..	..	..	*43	14	..	..	*29
..	..	..	..	45	44	..	..	1
..	..	..	..	45	44	..	..	1
..	..	..	..	45	44	..	..	1
..	..	..	..	45	44	..	..	1

Table 30

Net installed capacity of electricity plants by type

Megawatts

Country or area	Year	Total producers						
		Total	Combustible fuels	Hydro	Nuclear	Wind	Solar	Other
Canada	2011	132,959	38,939	75,573	12,665	5,265	497	20
	2012	130,554	34,660	75,537	13,370	6,201	766	20
	2013	133,816	35,215	75,537	14,033	7,801	1,210	20
	2014	137,344	36,217	75,537	14,033	9,694	1,843	20
Cayman Islands	2011	151	151	..	..	..	..	..
	2012	150	150	..	..	..	..	..
	2013	150	150	..	..	..	..	..
	2014	132	132	..	..	..	..	..
Costa Rica	2011	2,889	862	1,682	..	*133	*1	211
	2012	2,724	656	1,700	..	*149	*1	218
	2013	2,810	634	1,729	..	*229	1	218
	2014	3,215	618	2,045	..	305	1	246
Cuba	2011	5,914	5,849	65	..	..	..	..
	2012	5,699	5,637	62	..	..	..	..
	2013	6,055	5,992	63	..	..	..	..
	2014	6,169	6,106	63	..	..	..	..
Curaçao	2011	..	..	..	..	..	..	..
	2012	*265	*235	..	..	*30	..	..
	2013	*265	*235	..	..	*30	..	..
	2014	*265	*235	..	..	*30	..	..
Dominica	2011	27	20	7	..	..	..	..
	2012	27	20	7	..	..	..	..
	2013	27	20	7	..	..	..	..
	2014	27	20	7	..	..	..	..
Dominican Republic	2011	3,440	2,883	523	..	33	..	..
	2012	3,678	2,986	605	..	85	2	..
	2013	4,120	3,421	606	..	85	7	..
	2014	4,042	3,328	616	..	84	14	..
El Salvador	2011	1,477	801	472	..	..	..	204
	2012	1,478	.801	473	..	..	..	204
	2013	1,537	860	473	..	..	..	204
	2014	1,563	886	473	..	..	..	204
Greenland	2011	*174	*106	*68	..	..	..	..
	2012	*172	*96	76	..	..	..	..
	2013	187	*96	91	..	..	..	..
	2014	187	96	91	..	..	..	..
Grenada	2011	50	50	..	..	..	..	..
	2012	49	49	..	..	..	..	..
	2013	49	49	..	..	..	..	..
	2014	49	49	..	..	..	..	..
Guadeloupe	2011	467	359	8	..	21	*63	16
	2012	493	380	9	..	27	64	14
	2013	496	380	9	..	27	67	14
	2014	496	380	9	..	27	67	14
Guatemala	2011	2,510	1,570	891	..	..	..	49
	2012	2,796	1,773	991	..	..	..	32
	2013	2,974	1,923	1,002	..	..	..	49
	2014	2,610	1,580	991	..	..	5	34
Haiti	2011	267	207	61	..	..	..	..
	2012	267	207	61	..	..	..	..
	2013	*340	*262	77	..	..	..	..
	2014	*320	*259	61	..	..	..	..
Honduras	2011	1,722	1,089	531	..	102	..	..
	2012	1,783	1,143	538	..	102	..	..
	2013	1,806	1,146	558	..	102	..	..
	2014	1,915	1,139	624	..	152	..	..

Net installed capacity of electricity plants by type

Megawatts

Autoproducers				Main activity producers				
Total	Combustible fuels	Hydro	Other	Total	Combustible fuels	Hydro	Nuclear	Other
10,494	5,407	5,087	..	122,465	33,532	70,486	12,665	5,782
10,546	5,407	5,139	..	120,008	29,253	70,398	13,370	6,987
9,254	3,878	5,312	64	124,562	31,337	70,225	14,033	8,967
9,494	4,116	5,305	73	127,850	32,101	70,232	14,033	11,484
..	..	..	..	151	151	..	..	..
..	..	..	..	150	150	..	..	..
..	..	..	..	150	150	..	..	..
..	..	..	..	132	132	..	..	..
*368	28	*210	*130	2,521	834	1,472	..	215
*370	21	*210	*139	2,354	635	1,490	..	229
*370	*21	*210	*139	2,440	613	1,519	..	309
370	*21	210	*139	2,845	597	1,835	..	413
544	544	..	..	5,370	5,305	65	..	..
519	519	..	..	5,180	5,118	62	..	..
470	470	..	..	5,585	5,522	63	..	..
498	498	..	..	5,671	5,608	63	..	..
..	..	..	..	..	..	..	..	..
*133	*133	..	..	*132	*102	..	..	*30
*133	*133	..	..	*132	*102	..	..	*30
*133	*133	..	..	*132	*102	..	..	*30
..	..	..	..	27	20	7	..	..
..	..	..	..	27	20	7	..	..
..	..	..	..	27	20	7	..	..
..	..	..	..	27	20	7	..	..
175	175	..	..	3,265	2,709	523	..	33
176	175	..	2	3,501	2,811	605	..	85
407	400	..	7	3,713	3,021	606	..	85
426	412	..	14	3,616	2,916	616	..	84
193	193	0	..	1,284	608	472	..	204
193	193	0	..	1,285	608	473	..	204
*193	*193	0	..	1,344	667	473	..	204
*193	*193	0	..	1,370	693	473	..	204
*16	*16	..	..	*158	*90	*68	..	..
*16	*16	..	..	*156	*80	76	..	..
*16	*16	..	..	171	*80	91	..	..
16	16	..	..	171	80	91	..	..
..	..	..	..	50	50	..	..	..
..	..	..	..	49	49	..	..	..
..	..	..	..	49	49	..	..	..
..	..	..	..	49	49	..	..	..
..	..	..	..	467	359	8	..	*100
..	..	..	..	493	380	9	..	105
..	..	..	..	496	380	9	..	108
..	..	..	..	496	380	9	..	108
384	380	4	..	2,126	1,190	887	..	49
434	430	4	..	2,362	1,344	987	..	32
599	594	4	..	2,375	1,328	998	..	49
474	469	4	..	2,137	1,111	987	..	39
74	74	..	..	193	133	61	..	..
74	74	..	..	193	133	61	..	..
*94	*94	..	..	*246	*168	77	..	..
*93	*93	..	..	*227	*166	61	..	..
128	128	..	..	1,594	961	531	..	102
159	159	..	..	1,624	984	538	..	102
158	158	..	..	1,648	988	558	..	102
181	181	..	..	1,734	958	624	..	152

Table 30

Net installed capacity of electricity plants by type

Megawatts

Country or area	Year	Total producers						
		Total	Combustible fuels	Hydro	Nuclear	Wind	Solar	Other
Jamaica	2011	840	776	23	..	42	..	..
	2012	912	848	23	..	42	..	..
	2013	938	874	23	..	42	..	..
	2014	945	874	29	..	42	..	..
Martinique	2011	*415	*400	..	..	1	*14	..
	2012	*461	*400	..	..	1	60	..
	2013	*463	*400	..	..	1	62	..
	2014	492	428	..	..	1	63	..
Mexico	2011	62,046	47,586	11,571	1,365	601	36	887
	2012	64,077	48,364	11,626	1,400	1,815	48	824
	2013	64,092	48,047	11,633	1,400	2,122	67	823
	2014	66,238	48,893	12,464	1,400	2,569	99	813
Montserrat	2011	*5	*5	..	..	..	..	..
	2012	*5	*5	..	..	..	..	..
	2013	*5	*5	..	..	..	..	..
	2014	*5	*5	..	..	..	..	..
Neth. Antilles (former)	2011	*325	*325	..	..	..	..	..
	2012	..	..	..	..	..	..	..
	2013	..	..	..	..	..	..	..
	2014	..	..	..	..	..	..	..
Nicaragua	2011	1,110	853	106	..	63	..	88
	2012	1,288	870	107	..	146	..	165
	2013	1,292	869	121	..	147	..	155
	2014	1,329	867	121	..	186	..	155
Panama	2011	2,391	1,040	1,351	..	0	..	..
	2012	2,422	954	1,468	..	0	..	..
	2013	2,546	1,032	1,494	..	20	..	..
	2014	2,829	1,148	1,623	..	55	2	..
Puerto Rico	2011	5,900	5,800	100	..	0	0	..
	2012	5,903	5,800	100	..	1	*2	..
	2013	6,024	5,800	100	..	102	22	..
	2014	*6,036	*5,800	*100	..	*102	*34	..
Sint Maarten (Dutch part)	2011	..	..	..	..	..	..	..
	2012	97	97	..	..	..	..	..
	2013	97	97	..	..	..	..	..
	2014	97	97	..	..	..	..	..
St. Kitts-Nevis	2011	*57	*55	..	..	2	..	..
	2012	*62	*59	..	..	2	0	..
	2013	*62	*59	..	..	2	1	..
	2014	*63	*60	..	..	*2	1	..
St. Lucia	2011	76	76	..	..	..	..	..
	2012	86	86	..	..	..	..	..
	2013	86	86	..	..	..	..	..
	2014	88	88	..	..	..	..	..
St. Pierre-Miquelon	2011	27	26	..	..	1	..	..
	2012	27	26	..	..	1	..	..
	2013	28	27	..	..	1	..	..
	2014	27	26	..	..	1	..	..
St. Vincent-Grenadines	2011	*58	*46	*13	..	..	..	..
	2012	*58	*46	*13	..	..	..	..
	2013	*58	*46	*13	..	..	..	..
	2014	*58	*46	*13	..	..	..	..
Trinidad and Tobago	2011	2,295	2,295	..	..	..	..	..
	2012	2,577	2,577	..	..	..	..	..
	2013	2,598	2,598	..	..	..	..	..
	2014	2,598	2,598	..	..	..	..	..

Net installed capacity of electricity plants by type

Megawatts

Autoproducers				Main activity producers				
Total	Combustible fuels	Hydro	Other	Total	Combustible fuels	Hydro	Nuclear	Other
229	190	..	39	611	586	23	..	3
301	262	..	39	611	586	23	..	3
301	262	..	39	637	612	23	..	3
301	262	..	39	644	612	29	..	3
4	4	..	..	*411	*396	..	..	*15
4	4	..	..	*457	*396	..	..	61
4	4	..	..	*459	*396	..	..	63
4	4	..	..	488	424	..	..	64
9,581	8,913	118	550	52,465	38,673	11,453	1,365	974
11,741	10,349	128	1,264	52,336	38,015	11,498	1,400	1,423
10,595	8,886	124	1,585	53,497	39,161	11,509	1,400	1,427
11,859	9,599	195	2,065	54,379	39,294	12,269	1,400	1,416
..	..	..	..	*5	*5	..	..	..
..	..	..	..	*5	*5	..	..	..
..	..	..	..	*5	*5	..	..	..
..	..	..	..	*5	*5	..	..	..
*133	*133	..	..	*192	*192	..	..	..
..	..	..	..	..	..	..	..	..
..	..	..	..	..	..	..	..	..
..	..	..	..	..	..	..	..	..
123	122	1	..	987	731	105	..	151
135	134	1	..	1,153	736	106	..	311
135	134	1	..	1,157	735	120	..	302
135	134	1	..	1,194	733	120	..	341
224	159	65	..	2,167	881	1,286	..	0
224	159	65	..	2,198	795	1,403	..	0
213	153	60	..	2,333	879	1,434	..	20
269	209	60	..	2,560	939	1,563	..	57
*61	*61	..	..	5,839	5,739	100	..	0
*61	*61	..	..	5,842	5,739	100	..	*3
*61	*61	..	..	5,963	5,739	100	..	124
*61	*61	..	..	*5,975	*5,739	*100	..	*136
..	..	..	..	..	..	..	..	..
..	..	..	..	97	97	..	..	..
..	..	..	..	97	97	..	..	..
..	..	..	..	97	97	..	..	..
*4	*4	..	..	*53	*51	..	..	2
*4	*4	..	..	*58	*55	..	..	2
*4	*4	..	..	*58	*55	..	..	3
*4	*4	..	..	*59	*56	..	..	*3
..	..	..	..	76	76	..	..	..
..	..	..	..	86	86	..	..	..
..	..	..	..	86	86	..	..	..
..	..	..	..	88	88	..	..	..
..	..	..	..	27	26	..	..	1
..	..	..	..	27	26	..	..	1
..	..	..	..	28	27	..	..	1
..	..	..	..	27	26	..	..	1
..	..	..	..	*58	*46	*13	..	..
..	..	..	..	*58	*46	*13	..	..
..	..	..	..	*58	*46	*13	..	..
..	..	..	..	*58	*46	*13	..	..
*196	*196	..	..	2,099	2,099	..	..	..
*196	*196	..	..	2,381	2,381	..	..	..
*196	*196	..	..	2,402	2,402	..	..	..
*196	*196	..	..	2,402	2,402	..	..	..

Table 30

Net installed capacity of electricity plants by type

Megawatts

Country or area	Year	Total producers						
		Total	Combustible fuels	Hydro	Nuclear	Wind	Solar	Other
Turks and Caicos Islands	2011	65	65	..	..	..	..	..
	2012	76	76	..	..	..	..	..
	2013	76	76	..	..	..	..	..
	2014	79	79	..	..	..	..	..
United States	2011	1,055,279	798,833	100,944	101,419	45,676	5,643	2,764
	2012	1,067,878	793,352	101,106	101,885	59,075	8,613	3,847
	2013	1,065,293	787,297	101,589	99,240	59,974	13,045	4,148
	2014	1,073,438	787,407	102,162	98,569	64,232	16,545	4,523
United States Virgin Is.	2011	345	345	..	..	..	*1	..
	2012	331	330	..	..	..	*1	..
	2013	317	316	..	..	..	*1	..
	2014	321	316	..	..	..	*5	..
America, South	**2011**	**238,265**	**93,730**	**139,689**	**3,025**	**1,758**	**7**	**55**
	2012	**247,467**	**100,225**	**141,700**	**3,025**	**2,328**	**130**	**60**
	2013	**263,171**	**113,043**	**143,975**	**3,008**	**2,847**	**233**	**64**
	2014	**271,601**	**114,244**	**147,318**	**3,008**	**6,533**	**429**	**69**
Argentina	2011	33,810	22,660	10,045	1,018	*82	*5	..
	2012	35,023	23,807	10,053	1,018	134	11	..
	2013	39,563	28,269	10,074	1,018	194	8	..
	2014	*35,945	*24,651	10,074	1,018	194	8	..
Bolivia (Plur. State of)	2011	1,684	1,188	494	..	..	*2	..
	2012	1,882	1,387	494	..	..	*2	..
	2013	*2,114	*1,618	*494	..	..	*2	..
	2014	*1,606	*1,143	*455	..	3	5	..
Brazil	2011	117,134	31,243	82,459	2,007	1,425	..	..
	2012	120,973	32,786	84,294	2,007	1,886	..	..
	2013	126,743	36,528	86,018	1,990	2,202	5	..
	2014	133,913	37,827	89,193	1,990	4,888	15	..
Chile	2011	17,549	11,419	5,946	..	184	0	0
	2012	18,153	11,957	5,992	..	202	2	0
	2013	18,600	12,105	6,094	..	301	100	0
	2014	23,400	16,049	6,378	..	731	242	..
Colombia	2011	15,041	5,158	9,810	..	18	..	55
	2012	15,033	5,088	9,870	..	18	..	57
	2013	15,177	5,131	9,967	..	18	..	61
	2014	14,823	5,043	9,696	..	18	..	66
Ecuador	2011	4,796	2,586	2,207	..	2	0	..
	2012	5,063	2,824	2,237	..	2	0	..
	2013	5,103	2,843	2,237	..	19	4	..
	2014	5,299	3,011	2,241	..	21	26	..
Falkland Is. (Malvinas)	2011	*12	*9	..	..	*3	..	..
	2012	*12	*9	..	..	*3	..	..
	2013	*12	*9	..	..	*3	..	..
	2014	*12	*9	..	..	*3	..	..
French Guiana	2011	247	129	118	..	..	0	..
	2012	281	129	118	..	..	34	..
	2013	292	140	119	..	..	33	..
	2014	292	140	119	..	..	33	..
Guyana	2011	349	349	..	..	..	..	..
	2012	363	363	..	..	..	..	..
	2013	*383	*383	..	..	..	..	..
	2014	*380	*380	..	..	..	..	..
Paraguay	2011	8,816	6	8,810	..	..	..	..
	2012	8,816	6	8,810	..	..	..	..
	2013	8,825	15	8,810	..	..	..	..
	2014	8,825	15	8,810	..	..	..	..

Net installed capacity of electricity plants by type

Megawatts

Autoproducers				Main activity producers				
Total	Combustible fuels	Hydro	Other	Total	Combustible fuels	Hydro	Nuclear	Other
..	..	..	..	65	65	..	..	..
..	..	..	..	76	76	..	..	..
..	..	..	..	76	76	..	..	..
..	..	..	..	79	79	..	..	..
34,815	29,323	930	4,562	1,020,464	769,510	100,014	101,419	49,521
36,461	29,267	664	6,530	1,031,417	764,085	100,442	101,885	65,005
37,517	29,245	678	7,594	1,027,776	758,052	100,911	99,240	69,573
37,087	29,202	306	7,579	1,036,351	758,205	101,856	98,569	77,721
*15	*15	..	*1	330	330	..	..	..
*1	0	..	*1	330	330	..	..	..
*1	0	..	*1	316	316	..	..	..
*5	0	..	*5	316	316	..	..	..
27,774	**22,809**	**4,945**	**20**	**210,491**	**70,921**	**134,744**	**3,025**	**1,801**
28,862	**23,802**	**5,028**	**32**	**218,605**	**76,423**	**136,671**	**3,025**	**2,486**
30,507	**24,982**	**5,396**	**128**	**232,664**	**88,061**	**138,579**	**3,008**	**3,016**
31,559	**26,109**	**5,395**	**55**	**240,042**	**88,136**	**141,923**	**3,008**	**6,975**
3,923	3,901	20	2	29,887	18,759	10,025	1,018	*85
4,014	3,992	20	2	31,008	19,815	10,033	1,018	143
4,063	4,041	20	2	35,500	24,228	10,054	1,018	200
*4,014	*3,992	20	2	*31,931	*20,659	10,054	1,018	200
121	93	26	*2	1,563	1,096	467	..	..
137	108	26	*2	1,746	1,279	467	..	..
*155	*126	*26	*2	*1,959	*1,492	*467	..	..
*119	*89	*24	5	*1,487	*1,053	*431	..	3
17,775	13,338	4,437	..	99,359	17,905	78,022	2,007	1,425
18,818	14,338	4,480	..	102,155	18,448	79,814	2,007	1,886
19,912	15,102	4,805	5	106,831	21,426	81,213	1,990	2,202
20,901	16,027	4,863	11	113,012	21,800	84,330	1,990	4,892
1,191	1,175	14	2	16,358	10,244	5,932	..	182
1,037	983	42	12	17,116	10,974	5,950	..	192
1,295	1,117	82	96	17,305	10,988	6,012	..	305
1,320	1,282	28	10	22,080	14,767	6,350	..	963
*614	*522	92	..	14,427	4,636	9,718	..	73
*620	*528	92	..	14,413	4,560	9,778	..	75
*620	*528	92	..	14,557	4,603	9,875	..	79
*620	*528	92	..	14,203	4,515	9,604	..	84
712	639	73	..	4,083	1,947	2,134	..	2
740	664	76	..	4,323	2,160	2,161	..	2
794	718	76	..	4,308	2,125	2,161	..	23
865	791	75	..	4,434	2,220	2,166	..	48
*2	*2	..	..	*10	*7	..	..	*3
*2	*2	..	..	*10	*7	..	..	*3
*2	*2	..	..	*10	*7	..	..	*3
*2	*2	..	..	*10	*7	..	..	*3
..	..	..	..	247	129	118	..	0
..	..	..	..	281	129	118	..	34
..	..	..	..	292	140	119	..	33
..	..	..	..	292	140	119	..	33
179	179	..	..	170	170	..	..	..
180	180	..	..	182	182	..	..	..
*190	*190	..	..	*192	*192	..	..	..
*190	*190	..	..	*190	*190	..	..	..
..	..	..	..	8,816	6	8,810	..	..
..	..	..	..	8,816	6	8,810	..	..
..	..	..	..	8,825	15	8,810	..	..
..	..	..	..	8,825	15	8,810	..	..

Table 30

Net installed capacity of electricity plants by type

Megawatts

Country or area	Year	Total producers						
		Total	Combustible fuels	Hydro	Nuclear	Wind	Solar	Other
Peru	2011	8,691	5,240	3,451	..	1	0	..
	2012	9,699	6,134	3,484	..	1	80	..
	2013	11,051	7,414	3,556	..	1	80	..
	2014	11,203	7,302	3,662	..	143	96	..
Suriname	2011	427	238	189	..	..	..	..
	2012	427	238	189	..	..	..	..
	2013	427	238	189	..	..	..	..
	2014	427	238	189	..	..	..	..
Uruguay	2011	2,701	1,119	1,538	..	44	0	..
	2012	2,911	1,320	1,538	..	53	1	..
	2013	3,288	1,689	1,538	..	59	2	..
	2014	3,713	1,690	1,538	..	481	4	..
Venezuela (Bolivar. Rep.)	2011	27,008	12,386	14,622	..	..	..	..
	2012	28,832	14,177	14,622	..	30	..	3
	2013	31,595	16,662	14,880	..	50	..	3
	2014	31,765	16,748	14,963	..	51	..	3
Asia	**2011**	**2,401,716**	**1,800,504**	**428,737**	**92,385**	**67,847**	**8,411**	**3,833**
	2012	**2,557,859**	**1,908,289**	**455,667**	**91,536**	**86,240**	**11,830**	**4,297**
	2013	**2,734,113**	**2,004,352**	**491,623**	**91,742**	**105,383**	**36,296**	**4,717**
	2014	**2,936,177**	**2,117,840**	**525,721**	**97,962**	**130,004**	**59,723**	**4,927**
Afghanistan	2011	431	115	316	..	..	..	..
	2012	431	115	316	..	..	..	..
	2013	*431	*115	*316	..	..	..	..
	2014	*431	*115	*316	..	..	..	..
Armenia	2011	3,509	1,906	1,191	408	4	..	..
	2012	4,055	2,394	1,249	408	4	..	..
	2013	4,095	2,394	1,289	408	4	..	..
	2014	4,091	2,390	1,289	408	4	..	..
Azerbaijan	2011	6,350	5,352	998	..	0	0	..
	2012	6,420	5,397	1,023	..	0	..	..
	2013	7,353	6,264	1,084	..	3	2	..
	2014	7,355	6,270	1,078	..	3	4	..
Bahrain	2011	*5,868	*5,868	..	..	..	..	..
	2012	*5,868	*5,868	..	..	..	..	..
	2013	*5,868	*5,868	..	..	..	..	..
	2014	*6,050	*6,050	..	..	..	..	..
Bangladesh	2011	6,639	6,409	230	..	..	..	..
	2012	8,100	7,870	230	..	..	..	..
	2013	8,562	8,222	315	..	*1	*24	..
	2014	11,557	11,302	230	..	*1	*24	..
Bhutan	2011	1,505	17	1,488	..	..	0	..
	2012	1,505	17	1,488	..	..	0	..
	2013	1,506	17	1,488	..	..	0	..
	2014	1,632	17	1,614	..	..	0	..
Brunei Darussalam	2011	*803	*802	..	..	..	*1	..
	2012	*826	*825	..	..	..	*1	..
	2013	*826	*825	..	..	..	*1	..
	2014	*844	*843	..	..	..	*1	..
Cambodia	2011	571	362	207	..	..	2	..
	2012	584	357	225	..	..	*2	..
	2013	1,157	472	683	..	..	*2	..
	2014	1,513	582	929	..	..	*2	..
China	2011	1,136,211	834,799	240,364	12,570	46,230	2,220	28
	2012	1,222,980	888,257	257,090	12,570	61,420	3,410	233
	2013	1,341,077	945,120	288,777	14,660	76,520	15,890	110
	2014	1,461,939	1,005,720	313,981	19,880	95,810	26,520	*28

Net installed capacity of electricity plants by type

Megawatts

Autoproducers				Main activity producers				
Total	Combustible fuels	Hydro	Other	Total	Combustible fuels	Hydro	Nuclear	Other
1,377	1,283	94	..	7,314	3,957	3,357	..	1
1,432	1,329	103	..	8,267	4,806	3,381	..	81
1,416	1,311	106	..	9,635	6,104	3,451	..	81
1,464	1,360	104	..	9,739	5,942	3,558	..	239
312	123	189	..	115	115	..	..	..
312	123	189	..	115	115	..	..	..
312	123	189	..	115	115	..	..	..
312	123	189	..	115	115	..	..	..
264	250	..	14	2,437	869	1,538	..	30
267	251	..	16	2,644	1,069	1,538	..	37
444	421	..	23	2,844	1,268	1,538	..	38
449	422	..	27	3,264	1,268	1,538	..	458
*1,303	*1,303	..	..	25,705	11,083	14,622	..	..
*1,303	*1,303	..	..	27,529	12,874	14,622	..	33
1,303	1,303	..	..	30,292	15,359	14,880	..	53
1,303	1,303	..	..	30,462	15,445	14,963	..	53
258,656	**237,719**	**13,039**	**7,898**	**2,143,060**	**1,562,784**	**415,698**	**92,385**	**72,192**
269,032	**245,711**	**13,445**	**9,876**	**2,288,827**	**1,662,578**	**442,222**	**91,536**	**92,491**
285,908	**252,811**	**14,174**	**18,923**	**2,448,205**	**1,751,541**	**477,449**	**91,742**	**127,473**
314,473	**269,669**	**15,226**	**29,577**	**2,621,704**	**1,848,171**	**510,495**	**97,962**	**165,077**
182	55	127	..	249	60	189	..	..
182	55	127	..	249	60	189	..	..
*182	*55	*127	..	*249	*60	*189	..	..
*182	*55	*127	..	*249	*60	*189	..	..
..	..	..	..	3,509	1,906	1,191	408	4
..	..	..	..	4,055	2,394	1,249	408	4
..	..	..	..	4,095	2,394	1,289	408	4
..	..	..	..	4,091	2,390	1,289	408	4
100	100	..	..	6,250	5,252	998	..	0
126	126	..	..	6,294	5,271	1,023	..	0
236	236	..	..	7,117	6,028	1,084	..	5
242	242	..	..	7,113	6,028	1,078	..	7
*3,000	*3,000	..	..	*2,868	*2,868	..	..	..
*3,000	*3,000	..	..	*2,868	*2,868	..	..	..
*3,000	*3,000	..	..	*2,868	*2,868	..	..	..
*3,050	*3,050	..	..	*3,000	*3,000	..	..	..
1,911	1,911	..	..	4,728	4,498	230	..	..
2,375	2,375	..	..	5,725	5,495	230	..	..
2,222	2,222	..	..	6,340	6,000	315	..	*25
2,357	2,357	..	..	9,200	8,945	230	..	*25
..	..	..	..	1,505	17	1,488	..	0
..	..	..	..	1,505	17	1,488	..	0
..	..	..	..	1,506	17	1,488	..	0
..	..	..	..	1,632	17	1,614	..	0
*72	*72	..	..	*731	*730	..	..	*1
*75	*75	..	..	*751	*750	..	..	*1
*75	*75	..	..	*751	*750	..	..	*1
*78	*78	..	..	*766	*765	..	..	*1
..	..	..	..	571	362	207	..	2
..	..	..	..	584	357	225	..	*2
..	..	..	..	1,157	472	683	..	*2
..	..	..	..	1,513	582	929	..	*2
*73,843	*66,459	*7,384	..	1,062,368	768,340	232,980	12,570	48,478
*76,197	*68,577	*7,620	..	1,146,783	819,680	249,470	12,570	65,063
*83,367	*75,030	*8,337	..	1,257,710	870,090	280,440	14,660	92,520
*91,211	*82,090	*9,121	..	1,370,728	923,630	304,860	19,880	122,358

Table 30

Net installed capacity of electricity plants by type

Megawatts

Country or area	Year	Total producers						
		Total	Combustible fuels	Hydro	Nuclear	Wind	Solar	Other
China, Hong Kong SAR	2011	12,624	12,624	..	..	..	..	..
	2012	12,625	12,625	..	..	..	..	..
	2013	12,625	12,625	..	..	..	..	..
	2014	12,625	12,625	..	..	..	..	..
China, Macao SAR	2011	497	497	..	..	..	..	..
	2012	497	497	..	..	..	..	..
	2013	497	497	..	..	..	..	..
	2014	497	497	..	..	..	..	..
Cyprus	2011	1,736	1,591	..	..	134	11	..
	2012	1,726	1,561	..	..	147	18	..
	2013	1,700	1,517	..	..	147	36	..
	2014	1,734	1,513	..	..	147	74	..
Georgia	2011	*4,350	*1,688	*2,662	..	..	..	..
	2012	*4,350	*1,688	*2,662	..	..	..	..
	2013	*4,350	*1,688	*2,662	..	..	..	..
	2014	*4,350	*1,688	*2,662	..	..	..	..
India	2011	252,471	192,556	39,051	4,780	16,084	0	..
	2012	278,275	215,522	39,552	4,780	18,421	0	..
	2013	*294,873	*226,449	*40,061	4,780	*21,264	*2,319	..
	2014	*316,379	*241,396	*40,061	5,780	25,244	3,898	..
Indonesia	2011	47,756	42,163	4,406	..	9	1	1,176
	2012	53,125	46,603	5,067	..	9	4	1,441
	2013	55,426	48,875	5,054	..	5	9	1,484
	2014	60,588	53,692	5,306	..	5	9	1,576
Iran (Islamic Rep. of)	2011	70,356	60,485	8,746	1,020	98	0	7
	2012	74,475	63,596	9,746	1,020	106	0	7
	2013	75,816	64,413	10,266	1,020	110	0	7
	2014	*73,001	*63,081	*8,770	*1,020	*120	0	*10
Iraq	2011	16,952	14,439	2,513	..	..	..	..
	2012	16,952	14,439	2,513	..	..	..	..
	2013	27,110	24,597	2,513	..	..	..	..
	2014	25,589	23,076	2,513	..	..	..	..
Israel	2011	13,990	13,662	7	..	6	190	125
	2012	14,413	14,163	7	..	6	237	0
	2013	14,988	14,491	10	..	6	481	0
	2014	16,223	15,526	10	..	6	681	0
Japan	2011	292,058	186,810	48,418	48,960	2,419	4,914	537
	2012	295,193	190,405	48,934	46,148	2,562	6,632	512
	2013	302,711	192,759	48,932	44,264	2,645	13,599	512
	2014	315,318	194,857	49,597	44,264	2,753	23,339	508
Jordan	2011	3,511	3,498	12	..	1	..	..
	2012	3,511	3,498	12	..	1	..	..
	2013	3,385	3,372	12	..	1	..	..
	2014	4,199	4,186	12	..	1	..	..
Kazakhstan	2011	*20,217	*18,000	*2,217	..	..	..	..
	2012	*21,219	*19,000	*2,217	..	*2	..	..
	2013	*22,222	*20,000	*2,217	..	*4	*1	..
	2014	*25,011	*22,500	*2,500	..	*10	*1	..
Korea, Dem.Ppl's.Rep.	2011	*9,500	*4,500	*5,000	..	..	..	..
	2012	*9,500	*4,500	*5,000	..	..	..	..
	2013	*9,500	*4,500	*5,000	..	..	..	..
	2014	*9,500	*4,500	*5,000	..	..	..	..
Korea, Republic of	2011	84,654	58,303	6,418	18,716	425	730	62
	2012	87,830	59,086	6,447	20,716	464	1,024	93
	2013	91,485	61,795	6,454	20,716	576	1,555	389
	2014	99,830	69,122	6,467	20,716	612	2,481	432

Net installed capacity of electricity plants by type

Megawatts

Autoproducers				Main activity producers				
Total	Combustible fuels	Hydro	Other	Total	Combustible fuels	Hydro	Nuclear	Other
..	..	..	..	12,624	12,624	..	..	..
..	..	..	..	12,625	12,625	..	..	..
..	..	..	..	12,625	12,625	..	..	..
..	..	..	..	12,625	12,625	..	..	..
*25	*25	..	..	472	472	..	..	..
*25	*25	..	..	472	472	..	..	..
*25	*25	..	..	472	472	..	..	..
*25	*25	..	..	472	472	..	..	..
37	36	..	*1	1,699	1,555	..	..	144
37	36	..	*1	1,689	1,525	..	..	164
37	36	..	*1	1,663	1,481	..	..	182
42	32	..	*10	1,692	1,481	..	..	211
..	..	..	..	*4,350	*1,688	*2,662	..	..
..	..	..	..	*4,350	*1,688	*2,662	..	..
..	..	..	..	*4,350	*1,688	*2,662	..	..
..	..	..	..	*4,350	*1,688	*2,662	..	..
*36,510	*36,449	*61	..	215,961	156,107	38,990	4,780	16,084
*36,510	*36,449	*61	..	241,765	179,073	39,491	4,780	18,421
38,087	36,449	*61	*1,577	*256,786	*190,000	*40,000	4,780	*22,006
*44,657	*42,651	*61	*1,945	*271,722	*198,745	*40,000	5,780	27,197
*6,924	*6,420	*500	*4	40,832	35,743	3,906	..	1,182
*8,102	*7,583	*515	*4	45,023	39,020	4,552	..	1,450
*8,937	*8,434	*501	*2	46,489	40,441	4,553	..	1,495
*8,937	*8,434	*501	*2	51,651	45,258	4,805	..	1,588
5,134	5,134	..	..	65,222	55,351	8,746	1,020	105
5,581	5,581	..	..	68,894	58,015	9,746	1,020	113
5,581	5,581	..	..	70,235	58,832	10,266	1,020	117
*5,581	*5,581	..	..	*67,420	*57,500	*8,770	*1,020	*130
..	..	..	..	16,952	14,439	2,513	..	..
..	..	..	..	16,952	14,439	2,513	..	..
..	..	..	..	27,110	24,597	2,513	..	..
..	..	..	..	25,589	23,076	2,513	..	..
685	393	2	290	13,305	13,269	5	..	31
667	428	2	237	13,746	13,735	5	..	6
894	411	2	481	14,094	14,080	8	..	6
1,145	462	2	681	15,078	15,064	8	..	6
62,150	50,678	4,250	7,222	229,908	136,132	44,168	48,960	648
63,973	50,610	4,282	9,081	231,220	139,795	44,652	46,148	625
71,244	50,858	4,256	16,130	231,467	141,901	44,676	44,264	626
81,290	51,080	4,194	26,016	234,028	143,777	45,403	44,264	584
*199	*199	..	..	3,312	3,299	12	..	1
*199	*199	..	..	3,312	3,299	12	..	1
*199	*199	..	..	3,186	3,173	12	..	1
*199	*199	..	..	4,000	3,987	12	..	1
..	..	..	..	*20,217	*18,000	*2,217	..	..
..	..	..	..	*21,219	*19,000	*2,217	..	*2
*1	..	..	*1	*22,221	*20,000	*2,217	..	*4
*1	..	..	*1	*25,010	*22,500	*2,500	..	*10
..	..	..	..	*9,500	*4,500	*5,000	..	..
..	..	..	..	*9,500	*4,500	*5,000	..	..
..	..	..	..	*9,500	*4,500	*5,000	..	..
..	..	..	..	*9,500	*4,500	*5,000	..	..
5,702	5,503	..	199	78,952	52,800	6,418	18,716	1,018
6,817	6,452	..	365	81,013	52,634	6,447	20,716	1,216
6,338	5,835	..	503	85,147	55,960	6,454	20,716	2,017
7,715	7,001	..	714	92,115	62,121	6,467	20,716	2,811

Table 30

Net installed capacity of electricity plants by type

Megawatts

Country or area	Year	Total producers						
		Total	Combustible fuels	Hydro	Nuclear	Wind	Solar	Other
Kuwait	2011	14,803	14,803	..	..	..	..	..
	2012	14,803	14,803	..	..	..	..	..
	2013	15,719	15,719	..	..	..	..	..
	2014	15,819	15,819	..	..	..	..	..
Kyrgyzstan	2011	*3,859	*795	*3,064	..	..	..	..
	2012	*3,859	*795	*3,064	..	..	..	..
	2013	*3,861	*797	*3,064	..	..	..	..
	2014	*3,864	*800	*3,064	..	..	..	..
Lao People's Dem. Rep.	2011	2,796	*50	2,570	..	..	*176	..
	2012	3,199	*50	2,973	..	..	*176	..
	2013	3,206	*50	2,980	..	..	*176	..
	2014	3,284	*50	3,058	..	..	*176	..
Lebanon	2011	3,262	2,988	*274	..	..	..	..
	2012	3,262	2,988	*274	..	..	..	..
	2013	3,262	2,988	*274	..	..	..	..
	2014	3,262	2,988	*274	..	..	..	..
Malaysia	2011	28,750	25,734	3,015	..	..	*1	..
	2012	29,198	25,826	3,317	..	..	*55	..
	2013	31,817	27,725	3,931	..	..	*161	..
	2014	29,974	25,040	4,773	..	..	161	..
Maldives	2011	81	78	..	..	..	*3	..
	2012	84	81	..	..	..	*3	..
	2013	85	81	..	..	..	*5	..
	2014	271	266	..	..	..	5	..
Mongolia	2011	*1,340	*1,340	..	..	..	..	..
	2012	*1,340	*1,340	..	..	..	..	..
	2013	*1,340	*1,340	..	..	..	..	..
	2014	*1,370	*1,370	..	..	..	..	..
Myanmar	2011	3,588	895	2,693	..	..	..	..
	2012	3,726	913	2,813	..	..	..	..
	2013	4,146	1,141	3,005	..	..	..	..
	2014	4,805	1,620	3,185	..	..	..	..
Nepal	2011	*762	53	*709	..	..	..	..
	2012	*762	53	*709	..	..	..	..
	2013	*772	53	*718	..	..	..	..
	2014	*829	53	*776	..	..	..	..
Oman	2011	4,861	4,861	..	..	..	..	..
	2012	5,808	5,808	..	..	..	..	..
	2013	6,598	6,598	..	..	..	..	..
	2014	8,214	8,214	..	..	..	..	..
Other Asia	2011	48,795	38,367	4,643	5,144	523	118	..
	2012	48,424	37,803	4,683	5,144	571	223	..
	2013	48,860	38,026	4,683	5,144	614	392	..
	2014	48,476	37,391	4,683	5,144	637	620	..
Pakistan	2011	22,797	15,454	6,556	787	..	..	..
	2012	22,812	*15,289	6,773	750	..	..	..
	2013	23,686	*15,888	6,893	750	*155	..	..
	2014	*23,686	*15,888	*6,893	*750	*155	..	..
Philippines	2011	16,189	10,782	3,507	..	115	2	1,783
	2012	17,052	11,514	3,537	..	150	3	1,848
	2013	17,351	11,793	3,537	..	*150	*3	1,868
	2014	17,970	12,056	3,559	..	*400	*37	1,918
Qatar	2011	8,751	8,751	..	..	..	..	..
	2012	8,751	8,751	..	..	..	..	..
	2013	8,751	8,751	..	..	..	..	..
	2014	8,751	8,751	..	..	..	..	..

Net installed capacity of electricity plants by type

Megawatts

Autoproducers				Main activity producers				
Total	Combustible fuels	Hydro	Other	Total	Combustible fuels	Hydro	Nuclear	Other
*100	*100	..	..	14,703	14,703	..	..	..
*100	*100	..	..	14,703	14,703	..	..	..
*100	*100	..	..	15,619	15,619	..	..	..
*100	*100	..	..	15,719	15,719	..	..	..
..	..	..	..	*3,859	*795	*3,064	..	..
..	..	..	..	*3,859	*795	*3,064	..	..
*2	*2	..	..	*3,859	*795	*3,064	..	..
*5	*5	..	..	*3,859	*795	*3,064	..	..
*176	..	..	*176	2,620	*50	2,570	..	..
*176	..	..	*176	3,023	*50	2,973	..	..
*176	..	..	*176	3,030	*50	2,980	..	..
*176	..	..	*176	3,108	*50	3,058	..	..
*950	*950	..	..	2,312	2,038	*274	..	..
*950	*950	..	..	2,312	2,038	*274	..	..
950	950	..	..	2,312	2,038	*274	..	..
*950	*950	..	..	2,312	2,038	*274	..	..
2,998	2,998	0	0	25,752	22,736	3,015	..	*1
3,174	3,174	0	0	26,024	22,652	3,317	..	*55
3,037	3,037	0	0	28,780	24,688	3,931	..	*161
2,351	2,345	5	1	27,623	22,695	4,768	..	160
..	..	..	..	81	78	..	..	*3
..	..	..	..	84	81	..	..	*3
..	..	..	..	85	81	..	..	*5
..	..	..	..	271	266	..	..	5
*150	*150	..	..	*1,190	*1,190	..	..	..
*150	*150	..	..	*1,190	*1,190	..	..	..
*150	*150	..	..	*1,190	*1,190	..	..	..
*180	*180	..	..	*1,190	*1,190	..	..	..
..	..	..	..	3,588	895	2,693	..	..
120	..	120	..	3,606	913	2,693	..	..
393	221	172	..	3,753	920	2,833	..	..
628	456	172	..	4,177	1,164	3,013	..	..
*155	..	*155	..	*607	53	*554	..	..
*155	..	*155	..	*607	53	*554	..	..
*155	..	*155	..	*617	53	*563	..	..
*175	..	*175	..	*654	53	*601	..	..
..	..	..	..	4,861	4,861	..	..	..
..	..	..	..	5,808	5,808	..	..	..
..	..	..	..	6,598	6,598	..	..	..
..	..	..	..	8,214	8,214	..	..	..
7,204	7,204	..	..	41,591	31,164	4,643	5,144	641
7,233	7,233	..	..	41,191	30,570	4,683	5,144	794
7,443	7,443	..	..	41,417	30,583	4,683	5,144	1,006
7,380	7,380	..	..	41,096	30,011	4,683	5,144	1,257
8,353	8,353	..	..	14,444	7,101	6,556	787	..
*8,183	*8,183	..	..	14,629	*7,106	6,773	750	..
*8,558	*8,558	..	..	15,128	*7,330	6,893	750	*155
*8,558	*8,558	..	..	*15,128	*7,330	*6,893	*750	*155
*26	*10	*16	..	16,163	10,772	3,491	..	1,900
*26	*10	*16	..	17,026	11,504	3,521	..	2,001
*26	*10	*16	..	17,325	11,783	3,521	..	2,021
*26	*10	*16	..	17,944	12,046	3,543	..	2,355
6,561	6,561	..	..	2,190	2,190	..	..	..
6,561	6,561	..	..	2,190	2,190	..	..	..
6,561	6,561	..	..	2,190	2,190	..	..	..
6,561	6,561	..	..	2,190	2,190	..	..	..

Table 30

Net installed capacity of electricity plants by type

Megawatts

Country or area	Year	Total producers: Total	Combustible fuels	Hydro	Nuclear	Wind	Solar	Other
Saudi Arabia	2011	51,149	51,149	..	..	..	0	..
	2012	62,133	62,132	..	..	..	*1	..
	2013	*58,472	*58,471	..	..	..	*1	..
	2014	*65,507	*65,506	..	..	..	*1	..
Singapore	2011	10,131	10,131	..	..	..	..	..
	2012	10,296	10,296	..	..	..	..	..
	2013	11,425	11,425	..	..	..	..	..
	2014	13,047	13,047	..	..	..	..	..
Sri Lanka	2011	3,155	1,712	1,401	..	36	*6	..
	2012	3,383	1,716	1,585	..	76	*6	..
	2013	3,309	1,601	1,617	..	81	*10	..
	2014	4,058	2,243	1,664	..	131	19	..
State of Palestine	2011	140	140	..	..	..	..	..
	2012	140	140	..	..	..	..	..
	2013	140	140	..	..	..	..	..
	2014	140	140	..	..	..	..	..
Syrian Arab Republic	2011	8,206	7,356	850	..	..	..	..
	2012	*8,206	*7,356	*850	..	..	..	..
	2013	8,206	7,356	850	..	..	..	..
	2014	*9,603	*8,109	1,494	..	..	..	..
Tajikistan	2011	5,079	324	4,755	..	..	..	..
	2012	6,168	314	5,854	..	..	..	..
	2013	5,190	318	4,872	..	..	..	..
	2014	5,190	318	4,872	..	..	..	..
Thailand	2011	48,750	45,233	*3,488	..	*3	*26	0
	2012	49,806	46,289	*3,488	..	*3	*26	0
	2013	52,538	47,250	3,418	..	*300	1,570	0
	2014	53,472	48,238	3,444	..	220	1,570	0
Timor-Leste	2011	*50	*50	..	..	..	..	..
	2012	51	51	..	..	..	..	..
	2013	138	138	..	..	..	..	..
	2014	286	286	..	..	..	..	..
Turkey	2011	52,911	33,931	17,137	..	1,729	..	114
	2012	57,059	35,027	19,609	..	2,261	..	162
	2013	64,007	38,612	22,289	..	2,759	..	347
	2014	72,893	44,576	24,187	..	3,635	40	455
Turkmenistan	2011	*3,201	*3,200	*1	..	..	..	..
	2012	*3,401	*3,400	*1	..	..	..	..
	2013	*3,801	*3,800	*1	..	..	..	..
	2014	*4,001	*4,000	*1	..	..	..	..
United Arab Emirates	2011	26,086	26,076	..	..	..	10	..
	2012	27,180	27,170	..	..	..	10	..
	2013	27,374	27,314	..	..	..	60	..
	2014	28,829	28,769	..	..	..	60	..
Uzbekistan	2011	*12,400	*10,569	*1,831	..	..	..	..
	2012	*12,722	*10,867	*1,855	..	..	..	..
	2013	*12,722	*10,867	*1,855	..	..	..	..
	2014	*12,722	*10,867	*1,855	..	..	..	..
Viet Nam	2011	*25,731	*17,700	*8,000	..	*31	..	..
	2012	*28,240	*17,700	*10,503	..	*37	..	..
	2013	*28,240	*17,700	*10,503	..	*37	..	..
	2014	*34,080	*18,368	*15,603	..	*109	..	..
Yemen	2011	1,535	1,535	..	..	..	..	..
	2012	1,535	1,535	..	..	..	..	..
	2013	1,535	1,535	..	..	..	..	..
	2014	1,519	1,519	..	..	..	..	..

Net installed capacity of electricity plants by type

Megawatts

Autoproducers				Main activity producers				
Total	Combustible fuels	Hydro	Other	Total	Combustible fuels	Hydro	Nuclear	Other
*12,230	*12,230	..	..	38,919	38,919	..	..	0
*14,857	*14,857	..	..	47,276	47,275	..	..	*1
*14,040	*14,040	..	..	*44,432	*44,431	..	..	*1
*15,729	*15,729	..	..	*49,778	*49,777	..	..	*1
*306	*306	..	..	9,825	9,825	..	..	..
*314	*314	..	..	9,982	9,982	..	..	..
*502	*502	..	..	10,923	10,923	..	..	..
*502	*502	..	..	12,545	12,545	..	..	..
*15	*10	..	*5	3,140	1,702	1,401	..	37
*15	*10	..	*5	3,368	1,706	1,585	..	77
*15	*10	..	*5	3,294	1,591	1,617	..	86
*15	*10	..	*5	4,043	2,233	1,664	..	145
..	..	..	..	140	140	..	..	..
..	..	..	..	140	140	..	..	..
..	..	..	..	140	140	..	..	..
..	..	..	..	140	140	..	..	..
996	996	..	..	7,210	6,360	850	..	..
*996	*996	..	..	*7,210	*6,360	*850	..	..
996	996	..	..	7,210	6,360	850	..	..
*1,095	*1,095	..	..	*8,508	*7,014	1,494	..	..
..	..	..	..	5,079	324	4,755	..	..
..	..	..	..	6,168	314	5,854	..	..
..	..	..	..	5,190	318	4,872	..	..
..	..	..	..	5,190	318	4,872	..	..
*15,971	*15,971	..	..	32,779	29,262	*3,488	..	*29
*15,971	*15,971	..	..	33,835	30,318	*3,488	..	*29
*15,974	*15,974	..	..	36,564	31,276	3,418	..	1,870
*15,974	*15,974	..	..	37,498	32,264	3,444	..	1,790
..	..	..	..	*50	*50	..	..	..
..	..	..	..	51	51	..	..	..
..	..	..	..	138	138	..	..	..
..	..	..	..	286	286	..	..	..
3,018	2,473	544	1	49,893	31,458	16,593	..	1,842
3,201	2,656	544	1	53,858	32,371	19,065	..	2,422
3,421	2,836	544	41	60,586	35,776	21,745	..	3,065
*3,400	*2,836	*544	20	69,493	41,740	23,643	..	4,110
..	..	..	..	*3,201	*3,200	*1	..	..
..	..	..	..	*3,401	*3,400	*1	..	..
..	..	..	..	*3,801	*3,800	*1	..	..
..	..	..	..	*4,001	*4,000	*1	..	..
..	..	..	..	26,086	26,076	..	..	10
..	..	..	..	27,180	27,170	..	..	10
..	..	..	..	27,374	27,314	..	..	60
..	..	..	..	28,829	28,769	..	..	60
*155	*155	..	..	*12,245	*10,414	*1,831	..	..
*157	*157	..	..	*12,565	*10,710	*1,855	..	..
*157	*157	..	..	*12,565	*10,710	*1,855	..	..
*157	*157	..	..	*12,565	*10,710	*1,855	..	..
2,700	2,700	0	0	*23,031	*15,000	*8,000	..	*31
2,709	2,700	3	6	*25,531	*15,000	*10,500	..	*31
*2,709	*2,700	*3	*6	*25,531	*15,000	*10,500	..	*31
*3,682	*3,368	*308	*6	*30,398	*15,000	*15,295	..	*103
*118	*118	..	..	1,417	1,417	..	..	..
*118	*118	..	..	1,417	1,417	..	..	..
*118	*118	..	..	1,417	1,417	..	..	..
*117	*117	..	..	1,402	1,402	..	..	..

Table 30

Net installed capacity of electricity plants by type

Megawatts

Country or area	Year	Total producers: Total	Combustible fuels	Hydro	Nuclear	Wind	Solar	Other
Europe	**2011**	**1,285,393**	**704,749**	**255,179**	**173,500**	**95,072**	**52,854**	**4,038**
	2012	**1,322,988**	**716,145**	**258,201**	**165,600**	**107,520**	**71,596**	**3,926**
	2013	**1,345,300**	**712,033**	**260,874**	**165,388**	**119,529**	**83,472**	**4,004**
	2014	**1,383,760**	**729,853**	**262,073**	**165,962**	**130,793**	**90,984**	**4,096**
Albania	2011	1,613	123	1,476	..	..	14	..
	2012	1,767	123	1,629	..	..	15	..
	2013	1,918	123	1,781	..	..	14	..
	2014	1,865	123	1,726	..	..	16	..
Andorra	2011	*32	*5	*27	..	..	..	..
	2012	*32	*5	*27	..	..	..	..
	2013	*32	*5	*27	..	..	..	..
	2014	*32	*5	*27	..	..	..	..
Austria	2011	22,556	8,178	12,980	..	1,080	317	1
	2012	22,917	8,161	13,076	..	1,316	363	1
	2013	23,591	8,170	13,149	..	1,645	626	1
	2014	24,025	7,859	13,293	..	2,086	786	1
Belarus	2011	8,375	8,356	15	..	3	0	..
	2012	8,975	8,939	33	..	3	0	..
	2013	9,181	9,144	33	..	*3	0	..
	2014	10,037	9,999	33	..	*3	*1	..
Belgium	2011	20,098	10,285	1,426	5,927	1,069	1,391	0
	2012	20,773	9,468	1,427	5,927	1,370	2,581	0
	2013	20,984	8,910	1,429	5,927	1,792	2,922	4
	2014	20,918	8,605	1,429	5,927	1,930	3,023	4
Bosnia and Herzegovina	2011	*4,304	*2,187	*2,117	..	..	..	..
	2012	*4,304	*2,187	*2,117	..	..	..	..
	2013	*4,304	*2,187	*2,117	..	..	..	..
	2014	*4,304	*2,187	*2,117	..	..	..	..
Bulgaria	2011	10,236	4,541	3,108	1,892	541	154	..
	2012	11,637	4,912	3,129	1,906	677	1,013	..
	2013	11,604	4,701	3,202	1,982	683	1,036	..
	2014	11,390	4,470	3,219	1,975	700	1,026	..
Croatia	2011	4,015	1,787	2,110	..	118	0	..
	2012	4,222	1,897	2,141	..	180	4	..
	2013	4,312	1,849	2,190	..	254	19	..
	2014	4,428	1,933	2,114	..	347	34	..
Czechia	2011	20,181	11,888	2,197	3,970	213	1,913	..
	2012	20,447	11,915	2,212	4,040	258	2,022	..
	2013	21,079	12,211	2,252	4,290	262	2,064	..
	2014	21,970	13,082	2,252	4,290	278	2,068	..
Denmark	2011	13,587	9,609	9	..	3,952	17	..
	2012	14,075	9,500	9	..	4,164	402	..
	2013	13,810	8,410	9	..	4,820	571	..
	2014	13,656	8,152	9	..	4,888	607	..
Estonia	2011	2,825	2,640	5	..	180	..	..
	2012	2,923	2,649	8	..	266	..	..
	2013	2,910	2,654	8	..	248	..	..
	2014	3,096	2,750	5	..	341	..	..
Faeroe Islands	2011	103	65	32	..	*7	..	..
	2012	111	65	*39	..	*7	..	..
	2013	*115	*65	*39	..	*11	..	..
	2014	*124	*68	*39	..	*16	..	..
Finland	2011	15,677	9,559	3,196	2,716	199	7	..
	2012	15,757	9,564	3,196	2,732	257	8	..
	2013	16,653	10,221	3,224	2,752	447	9	..
	2014	16,245	9,607	3,248	2,752	627	11	..

Net installed capacity of electricity plants by type

Megawatts

Autoproducers				Main activity producers				
Total	Combustible fuels	Hydro	Other	Total	Combustible fuels	Hydro	Nuclear	Other
89,104	**75,531**	**4,248**	**9,325**	**1,196,289**	**629,218**	**250,932**	**173,500**	**142,639**
91,363	**73,829**	**4,081**	**13,453**	**1,231,625**	**642,316**	**254,120**	**165,600**	**169,589**
98,691	**77,261**	**4,085**	**17,345**	**1,246,609**	**634,772**	**256,789**	**165,388**	**189,660**
107,481	**81,552**	**4,104**	**21,825**	**1,276,280**	**648,301**	**257,969**	**165,962**	**204,048**
25	25	..	..	1,588	98	1,476	..	14
25	25	..	..	1,742	98	1,629	..	15
25	25	..	..	1,893	98	1,781	..	14
25	25	..	..	1,840	98	1,726	..	16
..	..	..	..	*32	*5	*27	..	..
..	..	..	..	*32	*5	*27	..	..
..	..	..	..	*32	*5	*27	..	..
..	..	..	..	*32	*5	*27	..	..
2,374	1,839	535	0	20,182	6,339	12,445	..	1,398
2,349	1,814	535	0	20,568	6,347	12,541	..	1,680
2,379	1,844	535	0	21,212	6,326	12,614	..	2,272
2,257	1,723	533	1	21,768	6,136	12,760	..	2,872
470	462	6	2	7,906	7,895	9	..	2
583	574	7	2	8,393	8,365	26	..	2
708	699	*7	*2	8,473	8,445	26	..	2
710	701	*7	*2	9,327	9,298	26	..	2
2,545	1,157	..	1,388	17,553	9,128	1,426	5,927	1,072
3,869	1,286	..	2,583	16,904	8,182	1,427	5,927	1,368
4,276	1,351	..	2,925	16,708	7,559	1,429	5,927	1,793
4,424	1,396	..	3,028	16,494	7,209	1,429	5,927	1,929
*230	*230	..	..	*4,074	*1,957	*2,117	..	..
*230	*230	..	..	*4,074	*1,957	*2,117	..	..
*230	*230	..	..	*4,074	*1,957	*2,117	..	..
*230	*230	..	..	*4,074	*1,957	*2,117	..	..
46	46	..	..	10,190	4,495	3,108	1,892	695
63	63	..	..	11,574	4,849	3,129	1,906	1,690
65	65	..	..	11,539	4,636	3,202	1,982	1,719
60	60	..	..	11,330	4,410	3,219	1,975	1,726
*214	*210	*4	..	3,801	1,577	2,106	..	118
201	199	2	..	4,021	1,698	2,139	..	184
157	155	2	..	4,155	1,694	2,188	..	273
*157	*155	*2	..	4,271	1,778	2,112	..	381
1,887	1,713	174	..	18,294	10,175	2,023	3,970	2,126
1,893	1,710	183	..	18,554	10,205	2,029	4,040	2,280
1,901	1,713	188	..	19,178	10,498	2,064	4,290	2,326
1,880	1,690	190	..	20,090	11,392	2,062	4,290	2,346
649	632	..	17	12,938	8,977	9	..	3,952
1,032	630	..	402	13,043	8,870	9	..	4,164
1,145	574	..	571	12,665	7,836	9	..	4,820
1,180	573	..	607	12,476	7,579	9	..	4,888
19	19	0	0	2,806	2,621	5	..	180
19	19	0	0	2,904	2,630	8	..	266
18	17	0	1	2,892	2,637	8	..	247
24	17	0	7	3,072	2,733	5	..	334
..	..	..	..	103	65	32	..	*7
..	..	..	..	111	65	*39	..	*7
..	..	..	..	*115	*65	*39	..	*11
..	..	..	..	*124	*68	*39	..	*16
2,038	1,836	195	7	13,639	7,723	3,001	2,716	199
2,008	1,833	167	8	13,749	7,731	3,029	2,732	257
2,043	1,848	186	9	14,610	8,373	3,038	2,752	447
2,045	1,848	186	11	14,200	7,759	3,062	2,752	627

Table 30

Net installed capacity of electricity plants by type

Megawatts

Country or area	Year	Total producers: Total	Combustible fuels	Hydro	Nuclear	Wind	Solar	Other
France	2011	127,256	27,792	25,347	63,130	6,679	2,796	1,512
	2012	129,254	27,764	25,366	63,130	7,517	3,965	1,512
	2013	128,432	25,576	25,360	63,130	8,202	4,652	1,512
	2014	129,069	24,411	25,294	63,130	9,068	5,654	1,512
Germany	2011	175,896	89,421	11,436	20,467	29,060	25,039	473
	2012	177,291	89,649	11,257	12,068	31,304	32,643	370
	2013	186,117	91,368	11,240	12,068	34,660	36,337	444
	2014	198,416	97,203	11,234	12,074	39,193	38,236	476
Gibraltar	2011	43	43	..	..	..	..	..
	2012	43	43	..	..	..	..	..
	2013	43	43	..	..	..	..	..
	2014	43	43	..	..	..	..	..
Greece	2011	16,524	11,048	3,224	..	1,640	612	..
	2012	17,751	11,226	3,236	..	1,753	1,536	..
	2013	18,855	11,229	3,238	..	1,809	2,579	..
	2014	18,895	10,932	3,389	..	1,978	2,596	..
Guernsey	2011	115	115	..	..	..	..	..
	2012	*115	*115	..	..	..	..	..
	2013	*115	*115	..	..	..	..	..
	2014	*115	*115	..	..	..	..	..
Hungary	2011	9,654	7,264	55	2,000	331	4	..
	2012	9,400	7,007	56	2,000	325	12	..
	2013	8,418	5,987	57	2,000	329	35	10
	2014	8,809	6,333	57	2,000	329	77	13
Iceland	2011	2,669	120	1,884	..	..	..	665
	2012	2,656	114	1,877	..	0	..	665
	2013	2,765	114	1,984	..	2	..	665
	2014	2,766	114	1,984	..	3	..	665
Ireland	2011	8,588	6,427	529	..	1,631	1	..
	2012	8,589	6,295	529	..	1,764	1	..
	2013	8,798	6,327	529	..	1,941	1	..
	2014	9,078	6,337	529	..	2,211	1	..
Isle of Man	2011	180	179	1	..	..	..	..
	2012	180	179	1	..	..	..	..
	2013	186	185	1	..	..	..	..
	2014	186	185	1	..	..	..	..
Italy	2011	118,443	75,977	21,737	..	6,918	12,773	1,038
	2012	124,234	76,793	21,880	..	8,102	16,420	1,039
	2013	124,750	74,733	22,009	..	8,542	18,420	1,046
	2014	121,762	71,272	22,098	..	8,683	18,609	1,100
Jersey	2011	*219	*219	..	..	..	..	..
	2012	*219	*219	..	..	..	..	..
	2013	*240	*240	..	..	..	..	..
	2014	*240	*240	..	..	..	..	..
Latvia	2011	2,604	986	1,581	..	37	..	0
	2012	2,702	1,062	1,579	..	59	..	2
	2013	2,963	1,305	1,591	..	67	..	0
	2014	2,967	1,306	1,592	..	69	..	0
Liechtenstein	2011	*12	*1	*10	..	..	*1	..
	2012	*12	*1	*10	..	..	*1	..
	2013	*13	*1	*10	..	..	*2	..
	2014	*13	1	*10	..	..	*2	..
Lithuania	2011	3,940	2,837	876	0	202	0	25
	2012	4,484	3,301	876	0	275	7	25
	2013	4,568	3,320	876	0	279	68	25
	2014	4,285	3,026	877	0	288	69	25

Net installed capacity of electricity plants by type

Megawatts

Autoproducers				Main activity producers				
Total	Combustible fuels	Hydro	Other	Total	Combustible fuels	Hydro	Nuclear	Other
8,046	4,138	347	3,561	119,210	23,654	25,000	63,130	7,426
9,046	4,273	366	4,407	120,208	23,491	25,000	63,130	8,587
8,401	3,414	207	4,780	120,031	22,162	25,153	63,130	9,586
8,234	2,756	210	5,268	120,835	21,655	25,084	63,130	10,966
9,755	9,371	69	315	166,141	80,050	11,367	20,467	54,257
7,019	6,737	72	210	170,272	82,912	11,185	12,068	64,107
9,619	9,296	42	281	176,498	82,072	11,198	12,068	71,160
9,807	9,456	44	307	188,609	87,747	11,190	12,074	77,598
..	..	..	..	43	43	..	..	..
..	..	..	..	43	43	..	..	..
..	..	..	..	43	43	..	..	..
..	..	..	..	43	43	..	..	..
540	540	..	..	15,984	10,508	3,224	..	2,252
522	522	..	..	17,229	10,704	3,236	..	3,289
527	527	..	..	18,328	10,702	3,238	..	4,388
505	505	..	..	18,390	10,427	3,389	..	4,574
..	..	..	..	115	115	..	..	..
..	..	..	..	*115	*115	..	..	..
..	..	..	..	*115	*115	..	..	..
..	..	..	..	*115	*115	..	..	..
171	167	..	4	9,483	7,097	55	2,000	331
113	101	..	12	9,287	6,906	56	2,000	325
201	156	..	45	8,217	5,831	57	2,000	329
262	172	..	90	8,547	6,161	57	2,000	329
0	0	0	..	2,669	120	1,884	..	665
0	0	0	..	2,656	114	1,877	..	665
0	0	0	..	2,765	114	1,984	..	667
0	0	0	..	2,766	114	1,984	..	668
328	327	..	1	8,260	6,100	529	..	1,631
333	332	..	1	8,256	5,963	529	..	1,764
338	337	..	1	8,460	5,990	529	..	1,941
343	342	..	1	8,735	5,995	529	..	2,211
..	..	..	..	180	179	1	..	..
..	..	..	..	180	179	1	..	..
..	..	..	..	186	185	1	..	..
..	..	..	..	186	185	1	..	..
5,776	5,476	169	131	112,667	70,501	21,568	..	20,598
4,195	3,941	128	126	120,039	72,852	21,752	..	25,435
4,302	4,069	119	114	120,448	70,664	21,890	..	27,894
4,074	3,853	119	102	117,688	67,419	21,979	..	28,290
*7	*7	..	..	*212	*212	..	..	..
*7	*7	..	..	*212	*212	..	..	..
*28	*28	..	..	*212	*212	..	..	..
*28	*28	..	..	*212	*212	..	..	..
27	21	5	1	2,577	965	1,576	..	36
42	39	3	0	2,660	1,023	1,576	..	61
52	50	2	0	2,911	1,255	1,589	..	67
43	41	2	0	2,924	1,265	1,590	..	69
..	..	..	..	*12	*1	*10	..	*1
..	..	..	..	*12	*1	*10	..	*1
..	..	..	..	*13	*1	*10	..	*2
..	..	..	..	*13	1	*10	..	*2
274	249	..	25	3,666	2,588	876	0	202
272	247	..	25	4,212	3,054	876	0	282
270	245	..	25	4,298	3,075	876	0	347
273	248	..	25	4,012	2,778	877	0	357

Table 30

Net installed capacity of electricity plants by type

Megawatts

Country or area	Year	Total producers						
		Total	Combustible fuels	Hydro	Nuclear	Wind	Solar	Other
Luxembourg	2011	1,741	521	1,134	..	45	41	..
	2012	1,789	522	1,134	..	58	75	..
	2013	1,813	526	1,134	..	58	95	..
	2014	2,022	524	1,330	..	58	110	..
Malta	2011	581	574	..	..	..	7	..
	2012	644	623	..	..	..	21	..
	2013	671	623	..	..	..	48	..
	2014	698	623	..	..	..	75	..
Montenegro	2011	868	210	658	..	..	..	..
	2012	868	210	658	..	..	..	..
	2013	868	210	658	..	..	..	..
	2014	870	210	658	..	..	2	..
Netherlands	2011	28,053	24,970	37	510	2,316	149	71
	2012	29,924	26,518	37	510	2,433	369	57
	2013	30,539	26,520	37	485	2,713	746	38
	2014	31,762	27,286	37	485	2,865	1,048	41
Norway	2011	32,082	1,566	29,969	..	512	..	35
	2012	32,860	1,611	30,509	..	705	..	35
	2013	33,486	1,600	31,033	..	818	..	35
	2014	33,651	1,600	31,153	..	863	..	35
Poland	2011	34,554	30,405	2,346	..	1,800	1	2
	2012	35,283	30,365	2,351	..	2,564	1	2
	2013	35,815	30,027	2,355	..	3,429	2	2
	2014	35,989	29,760	2,364	..	3,836	27	2
Portugal	2011	19,924	9,936	5,535	..	4,256	172	25
	2012	19,747	9,360	5,712	..	4,412	238	25
	2013	18,900	8,308	5,661	..	4,610	296	25
	2014	19,125	8,113	5,715	..	4,856	415	26
Republic of Moldova	2011	503	487	16	..	0	..	..
	2012	487	471	16	..	0	..	..
	2013	488	471	16	..	1	..	..
	2014	421	403	16	..	1	1	..
Romania	2011	20,499	11,616	6,483	1,411	988	1	..
	2012	21,774	11,945	6,548	1,411	1,822	48	..
	2013	23,033	11,393	6,610	1,411	2,773	846	..
	2014	24,054	11,323	6,613	1,411	3,244	1,463	..
Russian Federation	2011	222,984	151,114	47,479	24,300	10	..	81
	2012	232,685	157,845	49,445	25,304	10	..	81
	2013	239,423	163,926	50,104	25,304	10	..	79
	2014	259,020	182,465	50,845	25,304	101	227	78
Serbia	2011	*8,714	*5,559	*2,835	..	320	..	..
	2012	*8,714	*5,559	*2,835	..	320	..	..
	2013	*8,714	*5,559	*2,835	..	*320	..	..
	2014	8,385	5,175	2,835	..	353	22	..
Slovakia	2011	8,363	3,382	2,523	1,940	3	496	19
	2012	8,412	3,416	2,522	1,939	3	513	19
	2013	8,458	3,432	2,523	1,940	5	533	25
	2014	8,092	3,068	2,523	1,940	3	533	25
Slovenia	2011	3,268	1,270	1,253	688	0	57	..
	2012	3,351	1,267	1,254	688	0	142	..
	2013	3,434	1,256	1,299	688	4	187	..
	2014	3,454	1,242	1,296	688	4	224	..
Spain	2011	110,110	56,786	18,740	7,450	21,533	5,511	*90
	2012	112,475	56,736	18,750	7,450	22,793	6,656	*90
	2013	113,302	56,786	19,385	6,984	22,962	7,095	*90
	2014	113,774	56,786	19,423	7,399	22,979	7,097	*90

Net installed capacity of electricity plants by type

Megawatts

Autoproducers				Main activity producers				
Total	Combustible fuels	Hydro	Other	Total	Combustible fuels	Hydro	Nuclear	Other
101	58	2	41	1,640	463	1,132	..	45
136	59	2	75	1,653	463	1,132	..	58
156	59	2	95	1,657	467	1,132	..	58
171	59	2	110	1,851	465	1,328	..	58
5	3	..	*2	576	571	..	..	5
*8	3	..	*5	636	620	..	..	16
*23	3	..	*20	648	620	..	..	28
*23	3	..	*20	675	620	..	..	55
..	..	..	..	868	210	658	..	..
..	..	..	..	868	210	658	..	..
..	..	..	..	868	210	658	..	..
2	..	..	2	868	210	658	..	..
6,031	5,361	..	670	22,022	19,609	37	510	1,866
6,210	5,315	..	895	23,714	21,203	37	510	1,964
6,403	5,130	..	1,273	24,136	21,390	37	485	2,224
6,793	5,194	..	1,599	24,969	22,092	37	485	2,355
2,074	993	1,059	22	30,008	573	28,910	..	525
2,074	993	1,059	22	30,786	618	29,450	..	718
2,074	993	1,059	22	31,412	607	29,974	..	831
2,074	993	1,059	22	31,577	607	30,094	..	876
1,731	1,727	1	3	32,823	28,678	2,345	..	1,800
1,769	1,765	1	3	33,514	28,600	2,350	..	2,564
1,806	1,801	1	4	34,009	28,226	2,354	..	3,429
1,848	1,818	1	29	34,141	27,942	2,363	..	3,836
1,955	1,885	6	64	17,969	8,051	5,529	..	4,389
2,112	2,003	6	103	17,635	7,357	5,706	..	4,572
2,059	1,899	6	154	16,841	6,409	5,655	..	4,777
2,105	1,908	6	191	17,020	6,205	5,709	..	5,106
157	157	..	..	346	330	16	..	0
141	141	..	..	346	330	16	..	0
141	141	..	..	347	330	16	..	1
74	73	..	1	347	330	16	..	1
765	668	72	25	19,734	10,948	6,411	1,411	964
2,040	1,520	93	427	19,734	10,425	6,455	1,411	1,443
3,059	2,329	101	629	19,974	9,064	6,509	1,411	2,990
3,158	2,342	90	726	20,896	8,981	6,523	1,411	3,981
17,530	16,971	559	..	205,454	134,143	46,920	24,300	91
18,365	18,061	304	..	214,320	139,784	49,141	25,304	91
19,634	19,182	452	..	219,789	144,744	49,652	25,304	89
25,360	24,873	487	..	233,660	157,592	50,358	25,304	406
*35	*35	..	..	*8,679	*5,524	*2,835	..	320
*35	*35	..	..	*8,679	*5,524	*2,835	..	320
*35	*35	..	..	*8,679	*5,524	*2,835	..	*320
*35	*35	..	..	8,350	5,140	2,835	..	375
1,001	562	29	410	7,362	2,820	2,494	1,940	108
1,042	583	29	430	7,370	2,833	2,493	1,939	105
1,073	585	30	458	7,385	2,847	2,493	1,940	105
1,145	657	30	458	6,947	2,411	2,493	1,940	103
228	57	116	55	3,040	1,213	1,137	688	2
293	40	116	137	3,058	1,227	1,138	688	5
339	42	116	181	3,095	1,214	1,183	688	10
384	50	116	218	3,070	1,192	1,180	688	10
*7,304	*7,000	*200	*104	102,806	49,786	18,540	7,450	27,030
*7,304	*7,000	*200	*104	105,171	49,736	18,550	7,450	29,435
*7,304	*7,000	*200	*104	105,998	49,786	19,185	6,984	30,043
*7,304	*7,000	*200	*104	106,470	49,786	19,223	7,399	30,062

Table 30

Net installed capacity of electricity plants by type

Megawatts

Country or area	Year	Total producers						
		Total	Combustible fuels	Hydro	Nuclear	Wind	Solar	Other
Sweden	2011	35,227	6,546	16,577	9,323	2,769	12	..
	2012	37,843	8,362	16,414	9,436	3,607	24	..
	2013	37,915	7,776	16,494	9,408	4,194	43	..
	2014	39,930	9,270	15,996	9,507	5,097	60	..
Switzerland	2011	18,234	918	13,769	3,278	46	223	..
	2012	18,584	1,018	13,802	3,278	49	437	..
	2013	18,932	1,021	13,817	3,278	60	756	..
	2014	19,165	993	13,743	3,308	60	1,061	..
T.F.Yug.Rep. Macedonia	2011	1,647	1,089	556	..	..	2	..
	2012	1,709	1,110	595	..	..	4	..
	2013	1,734	1,110	617	..	..	7	..
	2014	1,764	1,080	632	..	37	15	..
Ukraine	2011	54,624	35,026	5,469	13,835	146	148	..
	2012	55,001	35,129	5,470	13,835	248	319	..
	2013	55,914	35,616	5,489	13,835	362	612	..
	2014	55,843	35,335	5,851	13,835	411	411	..
United Kingdom	2011	93,671	71,121	4,423	10,663	6,468	995	1
	2012	95,968	70,925	4,439	9,946	8,899	1,756	3
	2013	95,105	66,679	4,452	9,906	11,214	2,851	3
	2014	97,009	64,238	4,467	9,937	12,987	5,377	3
Oceania	**2011**	**74,820**	**55,243**	**14,564**	**..**	**2,800**	**1,406**	**807**
	2012	**76,529**	**55,460**	**14,578**	**..**	**3,235**	**2,449**	**807**
	2013	**77,183**	**55,289**	**13,834**	**..**	**3,895**	**3,276**	**889**
	2014	**79,399**	**55,921**	**13,855**	**..**	**4,530**	**4,038**	**1,055**
American Samoa	2011	41	41	..	..	..	0	..
	2012	*42	*41	..	..	..	1	..
	2013	*42	*41	..	..	..	*1	..
	2014	42	41	..	..	..	1	..
Australia	2011	62,357	50,044	8,788	..	2,127	1,397	1
	2012	64,164	50,377	8,790	..	2,561	2,435	1
	2013	64,688	50,171	8,037	..	3,221	3,258	1
	2014	66,557	50,704	8,048	..	3,797	4,007	1
Cook Islands	2011	*8	*8	..	..	..	..	..
	2012	*8	*8	..	..	..	..	..
	2013	*8	*8	..	..	..	..	..
	2014	*8	*8	..	..	..	..	..
Fiji	2011	*260	*120	*129	..	10	..	..
	2012	285	*140	135	..	10	..	..
	2013	309	164	135	..	10	..	..
	2014	309	164	135	..	10	..	..
French Polynesia	2011	*227	*177	*47	..	..	*3	..
	2012	*227	*177	*47	..	..	*3	..
	2013	*227	*177	*48	..	..	*3	..
	2014	*227	*177	*48	..	..	*3	..
Guam	2011	552	552	..	..	..	..	..
	2012	552	552	..	..	..	..	..
	2013	552	552	..	..	..	..	..
	2014	*552	*552	..	..	..	..	..
Kiribati	2011	*6	*5	..	..	..	*1	..
	2012	*6	*5	..	..	..	*1	..
	2013	*6	*5	..	..	..	*1	..
	2014	*6	*5	..	..	..	*1	..
Marshall Islands	2011	29	29	..	..	..	..	..
	2012	33	33	..	..	..	0	..
	2013	33	33	..	..	..	0	..
	2014	34	33	..	..	..	1	..

Net installed capacity of electricity plants by type

Megawatts

Autoproducers				Main activity producers				
Total	Combustible fuels	Hydro	Other	Total	Combustible fuels	Hydro	Nuclear	Other
1,402	1,399	3	..	33,825	5,147	16,574	9,323	2,781
1,241	1,238	3	..	36,602	7,124	16,411	9,436	3,631
1,203	1,194	9	..	36,712	6,582	16,485	9,408	4,237
*1,203	*1,194	9	..	38,727	8,076	15,987	9,507	5,157
1,390	678	489	223	16,844	240	13,280	3,278	46
1,746	727	582	437	16,838	291	13,220	3,278	49
2,071	730	585	756	16,861	291	13,232	3,278	60
2,352	730	561	1,061	16,813	263	13,182	3,308	60
3	3	..	..	1,644	1,086	556	..	2
3	3	..	..	1,706	1,107	595	..	4
3	3	..	..	1,731	1,107	617	..	7
3	3	..	..	1,761	1,077	632	..	52
3,138	3,132	6	..	51,486	31,894	5,463	13,835	294
3,356	3,351	5	..	51,645	31,778	5,465	13,835	567
3,336	3,326	5	5	52,578	32,290	5,484	13,835	969
2,933	2,925	4	4	52,910	32,410	5,847	13,835	818
8,833	6,377	202	2,254	84,838	64,744	4,221	10,663	5,210
9,667	6,410	218	3,039	86,301	64,515	4,221	9,946	7,619
11,287	6,166	231	4,890	83,818	60,513	4,221	9,906	9,178
13,953	5,876	246	7,831	83,056	58,362	4,221	9,937	10,536
6,696	**5,117**	**102**	**1,478**	**68,124**	**50,127**	**14,462**	**..**	**3,535**
7,435	**4,814**	**102**	**2,520**	**69,094**	**50,647**	**14,476**	**..**	**3,971**
9,128	**5,681**	**102**	**3,346**	**68,055**	**49,608**	**13,732**	**..**	**4,714**
9,947	**5,739**	**102**	**4,107**	**69,452**	**50,182**	**13,753**	**..**	**5,516**
..	..	..	..	41	41	..	..	0
..	..	..	..	*42	*41	..	..	1
..	..	..	..	*42	*41	..	..	*1
..	..	..	..	42	41	..	..	1
5,867	4,473	..	1,394	56,490	45,571	8,788	..	2,131
6,598	4,166	..	2,432	57,566	46,211	8,790	..	2,565
8,165	4,910	..	3,255	56,523	45,261	8,037	..	3,225
8,902	4,898	..	4,004	57,655	45,806	8,048	..	3,801
..	..	..	..	*8	*8	..	..	..
..	..	..	..	*8	*8	..	..	..
..	..	..	..	*8	*8	..	..	..
..	..	..	..	*8	*8	..	..	..
*50	*50	..	..	*210	*70	*129	..	10
*50	*50	..	..	235	*90	135	..	10
*50	*50	..	..	259	114	135	..	10
*50	*50	..	..	259	114	135	..	10
..	..	..	..	*227	*177	*47	..	*3
..	..	..	..	*227	*177	*47	..	*3
..	..	..	..	*227	*177	*48	..	*3
..	..	..	..	*227	*177	*48	..	*3
..	..	..	..	552	552	..	..	..
..	..	..	..	552	552	..	..	..
..	..	..	..	552	552	..	..	..
..	..	..	..	*552	*552	..	..	..
..	..	..	..	*6	*5	..	..	*1
..	..	..	..	*6	*5	..	..	*1
..	..	..	..	*6	*5	..	..	*1
..	..	..	..	*6	*5	..	..	*1
..	..	..	..	29	29	..	..	..
..	..	..	..	33	33	..	..	0
..	..	..	..	33	33	..	..	0
..	..	..	..	34	33	..	..	1

Table 30

Net installed capacity of electricity plants by type

Megawatts

Country or area	Year	Total producers						
		Total	Combustible fuels	Hydro	Nuclear	Wind	Solar	Other
Micronesia (Fed. States of)	2011	33	32	..	..	..	*2	..
	2012	33	32	..	..	..	*2	..
	2013	33	32	..	..	..	1	..
	2014	34	32	2	..	..	1	..
Nauru	2011	5	5	..	..	..	0	..
	2012	5	5	..	..	..	0	..
	2013	*5	*5	..	..	..	0	..
	2014	*6	*6	..	..	..	0	..
New Caledonia	2011	498	380	78	..	37	4	..
	2012	503	384	78	..	38	4	..
	2013	691	572	78	..	38	4	..
	2014	683	565	78	..	37	4	..
New Zealand	2011	9,765	3,138	5,254	..	623	..	750
	2012	9,619	2,988	5,254	..	623	4	750
	2013	9,463	2,738	5,263	..	623	7	832
	2014	9,704	2,741	5,263	..	683	19	998
Niue	2011	2	2	..	..	..	0	..
	2012	2	2	..	..	..	0	..
	2013	2	2	..	..	..	0	..
	2014	2	2	..	..	..	0	..
Northern Mariana Islands	2011	*82	*82	..	..	..	..	..
	2012	*88	*88	..	..	..	..	..
	2013	*91	*91	..	..	..	..	..
	2014	95	95	..	..	..	..	..
Palau	2011	*30	*30	..	..	..	..	..
	2012	*30	*30	..	..	..	..	..
	2013	*30	*30	..	..	..	..	..
	2014	30	30	..	..	..	..	..
Papua New Guinea	2011	*788	*478	254	..	..	..	56
	2012	*795	*478	261	..	..	..	56
	2013	*865	*548	261	..	..	..	56
	2014	*975	*650	*269	..	..	..	*56
Samoa	2011	*42	*30	*12	..	..	..	..
	2012	*42	*30	*12	..	..	..	..
	2013	*42	*30	*12	..	..	..	..
	2014	38	26	12	..	..	..	..
Solomon Islands	2011	*36	*36	0	..	..	..	..
	2012	*36	*36	0	..	..	..	..
	2013	*36	*36	0	..	..	..	..
	2014	*36	*36	0	..	..	..	..
Tonga	2011	*14	*14	..	..	..	..	..
	2012	*14	*14	..	..	..	..	..
	2013	*15	*14	..	..	..	*1	..
	2014	*16	*14	..	..	..	*2	..
Tuvalu	2011	*3	*3	..	..	..	..	..
	2012	*3	*3	..	..	..	..	..
	2013	3	3	..	..	..	..	..
	2014	3	3	..	..	..	..	..
Vanuatu	2011	31	27	1	..	3	0	..
	2012	31	27	1	..	3	0	..
	2013	31	27	1	..	3	0	..
	2014	31	27	1	..	3	0	..
Wallis and Futuna Is.	2011	11	11	0	..	..	0	..
	2012	11	11	0	..	..	0	..
	2013	*11	*11	0	..	..	0	..
	2014	*11	*11	0	..	..	0	..

Net installed capacity of electricity plants by type

Megawatts

Autoproducers				Main activity producers				
Total	Combustible fuels	Hydro	Other	Total	Combustible fuels	Hydro	Nuclear	Other
..	..	..	..	33	32	..	..	*2
..	..	..	..	33	32	..	..	*2
..	..	..	..	33	32	..	..	1
..	..	..	..	34	32	2	..	1
..	..	..	..	5	5	..	..	0
..	..	..	..	5	5	..	..	0
..	..	..	..	*5	*5	..	..	0
..	..	..	..	*6	*6	..	..	0
*4	*3	..	*1	494	377	78	..	40
*8	*7	..	*1	495	377	78	..	41
*61	*60	..	*1	630	512	78	..	41
*61	*60	..	*1	622	505	78	..	40
251	221	3	27	9,514	2,917	5,251	..	1,346
255	221	3	31	9,364	2,767	5,251	..	1,346
258	221	3	34	9,205	2,517	5,260	..	1,428
270	221	3	46	9,434	2,520	5,260	..	1,654
..	..	..	..	2	2	..	..	0
..	..	..	..	2	2	..	..	0
..	..	..	..	2	2	..	..	0
..	..	..	..	2	2	..	..	0
..	..	..	..	*82	*82	..	..	..
..	..	..	..	*88	*88	..	..	..
..	..	..	..	*91	*91	..	..	..
..	..	..	..	95	95	..	..	..
*2	*2	..	..	*28	*28	..	..	..
*2	*2	..	..	*28	*28	..	..	..
*2	*2	..	..	*28	*28	..	..	..
*2	*2	..	..	28	28	..	..	..
*515	*360	99	56	273	118	155	..	..
*515	*360	99	56	280	*118	162	..	..
*585	*430	99	56	280	*118	162	..	..
*655	*500	*99	*56	*320	*150	*170	..	..
*3	*3	..	..	*39	*27	*12	..	..
*3	*3	..	..	*39	*27	*12	..	..
*3	*3	..	..	*39	*27	*12	..	..
*3	*3	..	..	35	23	12	..	..
*5	*5	..	..	*31	*31	0	..	..
*5	*5	..	..	*31	*31	0	..	..
*5	*5	..	..	*31	*31	0	..	..
*5	*5	..	..	*31	*31	0	..	..
..	..	..	..	*14	*14	..	..	..
..	..	..	..	*14	*14	..	..	..
..	..	..	..	*15	*14	..	..	*1
..	..	..	..	*16	*14	..	..	*2
..	..	..	..	*3	*3	..	..	..
..	..	..	..	*3	*3	..	..	..
..	..	..	..	3	3	..	..	..
..	..	..	..	3	3	..	..	..
..	..	..	..	31	27	1	..	3
..	..	..	..	31	27	1	..	3
..	..	..	..	31	27	1	..	3
..	..	..	..	31	27	1	..	3
..	..	..	..	11	11	0	..	0
..	..	..	..	11	11	0	..	0
..	..	..	..	*11	*11	0	..	0
..	..	..	..	*11	*11	0	..	0

Table 31

Utilization of installed electricity capacity - by type

Kilowatt-hours per kilowatt divided by 8760

Table Notes

Data on utilization are derived by dividing the data on electricity production shown in Table 32 by the net installed capacity shown in Table 30, and then by the total number of hours in a year to yield a percentage of theoretical maximal utilization.

For this table, **Other** under total producers refers to geothermal, tide, wave, marine and other non-specified sources. **Other** under autoproducers refers to nuclear (if any), wind, solar, tide, wave, marine and other non-specified sources. **Other** under main activity producers refers to wind, solar, tide, wave, marine and other non-specified sources.

For Austria data are given in gross maximum capacity.

Please refer to the Definitions Section on pages ix to xv for the appropriate product description /classification.

Figure 74: Utilization of installed electric generating capacity by source and region, 2014

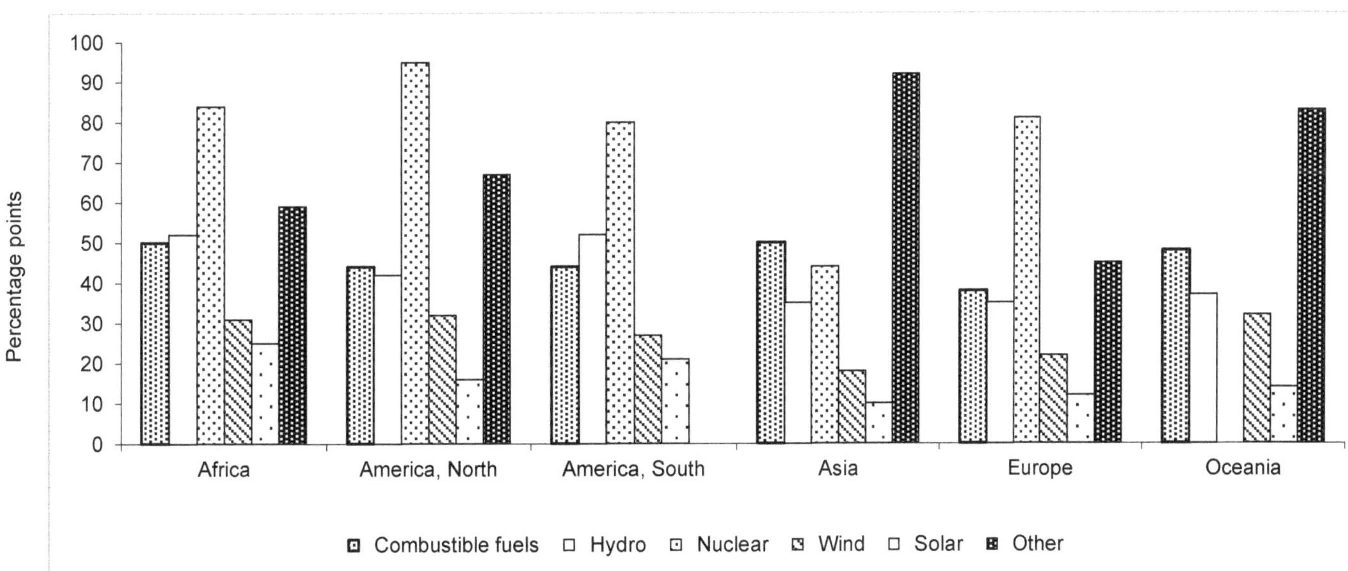

Table 31

Utilization of installed electricity capacity - by type

Kilowatt-hours per kilowatt divided by 8760

Country or area	Year	Total producers						
		Total	Combustible fuels	Hydro	Nuclear	Wind	Solar	Other
World	**2011**	**47**	**48**	**38**	**76**	**22**	**10**	**86**
	2012	**46**	**47**	**39**	**74**	**22**	**11**	**85**
	2013	**46**	**47**	**39**	**75**	**24**	**11**	**66**
	2014	**44**	**46**	**39**	**76**	**23**	**12**	**69**
Africa	**2011**	**54**	**55**	**50**	**82**	**26**	**25**	**84**
	2012	**54**	**55**	**51**	**72**	**24**	**17**	**85**
	2013	**54**	**54**	**52**	**89**	**34**	**17**	**84**
	2014	**52**	**52**	**52**	**84**	**31**	**25**	**59**
Algeria	2011	51	51	21	..	..	..	..
	2012	50	51	24	..	..	..	..
	2013	45	46	13	..	..	..	..
	2014	46	47	10	..	..	..	..
Angola	2011	42	24	60	..	..	..	..
	2012	46	36	57	..	..	..	..
	2013	61	51	72	..	..	..	..
	2014	52	66	44	..	..	..	..
Benin	2011	18	18	..	..	..	..	..
	2012	30	30	..	..	..	..	..
	2013	9	10	..	..	..	..	..
	2014	10	10	..	..	..	..	..
Botswana	2011	33	33	..	..	..	..	..
	2012	19	19	..	..	..	11	..
	2013	12	12	..	..	..	11	..
	2014	43	44	..	..	..	18	..
Burkina Faso	2011	23	22	29	..	..	..	..
	2012	24	23	34	..	..	..	..
	2013	28	27	38	..	..	..	..
	2014	33	33	32	..	..	..	..
Burundi	2011	44	29	46	..	..	..	..
	2012	44	7	50	..	..	..	..
	2013	47	34	49	..	..	..	..
	2014	48	43	50	..	..	..	..
Cabo Verde	2011	39	39	..	..	85	16	..
	2012	32	34	..	..	27	13	..
	2013	32	34	..	..	33	11	..
	2014	30	32	..	..	29	8	..
Cameroon	2011	52	31	69	..	..	..	..
	2012	53	31	72	..	..	..	..
	2013	50	27	76	..	..	..	..
	2014	51	26	79	..	..	..	..
Central African Rep.	2011	45	16	67	..	..	..	..
	2012	46	16	68	..	..	..	..
	2013	47	16	71	..	..	..	..
	2014	47	16	72	..	..	..	..
Chad	2011	61	61	..	..	..	..	..
	2012	57	57	..	..	..	..	..
	2013	55	55	..	..	..	..	..
	2014	55	55	..	..	..	..	..
Comoros	2011	28	28	..	..	..	..	..
	2012	29	29	..	..	..	..	..
	2013	27	27	..	..	..	..	..
	2014	25	25	..	..	..	..	..
Congo	2011	40	38	41	..	..	..	..
	2012	50	48	51	..	..	..	..
	2013	50	50	50	..	..	..	..
	2014	51	53	49	..	..	..	..

Utilization of installed electricity capacity - by type

Kilowatt-hours per kilowatt divided by 8760

Autoproducers				Main activity producers				
Total	Combustible fuels	Hydro	Other	Total	Combustible fuels	Hydro	Nuclear	Other
43	**45**	**37**	**14**	**47**	**48**	**38**	**76**	**23**
43	**45**	**37**	**14**	**46**	**47**	**39**	**74**	**23**
43	**47**	**37**	**14**	**46**	**47**	**39**	**75**	**23**
40	**45**	**35**	**14**	**45**	**46**	**39**	**76**	**22**
39	**41**	**15**	**17**	**55**	**56**	**52**	**82**	**33**
37	**39**	**9**	**18**	**56**	**57**	**52**	**72**	**30**
36	**37**	**9**	**18**	**55**	**55**	**53**	**89**	**37**
33	**34**	**11**	**14**	**54**	**54**	**53**	**84**	**35**
23	23	..	..	54	55	21	..	..
14	14	..	..	59	60	24	..	..
14	14	..	..	53	54	13	..	..
13	13	..	..	54	55	10	..	..
29	29	..	..	43	24	60	..	..
31	31	..	..	48	37	57	..	..
42	42	..	..	63	53	72	..	..
48	48	..	..	53	69	44	..	..
40	40	..	..	16	16	..	..	..
41	41	..	..	28	29	..	..	..
44	44	..	..	8	8	..	..	..
47	47	..	..	9	9	..	..	..
..	..	..	..	38	38	..	..	..
..	..	..	..	21	22	..	..	11
..	..	..	..	12	12	..	..	11
..	..	..	..	45	45	..	..	18
16	16	..	..	25	24	29	..	..
28	28	..	..	23	21	34	..	..
37	37	..	..	26	24	38	..	..
47	47	..	..	30	30	32	..	..
..	..	..	..	44	29	46	..	..
..	..	..	..	44	7	50	..	..
..	..	..	..	47	34	49	..	..
..	..	..	..	48	43	50	..	..
17	17	..	..	39	39	..	..	33
18	18	..	..	32	34	..	..	24
17	17	..	..	32	34	..	..	28
15	15	..	..	30	32	..	..	24
45	45	..	..	54	15	69	..	..
49	49	..	..	54	13	72	..	..
61	61	..	..	47	8	76	..	..
55	55	..	..	50	9	79	..	..
..	..	..	..	45	16	67	..	..
..	..	..	..	46	16	68	..	..
..	..	..	..	47	16	71	..	..
..	..	..	..	47	16	72	..	..
..	..	..	..	61	61	..	..	..
..	..	..	..	57	57	..	..	..
..	..	..	..	55	55	..	..	..
..	..	..	..	55	55	..	..	..
..	..	..	..	28	28	..	..	..
..	..	..	..	29	29	..	..	..
..	..	..	..	27	27	..	..	..
..	..	..	..	25	25	..	..	..
47	47	..	..	37	4	41	..	..
51	51	..	..	49	0	51	..	..
53	53	..	..	48	0	50	..	..
56	56	..	..	47	0	49	..	..

Table 31

Utilization of installed electricity capacity - by type

Kilowatt-hours per kilowatt divided by 8760

Country or area	Year	Total producers						
		Total	Combustible fuels	Hydro	Nuclear	Wind	Solar	Other
Côte d'Ivoire	2011	49	59	35	..	..	..	..
	2012	57	57	55	..	..	..	..
	2013	57	60	49	..	..	..	..
	2014	57	59	51	..	..	..	..
Dem. Rep. of the Congo	2011	36	30	36	..	..	..	..
	2012	35	3	35	..	..	..	..
	2013	36	3	37	..	..	..	..
	2014	39	4	39	..	..	..	..
Djibouti	2011	34	34	..	..	..	..	..
	2012	35	35	..	..	..	..	..
	2013	35	35	..	..	..	..	..
	2014	35	35	..	..	..	..	..
Egypt	2011	61	64	49	..	32	39	..
	2012	60	61	53	..	26	19	..
	2013	59	60	54	..	28	9	..
	2014	56	57	57	..	27	20	..
Equatorial Guinea	2011	15	15	15	..	..	..	..
	2012	15	15	15	..	..	..	..
	2013	15	15	15	..	..	..	..
	2014	15	15	15	..	..	..	..
Eritrea	2011	27	28	..	..	..	23	..
	2012	29	29	..	..	..	23	..
	2013	30	30	..	..	..	23	..
	2014	30	30	..	..	..	23	..
Ethiopia	2011	31	2	37	..	2	..	13
	2012	38	1	43	..	13	..	26
	2013	43	0	49	..	24	..	28
	2014	47	1	54	..	26	..	31
Gabon	2011	52	50	56	..	..	..	..
	2012	53	48	61	..	..	34	..
	2013	50	44	63	..	..	23	..
	2014	53	53	53	..	..	23	..
Gambia	2011	39	39	..	..	..	..	..
	2012	41	41	..	..	..	..	..
	2013	33	33	..	..	..	..	..
	2014	26	26	..	..	..	..	..
Ghana	2011	46	27	73	..	..	..	..
	2012	48	27	78	..	..	..	..
	2013	49	38	59	..	..	..	..
	2014	49	37	61	..	..	..	..
Guinea	2011	26	23	32	..	..	..	..
	2012	31	27	38	..	..	..	..
	2013	22	16	37	..	..	..	..
	2014	22	16	38	..	..	..	..
Guinea-Bissau	2011	14	14	..	..	..	..	..
	2012	14	14	..	..	..	..	..
	2013	15	15	..	..	..	..	..
	2014	14	14	..	..	..	..	..
Kenya	2011	56	54	50	..	33	..	86
	2012	56	41	60	..	35	..	87
	2013	56	35	66	..	34	..	86
	2014	50	41	51	..	37	..	60
Lesotho	2011	70	..	70	..	..	..	..
	2012	69	..	69	..	..	..	..
	2013	74	..	74	..	..	..	..
	2014	74	..	74	..	..	..	..

Utilization of installed electricity capacity - by type

Kilowatt-hours per kilowatt divided by 8760

Autoproducers				Main activity producers				
Total	Combustible fuels	Hydro	Other	Total	Combustible fuels	Hydro	Nuclear	Other
32	32	..	..	50	59	35	..	..
34	34	..	..	57	58	55	..	..
35	35	..	..	58	60	49	..	..
36	36	..	..	57	60	51	..	..
7	228	6	..	45	3	45	..	..
0	0	0	..	45	3	46	..	..
0	0	0	..	46	3	47	..	..
0	20	0	..	50	2	50	..	..
..	..	..	..	34	34	..	..	..
..	..	..	..	35	35	..	..	..
..	..	..	..	35	35	..	..	..
..	..	..	..	35	35	..	..	..
62	62	..	..	61	64	49	..	33
90	90	..	..	59	60	53	..	25
91	91	..	..	58	59	54	..	24
91	91	..	..	56	56	57	..	26
..	..	..	..	15	15	15	..	..
..	..	..	..	15	15	15	..	..
..	..	..	..	15	15	15	..	..
..	..	..	..	15	15	15	..	..
21	21	..	..	28	28	..	..	23
21	21	..	..	30	30	..	..	23
21	21	..	..	30	30	..	..	23
21	21	..	..	31	31	..	..	23
..	..	..	..	31	2	37	..	2
..	..	..	..	38	1	43	..	13
..	..	..	..	43	0	49	..	24
..	..	..	..	47	1	54	..	26
32	32	..	..	56	56	56	..	..
30	30	..	34	57	54	61	..	..
31	31	..	23	53	48	63	..	..
31	32	..	23	57	59	53	..	..
10	10	..	..	65	65	..	..	..
10	10	..	..	67	67	..	..	..
10	10	..	..	46	46	..	..	..
10	10	..	..	32	33	..	..	..
9	9	..	..	49	29	73	..	..
10	10	..	..	51	30	78	..	..
10	10	..	..	52	42	59	..	..
10	10	..	..	52	42	61	..	..
22	25	..	..	31	20	37	..	..
22	25	..	..	39	31	44	..	..
22	25	..	..	22	9	43	..	..
22	25	..	..	22	9	44	..	..
43	43	..	..	12	12	..	..	..
45	45	..	..	12	12	..	..	..
47	47	..	..	12	12	..	..	..
47	47	..	..	12	12	..	..	..
..	..	..	..	56	54	50	..	85
..	..	..	..	56	41	60	..	86
..	..	..	..	56	35	66	..	85
..	..	..	..	50	41	51	..	59
..	..	..	..	70	..	70	..	..
..	..	..	..	69	..	69	..	..
..	..	..	..	74	..	74	..	..
..	..	..	..	74	..	74	..	..

Table 31

Utilization of installed electricity capacity - by type

Kilowatt-hours per kilowatt divided by 8760

Country or area	Year	Total producers						
		Total	Combustible fuels	Hydro	Nuclear	Wind	Solar	Other
Liberia	2011	38	38	..	..	..	..	..
	2012	38	38	..	..	..	..	..
	2013	38	38	..	..	..	..	..
	2014	38	38	..	..	..	..	..
Libya	2011	33	33	..	..	..	..	..
	2012	44	44	..	..	..	..	..
	2013	46	46	..	..	..	..	..
	2014	46	46	..	..	..	..	..
Madagascar	2011	32	22	62	..	..	10	..
	2012	34	23	67	..	..	10	..
	2013	36	24	72	..	..	11	..
	2014	37	23	79	..	..	11	..
Malawi	2011	45	10	71	..	..	..	..
	2012	46	10	72	..	..	..	..
	2013	48	10	75	..	..	..	..
	2014	48	10	75	..	..	..	..
Mali	2011	24	26	23	..	..	..	..
	2012	23	20	24	..	..	..	..
	2013	21	17	25	..	..	..	..
	2014	22	27	17	..	..	..	..
Mauritania	2011	43	43	..	..	..	..	..
	2012	48	48	..	..	..	..	..
	2013	34	34	..	..	..	..	..
	2014	33	33	..	..	..	..	..
Mauritius	2011	42	45	11	..	25	..	..
	2012	41	43	14	..	32	7	..
	2013	42	44	18	..	32	13	..
	2014	43	46	17	..	37	16	..
Mayotte	2011	33	36	..	..	..	13	..
	2012	34	37	..	..	..	14	..
	2013	36	39	..	..	..	15	..
	2014	38	41	..	..	..	15	..
Morocco	2011	41	52	14	..	32	..	..
	2012	43	55	12	..	33	..	..
	2013	45	54	18	..	68	..	..
	2014	46	58	13	..	44	..	..
Mozambique	2011	78	1	86	..	..	..	..
	2012	70	1	77	..	..	..	..
	2013	67	27	74	..	..	..	..
	2014	75	44	80	..	..	10	..
Namibia	2011	47	1	73	..	..	..	..
	2012	35	5	49	..	..	..	..
	2013	40	5	56	..	..	..	..
	2014	35	1	51	..	..	..	..
Niger	2011	27	28	..	..	..	0	..
	2012	33	33	..	..	..	0	..
	2013	28	29	..	..	..	0	..
	2014	28	29	..	..	..	0	..
Nigeria	2011	35	35	35	..	..	..	..
	2012	37	38	33	..	..	..	..
	2013	34	34	32	..	..	..	..
	2014	33	34	31	..	..	..	..
Réunion	2011	41	53	31	..	9	13	..
	2012	37	43	42	..	14	14	..
	2013	37	41	48	..	12	16	..
	2014	39	49	36	..	12	16	..

Utilization of installed electricity capacity - by type

Kilowatt-hours per kilowatt divided by 8760

Autoproducers				Main activity producers				
Total	Combustible fuels	Hydro	Other	Total	Combustible fuels	Hydro	Nuclear	Other
33	33	..	..	54	54	..	..	..
33	33	..	..	54	54	..	..	..
33	33	..	..	54	54	..	..	..
33	33	..	..	54	54	..	..	..
..	..	..	..	33	33	..	..	..
..	..	..	..	44	44	..	..	..
..	..	..	..	46	46	..	..	..
..	..	..	..	46	46	..	..	..
61	61	..	..	30	19	62	..	10
61	61	..	..	32	20	67	..	10
61	61	..	..	34	20	72	..	11
61	61	..	..	36	20	79	..	11
10	10	11	..	72	52	72	..	..
10	10	11	..	73	52	73	..	..
10	10	11	..	76	52	76	..	..
10	10	13	..	76	52	76	..	..
6	6	..	..	27	36	23	..	..
6	6	..	..	25	27	24	..	..
6	6	..	..	23	21	25	..	..
5	5	..	..	24	36	17	..	..
50	50	..	..	39	39	..	..	..
53	53	..	..	45	45	..	..	..
52	52	..	..	28	28	..	..	..
55	55	..	..	26	26	..	..	..
64	64	..	..	29	31	11	..	25
69	69	..	7	26	27	14	..	32
70	71	..	12	27	28	18	..	32
69	72	..	16	27	29	17	..	37
..	..	..	..	33	36	..	..	13
..	..	..	..	34	37	..	..	14
..	..	..	..	36	39	..	..	15
..	..	..	..	38	41	..	..	15
7	7	..	..	43	56	14	..	32
5	5	..	..	45	58	12	..	33
40	40	..	..	46	55	18	..	68
41	41	..	..	46	59	13	..	44
79	..	79	..	77	1	86	..	..
60	..	60	..	70	1	78	..	..
60	..	60	..	68	27	74	..	..
62	..	63	10	76	44	81	..	..
..	..	..	..	47	1	73	..	..
..	..	..	..	35	5	49	..	..
..	..	..	..	40	5	56	..	..
..	..	..	..	35	1	51	..	..
20	21	..	0	29	29	..	..	..
20	21	..	0	37	37	..	..	..
11	12	..	0	34	34	..	..	..
15	17	..	0	33	33	..	..	..
37	38	34	..	34	33	35	..	..
40	40	33	..	36	37	33	..	..
28	28	31	..	35	36	32	..	..
21	21	31	..	38	41	31	..	..
..	..	..	..	41	53	31	..	12
..	..	..	..	37	43	42	..	14
..	..	..	..	37	41	48	..	15
..	..	..	..	39	49	36	..	15

Table 31

Utilization of installed electricity capacity - by type

Kilowatt-hours per kilowatt divided by 8760

Country or area	Year	Total producers						
		Total	Combustible fuels	Hydro	Nuclear	Wind	Solar	Other
Rwanda	2011	44	69	30	..	..	14	..
	2012	45	67	33	..	..	11	..
	2013	43	76	25	..	..	4	..
	2014	46	79	28	..	..	6	..
Sao Tome and Principe	2011	37	39	30	..	..	..	..
	2012	42	43	35	..	..	..	..
	2013	41	43	32	..	..	..	..
	2014	41	43	32	..	..	..	..
Senegal	2011	35	35	..	..	..	..	..
	2012	39	39	..	..	..	..	..
	2013	45	45	..	..	..	..	..
	2014	44	45	..	..	..	2	..
Seychelles	2011	47	47	..	..	..	..	..
	2012	48	48	..	..	..	..	..
	2013	47	50	..	..	13	..	..
	2014	48	50	..	..	13	..	..
Sierra Leone	2011	26	10	35	..	..	..	..
	2012	27	26	27	..	..	..	..
	2013	25	25	24	..	..	..	..
	2014	27	27	27	..	..	..	..
Somalia	2011	48	48	..	..	..	..	..
	2012	49	49	..	..	..	..	..
	2013	50	50	..	..	..	..	..
	2014	50	50	..	..	..	..	..
South Africa	2011	67	66	80	82	25	2	..
	2012	66	65	67	72	25	34	..
	2013	66	65	65	89	25	35	..
	2014	61	61	64	84	27	35	..
South Sudan	2011	..	..	..	..	..	..	..
	2012	62	63	..	..	..	11	..
	2013	66	67	..	..	..	11	..
	2014	68	69	..	..	..	11	..
St. Helena and Depend.	2011	20	19	..	..	27	..	..
	2012	14	13	..	..	21	5	..
	2013	15	14	..	..	24	12	..
	2014	14	14	..	..	12	9	..
Sudan	2011	..	..	..	..	..	..	..
	2012	38	26	48	..	..	..	..
	2013	39	20	52	..	..	..	..
	2014	42	22	56	..	..	..	..
Sudan (former)	2011	35	20	46	..	..	..	..
	2012	..	..	..	..	..	..	..
	2013	..	..	..	..	..	..	..
	2014	..	..	..	..	..	..	..
Swaziland	2011	47	39	59	..	..	..	..
	2012	43	40	48	..	..	..	..
	2013	41	40	43	..	..	..	..
	2014	38	30	47	..	..	..	..
Togo	2011	18	10	34	..	..	..	..
	2012	20	17	28	..	..	..	..
	2013	18	20	13	..	..	..	..
	2014	18	17	20	..	..	..	..
Tunisia	2011	45	45	10	..	23	16	..
	2012	48	50	20	..	13	12	..
	2013	46	47	11	..	20	18	..
	2014	43	44	10	..	25	36	..

Utilization of installed electricity capacity - by type

Kilowatt-hours per kilowatt divided by 8760

Autoproducers				Main activity producers				
Total	Combustible fuels	Hydro	Other	Total	Combustible fuels	Hydro	Nuclear	Other
..	..	..	..	44	69	30	..	14
..	..	..	..	45	67	33	..	11
..	..	..	..	43	76	25	..	4
..	..	..	..	46	79	28	..	6
..	..	..	..	37	39	30	..	..
..	..	..	..	42	43	35	..	..
..	..	..	..	41	43	32	..	..
..	..	..	..	41	43	32	..	..
40	40	..	..	32	32	..	..	..
47	47	..	..	33	33	..	..	..
21	19	..	..	64	64	..	..	..
16	15	..	2	67	67	..	..	..
..	..	..	..	47	47	..	..	..
..	..	..	..	48	48	..	..	..
..	..	..	..	47	50	..	..	13
..	..	..	..	48	50	..	..	15
..	..	..	..	26	10	35	..	..
..	..	..	..	27	26	27	..	..
..	..	..	..	25	25	24	..	..
..	..	..	..	27	27	27	..	..
..	..	..	..	48	48	..	..	..
..	..	..	..	49	49	..	..	..
..	..	..	..	50	50	..	..	..
..	..	..	..	50	50	..	..	..
61	61	49	..	67	66	82	82	19
58	59	41	..	66	66	69	72	29
63	63	39	..	66	65	66	89	33
65	66	35	..	61	61	66	84	30
..	..	..	..	..	..	..	..	..
65	66	..	11	59	60	..	..	11
68	70	..	11	63	64	..	..	11
71	72	..	11	65	66	..	..	11
..	..	..	..	20	19	..	..	27
..	..	..	..	14	13	..	..	20
..	..	..	..	15	14	..	..	23
..	..	..	..	14	14	..	..	12
..	..	..	..	..	..	..	..	..
..	..	..	..	37	25	48	..	..
..	..	..	..	38	17	52	..	..
..	..	..	..	42	22	56	..	..
..	..	..	..	35	20	46	..	..
..	..	..	..	..	..	..	..	..
..	..	..	..	..	..	..	..	..
..	..	..	..	..	..	..	..	..
41	43	..	..	55	0	63	..	..
42	43	..	..	46	9	52	..	..
42	43	..	..	41	10	46	..	..
32	34	..	..	44	0	50	..	..
16	16	..	..	18	10	34	..	..
16	16	..	..	20	17	28	..	..
10	10	..	..	18	21	13	..	..
16	16	..	..	18	17	20	..	..
41	30	..	16	45	46	10	..	23
45	30	..	12	48	50	20	..	13
41	30	..	18	46	48	11	..	20
43	32	..	36	43	45	10	..	25

Table 31

Utilization of installed electricity capacity - by type

Kilowatt-hours per kilowatt divided by 8760

Country or area	Year	Total producers						
		Total	Combustible fuels	Hydro	Nuclear	Wind	Solar	Other
Uganda	2011	47	55	44	..	..	..	..
	2012	41	77	33	..	..	..	..
	2013	39	77	32	..	..	..	..
	2014	42	45	41	..	..	..	..
United Rep. of Tanzania	2011	52	65	40	..	..	21	..
	2012	57	80	36	..	..	25	..
	2013	61	88	35	..	..	24	..
	2014	64	75	53	..	..	29	..
Zambia	2011	67	2	70	..	..	..	..
	2012	72	2	75	..	..	..	..
	2013	74	2	80	..	..	..	..
	2014	67	24	71	..	..	..	..
Zimbabwe	2011	50	33	78	..	..	..	..
	2012	49	32	81	..	..	..	..
	2013	51	38	75	..	..	..	..
	2014	57	42	82	..	..	..	..
America, North	**2011**	**48**	**44**	**46**	**91**	**29**	**12**	**77**
	2012	**47**	**44**	**43**	**89**	**27**	**13**	**68**
	2013	**47**	**44**	**43**	**93**	**31**	**14**	**72**
	2014	**47**	**44**	**42**	**95**	**32**	**16**	**67**
Anguilla	2011	39	39	..	..	..	..	..
	2012	36	36	..	..	..	..	..
	2013	36	36	..	..	..	..	..
	2014	36	36	..	..	..	..	..
Antigua and Barbuda	2011	44	44	..	..	..	..	..
	2012	44	44	..	..	..	..	..
	2013	45	45	..	..	..	..	..
	2014	46	46	..	..	..	..	..
Aruba	2011	33	32	..	..	43	..	..
	2012	33	31	..	..	52	..	..
	2013	37	35	..	..	53	..	..
	2014	36	34	..	..	53	..	..
Bahamas	2011	34	34	..	..	..	..	..
	2012	34	34	..	..	..	..	..
	2013	36	36	..	..	..	..	..
	2014	37	37	..	..	..	..	..
Barbados	2011	49	49	..	..	..	..	..
	2012	53	53	..	..	..	..	..
	2013	48	48	..	..	..	..	..
	2014	46	46	..	..	..	..	..
Belize	2011	31	18	53	..	..	..	..
	2012	34	27	45	..	..	..	..
	2013	30	21	43	..	..	..	..
	2014	30	23	43	..	..	..	..
Bermuda	2011	49	49	..	..	..	..	..
	2012	46	46	..	..	..	..	..
	2013	45	45	..	..	..	..	..
	2014	47	47	..	..	..	..	..
Bonaire, St Eustatius, Saba	2011	..	..	..	..	..	..	..
	2012	32	97	..	..	..	..	..
	2013	33	99	..	..	..	..	..
	2014	33	99	..	..	1	..	..
British Virgin Islands	2011	34	34	..	..	..	..	..
	2012	33	34	..	..	..	..	..
	2013	33	34	..	..	..	..	..
	2014	34	34	..	..	..	..	..

Utilization of installed electricity capacity - by type

Kilowatt-hours per kilowatt divided by 8760

Autoproducers				Main activity producers				
Total	Combustible fuels	Hydro	Other	Total	Combustible fuels	Hydro	Nuclear	Other
37	22	63	..	47	58	43	..	..
38	22	64	..	41	84	33	..	..
22	10	64	..	40	101	31	..	..
16	8	64	..	44	63	41	..	..
4	..	..	21	53	67	40	..	..
6	..	..	25	59	83	36	..	..
6	..	..	24	62	91	35	..	..
8	..	..	29	65	78	53	..	..
..	..	..	..	67	2	70	..	..
..	..	..	..	72	2	75	..	..
..	..	..	..	74	2	80	..	..
..	..	..	..	67	24	71	..	..
40	40	..	..	50	33	78	..	..
47	47	..	..	49	31	81	..	..
47	47	..	..	51	38	75	..	..
41	41	..	..	58	43	82	..	..
52	**55**	**61**	**16**	**48**	**44**	**45**	**91**	**32**
53	**57**	**67**	**18**	**47**	**44**	**42**	**89**	**29**
53	**59**	**63**	**16**	**47**	**43**	**42**	**93**	**33**
52	**58**	**62**	**18**	**47**	**43**	**41**	**95**	**32**
..	..	..	..	39	39	..	..	..
..	..	..	..	36	36	..	..	..
..	..	..	..	36	36	..	..	..
..	..	..	..	36	36	..	..	..
11	11	..	..	45	45	..	..	..
11	11	..	..	45	45	..	..	..
11	11	..	..	45	45	..	..	..
11	11	..	..	46	46	..	..	..
..	..	..	..	33	32	..	..	43
..	..	..	..	33	31	..	..	52
..	..	..	..	37	35	..	..	53
..	..	..	..	36	34	..	..	53
11	11	..	..	36	36	..	..	..
11	11	..	..	36	36	..	..	..
15	15	..	..	39	39	..	..	..
22	22	..	..	38	38	..	..	..
19	19	..	..	50	50	..	..	..
23	23	..	..	54	54	..	..	..
17	17	..	..	48	48	..	..	..
3	3	..	..	48	48	..	..	..
29	29	..	..	31	14	53	..	..
31	31	..	..	34	25	45	..	..
32	32	..	..	29	18	43	..	..
23	23	..	..	32	23	43	..	..
36	36	..	..	49	49	..	..	..
36	36	..	..	46	46	..	..	..
36	36	..	..	45	45	..	..	..
36	36	..	..	48	48	..	..	..
..	..	..	..	..	..	..	..	..
..	..	..	..	32	97	..	..	..
..	..	..	..	33	99	..	..	..
..	..	..	..	33	99	..	..	1
..	..	..	..	34	34	..	..	..
..	..	..	..	33	34	..	..	..
..	..	..	..	33	34	..	..	..
..	..	..	..	34	34	..	..	..

Table 31

Utilization of installed electricity capacity - by type

Kilowatt-hours per kilowatt divided by 8760

Country or area	Year	Total producers						
		Total	Combustible fuels	Hydro	Nuclear	Wind	Solar	Other
Canada	2011	54	44	57	84	22	6	15
	2012	55	48	57	81	21	5	15
	2013	56	46	59	84	26	14	1,499
	2014	55	44	58	88	27	11	1,300
Cayman Islands	2011	47	47	..	..	..	..	..
	2012	46	46	..	..	..	..	..
	2013	47	47	..	..	..	..	..
	2014	54	54	..	..	..	..	..
Costa Rica	2011	39	13	48	..	36	5	69
	2012	43	18	49	..	40	15	74
	2013	42	25	45	..	24	34	80
	2014	36	23	37	..	28	34	71
Cuba	2011	34	34	17	..	..	..	..
	2012	37	37	20	..	..	..	..
	2013	36	36	23	..	..	..	..
	2014	36	36	19	..	..	..	..
Curaçao	2011	..	..	..	..	..	..	..
	2012	40	43	..	..	12	..	..
	2013	39	42	..	..	12	..	..
	2014	38	42	..	..	12	..	..
Dominica	2011	43	37	62	..	..	..	..
	2012	43	42	46	..	..	..	..
	2013	43	36	63	..	..	..	..
	2014	44	41	53	..	..	..	..
Dominican Republic	2011	55	60	33	..	5	..	..
	2012	55	61	34	..	13	..	..
	2013	52	56	35	..	32	..	..
	2014	51	56	23	..	33	..	..
El Salvador	2011	47	36	49	..	..	..	86
	2012	48	41	45	..	..	..	86
	2013	47	39	43	..	..	..	87
	2014	45	38	41	..	..	..	87
Greenland	2011	37	24	56	..	..	..	..
	2012	38	27	52	..	..	..	..
	2013	35	23	47	..	..	..	..
	2014	35	22	50	..	..	..	..
Grenada	2011	47	47	..	..	..	..	..
	2012	47	47	..	..	..	..	..
	2013	46	46	..	..	..	..	..
	2014	47	47	..	..	..	..	..
Guadeloupe	2011	43	49	20	..	29	17	36
	2012	42	48	18	..	22	16	42
	2013	42	47	25	..	24	17	68
	2014	42	47	29	..	22	16	63
Guatemala	2011	40	32	53	..	..	..	57
	2012	38	30	51	..	..	..	89
	2013	38	31	53	..	..	..	49
	2014	47	40	56	..	..	16	83
Haiti	2011	42	47	23	..	..	..	..
	2012	47	49	41	..	..	..	..
	2013	35	40	21	..	..	..	..
	2014	37	42	17	..	..	..	..
Honduras	2011	51	49	61	..	13	..	..
	2012	50	46	59	..	38	..	..
	2013	51	50	56	..	35	..	..
	2014	48	50	48	..	30	..	..

Utilization of installed electricity capacity - by type

Kilowatt-hours per kilowatt divided by 8760

Autoproducers				Main activity producers				
Total	Combustible fuels	Hydro	Other	Total	Combustible fuels	Hydro	Nuclear	Other
57	46	68	..	54	44	56	84	21
60	48	71	..	55	48	56	81	19
69	73	67	3	55	43	59	84	28
68	71	67	20	54	41	57	88	26
..	..	..	..	47	47	..	..	..
..	..	..	..	46	46	..	..	..
..	..	..	..	47	47	..	..	..
..	..	..	..	54	54	..	..	..
62	57	71	48	35	12	45	..	61
67	97	74	53	39	15	45	..	64
6	99	..	..	47	22	51	..	74
6	98	..	..	40	20	42	..	63
17	17	..	..	36	36	17	..	..
18	18	..	..	39	39	20	..	..
21	21	..	..	37	37	23	..	..
19	19	..	..	37	38	19	..	..
..	..	..	..	..	..	..	..	..
42	42	..	..	37	45	..	..	12
41	41	..	..	37	44	..	..	12
41	41	..	..	36	43	..	..	12
..	..	..	..	43	37	62	..	..
..	..	..	..	43	42	46	..	..
..	..	..	..	43	36	63	..	..
..	..	..	..	44	41	53	..	..
191	191	..	..	48	52	33	..	5
198	200	..	..	48	52	34	..	13
89	91	..	..	48	51	35	..	32
80	82	..	..	47	53	23	..	33
35	35	..	..	49	37	49	..	86
41	41	..	..	49	41	45	..	86
39	39	..	..	48	39	43	..	87
39	39	..	..	46	38	41	..	87
81	81	..	..	33	14	56	..	..
82	82	..	..	34	16	52	..	..
68	68	..	..	32	14	47	..	..
65	65	..	..	33	13	50	..	..
..	..	..	..	47	47	..	..	..
..	..	..	..	47	47	..	..	..
..	..	..	..	46	46	..	..	..
..	..	..	..	47	47	..	..	..
..	..	..	..	43	49	20	..	23
..	..	..	..	42	48	18	..	21
..	..	..	..	42	47	25	..	25
..	..	..	..	42	47	29	..	24
18	18	64	..	44	36	53	..	57
17	16	59	..	42	34	51	..	89
16	16	73	..	44	37	53	..	49
21	20	73	..	52	48	56	..	74
5	5	..	..	56	71	23	..	..
7	7	..	..	62	72	41	..	..
6	6	..	..	47	59	21	..	..
11	11	..	..	47	59	17	..	..
43	43	..	..	51	50	61	..	13
37	37	..	..	51	48	59	..	38
38	38	..	..	52	52	56	..	35
22	22	..	..	51	56	48	..	30

Table 31

Utilization of installed electricity capacity - by type

Kilowatt-hours per kilowatt divided by 8760

Country or area	Year	Total producers						
		Total	Combustible fuels	Hydro	Nuclear	Wind	Solar	Other
Jamaica	2011	57	58	76	..	25	..	..
	2012	52	52	75	..	30	..	..
	2013	52	52	62	..	31	..	..
	2014	50	51	53	..	33	..	..
Martinique	2011	49	49	..	..	17	35	..
	2012	44	50	..	..	17	8	..
	2013	43	49	..	..	17	8	..
	2014	41	46	..	..	17	8	..
Mexico	2011	56	60	36	84	31	13	84
	2012	55	61	31	72	23	16	81
	2013	53	59	27	96	23	18	84
	2014	52	56	36	79	29	25	84
Montserrat	2011	58	58	..	..	..	..	..
	2012	58	58	..	..	..	..	..
	2013	58	58	..	..	..	..	..
	2014	50	50	..	..	..	..	..
Neth. Antilles (former)	2011	47	47	..	..	..	..	..
	2012	..	..	..	..	..	..	..
	2013	..	..	..	..	..	..	..
	2014	..	..	..	..	..	..	..
Nicaragua	2011	39	39	48	..	38	..	35
	2012	36	36	45	..	26	..	36
	2013	37	32	43	..	44	..	50
	2014	38	33	37	..	52	..	49
Panama	2011	38	41	35	..	..	..	..
	2012	41	39	42	..	..	..	..
	2013	40	42	39	..	..	..	..
	2014	37	41	35	..	24	6	..
Puerto Rico	2011	44	44	17	..	..	..	..
	2012	44	44	14	..	..	12	..
	2013	41	42	10	..	13	13	..
	2014	40	41	9	..	25	16	..
Sint Maarten (Dutch part)	2011	..	..	..	..	..	..	..
	2012	47	47	..	..	..	..	..
	2013	48	48	..	..	..	..	..
	2014	49	49	..	..	..	..	..
St. Kitts-Nevis	2011	44	45	..	..	29	..	..
	2012	40	40	..	..	40	..	..
	2013	40	41	..	..	42	15	..
	2014	40	40	..	..	42	15	..
St. Lucia	2011	58	58	..	..	..	..	..
	2012	51	51	..	..	..	..	..
	2013	51	51	..	..	..	..	..
	2014	49	49	..	..	..	..	..
St. Pierre-Miquelon	2011	19	19	..	..	21	..	..
	2012	19	19	..	..	17	..	..
	2013	19	19	..	..	17	..	..
	2014	20	21	..	..	12	..	..
St. Vincent-Grenadines	2011	28	29	28	..	..	..	..
	2012	28	29	28	..	..	..	..
	2013	27	27	27	..	..	..	..
	2014	33	33	33	..	..	..	..
Trinidad and Tobago	2011	44	44	..	..	..	..	..
	2012	40	40	..	..	..	..	..
	2013	42	42	..	..	..	..	..
	2014	43	43	..	..	..	..	..

Utilization of installed electricity capacity - by type

Kilowatt-hours per kilowatt divided by 8760

Autoproducers				Main activity producers				
Total	Combustible fuels	Hydro	Other	Total	Combustible fuels	Hydro	Nuclear	Other
70	79	..	27	52	51	76	..	..
58	62	..	32	49	48	75	..	..
73	78	..	34	42	41	62	..	..
63	67	..	35	44	43	53	..	..
86	86	..	..	48	49	..	..	34
86	86	..	..	44	49	..	..	8
86	86	..	..	43	49	..	..	8
86	86	..	..	41	45	..	..	8
43	44	44	28	58	63	36	84	80
44	47	50	18	57	64	31	72	61
41	46	51	18	55	62	27	96	63
42	45	44	25	54	59	35	79	65
..	..	..	..	58	58	..	..	..
..	..	..	..	58	58	..	..	..
..	..	..	..	58	58	..	..	..
..	..	..	..	50	50	..	..	..
60	60	..	..	37	37	..	..	..
..	..	..	..	..	..	..	..	..
..	..	..	..	..	..	..	..	..
..	..	..	..	..	..	..	..	..
39	39	..	..	39	39	48	..	37
42	43	13	..	35	35	45	..	31
45	45	13	..	36	30	43	..	47
46	46	13	..	37	31	37	..	50
1	2	..	..	41	48	36	..	..
1	2	..	..	45	46	44	..	..
2	2	..	..	44	49	41	..	..
1	2	..	..	41	50	37	..	23
47	47	..	..	44	44	17	..	..
47	47	..	..	44	44	14	..	7
47	47	..	..	41	42	10	..	13
47	47	..	..	40	41	9	..	22
..	..	..	..	..	..	..	..	..
..	..	..	..	47	47	..	..	..
..	..	..	..	48	48	..	..	..
..	..	..	..	49	49	..	..	..
2	2	..	..	48	48	..	..	29
2	2	..	..	43	43	..	..	40
3	3	..	..	43	43	..	..	35
3	3	..	..	42	43	..	..	35
..	..	..	..	58	58	..	..	..
..	..	..	..	51	51	..	..	..
..	..	..	..	51	51	..	..	..
..	..	..	..	49	49	..	..	..
..	..	..	..	19	19	..	..	21
..	..	..	..	19	19	..	..	17
..	..	..	..	19	19	..	..	17
..	..	..	..	20	21	..	..	12
..	..	..	..	28	29	28	..	..
..	..	..	..	28	29	28	..	..
..	..	..	..	27	27	27	..	..
..	..	..	..	33	33	33	..	..
4	4	..	..	47	47	..	..	..
4	4	..	..	43	43	..	..	..
5	5	..	..	45	45	..	..	..
5	5	..	..	47	47	..	..	..

Table 31

Utilization of installed electricity capacity - by type

Kilowatt-hours per kilowatt divided by 8760

Country or area	Year	Total producers						
		Total	Combustible fuels	Hydro	Nuclear	Wind	Solar	Other
Turks and Caicos Islands	2011	38	38	..	..	..	..	..
	2012	34	34	..	..	..	..	..
	2013	34	34	..	..	..	..	..
	2014	33	33	..	..	..	..	..
United States	2011	47	43	39	92	30	12	78
	2012	46	43	34	90	27	13	66
	2013	46	43	33	95	32	14	63
	2014	46	43	31	96	33	17	58
United States Virgin Is.	2011	31	31	..	..	..	15	..
	2012	30	30	..	..	..	15	..
	2013	30	30	..	..	..	15	..
	2014	28	28	..	..	..	15	..
America, South	**2011**	**51**	**38**	**58**	**83**	**21**	**55**	**0**
	2012	**50**	**42**	**56**	**85**	**29**	**11**	**0**
	2013	**49**	**43**	**54**	**82**	**32**	**14**	**0**
	2014	**49**	**44**	**52**	**80**	**27**	**21**	**0**
Argentina	2011	44	46	36	71	4	5	..
	2012	44	47	34	72	31	8	..
	2013	41	41	38	70	27	21	..
	2014	45	43	47	65	43	23	..
Bolivia (Plur. State of)	2011	49	47	54	..	..	17	..
	2012	46	44	54	..	..	17	..
	2013	44	39	59	..	..	17	..
	2014	62	65	56	..	30	14	..
Brazil	2011	52	31	59	89	22	..	..
	2012	52	40	56	91	31	..	..
	2013	51	49	52	89	34	16	..
	2014	50	57	48	88	29	12	..
Chile	2011	43	44	40	..	21	..	..
	2012	44	47	38	..	23	0	..
	2013	45	50	37	..	21	1	..
	2014	36	34	41	..	23	23	..
Colombia	2011	50	39	57	..	26	..	..
	2012	52	46	55	..	35	..	..
	2013	53	59	51	..	37	..	..
	2014	54	57	53	..	37	..	..
Ecuador	2011	49	42	58	..	16	29	..
	2012	52	43	62	..	11	43	..
	2013	52	49	56	..	34	11	..
	2014	52	48	58	..	43	7	..
Falkland Is. (Malvinas)	2011	17	15	..	..	22	..	..
	2012	17	15	..	..	22	..	..
	2013	17	15	..	..	22	..	..
	2014	17	15	..	..	22	..	..
French Guiana	2011	40	33	45	..	..	..	..
	2012	36	24	54	..	..	17	..
	2013	35	29	47	..	..	15	..
	2014	36	29	48	..	..	18	..
Guyana	2011	30	30	..	..	..	..	..
	2012	30	30	..	..	..	..	..
	2013	29	29	..	..	..	..	..
	2014	29	29	..	..	..	..	..
Paraguay	2011	75	2	75	..	..	..	..
	2012	78	6	78	..	..	..	..
	2013	78	2	78	..	..	..	..
	2014	72	5	72	..	..	..	..

Utilization of installed electricity capacity - by type

Kilowatt-hours per kilowatt divided by 8760

Autoproducers				Main activity producers				
Total	Combustible fuels	Hydro	Other	Total	Combustible fuels	Hydro	Nuclear	Other
..	..	..	..	38	38	..	..	..
..	..	..	..	34	34	..	..	..
..	..	..	..	34	34	..	..	..
..	..	..	..	33	33	..	..	..
54	61	23	13	47	43	39	92	32
55	63	41	17	46	43	34	90	29
54	64	60	16	46	43	32	95	32
53	63	50	16	46	43	31	96	32
30	31	..	15	31	31	..	..	..
15	..	..	15	30	30	..	..	..
15	..	..	15	30	30	..	..	..
15	..	..	15	28	28	..	..	..
46	**43**	**58**	**5**	**51**	**37**	**58**	**83**	**21**
45	**43**	**52**	**9**	**51**	**41**	**56**	**85**	**28**
46	**45**	**51**	**3**	**49**	**42**	**54**	**82**	**31**
47	**46**	**51**	**11**	**49**	**44**	**52**	**80**	**27**
44	44	33	..	44	47	36	71	4
43	43	33	..	44	48	34	72	30
41	41	30	..	41	41	38	70	27
42	42	29	..	45	44	47	65	43
11	15	..	17	52	50	57	..	..
10	13	..	17	49	46	57	..	..
4	5	..	17	47	42	62	..	..
18	24	..	14	66	68	60	..	30
49	46	58	..	52	20	59	89	22
47	46	52	..	53	36	56	91	31
49	48	52	11	52	50	52	89	34
51	51	52	11	50	62	48	88	29
52	52	46	34	42	43	40	..	21
64	65	49	21	43	45	38	..	23
57	64	24	2	44	48	37	..	20
55	55	65	32	35	33	41	..	23
67	75	19	..	50	35	57	..	6
70	79	16	..	51	43	55	..	8
64	72	15	..	53	57	51	..	8
69	78	17	..	53	54	53	..	8
55	52	79	..	48	38	57	..	16
60	58	77	..	50	38	62	..	12
56	54	77	..	51	47	56	..	30
51	48	80	..	53	49	58	..	23
11	11	..	..	18	16	..	..	22
11	11	..	..	18	16	..	..	22
11	11	..	..	18	16	..	..	22
11	11	..	..	18	16	..	..	22
..	..	..	..	40	33	45	..	..
..	..	..	..	36	24	54	..	17
..	..	..	..	35	29	47	..	15
..	..	..	..	36	29	48	..	18
20	20	..	..	41	41	..	..	..
18	18	..	..	41	41	..	..	..
18	18	..	..	39	39	..	..	..
12	12	..	..	47	47	..	..	..
..	..	..	..	75	2	75	..	..
..	..	..	..	78	6	78	..	..
..	..	..	..	78	2	78	..	..
..	..	..	..	72	5	72	..	..

Table 31

Utilization of installed electricity capacity - by type

Kilowatt-hours per kilowatt divided by 8760

Country or area	Year	Total producers						
		Total	Combustible fuels	Hydro	Nuclear	Wind	Solar	Other
Peru	2011	51	38	71	..	20	..	..
	2012	47	33	72	..	20	9	..
	2013	45	32	72	..	20	28	..
	2014	46	36	69	..	20	24	..
Suriname	2011	52	48	57	..	..	..	..
	2012	53	47	60	..	..	..	..
	2013	58	82	29	..	..	..	..
	2014	58	66	48	..	..	..	..
Uruguay	2011	44	38	48	..	29	..	..
	2012	42	44	40	..	26	..	..
	2013	40	22	61	..	27	..	..
	2014	40	18	72	..	17	10	..
Venezuela (Bolivar. Rep.)	2011	52	35	65	..	..	..	..
	2012	48	32	64	..	..	..	..
	2013	44	27	64	..	..	..	..
	2014	46	28	67	..	..	..	..
Asia	**2011**	**47**	**51**	**32**	**53**	**18**	**9**	**139**
	2012	**46**	**50**	**35**	**43**	**18**	**9**	**140**
	2013	**46**	**51**	**35**	**43**	**21**	**9**	**73**
	2014	**45**	**50**	**35**	**44**	**18**	**10**	**92**
Afghanistan	2011	27	17	30	..	..	..	..
	2012	27	14	32	..	..	..	..
	2013	27	22	29	..	..	..	..
	2014	28	15	32	..	..	..	..
Armenia	2011	24	14	24	71	17	..	..
	2012	23	16	21	65	11	..	..
	2013	21	15	19	66	11	..	..
	2014	22	16	18	69	11	..	..
Azerbaijan	2011	36	38	31	..	..	..	..
	2012	41	45	20	..	..	..	..
	2013	36	40	16	..	4	6	..
	2014	38	43	14	..	8	9	..
Bahrain	2011	47	47	..	..	..	..	..
	2012	48	48	..	..	..	..	..
	2013	50	50	..	..	..	..	..
	2014	51	51	..	..	..	..	..
Bangladesh	2011	76	77	43	..	..	..	..
	2012	68	69	39	..	..	..	..
	2013	71	72	32	..	46	69	..
	2014	55	56	29	..	46	69	..
Bhutan	2011	54	0	54	..	..	..	..
	2012	52	0	52	..	..	..	..
	2013	58	0	59	..	..	..	..
	2014	49	0	50	..	..	..	..
Brunei Darussalam	2011	53	53	..	..	..	23	..
	2012	54	54	..	..	..	23	..
	2013	61	61	..	..	..	23	..
	2014	61	61	..	..	..	23	..
Cambodia	2011	21	32	2	..	..	17	..
	2012	28	29	26	..	..	17	..
	2013	18	18	17	..	..	17	..
	2014	23	24	23	..	..	17	..
China	2011	47	52	33	78	17	..	9,659
	2012	47	50	39	88	18	..	1,434
	2013	46	51	36	87	21	4	619
	2014	44	48	39	76	19	7	5,215

Utilization of installed electricity capacity - by type

Kilowatt-hours per kilowatt divided by 8760

Autoproducers				Main activity producers				
Total	Combustible fuels	Hydro	Other	Total	Combustible fuels	Hydro	Nuclear	Other
21	18	64	..	57	44	72	..	20
21	18	60	..	52	38	73	..	9
21	18	66	..	48	35	72	..	28
21	18	61	..	50	40	69	..	22
59	61	57	..	34	34	..	..	..
53	42	60	..	52	52	..	..	..
55	96	29	..	67	67	..	..	..
58	74	48	..	58	58	..	..	..
35	37	..	..	45	39	48	..	42
37	39	..	1	42	45	40	..	36
24	25	..	2	43	21	61	..	42
33	34	..	3	41	12	72	..	18
13	13	..	..	54	38	65	..	..
13	13	..	..	50	34	64	..	..
13	13	..	..	46	28	64	..	..
13	13	..	..	47	29	67	..	..
40	**42**	**19**	**15**	**48**	**53**	**32**	**53**	**24**
40	**42**	**18**	**14**	**47**	**51**	**35**	**43**	**23**
43	**47**	**18**	**14**	**46**	**52**	**35**	**43**	**20**
39	**43**	**17**	**13**	**45**	**51**	**35**	**44**	**18**
18	1	25	..	33	33	33	..	..
19	1	27	..	33	26	35	..	..
17	1	24	..	34	41	32	..	..
19	1	27	..	34	28	36	..	..
..	..	..	..	24	14	24	71	17
..	..	..	..	23	16	21	65	11
..	..	..	..	21	15	19	66	11
..	..	..	..	22	16	18	69	11
34	34	..	..	37	38	31	..	..
148	148	..	..	39	42	20	..	..
87	87	..	..	35	38	16	..	5
95	95	..	..	36	41	14	..	8
40	40	..	..	55	55	..	..	..
40	40	..	..	57	57	..	..	..
42	42	..	..	59	59	..	..	..
41	41	..	..	62	62	..	..	..
77	77	..	..	75	77	43	..	..
63	63	..	..	70	72	39	..	..
73	73	..	..	70	72	32	..	68
75	75	..	..	50	51	29	..	68
..	..	..	..	54	0	54	..	..
..	..	..	..	52	0	52	..	..
..	..	..	..	58	0	59	..	..
..	..	..	..	49	0	50	..	..
52	52	..	..	53	53	..	..	23
62	62	..	..	54	54	..	..	23
70	70	..	..	60	60	..	..	23
73	73	..	..	60	60	..	..	23
..	..	..	..	21	32	2	..	17
..	..	..	..	28	29	26	..	17
..	..	..	..	18	18	17	..	17
..	..	..	..	23	24	23	..	17
48	53	..	..	47	52	34	78	22
47	53	..	..	47	50	40	88	22
47	52	..	..	46	51	37	87	19
43	48	..	..	44	48	40	76	17

Table 31

Utilization of installed electricity capacity - by type

Kilowatt-hours per kilowatt divided by 8760

Country or area	Year	Total producers						
		Total	Combustible fuels	Hydro	Nuclear	Wind	Solar	Other
China, Hong Kong SAR	2011	35	35	..	..	..	..	..
	2012	35	35	..	..	..	..	..
	2013	35	35	..	..	..	..	..
	2014	36	36	..	..	..	..	..
China, Macao SAR	2011	20	20	..	..	..	..	..
	2012	13	13	..	..	..	..	..
	2013	9	9	..	..	..	..	..
	2014	15	15	..	..	..	..	..
Cyprus	2011	32	34	..	..	10	12	..
	2012	31	33	..	..	14	14	..
	2013	29	30	..	..	18	15	..
	2014	29	31	..	..	14	13	..
Georgia	2011	27	15	34	..	..	..	..
	2012	25	17	31	..	..	..	..
	2013	26	12	35	..	..	..	..
	2014	27	14	36	..	..	..	..
India	2011	44	48	30	77	17	..	..
	2012	42	46	33	79	18	..	..
	2013	46	49	40	82	18	17	..
	2014	47	53	37	71	12	15	..
Indonesia	2011	44	43	32	..	6	11	91
	2012	42	43	29	..	6	8	75
	2013	45	44	38	..	3	7	72
	2014	42	41	40	..	0	8	72
Iran (Islamic Rep. of)	2011	39	43	16	4	25	6	36
	2012	39	43	15	21	22	7	38
	2013	40	43	16	51	39	8	34
	2014	43	46	18	50	34	..	..
Iraq	2011	37	40	19	..	..	..	..
	2012	31	33	20	..	..	..	..
	2013	25	25	22	..	..	..	..
	2014	30	32	13	..	..	..	..
Israel	2011	49	50	47	..	13	12	7
	2012	50	50	54	..	11	18	..
	2013	47	48	32	..	11	12	..
	2014	43	44	15	..	11	14	..
Japan	2011	42	54	22	24	22	12	57
	2012	41	57	20	4	21	12	58
	2013	40	56	20	2	18	12	58
	2014	38	54	20	0	21	12	58
Jordan	2011	48	48	52	..	24	..	..
	2012	54	54	58	..	24	..	..
	2013	58	58	52	..	24	..	..
	2014	50	50	55	..	20	..	..
Kazakhstan	2011	49	50	41	..	..	..	..
	2012	50	51	39	..	17	..	..
	2013	53	54	40	..	14	11	..
	2014	48	49	38	..	15	11	..
Korea, Dem.Ppl's.Rep.	2011	23	15	30	..	..	..	..
	2012	23	15	31	..	..	..	..
	2013	22	11	32	..	..	..	..
	2014	22	12	30	..	..	..	..
Korea, Republic of	2011	71	70	14	94	23	14	54
	2012	69	72	14	83	23	12	48
	2013	68	72	15	76	23	12	31
	2014	63	63	14	86	21	12	38

Utilization of installed electricity capacity - by type

Kilowatt-hours per kilowatt divided by 8760

Autoproducers				Main activity producers				
Total	Combustible fuels	Hydro	Other	Total	Combustible fuels	Hydro	Nuclear	Other
..	..	..	..	35	35	..	..	..
..	..	..	..	35	35	..	..	..
..	..	..	..	35	35	..	..	..
..	..	..	..	36	36	..	..	..
70	70	..	..	18	18	..	..	..
77	77	..	..	9	9	..	..	..
85	85	..	..	5	5	..	..	..
94	94	..	..	11	11	..	..	..
21	21	..	23	33	35	..	..	10
19	18	..	23	31	33	..	..	14
19	18	..	23	29	30	..	..	17
21	21	..	23	29	31	..	..	13
..	..	..	..	27	15	34	..	..
..	..	..	..	25	17	31	..	..
..	..	..	..	26	12	35	..	..
..	..	..	..	27	14	36	..	..
40	40	43	..	45	50	30	77	17
40	40	43	..	43	47	33	79	18
57	59	24	18	45	48	40	82	18
41	43	20	12	48	55	37	71	13
56	56	48	14	42	41	30	..	90
57	58	50	14	39	40	26	..	74
60	59	89	8	42	41	33	..	72
58	56	91	0	39	38	35	..	72
22	22	..	..	40	45	16	4	26
22	22	..	..	40	45	15	21	23
13	13	..	..	42	46	16	51	39
13	13	..	..	45	50	18	50	31
..	..	..	..	37	40	19	..	..
..	..	..	..	31	33	20	..	..
..	..	..	..	25	25	22	..	..
..	..	..	..	30	32	13	..	..
30	45	29	11	50	50	55	..	4
26	30	23	18	51	51	66	..	11
29	50	29	12	48	48	33	..	11
33	62	29	14	44	44	11	..	11
30	31	47	15	46	62	19	24	46
29	30	43	15	44	67	17	4	47
27	30	44	13	44	66	18	2	47
25	29	45	13	42	63	18	0	50
15	15	..	..	50	50	52	..	24
14	14	..	..	56	56	58	..	24
16	16	..	..	61	61	52	..	24
22	22	..	..	51	51	55	..	20
..	..	..	..	49	50	41	..	..
..	..	..	..	50	51	39	..	17
11	..	..	11	53	54	40	..	14
11	..	..	11	48	49	38	..	15
..	..	..	..	23	15	30	..	..
..	..	..	..	23	15	31	..	..
..	..	..	..	22	11	32	..	..
..	..	..	..	22	12	30	..	..
59	61	..	13	71	71	14	94	21
52	54	..	7	71	74	14	83	21
51	54	..	7	69	74	15	76	20
49	53	..	7	64	64	14	86	19

Table 31

Utilization of installed electricity capacity - by type

Kilowatt-hours per kilowatt divided by 8760

Country or area	Year	Total producers						
		Total	Combustible fuels	Hydro	Nuclear	Wind	Solar	Other
Kuwait	2011	44	44	..	..	..	..	..
	2012	48	48	..	..	..	..	..
	2013	44	44	..	..	..	..	..
	2014	47	47	..	..	..	..	..
Kyrgyzstan	2011	45	15	53	..	..	..	..
	2012	45	14	53	..	..	..	..
	2013	41	13	49	..	..	..	..
	2014	43	18	50	..	..	..	..
Lao People's Dem. Rep.	2011	16	68	15	..	..	13	..
	2012	14	68	13	..	..	13	..
	2013	16	68	15	..	..	13	..
	2014	15	68	15	..	..	13	..
Lebanon	2011	57	59	34	..	..	..	..
	2012	52	53	42	..	..	..	..
	2013	61	62	53	..	..	..	..
	2014	63	68	8	..	..	..	..
Malaysia	2011	50	52	29	..	..	10	..
	2012	53	55	31	..	..	10	..
	2013	50	53	31	..	..	10	..
	2014	56	61	32	..	..	16	..
Maldives	2011	39	40	..	..	..	4	..
	2012	41	42	..	..	..	7	..
	2013	41	43	..	..	..	5	..
	2014	14	14	..	..	..	5	..
Mongolia	2011	41	41	..	..	..	..	..
	2012	44	44	..	..	..	..	..
	2013	53	53	..	..	..	..	..
	2014	56	56	..	..	..	..	..
Myanmar	2011	33	37	32	..	..	..	..
	2012	34	40	32	..	..	..	..
	2013	34	34	34	..	..	..	..
	2014	34	38	32	..	..	..	..
Nepal	2011	52	0	56	..	..	..	..
	2012	53	2	57	..	..	..	..
	2013	46	2	50	..	..	..	..
	2014	48	2	51	..	..	..	..
Oman	2011	51	51	..	..	..	..	..
	2012	49	49	..	..	..	..	..
	2013	45	45	..	..	..	..	..
	2014	40	40	..	..	..	..	..
Other Asia	2011	59	60	17	93	30	3	..
	2012	59	60	21	90	25	14	..
	2013	59	60	21	92	30	10	..
	2014	61	64	18	94	27	10	..
Pakistan	2011	48	45	50	76	..	..	..
	2012	48	46	50	69	..	..	..
	2013	50	48	53	77	29	..	..
	2014	51	49	52	77	29	..	..
Philippines	2011	49	52	32	..	9	7	64
	2012	49	52	33	..	6	5	63
	2013	50	54	32	..	5	5	59
	2014	49	55	29	..	4	5	61
Qatar	2011	40	40	..	..	..	..	..
	2012	45	45	..	..	..	..	..
	2013	45	45	..	..	..	..	..
	2014	50	50	..	..	..	..	..

Utilization of installed electricity capacity - by type

Kilowatt-hours per kilowatt divided by 8760

Autoproducers				Main activity producers				
Total	Combustible fuels	Hydro	Other	Total	Combustible fuels	Hydro	Nuclear	Other
..	..	..	..	45	45	..	..	..
..	..	..	..	49	49	..	..	..
..	..	..	..	45	45	..	..	..
..	..	..	..	47	47	..	..	..
..	..	..	..	45	15	53	..	..
..	..	..	..	45	14	53	..	..
40	40	..	..	41	13	49	..	..
46	46	..	..	43	18	50	..	..
13	..	..	13	16	68	15	..	..
13	..	..	13	14	68	13	..	..
13	..	..	13	16	68	15	..	..
13	..	..	13	15	68	15	..	..
57	57	..	..	57	61	34	..	..
72	72	..	..	44	44	42	..	..
84	84	..	..	52	52	53	..	..
67	67	..	..	61	68	8	..	..
20	20	..	..	53	56	29	..	10
22	22	..	..	56	60	31	..	10
19	19	..	..	53	57	31	..	10
23	23	..	..	59	65	32	..	16
..	..	..	..	39	40	..	..	4
..	..	..	..	41	42	..	..	7
..	..	..	..	41	43	..	..	5
..	..	..	..	14	14	..	..	5
21	21	..	..	44	44	..	..	..
28	28	..	..	46	46	..	..	..
91	91	..	..	48	48	..	..	..
86	86	..	..	52	52	..	..	..
..	..	..	..	33	37	32	..	..
4	..	4	..	35	40	33	..	..
27	26	30	..	34	36	34	..	..
54	57	45	..	31	30	31	..	..
4	..	4	..	65	0	71	..	..
6	..	6	..	65	2	71	..	..
10	..	10	..	55	2	60	..	..
9	..	9	..	59	2	64	..	..
..	..	..	..	51	51	..	..	..
..	..	..	..	49	49	..	..	..
..	..	..	..	45	45	..	..	..
..	..	..	..	40	40	..	..	..
58	58	..	..	59	60	17	93	25
54	54	..	..	60	62	21	90	22
57	57	..	..	59	61	21	92	22
59	59	..	..	62	65	18	94	19
54	54	..	..	44	35	50	76	..
56	56	..	..	44	35	50	69	..
58	58	..	..	46	37	53	77	29
59	59	..	..	46	38	52	77	29
15	19	12	..	49	52	32	..	60
15	19	12	..	49	52	33	..	59
15	19	12	..	50	54	32	..	55
15	19	12	..	49	55	29	..	51
49	49	..	..	12	12	..	..	..
55	55	..	..	16	16	..	..	..
55	55	..	..	16	16	..	..	..
61	61	..	..	18	18	..	..	..

Table 31

Utilization of installed electricity capacity - by type

Kilowatt-hours per kilowatt divided by 8760

Country or area	Year	Total producers						
		Total	Combustible fuels	Hydro	Nuclear	Wind	Solar	Other
Saudi Arabia	2011	56	56	..	..	..	..	..
	2012	50	50	..	..	..	11	..
	2013	55	55	..	..	..	11	..
	2014	54	54	..	..	..	11	..
Singapore	2011	52	52	..	..	..	..	..
	2012	52	52	..	..	..	..	..
	2013	48	48	..	..	..	..	..
	2014	43	43	..	..	..	..	..
Sri Lanka	2011	42	46	37	..	29	16	..
	2012	40	56	24	..	22	17	..
	2013	42	35	49	..	33	11	..
	2014	36	41	31	..	24	6	..
State of Palestine	2011	46	46	..	..	..	..	..
	2012	38	38	..	..	..	..	..
	2013	44	44	..	..	..	..	..
	2014	27	27	..	..	..	..	..
Syrian Arab Republic	2011	58	60	44	..	..	..	..
	2012	43	43	43	..	..	..	..
	2013	36	36	40	..	..	..	..
	2014	26	26	23	..	..	..	..
Tajikistan	2011	36	1	39	..	..	..	..
	2012	31	3	33	..	..	..	..
	2013	38	2	40	..	..	..	..
	2014	36	17	37	..	..	..	..
Thailand	2011	35	36	26	..	11	10	76
	2012	38	39	28	..	11	10	38
	2013	38	41	19	..	12	8	38
	2014	39	41	18	..	16	10	38
Timor-Leste	2011	32	32	..	..	..	..	..
	2012	28	28	..	..	..	..	..
	2013	24	24	..	..	..	..	..
	2014	14	14	..	..	..	..	..
Turkey	2011	49	58	35	..	31	..	69
	2012	48	57	34	..	30	..	63
	2013	43	51	30	..	31	..	51
	2014	39	51	19	..	27	5	65
Turkmenistan	2011	61	61	0	..	..	..	..
	2012	60	60	0	..	..	..	..
	2013	57	57	..	..	..	..	..
	2014	58	58	..	..	..	..	..
United Arab Emirates	2011	43	43	..	..	..	..	..
	2012	45	45	..	..	..	..	..
	2013	46	46	..	..	..	..	..
	2014	46	46	..	..	..	..	..
Uzbekistan	2011	48	46	64	..	..	..	..
	2012	47	43	69	..	..	..	..
	2013	49	45	71	..	..	..	..
	2014	50	46	73	..	..	..	..
Viet Nam	2011	47	41	59	..	32	..	..
	2012	48	42	58	..	28	..	..
	2013	52	45	63	..	28	..	..
	2014	49	52	45	..	31	..	..
Yemen	2011	46	46	..	..	..	..	..
	2012	53	53	..	..	..	..	..
	2013	63	63	..	..	..	..	..
	2014	57	57	..	..	..	..	..

Utilization of installed electricity capacity - by type

Kilowatt-hours per kilowatt divided by 8760

Autoproducers				Main activity producers				
Total	Combustible fuels	Hydro	Other	Total	Combustible fuels	Hydro	Nuclear	Other
56	56	..	..	56	56	..	..	..
50	50	..	..	50	50	..	..	11
69	69	..	..	51	51	..	..	11
71	71	..	..	49	49	..	..	11
49	49	..	..	52	52	..	..	..
65	65	..	..	52	52	..	..	..
64	64	..	..	47	47	..	..	..
68	68	..	..	42	42	..	..	..
14	0	..	17	42	46	37	..	28
14	0	..	17	40	56	24	..	22
15	0	..	21	42	35	49	..	31
15	..	..	21	36	41	31	..	22
..	..	..	..	46	46	..	..	..
..	..	..	..	38	38	..	..	..
..	..	..	..	44	44	..	..	..
..	..	..	..	27	27	..	..	..
12	12	..	..	65	68	44	..	..
9	9	..	..	48	49	43	..	..
6	6	..	..	40	40	40	..	..
5	5	..	..	28	30	23	..	..
..	..	..	..	36	1	39	..	..
..	..	..	..	31	3	33	..	..
..	..	..	..	38	2	40	..	..
..	..	..	..	36	17	37	..	..
11	11	..	..	48	50	26	..	11
11	11	..	..	51	54	28	..	11
21	20	..	..	46	51	19	..	3
25	24	..	..	44	49	18	..	4
..	..	..	..	32	32	..	..	..
..	..	..	..	28	28	..	..	..
..	..	..	..	24	24	..	..	..
..	..	..	..	14	14	..	..	..
46	51	26	23	50	58	35	..	34
47	50	35	23	48	58	34	..	32
44	47	23	54	43	51	31	..	33
15	16	8	82	41	54	19	..	31
..	..	..	..	61	61	0	..	..
..	..	..	..	60	60	0	..	..
..	..	..	..	57	57	..	..	..
..	..	..	..	58	58	..	..	..
..	..	..	..	43	43	..	..	..
..	..	..	..	45	45	..	..	..
..	..	..	..	46	46	..	..	..
..	..	..	..	46	46	..	..	..
15	15	..	..	49	46	64	..	..
14	14	..	..	48	44	69	..	..
14	14	..	..	49	45	71	..	..
15	15	..	..	50	46	73	..	..
15	15	..	..	50	46	59	..	32
16	16	4	4	52	47	58	..	33
17	17	4	4	56	50	63	..	33
13	14	3	4	53	61	46	..	33
26	26	..	..	48	48	..	..	..
33	33	..	..	54	54	..	..	..
24	24	..	..	67	67	..	..	..
23	23	..	..	60	60	..	..	..

Table 31

Utilization of installed electricity capacity - by type

Kilowatt-hours per kilowatt divided by 8760

Country or area	Year	Total producers						
		Total	Combustible fuels	Hydro	Nuclear	Wind	Solar	Other
Europe	**2011**	**43**	**44**	**32**	**79**	**22**	**10**	**43**
	2012	**42**	**42**	**34**	**81**	**22**	**11**	**46**
	2013	**41**	**41**	**36**	**80**	**23**	**12**	**44**
	2014	**39**	**38**	**35**	**81**	**22**	**12**	**45**
Albania	2011	30	6	32	..	..	..	..
	2012	31	..	33	..	..	..	..
	2013	41	..	45	..	..	..	..
	2014	29	..	31	..	..	..	..
Andorra	2011	33	28	34	..	..	..	..
	2012	32	27	33	..	..	..	..
	2013	42	29	44	..	..	..	..
	2014	46	31	49	..	..	..	..
Austria	2011	33	36	33	..	20	6	11
	2012	36	31	42	..	21	11	11
	2013	33	26	40	..	22	11	0
	2014	31	23	38	..	21	11	0
Belarus	2011	44	44	31	..	3	..	..
	2012	39	39	24	..	20	..	..
	2013	39	39	47	..	27	..	..
	2014	40	40	42	..	37	16	..
Belgium	2011	51	41	11	93	25	10	..
	2012	46	43	13	78	23	10	..
	2013	45	42	14	82	23	10	31
	2014	40	39	12	65	27	11	23
Bosnia and Herzegovina	2011	41	57	24	..	..	..	..
	2012	37	52	23	..	..	..	..
	2013	46	53	39	..	..	..	..
	2014	43	53	32	..	..	..	..
Bulgaria	2011	57	75	14	98	18	7	..
	2012	46	59	15	95	21	9	..
	2013	43	54	17	82	23	15	..
	2014	48	61	18	92	22	14	..
Croatia	2011	32	38	28	..	19	..	..
	2012	29	33	27	..	21	6	..
	2013	37	30	45	..	23	7	..
	2014	35	22	49	..	24	12	..
Czechia	2011	50	52	14	81	21	13	..
	2012	49	50	15	86	18	12	..
	2013	47	47	18	82	21	11	..
	2014	45	44	15	81	20	12	..
Denmark	2011	30	30	22	..	28	10	..
	2012	25	24	22	..	28	3	..
	2013	29	31	16	..	26	10	..
	2014	27	26	19	..	31	11	..
Estonia	2011	52	54	68	..	23	..	..
	2012	47	50	60	..	19	..	..
	2013	52	55	37	..	24	..	..
	2014	46	49	62	..	20	..	..
Faeroe Islands	2011	30	29	33	..	26	..	..
	2012	30	32	29	..	19	..	..
	2013	29	32	26	..	23	..	..
	2014	28	25	35	..	24	..	..
Finland	2011	54	44	44	97	28	16	..
	2012	51	36	60	96	22	16	..
	2013	49	38	45	98	20	15	..
	2014	48	35	47	98	20	8	..

Utilization of installed electricity capacity - by type

Kilowatt-hours per kilowatt divided by 8760

Autoproducers				Main activity producers				
Total	Combustible fuels	Hydro	Other	Total	Combustible fuels	Hydro	Nuclear	Other
45	**50**	**41**	**13**	**43**	**43**	**32**	**79**	**19**
44	**50**	**43**	**13**	**42**	**42**	**34**	**81**	**19**
40	**46**	**45**	**13**	**41**	**40**	**35**	**80**	**19**
36	**42**	**45**	**13**	**40**	**37**	**35**	**81**	**19**
..	..	..	..	30	7	32	..	..
..	..	..	..	31	..	33	..	..
..	..	..	..	42	..	45	..	..
..	..	..	..	29	..	31	..	..
..	..	..	..	33	28	34	..	..
..	..	..	..	32	27	33	..	..
..	..	..	..	42	29	44	..	..
..	..	..	..	46	31	49	..	..
45	53	21	..	32	31	34	..	17
42	50	11	..	36	25	43	..	19
39	47	12	..	32	20	41	..	19
39	47	12	..	30	16	40	..	18
63	64	8	6	43	43	46	..	..
60	60	10	12	38	38	28	..	30
48	48	12	12	38	38	57	..	46
49	50	12	24	39	39	49	..	42
31	55	..	10	54	39	11	93	25
25	54	..	10	50	41	13	78	23
24	52	..	10	51	40	14	82	23
25	53	..	11	44	37	12	65	27
16	16	..	..	42	62	24	..	..
17	17	..	..	38	56	23	..	..
18	18	..	..	48	57	39	..	..
20	20	..	..	44	57	32	..	..
36	31	..	..	57	75	14	98	16
49	46	..	..	46	59	15	95	14
47	44	..	..	43	54	17	82	18
48	45	..	..	48	61	18	92	17
28	28	11	..	33	40	28	..	19
22	22	29	..	29	34	27	..	21
24	24	29	..	38	30	46	..	22
25	25	40	..	35	21	49	..	23
49	51	33	..	50	52	12	81	14
48	50	33	..	49	50	13	86	13
49	51	38	..	47	46	17	82	12
51	53	34	..	44	42	13	81	13
38	39	..	10	29	30	22	..	28
23	35	..	3	25	24	22	..	28
24	38	..	10	29	31	16	..	26
25	40	..	11	27	25	19	..	31
63	63	..	..	52	54	68	..	23
59	58	..	..	47	49	60	..	19
58	58	..	57	52	55	37	..	24
27	28	..	23	46	49	59	..	20
..	..	..	..	30	29	33	..	26
..	..	..	..	30	32	29	..	19
..	..	..	..	29	32	26	..	23
..	..	..	..	28	25	35	..	24
60	59	56	36	53	41	44	97	31
58	55	85	14	50	31	59	96	23
57	55	60	8	48	34	45	98	20
56	55	57	8	47	31	46	98	20

Table 31

Utilization of installed electricity capacity - by type

Kilowatt-hours per kilowatt divided by 8760

Country or area	Year	Total producers						
		Total	Combustible fuels	Hydro	Nuclear	Wind	Solar	Other
France	2011	50	22	22	80	21	8	4
	2012	50	23	29	77	23	12	8
	2013	51	23	34	77	22	12	8
	2014	50	16	31	79	22	12	8
Germany	2011	40	52	23	60	19	9	56
	2012	41	54	28	94	18	9	63
	2013	39	53	29	92	17	10	47
	2014	36	48	26	92	17	11	51
Gibraltar	2011	45	45	..	..	..	..	..
	2012	47	47	..	..	..	..	..
	2013	50	50	..	..	..	..	..
	2014	53	53	..	..	..	..	..
Greece	2011	41	53	15	..	23	11	..
	2012	39	52	16	..	25	13	..
	2013	35	44	23	..	26	16	..
	2014	30	40	16	..	21	17	..
Guernsey	2011	8	8	..	..	..	..	..
	2012	7	7	..	..	..	..	..
	2013	29	29	..	..	..	..	..
	2014	22	22	..	..	..	..	..
Hungary	2011	43	31	46	90	22	3	..
	2012	42	29	43	90	27	8	..
	2013	41	27	43	88	25	8	37
	2014	38	23	60	89	23	8	48
Iceland	2011	74	0	76	..	..	..	81
	2012	75	0	75	..	..	..	89
	2013	75	1	74	..	17	..	90
	2014	75	0	74	..	30	..	90
Ireland	2011	37	40	15	..	31	11	..
	2012	37	41	22	..	26	11	..
	2013	34	37	20	..	27	11	..
	2014	33	36	21	..	27	11	..
Isle of Man	2011	32	32	35	..	..	..	..
	2012	19	19	50	..	..	..	..
	2013	31	31	46	..	..	..	..
	2014	31	31	46	..	..	..	..
Italy	2011	29	34	25	..	16	10	71
	2012	27	32	23	..	19	13	70
	2013	27	29	28	..	20	13	70
	2014	26	28	31	..	20	14	68
Jersey	2011	2	2	..	..	..	..	..
	2012	3	3	..	..	..	..	..
	2013	16	16	..	..	..	..	..
	2014	6	6	..	..	..	..	..
Latvia	2011	27	36	21	..	22	..	..
	2012	26	25	27	..	22	..	..
	2013	24	28	21	..	20	..	..
	2014	20	26	14	..	23	..	..
Liechtenstein	2011	69	45	71	..	..	71	..
	2012	82	42	84	..	..	99	..
	2013	78	39	79	..	..	97	..
	2014	35	29	25	..	..	89	..
Lithuania	2011	14	12	14	..	27	..	..
	2012	13	12	12	..	22	3	..
	2013	12	10	14	..	25	8	..
	2014	12	9	14	..	25	12	..

Utilization of installed electricity capacity - by type

Kilowatt-hours per kilowatt divided by 8760

Autoproducers				Main activity producers				
Total	Combustible fuels	Hydro	Other	Total	Combustible fuels	Hydro	Nuclear	Other
22	37	16	7	52	20	23	80	19
28	46	20	10	52	19	29	77	21
26	44	41	12	53	19	34	77	20
26	52	43	11	51	11	31	79	20
59	59	54	59	39	52	23	60	15
72	72	55	72	39	52	28	94	14
53	53	49	51	38	53	29	92	13
53	53	42	53	35	48	26	92	14
..	..	..	..	45	45	..	..	..
..	..	..	..	47	47	..	..	..
..	..	..	..	50	50	..	..	..
..	..	..	..	53	53	..	..	..
54	54	..	..	41	53	15	..	20
51	51	..	..	39	52	16	..	19
41	41	..	..	34	44	23	..	20
41	41	..	..	30	40	16	..	19
..	..	..	..	8	8	..	..	..
..	..	..	..	7	7	..	..	..
..	..	..	..	29	29	..	..	..
..	..	..	..	22	22	..	..	..
46	48	..	3	43	30	46	90	22
52	57	..	8	42	29	43	90	27
35	41	..	14	41	26	43	88	25
28	35	..	14	38	22	60	89	23
..	..	..	..	74	0	76	..	81
..	..	..	..	75	0	75	..	89
..	..	..	..	75	1	74	..	90
..	..	..	..	75	0	74	..	90
68	69	..	11	35	38	15	..	31
73	73	..	11	35	39	22	..	26
69	70	..	11	33	35	20	..	27
69	69	..	11	32	34	21	..	27
..	..	..	..	32	32	35	..	..
..	..	..	..	19	19	50	..	..
..	..	..	..	31	31	46	..	..
..	..	..	..	31	31	46	..	..
47	46	59	45	28	33	25	..	15
44	44	51	42	27	32	23	..	17
43	42	60	32	26	29	28	..	17
44	44	71	29	26	27	31	..	18
54	54	..	..	1	1	..	..	..
71	71	..	..	1	1	..	..	..
74	74	..	..	8	8	..	..	..
15	15	..	..	5	5	..	..	..
37	43	21	0	27	36	21	..	23
50	52	27	..	26	24	27	..	21
56	57	17	..	23	27	21	..	20
68	68	40	..	19	25	14	..	23
..	..	..	..	69	45	71	..	71
..	..	..	..	82	42	84	..	99
..	..	..	..	78	39	79	..	97
..	..	..	..	35	29	25	..	89
25	16	..	..	13	12	14	..	27
33	26	..	..	12	10	12	..	22
26	18	..	..	11	9	14	..	21
31	23	..	..	10	8	14	..	23

Table 31

Utilization of installed electricity capacity - by type

Kilowatt-hours per kilowatt divided by 8760

Country or area	Year	Total producers						
		Total	Combustible fuels	Hydro	Nuclear	Wind	Solar	Other
Luxembourg	2011	24	55	11	..	16	7	..
	2012	24	56	12	..	15	6	..
	2013	18	34	12	..	16	9	..
	2014	17	35	10	..	16	10	..
Malta	2011	43	43	..	..	..	21	..
	2012	41	42	..	..	..	16	..
	2013	38	41	..	..	..	7	..
	2014	37	40	..	..	..	10	..
Montenegro	2011	37	87	21	..	..	..	..
	2012	37	74	26	..	..	..	..
	2013	52	78	43	..	..	..	..
	2014	40	72	30	..	..	11	..
Netherlands	2011	46	48	18	93	25	8	24
	2012	39	40	32	88	23	7	29
	2013	38	40	35	68	24	7	37
	2014	37	39	35	96	23	9	40
Norway	2011	45	35	46	..	29	..	8
	2012	51	24	53	..	25	..	0
	2013	46	22	47	..	26	..	0
	2014	48	23	50	..	29	..	0
Poland	2011	54	59	13	..	20	..	46
	2012	52	58	12	..	21	11	46
	2013	52	59	15	..	20	6	46
	2014	50	57	13	..	23	3	46
Portugal	2011	30	35	25	..	25	19	96
	2012	27	36	13	..	27	19	67
	2013	31	33	30	..	30	18	90
	2014	32	33	33	..	28	17	90
Republic of Moldova	2011	23	22	54	..	..	..	..
	2012	22	22	24	..	..	..	..
	2013	21	21	32	..	11	..	..
	2014	26	26	42	..	11	11	..
Romania	2011	35	34	26	95	16	11	..
	2012	31	31	22	93	17	2	..
	2013	29	27	26	94	19	6	..
	2014	31	27	33	94	22	13	..
Russian Federation	2011	54	54	40	81	6	..	74
	2012	53	52	39	80	6	..	67
	2013	50	49	42	78	6	..	64
	2014	47	44	40	82	11	8	67
Serbia	2011	51	60	37	..	..	..	..
	2012	48	55	40	..	..	..	..
	2013	52	60	44	..	..	..	..
	2014	46	49	47	..	..	3	..
Slovakia	2011	39	29	19	91	19	9	47
	2012	39	28	20	91	23	9	41
	2013	39	24	23	93	14	13	32
	2014	39	25	20	91	23	13	54
Slovenia	2011	56	55	34	103	..	13	..
	2012	54	54	37	92	..	13	..
	2013	54	51	43	88	11	13	..
	2014	58	41	56	106	11	13	..
Spain	2011	30	30	20	88	23	19	22
	2012	30	30	15	94	25	21	22
	2013	29	24	24	93	28	21	..
	2014	28	23	25	88	26	22	..

Utilization of installed electricity capacity - by type

Kilowatt-hours per kilowatt divided by 8760

Autoproducers				Main activity producers				
Total	Combustible fuels	Hydro	Other	Total	Combustible fuels	Hydro	Nuclear	Other
24	35	23	7	24	57	11	..	16
19	35	29	6	25	58	12	..	15
19	36	34	9	18	34	12	..	16
18	33	34	10	17	36	10	..	16
23	19	..	29	43	43	..	..	18
37	34	..	39	41	42	..	..	9
17	23	..	17	39	41	..	..	0
37	27	..	39	37	40	..	..	..
..	..	..	..	37	87	21	..	..
..	..	..	..	37	74	26	..	..
..	..	..	..	52	78	43	..	..
11	..	..	11	40	72	30	..	..
47	50	..	20	46	47	18	93	26
46	51	..	16	38	38	32	88	24
44	51	..	14	36	37	35	68	24
37	44	..	14	37	37	35	96	23
52	50	55	..	45	7	46	..	28
47	33	62	..	52	8	53	..	25
47	31	60	..	46	9	47	..	26
42	30	51	..	49	10	50	..	29
55	54	23	30	54	59	13	..	20
55	54	23	34	52	58	12	..	21
56	56	34	26	52	59	15	..	20
55	55	34	6	50	57	13	..	23
49	50	27	14	28	32	25	..	25
45	47	19	16	25	33	13	..	27
50	53	38	15	29	27	30	..	30
46	49	44	17	30	28	33	..	28
2	2	..	..	32	31	54	..	..
3	3	..	..	30	30	24	..	..
4	4	..	..	28	28	32	..	11
11	11	..	11	29	29	42	..	11
41	45	20	11	34	33	26	95	16
24	29	19	5	32	31	22	93	19
24	29	28	6	30	27	26	94	18
22	25	41	11	33	28	33	94	20
42	42	23	..	55	55	41	81	66
42	42	38	..	53	54	39	80	60
37	38	37	..	52	51	42	78	58
30	30	36	..	49	46	40	82	20
75	75	..	..	50	60	37	..	..
56	56	..	..	48	55	40	..	..
66	66	..	..	52	60	44	..	..
94	94	..	..	46	49	47	..	0
28	41	40	11	41	27	19	91	9
28	40	35	11	40	25	20	91	10
26	36	40	13	41	22	23	93	13
30	40	44	15	40	21	20	91	13
22	50	12	13	59	55	36	103	11
21	72	12	13	57	53	40	92	14
21	67	16	13	57	51	46	88	11
22	62	22	13	62	40	59	106	13
64	65	49	23	28	25	20	88	22
65	66	26	23	28	25	15	94	24
60	61	54	4	27	19	24	93	26
50	51	61	5	26	19	25	88	25

Table 31

Utilization of installed electricity capacity - by type

Kilowatt-hours per kilowatt divided by 8760

Country or area	Year	Total producers						
		Total	Combustible fuels	Hydro	Nuclear	Wind	Solar	Other
Sweden	2011	49	30	46	74	25	10	..
	2012	50	22	55	77	23	9	..
	2013	46	23	43	81	27	9	..
	2014	44	17	46	78	25	9	..
Switzerland	2011	40	44	28	93	17	8	..
	2012	43	42	33	89	21	8	..
	2013	42	42	33	91	17	8	..
	2014	43	41	33	95	19	9	..
T.F.Yug.Rep. Macedonia	2011	47	56	29	..	..	6	..
	2012	42	54	20	..	..	9	..
	2013	40	46	29	..	..	15	..
	2014	34	43	22	..	..	11	..
Ukraine	2011	41	31	23	74	7	2	..
	2012	41	32	23	74	13	12	..
	2013	40	31	30	69	20	11	..
	2014	37	27	18	73	31	12	..
United Kingdom	2011	45	44	22	74	28	3	11
	2012	43	42	21	81	25	9	15
	2013	43	43	20	81	29	8	23
	2014	40	41	22	73	28	9	8
Oceania	**2011**	**47**	**51**	**34**	**..**	**33**	**13**	**93**
	2012	**46**	**51**	**31**	**..**	**32**	**12**	**94**
	2013	**45**	**50**	**36**	**..**	**29**	**13**	**88**
	2014	**44**	**48**	**37**	**..**	**32**	**14**	**83**
American Samoa	2011	44	44	..	..	..	..	..
	2012	43	44	..	..	..	13	..
	2013	43	44	..	..	..	13	..
	2014	43	44	..	..	..	13	..
Australia	2011	46	52	22	..	33	13	11
	2012	45	52	18	..	31	12	11
	2013	44	50	26	..	28	13	11
	2014	43	48	26	..	31	14	11
Cook Islands	2011	48	48	..	..	..	..	..
	2012	47	47	..	..	..	..	..
	2013	46	46	..	..	..	..	..
	2014	45	45	..	..	..	..	..
Fiji	2011	35	32	40	..	6	..	..
	2012	32	22	44	..	8	..	..
	2013	32	23	45	..	6	..	..
	2014	32	32	34	..	5	..	..
French Polynesia	2011	38	34	54	..	..	14	..
	2012	39	36	50	..	..	25	..
	2013	37	35	47	..	..	30	..
	2014	36	32	52	..	..	42	..
Guam	2011	38	38	..	..	..	..	..
	2012	37	37	..	..	..	..	..
	2013	36	36	..	..	..	..	..
	2014	36	36	..	..	..	..	..
Kiribati	2011	46	50	..	..	..	..	..
	2012	46	50	..	..	..	..	..
	2013	46	50	..	..	..	..	..
	2014	46	51	..	..	..	..	..
Marshall Islands	2011	31	31	..	..	..	..	..
	2012	27	27	..	..	..	8	..
	2013	29	29	..	..	..	8	..
	2014	29	29	..	..	..	5	..

Utilization of installed electricity capacity - by type

Kilowatt-hours per kilowatt divided by 8760

Autoproducers				Main activity producers				
Total	Combustible fuels	Hydro	Other	Total	Combustible fuels	Hydro	Nuclear	Other
49	49	57	..	49	25	46	74	25
58	58	65	..	50	16	55	77	23
56	56	13	..	46	16	43	81	27
55	55	15	..	44	11	46	78	25
44	47	58	8	40	37	27	93	17
40	46	55	8	43	32	32	89	21
35	47	55	8	43	27	32	91	17
33	49	59	9	44	20	32	95	19
8	8	..	..	47	56	29	..	6
8	8	..	..	42	54	20	..	9
6	6	..	..	40	46	29	..	15
3	3	..	..	34	43	22	..	3
13	13	40	..	42	32	23	74	5
13	13	46	..	43	34	23	74	13
11	11	50	2	42	33	30	69	14
11	11	54	9	39	28	18	73	22
52	65	61	15	44	42	20	74	28
49	64	58	15	43	40	19	81	26
39	58	54	15	44	41	18	81	30
35	63	58	14	41	39	20	73	29
41	**48**	**41**	**16**	**48**	**52**	**34**	**..**	**46**
37	**49**	**41**	**14**	**47**	**51**	**31**	**..**	**44**
35	**47**	**42**	**15**	**46**	**50**	**36**	**..**	**40**
34	**47**	**42**	**15**	**45**	**48**	**37**	**..**	**41**
..	..	..	..	44	44	..	..	..
..	..	..	..	43	44	..	..	13
..	..	..	..	43	44	..	..	13
..	..	..	..	43	44	..	..	13
39	47	..	12	47	53	22	..	33
34	47	..	12	46	52	18	..	31
33	45	..	13	46	50	26	..	28
31	45	..	14	44	49	26	..	31
..	..	..	..	48	48	..	..	..
..	..	..	..	47	47	..	..	..
..	..	..	..	46	46	..	..	..
..	..	..	..	45	45	..	..	..
10	10	..	..	41	48	40	..	6
11	11	..	..	37	28	44	..	8
11	11	..	..	36	28	45	..	6
11	11	..	..	36	40	34	..	5
..	..	..	..	38	34	54	..	14
..	..	..	..	39	36	50	..	25
..	..	..	..	37	35	47	..	30
..	..	..	..	36	32	52	..	42
..	..	..	..	38	38	..	..	..
..	..	..	..	37	37	..	..	..
..	..	..	..	36	36	..	..	..
..	..	..	..	36	36	..	..	..
..	..	..	..	46	50	..	..	..
..	..	..	..	46	50	..	..	..
..	..	..	..	46	50	..	..	..
..	..	..	..	46	51	..	..	..
..	..	..	..	31	31	..	..	..
..	..	..	..	27	27	..	..	8
..	..	..	..	29	29	..	..	8
..	..	..	..	29	29	..	..	5

Table 31

Utilization of installed electricity capacity - by type

Kilowatt-hours per kilowatt divided by 8760

Country or area	Year	Total producers						
		Total	Combustible fuels	Hydro	Nuclear	Wind	Solar	Other
Micronesia (Fed. States of)	2011	23	23	..	..	..	15	..
	2012	23	24	..	..	..	15	..
	2013	24	24	..	..	..	23	..
	2014	23	24	..	..	..	23	..
Nauru	2011	49	50	..	..	..	29	..
	2012	50	51	..	..	..	16	..
	2013	52	52	..	..	..	16	..
	2014	47	48	..	..	..	9	..
New Caledonia	2011	52	55	55	..	14	12	..
	2012	52	54	59	..	17	15	..
	2013	41	40	67	..	16	14	..
	2014	44	46	42	..	18	14	..
New Zealand	2011	52	41	55	..	36	..	94
	2012	53	50	50	..	38	14	95
	2013	52	49	50	..	37	11	89
	2014	51	40	53	..	37	10	83
Niue	2011	17	18	..	..	..	15	..
	2012	18	18	..	..	..	14	..
	2013	18	18	..	..	..	15	..
	2014	15	17	..	..	..	2	..
Northern Mariana Islands	2011	57	57	..	..	..	..	..
	2012	53	53	..	..	..	..	..
	2013	52	52	..	..	..	..	..
	2014	51	51	..	..	..	..	..
Palau	2011	29	29	..	..	..	..	..
	2012	29	29	..	..	..	..	..
	2013	29	29	..	..	..	..	..
	2014	30	30	..	..	..	..	..
Papua New Guinea	2011	53	54	44	..	..	..	86
	2012	54	56	43	..	..	..	86
	2013	55	57	44	..	..	..	86
	2014	49	48	42	..	..	..	86
Samoa	2011	31	30	33	..	..	..	..
	2012	32	31	35	..	..	..	..
	2013	33	33	34	..	..	..	..
	2014	37	38	34	..	..	..	..
Solomon Islands	2011	26	26	..	..	..	..	..
	2012	26	26	..	..	..	..	..
	2013	28	28	..	..	..	..	..
	2014	29	29	..	..	..	..	..
Tonga	2011	43	43	..	..	..	..	..
	2012	43	43	..	..	..	..	..
	2013	41	43	..	..	..	16	..
	2014	40	43	..	..	..	18	..
Tuvalu	2011	22	22	..	..	..	..	..
	2012	22	22	..	..	..	..	..
	2013	21	21	..	..	..	..	..
	2014	21	21	..	..	..	..	..
Vanuatu	2011	25	24	54	..	19	16	..
	2012	24	24	69	..	11	21	..
	2013	25	24	67	..	11	21	..
	2014	24	24	67	..	11	21	..
Wallis and Futuna Is.	2011	20	20	..	..	..	..	..
	2012	19	20	..	..	..	..	..
	2013	19	19	..	..	..	..	..
	2014	19	19	..	..	..	..	..

Utilization of installed electricity capacity - by type

Kilowatt-hours per kilowatt divided by 8760

Autoproducers				Main activity producers				
Total	Combustible fuels	Hydro	Other	Total	Combustible fuels	Hydro	Nuclear	Other
..	..	..	..	23	23	..	..	15
..	..	..	..	23	24	..	..	15
..	..	..	..	24	24	..	..	23
..	..	..	..	23	24	..	..	23
..	..	..	..	49	50	..	..	29
..	..	..	..	50	51	..	..	16
..	..	..	..	52	52	..	..	16
..	..	..	..	47	48	..	..	9
35	38	..	14	52	56	55	..	14
36	38	..	14	52	55	59	..	17
39	39	..	14	42	40	67	..	16
46	46	..	14	44	46	42	..	18
58	60	46	43	52	39	55	..	68
62	65	49	37	52	49	50	..	70
62	67	49	33	52	47	50	..	67
59	66	49	27	51	38	53	..	65
..	..	..	..	17	18	..	..	15
..	..	..	..	18	18	..	..	14
..	..	..	..	18	18	..	..	15
..	..	..	..	15	17	..	..	2
..	..	..	..	57	57	..	..	..
..	..	..	..	53	53	..	..	..
..	..	..	..	52	52	..	..	..
..	..	..	..	51	51	..	..	..
34	34	..	..	29	29	..	..	..
34	34	..	..	29	29	..	..	..
34	34	..	..	29	29	..	..	..
35	35	..	..	30	30	..	..	..
62	64	40	86	37	26	46	..	..
64	66	40	86	37	26	45	..	..
62	64	42	86	40	33	45	..	..
56	55	42	86	35	26	43	..	..
23	23	..	..	32	31	33	..	..
23	23	..	..	33	32	35	..	..
23	23	..	..	34	34	34	..	..
23	23	..	..	38	39	34	..	..
18	18	..	..	27	28	..	..	..
18	18	..	..	28	28	..	..	..
18	18	..	..	30	30	..	..	..
18	18	..	..	31	31	..	..	..
..	..	..	..	43	43	..	..	..
..	..	..	..	43	43	..	..	..
..	..	..	..	41	43	..	..	16
..	..	..	..	40	43	..	..	18
..	..	..	..	22	22	..	..	..
..	..	..	..	22	22	..	..	..
..	..	..	..	21	21	..	..	..
..	..	..	..	21	21	..	..	..
..	..	..	..	25	24	54	..	19
..	..	..	..	24	24	69	..	11
..	..	..	..	25	24	67	..	12
..	..	..	..	24	24	67	..	12
..	..	..	..	20	20	..	..	..
..	..	..	..	19	20	..	..	..
..	..	..	..	19	19	..	..	..
..	..	..	..	19	19	..	..	..

Table 32

Production of electricity - by type

Gigawatt-hours

Table Notes

For this table, **Other** under total producers refers to geothermal, tide, wave, marine, electricity from chemical heat, and other non-specified sources. **Other** under autoproducers refers to nuclear (if any), wind, solar, tide, wave, marine, electricity from chemical heat, and other non-specified sources. **Other** under main activity producers refers to wind, solar, tide, wave, marine, electricity from chemical heat, and other non-specified sources.

For India, thermal production includes also wind electricity.

Please refer to the Definitions Section on pages ix to xv for the appropriate product description /classification.

Figure 75: World electricity generation by source, 1990-2014

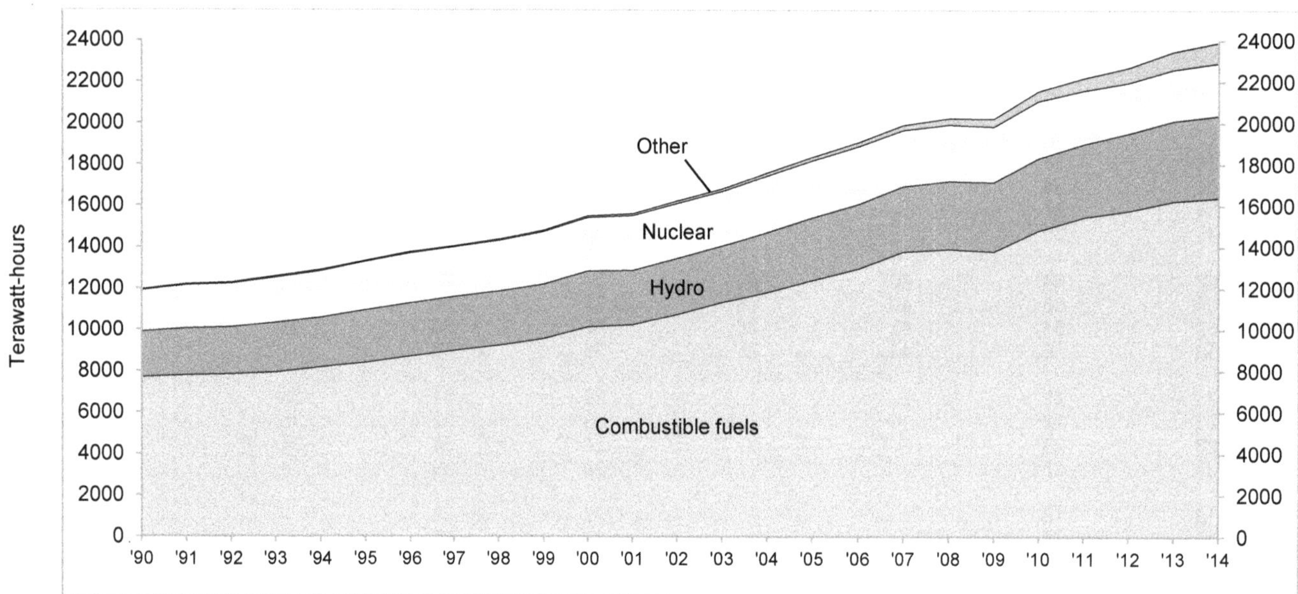

Figure 76: World production of electricity by source and region, 2014

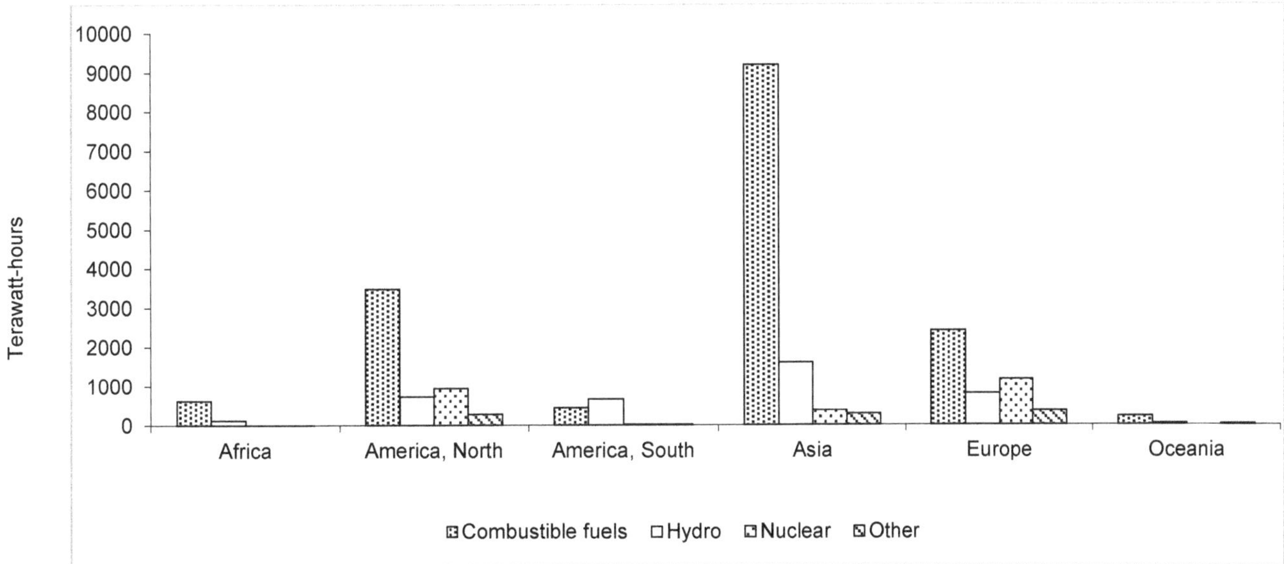

Figure 77: Major countries producing electricity from combustible fuels, 2014

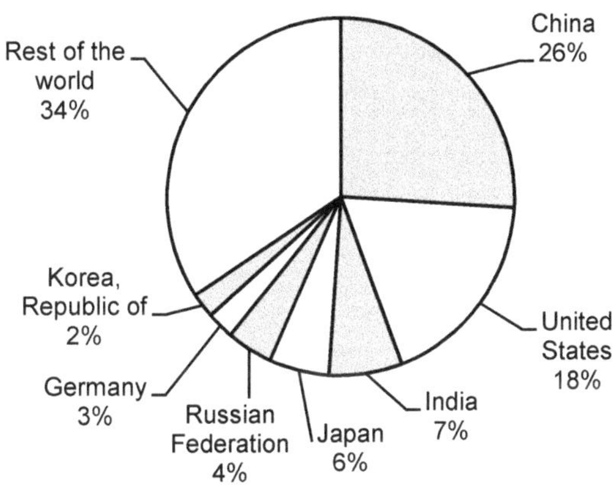

Figure 78: Major countries producing hydro electricity, 2014

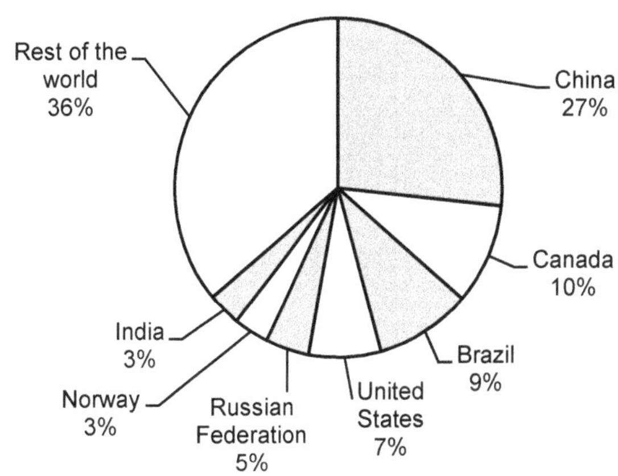

Figure 79: Major countries producing nuclear electricity, 2014

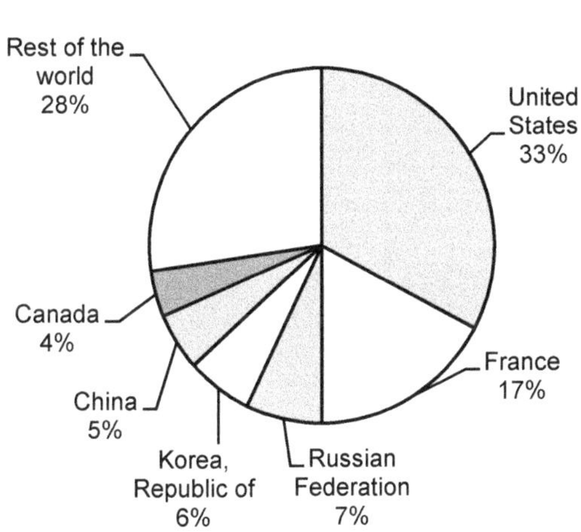

Figure 80: Major countries producing electricity from other sources, 2014

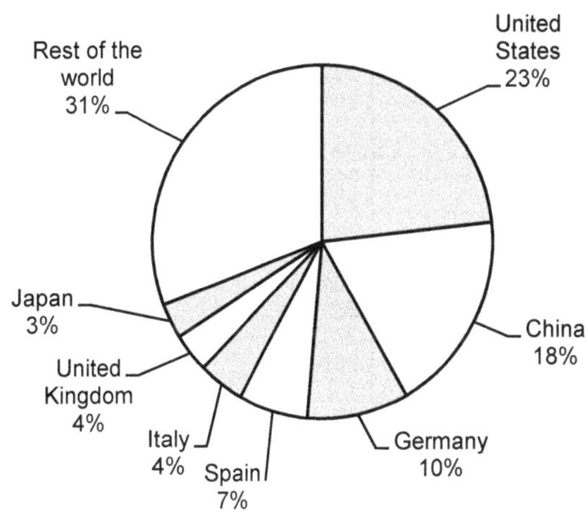

Table 32

Production of electricity - by type

Gigawatt-hours

Country or area	Year	Total producers: Total	Combustible fuels	Hydro	Nuclear	Wind	Solar	Other
World	**2011**	**22,173,726**	**15,441,720**	**3,552,328**	**2,582,635**	**434,319**	**62,836**	**99,888**
	2012	**22,681,641**	**15,753,928**	**3,741,121**	**2,460,301**	**521,800**	**94,721**	**109,770**
	2013	**23,453,355**	**16,222,806**	**3,877,281**	**2,479,106**	**645,870**	**136,537**	**91,755**
	2014	**23,906,233**	**16,396,065**	**3,979,941**	**2,535,326**	**707,052**	**184,113**	**103,736**
Africa	**2011**	**699,697**	**567,204**	**114,251**	**13,502**	**2,441**	**658**	**1,642**
	2012	**725,089**	**592,994**	**115,368**	**11,954**	**2,512**	**516**	**1,745**
	2013	**746,422**	**604,442**	**121,471**	**14,106**	**3,681**	**647**	**2,074**
	2014	**771,383**	**620,149**	**127,230**	**13,794**	**5,318**	**1,707**	**3,185**
Algeria	2011	51,224	50,722	502	..	..	..	..
	2012	57,397	56,775	622	..	..	..	..
	2013	59,890	59,560	330	..	..	..	..
	2014	64,242	63,988	254	..	..	..	..
Angola	2011	5,651	1,644	4,007	..	..	..	..
	2012	6,204	2,432	3,772	..	..	..	..
	2013	8,216	3,449	4,767	..	..	..	..
	2014	9,480	4,439	5,041	..	..	..	..
Benin	2011	155	155	..	..	..	..	..
	2012	162	162	..	..	..	..	..
	2013	173	173	..	..	..	..	..
	2014	184	184	..	..	..	..	..
Botswana	2011	437	437	..	..	..	..	..
	2012	251	250	..	..	..	1	..
	2013	761	760	..	..	..	1	..
	2014	2,365	2,363	..	..	..	*2	..
Burkina Faso	2011	530	448	82	..	..	..	..
	2012	626	529	97	..	..	..	..
	2013	*731	*625	106	..	..	..	..
	2014	*870	*779	90	..	..	..	..
Burundi	2011	141	13	128	..	..	..	..
	2012	142	3	139	..	..	..	..
	2013	159	21	138	..	..	..	..
	2014	174	34	140	..	..	..	..
Cabo Verde	2011	365	341	..	..	16	9	..
	2012	375	306	..	..	61	7	..
	2013	404	321	..	..	76	7	..
	2014	409	338	..	..	66	5	..
Cameroon	2011	5,943	1,546	4,397	..	..	..	..
	2012	6,263	1,664	4,599	..	..	..	..
	2013	6,849	1,994	4,855	..	..	..	..
	2014	6,922	1,854	5,068	..	..	..	..
Central African Rep.	2011	*174	*27	*147	..	..	..	..
	2012	*177	*27	*150	..	..	..	..
	2013	*181	*26	*155	..	..	..	..
	2014	*183	*26	*157	..	..	..	..
Chad	2011	*218	*218	..	..	..	..	..
	2012	205	205	..	..	..	..	..
	2013	*225	*225	..	..	..	..	..
	2014	*225	*225	..	..	..	..	..
Comoros	2011	*57	*57	..	..	..	..	..
	2012	*58	*58	..	..	..	..	..
	2013	*56	*56	..	..	..	..	..
	2014	*55	*55	..	..	..	..	..
Congo	2011	1,293	502	791	..	..	..	..
	2012	1,700	713	987	..	..	..	..
	2013	1,709	738	971	..	..	..	..
	2014	1,740	788	952	..	..	..	..

Production of electricity - by type

Gigawatt-hours

Autoproducers				Main activity producers				
Total	Combustible fuels	Hydro	Other	Total	Combustible fuels	Hydro	Nuclear	Other
1,700,294	**1,571,855**	**96,854**	**31,584**	**20,473,433**	**13,869,865**	**3,455,473**	**2,582,635**	**565,459**
1,748,861	**1,608,415**	**96,558**	**43,888**	**20,932,780**	**14,145,512**	**3,644,563**	**2,460,301**	**682,403**
1,891,894	**1,730,738**	**98,812**	**62,343**	**21,561,461**	**14,492,068**	**3,778,469**	**2,479,106**	**811,818**
1,901,338	**1,722,393**	**97,028**	**81,917**	**22,004,895**	**14,673,672**	**3,882,913**	**2,535,326**	**912,985**
32,833	**31,668**	**965**	**201**	**666,864**	**535,536**	**113,286**	**13,502**	**4,540**
36,032	**35,230**	**567**	**234**	**689,057**	**557,763**	**114,801**	**11,954**	**4,538**
36,336	**35,467**	**556**	**312**	**710,086**	**568,975**	**120,915**	**14,106**	**6,090**
37,505	**36,420**	**761**	**325**	**733,878**	**583,729**	**126,469**	**13,794**	**9,886**
2,352	2,352	..	..	48,872	48,370	502	..	..
3,311	3,311	..	..	54,086	53,464	622	..	..
3,742	3,742	..	..	56,148	55,818	330	..	..
3,742	3,742	..	..	60,500	60,246	254	..	..
325	325	..	..	5,326	1,319	4,007	..	..
357	357	..	..	5,847	2,075	3,772	..	..
473	473	..	..	7,743	2,976	4,767	..	..
546	546	..	..	8,934	3,893	5,041	..	..
28	28	..	..	127	127	..	..	..
29	29	..	..	133	133	..	..	..
31	31	..	..	142	142	..	..	..
33	33	..	..	151	151	..	..	..
..	..	..	..	437	437	..	..	..
..	..	..	..	251	250	..	..	1
..	..	..	..	761	760	..	..	1
..	..	..	..	2,365	2,363	..	..	*2
88	88	..	..	442	360	82	..	..
147	147	..	..	479	382	97	..	..
*198	*198	..	..	*533	*427	106	..	..
*249	*249	..	..	*621	*530	90	..	..
..	..	..	..	141	13	128	..	..
..	..	..	..	142	3	139	..	..
..	..	..	..	159	21	138	..	..
..	..	..	..	174	34	140	..	..
1	1	..	..	365	340	..	..	25
1	1	..	..	374	306	..	..	69
1	1	..	..	404	321	..	..	83
1	1	..	..	408	338	..	..	71
1,194	1,194	..	..	4,749	352	4,397	..	..
1,295	1,295	..	..	4,968	369	4,599	..	..
1,607	1,607	..	..	5,242	387	4,855	..	..
1,449	1,449	..	..	5,473	405	5,068	..	..
..	..	..	..	*174	*27	*147	..	..
..	..	..	..	*177	*27	*150	..	..
..	..	..	..	*181	*26	*155	..	..
..	..	..	..	*183	*26	*157	..	..
..	..	..	..	*218	*218	..	..	..
..	..	..	..	205	205	..	..	..
..	..	..	..	*225	*225	..	..	..
..	..	..	..	*225	*225	..	..	..
..	..	..	..	*57	*57	..	..	..
..	..	..	..	*58	*58	..	..	..
..	..	..	..	*56	*56	..	..	..
..	..	..	..	*55	*55	..	..	..
493	493	..	..	800	9	791	..	..
713	713	..	..	987	0	987	..	..
738	738	..	..	971	0	971	..	..
788	788	..	..	952	0	952	..	..

Table 32

Production of electricity - by type

Gigawatt-hours

Country or area	Year	Total producers						
		Total	Combustible fuels	Hydro	Nuclear	Wind	Solar	Other
Côte d'Ivoire	2011	6,099	4,325	1,774	..	..	..	..
	2012	7,016	5,227	1,789	..	..	..	..
	2013	7,650	6,044	1,606	..	..	..	..
	2014	8,286	6,373	1,913	..	..	..	..
Dem. Rep. of the Congo	2011	7,899	88	7,811	..	..	..	..
	2012	7,595	9	7,586	..	..	..	..
	2013	8,240	9	8,231	..	..	..	..
	2014	8,831	11	8,820	..	..	..	..
Djibouti	2011	*387	*387	..	..	..	..	..
	2012	393	393	..	..	..	..	..
	2013	402	402	..	..	..	..	..
	2014	402	402	..	..	..	..	..
Egypt	2011	161,162	146,224	12,934	..	1,525	479	..
	2012	164,628	150,010	13,121	..	1,260	237	..
	2013	168,050	153,252	13,352	..	1,332	114	..
	2014	177,249	161,708	13,979	..	1,315	247	..
Equatorial Guinea	2011	*281	*242	*39	..	..	..	..
	2012	*439	*242	*197	..	..	..	..
	2013	*439	*242	*197	..	..	..	..
	2014	*439	*242	*197	..	..	..	..
Eritrea	2011	337	335	..	..	..	2	..
	2012	359	357	..	..	..	2	..
	2013	363	361	..	..	..	2	..
	2014	370	368	..	..	..	2	..
Ethiopia	2011	6,336	37	6,262	..	29	..	8
	2012	7,608	12	7,388	..	192	..	16
	2013	8,719	8	8,338	..	356	..	17
	2014	9,615	9	9,195	..	392	..	19
Gabon	2011	2,033	1,193	840	..	..	..	..
	2012	2,151	1,240	908	..	..	3	..
	2013	2,264	1,330	932	..	..	2	..
	2014	2,367	1,570	795	..	..	2	..
Gambia	2011	*256	*256	..	..	..	..	..
	2012	*266	*266	..	..	..	..	..
	2013	*261	*261	..	..	..	..	..
	2014	*261	*261	..	..	..	..	..
Ghana	2011	11,366	3,805	7,561	..	..	..	..
	2012	12,204	4,133	8,071	..	..	..	..
	2013	13,047	4,814	8,233	..	..	..	..
	2014	13,139	4,752	8,387	..	..	..	..
Guinea	2011	890	*476	414	..	..	..	..
	2012	1,071	*580	491	..	..	..	..
	2013	995	*513	482	..	..	..	..
	2014	*1,015	*525	*490	..	..	..	..
Guinea-Bissau	2011	*33	*33	..	..	..	..	..
	2012	*34	*34	..	..	..	..	..
	2013	*35	*35	..	..	..	..	..
	2014	*35	*35	..	..	..	..	..
Kenya	2011	7,560	2,882	3,217	..	18	..	1,444
	2012	7,852	2,305	4,016	..	14	..	1,516
	2013	8,448	2,218	4,435	..	15	..	1,781
	2014	9,139	2,635	3,569	..	17	..	2,917
Lesotho	2011	490	..	490	..	..	..	..
	2012	485	..	485	..	..	..	..
	2013	515	..	515	..	..	..	..
	2014	515	..	515	..	..	..	..

Production of electricity - by type

Gigawatt-hours

Autoproducers				Main activity producers				
Total	Combustible fuels	Hydro	Other	Total	Combustible fuels	Hydro	Nuclear	Other
65	65	..	..	6,034	4,260	1,774	..	..
68	68	..	..	6,948	5,159	1,789	..	..
70	70	..	..	7,580	5,974	1,606	..	..
72	72	..	..	8,214	6,301	1,913	..	..
364	80	284	..	7,535	8	7,527	..	..
0	0	0	..	7,595	9	7,586	..	..
0	0	0	..	8,240	9	8,231	..	..
7	7	0	..	8,824	4	8,820	..	..
..	..	..	..	*387	*387	..	..	..
..	..	..	..	393	393	..	..	..
..	..	..	..	402	402	..	..	..
..	..	..	..	402	402	..	..	..
3,756	3,756	..	..	157,406	142,468	12,934	..	2,004
5,405	5,405	..	..	159,223	144,605	13,121	..	1,497
5,502	5,502	..	..	162,548	147,750	13,352	..	1,446
*5,502	*5,502	..	..	171,747	156,206	13,979	..	1,562
..	..	..	..	*281	*242	*39	..	..
..	..	..	..	*439	*242	*197	..	..
..	..	..	..	*439	*242	*197	..	..
..	..	..	..	*439	*242	*197	..	..
11	11	..	..	326	324	..	..	2
11	11	..	..	348	346	..	..	2
11	11	..	..	352	350	..	..	2
11	11	..	..	359	357	..	..	2
..	..	..	..	6,336	37	6,262	..	37
..	..	..	..	7,608	12	7,388	..	208
..	..	..	..	8,719	8	8,338	..	373
..	..	..	..	9,615	9	9,195	..	411
195	195	..	..	1,838	998	840	..	..
185	182	..	3	1,966	1,058	908	..	..
193	191	..	2	2,071	1,139	932	..	..
194	192	..	2	2,173	1,378	795	..	..
*30	*30	..	..	*226	*226	..	..	..
*31	*31	..	..	*236	*236	..	..	..
*31	*31	..	..	*230	*230	..	..	..
*31	*31	..	..	*230	*230	..	..	..
*166	*166	..	..	11,200	3,639	7,561	..	..
*180	*180	..	..	12,024	3,953	8,071	..	..
*180	*180	..	..	12,867	4,634	8,233	..	..
*180	*180	..	..	12,959	4,572	8,387	..	..
*342	*342	..	..	548	134	414	..	..
*342	*342	..	..	729	238	491	..	..
*342	*342	..	..	653	171	482	..	..
*350	*350	..	..	*665	*175	*490	..	..
*7	*7	..	..	*26	*26	..	..	..
*8	*8	..	..	*26	*26	..	..	..
*8	*8	..	..	*27	*27	..	..	..
*8	*8	..	..	*27	*27	..	..	..
..	..	..	..	7,560	2,882	3,217	..	1,461
..	..	..	..	7,852	2,305	4,016	..	1,530
..	..	..	..	8,448	2,218	4,435	..	1,796
..	..	..	..	9,139	2,635	3,569	..	2,934
..	..	..	..	490	..	490	..	..
..	..	..	..	485	..	485	..	..
..	..	..	..	515	..	515	..	..
..	..	..	..	515	..	515	..	..

Table 32

Production of electricity - by type

Gigawatt-hours

Country or area	Year	Total producers: Total	Combustible fuels	Hydro	Nuclear	Wind	Solar	Other
Liberia	2011	*300	*300	..	..	..	..	..
	2012	*300	*300	..	..	..	..	..
	2013	*300	*300	..	..	..	..	..
	2014	*300	*300	..	..	..	..	..
Libya	2011	25,999	25,999	..	..	..	..	..
	2012	34,275	34,275	..	..	..	..	..
	2013	37,913	37,913	..	..	..	..	..
	2014	37,731	37,731	..	..	..	..	..
Madagascar	2011	1,438	747	690	..	..	0	..
	2012	1,520	765	755	..	..	0	..
	2013	1,594	784	809	..	..	*1	..
	2014	1,659	774	884	..	..	*1	..
Malawi	2011	1,979	*181	1,798	..	..	..	..
	2012	2,027	*181	1,846	..	..	..	..
	2013	2,098	*181	1,917	..	..	..	..
	2014	*2,098	*181	*1,917	..	..	..	..
Mali	2011	1,381	654	727	..	..	..	..
	2012	1,363	579	783	..	..	..	..
	2013	1,510	532	978	..	..	..	..
	2014	*1,679	*959	*720	..	..	..	..
Mauritania	2011	744	744	..	..	..	..	..
	2012	824	824	..	..	..	..	..
	2013	790	790	..	..	..	..	..
	2014	910	910	..	..	..	..	..
Mauritius	2011	2,739	2,679	57	..	3	..	..
	2012	2,798	2,719	74	..	4	1	..
	2013	2,882	2,781	95	..	4	3	..
	2014	2,936	2,817	91	..	3	25	..
Mayotte	2011	262	248	..	..	..	14	..
	2012	271	256	..	..	..	15	..
	2013	285	268	..	..	..	17	..
	2014	299	283	..	..	..	17	..
Morocco	2011	24,364	21,533	2,139	..	692	..	..
	2012	26,495	23,951	1,816	..	728	..	..
	2013	27,929	23,663	2,785	..	1,481	..	..
	2014	29,142	25,185	2,033	..	1,924	..	..
Mozambique	2011	16,830	20	16,810	..	..	..	..
	2012	15,166	21	15,145	..	..	..	..
	2013	15,645	840	14,805	..	..	..	..
	2014	17,739	1,379	16,359	..	..	1	..
Namibia	2011	1,607	9	1,598	..	..	..	..
	2012	1,538	67	1,471	..	..	..	..
	2013	1,712	72	1,640	..	..	..	..
	2014	1,498	13	1,485	..	..	..	..
Niger	2011	324	324	..	..	..	0	..
	2012	392	392	..	..	..	0	..
	2013	438	438	..	..	..	0	..
	2014	439	439	..	..	..	0	..
Nigeria	2011	27,034	21,151	5,883	..	..	..	..
	2012	28,706	23,047	5,659	..	..	..	..
	2013	28,883	23,557	5,326	..	..	..	..
	2014	30,390	25,044	5,346	..	..	..	..
Réunion	2011	2,750	2,195	401	..	12	142	..
	2012	2,812	2,114	488	..	19	191	..
	2013	2,813	2,017	557	..	15	224	..
	2014	2,857	2,180	426	..	16	236	..

Production of electricity - by type

Gigawatt-hours

Autoproducers				Main activity producers				
Total	Combustible fuels	Hydro	Other	Total	Combustible fuels	Hydro	Nuclear	Other
*200	*200	..	..	*100	*100	..	..	..
*200	*200	..	..	*100	*100	..	..	..
*200	*200	..	..	*100	*100	..	..	..
*200	*200	..	..	*100	*100	..	..	..
..	..	..	..	25,999	25,999	..	..	..
..	..	..	..	34,275	34,275	..	..	..
..	..	..	..	37,913	37,913	..	..	..
..	..	..	..	37,731	37,731	..	..	..
*170	*170	..	..	1,268	577	690	..	0
*170	*170	..	..	1,350	595	755	..	0
*170	*170	..	..	1,424	614	809	..	*1
*170	*170	..	..	1,489	604	884	..	*1
*181	*176	*5	..	1,798	*5	1,793	..	..
*181	*176	*5	..	1,846	*5	1,841	..	..
*181	*176	*5	..	1,917	*5	1,912	..	..
*181	*176	*5	..	*1,917	*5	*1,912	..	..
*52	*52	..	..	1,329	602	727	..	..
*52	*52	..	..	1,311	527	783	..	..
52	52	..	..	1,458	480	978	..	..
*52	*52	..	..	*1,627	*907	*720	..	..
309	309	..	..	436	436	..	..	..
323	323	..	..	501	501	..	..	..
320	320	..	..	470	470	..	..	..
380	380	..	..	530	530	..	..	..
1,609	1,609	..	..	1,130	1,070	57	..	3
1,652	1,651	..	1	1,146	1,068	74	..	4
1,706	1,703	..	3	1,176	1,078	95	..	4
1,761	1,736	..	25	1,175	1,081	91	..	3
..	..	..	..	262	248	..	..	14
..	..	..	..	271	256	..	..	15
..	..	..	..	285	268	..	..	17
..	..	..	..	299	283	..	..	17
217	217	..	..	24,147	21,316	2,139	..	692
139	139	..	..	26,356	23,812	1,816	..	728
1,180	1,180	..	..	26,749	22,483	2,785	..	1,481
1,197	1,197	..	..	27,945	23,988	2,033	..	1,924
348	..	348	..	16,482	20	16,462	..	..
263	..	263	..	14,903	21	14,882	..	..
263	..	263	..	15,382	840	14,542	..	..
468	..	467	1	17,271	1,379	15,892	..	..
..	..	..	..	1,607	9	1,598	..	..
..	..	..	..	1,538	67	1,471	..	..
..	..	..	..	1,712	72	1,640	..	..
..	..	..	..	1,498	13	1,485	..	..
54	54	..	0	270	270	..	..	..
54	54	..	0	338	338	..	..	..
39	39	..	0	399	399	..	..	..
*56	*56	..	0	383	383	..	..	..
7,002	6,882	120	..	20,032	14,269	5,763	..	..
7,435	7,320	115	..	21,271	15,727	5,544	..	..
5,257	5,148	109	..	23,626	18,409	5,217	..	..
5,582	5,473	109	..	24,808	19,571	5,237	..	..
..	..	..	..	2,750	2,195	401	..	154
..	..	..	..	2,812	2,114	488	..	209
..	..	..	..	2,813	2,017	557	..	239
..	..	..	..	2,857	2,180	426	..	252

Table 32

Production of electricity - by type

Gigawatt-hours

Country or area	Year	Total producers						
		Total	Combustible fuels	Hydro	Nuclear	Wind	Solar	Other
Rwanda	2011	346	196	149	..	..	0	..
	2012	393	211	182	..	..	0	..
	2013	413	265	148	..	..	0	..
	2014	476	290	186	..	..	0	..
Sao Tome and Principe	2011	*60	*54	*6	..	..	..	..
	2012	*67	*60	*7	..	..	..	..
	2013	*67	*60	*7	..	..	..	..
	2014	*67	*60	*7	..	..	..	..
Senegal	2011	2,560	2,560	..	..	..	..	..
	2012	2,917	2,917	..	..	..	..	..
	2013	3,403	3,335	..	..	..	4	64
	2014	3,406	3,335	..	..	..	4	67
Seychelles	2011	324	324	..	..	..	..	..
	2012	336	336	..	..	..	..	..
	2013	354	347	..	..	7	..	..
	2014	363	355	..	..	7	1	..
Sierra Leone	2011	176	22	153	..	..	..	..
	2012	179	61	118	..	..	..	..
	2013	164	59	105	..	..	..	..
	2014	181	63	118	..	..	..	..
Somalia	2011	334	334	..	..	..	..	..
	2012	340	340	..	..	..	..	..
	2013	350	350	..	..	..	..	..
	2014	350	350	..	..	..	..	..
South Africa	2011	262,538	243,979	5,019	13,502	37	1	..
	2012	257,919	241,678	4,211	11,954	37	39	..
	2013	256,073	237,647	4,040	14,106	37	243	..
	2014	252,578	232,512	4,082	13,794	1,070	1,120	..
South Sudan	2011	..	..	..	..	..	..	..
	2012	445	443	..	..	..	2	..
	2013	472	470	..	..	..	2	..
	2014	488	486	..	..	..	2	..
St. Helena and Depend.	2011	9	8	..	..	1	..	..
	2012	10	9	..	..	1	0	..
	2013	11	10	..	..	1	0	..
	2014	*11	*10	..	..	*1	0	..
Sudan	2011	..	..	..	..	..	..	..
	2012	9,510	2,891	6,619	..	..	..	..
	2013	10,607	2,290	8,317	..	..	..	..
	2014	11,376	2,463	8,913	..	..	..	..
Sudan (former)	2011	8,455	1,988	6,467	..	..	..	..
	2012	..	..	..	..	..	..	..
	2013	..	..	..	..	..	..	..
	2014	..	..	..	..	..	..	..
Swaziland	2011	654	*321	333	..	..	..	..
	2012	*622	*350	272	..	..	..	..
	2013	*590	*350	240	..	..	..	..
	2014	567	265	302	..	..	..	..
Togo	2011	320	119	202	..	..	..	..
	2012	399	234	165	..	..	..	..
	2013	360	286	74	..	..	..	..
	2014	358	238	120	..	..	..	..
Tunisia	2011	16,497	16,142	54	..	109	2	190
	2012	18,059	17,537	110	..	196	3	213
	2013	18,382	17,740	60	..	358	12	212
	2014	19,024	18,254	56	..	507	25	182

Production of electricity - by type

Gigawatt-hours

Autoproducers				Main activity producers				
Total	Combustible fuels	Hydro	Other	Total	Combustible fuels	Hydro	Nuclear	Other
..	..	..	..	346	196	149	..	0
..	..	..	..	393	211	182	..	0
..	..	..	..	413	265	148	..	0
..	..	..	..	476	290	186	..	0
..	..	..	..	*60	*54	*6	..	..
..	..	..	..	*67	*60	*7	..	..
..	..	..	..	*67	*60	*7	..	..
..	..	..	..	*67	*60	*7	..	..
1,203	1,203	..	..	1,357	1,357	..	..	..
1,496	1,496	..	..	1,421	1,421	..	..	..
673	605	..	68	2,730	2,730	..	..	..
548	477	..	71	2,858	2,858	..	..	..
..	..	..	..	324	324	..	..	..
..	..	..	..	336	336	..	..	..
..	..	..	..	354	347	..	..	7
..	..	..	..	363	355	..	..	8
..	..	..	..	176	22	153	..	..
..	..	..	..	179	61	118	..	..
..	..	..	..	164	59	105	..	..
..	..	..	..	181	63	118	..	..
..	..	..	..	334	334	..	..	..
..	..	..	..	340	340	..	..	..
..	..	..	..	350	350	..	..	..
..	..	..	..	350	350	..	..	..
10,792	10,639	153	..	251,746	233,340	4,866	13,502	38
10,403	10,275	128	..	247,516	231,403	4,083	11,954	76
11,222	11,099	123	..	244,851	226,548	3,917	14,106	280
12,177	12,053	124	..	240,401	220,459	3,958	13,794	2,190
..	..	..	..	..	..	..	..	..
234	233	..	1	211	210	..	..	1
246	245	..	1	226	225	..	..	1
254	253	..	1	234	233	..	..	1
..	..	..	..	9	8	..	..	1
..	..	..	..	10	9	..	..	1
..	..	..	..	11	10	..	..	1
..	..	..	..	*11	*10	..	..	*1
..	..	..	..	..	..	..	..	..
*74	*74	..	..	9,436	2,817	6,619	..	..
*320	*320	..	..	10,287	1,970	8,317	..	..
..	..	..	..	11,376	2,463	8,913	..	..
..	..	..	..	8,455	1,988	6,467	..	..
..	..	..	..	..	..	..	..	..
..	..	..	..	..	..	..	..	..
..	..	..	..	..	..	..	..	..
*321	*321	..	..	333	0	333	..	..
*342	*342	..	..	280	8	272	..	..
*342	*342	..	..	248	*8	240	..	..
265	265	..	..	302	0	302	..	..
11	11	..	..	309	108	202	..	..
11	11	..	..	388	223	165	..	..
7	7	..	..	353	279	74	..	..
11	11	..	..	347	227	120	..	..
696	504	..	192	15,801	15,638	54	..	109
634	418	..	216	17,425	17,119	110	..	196
745	521	..	224	17,637	17,219	60	..	358
767	560	..	207	18,257	17,694	56	..	507

Table 32

Production of electricity - by type

Gigawatt-hours

Country or area	Year	Total producers						
		Total	Combustible fuels	Hydro	Nuclear	Wind	Solar	Other
Uganda	2011	*2,588	897	*1,691	..	..	..	..
	2012	*2,640	*915	*1,725	..	..	..	..
	2013	*2,861	*915	*1,946	..	..	..	..
	2014	3,258	697	2,561	..	..	..	..
United Rep. of Tanzania	2011	5,093	3,091	1,993	..	..	9	..
	2012	5,589	3,809	1,767	..	..	13	..
	2013	5,941	4,209	1,717	..	..	15	..
	2014	6,219	3,611	2,590	..	..	18	..
Zambia	2011	11,498	15	11,483	..	..	..	..
	2012	12,368	18	12,350	..	..	..	..
	2013	13,300	19	13,281	..	..	..	..
	2014	14,452	410	14,042	..	..	..	..
Zimbabwe	2011	9,177	3,975	5,202	..	..	..	..
	2012	9,149	3,761	5,387	..	..	..	..
	2013	9,499	4,517	4,982	..	..	..	..
	2014	10,023	4,592	5,431	..	..	..	..
America, North	**2011**	**5,410,924**	**3,536,725**	**780,100**	**925,083**	**133,711**	**6,592**	**28,714**
	2012	**5,363,980**	**3,522,469**	**735,658**	**904,761**	**158,551**	**10,671**	**31,870**
	2013	**5,400,500**	**3,481,498**	**734,652**	**937,243**	**193,932**	**17,650**	**35,524**
	2014	**5,433,273**	**3,480,694**	**726,681**	**947,939**	**215,770**	**26,785**	**35,405**
Anguilla	2011	95	95	..	..	..	..	..
	2012	90	90	..	..	..	..	..
	2013	89	89	..	..	..	..	..
	2014	90	90	..	..	..	..	..
Antigua and Barbuda	2011	*325	*325	..	..	..	..	..
	2012	*327	*327	..	..	..	..	..
	2013	*330	*330	..	..	..	..	..
	2014	*336	*336	..	..	..	..	..
Aruba	2011	931	818	..	..	113	..	..
	2012	915	778	..	..	136	..	..
	2013	*928	*790	..	..	*138	..	..
	2014	*917	*778	..	..	*138	..	..
Bahamas	2011	1,754	1,754	..	..	..	..	..
	2012	1,750	1,750	..	..	..	..	..
	2013	1,885	1,885	..	..	..	..	..
	2014	1,853	1,853	..	..	..	..	..
Barbados	2011	1,049	1,049	..	..	..	..	..
	2012	1,129	1,129	..	..	..	..	..
	2013	1,016	1,016	..	..	..	..	..
	2014	1,010	1,010	..	..	..	..	..
Belize	2011	*388	*144	*244	..	..	..	..
	2012	*424	*215	*208	..	..	..	..
	2013	*371	*164	*207	..	..	..	..
	2014	*404	*199	*205	..	..	..	..
Bermuda	2011	*726	*726	..	..	..	..	..
	2012	*692	*692	..	..	..	..	..
	2013	*673	*673	..	..	..	..	..
	2014	*710	*710	..	..	..	..	..
Bonaire, St Eustatius, Saba	2011	..	..	..	..	..	..	..
	2012	*119	*119	..	..	..	..	..
	2013	*122	*122	..	..	..	..	..
	2014	*124	*122	..	..	2	..	..
British Virgin Islands	2011	131	131	..	..	..	..	..
	2012	*130	*130	..	..	..	..	..
	2013	*130	*130	..	..	..	..	..
	2014	*132	*132	..	..	..	..	..

Production of electricity - by type

Gigawatt-hours

Autoproducers				Main activity producers				
Total	Combustible fuels	Hydro	Other	Total	Combustible fuels	Hydro	Nuclear	Other
*85	*30	*55	..	*2,503	867	*1,636	..	..
*87	*31	*56	..	*2,553	*884	*1,669	..	..
*87	*31	*56	..	*2,774	*884	*1,890	..	..
*96	*40	*56	..	3,162	657	2,505	..	..
9	..	..	9	5,084	3,091	1,993	..	..
13	..	..	13	5,576	3,809	1,767	..	..
15	..	..	15	5,926	4,209	1,717	..	..
18	..	..	18	6,201	3,611	2,590	..	..
..	..	..	..	11,498	15	11,483	..	..
..	..	..	..	12,368	18	12,350	..	..
..	..	..	..	13,300	19	13,281	..	..
..	..	..	..	14,452	410	14,042	..	..
158	158	..	..	9,019	3,817	5,202	..	..
186	186	..	..	8,963	3,575	5,387	..	..
184	184	..	..	9,315	4,333	4,982	..	..
160	160	..	..	9,863	4,432	5,431	..	..
263,841	**222,444**	**34,133**	**7,263**	**5,147,084**	**3,314,281**	**745,966**	**925,083**	**161,753**
285,226	**236,562**	**36,380**	**12,284**	**5,078,754**	**3,285,907**	**699,278**	**904,761**	**188,808**
282,602	**233,728**	**35,367**	**13,507**	**5,117,898**	**3,247,770**	**699,285**	**937,243**	**233,600**
281,874	**233,039**	**33,162**	**15,673**	**5,151,399**	**3,247,655**	**693,519**	**947,939**	**262,287**
..	..	..	..	95	95	..	..	..
..	..	..	..	90	90	..	..	..
..	..	..	..	89	89	..	..	..
..	..	..	..	90	90	..	..	..
*1	*1	..	..	*324	*324	..	..	..
*1	*1	..	..	*326	*326	..	..	..
*1	*1	..	..	*329	*329	..	..	..
*1	*1	..	..	*335	*335	..	..	..
..	..	..	..	931	818	..	..	113
..	..	..	..	915	778	..	..	136
..	..	..	..	*928	*790	..	..	*138
..	..	..	..	*917	*778	..	..	*138
51	51	..	..	1,703	1,703	..	..	..
51	51	..	..	1,699	1,699	..	..	..
74	74	..	..	1,811	1,811	..	..	..
*74	*74	..	..	1,779	1,779	..	..	..
*5	*5	..	..	1,044	1,044	..	..	..
*4	*4	..	..	1,125	1,125	..	..	..
*3	*3	..	..	1,013	1,013	..	..	..
*3	*3	..	..	1,007	1,007	..	..	..
66	66	..	..	*322	*78	*244	..	..
72	72	..	..	*352	*143	*208	..	..
63	63	..	..	*308	*101	*207	..	..
69	69	..	..	*335	*130	*205	..	..
12	12	..	..	*714	*714	..	..	..
*12	*12	..	..	*680	*680	..	..	..
*12	*12	..	..	*661	*661	..	..	..
*12	*12	..	..	*698	*698	..	..	..
..	..	..	..	..	..	..	..	..
..	..	..	..	*119	*119	..	..	..
..	..	..	..	*122	*122	..	..	..
..	..	..	..	*124	*122	..	..	2
..	..	..	..	131	131	..	..	..
..	..	..	..	*130	*130	..	..	..
..	..	..	..	*130	*130	..	..	..
..	..	..	..	*132	*132	..	..	..

Table 32

Production of electricity - by type

Gigawatt-hours

Country or area	Year	Total producers						
		Total	Combustible fuels	Hydro	Nuclear	Wind	Solar	Other
Canada	2011	629,901	150,042	375,797	93,589	10,187	260	26
	2012	632,921	146,063	380,340	94,862	11,310	319	27
	2013	660,795	143,406	391,861	103,439	17,963	1,499	2,627
	2014	656,225	139,402	382,574	107,678	22,538	1,756	2,277
Cayman Islands	2011	626	626	..	..	..	..	..
	2012	607	607	..	..	..	..	..
	2013	615	615	..	..	..	..	..
	2014	621	621	..	..	..	..	..
Costa Rica	2011	9,831	1,002	7,135	..	414	0	1,280
	2012	10,174	1,009	7,233	..	528	1	1,403
	2013	10,235	1,379	6,851	..	485	3	1,517
	2014	10,217	1,224	6,717	..	735	3	1,538
Cuba	2011	17,759	17,660	99	..	..	..	..
	2012	18,432	18,321	111	..	..	..	..
	2013	19,140	19,012	127	..	..	..	..
	2014	19,366	19,262	104	..	..	..	..
Curaçao	2011	..	..	..	..	..	..	..
	2012	917	885	..	..	32	..	..
	2013	904	872	..	..	32	..	..
	2014	891	859	..	..	32	..	..
Dominica	2011	100	65	36	..	..	..	..
	2012	102	75	27	..	..	..	..
	2013	101	64	37	..	..	..	..
	2014	102	72	31	..	..	..	..
Dominican Republic	2011	16,701	15,174	1,513	..	14	..	..
	2012	17,746	15,867	1,783	..	96	..	..
	2013	18,799	16,685	1,872	..	243	..	..
	2014	17,962	16,452	1,267	..	243	..	..
El Salvador	2011	6,083	2,538	2,011	..	..	..	1,534
	2012	6,237	2,855	1,847	..	..	..	1,535
	2013	6,272	2,922	1,790	..	..	..	1,560
	2014	6,223	2,947	1,718	..	..	..	1,558
Greenland	2011	565	227	338	..	..	..	..
	2012	574	229	345	..	..	..	..
	2013	568	191	378	..	..	..	..
	2014	580	184	396	..	..	..	..
Grenada	2011	204	204	..	..	..	..	..
	2012	200	200	..	..	..	..	..
	2013	197	197	..	..	..	..	..
	2014	200	200	..	..	..	..	..
Guadeloupe	2011	*1,767	*1,555	*14	..	*53	*94	*51
	2012	1,817	1,610	14	..	52	91	50
	2013	1,818	1,562	19	..	57	99	81
	2014	1,809	1,563	22	..	54	95	75
Guatemala	2011	8,798	4,376	4,175	..	..	..	246
	2012	9,335	4,631	4,459	..	..	..	246
	2013	10,024	5,152	4,660	..	..	..	212
	2014	10,638	5,531	4,853	..	..	7	247
Haiti	2011	974	853	121	..	..	..	..
	2012	1,100	879	221	..	..	..	..
	2013	1,055	914	141	..	..	..	..
	2014	1,033	943	90	..	..	..	..
Honduras	2011	7,653	4,721	2,815	..	117	..	..
	2012	7,740	4,615	2,787	..	338	..	..
	2013	8,074	5,025	2,739	..	310	..	..
	2014	8,038	5,038	2,602	..	398	..	..

Production of electricity - by type

Gigawatt-hours

Autoproducers				Main activity producers				
Total	Combustible fuels	Hydro	Other	Total	Combustible fuels	Hydro	Nuclear	Other
52,204	21,688	30,515	1	577,697	128,354	345,282	93,589	10,472
54,971	22,933	32,034	4	577,950	123,130	348,306	94,862	11,652
56,197	24,933	31,246	18	604,598	118,473	360,615	103,439	22,071
56,693	25,512	31,053	128	599,532	113,890	351,521	107,678	26,443
..	..	..	..	626	626	..	..	..
..	..	..	..	607	607	..	..	..
..	..	..	..	615	615	..	..	..
..	..	..	..	621	621	..	..	..
1,995	139	1,304	552	7,837	863	5,831	..	1,143
2,186	179	1,357	651	7,988	830	5,877	..	1,281
183	183	..	..	10,052	1,196	6,851	..	2,005
181	181	..	..	10,036	1,043	6,717	..	2,276
815	815	..	..	16,944	16,845	99	..	..
836	836	..	..	17,596	17,485	111	..	..
877	877	..	..	18,263	18,135	127	..	..
838	838	..	..	18,529	18,424	104	..	..
..	..	..	..	..	..	..	..	..
486	486	..	..	431	399	..	..	32
479	479	..	..	425	393	..	..	32
472	472	..	..	419	387	..	..	32
..	..	..	..	100	65	36	..	..
..	..	..	..	102	75	27	..	..
..	..	..	..	101	64	37	..	..
..	..	..	..	102	72	31	..	..
2,919	2,919	..	..	13,782	12,254	1,513	..	14
3,055	3,055	..	..	14,690	12,812	1,783	..	96
3,190	3,190	..	..	15,610	13,495	1,872	..	243
2,970	2,970	..	..	14,992	13,482	1,267	..	243
593	593	..	..	5,490	1,945	2,011	..	1,534
696	696	..	..	5,541	2,159	1,847	..	1,535
660	660	..	..	5,612	2,262	1,790	..	1,560
654	654	..	..	5,569	2,293	1,718	..	1,558
114	114	..	..	452	114	338	..	..
115	115	..	..	459	115	345	..	..
95	95	..	..	473	95	378	..	..
92	92	..	..	488	92	396	..	..
..	..	..	..	204	204	..	..	..
..	..	..	..	200	200	..	..	..
..	..	..	..	197	197	..	..	..
..	..	..	..	200	200	..	..	..
..	..	..	..	*1,767	*1,555	*14	..	*198
..	..	..	..	1,817	1,610	14	..	193
..	..	..	..	1,818	1,562	19	..	237
..	..	..	..	1,809	1,563	22	..	224
608	586	22	..	8,189	3,790	4,153	..	246
634	611	23	..	8,702	4,020	4,436	..	246
855	827	28	..	9,170	4,326	4,632	..	212
856	827	28	..	9,782	4,703	4,825	..	254
31	31	..	..	943	822	121	..	..
48	48	..	..	1,052	831	221	..	..
50	50	..	..	1,005	864	141	..	..
90	90	..	..	943	853	90	..	..
483	483	..	..	7,170	4,238	2,815	..	117
509	509	..	..	7,231	4,106	2,787	..	338
531	531	..	..	7,543	4,494	2,739	..	310
351	351	..	..	7,687	4,687	2,602	..	398

Table 32

Production of electricity - by type

Gigawatt-hours

Country or area	Year	Total producers						
		Total	Combustible fuels	Hydro	Nuclear	Wind	Solar	Other
Jamaica	2011	4,171	3,926	152	..	92	..	..
	2012	4,156	3,898	151	..	108	..	..
	2013	4,256	4,017	124	..	115	..	..
	2014	4,123	3,869	135	..	119	..	..
Martinique	2011	*1,764	*1,720	..	..	*2	*43	..
	2012	*1,780	*1,736	..	..	*2	*43	..
	2013	*1,764	*1,719	..	..	*2	*44	..
	2014	*1,764	*1,719	..	..	*2	*44	..
Mexico	2011	302,785	248,253	36,247	10,089	1,648	41	6,507
	2012	307,268	257,041	31,883	8,770	3,688	69	5,817
	2013	297,326	247,163	28,002	11,800	4,185	106	6,070
	2014	301,496	240,279	38,893	9,677	6,426	221	6,000
Montserrat	2011	*23	*23	..	..	..	..	..
	2012	*23	*23	..	..	..	..	..
	2013	*23	*23	..	..	..	..	..
	2014	24	24	..	..	..	..	..
Neth. Antilles (former)	2011	1,325	1,325	..	..	..	..	..
	2012	..	..	..	..	..	..	..
	2013	..	..	..	..	..	..	..
	2014	..	..	..	..	..	..	..
Nicaragua	2011	3,824	2,896	444	..	211	..	273
	2012	4,031	2,759	419	..	330	..	523
	2013	4,163	2,466	456	..	562	..	679
	2014	4,444	2,541	395	..	846	..	662
Panama	2011	7,857	3,759	4,098	..	..	..	..
	2012	8,606	3,218	5,388	..	..	..	..
	2013	8,962	3,808	5,154	..	..	..	..
	2014	9,287	4,136	5,034	..	116	1	..
Puerto Rico	2011	22,507	22,358	150	..	..	..	..
	2012	22,556	22,429	125	..	..	2	..
	2013	21,743	21,509	91	..	118	26	..
	2014	21,317	20,969	81	..	219	48	..
Sint Maarten (Dutch part)	2011	..	..	..	..	..	..	..
	2012	*402	*402	..	..	..	..	..
	2013	410	410	..	..	..	..	..
	2014	*418	*418	..	..	..	..	..
St. Kitts-Nevis	2011	223	217	..	..	6	..	..
	2012	218	210	..	..	8	0	..
	2013	*220	*211	..	..	*8	*1	..
	2014	*220	*211	..	..	*8	*1	..
St. Lucia	2011	385	385	..	..	..	..	..
	2012	385	385	..	..	..	..	..
	2013	383	383	..	..	..	..	..
	2014	379	379	..	..	..	..	..
St. Pierre-Miquelon	2011	45	44	..	..	1	..	..
	2012	43	43	..	..	1	..	..
	2013	47	46	..	..	*1	..	..
	2014	48	47	..	..	*1	..	..
St. Vincent-Grenadines	2011	*145	*114	*31	..	..	..	..
	2012	*145	*114	*31	..	..	..	..
	2013	*139	*108	*31	..	..	..	..
	2014	*170	*132	*37	..	..	..	..
Trinidad and Tobago	2011	8,772	8,772	..	..	..	..	..
	2012	9,132	9,132	..	..	..	..	..
	2013	9,505	9,505	..	..	..	..	..
	2014	9,891	9,891	..	..	..	..	..

Production of electricity - by type

Gigawatt-hours

Autoproducers				Main activity producers				
Total	Combustible fuels	Hydro	Other	Total	Combustible fuels	Hydro	Nuclear	Other
1,411	1,319	..	92	2,760	2,607	152	..	..
1,542	1,434	..	108	2,615	2,464	151	..	..
1,914	1,799	..	115	2,343	2,218	124	..	..
1,661	1,542	..	119	2,462	2,327	135	..	..
*30	*30	..	..	*1,734	*1,690	..	..	*45
*30	*30	..	..	*1,750	*1,706	..	..	*45
*30	*30	..	..	*1,734	*1,689	..	..	*46
*30	*30	..	..	*1,734	*1,689	..	..	*46
36,426	34,643	451	1,332	266,359	213,610	35,796	10,089	6,864
45,027	42,450	566	2,011	262,241	214,591	31,317	8,770	7,563
38,505	35,483	558	2,464	258,821	211,680	27,444	11,800	7,897
43,238	37,933	748	4,557	258,258	202,346	38,145	9,677	8,090
..	..	..	..	*23	*23	..	..	..
..	..	..	..	*23	*23	..	..	..
..	..	..	..	*23	*23	..	..	..
..	..	..	..	24	24	..	..	..
702	702	..	..	623	623	..	..	..
..	..	..	..	..	..	..	..	..
..	..	..	..	..	..	..	..	..
..	..	..	..	..	..	..	..	..
419	419	..	..	3,405	2,477	444	..	484
502	501	1	..	3,529	2,258	418	..	853
535	534	1	..	3,628	1,932	455	..	1,241
545	544	1	..	3,899	1,997	394	..	1,508
23	23	..	..	7,834	3,736	4,098	..	..
26	26	..	..	8,580	3,192	5,388	..	..
33	33	..	..	8,929	3,775	5,154	..	..
31	31	..	..	9,256	4,105	5,034	..	118
*250	*250	..	..	22,257	22,108	150	..	..
*250	*250	..	..	22,306	22,179	125	..	2
*250	*250	..	..	21,493	21,259	91	..	144
*250	*250	..	..	21,067	20,719	81	..	267
..	..	..	..	..	..	..	..	..
..	..	..	..	*402	*402	..	..	..
..	..	..	..	410	410	..	..	..
..	..	..	..	*418	*418	..	..	..
1	1	..	..	222	216	..	..	6
1	1	..	..	217	209	..	..	8
*1	*1	..	..	*219	*210	..	..	*9
*1	*1	..	..	*219	*210	..	..	*9
..	..	..	..	385	385	..	..	..
..	..	..	..	385	385	..	..	..
..	..	..	..	383	383	..	..	..
..	..	..	..	379	379	..	..	..
..	..	..	..	45	44	..	..	1
..	..	..	..	43	43	..	..	1
..	..	..	..	47	46	..	..	*1
..	..	..	..	48	47	..	..	*1
..	..	..	..	*145	*114	*31	..	..
..	..	..	..	*145	*114	*31	..	..
..	..	..	..	*139	*108	*31	..	..
..	..	..	..	*170	*132	*37	..	..
72	72	..	..	8,700	8,700	..	..	..
75	75	..	..	9,057	9,057	..	..	..
78	78	..	..	9,427	9,427	..	..	..
78	78	..	..	9,813	9,813	..	..	..

Table 32

Production of electricity - by type

Gigawatt-hours

Country or area	Year	Total producers: Total	Combustible fuels	Hydro	Nuclear	Wind	Solar	Other
Turks and Caicos Islands	2011	*214	*214	..	..	..	..	..
	2012	*227	*227	..	..	..	..	..
	2013	*227	*227	..	..	..	..	..
	2014	*230	*230	..	..	..	..	..
United States	2011	4,349,571	3,037,683	344,679	821,405	120,854	6,153	18,797
	2012	4,290,660	3,016,908	298,287	801,129	141,922	10,145	22,269
	2013	4,306,371	2,985,891	290,113	822,004	169,713	15,872	22,778
	2014	4,339,210	2,995,556	281,527	830,584	183,892	24,603	23,048
United States Virgin Is.	2011	951	951	..	..	..	*1	..
	2012	873	872	..	..	..	*1	..
	2013	821	820	..	..	..	*1	..
	2014	*772	*766	..	..	..	*6	..
America, South	**2011**	**1,054,192**	**314,504**	**714,390**	**22,030**	**3,231**	**36**	**0**
	2012	**1,092,648**	**365,430**	**698,652**	**22,433**	**6,010**	**123**	**0**
	2013	**1,129,525**	**422,046**	**677,521**	**21,657**	**7,859**	**278**	**164**
	2014	**1,154,584**	**444,806**	**672,033**	**21,134**	**15,524**	**797**	**291**
Argentina	2011	129,892	91,592	31,901	6,371	26	2	..
	2012	135,207	98,719	29,716	6,395	369	8	..
	2013	140,863	100,758	33,422	6,207	461	15	..
	2014	141,586	93,736	41,348	5,756	730	16	..
Bolivia (Plur. State of)	2011	7,220	4,870	2,347	..	..	3	..
	2012	7,661	5,306	2,352	..	..	3	..
	2013	8,065	5,526	2,535	..	1	3	..
	2014	8,755	6,490	2,251	..	8	6	..
Brazil	2011	531,759	85,062	428,333	15,659	2,705	..	..
	2012	552,498	116,068	415,342	16,038	5,050	..	..
	2013	570,838	157,810	390,992	15,450	6,579	7	..
	2014	590,541	189,497	373,439	15,378	12,211	16	..
Chile	2011	65,713	44,366	21,009	..	338	0	0
	2012	69,751	49,184	20,158	..	409	0	0
	2013	73,065	52,602	19,737	..	554	8	164
	2014	73,719	48,397	23,099	..	1,443	489	291
Colombia	2011	66,472	17,514	48,917	..	41	..	..
	2012	68,344	20,689	47,600	..	55	..	..
	2013	70,858	26,423	44,377	..	58	..	..
	2014	70,059	25,149	44,852	..	58	..	..
Ecuador	2011	20,544	9,408	11,133	..	3	0	..
	2012	22,848	10,608	12,238	..	2	0	..
	2013	23,260	12,161	11,039	..	57	4	..
	2014	24,307	12,753	11,458	..	80	17	..
Falkland Is. (Malvinas)	2011	*18	*12	..	..	*6	..	..
	2012	*18	*12	..	..	*6	..	..
	2013	*18	*12	..	..	*6	..	..
	2014	*18	*12	..	..	*6	..	..
French Guiana	2011	873	373	469	..	..	31	..
	2012	874	266	556	..	..	52	..
	2013	883	349	490	..	..	44	..
	2014	910	359	500	..	..	51	..
Guyana	2011	910	910	..	..	..	..	..
	2012	942	942	..	..	..	..	..
	2013	962	962	..	..	..	..	..
	2014	979	979	..	..	..	..	..
Paraguay	2011	57,625	1	57,624	..	..	..	..
	2012	60,235	3	60,232	..	..	..	..
	2013	60,381	3	60,378	..	..	..	..
	2014	55,282	6	55,276	..	..	..	..

Production of electricity - by type

Gigawatt-hours

Autoproducers				Main activity producers				
Total	Combustible fuels	Hydro	Other	Total	Combustible fuels	Hydro	Nuclear	Other
..	..	..	..	*214	*214	..	..	..
..	..	..	..	*227	*227	..	..	..
..	..	..	..	*227	*227	..	..	..
..	..	..	..	*230	*230	..	..	..
164,569	157,442	1,841	5,286	4,185,002	2,880,241	342,838	821,405	140,518
174,097	162,188	2,400	9,509	4,116,563	2,854,720	295,887	801,129	164,827
177,985	163,542	3,534	10,909	4,128,386	2,822,349	286,579	822,004	197,454
172,680	160,485	1,332	10,863	4,166,530	2,835,071	280,195	830,584	220,680
*41	*40	..	*1	911	911	..	..	..
*1	0	..	*1	872	872	..	..	..
*1	0	..	*1	820	820	..	..	..
*6	0	..	*6	*766	*766	..	..	..
111,471	**86,499**	**24,964**	**9**	**942,721**	**228,006**	**689,427**	**22,030**	**3,259**
113,549	**90,640**	**22,882**	**26**	**979,099**	**274,789**	**675,770**	**22,433**	**6,107**
121,664	**97,627**	**24,007**	**30**	**1,007,861**	**324,419**	**653,514**	**21,657**	**8,271**
130,024	**105,660**	**24,314**	**51**	**1,024,560**	**339,146**	**647,719**	**21,134**	**16,561**
15,052	14,994	58	..	114,840	76,598	31,843	6,371	28
15,229	15,171	58	..	119,978	83,548	29,658	6,395	377
14,716	14,663	53	..	126,147	86,095	33,369	6,207	476
14,703	14,653	50	..	126,883	79,083	41,298	5,756	746
122	119	..	3	7,098	4,751	2,347	..	..
123	120	..	3	7,538	5,186	2,352	..	..
59	56	..	3	8,006	5,470	2,535	..	1
192	186	..	6	8,563	6,304	2,251	..	8
77,033	54,321	22,712	..	454,726	30,741	405,621	15,659	2,705
78,029	57,566	20,463	..	474,469	58,502	394,879	16,038	5,050
86,163	64,105	22,053	5	484,675	93,705	368,939	15,450	6,581
94,031	71,932	22,088	11	496,510	117,565	351,351	15,378	12,216
5,427	5,365	56	6	60,286	39,001	20,953	..	332
5,778	5,577	179	22	63,973	43,607	19,979	..	387
6,422	6,228	175	19	66,643	46,374	19,562	..	707
6,310	6,123	159	28	67,409	42,274	22,940	..	2,195
3,591	3,435	156	..	62,881	14,079	48,761	..	41
3,790	3,658	132	..	64,554	17,031	47,468	..	55
3,464	3,343	121	..	67,394	23,080	44,256	..	58
3,743	3,604	139	..	66,316	21,545	44,713	..	58
3,441	2,936	505	..	17,103	6,472	10,628	..	3
3,866	3,356	511	..	18,982	7,252	11,727	..	3
3,892	3,378	514	..	19,369	8,783	10,525	..	60
3,831	3,308	522	..	20,477	9,445	10,936	..	96
*2	*2	..	..	*16	*10	..	..	*6
*2	*2	..	..	*16	*10	..	..	*6
*2	*2	..	..	*16	*10	..	..	*6
*2	*2	..	..	*16	*10	..	..	*6
..	..	..	..	873	373	469	..	31
..	..	..	..	874	266	556	..	52
..	..	..	..	883	349	490	..	44
..	..	..	..	910	359	500	..	51
306	306	..	..	605	605	..	..	..
285	285	..	..	658	658	..	..	..
302	302	..	..	660	660	..	..	..
202	202	..	..	778	778	..	..	..
..	..	..	..	57,625	1	57,624	..	..
..	..	..	..	60,235	3	60,232	..	..
..	..	..	..	60,381	3	60,378	..	..
..	..	..	..	55,282	6	55,276	..	..

Table 32

Production of electricity - by type

Gigawatt-hours

Country or area	Year	Total producers						
		Total	Combustible fuels	Hydro	Nuclear	Wind	Solar	Other
Peru	2011	38,811	17,248	21,562	..	1	0	..
	2012	40,044	17,951	22,032	..	1	60	..
	2013	43,330	20,812	22,320	..	1	197	..
	2014	45,508	22,882	22,171	..	256	199	..
Suriname	2011	1,951	1,005	946	..	..	..	..
	2012	1,972	974	998	..	..	..	..
	2013	2,184	1,703	481	..	..	..	..
	2014	2,180	1,385	795	..	..	..	..
Uruguay	2011	10,345	3,755	6,479	..	111	..	..
	2012	10,601	5,062	5,421	..	118	..	..
	2013	11,658	3,309	8,206	..	143	..	..
	2014	13,008	2,623	9,649	..	733	3	..
Venezuela (Bolivar. Rep.)	2011	122,059	38,389	83,670	..	..	..	..
	2012	121,653	39,645	82,008	..	..	..	..
	2013	123,160	39,615	83,545	..	..	..	..
	2014	127,732	40,538	87,194	..	..	..	..
Asia	**2011**	**9,845,786**	**8,069,451**	**1,191,663**	**425,377**	**105,822**	**6,548**	**46,925**
	2012	**10,298,780**	**8,363,089**	**1,390,201**	**345,675**	**137,834**	**9,019**	**52,962**
	2013	**11,021,968**	**8,936,754**	**1,488,803**	**347,563**	**191,139**	**27,348**	**30,361**
	2014	**11,477,676**	**9,204,823**	**1,601,514**	**379,463**	**201,338**	**50,762**	**39,777**
Afghanistan	2011	1,005	176	829	..	..	..	..
	2012	1,022	144	878	..	..	..	..
	2013	*1,022	*218	*804	..	..	..	..
	2014	*1,049	*154	*895	..	..	..	..
Armenia	2011	7,433	2,390	2,489	2,548	6	..	..
	2012	8,036	3,399	2,311	2,322	4	..	..
	2013	7,710	3,173	2,173	2,360	4	..	..
	2014	7,750	3,289	1,992	2,465	4	..	..
Azerbaijan	2011	20,294	17,618	2,676	..	0	..	..
	2012	22,988	21,167	1,821	..	0	..	..
	2013	23,354	21,863	1,489	..	1	1	..
	2014	24,728	23,423	1,300	..	2	3	..
Bahrain	2011	24,338	24,338	..	..	..	..	..
	2012	24,770	24,770	..	..	..	..	..
	2013	25,909	25,909	..	..	..	..	..
	2014	27,246	27,246	..	..	..	..	..
Bangladesh	2011	44,153	43,281	872	..	..	..	..
	2012	48,562	47,785	777	..	..	..	..
	2013	53,043	52,000	894	..	4	145	..
	2014	55,845	55,108	588	..	4	145	..
Bhutan	2011	7,068	0	7,067	..	..	..	..
	2012	6,827	0	6,827	..	..	..	..
	2013	7,640	1	7,640	..	..	..	..
	2014	7,004	1	7,003	..	..	..	..
Brunei Darussalam	2011	3,725	3,723	..	..	..	2	..
	2012	3,930	3,928	..	..	..	2	..
	2013	4,402	4,400	..	..	..	2	..
	2014	4,506	4,504	..	..	..	2	..
Cambodia	2011	1,053	1,005	45	..	..	3	..
	2012	1,434	914	517	..	..	3	..
	2013	1,778	759	1,016	..	..	3	..
	2014	3,059	1,204	1,852	..	..	3	..
China	2011	4,713,019	3,833,702	698,945	86,350	70,331	..	23,691
	2012	4,987,553	3,892,814	872,107	97,394	95,978	..	29,260
	2013	5,431,637	4,247,009	920,291	111,613	141,197	5,564	5,963
	2014	5,649,583	4,268,649	1,064,337	132,538	156,078	15,189	12,792

Production of electricity - by type

Gigawatt-hours

Autoproducers				Main activity producers				
Total	Combustible fuels	Hydro	Other	Total	Combustible fuels	Hydro	Nuclear	Other
2,558	2,028	530	..	36,253	15,220	21,032	..	1
2,675	2,134	541	..	37,369	15,817	21,491	..	61
2,666	2,055	610	..	40,665	18,757	21,709	..	198
2,664	2,104	560	..	42,844	20,778	21,611	..	455
1,604	658	946	..	347	347	..	..	..
1,446	448	998	..	526	526	..	..	..
1,514	1,033	481	..	670	670	..	..	..
1,591	796	795	..	589	589	..	..	..
810	810	..	..	9,535	2,945	6,479	..	111
867	865	..	1	9,734	4,197	5,421	..	117
928	925	..	3	10,730	2,384	8,206	..	140
1,280	1,274	..	6	11,728	1,349	9,649	..	730
1,526	1,526	..	..	120,533	36,863	83,670	..	..
1,459	1,459	..	..	120,194	38,186	82,008	..	..
1,537	1,537	..	..	121,623	38,078	83,545	..	..
1,476	1,476	..	..	126,256	39,062	87,194	..	..
913,540	**881,628**	**21,276**	**10,636**	**8,932,247**	**7,187,823**	**1,170,387**	**425,377**	**148,659**
936,206	**902,707**	**20,900**	**12,599**	**9,362,573**	**7,460,382**	**1,369,301**	**345,675**	**187,216**
1,077,711	**1,032,082**	**22,335**	**23,293**	**9,944,257**	**7,904,672**	**1,466,468**	**347,563**	**225,554**
1,079,808	**1,022,891**	**22,404**	**34,513**	**10,397,868**	**8,181,931**	**1,579,110**	**379,463**	**257,364**
286	5	281	..	719	171	548	..	..
301	5	296	..	721	139	582	..	..
*276	*5	*271	..	*746	*213	*533	..	..
*307	*5	*302	..	*742	*149	*593	..	..
..	..	..	..	7,433	2,390	2,489	2,548	6
..	..	..	..	8,036	3,399	2,311	2,322	4
..	..	..	..	7,710	3,173	2,173	2,360	4
..	..	..	..	7,750	3,289	1,992	2,465	4
301	301	..	..	19,993	17,317	2,676	..	0
1,630	1,630	..	..	21,358	19,537	1,821	..	0
1,798	1,798	..	..	21,556	20,065	1,489	..	2
2,022	2,022	..	..	22,706	21,401	1,300	..	5
10,632	10,632	..	..	13,706	13,706	..	..	..
10,510	10,510	..	..	14,260	14,260	..	..	..
11,166	11,166	..	..	14,743	14,743	..	..	..
10,989	10,989	..	..	16,257	16,257	..	..	..
12,951	12,951	..	..	31,202	30,330	872	..	..
13,208	13,208	..	..	35,354	34,577	777	..	..
14,277	14,277	..	..	38,766	37,723	894	..	149
15,418	15,418	..	..	40,427	39,690	588	..	149
..	..	..	..	7,068	0	7,067	..	..
..	..	..	..	6,827	0	6,827	..	..
..	..	..	..	7,640	1	7,640	..	..
..	..	..	..	7,004	1	7,003	..	..
330	330	..	..	3,395	3,393	..	..	2
406	406	..	..	3,524	3,522	..	..	2
460	460	..	..	3,942	3,940	..	..	2
496	496	..	..	4,010	4,008	..	..	2
..	..	..	..	1,053	1,005	45	..	3
..	..	..	..	1,434	914	517	..	3
..	..	..	..	1,778	759	1,016	..	3
..	..	..	..	3,059	1,204	1,852	..	3
310,913	310,913	..	..	4,402,106	3,522,789	698,945	86,350	94,022
315,707	315,707	..	..	4,671,846	3,577,107	872,107	97,394	125,238
344,432	344,432	..	..	5,087,205	3,902,577	920,291	111,613	152,724
346,187	346,187	..	..	5,303,396	3,922,462	1,064,337	132,538	184,059

Table 32

Production of electricity - by type

Gigawatt-hours

Country or area	Year	Total producers						
		Total	Combustible fuels	Hydro	Nuclear	Wind	Solar	Other
China, Hong Kong SAR	2011	39,026	39,026	..	..	..	..	..
	2012	38,752	38,752	..	..	..	..	..
	2013	39,063	39,063	..	..	..	..	..
	2014	39,803	39,803	..	..	..	..	..
China, Macao SAR	2011	886	886	..	..	..	..	..
	2012	561	561	..	..	..	..	..
	2013	413	413	..	..	..	..	..
	2014	641	641	..	..	..	..	..
Cyprus	2011	4,929	4,803	..	..	114	12	..
	2012	4,717	4,510	..	..	185	22	..
	2013	4,290	4,012	..	..	231	47	..
	2014	4,350	4,084	..	..	182	84	..
Georgia	2011	10,104	2,212	7,892	..	..	..	..
	2012	9,698	2,477	7,221	..	..	..	..
	2013	10,059	1,788	8,271	..	..	..	..
	2014	10,370	2,036	8,334	..	..	..	..
India	2011	973,006	813,625	103,741	32,287	23,353	..	..
	2012	1,035,264	860,173	113,855	32,871	28,365	..	..
	2013	1,193,480	980,599	141,637	34,228	33,583	3,433	..
	2014	1,308,873	1,111,341	129,353	36,102	27,057	5,020	..
Indonesia	2011	182,389	160,593	12,419	..	5	1	9,371
	2012	196,181	173,957	12,799	..	5	3	9,417
	2013	216,186	189,835	16,930	..	1	6	9,414
	2014	220,794	191,991	18,800	..	0	7	9,996
Iran (Islamic Rep. of)	2011	240,052	227,428	12,058	327	217	0	22
	2012	254,276	239,752	12,447	1,847	207	0	23
	2013	262,434	242,909	14,582	4,546	376	0	21
	2014	274,609	255,917	13,862	4,472	358	..	..
Iraq	2011	54,240	50,100	4,140	..	..	..	..
	2012	46,018	41,626	4,392	..	..	..	..
	2013	58,422	53,665	4,757	..	..	..	..
	2014	67,768	64,837	2,931	..	..	..	..
Israel	2011	59,679	59,371	29	..	7	192	80
	2012	63,003	62,595	33	..	6	369	0
	2013	61,322	60,794	28	..	6	494	0
	2014	60,813	59,954	13	..	6	840	..
Japan	2011	1,082,230	876,365	91,709	101,761	4,559	5,160	2,676
	2012	1,064,065	950,187	83,645	15,939	4,722	6,963	2,609
	2013	1,065,623	950,236	84,923	9,303	4,286	14,279	2,596
	2014	1,040,676	921,613	86,942	0	5,038	24,506	2,577
Jordan	2011	14,647	14,589	55	..	3	..	..
	2012	16,596	16,532	61	..	3	..	..
	2013	17,263	17,205	55	..	3	..	..
	2014	18,270	18,210	58	..	2	..	..
Kazakhstan	2011	86,586	78,703	7,883	..	..	..	..
	2012	92,817	85,177	7,637	..	3	..	..
	2013	103,086	95,349	7,731	..	5	1	..
	2014	105,068	96,791	8,263	..	13	1	..
Korea, Dem.Ppl's.Rep.	2011	19,226	6,026	13,200	..	..	..	..
	2012	19,236	5,736	13,500	..	..	..	..
	2013	18,332	4,432	13,900	..	..	..	..
	2014	17,909	4,909	13,000	..	..	..	..
Korea, Republic of	2011	523,286	358,505	7,831	154,723	863	917	447
	2012	534,618	374,116	7,652	150,327	917	1,103	503
	2013	541,996	390,858	8,394	138,784	1,149	1,605	1,206
	2014	550,933	381,510	7,820	156,407	1,146	2,557	1,493

Production of electricity - by type

Gigawatt-hours

Autoproducers				Main activity producers				
Total	Combustible fuels	Hydro	Other	Total	Combustible fuels	Hydro	Nuclear	Other
..	..	..	..	39,026	39,026	..	..	..
..	..	..	..	38,752	38,752	..	..	..
..	..	..	..	39,063	39,063	..	..	..
..	..	..	..	39,803	39,803	..	..	..
153	153	..	..	733	733	..	..	..
169	169	..	..	392	392	..	..	..
186	186	..	..	227	227	..	..	..
206	206	..	..	435	435	..	..	..
68	66	..	2	4,861	4,737	..	..	124
60	58	..	2	4,657	4,452	..	..	205
60	58	..	2	4,230	3,954	..	..	276
78	58	..	20	4,272	4,026	..	..	246
..	..	..	..	10,104	2,212	7,892	..	..
..	..	..	..	9,698	2,477	7,221	..	..
..	..	..	..	10,059	1,788	8,271	..	..
..	..	..	..	10,370	2,036	8,334	..	..
128,401	128,172	*229	..	844,605	685,453	103,512	32,287	23,353
*128,401	*128,172	*229	..	906,863	732,001	113,626	32,871	28,365
190,737	188,122	129	2,486	1,002,743	792,477	141,508	34,228	34,530
*162,057	*159,837	*109	*2,111	1,146,816	951,504	129,244	36,102	29,966
33,767	31,659	2,103	5	148,622	128,934	10,316	..	9,372
40,566	38,287	2,274	5	155,615	135,670	10,525	..	9,420
47,150	43,233	3,916	1	169,036	146,602	13,014	..	9,420
45,497	41,513	3,984	0	175,297	150,478	14,816	..	10,003
9,836	9,836	..	..	230,216	217,592	12,058	327	239
10,740	10,740	..	..	243,536	229,012	12,447	1,847	230
6,549	6,549	..	..	255,885	236,360	14,582	4,546	397
6,270	6,270	..	..	268,339	249,647	13,862	4,472	358
..	..	..	..	54,240	50,100	4,140	..	..
..	..	..	..	46,018	41,626	4,392	..	..
..	..	..	..	58,422	53,665	4,757	..	..
..	..	..	..	67,768	64,837	2,931	..	..
1,824	1,552	5	267	57,855	57,819	24	..	12
1,515	1,142	4	369	61,488	61,453	29	..	6
2,296	1,797	5	494	59,026	58,997	23	..	6
3,335	2,490	5	840	57,478	57,464	8	..	6
165,684	138,579	17,331	9,774	916,546	737,786	74,378	101,761	2,621
163,002	135,020	16,285	11,697	901,063	815,167	67,360	15,939	2,597
167,907	132,992	16,322	18,593	897,716	817,244	68,601	9,303	2,568
176,472	130,206	16,687	29,579	864,204	791,407	70,255	0	2,542
257	257	..	..	14,390	14,332	55	..	3
241	241	..	..	16,355	16,291	61	..	3
286	286	..	..	16,977	16,919	55	..	3
383	383	..	..	17,887	17,827	58	..	2
..	..	..	..	86,586	78,703	7,883	..	..
..	..	..	..	92,817	85,177	7,637	..	3
1	..	..	1	103,085	95,349	7,731	..	5
1	..	..	1	105,067	96,791	8,263	..	13
..	..	..	..	19,226	6,026	13,200	..	..
..	..	..	..	19,236	5,736	13,500	..	..
..	..	..	..	18,332	4,432	13,900	..	..
..	..	..	..	17,909	4,909	13,000	..	..
29,690	29,305	..	385	493,596	329,200	7,831	154,723	1,842
30,958	30,637	..	321	503,660	343,479	7,652	150,327	2,202
28,163	27,712	..	451	513,833	363,146	8,394	138,784	3,509
32,971	32,466	..	505	517,962	349,044	7,820	156,407	4,691

Table 32

Production of electricity - by type

Gigawatt-hours

Country or area	Year	Total producers						
		Total	Combustible fuels	Hydro	Nuclear	Wind	Solar	Other
Kuwait	2011	57,457	57,457	..	..	..	..	..
	2012	62,655	62,655	..	..	..	..	..
	2013	60,982	60,982	..	..	..	..	..
	2014	65,140	65,140	..	..	..	..	..
Kyrgyzstan	2011	15,158	1,019	14,139	..	..	..	..
	2012	15,168	989	14,179	..	..	..	..
	2013	14,011	914	13,097	..	..	..	..
	2014	14,572	1,274	13,298	..	..	..	..
Lao People's Dem. Rep.	2011	3,802	*300	3,309	..	..	*193	..
	2012	*3,802	*300	*3,309	..	..	*193	..
	2013	4,471	*300	3,976	..	..	*195	..
	2014	4,380	*300	3,885	..	..	*195	..
Lebanon	2011	16,365	15,560	805	..	..	..	..
	2012	14,826	13,819	1,007	..	..	..	..
	2013	17,487	16,218	1,269	..	..	..	..
	2014	17,952	17,759	193	..	..	..	..
Malaysia	2011	124,894	117,269	7,624	..	..	1	..
	2012	134,421	125,319	9,054	..	..	48	..
	2013	138,331	127,604	10,586	..	..	141	..
	2014	147,461	133,846	13,388	..	..	227	..
Maldives	2011	275	274	..	..	..	1	..
	2012	299	297	..	..	..	2	..
	2013	305	303	..	..	..	*2	..
	2014	329	327	..	..	..	*2	..
Mongolia	2011	4,812	4,812	..	..	..	..	..
	2012	5,182	5,182	..	..	..	..	..
	2013	6,215	6,215	..	..	..	..	..
	2014	6,725	6,725	..	..	..	..	..
Myanmar	2011	10,425	2,907	7,518	..	..	..	..
	2012	10,965	3,199	7,766	..	..	..	..
	2013	12,247	3,424	8,823	..	..	..	..
	2014	14,156	5,327	8,829	..	..	..	..
Nepal	2011	3,492	2	3,490	..	..	..	..
	2012	3,543	10	3,533	..	..	..	..
	2013	3,130	10	3,120	..	..	..	..
	2014	3,508	10	3,498	..	..	..	..
Oman	2011	21,874	21,874	..	..	..	..	..
	2012	25,017	25,017	..	..	..	..	..
	2013	26,240	26,240	..	..	..	..	..
	2014	29,128	29,128	..	..	..	..	..
Other Asia	2011	252,016	201,594	6,902	42,116	1,371	33	..
	2012	250,336	199,770	8,606	40,422	1,262	276	..
	2013	252,351	200,124	8,610	41,639	1,640	338	..
	2014	260,025	208,145	7,439	42,389	1,500	552	..
Pakistan	2011	95,090	61,308	28,517	5,265	..	..	..
	2012	*96,126	*61,716	29,857	4,553	..	..	..
	2013	*104,076	*66,716	31,873	5,090	397	..	..
	2014	*105,305	*68,390	31,428	5,090	397	..	..
Philippines	2011	69,210	49,463	9,715	..	88	1	9,942
	2012	72,956	52,360	10,269	..	75	1	10,250
	2013	75,300	55,592	10,036	..	66	1	9,605
	2014	77,295	57,664	9,154	..	152	17	10,308
Qatar	2011	30,730	30,730	..	..	..	..	..
	2012	34,787	34,787	..	..	..	..	..
	2013	34,668	34,668	..	..	..	..	..
	2014	38,692	38,692	..	..	..	..	..

Production of electricity - by type

Gigawatt-hours

Autoproducers				Main activity producers				
Total	Combustible fuels	Hydro	Other	Total	Combustible fuels	Hydro	Nuclear	Other
..	..	..	..	57,457	57,457	..	..	..
..	..	..	..	62,655	62,655	..	..	..
..	..	..	..	60,982	60,982	..	..	..
..	..	..	..	65,140	65,140	..	..	..
..	..	..	..	15,158	1,019	14,139	..	..
..	..	..	..	15,168	989	14,179	..	..
7	7	..	..	14,004	907	13,097	..	..
20	20	..	..	14,552	1,254	13,298	..	..
*193	..	..	*193	3,609	*300	3,309	..	..
*193	..	..	*193	*3,609	*300	*3,309	..	..
*195	..	..	*195	4,276	*300	3,976	..	..
*195	..	..	*195	4,185	*300	3,885	..	..
4,750	4,750	..	..	11,615	10,810	805	..	..
6,000	6,000	..	..	8,826	7,819	1,007	..	..
7,000	7,000	..	..	10,487	9,218	1,269	..	..
5,565	5,565	..	..	12,387	12,194	193	..	..
5,133	5,133	..	..	119,761	112,136	7,624	..	1
6,154	6,154	..	..	128,267	119,165	9,054	..	48
5,020	5,020	..	..	133,311	122,584	10,586	..	141
4,673	4,673	..	..	142,788	129,173	13,388	..	227
..	..	..	..	275	274	..	..	1
..	..	..	..	299	297	..	..	2
..	..	..	..	305	303	..	..	*2
..	..	..	..	329	327	..	..	*2
*276	*276	..	..	4,536	4,536	..	..	..
*366	*366	..	..	4,816	4,816	..	..	..
*1,196	*1,196	..	..	5,020	5,020	..	..	..
*1,349	*1,349	..	..	5,376	5,376	..	..	..
..	..	..	..	10,425	2,907	7,518	..	..
42	..	42	..	10,923	3,199	7,725	..	..
945	495	450	..	11,302	2,929	8,373	..	..
2,959	2,281	678	..	11,197	3,046	8,151	..	..
59	..	59	..	3,433	2	3,431	..	..
84	..	84	..	3,459	10	3,449	..	..
137	..	137	..	2,993	10	2,983	..	..
*140	..	*140	..	3,368	10	3,358	..	..
..	..	..	..	21,874	21,874	..	..	..
..	..	..	..	25,017	25,017	..	..	..
..	..	..	..	26,240	26,240	..	..	..
..	..	..	..	29,128	29,128	..	..	..
36,489	36,489	..	..	215,527	165,105	6,902	42,116	1,404
34,223	34,223	..	..	216,113	165,547	8,606	40,422	1,538
37,192	37,192	..	..	215,159	162,932	8,610	41,639	1,978
38,071	38,071	..	..	221,954	170,074	7,439	42,389	2,052
39,674	39,674	..	..	55,416	21,634	28,517	5,265	..
*40,174	*40,174	..	..	55,952	*21,542	29,857	4,553	..
*43,174	*43,174	..	..	60,902	*23,542	31,873	5,090	397
*44,257	*44,257	..	..	61,048	*24,133	31,428	5,090	397
*34	*17	*17	..	69,176	49,446	9,698	..	10,032
*34	*17	*17	..	72,922	52,343	10,252	..	10,327
*34	*17	*17	..	75,266	55,575	10,019	..	9,672
*34	*17	*17	..	77,261	57,647	9,137	..	10,477
28,370	28,370	..	..	2,360	2,360	..	..	..
31,799	31,799	..	..	2,988	2,988	..	..	..
31,578	31,578	..	..	3,090	3,090	..	..	..
35,190	35,190	..	..	3,502	3,502	..	..	..

Table 32

Production of electricity - by type

Gigawatt-hours

Country or area	Year	Total producers						
		Total	Combustible fuels	Hydro	Nuclear	Wind	Solar	Other
Saudi Arabia	2011	250,077	250,077	..	..	..	0	..
	2012	271,680	271,679	..	..	..	1	..
	2013	284,017	284,016	..	..	..	1	..
	2014	311,806	311,805	..	..	..	1	..
Singapore	2011	46,000	46,000	..	..	..	..	..
	2012	46,971	46,971	..	..	..	..	..
	2013	47,964	47,964	..	..	..	..	..
	2014	49,310	49,310	..	..	..	..	..
Sri Lanka	2011	11,600	6,916	4,584	..	92	9	..
	2012	11,897	8,439	3,302	..	147	10	..
	2013	12,044	4,870	6,929	..	236	*10	..
	2014	12,832	7,986	4,563	..	273	*9	..
State of Palestine	2011	569	569	..	..	..	..	..
	2012	461	461	..	..	..	..	..
	2013	536	536	..	..	..	..	..
	2014	337	337	..	..	..	..	..
Syrian Arab Republic	2011	42,025	38,742	3,283	..	..	..	..
	2012	31,188	27,951	3,237	..	..	..	..
	2013	25,933	22,933	3,000	..	..	..	..
	2014	21,726	18,726	3,000	..	..	..	..
Tajikistan	2011	16,234	38	16,196	..	..	..	..
	2012	16,998	74	16,924	..	..	..	..
	2013	17,115	44	17,071	..	..	..	..
	2014	16,472	472	16,000	..	..	..	..
Thailand	2011	151,595	143,632	7,935	..	*3	*23	2
	2012	166,471	158,013	8,431	..	*3	*23	1
	2013	175,531	168,397	5,748	..	305	1,080	1
	2014	180,862	173,631	5,540	..	305	1,385	1
Timor-Leste	2011	140	140	..	..	..	..	..
	2012	126	126	..	..	..	..	..
	2013	295	295	..	..	..	..	..
	2014	349	349	..	..	..	..	..
Turkey	2011	229,393	171,638	52,338	..	4,723	..	694
	2012	239,496	174,872	57,865	..	5,860	..	899
	2013	240,154	171,622	59,420	..	7,557	..	1,555
	2014	251,963	200,171	40,645	..	8,520	17	2,610
Turkmenistan	2011	17,220	17,220	0	..	..	..	..
	2012	17,750	17,750	0	..	..	..	..
	2013	18,870	18,870	..	..	..	..	..
	2014	20,400	20,400	..	..	..	..	..
United Arab Emirates	2011	99,137	99,137	..	..	..	..	..
	2012	106,222	106,222	..	..	..	..	..
	2013	109,979	109,979	..	..	..	..	..
	2014	116,528	116,528	..	..	..	..	..
Uzbekistan	2011	52,400	42,160	10,240	..	..	..	..
	2012	52,500	41,290	11,210	..	..	..	..
	2013	54,200	42,640	11,560	..	..	..	..
	2014	55,400	43,570	11,830	..	..	..	..
Viet Nam	2011	105,219	63,944	41,188	..	87	..	..
	2012	118,940	65,676	53,172	..	*92	..	..
	2013	128,549	70,287	58,170	..	*92	..	..
	2014	145,730	83,950	61,480	..	*300	..	..
Yemen	2011	6,204	6,204	..	..	..	..	..
	2012	7,073	7,073	..	..	..	..	..
	2013	8,501	8,501	..	..	..	..	..
	2014	7,646	7,646	..	..	..	..	..

Production of electricity - by type

Gigawatt-hours

Autoproducers				Main activity producers				
Total	Combustible fuels	Hydro	Other	Total	Combustible fuels	Hydro	Nuclear	Other
59,797	59,797	..	..	190,280	190,280	..	..	0
64,548	64,548	..	..	207,132	207,131	..	..	1
85,117	85,117	..	..	198,900	198,899	..	..	1
97,218	97,218	..	..	214,588	214,587	..	..	1
1,315	1,315	..	..	44,685	44,685	..	..	..
1,775	1,775	..	..	45,197	45,197	..	..	..
2,804	2,804	..	..	45,160	45,160	..	..	..
2,974	2,974	..	..	46,336	46,336	..	..	..
18	0	11	8	11,582	6,916	4,573	..	93
18	0	11	8	11,879	8,439	3,291	..	149
*20	0	*11	*9	12,024	4,870	6,918	..	236
*20	..	*11	*9	12,812	7,986	4,552	..	274
..	..	..	..	569	569	..	..	..
..	..	..	..	461	461	..	..	..
..	..	..	..	536	536	..	..	..
..	..	..	..	337	337	..	..	..
1,060	1,060	..	..	40,965	37,682	3,283	..	..
752	752	..	..	30,436	27,199	3,237	..	..
498	498	..	..	25,435	22,435	3,000	..	..
497	497	..	..	21,229	18,229	3,000	..	..
..	..	..	..	16,234	38	16,196	..	..
..	..	..	..	16,998	74	16,924	..	..
..	..	..	..	17,115	44	17,071	..	..
..	..	..	..	16,472	472	16,000	..	..
14,936	14,936	..	..	136,659	128,696	7,935	..	*28
15,134	15,134	..	..	151,337	142,879	8,431	..	*27
29,532	28,668	..	864	145,999	139,729	5,748	..	522
34,890	33,782	..	1,108	145,972	139,849	5,540	..	583
..	..	..	..	140	140	..	..	..
..	..	..	..	126	126	..	..	..
..	..	..	..	295	295	..	..	..
..	..	..	..	349	349	..	..	..
12,243	11,001	1,240	2	217,150	160,637	51,098	..	5,415
13,254	11,594	1,658	2	226,242	163,278	56,207	..	6,757
13,071	11,800	1,076	195	227,083	159,822	58,344	..	8,917
4,441	3,907	391	143	247,522	196,264	40,254	..	11,004
..	..	..	..	17,220	17,220	0	..	..
..	..	..	..	17,750	17,750	0	..	..
..	..	..	..	18,870	18,870	..	..	..
..	..	..	..	20,400	20,400	..	..	..
..	..	..	..	99,137	99,137	..	..	..
..	..	..	..	106,222	106,222	..	..	..
..	..	..	..	109,979	109,979	..	..	..
..	..	..	..	116,528	116,528	..	..	..
197	197	..	..	52,203	41,963	10,240	..	..
193	193	..	..	52,307	41,097	11,210	..	..
199	199	..	..	54,001	42,441	11,560	..	..
204	204	..	..	55,196	43,366	11,830	..	..
3,633	3,633	..	..	101,586	60,311	41,188	..	87
*3,703	*3,700	*1	*2	115,237	61,976	53,171	..	*90
*4,003	*4,000	*1	*2	124,546	66,287	58,169	..	*90
*4,182	*4,100	*80	*2	141,548	79,850	61,400	..	*298
269	269	..	..	5,935	5,935	..	..	..
346	346	..	..	6,727	6,727	..	..	..
245	245	..	..	8,256	8,256	..	..	..
240	240	..	..	7,406	7,406	..	..	..

Table 32

Production of electricity - by type

Gigawatt-hours

Country or area	Year	Total producers						
		Total	Combustible fuels	Hydro	Nuclear	Wind	Solar	Other
Europe	**2011**	**4,854,141**	**2,705,057**	**707,938**	**1,196,643**	**181,022**	**47,463**	**16,018**
	2012	**4,895,081**	**2,661,372**	**762,097**	**1,175,478**	**207,780**	**71,815**	**16,539**
	2013	**4,850,703**	**2,538,127**	**811,304**	**1,158,537**	**239,214**	**86,764**	**16,758**
	2014	**4,766,092**	**2,412,213**	**807,781**	**1,172,996**	**256,572**	**99,168**	**17,361**
Albania	2011	4,217	59	4,158	..	..	..	..
	2012	4,725	..	4,725	..	..	..	..
	2013	6,956	..	6,956	..	..	..	..
	2014	4,724	..	4,724	..	..	..	..
Andorra	2011	91	12	79	..	..	..	..
	2012	88	12	76	..	..	..	..
	2013	115	13	102	..	..	..	..
	2014	127	13	113	..	..	..	..
Austria	2011	65,813	25,909	37,782	..	1,936	174	12
	2012	72,617	22,101	47,705	..	2,462	337	12
	2013	68,277	18,752	45,777	..	3,152	582	14
	2014	65,421	15,949	44,828	..	3,846	785	13
Belarus	2011	32,200	32,157	42	..	1	..	..
	2012	30,799	30,723	70	..	6	..	..
	2013	31,507	31,361	138	..	8	..	..
	2014	34,735	34,602	121	..	11	1	..
Belgium	2011	90,241	36,969	1,423	48,234	2,312	1,169	134
	2012	82,923	35,843	1,659	40,295	2,751	2,148	227
	2013	83,526	32,467	1,723	42,644	3,687	2,644	361
	2014	72,688	29,582	1,507	33,703	4,614	2,883	399
Bosnia and Herzegovina	2011	15,280	10,893	4,387	..	..	..	..
	2012	14,082	9,867	4,215	..	..	..	..
	2013	17,451	10,215	7,236	..	..	..	..
	2014	16,160	10,225	5,935	..	..	..	..
Bulgaria	2011	50,797	29,807	3,691	16,314	861	101	23
	2012	47,329	25,518	3,976	15,785	1,221	814	15
	2013	43,784	22,067	4,795	14,171	1,374	1,361	16
	2014	47,485	23,858	5,163	15,867	1,331	1,252	14
Croatia	2011	11,373	6,010	5,162	..	201	0	..
	2012	10,755	5,425	4,999	..	329	2	..
	2013	14,052	4,797	8,727	..	517	11	..
	2014	13,554	3,664	9,125	..	730	35	..
Czechia	2011	87,561	54,035	2,664	28,283	397	2,182	..
	2012	87,573	51,824	2,860	30,324	416	2,149	..
	2013	87,065	50,167	3,639	30,745	481	2,033	..
	2014	86,024	50,138	2,961	30,325	477	2,123	..
Denmark	2011	35,229	25,423	17	..	9,774	15	..
	2012	30,701	20,310	17	..	10,270	104	..
	2013	34,760	23,106	13	..	11,123	518	..
	2014	32,183	18,493	15	..	13,079	596	..
Estonia	2011	12,893	12,495	30	..	368	..	..
	2012	11,967	11,491	42	..	434	..	..
	2013	13,275	12,720	26	..	529	..	..
	2014	12,446	11,815	27	..	604	..	..
Faeroe Islands	2011	274	167	93	..	15	..	..
	2012	292	181	100	..	11	..	..
	2013	293	180	91	..	22	..	..
	2014	305	150	121	..	34	..	..
Finland	2011	73,506	37,083	12,445	23,187	481	10	300
	2012	70,416	29,746	16,859	22,987	494	11	319
	2013	71,263	33,744	12,838	23,606	774	12	289
	2014	68,093	29,711	13,397	23,580	1,107	8	290

Production of electricity - by type

Gigawatt-hours

Autoproducers				Main activity producers				
Total	Combustible fuels	Hydro	Other	Total	Combustible fuels	Hydro	Nuclear	Other
354,664	**328,084**	**15,154**	**11,426**	**4,499,477**	**2,376,973**	**692,784**	**1,196,643**	**233,077**
353,849	**322,717**	**15,465**	**15,667**	**4,541,231**	**2,338,655**	**746,632**	**1,175,478**	**280,466**
345,342	**308,310**	**16,174**	**20,858**	**4,505,362**	**2,229,817**	**795,130**	**1,158,537**	**321,878**
342,868	**300,880**	**16,015**	**25,973**	**4,423,223**	**2,111,332**	**791,766**	**1,172,996**	**347,128**
..	..	..	..	4,217	59	4,158	..	..
..	..	..	..	4,725	..	4,725	..	..
..	..	..	..	6,956	..	6,956	..	..
..	..	..	..	4,724	..	4,724	..	..
..	..	..	..	91	12	79	..	..
..	..	..	..	88	12	76	..	..
..	..	..	..	115	13	102	..	..
..	..	..	..	127	13	113	..	..
9,454	8,477	966	11	56,359	17,432	36,816	..	2,111
8,572	8,023	538	11	64,045	14,078	47,167	..	2,800
8,099	7,533	552	14	60,178	11,219	45,225	..	3,734
7,691	7,101	577	13	57,730	8,848	44,251	..	4,631
2,596	2,591	4	1	29,604	29,566	38	..	..
3,050	3,042	6	2	27,749	27,681	64	..	4
2,976	2,967	7	2	28,531	28,394	131	..	6
3,072	3,061	7	4	31,663	31,541	114	..	8
6,874	5,561	..	1,313	83,367	31,408	1,423	48,234	2,302
8,496	6,103	..	2,393	74,427	29,740	1,659	40,295	2,733
9,124	6,110	..	3,014	74,402	26,357	1,723	42,644	3,678
9,789	6,488	..	3,301	62,899	23,094	1,507	33,703	4,595
325	325	..	..	14,955	10,568	4,387	..	..
343	343	..	..	13,739	9,524	4,215	..	..
369	369	..	..	17,082	9,846	7,236	..	..
403	403	..	..	15,757	9,822	5,935	..	..
147	124	..	23	50,650	29,683	3,691	16,314	962
270	255	..	15	47,059	25,263	3,976	15,785	2,035
268	252	..	16	43,516	21,815	4,795	14,171	2,735
253	239	..	14	47,232	23,619	5,163	15,867	2,583
516	512	4	..	10,857	5,498	5,158	..	201
388	383	5	..	10,367	5,042	4,994	..	331
332	327	5	..	13,720	4,470	8,722	..	528
345	338	7	..	13,209	3,326	9,118	..	765
8,129	7,626	503	..	79,432	46,409	2,161	28,283	2,579
7,985	7,462	523	..	79,588	44,362	2,337	30,324	2,565
8,218	7,597	621	..	78,847	42,570	3,018	30,745	2,514
8,478	7,914	564	..	77,546	42,224	2,397	30,325	2,600
2,161	2,146	..	15	33,068	23,277	17	..	9,774
2,047	1,943	..	104	28,654	18,367	17	..	10,270
2,437	1,919	..	518	32,323	21,187	13	..	11,123
2,606	2,010	..	596	29,577	16,483	15	..	13,079
105	105	..	0	12,788	12,390	30	..	368
98	97	..	1	11,869	11,394	42	..	433
91	86	..	5	13,184	12,634	26	..	524
57	42	1	14	12,389	11,773	26	..	590
..	..	..	..	274	167	93	..	15
..	..	..	..	292	181	100	..	11
..	..	..	..	293	180	91	..	22
..	..	..	..	305	150	121	..	34
10,654	9,510	949	195	62,852	27,573	11,496	23,187	596
10,255	8,791	1,244	220	60,161	20,955	15,615	22,987	604
10,122	8,960	979	183	61,141	24,784	11,859	23,606	892
10,093	8,976	927	190	58,000	20,735	12,470	23,580	1,215

Table 32

Production of electricity - by type

Gigawatt-hours

Country or area	Year	Total producers						
		Total	Combustible fuels	Hydro	Nuclear	Wind	Solar	Other
France	2011	561,448	54,594	49,865	442,383	12,051	2,078	477
	2012	565,704	56,722	63,594	425,406	14,914	4,016	1,052
	2013	572,308	50,877	75,867	423,685	16,034	4,735	1,110
	2014	562,776	33,484	68,626	436,474	17,249	5,909	1,034
Germany	2011	613,068	410,773	23,511	107,971	48,883	19,599	2,331
	2012	629,812	423,424	27,849	99,460	50,670	26,380	2,029
	2013	638,729	428,094	28,782	97,290	51,708	31,010	1,845
	2014	627,795	409,688	25,444	97,129	57,357	36,056	2,121
Gibraltar	2011	171	171	..	..	..	..	..
	2012	176	176	..	..	..	..	..
	2013	189	189	..	..	..	..	..
	2014	198	198	..	..	..	..	..
Greece	2011	59,436	51,236	4,275	..	3,315	610	..
	2012	60,959	50,824	4,591	..	3,850	1,694	..
	2013	57,152	42,981	6,384	..	4,139	3,648	..
	2014	50,474	38,386	4,607	..	3,689	3,792	..
Guernsey	2011	85	85	..	..	..	..	..
	2012	68	68	..	..	..	..	..
	2013	289	289	..	..	..	..	..
	2014	227	227	..	..	..	..	..
Hungary	2011	36,019	19,485	222	15,685	626	1	..
	2012	34,635	17,851	213	15,793	770	8	..
	2013	30,291	13,933	213	15,370	718	25	32
	2014	29,371	12,652	302	15,649	657	56	55
Iceland	2011	17,211	2	12,507	..	..	..	4,702
	2012	17,549	3	12,337	..	..	..	5,209
	2013	18,116	5	12,863	..	3	..	5,245
	2014	18,122	3	12,873	..	8	..	5,238
Ireland	2011	27,474	22,386	707	..	4,380	1	..
	2012	27,600	22,575	1,014	..	4,010	1	..
	2013	26,142	20,655	944	..	4,542	1	..
	2014	26,314	20,185	988	..	5,140	1	..
Isle of Man	2011	499	496	3	..	..	..	..
	2012	296	292	4	..	..	..	..
	2013	*504	*500	*4	..	..	..	..
	2014	*504	*500	*4	..	..	..	..
Italy	2011	302,582	227,711	47,757	..	9,856	10,796	6,462
	2012	299,277	216,811	43,854	..	13,407	18,862	6,343
	2013	289,807	192,237	54,672	..	14,897	21,589	6,412
	2014	279,827	175,510	60,256	..	15,178	22,306	6,577
Jersey	2011	46	46	..	..	..	..	..
	2012	60	60	..	..	..	..	..
	2013	328	328	..	..	..	..	..
	2014	135	135	..	..	..	..	..
Latvia	2011	6,095	3,137	2,887	..	71	..	..
	2012	6,167	2,346	3,707	..	114	..	..
	2013	6,209	3,177	2,912	..	120	..	..
	2014	5,141	3,006	1,994	..	141	..	..
Liechtenstein	2011	72	4	62	..	..	6	..
	2012	86	4	74	..	..	9	..
	2013	85	3	69	..	..	13	..
	2014	40	3	22	..	..	16	..
Lithuania	2011	4,822	3,034	1,056	..	475	..	257
	2012	5,043	3,337	937	..	540	2	227
	2013	4,762	2,808	1,069	..	603	45	237
	2014	4,397	2,354	1,088	..	639	73	243

Production of electricity - by type

Gigawatt-hours

Autoproducers				Main activity producers				
Total	Combustible fuels	Hydro	Other	Total	Combustible fuels	Hydro	Nuclear	Other
15,852	13,255	476	2,121	545,596	41,339	49,389	442,383	12,485
21,882	17,220	654	4,008	543,822	39,502	62,940	425,406	15,974
18,786	13,225	743	4,818	553,522	37,652	75,124	423,685	17,061
18,629	12,619	800	5,210	544,147	20,865	67,826	436,474	18,982
50,614	48,649	327	1,638	562,454	362,124	23,184	107,971	69,175
44,094	42,421	350	1,323	585,718	381,003	27,499	99,460	77,756
44,892	43,466	182	1,244	593,837	384,628	28,600	97,290	83,319
45,450	43,874	162	1,414	582,345	365,814	25,282	97,129	94,120
..	..	..	..	171	171	..	..	..
..	..	..	..	176	176	..	..	..
..	..	..	..	189	189	..	..	..
..	..	..	..	198	198	..	..	..
2,553	2,553	..	..	56,883	48,683	4,275	..	3,925
2,351	2,351	..	..	58,608	48,473	4,591	..	5,544
1,885	1,885	..	..	55,267	41,096	6,384	..	7,787
1,792	1,792	..	..	48,682	36,594	4,607	..	7,481
..	..	..	..	85	85	..	..	..
..	..	..	..	68	68	..	..	..
..	..	..	..	289	289	..	..	..
..	..	..	..	227	227	..	..	..
696	695	..	1	35,323	18,790	222	15,685	626
516	508	..	8	34,119	17,343	213	15,793	770
612	555	..	57	29,679	13,378	213	15,370	718
642	531	..	111	28,729	12,121	302	15,649	657
..	..	..	..	17,211	2	12,507	..	4,702
..	..	..	..	17,549	3	12,337	..	5,209
..	..	..	..	18,116	5	12,863	..	5,248
..	..	..	..	18,122	3	12,873	..	5,246
1,964	1,963	..	1	25,510	20,423	707	..	4,380
2,126	2,125	..	1	25,474	20,450	1,014	..	4,010
2,056	2,055	..	1	24,086	18,600	944	..	4,542
2,073	2,072	..	1	24,241	18,113	988	..	5,140
..	..	..	..	499	496	3	..	..
..	..	..	..	296	292	4	..	..
..	..	..	..	*504	*500	*4	..	..
..	..	..	..	*504	*500	*4	..	..
23,563	22,183	868	512	279,019	205,528	46,889	..	26,602
16,057	15,025	573	459	283,220	201,786	43,281	..	38,153
16,073	15,124	627	322	273,734	177,113	54,045	..	42,576
15,742	14,749	738	255	264,085	160,761	59,518	..	43,806
33	33	..	..	13	13	..	..	..
43	43	..	..	17	17	..	..	..
181	181	..	..	147	147	..	..	..
37	37	..	..	99	99	..	..	..
88	79	9	0	6,007	3,058	2,878	..	71
183	176	7	0	5,984	2,170	3,700	..	114
257	251	3	3	5,952	2,926	2,909	..	117
257	246	7	4	4,884	2,760	1,987	..	137
..	..	..	..	72	4	62	..	6
..	..	..	..	86	4	74	..	9
..	..	..	..	85	3	69	..	13
..	..	..	..	40	3	22	..	16
600	343	..	257	4,222	2,691	1,056	..	475
787	560	..	227	4,256	2,777	937	..	542
621	384	..	237	4,141	2,424	1,069	..	648
736	493	..	243	3,661	1,861	1,088	..	712

Table 32

Production of electricity - by type

Gigawatt-hours

Country or area	Year	Total producers						
		Total	Combustible fuels	Hydro	Nuclear	Wind	Solar	Other
Luxembourg	2011	3,718	2,498	1,130	..	64	26	..
	2012	3,818	2,543	1,160	..	77	38	..
	2013	2,888	1,573	1,158	..	83	74	..
	2014	2,967	1,623	1,169	..	80	95	..
Malta	2011	2,187	2,174	..	..	..	13	..
	2012	2,307	2,277	..	..	..	30	..
	2013	2,251	2,222	..	..	..	29	..
	2014	2,245	2,177	..	..	..	68	..
Montenegro	2011	2,809	1,600	1,209	..	..	..	..
	2012	2,844	1,367	1,477	..	..	..	..
	2013	3,945	1,444	2,501	..	..	..	..
	2014	3,040	1,322	1,716	..	..	2	..
Netherlands	2011	113,963	104,412	57	4,141	5,101	104	148
	2012	103,298	93,925	104	3,915	4,981	226	147
	2013	101,736	92,494	114	2,891	5,627	487	123
	2014	103,418	92,488	112	4,091	5,797	785	145
Norway	2011	127,655	4,739	121,552	..	1,283	..	81
	2012	147,716	3,354	142,812	..	1,548	..	2
	2013	133,975	3,152	128,699	..	1,881	..	243
	2014	142,327	3,172	136,636	..	2,216	..	303
Poland	2011	163,548	157,477	2,761	..	3,205	..	105
	2012	162,139	154,845	2,465	..	4,747	1	81
	2013	164,557	155,444	2,997	..	6,004	1	111
	2014	159,059	148,500	2,734	..	7,676	7	142
Portugal	2011	52,463	30,693	12,115	..	9,161	280	214
	2012	46,614	29,151	6,660	..	10,259	393	151
	2013	51,672	24,110	14,868	..	12,014	479	201
	2014	52,802	23,440	16,412	..	12,111	627	212
Republic of Moldova	2011	1,016	940	76	..	0	..	..
	2012	932	898	34	..	0	..	..
	2013	905	859	45	..	1	..	..
	2014	963	902	59	..	1	1	..
Romania	2011	62,217	34,135	14,946	11,747	1,388	1	..
	2012	59,045	32,594	12,337	11,466	2,640	8	..
	2013	58,887	27,021	15,308	11,618	4,520	420	..
	2014	65,676	26,903	19,280	11,676	6,201	1,616	..
Russian Federation	2011	1,054,765	713,689	167,608	172,941	5	..	522
	2012	1,070,734	725,399	167,319	177,534	5	..	477
	2013	1,059,092	703,481	182,654	172,508	5	..	444
	2014	1,064,207	705,598	177,141	180,757	96	160	455
Serbia	2011	38,600	29,357	9,243	..	..	..	..
	2012	36,799	26,885	9,914	..	..	..	..
	2013	39,877	29,024	10,853	..	..	..	..
	2014	34,060	22,437	11,617	..	..	6	..
Slovakia	2011	28,656	8,619	4,146	15,411	5	397	78
	2012	28,664	8,231	4,439	15,495	6	424	69
	2013	28,832	7,283	5,166	15,720	6	588	69
	2014	27,401	6,719	4,462	15,499	6	597	118
Slovenia	2011	16,059	6,073	3,706	6,215	0	65	..
	2012	15,736	5,958	4,087	5,528	0	163	..
	2013	16,103	5,661	4,923	5,300	4	215	..
	2014	17,437	4,440	6,366	6,370	4	257	..
Spain	2011	293,848	150,730	32,911	57,718	42,918	9,400	171
	2012	297,559	150,312	24,162	61,470	49,472	11,968	175
	2013	285,632	119,111	41,052	56,726	55,646	13,097	..
	2014	278,750	112,789	42,970	57,305	52,013	13,673	..

Production of electricity - by type

Gigawatt-hours

Autoproducers				Main activity producers				
Total	Combustible fuels	Hydro	Other	Total	Combustible fuels	Hydro	Nuclear	Other
208	178	4	26	3,510	2,320	1,126	..	64
225	182	5	38	3,593	2,361	1,155	..	77
265	185	6	74	2,623	1,388	1,152	..	83
271	170	6	95	2,696	1,453	1,163	..	80
10	5	..	5	2,177	2,169	..	..	8
26	9	..	17	2,281	2,268	..	..	13
35	6	..	29	2,216	2,216	..	..	0
75	7	..	68	2,170	2,170	..	..	..
..	..	..	..	2,809	1,600	1,209	..	..
..	..	..	..	2,844	1,367	1,477	..	..
..	..	..	..	3,945	1,444	2,501	..	..
2	..	..	2	3,038	1,322	1,716	..	..
24,748	23,582	..	1,166	89,215	80,830	57	4,141	4,187
25,148	23,881	..	1,267	78,150	70,044	104	3,915	4,087
24,608	22,998	..	1,610	77,128	69,496	114	2,891	4,627
21,960	20,048	..	1,912	81,458	72,440	112	4,091	4,815
9,538	4,379	5,101	58	118,117	360	116,451	..	1,306
8,619	2,897	5,720	2	139,097	457	137,092	..	1,548
8,472	2,670	5,559	243	125,503	482	123,140	..	1,881
7,671	2,651	4,717	303	134,656	521	131,919	..	2,216
8,342	8,235	2	105	155,206	149,242	2,759	..	3,205
8,449	8,365	2	82	153,690	146,480	2,463	..	4,747
8,917	8,802	3	112	155,640	146,642	2,994	..	6,004
8,881	8,729	3	149	150,178	139,771	2,731	..	7,676
8,347	8,251	14	82	44,116	22,442	12,101	..	9,573
8,344	8,184	10	150	38,270	20,967	6,650	..	10,653
9,010	8,780	20	210	42,662	15,330	14,848	..	12,484
8,435	8,117	23	295	44,367	15,323	16,389	..	12,655
34	34	..	..	982	906	76	..	0
32	32	..	..	900	866	34	..	0
44	44	..	..	861	815	45	..	1
72	71	..	1	891	831	59	..	1
2,780	2,626	129	25	59,437	31,509	14,817	11,747	1,364
4,252	3,908	158	186	54,793	28,686	12,179	11,466	2,462
6,519	5,918	246	355	52,368	21,103	15,062	11,618	4,585
6,095	5,037	327	731	59,581	21,866	18,953	11,676	7,086
64,108	63,002	1,106	..	990,657	650,687	166,502	172,941	527
67,181	66,158	1,023	..	1,003,553	659,241	166,296	177,534	482
64,495	63,047	1,448	..	994,597	640,434	181,206	172,508	449
66,732	65,186	1,546	..	997,475	640,412	175,595	180,757	711
230	230	..	..	38,370	29,127	9,243	..	..
171	171	..	..	36,628	26,714	9,914	..	..
202	202	..	..	39,675	28,822	10,853	..	..
289	289	..	..	33,771	22,148	11,617	..	6
2,492	1,996	101	395	26,164	6,623	4,045	15,411	85
2,536	2,039	88	409	26,128	6,192	4,351	15,495	90
2,470	1,825	106	539	26,362	5,458	5,060	15,720	124
3,040	2,317	116	607	24,361	4,402	4,346	15,499	114
436	248	125	63	15,623	5,825	3,581	6,215	2
534	251	126	157	15,202	5,707	3,961	5,528	6
622	247	166	209	15,481	5,414	4,757	5,300	10
749	271	228	250	16,688	4,169	6,138	6,370	11
40,818	39,750	861	207	253,030	110,980	32,050	57,718	52,282
41,427	40,752	462	213	256,132	109,560	23,700	61,470	61,402
38,128	37,145	943	40	247,504	81,966	40,109	56,726	68,703
32,111	31,006	1,064	41	246,639	81,783	41,906	57,305	65,645

Table 32

Production of electricity - by type

Gigawatt-hours

Country or area	Year	Total producers						
		Total	Combustible fuels	Hydro	Nuclear	Wind	Solar	Other
Sweden	2011	150,376	17,256	66,556	60,475	6,078	11	..
	2012	166,562	16,283	79,058	64,037	7,165	19	..
	2013	153,166	15,336	61,496	66,457	9,842	35	..
	2014	153,662	13,632	63,872	64,877	11,234	47	..
Switzerland	2011	64,629	3,567	34,133	26,710	70	149	..
	2012	69,893	3,739	40,305	25,441	88	320	..
	2013	70,312	3,720	39,968	25,990	90	544	..
	2014	71,767	3,566	39,701	27,557	101	842	..
T.F.Yug.Rep. Macedonia	2011	6,759	5,325	1,433	..	..	1	..
	2012	6,262	5,218	1,041	..	..	3	..
	2013	6,094	4,501	1,584	..	..	9	..
	2014	5,303	4,082	1,207	..	..	14	..
Ukraine	2011	194,947	93,634	10,946	90,248	89	30	..
	2012	198,878	97,126	10,994	90,137	288	333	..
	2013	194,377	95,487	14,472	83,209	639	570	..
	2014	182,815	83,549	9,318	88,389	1,130	429	..
United Kingdom	2011	367,423	273,961	8,585	68,980	15,652	244	1
	2012	363,582	263,734	8,252	70,405	19,835	1,352	4
	2013	359,168	250,539	7,606	70,607	28,421	1,989	6
	2014	338,925	230,342	8,768	63,748	32,015	4,050	2
Oceania	**2011**	**308,986**	**248,779**	**43,987**	**..**	**8,092**	**1,540**	**6,589**
	2012	**306,065**	**248,574**	**39,145**	**..**	**9,113**	**2,578**	**6,655**
	2013	**304,237**	**239,939**	**43,529**	**..**	**10,045**	**3,850**	**6,874**
	2014	**303,224**	**233,381**	**44,703**	**..**	**12,531**	**4,895**	**7,716**
American Samoa	2011	155	155	..	..	..	..	..
	2012	157	156	..	..	..	1	..
	2013	*156	*155	..	..	..	*1	..
	2014	157	156	..	..	..	*1	..
Australia	2011	253,958	229,535	16,807	..	6,085	1,530	1
	2012	251,165	227,552	14,083	..	6,970	2,559	1
	2013	249,720	219,663	18,270	..	7,960	3,826	1
	2014	248,299	214,767	18,421	..	10,252	4,858	1
Cook Islands	2011	34	34	..	..	..	..	..
	2012	33	33	..	..	..	..	..
	2013	33	33	..	..	..	..	..
	2014	32	32	..	..	..	..	..
Fiji	2011	801	340	456	..	5	..	..
	2012	803	271	525	..	7	..	..
	2013	858	325	527	..	5	..	..
	2014	859	454	401	..	4	..	..
French Polynesia	2011	*761	*533	*225	..	..	*3	..
	2012	*768	*553	*209	..	..	*6	..
	2013	*744	*542	*195	..	..	*7	..
	2014	*721	*492	*219	..	..	*10	..
Guam	2011	*1,858	*1,858	..	..	..	..	..
	2012	*1,799	*1,799	..	..	..	..	..
	2013	1,736	1,736	..	..	..	..	..
	2014	*1,750	*1,750	..	..	..	..	..
Kiribati	2011	*24	*24	..	..	..	..	..
	2012	*24	*24	..	..	..	..	..
	2013	*24	*24	..	..	..	..	..
	2014	*24	*24	..	..	..	..	..
Marshall Islands	2011	*79	*79	..	..	..	..	..
	2012	78	78	..	..	..	0	..
	2013	84	83	..	..	..	0	..
	2014	*85	*85	..	..	..	0	..

Production of electricity - by type

Gigawatt-hours

Autoproducers				Main activity producers				
Total	Combustible fuels	Hydro	Other	Total	Combustible fuels	Hydro	Nuclear	Other
6,015	6,000	15	..	144,361	11,256	66,541	60,475	6,089
6,350	6,333	17	..	160,212	9,950	79,041	64,037	7,184
5,854	5,844	10	..	147,312	9,492	61,486	66,457	9,877
5,797	5,785	12	..	147,865	7,847	63,860	64,877	11,281
5,414	2,782	2,483	149	59,215	785	31,650	26,710	70
6,065	2,927	2,818	320	63,828	812	37,487	25,441	88
6,397	3,020	2,833	544	63,915	700	37,135	25,990	90
6,865	3,109	2,914	842	64,902	457	36,787	27,557	101
2	2	..	..	6,757	5,323	1,433	..	1
2	2	..	..	6,260	5,216	1,041	..	3
2	2	..	..	6,092	4,499	1,584	..	9
1	1	..	..	5,302	4,081	1,207	..	14
3,653	3,632	21	..	191,294	90,002	10,925	90,248	119
3,773	3,753	20	..	195,105	93,373	10,974	90,137	621
3,200	3,177	22	1	191,177	92,310	14,450	83,209	1,208
2,927	2,905	19	3	179,888	80,644	9,299	88,389	1,556
40,565	36,422	1,086	3,057	326,858	237,539	7,499	68,980	12,840
41,172	36,002	1,116	4,054	322,410	227,732	7,136	70,405	17,137
38,703	31,152	1,093	6,458	320,465	219,387	6,513	70,607	23,958
42,751	32,197	1,250	9,304	296,174	198,145	7,518	63,748	26,763
23,945	**21,532**	**363**	**2,050**	**285,041**	**227,246**	**43,624**	**..**	**14,171**
24,000	**20,558**	**364**	**3,078**	**282,065**	**228,016**	**38,781**	**..**	**15,269**
28,240	**23,524**	**373**	**4,343**	**275,997**	**216,415**	**43,156**	**..**	**16,426**
29,257	**23,503**	**373**	**5,382**	**273,967**	**209,878**	**44,330**	**..**	**19,760**
..	..	..	..	155	155	..	..	..
..	..	..	..	157	156	..	..	1
..	..	..	..	*156	*155	..	..	*1
..	..	..	..	157	156	..	..	*1
19,814	18,289	..	1,525	234,144	211,246	16,807	..	6,091
19,667	17,112	..	2,555	231,498	210,440	14,083	..	6,975
23,364	19,543	..	3,821	226,356	200,120	18,270	..	7,966
24,355	19,504	..	4,851	223,944	195,263	18,421	..	10,260
..	..	..	..	34	34	..	..	..
..	..	..	..	33	33	..	..	..
..	..	..	..	33	33	..	..	..
..	..	..	..	32	32	..	..	..
44	44	..	..	757	295	456	..	5
48	48	..	..	755	223	525	..	7
47	47	..	..	811	278	527	..	5
50	50	..	..	810	404	401	..	4
..	..	..	..	*761	*533	*225	..	*3
..	..	..	..	*768	*553	*209	..	*6
..	..	..	..	*744	*542	*195	..	*7
..	..	..	..	*721	*492	*219	..	*10
..	..	..	..	*1,858	*1,858	..	..	..
..	..	..	..	*1,799	*1,799	..	..	..
..	..	..	..	1,736	1,736	..	..	..
..	..	..	..	*1,750	*1,750	..	..	..
..	..	..	..	*24	*24	..	..	..
..	..	..	..	*24	*24	..	..	..
..	..	..	..	*24	*24	..	..	..
..	..	..	..	*24	*24	..	..	..
..	..	..	..	*79	*79	..	..	..
..	..	..	..	78	78	..	..	0
..	..	..	..	84	83	..	..	0
..	..	..	..	*85	*85	..	..	0

Table 32

Production of electricity - by type

Gigawatt-hours

Country or area	Year	Total producers						
		Total	Combustible fuels	Hydro	Nuclear	Wind	Solar	Other
Micronesia (Fed. States of)	2011	66	64	..	..	..	*2	..
	2012	*67	*65	..	..	..	*2	..
	2013	*68	*66	..	..	..	*2	..
	2014	70	68	..	..	..	*2	..
Nauru	2011	23	23	..	..	..	0	..
	2012	24	24	..	..	..	0	..
	2013	*24	*24	..	..	..	0	..
	2014	*25	*25	..	..	..	0	..
New Caledonia	2011	2,267	1,843	375	..	46	4	..
	2012	2,288	1,827	399	..	57	5	..
	2013	2,506	1,994	455	..	52	5	..
	2014	2,622	2,272	288	..	57	5	..
New Zealand	2011	44,465	11,233	25,114	..	1,952	..	6,166
	2012	44,260	13,046	22,901	..	2,076	5	6,232
	2013	43,265	11,739	23,044	..	2,024	7	6,451
	2014	43,553	9,694	24,336	..	2,214	16	7,293
Niue	2011	3	3	..	..	..	0	..
	2012	3	3	..	..	..	0	..
	2013	3	3	..	..	..	0	..
	2014	3	3	..	..	..	0	..
Northern Mariana Islands	2011	*407	*407	..	..	..	..	..
	2012	*411	*411	..	..	..	..	..
	2013	*415	*415	..	..	..	..	..
	2014	*419	*419	..	..	..	..	..
Palau	2011	*78	*78	..	..	..	..	..
	2012	*78	*78	..	..	..	..	..
	2013	*78	*78	..	..	..	..	..
	2014	*79	*79	..	..	..	..	..
Papua New Guinea	2011	3,666	2,275	969	..	..	..	422
	2012	3,764	2,359	983	..	..	..	422
	2013	4,169	2,752	995	..	..	..	422
	2014	*4,169	*2,752	*995	..	..	..	*422
Samoa	2011	114	79	35	..	..	..	..
	2012	117	80	37	..	..	..	..
	2013	*121	*85	*36	..	..	..	..
	2014	*121	*85	*36	..	..	..	..
Solomon Islands	2011	83	83	..	..	..	..	..
	2012	83	83	..	..	..	..	..
	2013	89	89	..	..	..	..	..
	2014	92	92	..	..	..	..	..
Tonga	2011	53	53	..	..	..	..	..
	2012	52	52	..	..	..	..	..
	2013	*53	*51	..	..	..	*2	..
	2014	*55	*52	..	..	..	*2	..
Tuvalu	2011	5	5	..	..	..	..	..
	2012	*5	*5	..	..	..	..	..
	2013	*5	*5	..	..	..	..	..
	2014	*5	*5	..	..	..	..	..
Vanuatu	2011	*68	*57	6	..	*5	0	..
	2012	67	57	7	..	3	0	..
	2013	*68	*58	*7	..	*3	0	..
	2014	*67	*56	*7	..	*3	0	..
Wallis and Futuna Is.	2011	20	20	..	..	..	..	..
	2012	19	19	..	..	..	..	..
	2013	19	19	..	..	..	..	..
	2014	19	19	..	..	..	..	..

Production of electricity - by type

Gigawatt-hours

Autoproducers				Main activity producers				
Total	Combustible fuels	Hydro	Other	Total	Combustible fuels	Hydro	Nuclear	Other
..	..	..	..	66	64	..	..	*2
..	..	..	..	*67	*65	..	..	*2
..	..	..	..	*68	*66	..	..	*2
..	..	..	..	70	68	..	..	*2
..	..	..	..	23	23	..	..	0
..	..	..	..	24	24	..	..	0
..	..	..	..	*24	*24	..	..	0
..	..	..	..	*25	*25	..	..	0
*11	*10	..	*1	2,256	1,833	375	..	49
24	23	..	*1	2,264	1,804	399	..	61
207	206	..	1	2,299	1,788	455	..	56
*242	*241	..	*1	2,380	2,031	288	..	61
1,276	1,162	12	102	43,189	10,071	25,102	..	8,016
1,374	1,261	13	100	42,886	11,785	22,888	..	8,213
1,409	1,297	13	99	41,856	10,442	23,031	..	8,383
1,398	1,277	13	108	42,155	8,417	24,323	..	9,415
..	..	..	..	3	3	..	..	0
..	..	..	..	3	3	..	..	0
..	..	..	..	3	3	..	..	0
..	..	..	..	3	3	..	..	0
..	..	..	..	*407	*407	..	..	..
..	..	..	..	*411	*411	..	..	..
..	..	..	..	*415	*415	..	..	..
..	..	..	..	*419	*419	..	..	..
*6	*6	..	..	*72	*72	..	..	..
*6	*6	..	..	*72	*72	..	..	..
*6	*6	..	..	*72	*72	..	..	..
*6	*6	..	..	*73	*73	..	..	..
2,781	2,008	351	422	885	267	618	..	..
2,868	2,095	351	422	896	264	632	..	..
3,194	2,412	360	422	975	340	635	..	..
*3,194	*2,412	*360	*422	*975	*340	*635	..	..
*5	*5	..	..	109	74	35	..	..
*5	*5	..	..	112	75	37	..	..
*5	*5	..	..	*116	*80	*36	..	..
*5	*5	..	..	*116	*80	*36	..	..
*8	*8	..	..	75	75	..	..	..
*8	*8	..	..	75	75	..	..	..
*8	*8	..	..	81	81	..	..	..
*8	*8	..	..	84	84	..	..	..
..	..	..	..	53	53	..	..	..
..	..	..	..	52	52	..	..	..
..	..	..	..	*53	*51	..	..	*2
..	..	..	..	*55	*52	..	..	*2
..	..	..	..	5	5	..	..	..
..	..	..	..	*5	*5	..	..	..
..	..	..	..	*5	*5	..	..	..
..	..	..	..	*5	*5	..	..	..
..	..	..	..	*68	*57	6	..	*5
..	..	..	..	67	57	7	..	3
..	..	..	..	*68	*58	*7	..	*3
..	..	..	..	*67	*56	*7	..	*3
..	..	..	..	20	20	..	..	..
..	..	..	..	19	19	..	..	..
..	..	..	..	19	19	..	..	..
..	..	..	..	19	19	..	..	..

Table 33

Electricity supply and consumption

Gigawatt-hours and kilowatt-hours per capita

Table Notes

Supply is calculated as **Production** plus **Imports** minus **Exports**.

EIOU and losses refers to energy industries own use of electricity, including for pumped storage, and losses.

Total Final Consumption (TFC) refers to final energy consumption of electricity.

Figures may not add up due to statistical differences, or for some countries quantities transformed into heat in heat pumps and/or electric boilers

Please refer to the Definitions Section on pages ix to xv for the appropriate product description /classification.

Figure 81: World electricity generation by region, 1990-2014

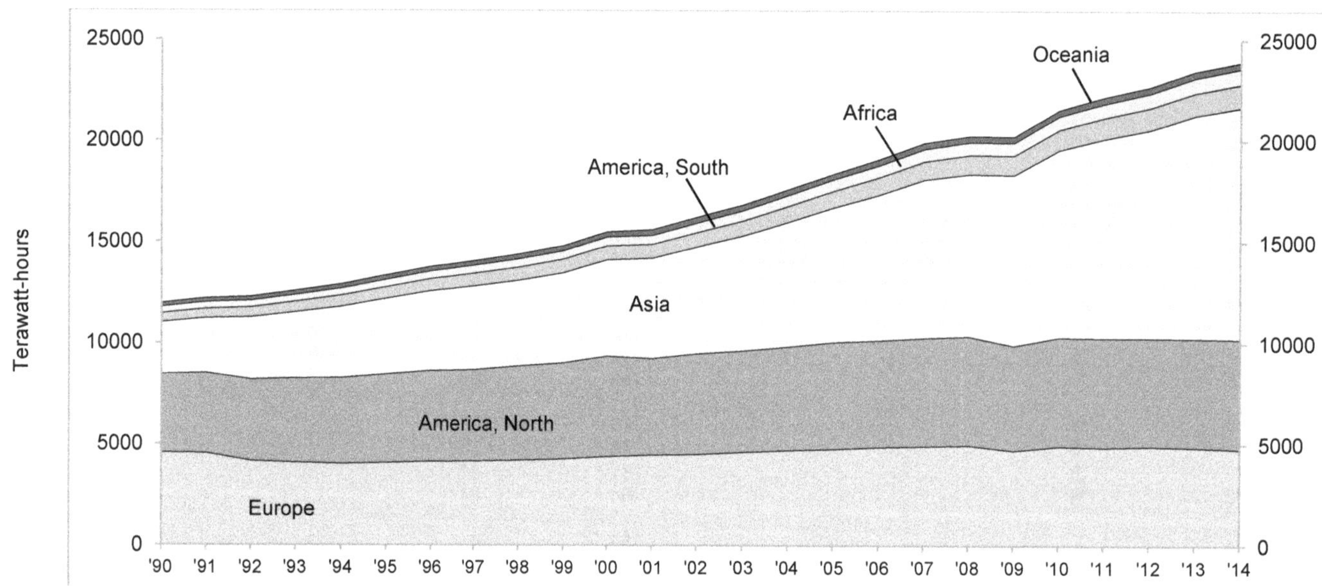

Figure 82: Major electricity producing countries, 2014

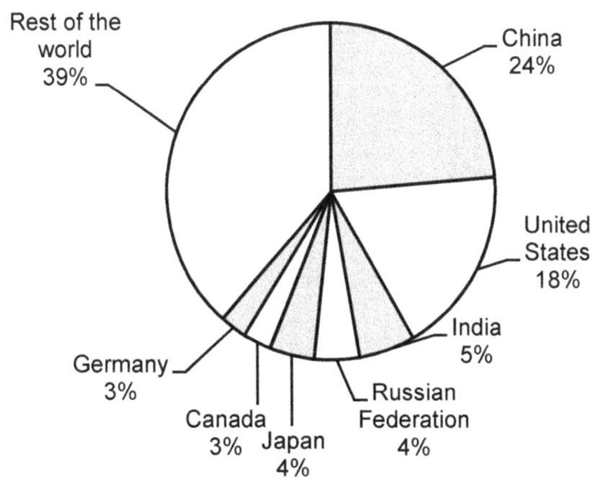

Figure 83: Major electricity consuming countries, 2014

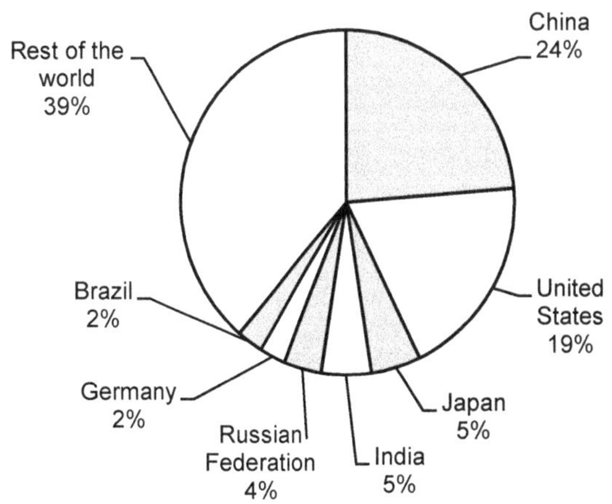

Table 33

Electricity supply and consumption

Gigawatt-hours and kilowatt-hours per capita

Country or area	Year	Production	Imports	Exports	Supply		Consumption	
					Total	Per Capita	EIOU+losses	TFC
World	**2011**	**22,173,726**	**659,021**	**641,699**	**22,191,049**	**3,164**	**3,776,828**	**18,427,954**
	2012	**22,681,641**	**685,934**	**666,678**	**22,700,898**	**3,198**	**3,851,713**	**18,850,149**
	2013	**23,453,355**	**690,064**	**654,668**	**23,488,750**	**3,271**	**3,987,142**	**19,514,321**
	2014	**23,906,233**	**732,499**	**688,470**	**23,950,262**	**3,296**	**4,024,070**	**19,788,948**
Africa	**2011**	**699,697**	**37,918**	**33,145**	**704,470**	**658**	**136,103**	**564,690**
	2012	**725,089**	**36,931**	**30,954**	**731,065**	**665**	**151,296**	**579,057**
	2013	**746,422**	**37,322**	**29,413**	**754,331**	**669**	**165,860**	**586,363**
	2014	**771,383**	**38,702**	**31,720**	**778,366**	**673**	**174,724**	**603,662**
Algeria	2011	51,224	657	799	51,082	1,391	15,213	35,867
	2012	57,397	936	985	57,348	1,532	16,571	40,777
	2013	59,890	295	384	59,801	1,566	16,646	43,158
	2014	64,242	686	877	64,051	1,645	18,318	45,733
Angola	2011	5,651	..	..	5,651	258	776	4,875
	2012	6,204	..	..	6,204	273	852	5,352
	2013	8,216	..	..	8,216	350	1,129	7,087
	2014	9,480	..	..	9,480	391	1,302	8,178
Benin	2011	155	924	..	1,079	110	213	866
	2012	162	967	..	1,129	112	223	906
	2013	173	1,033	..	1,206	117	238	968
	2014	184	1,101	..	1,285	121	254	1,031
Botswana	2011	437	3,180	..	3,617	1,731	500	3,118
	2012	251	3,371	..	3,622	1,698	424	3,198
	2013	761	2,981	..	3,743	1,720	432	3,310
	2014	2,365	1,684	..	4,049	1,824	547	3,449
Burkina Faso	2011	530	495	..	1,025	64	*175	*850
	2012	626	515	..	1,141	69	*196	*945
	2013	*731	532	..	*1,262	*74	*225	*1,037
	2014	*870	488	..	*1,358	*77	*235	*1,123
Burundi	2011	141	*99	..	240	24	8	*244
	2012	142	*99	..	240	24	8	*238
	2013	159	103	..	262	25	8	*259
	2014	174	91	..	265	24	8	*265
Cabo Verde	2011	365	..	..	365	738	102	263
	2012	375	..	..	375	749	108	266
	2013	404	..	..	404	797	112	291
	2014	409	..	..	409	795	113	295
Cameroon	2011	5,943	..	..	5,943	281	1,173	4,770
	2012	6,263	..	..	6,263	289	1,272	4,991
	2013	6,849	..	..	6,849	308	1,580	5,269
	2014	6,922	..	..	6,922	304	1,437	5,485
Central African Rep.	2011	*174	..	..	*174	*38	*18	*156
	2012	*177	..	..	*177	*38	*18	*159
	2013	*181	..	..	*181	*38	*21	*161
	2014	*183	..	..	*183	*38	*21	*163
Chad	2011	*218	..	..	*218	*18	*18	*200
	2012	205	..	..	205	16	19	*186
	2013	*225	..	..	*225	*17	*19	*206
	2014	*225	..	..	*225	*17	*19	*206
Comoros	2011	*57	..	..	*57	*80	*12	*46
	2012	*58	..	..	*58	*80	*12	*46
	2013	*56	..	..	*56	*75	*12	*44
	2014	*55	..	..	*55	*71	*11	*44

Electricity supply and consumption

Gigawatt-hours and kilowatt-hours per capita

Country or area	Year	Production	Imports	Exports	Supply		Consumption	
					Total	Per Capita	EIOU+losses	TFC
Congo	2011	1,293	27	0	1,320	316	816	606
	2012	1,700	25	0	1,725	402	948	792
	2013	1,709	56	19	1,746	397	953	802
	2014	1,740	18	22	1,736	385	970	798
Côte d'Ivoire	2011	6,099	22	615	5,506	267	1,670	3,836
	2012	7,016	54	645	6,425	304	1,717	4,708
	2013	7,650	0	820	6,830	316	2,063	4,767
	2014	8,286	0	877	7,409	334	1,601	5,808
Dem. Rep. of the Congo	2011	7,899	44	171	7,772	114	1,054	6,718
	2012	7,595	626	48	8,173	116	910	7,383
	2013	8,240	626	48	8,818	122	988	7,252
	2014	8,831	1,138	69	9,900	132	2,535	7,899
Djibouti	2011	*387	..	..	*387	*460	97	*290
	2012	393	..	..	393	461	*96	*297
	2013	402	..	..	402	465	*96	*306
	2014	402	..	..	402	459	*97	*305
Egypt	2011	161,162	102	1,679	159,585	1,905	21,568	135,838
	2012	164,628	77	474	164,231	1,917	23,710	140,257
	2013	168,050	61	460	167,651	1,914	29,726	137,591
	2014	177,249	81	470	176,860	1,974	30,304	146,556
Equatorial Guinea	2011	*281	..	..	*281	*374	*51	*230
	2012	*439	..	..	*439	*567	*79	*360
	2013	*439	..	..	*439	*551	*79	*360
	2014	*439	..	..	*439	*535	*79	*360
Eritrea	2011	337	..	..	337	70	66	271
	2012	359	..	..	359	73	74	285
	2013	363	..	..	363	73	68	295
	2014	370	..	..	370	72	67	303
Ethiopia	2011	6,336	..	332	6,004	67	1,634	4,370
	2012	7,608	..	563	7,045	76	1,769	5,276
	2013	8,719	..	954	7,765	82	1,678	6,087
	2014	9,615	..	1,052	8,563	88	2,034	6,529
Gabon	2011	2,033	..	..	2,033	1,289	505	1,528
	2012	2,151	..	..	2,151	1,333	532	1,619
	2013	2,264	184	..	2,448	1,483	606	1,842
	2014	2,367	388	..	2,755	1,632	684	2,071
Gambia	2011	*256	..	..	*256	*146	*54	*199
	2012	*266	..	..	*266	*147	*73	*208
	2013	*261	..	..	*261	*140	*64	*209
	2014	*261	..	..	*261	*135	*43	*209
Ghana	2011	11,366	81	691	10,756	431	2,279	7,924
	2012	12,204	128	667	11,665	457	2,210	8,890
	2013	13,047	27	530	12,544	479	2,360	9,472
	2014	13,139	51	522	12,668	473	2,524	9,522
Guinea	2011	890	..	..	890	79	*243	*647
	2012	1,071	..	..	1,071	92	*305	*766
	2013	995	..	..	995	83	*269	*726
	2014	*1,015	..	..	*1,015	*83	*274	*741
Guinea-Bissau	2011	*33	..	..	*33	*20	0	*33
	2012	*34	..	..	*34	*20	0	*34
	2013	*35	..	..	*35	*20	0	*35
	2014	*35	..	..	*35	*20	0	*35

Table 33

Electricity supply and consumption

Gigawatt-hours and kilowatt-hours per capita

Country or area	Year	Production	Imports	Exports	Supply		Consumption	
					Total	Per Capita	EIOU+losses	TFC
Kenya	2011	7,560	34	37	7,557	182	1,283	6,500
	2012	7,852	39	33	7,858	185	1,444	6,760
	2013	8,448	49	36	8,461	194	1,525	7,296
	2014	9,139	158	27	9,270	207	1,706	7,786
Lesotho	2011	490	262	39	713	351	67	645
	2012	485	310	21	774	376	98	676
	2013	515	285	2	798	383	88	711
	2014	515	271	3	783	371	98	*685
Liberia	2011	*300	..	..	*300	*74	*44	*256
	2012	*300	..	..	*300	*72	*44	*256
	2013	*300	..	..	*300	*70	*44	*256
	2014	*300	..	..	*300	*68	*44	*256
Libya	2011	25,999	56	121	25,934	4,124	10,126	15,808
	2012	34,275	61	14	34,322	5,462	21,328	12,994
	2013	37,913	64	1	37,976	6,061	23,910	14,066
	2014	37,731	88	..	37,819	6,042	26,929	10,890
Madagascar	2011	1,438	..	..	1,438	66	*62	1,053
	2012	1,520	..	..	1,520	68	*64	1,100
	2013	1,594	..	..	1,594	70	*65	1,222
	2014	1,659	..	..	1,659	70	*67	1,266
Malawi	2011	1,979	..	20	1,959	129	*281	1,678
	2012	2,027	..	18	2,009	128	*292	1,717
	2013	2,098	..	*18	2,080	128	*301	*1,779
	2014	*2,098	..	*18	*2,080	*125	*301	*1,779
Mali	2011	1,381	..	*368	1,013	65	*133	*1,201
	2012	1,363	..	*368	995	62	*135	*1,215
	2013	1,510	..	*300	1,210	73	131	1,281
	2014	*1,679	..	*300	*1,379	*81	*140	*1,340
Mauritania	2011	744	*115	..	859	233	152	698
	2012	824	*115	..	939	249	159	771
	2013	790	*115	..	905	234	199	706
	2014	910	135	..	1,045	263	*319	862
Mauritius	2011	2,739	..	..	2,739	2,185	502	2,227
	2012	2,798	..	..	2,798	2,223	501	2,294
	2013	2,882	..	..	2,882	2,281	499	2,384
	2014	2,936	..	..	2,936	2,314	481	2,452
Mayotte	2011	262	..	..	262	1,219	22	240
	2012	271	..	..	271	1,225	25	246
	2013	285	..	..	285	1,254	25	261
	2014	299	..	..	299	1,281	26	274
Morocco	2011	24,364	4,607	..	28,971	876	3,301	25,670
	2012	26,495	4,841	..	31,336	935	3,777	*27,559
	2013	27,929	5,551	151	33,329	980	5,483	27,846
	2014	29,142	6,138	128	35,152	1,019	6,346	28,806
Mozambique	2011	16,830	8,570	11,954	13,446	537	2,605	10,141
	2012	15,166	8,304	9,791	13,679	532	2,453	10,939
	2013	15,645	8,339	9,058	14,926	564	2,867	11,605
	2014	17,739	7,656	10,199	15,196	558	2,850	12,342
Namibia	2011	1,607	2,389	79	3,917	1,749	360	3,467
	2012	1,538	2,226	88	3,676	1,604	426	3,635
	2013	1,712	2,916	89	4,539	1,934	369	3,772
	2014	1,498	2,886	84	4,300	1,790	543	3,747

Electricity supply and consumption

Gigawatt-hours and kilowatt-hours per capita

Country or area	Year	Production	Imports	Exports	Supply		Consumption	
					Total	Per Capita	EIOU+losses	TFC
Niger	2011	324	607	4	927	55	181	747
	2012	392	637	4	1,024	58	202	826
	2013	438	602	3	1,037	56	194	836
	2014	439	574	0	1,013	53	*236	*797
Nigeria	2011	27,034	..	..	27,034	165	3,587	23,447
	2012	28,706	..	..	28,706	171	3,547	25,159
	2013	28,883	..	..	28,883	167	5,396	23,487
	2014	30,390	..	..	30,390	171	5,951	24,439
Réunion	2011	2,750	..	..	2,750	3,285	243	2,498
	2012	2,812	..	..	2,812	3,334	284	2,540
	2013	2,813	..	..	2,813	3,313	269	2,555
	2014	2,857	..	..	2,857	3,341	259	2,598
Rwanda	2011	346	78	5	418	40	*92	326
	2012	393	91	3	481	44	*101	380
	2013	413	94	0	507	46	*119	388
	2014	476	90	4	562	50	*124	438
Sao Tome and Principe	2011	*60	..	..	*60	*344	*16	*44
	2012	*67	..	..	*67	*375	*16	*51
	2013	*67	..	..	*67	*367	*16	*51
	2014	*67	..	..	*67	*360	*16	*51
Senegal	2011	2,560	257	..	2,817	211	299	2,032
	2012	2,917	290	..	3,207	233	343	2,313
	2013	3,403	308	..	3,711	261	632	3,079
	2014	3,406	318	..	3,724	254	514	3,210
Seychelles	2011	324	..	..	324	3,452	47	280
	2012	336	..	..	336	3,560	49	314
	2013	354	..	..	354	3,720	49	307
	2014	363	..	..	363	3,784	51	313
Sierra Leone	2011	176	..	..	176	30	90	86
	2012	179	..	..	179	30	79	100
	2013	164	..	..	164	27	95	69
	2014	181	..	..	181	29	81	86
Somalia	2011	334	..	..	334	34	*33	301
	2012	340	..	..	340	34	*32	308
	2013	350	..	..	350	34	*35	315
	2014	350	..	..	350	33	*35	315
South Africa	2011	262,538	11,890	14,964	259,464	4,967	52,223	207,241
	2012	257,919	10,006	15,035	252,890	4,786	52,777	200,113
	2013	256,073	9,428	13,929	251,572	4,710	52,264	199,308
	2014	252,578	11,177	13,836	249,919	4,631	51,826	198,093
South Sudan	2011	..	..	..	..	..	..	..
	2012	445	..	..	445	41	44	385
	2013	472	..	..	472	41	45	415
	2014	488	..	..	488	41	47	429
St. Helena and Depend.	2011	9	..	..	9	2,269	*1	8
	2012	10	..	..	10	2,362	*1	9
	2013	11	..	..	11	2,645	*2	9
	2014	*11	..	..	*11	*2,676	*2	*9
Sudan	2011	..	..	..	..	..	..	..
	2012	9,510	..	..	9,510	252	1,900	7,611
	2013	10,607	..	..	10,607	275	2,712	7,895
	2014	11,376	..	..	11,376	289	1,666	9,710

Table 33

Electricity supply and consumption

Gigawatt-hours and kilowatt-hours per capita

Country or area	Year	Production	Imports	Exports	Supply		Consumption	
					Total	Per Capita	EIOU+losses	TFC
Sudan (former)	2011	8,455	..	..	8,455	181	1,765	6,690
	2012	..	..	..	..	..	..	..
	2013	..	..	..	..	..	..	..
	2014	..	..	..	..	..	..	..
Swaziland	2011	654	806	..	1,460	1,204	177	1,283
	2012	*622	813	..	1,435	1,165	189	1,246
	2013	*590	822	..	1,412	1,129	184	1,227
	2014	567	860	..	1,427	1,124	456	1,018
Togo	2011	320	696	..	1,016	155	168	847
	2012	399	679	..	1,079	160	175	904
	2013	360	773	..	1,133	164	196	937
	2014	358	857	..	1,215	171	192	1,023
Tunisia	2011	16,497	131	161	16,467	1,531	3,210	13,239
	2012	18,059	384	426	18,017	1,656	3,628	14,386
	2013	18,382	177	233	18,326	1,665	3,589	14,729
	2014	19,024	536	625	18,935	1,701	3,723	15,190
Uganda	2011	*2,588	*36	89	*2,535	*74	*869	*1,681
	2012	*2,640	*37	*91	*2,586	*73	*896	*1,724
	2013	*2,861	47	105	*2,803	*77	*772	2,070
	2014	3,258	33	125	3,165	84	781	2,417
United Rep. of Tanzania	2011	5,093	55	..	5,148	109	1,168	3,995
	2012	5,589	61	..	5,650	116	1,061	4,441
	2013	5,941	59	..	6,000	119	1,155	4,836
	2014	6,219	59	..	6,278	121	1,116	4,976
Zambia	2011	11,498	120	29	11,589	808	3,018	8,571
	2012	12,368	163	980	11,551	781	1,232	10,319
	2013	13,300	73	1,083	12,290	806	1,444	10,846
	2014	14,452	13	1,256	13,209	840	2,490	10,719
Zimbabwe	2011	9,177	1,579	988	9,768	685	1,735	8,043
	2012	9,149	1,076	701	9,524	654	1,866	7,831
	2013	9,499	1,722	1,189	10,032	673	1,814	8,285
	2014	10,023	1,127	1,226	9,924	651	1,832	8,238
America, North	**2011**	**5,410,924**	**68,729**	**68,198**	**5,411,455**	**9,795**	**807,302**	**4,610,541**
	2012	**5,363,980**	**73,594**	**71,717**	**5,365,856**	**9,618**	**806,665**	**4,584,845**
	2013	**5,400,500**	**84,016**	**76,659**	**5,407,857**	**9,601**	**782,725**	**4,623,261**
	2014	**5,433,273**	**84,611**	**76,840**	**5,441,044**	**9,571**	**790,781**	**4,645,015**
Anguilla	2011	95	..	..	95	6,821	12	*84
	2012	90	..	..	90	6,333	11	*79
	2013	89	..	..	89	6,196	11	*77
	2014	90	..	..	90	6,189	11	*78
Antigua and Barbuda	2011	*325	..	..	*325	*3,687	*91	*230
	2012	*327	..	..	*327	*3,671	*92	*235
	2013	*330	..	..	*330	*3,667	*92	*237
	2014	*336	..	..	*336	*3,696	*94	*242
Aruba	2011	931	..	..	931	9,128	159	*772
	2012	915	..	..	915	8,931	148	*766
	2013	*928	..	..	*928	*9,018	149	*779
	2014	*917	..	..	*917	*8,861	*138	*779
Bahamas	2011	1,754	..	..	1,754	4,782	216	1,523
	2012	1,750	..	..	1,750	4,698	215	1,520
	2013	1,885	..	..	1,885	4,989	74	1,767
	2014	1,853	..	..	1,853	4,837	74	1,808

Electricity supply and consumption

Gigawatt-hours and kilowatt-hours per capita

Country or area	Year	Production	Imports	Exports	Supply		Consumption	
					Total	Per Capita	EIOU+losses	TFC
Barbados	2011	1,049	..	..	1,049	3,738	109	940
	2012	1,129	..	..	1,129	4,008	80	1,048
	2013	1,016	..	..	1,016	3,596	101	915
	2014	1,010	..	..	1,010	3,563	106	*903
Belize	2011	*388	171	..	*559	*1,698	64	495
	2012	*424	172	..	*596	*1,769	70	526
	2013	*371	234	..	*605	*1,758	80	536
	2014	*404	233	..	*637	*1,812	79	558
Bermuda	2011	*726	..	..	*726	*11,417	*96	636
	2012	*692	..	..	*692	*10,953	*87	606
	2013	*673	..	..	*673	*10,721	*87	586
	2014	*710	..	..	*710	*11,383	*133	577
Bonaire, St Eustatius, Saba	2011	..	..	..	..	..	..	..
	2012	*119	..	..	*119	*5,168	*30	*88
	2013	*122	..	..	*122	*5,146	*31	*91
	2014	*124	..	..	*124	*5,104	*32	*92
British Virgin Islands	2011	131	..	..	131	4,694	*21	*110
	2012	*130	..	..	*130	*4,560	*21	*109
	2013	*130	..	..	*130	*4,474	*21	*109
	2014	*132	..	..	*132	*4,462	*21	*111
Canada	2011	629,901	14,398	51,076	593,223	17,195	103,162	488,004
	2012	632,921	10,841	57,639	586,123	16,810	101,056	495,591
	2013	660,795	10,694	62,578	608,911	17,284	105,240	497,820
	2014	656,225	12,808	58,421	610,612	17,158	107,199	489,239
Cayman Islands	2011	626	..	..	626	11,064	53	573
	2012	607	..	..	607	10,552	39	568
	2013	615	..	..	615	10,537	40	575
	2014	621	..	..	621	10,490	37	584
Costa Rica	2011	9,831	283	320	9,795	2,129	1,151	8,673
	2012	10,174	419	402	10,191	2,190	1,192	9,019
	2013	10,235	531	484	10,282	2,185	1,192	9,103
	2014	10,217	759	555	10,421	2,190	1,223	9,207
Cuba	2011	17,759	..	..	17,759	1,568	3,734	17,760
	2012	18,432	..	..	18,432	1,625	3,901	18,432
	2013	19,140	..	..	19,140	1,684	3,961	18,135
	2014	19,366	..	..	19,366	1,702	4,017	16,405
Curaçao	2011	..	..	..	..	..	..	..
	2012	917	..	..	917	6,008	233	684
	2013	904	..	..	904	5,859	230	674
	2014	891	..	..	891	5,720	227	664
Dominica	2011	100	..	..	100	1,406	12	89
	2012	102	..	..	102	1,417	12	90
	2013	101	..	..	101	1,399	11	89
	2014	102	..	..	102	1,416	11	91
Dominican Republic	2011	16,701	..	..	16,701	1,666	2,509	14,192
	2012	17,746	..	..	17,746	1,747	2,661	15,085
	2013	18,799	..	..	18,799	1,828	3,096	15,704
	2014	17,962	..	..	17,962	1,726	2,775	15,189
El Salvador	2011	6,083	216	102	6,197	1,023	1,028	5,169
	2012	6,237	163	78	6,322	1,041	919	5,383
	2013	6,272	283	91	6,464	1,061	818	5,647
	2014	6,223	589	208	6,604	1,081	1,049	5,524

Table 33

Electricity supply and consumption

Gigawatt-hours and kilowatt-hours per capita

Country or area	Year	Production	Imports	Exports	Supply		Consumption	
					Total	Per Capita	EIOU+losses	TFC
Greenland	2011	565	..	..	565	10,016	229	320
	2012	574	..	..	574	10,186	264	314
	2013	568	..	..	568	10,097	248	318
	2014	580	..	..	580	10,317	411	331
Grenada	2011	204	..	..	204	1,942	23	181
	2012	200	..	..	200	1,893	21	179
	2013	197	..	..	197	1,857	21	176
	2014	200	..	..	200	1,880	21	179
Guadeloupe	2011	*1,767	..	..	*1,767	*3,852	*275	*1,492
	2012	1,817	..	..	1,817	3,942	310	1,507
	2013	1,818	..	..	1,818	3,922	308	1,510
	2014	1,809	..	..	1,809	3,881	282	1,508
Guatemala	2011	8,798	511	193	9,116	606	1,603	7,688
	2012	9,335	372	346	9,361	609	1,619	7,741
	2013	10,024	313	669	9,668	616	1,586	7,582
	2014	10,638	545	1,053	10,130	632	1,436	8,288
Haiti	2011	974	..	..	974	96	686	235
	2012	1,100	..	..	1,100	107	655	493
	2013	1,055	..	..	1,055	101	618	456
	2014	1,033	..	..	1,033	98	632	410
Honduras	2011	7,653	171	127	7,697	1,010	2,080	5,449
	2012	7,740	154	79	7,815	1,010	2,552	5,308
	2013	8,074	268	156	8,186	1,043	2,775	5,455
	2014	8,038	827	505	8,360	1,050	2,809	5,552
Jamaica	2011	4,171	..	..	4,171	1,515	997	3,176
	2012	4,156	..	..	4,156	1,504	1,076	3,081
	2013	4,256	..	..	4,256	1,535	1,212	*3,039
	2014	4,123	..	..	4,123	1,481	1,115	*3,008
Martinique	2011	*1,764	..	..	*1,764	*4,470	*196	*1,547
	2012	*1,780	..	..	*1,780	*4,507	*197	*1,558
	2013	*1,764	..	..	*1,764	*4,460	*195	*1,544
	2014	*1,764	..	..	*1,764	*4,454	*210	*1,544
Mexico	2011	302,785	596	1,293	302,088	2,510	73,430	233,817
	2012	307,268	2,177	1,116	308,329	2,526	72,268	247,710
	2013	297,326	1,210	1,240	297,296	2,403	57,789	241,470
	2014	301,496	2,124	2,653	300,967	2,400	57,327	252,220
Montserrat	2011	*23	..	..	*23	*4,606	*4	19
	2012	*23	..	..	*23	*4,573	*4	19
	2013	*23	..	..	*23	*4,543	*4	20
	2014	24	..	..	24	4,711	*5	20
Neth. Antilles (former)	2011	1,325	..	..	1,325	6,216	336	989
	2012	..	..	..	..	..	..	..
	2013	..	..	..	..	..	..	..
	2014	..	..	..	..	..	..	..
Nicaragua	2011	3,824	10	41	3,793	653	980	2,821
	2012	4,031	20	3	4,048	689	994	3,041
	2013	4,163	52	16	4,199	706	1,043	3,196
	2014	4,444	22	49	4,417	734	1,087	3,325
Panama	2011	7,857	72	8	7,921	2,151	1,088	6,695
	2012	8,606	19	59	8,566	2,288	1,185	7,176
	2013	8,962	75	71	8,966	2,356	1,214	7,511
	2014	9,287	193	99	9,381	2,426	1,339	7,832

Electricity supply and consumption

Gigawatt-hours and kilowatt-hours per capita

Country or area	Year	Production	Imports	Exports	Supply		Consumption	
					Total	Per Capita	EIOU+losses	TFC
Puerto Rico	2011	22,507	..	..	22,507	6,080	*2,281	19,977
	2012	22,556	..	..	22,556	6,103	*2,724	19,582
	2013	21,743	..	..	21,743	5,891	*1,787	19,706
	2014	21,317	..	..	21,317	5,782	1,365	19,702
Sint Maarten (Dutch part)	2011	..	..	..	..	..	..	..
	2012	*402	..	..	*402	*11,433	*36	*366
	2013	410	..	..	410	11,243	37	373
	2014	*418	..	..	*418	*11,086	*38	*380
St. Kitts-Nevis	2011	223	..	..	223	4,198	40	183
	2012	218	..	..	218	4,060	39	179
	2013	*220	..	..	*220	*4,051	*40	*180
	2014	*220	..	..	*220	*4,004	*40	*180
St. Lucia	2011	385	..	..	385	2,149	52	333
	2012	385	..	..	385	2,127	51	333
	2013	383	..	..	383	2,101	48	334
	2014	379	..	..	379	2,066	47	332
St. Pierre-Miquelon	2011	45	..	..	45	7,209	2	44
	2012	43	..	..	43	6,927	3	41
	2013	47	..	..	47	7,410	*3	*43
	2014	48	..	..	48	7,692	*3	*45
St. Vincent-Grenadines	2011	*145	..	..	*145	*1,329	*15	123
	2012	*145	..	..	*145	*1,329	13	123
	2013	*139	..	..	*139	*1,271	*14	125
	2014	*170	..	..	*170	*1,553	*18	*158
Trinidad and Tobago	2011	8,772	..	..	8,772	6,572	562	8,211
	2012	9,132	..	..	9,132	6,807	571	8,560
	2013	9,505	..	..	9,505	7,050	580	8,926
	2014	9,891	..	..	9,891	7,302	588	9,306
Turks and Caicos Islands	2011	*214	..	..	*214	*6,745	*18	*196
	2012	*227	..	..	*227	*7,000	*19	*208
	2013	*227	..	..	*227	*6,857	*19	*208
	2014	*230	..	..	*230	*6,817	*19	*211
United States	2011	4,349,571	52,301	15,038	4,386,834	14,043	609,833	3,777,001
	2012	4,290,660	59,257	11,995	4,337,922	13,780	611,148	3,726,774
	2013	4,306,371	70,355	11,353	4,365,373	13,765	597,809	3,767,564
	2014	4,339,210	66,511	13,298	4,392,423	13,750	604,630	3,787,793
United States Virgin Is.	2011	951	..	..	951	8,954	155	796
	2012	873	..	..	873	8,221	149	724
	2013	821	..	..	821	7,729	140	681
	2014	*772	..	..	*772	*7,272	*131	641
America, South	**2011**	**1,054,192**	**52,126**	**51,499**	**1,054,818**	**2,628**	**197,401**	**852,913**
	2012	**1,092,648**	**50,310**	**49,382**	**1,093,576**	**2,695**	**209,403**	**884,221**
	2013	**1,129,525**	**50,019**	**49,221**	**1,130,323**	**2,757**	**216,100**	**928,134**
	2014	**1,154,584**	**44,686**	**43,732**	**1,155,538**	**2,789**	**234,548**	**915,624**
Argentina	2011	129,892	10,929	1,262	139,559	3,350	23,566	115,990
	2012	135,207	8,116	506	142,817	3,393	23,080	119,850
	2013	140,863	8,303	247	148,919	3,501	24,792	124,127
	2014	141,586	10,024	166	151,444	3,524	24,691	126,757
Bolivia (Plur. State of)	2011	7,220	..	..	7,220	716	936	6,329
	2012	7,661	..	..	7,661	748	1,028	6,613
	2013	8,065	..	..	8,065	775	919	6,962
	2014	8,755	..	..	8,755	829	997	7,392

Table 33

Electricity supply and consumption

Gigawatt-hours and kilowatt-hours per capita

Country or area	Year	Production	Imports	Exports	Supply		Consumption	
					Total	Per Capita	EIOU+losses	TFC
Brazil	2011	531,759	38,430	2,544	567,645	2,831	110,896	456,748
	2012	552,498	40,722	467	592,753	2,929	120,705	472,048
	2013	570,838	40,334	0	611,172	2,992	124,717	486,454
	2014	590,541	33,778	3	624,316	3,030	122,918	501,399
Chile	2011	65,713	732	..	66,445	3,863	7,771	57,932
	2012	69,751	0	..	69,751	4,011	8,112	62,219
	2013	73,065	0	..	73,065	4,157	9,075	65,176
	2014	73,719	0	..	73,719	4,150	6,852	66,458
Colombia	2011	66,472	8	1,540	64,940	1,399	9,214	57,845
	2012	68,344	7	713	67,638	1,443	9,544	59,285
	2013	70,858	5	1,374	69,489	1,468	8,572	67,260
	2014	70,059	47	849	69,257	1,449	9,464	51,268
Ecuador	2011	20,544	1,295	14	21,824	1,438	3,610	18,185
	2012	22,848	238	12	23,074	1,496	3,565	19,492
	2013	23,260	662	29	23,894	1,526	3,430	20,430
	2014	24,307	837	47	25,097	1,578	3,656	21,491
Falkland Is. (Malvinas)	2011	*18	..	..	*18	*6,165	0	*18
	2012	*18	..	..	*18	*6,165	0	*18
	2013	*18	..	..	*18	*6,135	0	*18
	2014	*18	..	..	*18	*6,105	0	*18
French Guiana	2011	873	..	..	873	3,633	*96	749
	2012	874	..	..	874	3,536	*109	765
	2013	883	..	..	883	3,473	*114	769
	2014	910	..	..	910	3,480	*135	775
Guyana	2011	910	..	..	910	1,204	393	502
	2012	942	..	..	942	1,243	370	526
	2013	962	..	..	962	1,264	258	704
	2014	979	..	..	979	1,282	244	736
Paraguay	2011	57,625	..	46,120	11,505	1,828	3,935	7,571
	2012	60,235	..	47,663	12,572	1,971	4,350	8,222
	2013	60,381	..	47,365	13,016	2,013	3,998	9,018
	2014	55,282	..	41,400	13,882	2,119	4,088	9,795
Peru	2011	38,811	6	0	38,817	1,304	3,666	35,125
	2012	40,044	5	2	40,047	1,328	3,944	35,969
	2013	43,330	0	0	43,330	1,418	5,145	38,208
	2014	45,508	0	0	45,508	1,469	5,818	39,719
Suriname	2011	1,951	..	..	1,951	3,728	209	1,742
	2012	1,972	..	..	1,972	3,732	211	1,761
	2013	2,184	..	..	2,184	4,095	234	1,951
	2014	2,180	..	..	2,180	4,049	224	1,955
Uruguay	2011	10,345	477	19	10,803	3,191	1,558	9,307
	2012	10,601	742	20	11,323	3,333	1,640	9,580
	2013	11,658	0	206	11,451	3,360	1,712	9,850
	2014	13,008	0	1,267	11,742	3,434	1,720	10,132
Venezuela (Bolivar. Rep.)	2011	122,059	249	..	122,308	4,156	31,552	84,871
	2012	121,653	480	..	122,133	4,091	32,746	87,873
	2013	123,160	715	..	123,875	4,092	33,134	97,208
	2014	127,732	..	..	127,732	4,161	53,742	77,729
Asia	**2011**	**9,845,786**	**90,236**	**72,043**	**9,863,979**	**2,340**	**1,664,718**	**8,213,614**
	2012	**10,298,780**	**91,426**	**70,734**	**10,319,471**	**2,422**	**1,701,145**	**8,600,683**
	2013	**11,021,968**	**97,728**	**69,392**	**11,050,303**	**2,567**	**1,866,440**	**9,190,832**
	2014	**11,477,676**	**105,877**	**68,634**	**11,514,919**	**2,647**	**1,870,218**	**9,526,993**

Electricity supply and consumption

Gigawatt-hours and kilowatt-hours per capita

Country or area	Year	Production	Imports	Exports	Supply		Consumption	
					Total	Per Capita	EIOU+losses	TFC
Afghanistan	2011	1,005	2,732	..	3,737	130	690	2,300
	2012	1,022	3,071	..	4,093	138	610	3,265
	2013	*1,022	3,615	..	4,638	151	*756	3,022
	2014	*1,049	3,711	..	4,760	151	*773	*3,767
Armenia	2011	7,433	205	1,533	6,104	2,057	1,231	5,151
	2012	8,036	98	1,696	6,438	2,162	1,318	5,120
	2013	7,710	148	1,313	6,545	2,187	1,278	5,404
	2014	7,750	206	1,314	6,642	2,210	1,290	5,353
Azerbaijan	2011	20,294	128	805	19,617	2,126	6,220	13,267
	2012	22,988	141	680	22,449	2,398	6,974	15,394
	2013	23,354	127	495	22,986	2,420	6,921	15,982
	2014	24,728	124	489	24,363	2,530	7,368	16,907
Bahrain	2011	24,338	227	107	24,458	18,727	1,683	22,775
	2012	24,770	35	190	24,615	18,458	1,534	23,315
	2013	25,909	70	53	25,926	19,213	1,413	24,513
	2014	27,246	240	237	27,249	20,008	1,069	26,186
Bangladesh	2011	44,153	..	..	44,153	288	7,149	36,971
	2012	48,562	..	..	48,562	313	8,671	40,088
	2013	53,043	..	..	53,043	338	10,101	42,717
	2014	55,845	..	..	55,845	351	9,718	48,918
Bhutan	2011	7,068	20	5,268	1,820	2,485	129	1,638
	2012	6,827	37	4,890	1,973	2,653	143	1,770
	2013	7,640	108	5,648	2,100	2,783	197	1,841
	2014	7,004	187	4,992	2,199	2,875	137	2,004
Brunei Darussalam	2011	3,725	..	..	3,725	9,325	661	3,064
	2012	3,930	..	..	3,930	9,691	723	3,208
	2013	4,402	..	..	4,402	10,697	1,247	3,155
	2014	4,506	..	..	4,506	10,796	1,148	3,357
Cambodia	2011	1,053	1,644	..	2,697	185	327	2,370
	2012	1,434	1,891	..	3,325	224	290	3,035
	2013	1,778	2,050	..	3,828	254	522	3,306
	2014	3,059	1,803	..	4,862	317	753	4,109
China	2011	4,713,019	6,562	19,307	4,700,274	3,486	840,277	3,859,811
	2012	4,987,553	6,874	17,653	4,976,774	3,672	854,638	4,121,626
	2013	5,431,637	7,438	18,669	5,420,406	3,978	936,237	4,484,105
	2014	5,649,583	6,750	18,158	5,638,175	4,117	951,993	4,686,376
China, Hong Kong SAR	2011	39,026	10,735	2,957	46,804	6,644	4,739	42,065
	2012	38,752	11,156	1,838	48,070	6,769	5,039	43,031
	2013	39,063	9,969	1,650	47,382	6,614	4,826	42,556
	2014	39,803	10,288	1,226	48,865	6,762	4,986	43,880
China, Macao SAR	2011	886	3,165	..	4,051	7,410	197	3,855
	2012	561	3,855	..	4,416	7,917	213	4,203
	2013	413	4,059	..	4,472	7,872	183	4,288
	2014	641	4,099	..	4,740	8,202	210	4,530
Cyprus	2011	4,929	..	..	4,929	4,414	399	4,721
	2012	4,717	..	..	4,717	4,177	319	4,409
	2013	4,290	..	..	4,290	3,758	371	3,921
	2014	4,350	..	..	4,350	3,771	385	3,965
Georgia	2011	10,104	471	931	9,644	2,298	1,652	8,088
	2012	9,698	615	528	9,785	2,364	1,604	8,168
	2013	10,059	484	450	10,093	2,472	1,019	9,074
	2014	10,370	794	545	10,619	2,632	834	9,784

Table 33

Electricity supply and consumption

Gigawatt-hours and kilowatt-hours per capita

Country or area	Year	Production	Imports	Exports	Supply		Consumption	
					Total	Per Capita	EIOU+losses	TFC
India	2011	973,006	8,441	0	981,447	787	242,431	772,603
	2012	1,035,264	5,152	0	1,040,416	823	242,544	785,130
	2013	1,193,480	5,598	0	1,199,078	937	309,023	890,055
	2014	1,308,873	4,998	0	1,313,871	1,014	*279,116	948,327
Indonesia	2011	182,389	2,542	..	184,931	755	23,559	160,517
	2012	196,181	2,440	..	198,621	801	23,638	173,991
	2013	216,186	2,990	..	219,176	872	25,712	195,198
	2014	220,794	8,990	..	229,784	903	26,427	170,088
Iran (Islamic Rep. of)	2011	240,052	3,656	8,668	235,040	3,126	45,628	189,458
	2012	254,276	3,897	11,029	247,144	3,245	47,743	199,402
	2013	262,434	3,707	11,586	254,555	3,299	49,891	205,630
	2014	274,609	3,772	9,660	268,721	3,439	46,045	221,411
Iraq	2011	54,240	7,262	..	61,502	1,930	18,854	42,648
	2012	46,018	10,170	..	56,188	1,705	21,170	35,076
	2013	58,422	12,202	..	70,624	2,071	25,595	45,042
	2014	67,768	12,251	..	80,019	2,269	36,025	43,993
Israel	2011	59,679	..	4,222	55,457	7,332	7,158	48,956
	2012	63,003	..	4,434	58,569	7,612	7,407	51,804
	2013	61,322	..	4,675	56,647	7,246	5,185	51,640
	2014	60,813	..	4,844	55,969	7,049	4,610	51,318
Japan	2011	1,082,230	..	..	1,082,230	8,505	107,854	964,976
	2012	1,064,065	..	..	1,064,065	8,369	107,632	963,482
	2013	1,065,623	..	..	1,065,623	8,392	106,701	968,316
	2014	1,040,676	..	..	1,040,676	8,208	99,453	951,487
Jordan	2011	14,647	1,738	86	16,299	2,411	2,867	13,496
	2012	16,596	784	104	17,276	2,470	3,109	14,245
	2013	17,263	381	59	17,585	2,437	3,121	14,472
	2014	18,270	435	64	18,641	2,514	3,300	15,321
Kazakhstan	2011	86,586	3,406	1,809	88,183	5,327	23,188	64,752
	2012	92,817	4,252	2,933	94,136	5,596	26,123	68,248
	2013	103,086	2,147	2,996	102,237	5,979	36,122	65,727
	2014	105,068	1,749	2,917	103,900	5,981	34,636	68,900
Korea, Dem.Ppl's.Rep.	2011	19,226	..	..	19,226	781	4,864	14,362
	2012	19,236	..	..	19,236	777	4,867	14,369
	2013	18,332	..	..	18,332	736	4,638	13,694
	2014	17,909	..	..	17,909	716	4,531	13,378
Korea, Republic of	2011	523,286	..	..	523,286	10,602	52,168	470,613
	2012	534,618	..	..	534,618	10,777	53,516	481,405
	2013	541,996	..	..	541,996	10,873	55,282	487,125
	2014	550,933	..	..	550,933	11,002	58,052	486,834
Kuwait	2011	57,457	..	..	57,457	17,738	20,075	37,382
	2012	62,655	..	..	62,655	18,322	22,921	39,734
	2013	60,982	..	..	60,982	16,969	21,203	39,779
	2014	65,140	..	..	65,140	17,356	22,088	43,052
Kyrgyzstan	2011	15,158	174	2,848	12,484	2,248	3,676	8,712
	2012	15,168	177	1,840	13,505	2,391	3,666	9,591
	2013	14,011	..	377	13,634	2,373	3,095	10,267
	2014	14,572	286	72	14,786	2,530	3,802	10,984
Lao People's Dem. Rep.	2011	3,802	748	*2,020	2,530	397	*211	2,399
	2012	*3,802	*748	*2,020	*2,530	*391	*211	2,853
	2013	4,471	1,205	876	4,800	729	900	3,363
	2014	4,380	1,486	630	5,236	783	895	3,768

Electricity supply and consumption

Gigawatt-hours and kilowatt-hours per capita

Country or area	Year	Production	Imports	Exports	Supply		Consumption	
					Total	Per Capita	EIOU+losses	TFC
Lebanon	2011	16,365	840	..	17,205	3,747	1,868	15,337
	2012	14,826	323	..	15,149	3,076	1,372	13,776
	2013	17,487	522	..	18,009	3,406	1,651	16,358
	2014	17,952	136	..	18,088	3,223	1,879	16,210
Malaysia	2011	124,894	370	10	125,254	4,384	12,779	107,341
	2012	134,421	100	15	134,506	4,635	11,543	116,353
	2013	138,331	221	23	138,529	4,701	11,366	123,076
	2014	147,461	23	12	147,472	4,932	15,585	128,333
Maldives	2011	275	..	..	275	811	*25	*250
	2012	299	..	..	299	867	*28	*271
	2013	305	..	..	305	868	*31	*273
	2014	329	..	..	329	920	34	*295
Mongolia	2011	4,812	276	24	5,064	1,835	1,335	3,453
	2012	5,182	366	21	5,526	1,968	1,388	3,773
	2013	6,215	1,196	18	7,392	2,585	1,465	4,732
	2014	6,725	1,349	33	8,041	2,763	1,565	5,158
Myanmar	2011	10,425	..	..	10,425	200	2,708	7,717
	2012	10,965	..	..	10,965	209	2,710	8,255
	2013	12,247	..	..	12,247	231	2,634	9,613
	2014	14,156	..	..	14,156	265	2,901	11,255
Nepal	2011	3,492	746	4	4,234	156	1,134	3,099
	2012	3,543	793	4	4,332	158	1,090	3,240
	2013	3,130	1,072	4	4,198	151	785	3,413
	2014	3,508	1,319	3	4,823	171	*765	3,491
Oman	2011	21,874	..	..	21,874	6,814	3,362	18,513
	2012	25,017	..	..	25,017	7,057	4,062	20,956
	2013	26,240	..	..	26,240	6,716	3,450	22,790
	2014	29,128	..	..	29,128	6,876	3,955	25,173
Other Asia	2011	252,016	..	..	252,016	10,835	29,390	223,002
	2012	250,336	..	..	250,336	10,742	28,032	222,657
	2013	252,351	..	..	252,351	10,813	25,703	226,639
	2014	260,025	..	..	260,025	11,130	28,023	231,999
Pakistan	2011	95,090	274	..	95,364	549	18,604	76,761
	2012	*96,126	375	..	*96,501	*544	19,713	76,788
	2013	*104,076	419	..	*104,495	*577	20,689	83,409
	2014	*105,305	428	..	*105,733	*571	22,134	86,328
Philippines	2011	69,210	..	..	69,210	732	13,078	56,098
	2012	72,956	..	..	72,956	760	13,711	59,211
	2013	75,300	..	..	75,300	772	13,285	61,566
	2014	77,295	..	..	77,295	780	13,951	63,345
Qatar	2011	30,730	..	..	30,730	16,128	2,971	27,759
	2012	34,787	..	..	34,787	17,259	4,604	30,184
	2013	34,668	..	..	34,668	16,498	4,603	30,066
	2014	38,692	..	..	38,692	17,813	4,909	33,784
Saudi Arabia	2011	250,077	..	..	250,077	8,687	38,905	210,393
	2012	271,680	..	..	271,680	9,211	40,373	231,307
	2013	284,017	..	..	284,017	9,404	35,791	248,226
	2014	311,806	..	..	311,806	10,095	39,691	272,115
Singapore	2011	46,000	..	..	46,000	8,862	2,473	42,821
	2012	46,971	..	..	46,971	8,863	2,344	44,315
	2013	47,964	..	..	47,964	8,874	3,041	44,963
	2014	49,310	..	..	49,310	8,955	3,035	46,417

Table 33

Electricity supply and consumption

Gigawatt-hours and kilowatt-hours per capita

Country or area	Year	Production	Imports	Exports	Supply		Consumption	
					Total	Per Capita	EIOU+losses	TFC
Sri Lanka	2011	11,600	..	..	11,600	571	1,611	9,990
	2012	11,897	..	..	11,897	583	1,488	10,409
	2013	12,044	..	..	12,044	587	1,465	10,554
	2014	12,832	..	..	12,832	622	1,833	10,997
State of Palestine	2011	569	4,622	..	5,191	1,242	1,292	3,506
	2012	461	4,909	..	5,370	1,250	537	4,846
	2013	536	4,734	..	5,270	1,193	527	4,743
	2014	337	4,938	..	5,275	1,161	633	4,642
Syrian Arab Republic	2011	42,025	902	1,192	41,735	2,036	11,263	30,472
	2012	31,188	0	0	31,188	1,561	8,574	22,614
	2013	25,933	0	529	25,404	1,315	6,599	18,805
	2014	21,726	0	138	21,588	1,150	5,835	15,753
Tajikistan	2011	16,234	172	197	16,209	2,090	2,427	13,686
	2012	16,998	114	775	16,337	2,060	2,602	13,711
	2013	17,115	117	1,061	16,171	1,993	2,684	13,487
	2014	16,472	33	1,326	15,179	1,830	2,891	12,270
Thailand	2011	151,595	10,774	1,335	161,034	2,407	14,672	*146,362
	2012	166,471	10,527	1,912	175,086	2,607	15,858	*159,228
	2013	175,531	12,569	1,794	186,306	2,762	21,984	164,322
	2014	180,862	12,260	2,066	191,056	2,821	22,271	168,785
Timor-Leste	2011	140	..	..	140	130	*14	126
	2012	126	..	..	126	114	*13	113
	2013	295	..	..	295	261	*30	*265
	2014	349	..	..	349	302	*35	*314
Turkey	2011	229,393	4,556	3,645	230,304	3,133	46,311	183,993
	2012	239,496	5,827	2,954	242,369	3,238	49,473	192,896
	2013	240,154	7,429	1,227	246,356	3,232	50,188	196,168
	2014	251,963	7,953	2,696	257,220	3,318	51,778	205,442
Turkmenistan	2011	17,220	..	2,550	14,670	2,873	5,208	9,462
	2012	17,750	..	2,720	15,030	2,906	5,335	9,695
	2013	18,870	..	2,850	16,020	3,057	5,686	10,334
	2014	20,400	..	3,210	17,190	3,239	6,101	11,089
United Arab Emirates	2011	99,137	49	36	99,150	11,351	10,743	85,549
	2012	106,222	16	34	106,204	11,863	12,131	88,247
	2013	109,979	0	23	109,956	12,163	12,239	97,717
	2014	116,528	0	0	116,528	12,825	9,686	106,842
Uzbekistan	2011	52,400	12,500	12,411	52,489	1,864	9,113	43,376
	2012	52,500	12,524	12,435	52,589	1,839	9,128	43,461
	2013	54,200	12,930	12,838	54,292	1,870	9,424	44,868
	2014	55,400	13,216	13,122	55,494	1,883	9,632	45,862
Viet Nam	2011	105,219	301	79	105,441	1,180	13,415	93,523
	2012	118,940	159	29	119,070	1,318	*15,503	104,255
	2013	128,549	221	178	128,592	1,407	*16,049	115,283
	2014	145,730	2,053	880	146,903	1,589	*18,276	128,627
Yemen	2011	6,204	..	..	6,204	256	2,129	4,075
	2012	7,073	..	..	7,073	284	2,910	4,163
	2013	8,501	..	..	8,501	333	3,531	4,970
	2014	7,646	..	..	7,646	292	3,176	4,470
Europe	**2011**	**4,854,141**	**410,013**	**416,813**	**4,847,340**	**6,583**	**923,640**	**3,925,863**
	2012	**4,895,081**	**433,673**	**443,889**	**4,884,864**	**6,628**	**935,904**	**3,943,370**
	2013	**4,850,703**	**420,979**	**429,983**	**4,841,699**	**6,564**	**910,898**	**3,926,973**
	2014	**4,766,092**	**458,623**	**467,544**	**4,757,171**	**6,446**	**907,920**	**3,840,454**

Electricity supply and consumption

Gigawatt-hours and kilowatt-hours per capita

Country or area	Year	Production	Imports	Exports	Supply		Consumption	
					Total	Per Capita	EIOU+losses	TFC
Albania	2011	4,217	3,262	232	7,247	2,511	1,590	5,657
	2012	4,725	2,538	0	7,263	2,521	1,503	5,760
	2013	6,956	2,323	0	9,279	3,218	2,386	6,892
	2014	4,724	3,250	183	7,791	2,696	1,268	6,523
Andorra	2011	91	466	..	557	6,771	75	483
	2012	88	474	..	562	7,083	74	488
	2013	115	438	3	550	7,246	71	482
	2014	127	415	..	542	7,446	67	474
Austria	2011	65,813	24,972	16,777	74,008	8,786	13,815	60,193
	2012	72,617	23,264	20,455	75,426	8,920	14,339	61,087
	2013	68,277	24,960	17,689	75,548	8,902	14,537	61,011
	2014	65,421	26,712	17,437	74,696	8,770	14,228	60,468
Belarus	2011	32,200	9,289	3,704	37,785	3,983	7,889	29,896
	2012	30,799	10,398	2,797	38,400	4,046	8,022	30,378
	2013	31,507	9,382	3,012	37,877	3,988	7,942	29,935
	2014	34,735	7,806	4,488	38,053	4,005	7,833	30,220
Belgium	2011	90,241	13,189	10,652	92,778	8,430	12,667	80,592
	2012	82,923	16,848	6,912	92,859	8,381	11,695	81,713
	2013	83,526	17,243	7,603	93,166	8,353	10,479	81,744
	2014	72,688	21,791	4,188	90,291	8,043	9,663	80,561
Bosnia and Herzegovina	2011	15,280	4,171	5,660	13,791	3,599	3,003	10,788
	2012	14,082	4,481	4,525	14,038	3,667	2,941	11,097
	2013	17,451	3,167	6,862	13,756	3,598	2,823	10,933
	2014	16,160	3,162	5,998	13,324	3,490	2,737	10,587
Bulgaria	2011	50,797	1,449	12,110	40,136	5,457	11,690	28,421
	2012	47,329	2,353	10,661	39,021	5,343	11,140	27,845
	2013	43,784	3,351	9,532	37,603	5,185	10,070	27,532
	2014	47,485	4,319	13,774	38,030	5,281	10,298	27,674
Croatia	2011	11,373	13,985	6,830	18,528	4,307	2,793	15,735
	2012	10,755	13,174	5,743	18,186	4,242	2,836	15,350
	2013	14,052	11,260	7,391	17,921	4,195	2,849	15,072
	2014	13,554	10,898	6,945	17,507	4,114	2,674	14,833
Czechia	2011	87,561	10,457	27,501	70,517	6,694	13,757	56,751
	2012	87,573	11,587	28,707	70,453	6,681	13,791	56,654
	2013	87,065	10,571	27,458	70,178	6,655	13,479	56,691
	2014	86,024	11,842	28,142	69,724	6,614	13,505	56,203
Denmark	2011	35,229	11,694	10,374	36,549	6,554	4,806	31,659
	2012	30,701	15,920	10,706	35,915	6,412	4,542	31,181
	2013	34,760	11,459	10,377	35,842	6,373	4,473	31,213
	2014	32,183	12,702	9,847	35,038	6,205	4,287	30,625
Estonia	2011	12,893	1,690	5,252	9,331	7,026	2,704	6,627
	2012	11,967	2,710	4,950	9,727	7,346	2,749	6,978
	2013	13,275	2,712	6,300	9,687	7,338	2,867	6,820
	2014	12,446	3,730	6,484	9,692	7,364	2,786	6,906
Faeroe Islands	2011	274	..	..	274	5,647	*19	255
	2012	292	..	..	292	6,025	*30	261
	2013	293	..	..	293	6,058	*18	275
	2014	305	..	..	305	6,334	*22	284
Finland	2011	73,506	17,656	3,804	87,358	16,190	7,092	80,107
	2012	70,416	19,089	1,645	87,860	16,196	6,933	80,717
	2013	71,263	17,591	1,876	86,978	15,950	6,850	79,940
	2014	68,093	21,622	3,655	86,060	15,705	6,680	79,137

Table 33

Electricity supply and consumption

Gigawatt-hours and kilowatt-hours per capita

Country or area	Year	Production	Imports	Exports	Supply		Consumption	
					Total	Per Capita	EIOU+losses	TFC
France	2011	561,448	9,501	65,914	505,035	7,978	89,757	417,566
	2012	565,704	12,213	56,734	521,183	8,195	89,069	434,093
	2013	572,308	11,687	60,148	523,847	8,200	84,841	440,710
	2014	562,776	7,873	75,063	495,586	7,724	82,980	415,325
Germany	2011	613,068	51,003	54,768	609,303	7,576	83,757	525,546
	2012	629,812	46,268	66,810	609,270	7,571	83,437	525,834
	2013	638,729	39,222	71,415	606,536	7,528	83,335	523,201
	2014	627,795	40,435	74,320	593,910	7,364	81,074	512,835
Gibraltar	2011	171	..	..	171	5,504	8	163
	2012	176	..	..	176	5,605	9	167
	2013	189	..	..	189	5,958	10	179
	2014	198	..	..	198	6,188	10	188
Greece	2011	59,436	7,180	3,948	62,668	5,619	10,874	51,794
	2012	60,959	5,954	4,169	62,744	5,648	10,727	52,017
	2013	57,152	5,788	3,901	59,039	5,340	10,248	48,791
	2014	50,474	9,461	642	59,293	5,390	9,793	49,500
Guernsey	2011	85	309	..	393	6,261	*18	375
	2012	68	313	..	381	6,031	*18	*363
	2013	289	111	..	400	6,304	*22	*378
	2014	227	156	..	383	6,009	*21	*362
Hungary	2011	36,019	14,664	8,021	42,662	4,271	8,088	34,540
	2012	34,635	16,970	9,003	42,602	4,278	7,403	32,812
	2013	30,291	16,635	4,758	42,168	4,249	7,316	34,873
	2014	29,371	19,079	5,689	42,761	4,324	7,033	34,704
Iceland	2011	17,211	..	..	17,211	53,612	1,005	16,010
	2012	17,549	..	..	17,549	54,263	981	16,357
	2013	18,116	..	..	18,116	55,674	882	16,823
	2014	18,122	..	..	18,122	55,365	1,468	16,830
Ireland	2011	27,474	732	242	27,964	6,010	3,377	24,871
	2012	27,600	784	370	28,014	6,001	3,714	24,163
	2013	26,142	2,625	383	28,384	6,076	3,848	24,201
	2014	26,314	2,853	704	28,463	6,088	3,729	24,136
Isle of Man	2011	499	41	115	424	4,989	47	377
	2012	296	180	48	428	4,987	47	381
	2013	*504	*50	77	*477	*5,517	*104	373
	2014	*504	*50	93	*461	*5,291	*93	368
Italy	2011	302,582	47,519	1,787	348,314	5,833	46,491	301,828
	2012	299,277	45,407	2,304	342,380	5,728	45,644	296,742
	2013	289,807	44,338	2,200	331,945	5,551	44,547	287,398
	2014	279,827	46,747	3,031	323,543	5,409	42,046	281,498
Jersey	2011	46	700	..	746	7,638	67	677
	2012	60	681	..	741	7,550	85	670
	2013	328	553	..	881	8,922	71	690
	2014	135	581	..	717	7,227	65	650
Latvia	2011	6,095	4,009	2,764	7,340	3,557	1,147	6,191
	2012	6,167	4,935	3,244	7,858	3,857	1,007	6,848
	2013	6,209	5,005	3,650	7,564	3,760	984	6,576
	2014	5,141	5,340	3,023	7,458	3,749	874	6,582
Liechtenstein	2011	72	326	..	398	10,901	0	*398
	2012	86	318	..	404	10,981	0	*404
	2013	85	319	..	404	10,904	0	404
	2014	40	355	..	395	10,602	0	395

Electricity supply and consumption

Gigawatt-hours and kilowatt-hours per capita

Country or area	Year	Production	Imports	Exports	Supply		Consumption	
					Total	Per Capita	EIOU+losses	TFC
Lithuania	2011	4,822	8,086	1,347	11,561	3,765	2,977	8,580
	2012	5,043	8,060	1,441	11,662	3,866	2,738	8,921
	2013	4,762	8,073	1,127	11,708	3,950	2,750	8,955
	2014	4,397	8,521	898	12,020	4,121	2,780	9,237
Luxembourg	2011	3,718	7,096	2,614	8,200	15,770	1,674	6,529
	2012	3,818	6,732	2,622	7,928	14,889	1,670	6,270
	2013	2,888	6,852	1,907	7,833	14,376	1,618	6,102
	2014	2,967	6,961	2,067	7,861	14,124	1,645	6,228
Malta	2011	2,187	..	..	2,187	5,282	287	1,900
	2012	2,307	..	..	2,307	5,551	328	1,979
	2013	2,251	..	..	2,251	5,401	271	1,980
	2014	2,245	..	..	2,245	5,374	213	2,032
Montenegro	2011	2,809	1,383	431	3,761	6,037	804	3,414
	2012	2,844	964	228	3,580	5,738	836	3,220
	2013	3,945	388	647	3,686	5,901	758	2,706
	2014	3,040	903	642	3,301	5,279	692	2,610
Netherlands	2011	113,963	20,620	11,531	123,052	7,373	15,097	108,148
	2012	103,298	32,156	15,046	120,408	7,189	14,589	104,006
	2013	101,736	33,252	15,015	119,973	7,137	15,584	104,373
	2014	103,418	32,855	18,128	118,145	7,004	15,170	101,630
Norway	2011	127,655	11,255	14,329	124,581	25,148	16,528	107,389
	2012	147,716	4,190	22,006	129,900	25,885	18,270	110,778
	2013	133,975	10,135	15,141	128,969	25,370	16,270	111,699
	2014	142,327	6,347	21,932	126,742	24,620	17,466	108,409
Poland	2011	163,548	6,780	12,022	158,306	4,102	36,331	121,975
	2012	162,139	9,803	12,643	159,299	4,126	36,625	122,674
	2013	164,557	7,801	12,322	160,036	4,144	35,977	124,059
	2014	159,059	13,508	11,342	161,225	4,175	35,362	125,863
Portugal	2011	52,463	6,742	3,929	55,276	5,235	6,912	48,364
	2012	46,614	10,766	2,871	54,509	5,184	8,268	46,241
	2013	51,672	8,100	5,324	54,448	5,205	9,191	45,257
	2014	52,802	7,247	6,344	53,705	5,163	8,510	45,195
Republic of Moldova	2011	1,016	3,145	..	4,161	1,020	780	3,380
	2012	932	3,279	..	4,211	1,033	741	3,470
	2013	905	3,332	..	4,237	1,040	591	3,646
	2014	963	3,342	..	4,305	1,057	657	3,648
Romania	2011	62,217	3,410	5,316	60,311	2,999	17,687	42,714
	2012	59,045	3,903	3,650	59,298	2,973	17,240	42,386
	2013	58,887	2,737	4,753	56,871	2,873	16,554	40,628
	2014	65,676	2,811	9,937	58,550	2,979	16,593	41,905
Russian Federation	2011	1,054,765	1,558	24,111	1,032,212	7,208	303,388	728,824
	2012	1,070,734	2,661	19,143	1,054,252	7,358	312,553	740,285
	2013	1,059,092	4,706	18,382	1,045,416	7,292	301,325	744,091
	2014	1,064,207	6,623	14,671	1,056,159	7,364	318,329	737,830
Serbia	2011	38,600	6,701	6,979	38,322	4,247	10,331	27,991
	2012	36,799	5,781	5,392	37,188	4,140	10,021	27,167
	2013	39,877	4,077	6,614	37,340	4,178	10,437	26,903
	2014	34,060	7,008	5,445	35,623	4,006	9,465	26,158
Slovakia	2011	28,656	11,227	10,500	29,383	5,430	4,568	24,812
	2012	28,664	13,472	13,079	29,057	5,366	5,101	23,937
	2013	28,832	10,719	10,628	28,923	5,337	3,827	25,084
	2014	27,401	12,964	11,862	28,503	5,256	4,330	24,157

Table 33

Electricity supply and consumption

Gigawatt-hours and kilowatt-hours per capita

Country or area	Year	Production	Imports	Exports	Supply		Consumption	
					Total	Per Capita	EIOU+losses	TFC
Slovenia	2011	16,059	7,036	8,408	14,687	7,133	2,187	12,500
	2012	15,736	7,452	8,491	14,697	7,125	2,269	12,428
	2013	16,103	7,521	8,811	14,813	7,174	2,334	12,479
	2014	17,437	7,254	9,997	14,694	7,112	2,235	12,459
Spain	2011	293,848	7,932	14,023	287,757	6,161	45,138	243,509
	2012	297,559	7,787	18,986	286,360	6,140	46,940	240,248
	2013	285,632	9,887	16,638	278,881	6,003	47,930	230,087
	2014	278,750	12,310	15,716	275,344	5,952	48,374	226,897
Sweden	2011	150,376	12,481	19,714	143,143	15,128	17,183	124,628
	2012	166,562	11,682	31,255	146,989	15,402	17,910	127,286
	2013	153,166	12,674	22,676	143,164	14,875	16,689	125,016
	2014	153,662	13,852	29,475	138,039	14,226	14,044	122,191
Switzerland	2011	64,629	34,824	32,237	67,216	8,442	8,617	58,586
	2012	69,893	31,549	33,749	67,693	8,399	8,721	58,959
	2013	70,312	29,874	32,270	67,916	8,327	8,593	59,313
	2014	71,767	28,530	34,021	66,276	8,035	8,810	57,457
T.F.Yug.Rep. Macedonia	2011	6,759	2,749	73	9,435	4,567	2,057	7,378
	2012	6,262	2,741	72	8,931	4,316	1,926	7,005
	2013	6,094	2,491	62	8,522	4,112	1,722	6,801
	2014	5,303	3,074	115	8,263	3,981	1,534	6,729
Ukraine	2011	194,947	35	6,327	188,655	4,148	48,857	139,800
	2012	198,878	93	11,560	187,411	4,135	49,051	137,668
	2013	194,377	39	9,929	184,487	4,085	46,956	137,531
	2014	182,815	89	8,523	174,381	3,875	40,594	128,387
United Kingdom	2011	367,423	8,689	2,467	373,645	5,915	55,700	317,942
	2012	363,582	13,743	1,872	375,453	5,906	57,371	318,082
	2013	359,168	17,532	3,102	373,598	5,842	56,470	317,127
	2014	338,925	23,244	2,723	359,446	5,587	55,883	303,563
Oceania	**2011**	**308,986**	**..**	**..**	**308,986**	**8,348**	**47,663**	**260,333**
	2012	**306,065**	**..**	**..**	**306,065**	**8,139**	**47,299**	**257,973**
	2013	**304,237**	**..**	**..**	**304,237**	**7,968**	**45,118**	**258,757**
	2014	**303,224**	**..**	**..**	**303,224**	**7,824**	**45,879**	**257,201**
American Samoa	2011	155	..	..	155	2,802	19	136
	2012	157	..	..	157	2,836	22	134
	2013	*156	..	..	*156	*2,829	21	136
	2014	157	..	..	157	2,831	22	135
Australia	2011	253,958	..	..	253,958	11,266	41,629	212,237
	2012	251,165	..	..	251,165	10,962	41,144	209,917
	2013	249,720	..	..	249,720	10,731	39,333	210,180
	2014	248,299	..	..	248,299	10,511	40,243	208,055
Cook Islands	2011	34	..	..	34	1,646	0	34
	2012	33	..	..	33	1,603	0	33
	2013	33	..	..	33	1,576	0	33
	2014	32	..	..	32	1,534	0	32
Fiji	2011	801	..	..	801	924	*51	767
	2012	803	..	..	803	919	*72	732
	2013	858	..	..	858	974	*77	780
	2014	859	..	..	859	969	*93	783
French Polynesia	2011	*761	..	..	*761	*2,809	*75	*685
	2012	*768	..	..	*768	*2,804	*79	689
	2013	*744	..	..	*744	*2,689	*80	665
	2014	*721	..	..	*721	*2,577	*79	*642

Electricity supply and consumption

Gigawatt-hours and kilowatt-hours per capita

Country or area	Year	Production	Imports	Exports	Supply		Consumption	
					Total	Per Capita	EIOU+losses	TFC
Guam	2011	*1,858	..	..	*1,858	*11,551	*240	1,618
	2012	*1,799	..	..	*1,799	*11,050	*236	1,564
	2013	1,736	..	..	1,736	10,516	187	1,550
	2014	*1,750	..	..	*1,750	*10,445	*172	*1,578
Kiribati	2011	*24	..	..	*24	*230	*1	*23
	2012	*24	..	..	*24	*226	*1	*23
	2013	*24	..	..	*24	*222	*1	*23
	2014	*24	..	..	*24	*220	*1	*23
Marshall Islands	2011	*79	..	..	*79	*1,504	*10	*69
	2012	78	..	..	78	1,475	10	68
	2013	84	..	..	84	1,584	*11	*73
	2014	*85	..	..	*85	*1,611	*11	*74
Micronesia (Fed. States of)	2011	66	..	..	66	637	13	*53
	2012	*67	..	..	*67	*650	*13	*54
	2013	*68	..	..	*68	*656	*13	*55
	2014	70	..	..	70	669	*15	*55
Nauru	2011	23	..	..	23	2,300	*2	21
	2012	24	..	..	24	2,351	*2	*21
	2013	*24	..	..	*24	*2,400	*2	*22
	2014	*25	..	..	*25	*2,462	*2	*23
New Caledonia	2011	2,267	..	..	2,267	9,076	115	2,152
	2012	2,288	..	..	2,288	9,038	72	2,216
	2013	2,506	..	..	2,506	9,770	108	2,398
	2014	2,622	..	..	2,622	10,092	168	2,453
New Zealand	2011	44,465	..	..	44,465	10,095	4,949	38,601
	2012	44,260	..	..	44,260	9,978	5,061	38,514
	2013	43,265	..	..	43,265	9,689	4,884	38,224
	2014	43,553	..	..	43,553	9,688	4,756	38,636
Niue	2011	3	..	..	3	2,023	0	3
	2012	3	..	..	3	2,064	0	3
	2013	3	..	..	3	2,082	0	3
	2014	3	..	..	3	2,004	0	3
Northern Mariana Islands	2011	*407	..	..	*407	*7,638	*91	*316
	2012	*411	..	..	*411	*7,703	*91	*319
	2013	*415	..	..	*415	*7,700	*92	*322
	2014	*419	..	..	*419	*7,680	*93	*326
Palau	2011	*78	..	..	*78	*3,761	*16	71
	2012	*78	..	..	*78	*3,734	*16	*67
	2013	*78	..	..	*78	*3,705	*16	*64
	2014	*79	..	..	*79	*3,745	*16	*65
Papua New Guinea	2011	3,666	..	..	3,666	524	*407	3,259
	2012	3,764	..	..	3,764	526	*435	3,329
	2013	4,169	..	..	4,169	570	242	3,926
	2014	*4,169	..	..	*4,169	*559	*158	4,011
Samoa	2011	114	..	..	114	608	*16	89
	2012	117	..	..	117	620	*16	*93
	2013	*121	..	..	*121	*633	*23	*98
	2014	*121	..	..	*121	*629	*23	*98
Solomon Islands	2011	83	..	..	83	154	*12	*71
	2012	83	..	..	83	152	*12	*71
	2013	89	..	..	89	159	*12	*77
	2014	92	..	..	92	161	*12	*80

Electricity supply and consumption

Gigawatt-hours and kilowatt-hours per capita

Country or area	Year	Production	Imports	Exports	Supply		Consumption	
					Total	Per Capita	EIOU+losses	TFC
Tonga	2011	53	..	..	53	509	9	*45
	2012	52	..	..	52	500	8	*45
	2013	*53	..	..	*53	*507	7	*46
	2014	*55	..	..	*55	*517	7	*48
Tuvalu	2011	5	..	..	5	516	0	5
	2012	*5	..	..	*5	*517	0	*5
	2013	*5	..	..	*5	*516	0	*5
	2014	*5	..	..	*5	*516	0	*5
Vanuatu	2011	*68	..	..	*68	*280	5	63
	2012	67	..	..	67	272	5	61
	2013	*68	..	..	*68	*267	5	62
	2014	*67	..	..	*67	*257	5	61
Wallis and Futuna Is.	2011	20	..	..	20	1,469	4	*16
	2012	19	..	..	19	1,422	3	*16
	2013	19	..	..	19	1,420	3	*15
	2014	19	..	..	19	1,405	3	*16

Production of heat – by type

Terajoules

Table Notes

Direct use geothermal represents heat used directly for district heating, agriculture etc. without a commercial transaction taking place.

Direct use solar thermal represents heat used for the production of domestic hot water or for the seasonal heating of swimming pools (e.g. flat plate collectors, mainly of the thermosyphon type) without a commercial transaction taking place.

Please refer to the Definitions Section on pages ix to xv for the appropriate product description /classification.

Figure 84: Heat production by region, 1990-2014

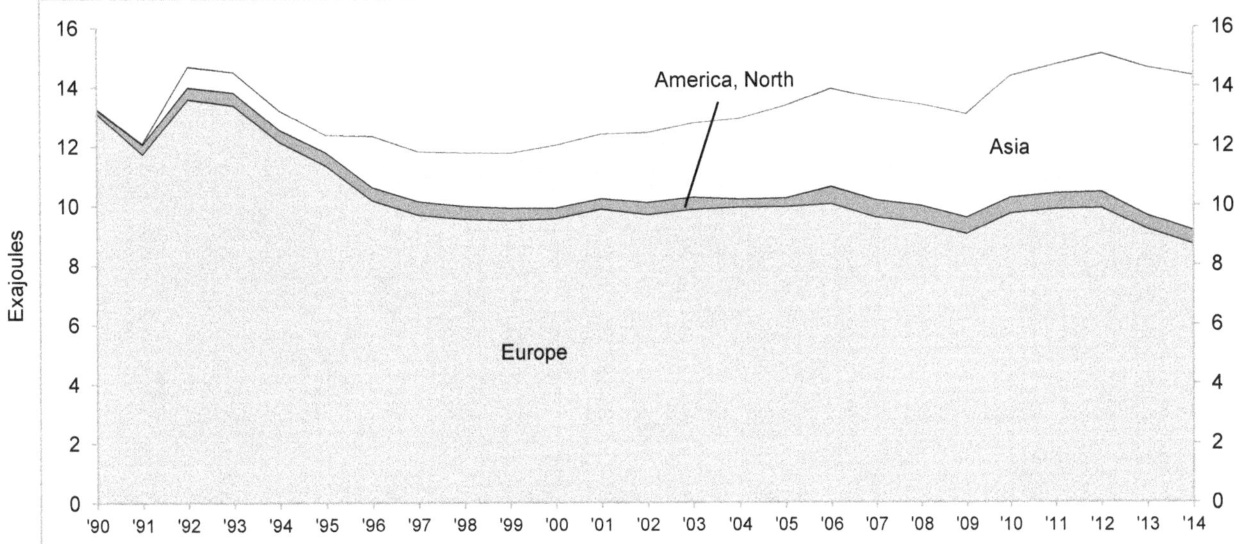

Figure 85: Major heat producing countries, 2014

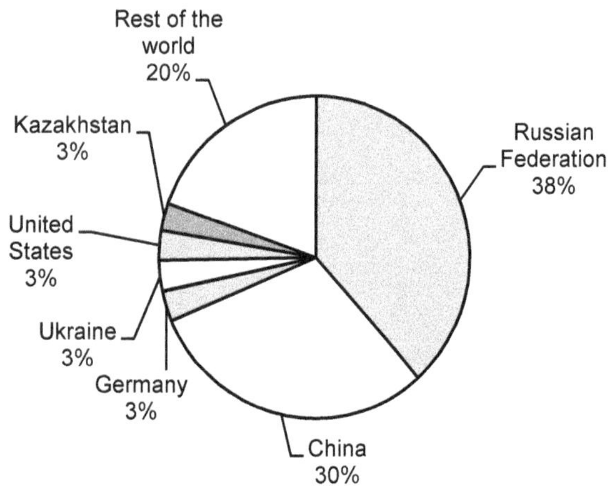

Figure 86: World heat production by type, 2014

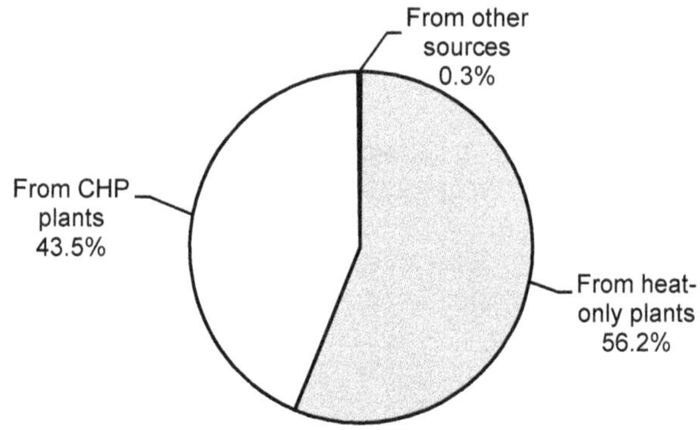

Production of heat - by type

Terajoules

Country or area	Year	Combust-ible fuels	Geo-thermal	Solar thermal	Nuclear	Chemical sources	Heat pumps and electric boilers	Other sources	Direct use geothermal	Direct use solar thermal
World	**2011**	**14,303,353**	**25,674**	**6,207**	**27,734**	**29,584**	**28,137**	**345,025**	**124,637**	**251,667**
	2012	**14,658,369**	**27,630**	**6,147**	**27,848**	**30,630**	**33,379**	**345,763**	**136,831**	**288,693**
	2013	**14,142,688**	**28,583**	**5,927**	**27,587**	**78,672**	**29,249**	**344,206**	**141,144**	**298,708**
	2014	**13,869,399**	**29,933**	**6,890**	**26,464**	**71,982**	**30,981**	**352,976**	**136,711**	**307,338**
Africa	**2011**	**12,421**	**..**	**716**	**..**	**3,865**	**..**	**..**	**..**	**4,382**
	2012	**12,923**	**..**	**749**	**..**	**4,364**	**..**	**..**	**..**	**4,822**
	2013	**12,895**	**..**	**786**	**..**	**3,748**	**..**	**..**	**..**	**5,531**
	2014	**11,265**	**..**	**815**	**..**	**3,355**	**..**	**..**	**..**	**6,136**
Mayotte	2011	..	..	..	..	..	..	..	..	4
	2012	..	..	..	..	..	..	..	..	*4
	2013	..	..	..	..	..	..	..	..	*5
	2014	..	..	..	..	..	..	..	..	*6
Namibia	2011	..	..	..	..	..	..	..	..	72
	2012	..	..	..	..	..	..	..	..	72
	2013	..	..	..	..	..	..	..	..	72
	2014	..	..	..	..	..	..	..	..	72
Réunion	2011	1,721	..	716	..	..	..	..	..	..
	2012	2,023	..	749	..	..	..	..	..	..
	2013	1,995	..	786	..	..	..	..	..	..
	2014	2,016	..	815	..	..	..	..	..	..
Senegal	2011	..	..	..	..	1,124	..	..	..	..
	2012	..	..	..	..	1,298	..	..	..	..
	2013	..	..	..	..	698	..	..	..	..
	2014	..	..	..	..	731	..	..	..	..
South Africa	2011	..	..	..	..	..	..	..	..	3,005
	2012	..	..	..	..	..	..	..	..	3,253
	2013	..	..	..	..	..	..	..	..	3,804
	2014	..	..	..	..	..	..	..	..	4,241
Swaziland	2011	*10,700	..	..	..	..	..	..	..	..
	2012	*10,900	..	..	..	..	..	..	..	..
	2013	*10,900	..	..	..	..	..	..	..	..
	2014	9,249	..	..	..	..	..	..	..	..
Tunisia	2011	..	..	..	..	2,741	..	..	..	1,301
	2012	..	..	..	..	3,066	..	..	..	1,493
	2013	..	..	..	..	3,050	..	..	..	1,650
	2014	..	..	..	..	2,624	..	..	..	1,817
America, North	**2011**	**536,236**	**..**	**..**	**..**	**..**	**..**	**..**	**9,285**	**67,766**
	2012	**530,927**	**..**	**..**	**..**	**..**	**..**	**..**	**17,687**	**94,312**
	2013	**483,191**	**..**	**..**	**..**	**..**	**..**	**..**	**18,268**	**98,358**
	2014	**467,289**	**..**	**..**	**..**	**..**	**..**	**..**	**10,969**	**102,538**
Canada	2011	26,773	..	..	..	..	..	..	..	1,770
	2012	28,988	..	..	..	..	..	..	..	1,553
	2013	28,679	..	..	..	..	..	..	..	1,630
	2014	37,679	..	..	..	..	..	..	..	1,702
Greenland	2011	826	..	..	..	..	..	..	..	..
	2012	1,140	..	..	..	..	..	..	..	..
	2013	1,099	..	..	..	..	..	..	..	..
	2014	1,003	..	..	..	..	..	..	..	..
Mexico	2011	..	..	..	..	..	..	..	..	5,665
	2012	..	..	..	..	..	..	..	..	6,427
	2013	..	..	..	..	..	..	..	..	7,236
	2014	..	..	..	..	..	..	..	..	8,064

Table 34

Production of heat - by type

Terajoules

Country or area	Year	Combust-ible fuels	Geo-thermal	Solar thermal	Nuclear	Chemical sources	Heat pumps and electric boilers	Other sources	Direct use geothermal	Direct use solar thermal
United States	2011	508,637	..	..	..	..	..	..	9,285	60,331
	2012	500,799	..	..	..	..	..	..	17,687	86,332
	2013	453,413	..	..	..	..	..	..	18,268	89,492
	2014	428,607	..	..	..	..	..	..	10,969	92,772
America, South	**2011**	**..**	**..**	**..**	**..**	**0**	**..**	**..**	**102**	**263**
	2012	**..**	**..**	**..**	**..**	**0**	**..**	**..**	**102**	**1,058**
	2013	**..**	**..**	**..**	**..**	**1,091**	**..**	**..**	**102**	**1,091**
	2014	**..**	**..**	**..**	**..**	**1,936**	**..**	**..**	**102**	**1,116**
Chile	2011	..	..	..	..	0	..	..	..	..
	2012	..	..	..	..	0	..	..	..	772
	2013	..	..	..	..	1,091	..	..	..	780
	2014	..	..	..	..	1,936	..	..	..	780
Ecuador	2011	..	..	..	..	..	..	..	102	..
	2012	..	..	..	..	..	..	..	102	..
	2013	..	..	..	..	..	..	..	102	..
	2014	..	..	..	..	..	..	..	102	..
Peru	2011	..	..	..	..	..	..	..	..	263
	2012	..	..	..	..	..	..	..	..	286
	2013	..	..	..	..	..	..	..	..	311
	2014	..	..	..	..	..	..	..	..	336
Asia	**2011**	**4,325,833**	**36**	**5,183**	**..**	**3,767**	**3,662**	**4,608**	**71,210**	**100,008**
	2012	**4,654,212**	**36**	**4,943**	**..**	**3,420**	**3,791**	**4,332**	**71,838**	**104,357**
	2013	**4,954,199**	**0**	**4,569**	**..**	**3,976**	**3,878**	**8,370**	**72,819**	**105,458**
	2014	**5,207,604**	**0**	**5,301**	**..**	**3,752**	**3,878**	**7,629**	**74,820**	**104,480**
Armenia	2011	186	..	..	..	..	..	..	..	..
	2012	90	..	..	..	..	..	..	..	..
	2013	127	..	..	..	..	..	..	..	..
	2014	*127	..	..	..	..	..	..	..	..
Azerbaijan	2011	4,536	..	..	..	..	..	..	..	..
	2012	5,004	..	..	..	..	..	..	..	..
	2013	5,436	..	..	..	..	..	..	..	..
	2014	6,778	..	..	..	..	..	..	..	..
China	2011	3,491,768	..	..	..	..	..	..	..	..
	2012	3,787,310	..	..	..	..	..	..	..	..
	2013	4,131,207	..	..	..	..	..	..	..	..
	2014	4,334,113	..	..	..	..	..	..	..	..
Cyprus	2011	31	..	..	..	..	..	..	44	2,625
	2012	31	..	..	..	..	..	..	62	2,700
	2013	42	..	..	..	..	..	..	62	2,751
	2014	45	..	..	..	..	..	..	65	2,796
Georgia	2011	0	36	..	..	..	..	..	377	0
	2012	0	36	..	..	..	..	..	380	6
	2013	0	0	..	..	..	..	..	520	47
	2014	0	0	..	..	..	..	..	546	77
Israel	2011	..	..	..	..	..	..	..	..	46,070
	2012	..	..	..	..	..	..	..	..	46,070
	2013	..	..	..	..	..	..	..	..	46,070
	2014	..	..	..	..	..	..	..	..	46,070
Japan	2011	14,202	..	..	..	..	3,662	4,608	7,535	18,726
	2012	13,883	..	..	..	..	3,791	4,332	7,409	17,098
	2013	13,697	..	..	..	..	3,878	4,377	7,343	16,905
	2014	13,697	..	..	..	..	3,878	4,377	7,242	15,391

Production of heat - by type

Terajoules

Country or area	Year	Combustible fuels	Geothermal	Solar thermal	Nuclear	Chemical sources	Heat pumps and electric boilers	Other sources	Direct use geothermal	Direct use solar thermal
Kazakhstan	2011	405,407	..	..	..	..	..	..	..	..
	2012	418,081	..	..	..	..	..	..	..	..
	2013	396,689	..	..	..	..	..	..	..	..
	2014	398,902	..	..	..	..	..	..	..	..
Korea, Republic of	2011	199,727	..	..	..	3,767	..	..	2,002	1,148
	2012	216,087	..	..	..	3,420	..	..	2,733	1,099
	2013	196,855	..	..	..	3,976	..	..	3,642	1,164
	2014	203,792	..	..	..	3,752	..	..	4,542	1,193
Kyrgyzstan	2011	13,389	..	..	..	..	..	..	..	..
	2012	13,534	..	..	..	..	..	..	..	..
	2013	12,525	..	..	..	..	..	..	..	..
	2014	13,858	..	..	..	..	..	..	..	..
Lebanon	2011	..	..	..	..	..	..	..	..	704
	2012	..	..	..	..	..	..	..	..	839
	2013	..	..	..	..	..	..	..	..	894
	2014	..	..	..	..	..	..	..	..	1,015
Mongolia	2011	36,354	..	..	..	..	..	..	..	..
	2012	39,029	..	..	..	..	..	..	..	..
	2013	39,824	..	..	..	..	..	..	..	..
	2014	42,003	..	..	..	..	..	..	..	..
Other Asia	2011	..	..	..	..	..	..	..	..	4,358
	2012	..	..	..	..	..	..	..	..	4,390
	2013	..	..	..	..	..	..	..	..	4,343
	2014	..	..	..	..	..	..	..	..	4,318
State of Palestine	2011	..	..	5,183	..	..	..	..	..	..
	2012	..	..	4,943	..	..	..	..	..	..
	2013	..	..	4,569	..	..	..	..	..	..
	2014	..	..	5,301	..	..	..	..	..	..
Tajikistan	2011	387	..	..	..	..	..	..	..	..
	2012	405	..	..	..	..	..	..	..	..
	2013	408	..	..	..	..	..	..	..	..
	2014	1,063	..	..	..	..	..	..	..	..
Turkey	2011	50,917	..	..	..	..	..	..	61,252	26,377
	2012	51,317	..	..	..	..	..	..	61,254	32,155
	2013	51,633	..	..	..	..	..	3,993	61,252	33,284
	2014	84,796	..	..	..	..	..	3,252	62,425	33,620
Turkmenistan	2011	8,311	..	..	..	..	..	..	..	..
	2012	8,567	..	..	..	..	..	..	..	..
	2013	9,108	..	..	..	..	..	..	..	..
	2014	9,846	..	..	..	..	..	..	..	..
Uzbekistan	2011	100,619	..	..	..	..	..	..	..	..
	2012	100,874	..	..	..	..	..	..	..	..
	2013	96,648	..	..	..	..	..	..	..	..
	2014	98,584	..	..	..	..	..	..	..	..
Europe	**2011**	**9,428,863**	**25,638**	**308**	**27,734**	**21,952**	**24,475**	**339,229**	**35,313**	**67,127**
	2012	**9,460,307**	**27,594**	**455**	**27,848**	**22,846**	**29,588**	**340,524**	**37,241**	**71,305**
	2013	**8,692,404**	**28,583**	**572**	**27,587**	**69,857**	**25,371**	**334,986**	**39,286**	**74,647**
	2014	**8,183,242**	**29,933**	**774**	**26,464**	**62,939**	**27,103**	**344,497**	**39,158**	**79,359**
Albania	2011	243	..	..	..	..	..	..	..	..
	2012	197	..	..	..	..	..	..	..	..
	2013	199	..	..	..	..	..	..	..	..
	2014	214	..	..	..	..	..	..	..	..

Table 34

Production of heat - by type

Terajoules

Country or area	Year	Combust-ible fuels	Geo-thermal	Solar thermal	Nuclear	Chemical sources	Heat pumps and electric boilers	Other sources	Direct use geothermal	Direct use solar thermal
Austria	2011	79,196	524	51	..	99	..	..	278	6,990
	2012	83,927	577	57	..	99	..	..	346	7,239
	2013	86,836	593	57	..	99	..	..	329	7,383
	2014	79,299	543	61	..	95	..	..	268	7,561
Belarus	2011	260,279	..	..	..	..	..	..	..	..
	2012	269,754	..	..	..	..	..	..	..	..
	2013	263,853	..	..	..	..	..	..	..	..
	2014	254,934	..	..	..	..	..	..	..	..
Belgium	2011	35,098	55	..	..	2,930	..	..	..	601
	2012	32,124	60	..	..	4,835	..	..	..	641
	2013	29,115	68	..	..	8,499	..	..	..	779
	2014	27,121	60	..	..	9,304	..	..	..	857
Bosnia and Herzegovina	2011	6,355	..	..	..	..	..	..	..	..
	2012	6,075	..	..	..	..	..	..	..	..
	2013	5,722	..	..	..	..	..	..	..	..
	2014	5,000	..	..	..	..	..	..	..	..
Bulgaria	2011	54,518	..	..	922	2,992	..	..	1,380	579
	2012	54,649	..	..	940	2,235	..	..	1,400	645
	2013	50,370	..	..	655	1,992	..	..	1,400	798
	2014	51,930	..	..	604	1,645	..	..	1,400	824
Croatia	2011	12,065	..	..	..	..	..	..	287	254
	2012	11,625	..	..	..	..	..	..	291	298
	2013	11,738	..	..	..	..	..	..	286	341
	2014	10,127	..	..	..	..	..	..	446	385
Czechia	2011	120,734	..	..	919	1,209	102	332	..	455
	2012	125,383	..	..	981	1,341	105	348	..	561
	2013	122,516	..	..	1,035	1,406	94	350	..	630
	2014	119,947	..	..	872	877	189	340	..	665
Denmark	2011	129,843	83	217	..	..	264	2,682	..	518
	2012	133,125	144	345	..	..	688	2,646	..	534
	2013	132,601	114	475	..	..	596	2,252	..	552
	2014	118,409	83	662	..	..	494	2,488	..	565
Estonia	2011	22,887	..	..	..	..	..	..	..	..
	2012	24,473	..	..	..	..	..	..	..	..
	2013	23,028	..	..	..	..	..	..	..	..
	2014	21,812	..	..	..	..	..	..	..	..
Finland	2011	180,873	..	..	..	2,925	1,400	1,204	..	43
	2012	191,553	..	..	..	3,698	1,866	890	..	47
	2013	179,584	..	..	..	4,475	1,792	1,043	..	52
	2014	174,077	..	..	..	4,526	2,236	960	..	57
France	2011	161,154	3,049	..	..	..	..	..	1,280	2,986
	2012	132,614	3,242	..	..	..	47	..	1,280	3,420
	2013	130,457	3,937	..	..	..	75	..	1,280	3,748
	2014	117,775	3,984	..	..	..	270	..	1,280	4,128
Germany	2011	460,483	51	..	..	..	..	6,703	2,477	23,184
	2012	474,421	85	11	..	..	..	6,296	2,682	24,109
	2013	480,642	207	1	..	..	..	7,461	2,852	24,408
	2014	430,982	304	11	..	..	..	7,043	3,487	26,221
Greece	2011	2,256	..	..	..	..	..	..	970	7,656
	2012	1,889	..	..	..	..	..	..	909	7,720
	2013	1,738	..	..	..	..	..	..	812	7,831
	2014	2,073	..	..	..	..	..	..	490	8,029

Production of heat - by type

Terajoules

Country or area	Year	Combust-ible fuels	Geo-thermal	Solar thermal	Nuclear	Chemical sources	Heat pumps and electric boilers	Other sources	Direct use geothermal	Direct use solar thermal
Hungary	2011	48,432	271	5	508	..	..	..	4,080	242
	2012	48,574	395	5	528	..	..	..	4,071	242
	2013	46,975	574	0	505	..	..	965	4,104	250
	2014	39,672	1,166	0	456	..	..	1,497	4,048	260
Iceland	2011	..	20,816	..	..	..	674	..	4,411	..
	2012	..	22,241	..	..	..	675	..	4,416	..
	2013	..	22,060	..	..	..	675	..	4,087	..
	2014	..	22,443	..	..	..	675	..	4,543	..
Ireland	2011	..	..	..	..	..	..	..	..	380
	2012	..	..	..	..	..	..	..	..	427
	2013	..	..	..	..	..	..	..	..	472
	2014	..	..	..	..	..	..	..	..	511
Italy	2011	218,519	589	1	..	..	..	..	5,243	5,877
	2012	206,335	650	1	..	..	..	..	4,950	6,503
	2013	215,757	650	2	..	..	..	..	4,987	7,040
	2014	205,188	764	2	..	..	..	..	4,660	7,517
Latvia	2011	24,985	..	..	..	..	8	7	..	..
	2012	26,840	..	..	..	..	10	7	..	..
	2013	26,227	..	..	..	..	9	13	..	..
	2014	25,740	..	..	..	..	7	..	..	..
Liechtenstein	2011	..	..	32	..	..	..	..	..	..
	2012	..	..	34	..	..	..	..	..	..
	2013	..	..	36	..	..	..	..	..	..
	2014	..	..	37	..	..	..	..	..	..
Lithuania	2011	35,583	67	..	..	10,225	12	..	..	..
	2012	36,489	79	..	..	9,877	10	..	..	..
	2013	34,217	35	..	..	9,363	10	..	..	..
	2014	32,420	39	..	..	10,832	9	..	..	..
Luxembourg	2011	3,150	..	..	..	..	..	..	..	63
	2012	3,076	..	..	..	..	..	..	..	88
	2013	3,247	..	..	..	..	..	..	..	104
	2014	3,587	..	..	..	..	..	..	..	128
Malta	2011	5	..	..	..	..	..	..	..	194
	2012	8	..	..	..	..	..	..	..	216
	2013	1	..	..	..	..	..	..	..	173
	2014	1	..	..	..	..	..	..	..	178
Netherlands	2011	153,951	..	..	..	..	..	..	316	1,040
	2012	149,117	..	..	..	..	..	..	495	1,070
	2013	150,090	..	..	..	..	..	..	993	1,106
	2014	147,157	..	..	..	..	..	..	1,502	1,128
Norway	2011	17,236	..	..	..	615	3,326	805	..	..
	2012	16,273	..	..	..	21	4,351	1,174	..	..
	2013	18,260	..	..	..	1,194	5,076	1,164	..	..
	2014	17,422	..	..	..	3,200	4,091	1,313	..	..
Poland	2011	297,957	..	..	..	913	4	101	531	434
	2012	308,168	..	..	..	685	5	191	661	544
	2013	302,848	..	..	..	891	5	128	778	639
	2014	275,602	..	..	..	1,205	4	76	847	720
Portugal	2011	20,600	..	..	..	44	..	..	65	2,491
	2012	21,419	..	..	..	55	..	..	65	2,820
	2013	25,523	..	..	..	46	..	..	52	3,046
	2014	21,367	..	..	..	55	..	..	55	3,218

Table 34

Production of heat - by type

Terajoules

Country or area	Year	Combust-ible fuels	Geo-thermal	Solar thermal	Nuclear	Chemical sources	Heat pumps and electric boilers	Other sources	Direct use geothermal	Direct use solar thermal
Republic of Moldova	2011	11,402	..	..	..	..	..	..	..	..
	2012	10,876	..	..	..	..	..	..	..	..
	2013	10,640	..	..	..	..	..	..	..	..
	2014	10,466	..	..	..	..	..	..	..	..
Romania	2011	98,853	7	..	..	..	..	..	923	2
	2012	89,338	4	..	..	..	..	..	902	4
	2013	84,803	9	..	..	..	..	..	1,049	9
	2014	77,756	217	..	..	..	..	..	835	23
Russian Federation	2011	6,038,583	..	..	15,482	..	..	327,324	..	..
	2012	6,063,487	..	..	15,109	..	..	328,963	..	..
	2013	5,374,829	211	..	15,136	..	..	321,571	..	..
	2014	5,187,219	191	..	14,982	..	..	330,772	..	..
Serbia	2011	41,440	..	..	..	..	..	..	265	..
	2012	35,203	..	..	..	..	..	..	260	..
	2013	34,313	..	..	..	..	..	..	187	..
	2014	30,694	..	..	..	..	..	..	234	..
Slovakia	2011	41,351	104	2	2,210	..	9	71	60	211
	2012	40,905	95	2	2,115	..	45	9	55	222
	2013	40,192	102	1	2,147	..	49	39	69	233
	2014	32,825	120	1	1,795	..	55	8	56	241
Slovenia	2011	9,698	22	..	..	..	..	..	1,169	376
	2012	9,477	22	..	..	..	..	..	1,133	412
	2013	9,405	23	..	..	..	..	..	1,250	440
	2014	8,164	19	..	..	..	..	..	1,275	455
Spain	2011	..	..	..	..	..	..	..	703	8,576
	2012	..	..	..	..	..	..	..	737	9,226
	2013	..	..	..	..	..	..	..	758	10,006
	2014	..	..	..	..	..	..	..	789	10,828
Sweden	2011	159,916	..	..	..	..	18,544	..	..	461
	2012	174,779	..	..	..	..	21,665	..	..	464
	2013	173,253	..	..	..	..	16,895	..	..	468
	2014	160,891	..	..	..	..	19,005	..	..	468
Switzerland	2011	18,399	..	..	1,290	..	132	..	10,418	1,655
	2012	18,866	..	..	1,370	..	121	..	12,184	1,853
	2013	19,674	..	..	1,270	..	95	..	13,630	2,042
	2014	17,313	..	..	1,190	..	68	..	12,616	2,212
T.F.Yug.Rep. Macedonia	2011	2,669	..	..	..	..	..	..	424	..
	2012	2,463	..	..	..	..	..	..	371	..
	2013	2,131	..	..	..	..	..	..	350	..
	2014	1,956	..	..	..	..	..	..	294	..
Ukraine	2011	602,027	..	..	6,403	..	..	..	..	..
	2012	592,624	..	..	6,805	..	..	..	..	..
	2013	537,503	..	..	6,839	41,892	..	..	..	..
	2014	406,078	..	..	6,565	31,200	..	..	..	..
United Kingdom	2011	58,123	..	..	..	..	..	..	33	1,859
	2012	64,181	..	..	..	..	..	..	33	2,000
	2013	64,117	..	..	..	..	..	..	33	2,097
	2014	68,024	..	..	..	..	..	..	33	2,180
Oceania	**2011**	**..**	**..**	**..**	**..**	**..**	**..**	**1,188**	**8,727**	**12,121**
	2012	**..**	**..**	**..**	**..**	**..**	**..**	**907**	**9,963**	**12,839**
	2013	**..**	**..**	**..**	**..**	**..**	**..**	**850**	**10,669**	**13,623**
	2014	**..**	**..**	**..**	**..**	**..**	**..**	**850**	**11,662**	**13,709**

Production of heat - by type

Terajoules

Country or area	Year	Combustible fuels	Geothermal	Solar thermal	Nuclear	Chemical sources	Heat pumps and electric boilers	Other sources	Direct use geothermal	Direct use solar thermal
Australia	2011	..	..	..	..	..	..	..	..	11,630
	2012	..	..	..	..	..	..	..	..	12,348
	2013	..	..	..	..	..	..	..	..	13,132
	2014	..	..	..	..	..	..	..	..	13,218
New Caledonia	2011	..	..	..	..	..	..	..	..	111
	2012	..	..	..	..	..	..	..	..	111
	2013	..	..	..	..	..	..	..	..	111
	2014	..	..	..	..	..	..	..	..	111
New Zealand	2011	..	..	..	..	..	..	1,188	8,727	364
	2012	..	..	..	..	..	..	907	9,963	364
	2013	..	..	..	..	..	..	850	10,669	364
	2014	..	..	..	..	..	..	850	11,662	364
Niue	2011	..	..	..	..	..	..	..	..	16
	2012	..	..	..	..	..	..	..	..	16
	2013	..	..	..	..	..	..	..	..	16
	2014	..	..	..	..	..	..	..	..	16

Table 35

Production of uranium (uranium content)

Metric tons

Table Notes

Data for 2011 to 2013 are based on country submissions and Table 1.21 of the OECD and IAEA publication *Uranium 2014: Resources, Production and Demand*, OECD Publishing, available at http://dx.doi.org/10.1787/uranium-2014-en.

2014 data are based on country submissions and the World Uranium Mining Production data on the World Nuclear Association website, available at http://www.world-nuclear.org/information-library/nuclear-fuel-cycle/mining-of-uranium/world-uranium-mining-production.aspx.

Production for Bulgaria, France, Germany and Hungary comes from mine rehabilitation efforts only.

Please refer to the Definitions Section on pages ix to xv for the appropriate product description /classification.

Production of uranium (uranium content)

Metric tons

Country or area	2011	2012	2013	2014
World	**54,766**	**59,197**	**60,541**	**56,714**
Africa	**9,616**	**11,369**	**11,791**	**8,869**
Malawi	842	1,103	1,347	1,065
Namibia	3,954	5,026	5,627	3,255
Niger	4,264	4,773	4,277	4,156
South Africa	556	467	540	393
America, North	**10,727**	**10,665**	**10,792**	**11,053**
Canada	9,145	8,998	9,000	9,134
United States	1,582	1,667	1,792	1,919
America, South	**415**	**383**	**234**	**67**
Brazil	415	383	234	67
Asia	**23,807**	**25,535**	**26,835**	**27,497**
China	1,400	1,450	1,450	1,500
India	400	385	400	385
Iran (Islamic Rep. of)	12	15	40	*40
Kazakhstan	19,450	21,240	22,500	23,127
Pakistan	45	45	45	45
Uzbekistan	2,500	2,400	2,400	2,400
Europe	**4,234**	**4,236**	**4,539**	**4,227**
Czechia	229	228	215	193
France	6	3	3	2
Germany	51	50	30	33
Hungary	2	1	3	*3
Romania	80	80	80	80
Russian Federation	2,993	2,862	3,133	2,990
Ukraine	873	1,012	1,075	926
Oceania	**5,967**	**7,009**	**6,350**	**5,001**
Australia	5,967	7,009	6,350	5,001

Selected energy resources and reserves

Million metric tons unless otherwise indicated

Table Notes

Based on country submissions and the publication *World Energy Resources - 2016*, World Energy Council (WEC), available at https://www.worldenergy.org/publications/2016/world-energy-resources-2016/.

Please refer to the Definitions Section on pages ix to xv for the appropriate product description /classification.

Table 36

Selected energy resources and reserves

Million metric tons unless otherwise indicated

Country or area	Bituminous coal / Anthracite			Sub-bituminous coal / Lignite			Peat	
	Proved amount in place	Proved recoverable reserves	Estimated additional amount in place	Proved amount in place	Proved recoverable reserves	Estimated additional amount in place	Proved amount in place	Proved recoverable reserves
Afghanistan	66	66	400	..	..	..	..	..
Albania	..	..	..	727	522	..	..	155
Algeria	223	59	164	..	..	..	..	..
Angola	..	..	..	..	..	..	..	..
Argentina	800	500	..	7,300	500	273	90	*80
Armenia	317	163	..	..	..	..	..	..
Australia	1,598,761	62,095	125,000	443,431	44,164	215,000	..	..
Austria	1	..	3	333	25	61	..	..
Azerbaijan	..	..	..	..	..	..	..	..
Bahrain	..	..	..	..	..	..	..	..
Bangladesh	3,260	293	..	3	..	..	*138	..
Barbados	..	..	..	..	..	..	..	..
Belarus	..	..	..	1,500	100	..	323	323
Belgium	4,100	410	1,400	..	..	..	..	..
Belize	..	..	..	..	..	..	..	..
Benin	..	..	..	..	..	..	..	..
Bhutan	..	..	..	..	..	..	..	..
Bolivia (Plur. State of)	1	1	..	..	..	..	..	..
Bosnia and Herzegovina	..	484	..	5,274	2,264	..	..	..
Botswana	21,240	40	205,253	..	..	..	..	..
Brazil	6,212	1,547	..	17,636	5,049	15,319	..	..
Brunei Darussalam	..	..	..	..	..	..	..	..
Bulgaria	4,112	192	1,200	4,574	2,174	2,618	5	5
Burkina Faso	..	..	..	..	..	..	..	..
Burundi	..	..	..	..	..	..	56	56
Cambodia	..	..	..	..	..	..	..	..
Cameroon	..	..	..	..	..	..	..	..
Canada	187,606	4,346	26,045	120,506	2,236	31,990	1,092	..
Central African Rep.	..	..	..	3	3	..	..	..
Chad	..	..	..	..	..	..	..	..
Chile	5,316	1,181	125	7	155	5,000	..	..
China	5,462,672	124,059	363,200	332,652	7,555	304,700	4,687	328
Colombia	14,809	4,881	13,173	411	380	3,176	..	..
Congo	..	..	..	..	..	..	..	..
Costa Rica	17	..	..	27	..	22	48	..
Côte d'Ivoire	..	..	..	..	..	..	..	..
Croatia	4	4	..	300	33	..	..	..
Cuba	..	..	..	..	..	..	..	..
Cyprus	..	..	..	..	..	..	..	..
Czechia	16,526	1,107	..	9,767	2,604	..	..	..
Dem. Rep. of the Congo	988	88	..	..	..	..	..	..
Denmark	..	..	..	63	..	..	..	..
Dominica	..	..	..	..	..	..	..	..
Dominican Republic	..	..	..	84	..	..	..	..
Ecuador	..	..	..	24	24	6	..	..
Egypt	182	16	..	27	22	52	..	..
El Salvador	..	..	..	..	..	..	..	..
Equatorial Guinea	..	..	..	..	..	..	..	..
Estonia	..	..	..	..	..	..	2,000	2,000
Ethiopia	..	..	..	23	11	..	..	..
Faeroe Islands	..	..	..	..	..	..	..	..
Falkland Is. (Malvinas)	..	..	..	..	..	..	..	..
Fiji	..	..	..	..	..	..	..	..
Finland	..	..	..	..	..	..	850	420
France	160	95	200	114	21	165	..	..
French Guiana	..	..	..	..	..	..	..	..
French Polynesia	..	..	..	..	..	..	..	..
Gabon	..	..	..	..	..	..	..	..
Georgia	901	201	..	..	..	..	..	..
Germany	82,982	21	8,065	76,800	36,300	34,100	157	96

Peat	Proved recoverable reserves				Uranium (metric tons)		Hydropower (Gross theoretical capability) (GWh/Year)
Estimated additional amount in place	Natural gas (billions cubic metres)	Crude oil and NGL	Oil shale	Bituminous sands	Reasonably assured resources	Inferred Resources	
..	*50	..	..	..	..	..	394,000
..	*1	*26	..	5	..	..	40,000
..	4,504	1,537	..	..	19,500	..	12,000
..	*308	1,709	..	74	..	..	150,000
50	332	328	57	..	8,600	11,000	354,000
..	*18	..	44	..	..	..	22,000
..	3,471	442	4,531	..	1,174,000	558,700	265,000
..	*10	*7	1	..	..	..	150,000
..	1,148	959	..	20	..	..	44,000
..	172	15	..	..	..	..	..
*138	232	*4	..	..	..	..	4,000
..	0	0	..	..	..	..	..
1,479	*3	27	1,000	..	..	..	8,000
..	0	..	..	..	..	..	1,000
..	..	0	..	..	..	..	1,000
..	*1	1	..	..	..	..	2,000
..	..	..	..	..	..	..	263,000
..	281	*22	..	..	..	..	178,000
..	..	..	..	..	..	..	70,000
..	..	..	..	..	23,100	59,100	..
487	424	1,890	11,734	..	155,100	121,000	3,040,000
..	276	*150	..	..	..	..	..
..	*6	*2	18	..	..	..	27,000
..	..	..	..	..	..	..	1,000
..	..	..	..	..	..	..	6,000
..	..	..	..	..	..	..	88,000
..	*150	*18	..	..	..	..	294,000
336,908	1,987	27,755	2,192	27,134	357,500	192,500	2,067,000
..	..	..	..	..	12,000	..	7,000
..	..	216	..	..	..	..	150
..	*40	*20	3	..	700	1,200	227,000
952	3,841	2,521	47,600	120	120,000	56,600	6,083,000
..	135	333	..	5	..	..	1,000,000
..	115	*190	..	81	..	..	50,000
22	..	..	..	..	..	..	224,000
..	28	*14	..	..	..	..	46,000
..	*25	*7	..	..	..	..	20,000
280	*71	*3	..	8	..	..	3,000
..	*142	..	..	..	..	..	59,000
..	4	*2	..	..	1,300	100	13,000
..	*1	*24	14,310	5	1,400	1,300	1,397,000
..	31	74	..	..	..	..	120
..	..	..	..	..	..	..	200
..	..	..	..	..	..	..	50,000
..	*11	1,175	..	7	..	..	169,000
..	1,846	458	816	8	..	1,900	125,000
..	..	..	..	..	..	..	7,000
..	*37	149	..	..	..	..	..
..	..	..	2,494	..	..	..	2,000
..	*25	0	..	..	..	..	650,000
..	..	..	..	..	..	..	1,000
28	..	..	..	..	..	..	..
..	..	..	..	..	..	..	3,000
2,200	..	..	..	..	1,100	..	31,000
..	*10	*11	1,002	..	11,500	100	270,000
..	..	..	..	..	..	..	2,000
..	..	..	..	..	..	..	1,000
..	*28	274	..	..	4,800	1,000	200,000
..	*9	*5	..	0	..	..	136,000
200	39	*31	286	..	3,000	4,000	120,000

Table 36

Selected energy resources and reserves

Million metric tons unless otherwise indicated

Country or area	Bituminous coal / Anthracite			Sub-bituminous coal / Lignite			Peat	
	Proved amount in place	Proved recoverable reserves	Estimated additional amount in place	Proved amount in place	Proved recoverable reserves	Estimated additional amount in place	Proved amount in place	Proved recoverable reserves
Ghana	..	..	..	..	..	..	..	..
Greece	..	..	..	6,430	2,876	..	4,000	..
Greenland	383	183	..	183	183	200	..	..
Grenada	..	..	..	..	..	..	..	..
Guatemala	..	..	..	..	..	..	..	..
Guinea	..	..	..	..	..	..	..	..
Guinea-Bissau	..	..	..	..	..	..	..	..
Guyana	..	..	..	..	..	..	..	..
Haiti	..	..	..	40	..	27	..	..
Honduras	..	..	..	21	..	..	..	..
Hungary	5,351	276	298	5,337	2,633	2,630	..	..
Iceland	..	..	..	..	..	..	..	..
India	260,544	85,562	157,435	42,645	4,714	32,893	..	..
Indonesia	109,825	17,394	6,035	40,639	8,274	38,704	49,000	..
Iran (Islamic Rep. of)	41,203	1,203	..	2,295	..	..	..	..
Iraq	..	..	..	..	..	..	..	..
Ireland	40	14	26	..	9	..	154	106
Israel	..	..	..	..	..	..	500	386
Italy	610	10	..	29	7	600	..	..
Jamaica	..	..	..	..	..	..	200	32
Japan	13,883	340	6,298	1,036	10	7,122	..	..
Jordan	..	..	..	..	..	..	..	..
Kazakhstan	148,695	25,605	..	..	12,100	..	..	..
Kenya	..	..	..	..	..	..	..	..
Korea, Dem.Ppl's.Rep.	10,600	600	2,700	300	300	2,200	..	..
Korea, Republic of	1,686	326	272	209	126	812	..	..
Kuwait	..	..	..	..	..	..	..	..
Kyrgyzstan	28,499	971	..	..	812	..	..	..
Lao People's Dem. Rep.	62	4	..	521	499	..	..	..
Latvia	..	..	..	..	..	..	190	94
Lebanon	..	..	..	..	..	..	..	..
Lesotho	..	..	..	..	..	..	..	..
Liberia	..	..	..	..	..	..	..	..
Libya	..	..	..	..	..	..	..	..
Lithuania	..	..	..	..	..	..	937	269
Madagascar	150	..	..	37	..	..	..	..
Malawi	802	2	..	..	2	..	..	..
Malaysia	1,209	141	78	451	39	575	..	..
Mali	..	..	..	3	..	3	..	..
Mauritania	..	..	..	..	..	..	..	..
Mauritius	..	..	..	..	..	..	..	..
Mexico	4,160	1,160	1,960	51	51	213	..	..
Mongolia	41,024	1,170	..	120,776	1,350	..	..	..
Montenegro	337	142	..	..	..	..	..	..
Morocco	96	14	..	40	..	2	..	..
Mozambique	23,636	1,792	155	..	..	..	..	..
Myanmar	252	3	120	5	3	80	..	..
Namibia	350	..	..	..	..	..	..	..
Nepal	8	1	..	1	1	7	..	..
Netherlands	3,247	497	2,750	..	..	..	120	120
New Caledonia	2	2	8	..	..	..	..	..
New Zealand	3,175	825	942	11,350	6,750	11,902	1,640	..
Nicaragua	..	..	..	..	..	..	..	..
Niger	90	70	..	6	6	..	..	..
Nigeria	2,144	287	21	377	57	1,000	..	..
Norway	93	2	..	64	5	100	745	350
Oman	..	..	..	..	..	..	..	..
Pakistan	5,996	207	5	179,596	2,857	68,068	..	..
Panama	..	..	..	..	..	..	..	..
Papua New Guinea	..	..	..	..	..	..	..	..

Peat	Proved recoverable reserves				Uranium (metric tons)		
Estimated additional amount in place	Natural gas (billions cubic metres)	Crude oil and NGL	Oil shale	Bituminous sands	Reasonably assured resources	Inferred Resources	Hydropower (Gross theoretical capability) (GWh/Year)
..	*23	*90	..	..	..	..	28,000
..	*1	*1	..	..	1,000	6,000	80,000
..	..	..	..	..	..	134,700	550,000
..	..	..	..	..	..	..	38
..	3	*11	..	..	..	..	59,000
..	..	232	..	..	..	..	26,000
..	..	..	..	..	..	..	1,000
..	..	..	..	..	..	..	81,000
..	..	..	..	..	..	..	4,000
..	..	..	..	..	..	..	16,000
238	*8	*4	8	..	..	8,600	10,000
..	..	..	..	..	..	..	184,000
..	1,489	764	..	..	77,000	27,900	2,638,000
..	2,839	498	..	67	8,400	2,200	2,147,000
..	34,020	*21,433	..	..	700	1,800	448,000
..	3,694	19,308	..	..	..	..	225,000
140	*10	..	..	..	..	..	1,000
1,000	181	*2	550	0	..	..	0
2,500	45	82	10,446	48	4,800	1,300	190,000
..	..	..	..	..	..	..	1,000
..	*21	*4	..	..	6,600	..	718,000
..	*6	0	5,242	5	0	33,800	4,000
..	936	3,932	400	6,689	285,600	417,300	170,000
..	..	..	..	..	..	..	24,000
..	..	..	..	..	..	..	..
..	6	..	..	..	..	..	52,000
..	1,784	13,981	..	..	..	..	..
..	*6	*5	..	..	..	..	163,000
..	..	..	..	..	..	..	233,000
760	..	..	..	..	..	..	7,000
..	..	..	..	..	..	..	2,000
..	..	..	..	..	..	..	5,000
..	..	..	..	..	..	..	28,000
..	1,505	6,297	..	..	..	..	..
..	..	*1	..	..	..	..	6,000
..	2	..	5	35	..	..	321,000
..	..	..	..	..	8,200	5,700	15,000
..	*1,169	471	..	..	..	..	230,000
..	..	..	..	..	..	..	12,000
..	*28	*3	..	..	..	..	..
..	..	..	..	..	..	..	150
..	324	1,496	..	0	2,800	0	430,000
..	..	*35	42	..	30,600	25,100	57,000
..	..	..	..	..	..	..	27,000
..	*1	0	8,167	..	..	..	12,000
..	*2,832	*2	..	..	..	..	103,000
..	528	*3	286	..	..	..	348,000
..	*62	..	..	..	248,200	155,500	23,000
..	0	..	..	..	..	..	733,000
..	675	*30	..	..	..	..	11,000
..	..	..	..	..	..	..	2,000
..	*28	*17	3	..	..	..	205,000
..	..	..	..	..	..	..	33,000
..	..	*20	..	..	325,000	104,900	3,000
..	5,111	5,003	..	91	..	..	43,000
7,015	1,857	994	..	..	..	..	600,000
..	688	720	..	..	..	..	..
..	543	*50	..	..	..	..	475,000
..	..	..	..	..	..	..	26,000
..	141	*24	..	..	..	..	175,000

Table 36

Selected energy resources and reserves

Million metric tons unless otherwise indicated

Country or area	Bituminous coal / Anthracite			Sub-bituminous coal / Lignite			Peat	
	Proved amount in place	Proved recoverable reserves	Estimated additional amount in place	Proved amount in place	Proved recoverable reserves	Estimated additional amount in place	Proved amount in place	Proved recoverable reserves
Paraguay	..	..	..	..	..	..	..	..
Peru	1,567	102	..	100	100	..	..	..
Philippines	1,223	211	8	1,017	105	37	..	..
Poland	178,913	16,203	27,405	227,886	5,429	11,837	890	..
Portugal	3	3	..	66	33	..	..	..
Qatar	..	..	..	..	..	..	..	..
Republic of Moldova	..	..	..	..	..	..	..	..
Romania	2,446	11	2,143	9,920	280	8,177	25	13
Russian Federation	2,727,915	69,634	200,000	1,379,623	90,730	200,000	17,680	11,554
Rwanda	..	..	..	..	..	..	..	..
Samoa	..	..	..	..	..	..	..	..
Saudi Arabia	..	..	..	..	..	..	..	..
Senegal	..	..	..	..	..	..	40	24
Serbia	855	402	..	20,186	7,112	..	..	..
Serbia and Montenegro	..	..	..	..	..	..	..	..
Sierra Leone	..	..	..	2	..	..	..	..
Slovakia	19	2	..	1,073	135	..	..	..
Slovenia	95	56	..	656	315	40	..	..
Solomon Islands	..	..	..	..	..	..	..	..
Somalia	..	..	..	..	..	..	..	..
South Africa	213,560	9,893	5,000	30	..	100	47	30
South Sudan	..	..	..	..	..	..	..	..
Spain	4,231	868	3,000	319	319	1,260	94	70
Sri Lanka	..	..	..	..	..	..	5	5
Sudan	..	..	..	..	..	..	..	..
Sudan (former)	..	..	..	..	..	..	..	..
Suriname	..	..	..	..	..	..	..	..
Swaziland	4,644	144	793	..	..	..	..	..
Sweden	5	1	..	4	1	20	700	70
Switzerland	..	..	..	..	..	..	..	..
Syrian Arab Republic	..	..	..	..	..	..	..	..
T.F.Yug.Rep. Macedonia	..	..	..	632	332	..	..	..
Tajikistan	4,075	375	..	..	..	..	..	..
Thailand	..	..	..	1,889	1,063	2,857	..	..
Timor-Leste	..	..	..	..	..	..	..	..
Togo	..	..	..	..	..	..	..	..
Trinidad and Tobago	..	..	..	..	..	..	..	..
Tunisia	..	..	..	..	..	..	..	..
Turkey	1,182	380	249	12,828	12,466	296	..	..
Turkmenistan	800	..	..	..	..	..	..	..
Uganda	800	..	..	..	..	..	..	..
Ukraine	81,045	32,039	5,170	7,717	2,336	6,174	2,160	684
United Arab Emirates	..	..	..	..	..	..	..	..
United Kingdom	186,770	70	190	1,000	500	..	..	..
United Rep. of Tanzania	1,410	269	*1,500	..	..	..	..	..
United States	6,680,329	222,641	445,346	1,398,360	30,483	667,415	26,000	13,000
Uruguay	..	..	..	..	..	..	..	..
Uzbekistan	10,852	1,375	..	..	1,853	..	..	..
Venezuela (Bolivar. Rep.)	6,712	731	4,528	..	..	..	..	..
Viet Nam	6,635	3,116	..	200,120	244	..	..	..
Yemen	..	..	..	..	..	..	..	..
Zambia	945	45	..	69	55	18	..	..
Zimbabwe	25,502	502	5,820	965	..	..	..	..

Peat	Proved recoverable reserves				Uranium (metric tons)		Hydropower (Gross theoretical capability) (GWh/Year)
Estimated additional amount in place	Natural gas (billions cubic metres)	Crude oil and NGL	Oil shale	Bituminous sands	Reasonably assured resources	Inferred Resources	
200	..	..	..	..	..	..	111,000
..	414	170	..	1	1,300	1,300	1,577,000
..	*99	*15	..	..	..	..	47,000
2,300	95	15	7	..	..	..	25,000
..	..	..	..	..	6,000	1,000	32,000
..	24,528	2,695	..	..	..	..	..
..	..	..	..	..	..	..	2,000
10	110	80	..	1	3,100	3,600	70,000
168,320	32,271	14,024	35,470	4,518	216,500	432,000	2,295,000
2,000	*57	..	..	..	..	..	2,000
..	..	..	..	..	..	..	140
..	8,325	36,618	..	..	..	..	..
52	10	0	..	..	..	..	11,000
..	31	*7	..	..	..	..	27,000
..	45	22	..	..	..	..	118,000
..	..	*60	..	..	..	..	11,000
..	*14	*1	..	..	0	9,000	10,000
..	0	0	..	..	1,700	7,500	19,000
..	..	..	..	..	..	..	3,000
..	*6	..	..	..	5,000	2,600	2,000
..	10	*2	19	..	175,300	179,200	73,000
..	..	472	..	..	..	..	..
..	*3	*19	40	..	14,000	0	162,000
46	..	..	..	..	..	..	21,000
..	..	202	..	..	..	..	..
..	85	682	..	..	..	..	48,000
..	..	*12	..	..	..	..	39,000
..	..	..	..	..	..	..	4,000
..	..	..	875	..	5,000	8,500	200,000
..	..	..	..	..	..	..	125,000
..	285	*340	..	..	..	..	11,000
..	..	..	..	..	..	..	9,000
..	*6	*2	..	..	..	..	527,000
..	220	49	916	..	..	..	18,000
..	..	*59	..	..	..	..	..
..	..	..	..	..	..	..	4,000
..	326	*113	..	1	..	..	..
..	*65	55	..	..	..	..	1,000
..	*7	*46	284	..	7,300	..	433,000
..	17,479	82	1,100	..	..	..	24,000
..	*14	*136	..	..	..	..	33,000
1,846	604	*54	600	..	84,800	81,300	45,000
..	6,091	12,976	..	..	..	..	..
..	206	374	501	12	..	..	35,000
..	*1,317	..	..	..	30,100	15,600	39,000
..	10,441	*6,857	536,931	3	207,400	..	2,040,000
46	..	..	..	..	..	..	32,000
..	1,086	81	1,200	..	59,400	31,900	88,000
..	5,617	*26,807	..	9,212	..	..	731,000
..	617	*599	..	..	1,000	5,400	300,000
..	266	409	..	..	..	..	..
..	..	..	..	..	9,900	5,700	53,000
..	..	..	..	..	1,400	..	44,000

www.ingramcontent.com/pod-product-compliance
Lightning Source LLC
Chambersburg PA
CBHW080934240926
51849CB00003BA/459
9789211616163